Settlement of Structures

Settlement
of
Structures

*Conference organised by the British Geotechnical
Society at The Lady Mitchell Hall, Cambridge held
in April 1974*

Pentech Press
London

Pentech Press Limited
8 John Street, London WC1N 2HY

ISBN 0 7273 1901 9

Printed and bound in Great Britain
by Butler & Tanner Limited, Frome and London

The importance of settlement observations was highlighted during discussions of the Earthworks and Foundations Committee of CIRIA in 1971.
CIRIA also felt the need for closer cooperation between geotechnical and
structural engineers on the all-important problem of soil structure interaction.
This feeling was shared by the Institution of Structural Engineers who had
set up their own Working Party to review the matter.

Arising from these discussions, CIRIA held a colloquium on Settlement of
Structures in December 1971. This was attended by members of the Earthworks and Foundations and Structural Design Committees of CIRIA,
members of the Institution of Structural Engineers Working Party, and
engineers from universities, Government Research Organisations, Consulting
Engineers and Contractors.

Following this colloquium it was decided that a conference on settlements
was needed, and after discussion the British Geotechnical Society agreed to
organise a conference in conjunction with the University of Cambridge.

A word of explanation is needed concerning the marked differences in the
lengths of papers. In order to encourage the maximum presentation of
settlement data the Committee solicited technical notes as well as full length
papers, intending that technical notes would give factual settlement data
and basic description of ground conditions without discussion of the results.
In the event papers were submitted which fell between 'papers' and 'technical
notes' both in terms of length and content. It was therefore decided not to
distinguish between the types of papers.

Finally it should be explained that papers about embankment settlements
were omitted in order to keep the overall size of the Proceedings within
reasonable bounds, on the grounds that there has been good recent coverage of
embankment problems, as noted at the beginning of the General Report for
Session II.

A. C. MEIGH
Chairman, Organising Committee

v

ORGANISING COMMITTEE

A. C. Meigh	Soil Mechanics Limited (Chairman)
J. B. Burland	Building Research Establishment
F. G. Butler	Ove Arup and Partners
A. D. M. Penman	Building Research Establishment
N. E. Simons	University of Surrey
C. P. Wroth	University of Cambridge

Conference Secretary

W. M. G. Bompas	University of Cambridge

CONTENTS

Session III—Heavily over-consolidated cohesive materials

Session IV—Rocks

Opening Address
Professor SIR JOHN BAKER, O.B.E., Sc.D., F.R.S., C.Eng.

After all these years I am still a little vague about what the opener should say, although I should not be because since my retirement most of my public appearances have, oddly enough, been in opening conferences on one or other aspect of soil mechanics. If I tend to reminisce, however, old men's stories have their uses if they remove some of those obscurities from the past which, in the future, might cause historians to speculate.

I do not really know how my very great involvement in soil mechanics came about. It may very well have stemmed from that day in 1928 when, having arrived at the Building Research Station to start some new activity on steel structures I found that the authorities had provided no accommodation for me. I bumped into Cooling who very kindly invited me to his office, and gave me a seat there until something more permanent could be found for me. Whatever the link, I have been very much attached to your subject ever since.

Resisting the temptation to reminisce wildly, I have one story that may be of some value to the historian. Cambridge is the most democratic institution in the world: we do all our business quite openly and we record all our decisions. They appear in a weekly newspaper, the *University Reporter*. If you were to refer to the *Reporter* of August 22, 1944, you would find among the list of appointments a name that is well known in soil mechanics' circles: it is the name A. W. Skempton. You would find from the *Reporter* that he was appointed a University Demonstrator in Engineering here to date from October 1, 1944. This was my first attempt to introduce teaching and research in soil mechanics into this University. Considering that I only arrived here to become Head of the Department of Engineering in 1943, I obviously was losing no time, or so I thought. Unfortunately, search as you will, you will not find the name Skempton appearing in our archives again, because he did not take up the appointment. I was extremely disappointed at losing what I considered, on good evidence, to be the brightest young man in the business, and I have no doubt that I blamed my old friend and collaborator, Sutton Pippard, who was Professor of Civil Engineering at the Imperial College, for perhaps saying to Skempton, 'Well now, if you really want to go into University work, I can offer you something very much better at Imperial College.' That is where Skempton moved a little later, with the brilliant results that you know so well.

Having denied the future researcher into the history of soil mechanics the opportunity of speculating about this very odd single appearance of this distinguished name, I have not denied him the opportunity of further speculation; speculating on whether, had Skempton stayed at Cambridge, Roscoe, on his return from the War, would have chosen a project in the soil mechanics field for his PhD course, a move which started all our work in Cambridge. Nor have I stopped him speculating that if Roscoe had made that choice, how he and his supervisor would have hit it off together. Perhaps, most pertinent of all, what would have been the development of soil mechanics at Imperial College?

This Conference is on the Settlement of Structures which reminds me that I spent twenty years developing a design method which was impelled by the aim to make soil mechanics and differential settlement irrelevant. My aim was to put you out of business almost before you began. If engineers would only construct the bottom parts of their structures as continuous ductile frame structures, how much happier they could all be. Happiness, this morning is not your aim; you are searching for knowledge and I won't stop you any longer—but I wish you a most successful conference.

SESSION I
Granular materials

I/1. The performance of two large oil tanks founded on compacted gravel at Fawley, Southampton, Hampshire

G. E. Bratchell and A. J. Leggatt,
Nachshen, Crofts and Leggatt, Consulting Engineers, Westminster, London
N. E. Simons,
Department of Civil Engineering, University of Surrey

Summary

The Paper describes briefly the new foundations, consisting of imported sandy gravel fill some 10 m thick, compacted by vibroflotation, for two tanks, nos. 281 and 282, at the Esso Refinery, Fawley. The effects of vibroflotation on S.P.T. N values, and dynamic and static cone penetration tests are presented. The results of settlement observations taken at various points on the tanks and in the ground during hydraulic testing are summarised, and the observed settlements within the gravel fill are compared with those computed on the basis of N values and static cone point resistances.

Introduction

The design and construction of new foundations for tanks 281 and 282 for the Esso Refinery, Fawley, Hampshire, involved a number of matters of geotechnical interest. A full account of the project covering the general civil engineering aspects has been published, Leggatt and Bratchell (1973).

The two tanks, 79 m in diameter, 20 m high, with floating roofs, were originally built on piled foundations and were ready for loading tests in the summer of 1968. During the hydraulic testing of the first tank, evidence of foundation failure was detected and subsequent investigation led to the foundations for both tanks being abandoned. After considering several alternatives, new foundations were constructed by digging out the soft

3

alluvium down to a bed of natural gravel, filling with imported gravel, compacting the fill by vibroflotation and then floating the tanks onto the new foundation.

The tanks and their foundations were tested by filling with seawater, and the performance of the compacted gravel fill was checked by taking settlement observations at various points on the perimeter and in the central area of the tanks, and on selected points underneath the fill itself.

Soil conditions

The site is on the foreshore of Southampton Water and the descending soil succession is:
(1) Sandy gravel fill, some 1.4 m thick,
(2) soft alluvium, approximately 9 m thick,
(3) medium dense sandy gravel, of average thickness 3 m,
(4) Barton Clay.
A simplified soil profile is shown in *Fig. 1*.

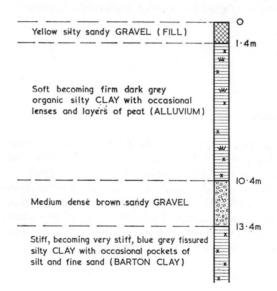

Fig. 1. Simplified soil profile

The alluvium was removed by dredging under water and was replaced by sandy gravel conforming to the grading limits in *Table 1*. The sandy gravel was compacted by vibroflotation using a triangular grid at 2.6 m centres. On top of the vibroflotated fill, a 0.9 m thickness of selected sand and gravel was

Table 1. GRADING LIMITS FOR IMPORTED SANDY GRAVEL FILL

BS Sieve	40 mm	20 mm	10 mm	5 mm	No. 7	No. 25	No. 200
Passing %	100	65–90	50–70	35–60	25–50	10–30	2

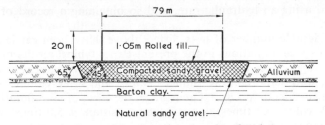

Fig. 2. Cross-section through tank and its foundations

rolled in four layers with a vibrating roller, and finally a 0.15 m thick layer of crushed stone was rolled down to form a suitable finish to receive the tank.

The underlying natural gravel had Standard Penetration Test N values generally in the range 25 to 40.

The Barton Clay can be described as a stiff becoming very stiff blue grey fissured silty clay.

A cross-section through the tank and its foundations is shown in *Fig. 2.*

Effects of vibroflotation on soil properties

The effects of vibroflotation were checked using Standard Penetration Tests, and Dynamic and Static Sounding Tests, with a series of each type of test being carried out before and after vibroflotation, the location of the post-vibroflotation tests being equidistant from the positions where the vibrators

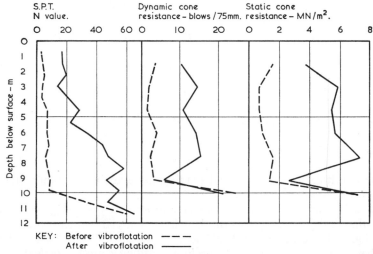

Fig. 3. Average values of S.P.T. tests, and dynamic and static cone resistances, before and after vibroflotation

were inserted. In the dynamic and static sounding test a 35 mm diameter 60° sliding cone is driven into the ground by a 55 kg hammer with a 0.75 m drop and the resistance to penetration is recorded. At intervals, a static penetration test was carried out, by forcing the sliding cone itself into the soil a distance

5

of 60 mm, using an hydraulic ram, and so obtaining a record of the cone resistance.

A considerable number of tests, listed above, have been carried out, but because of space limitations in the present Paper typical results only can be given. Average values of Standard Penetration Test N value, dynamic cone resistance, and static cone resistance, before and after vibroflotation, are plotted against depth in *Fig. 3*. On average, the effect of vibroflotation was to increase N values 5.2 times, dynamic cone resistances 5.0 times, and static cone resistances 5.6 times.

Settlement observations

Testing by filling with sea-water is the Esso basic practice since this tests both tank and foundation with well-defined overloads; in this case, an overload of about 20% of the normal loading was obtained. The total loading was reached after 13 days for tank 281 and after 16 days for tank 282, and the maximum load was held for 14 days for tank 281 and 10 days for tank 282.

During the entire tests, levels were taken directly on the outside perimeter, on three concentric circles within the tank by direct soundings through holes in the roof, on the ground near the tank, and on a number of direct-reading and remote-reading settlement measuring devices.

The tape soundings were obtained by lowering a weighted tape through

Table 2. OBSERVED AND DEDUCED SETTLEMENTS OF TANKS

	Settlement		
	In clay mm	In fill mm	Total mm
Perimeter tank 281	76	32	108
Centre area tank 281	184*	63*	247
Perimeter tank 282	57	32	89
Centre area tank 282	95*	57*	152

* Deduced values.

appropriate holes in the roof, and reading these against a manometer tube datum transfer arrangement set up on the roof.

The settlement measurement devices comprised vertical steel joists installed within the fill depth with plates on their lower ends so that they stood on the natural gravel layer. Some of the devices were situated beneath the tank itself and these were read by means of remote reading gauges attached to their heads. Four settlement devices were installed outside the tank perimeter and were read by direct levelling on top of the joists.

The dipping tape methods were found to be successful in practice and these readings, together with the levels on the periphery of the tank, and the direct levels on the external settlement devices, form the basis of the results. The readings from the remote reading gauges were inconsistent and must be regarded as unsatisfactory. When designing these gauges, it was appreciated that a high degree of operational reliability was required but, despite this, there were difficulties with installation and protective measures, and the results were disappointing. The measured and deduced results obtained are shown in *Table 2*. After unloading, there was a small amount of recovery averaging about 12.5 mm for tank 282, and about 32 mm for tank 281.

It can be seen that the deduced settlement at the centre of the tanks in the sandy gravel fill alone averages 60 mm, while the corresponding figure for the perimeter settlement is 32 mm.

Predicted settlements

Many methods have been proposed to predict the settlement of a structure founded on granular material, based on S.P.T. N values and on static cone point resistances. Those methods which the authors believe to be most widely used in practice, have been adopted in calculations of the settlement in the sandy gravel fill only, and these methods are summarised in *Table 3*, which

Table 3. RESULTS OF SETTLEMENT CALCULATIONS FOR COMPACTED SANDY GRAVEL FILL

Method of calculation	Calculated settlement mm	$\dfrac{\text{Calculated settlement}}{\text{Observed settlement}}$
Terzaghi and Peck (1967)	33	0.55
Terzaghi and Peck design chart, but making no correction for the presence of the water table, and then increasing the pressure to give 25 mm settlement by 50%, Meyerhof (1965)	11	0.18
'N' values corrected for effective over-burden pressure, Tomlinson (1969), Terzaghi and Peck design chart, but making no correction for the presence of the water table	7	0.12
Peck and Bazaraa (1969)	16	0.27
Alpan (1964)	10	0.17
Parry (1971)	8	0.13
de Beer and Martens (1957)	110	1.83
Schmertmann (1970)	14	0.23

Note: The first six methods listed above are based on the use of S.P.T. N values, and the last two on the use of static cone point resistances.

also lists the computed settlements. The intensity of applied loading has been taken as 188 kN/m², and the depth of the ground water table as 1 m. It should be stressed that the case record presented in this paper cannot be considered a classic example on which to base conclusions concerning the accuracy of the various methods of computation considered above, because of

(a) the large width of the loaded area, 79 m,
(b) the relatively limited thickness, 10 m, of sandy gravel compared with the foundation width.

It should also be noted that although the oil tanks are flexible structures, several of the methods of computation adopted, strictly apply to rigid footings.

Conclusions

The soil replacement solution for the foundations is regarded as being technically successful, since the permanent works have performed satisfactorily under full hydrotesting, and in subsequent use, justifying the design assumptions.

Markedly different computed settlements result from the different methods of settlement analysis which have been adopted. All methods based on S.P.T. values under-estimate the observed settlement. This is of interest particularly because it is known that the Terzaghi and Peck basic method is generally conservative. While the de Beer and Martens procedure generally over-estimates the observed settlement, a much greater over-estimate may in fact have been anticipated, since their procedure is based on maximum vertical strains occurring immediately under a foundation, while the actual point of maximum vertical strain develops at a greater depth, approximately equal to half the foundation width, Eggestad (1963), and Schmertmann (1970). At Fawley, the total thickness of compressible granular fill is only 0.13 times the foundation width.

It is possible that deformations occurring in the underlying Barton clay have allowed greater horizontal strains (and hence vertical settlements) to take place in the granular fill than otherwise would have been the case, if the granular fill had rested on a more rigid stratum. The surrounding soft alluvium may also have influenced the observed settlements.

Acknowledgements

The Authors wish to acknowledge the co-operation of the Esso Petroleum Co. Ltd., throughout the project and for their permission to publish this Paper.

REFERENCES

Alpan, I. (1964). 'Estimating the Settlements of Foundations on Sands', *Civ. Eng. and Pub. Wks Rev.* Vol. 59, 1415–1418

de Beer, E. and Martens, A. (1957). 'Method of Computation of an Upper Limit for the Influence of the Heterogeneity of Sand Layers in the Settlement of Bridges, *Proc. 4th Int. Conf. Soil Mech. & Found. Eng.*, London. Vol. 1, 275–282

Eggestad, A. (1963). 'Deformation Measurements below a Model Footing on the Surface of Dry Sand', *Proc. European Conf. Soil Mech. & Found. Eng.*, Wiesbaden. Vol. I, 233–239

Leggatt, A. J. and Bratchell, G. E. (1973). 'Submerged Foundations for 100 000 ton Oil Tanks', *Proc. I.C.E.*, Part 1, Vol. 54, 291–305

Meyerhof, G. G. (1965). 'Shallow Foundations', *J. Soil Mech. & Found. Div., Proc. A.S.C.E.*, Vol. 82, SMI

Parry, R. H. G. (1971). 'A Direct Method of Estimating Settlements in Sand from S.P.T. Values', *Proc. Sym. on the Interaction of Structure & Foundation*, Birmingham, 29–37

Peck, R. B. and Bazaraa, A. R. S. (1969). *Disc., J. Soil Mech. & Found. Div., A.S.C.E.*, No. SM3, 905–909

Schmertmann, J. H. (1970). 'Static Cone to Compute Static Settlement Over Sand', *J. Soil Mech. & Found. Div., Proc. A.S.C.E.*, No. SM3, 1011–1042

Terzaghi, K. and Peck, R. B. (1967). *Soil Mechanics in Engineering Practice*, 2nd edn., 491. New York, Wiley

Tomlinson, M. J. (1969). *Foundation Design and Construction.* 2nd edn., 103. London, Pitman

I/2. Settlement distribution in the subsoil underneath a nuclear reactor

H. Breth,
Professor of Soil Mechanics

G. Chambosse
Institute of Soil Mechanics and Foundation Engineering, Technische Hochschule, Darmstadt, West Germany

Summary

A nuclear reactor was built near the River Rhine in Biblis, Germany. The load attributable to the reactor amounted to 120 000 Mp (1 180 000 kN). The reactor was founded at a depth of 5 m on a plate of 60 m diameter. The subsoil, which was explored by way of borings reaching a depth of 100 m, consisted of a prestressed granular material (classification SW). The soil characteristics were obtained from penetration tests. Settlements of 50 mm were measured in the foundation level of the reactor.

Subsoil

The reactor is located in the Rhine Valley. The subsoil was explored with the help of borings reaching a depth of 100 m. Underneath the ground surface is a 2.5 m thick silt layer and underneath the silt layer is gravelly sand. Below, are fine and coarse sands down to 60 m and from 60 m to the bottom of the bore hole the subsoil consists of sandy silts and sandy clays. The density of the sands was examined in the bore holes by way of static penetration. In *Fig. 1* the measured peak pressures are shown besides the bore profile. In the soft silt layer close to the surface the peak pressure is low, amounting to roughly 10 kg/cm². Underneath the silt layer, the peak pressure rapidly increases in the gravel to 150–300 kg/cm². Beginning at a depth of 10 m, peak pressures of between 250 and 500 kg/cm² were measured in fine and medium sand. At a depth of 45 m the pressure increased to 750 kg/cm.² The

10

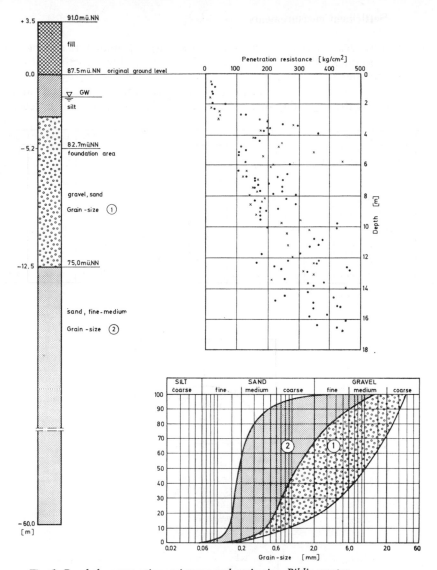

Fig. 1. Borehole, penetration resistance and grain size, Biblis reactor

density of the sandy gravel may be defined as dense, that of the fine and medium sand as dense to very dense. The results of the static penetration tests suggest a friction angle of 35° to 37.5°. The compression index of the sand was estimated at 2500 to 5000 kp/cm². The cohesive layers at a depth below 60 m are stiff to very stiff and of low plasticity (plasticity 8–29%, classification ML to CL). A one-dimensional compression test resulted in a compression index of 280 kg/cm² for an applied load of 6–8 kg/cm². The ground water is connected with the water level of the Rhine river and thus subjected to fluctuations. On average it is 1.5 m beneath the surface.

11

Settlement measurements

The cylindric reactor (with circular cross-section) is founded at a depth of 5 m on a 3 m thick plate 60 m in diameter. The weight of the reactor amounts to approximately 120 000 Mp (1 180 000 kN), leading to an average soil pressure of 4.25 kg/cm². The settlement of the reactor was measured at 18 points and by the beginning of 1973 the centre of the reactor had settled by

Point:	6	7	8	20	19	18	2	3	4	16	15	14	10	11	12	22	23	24	Load [Mp]	Date
Settlement [mm]	2	6	4	6	6	5	4	4	4	7	7	5	4	.8	7	5	7	6	7000	4.1970
	17	19	20	20	20	19	16	13	17	21	19	17	16	19	19	18	19	21	45000	11.1970
	19	21	22	22	22	21	18	15	19	23	22	18	17	20	21	19	21	23	50000	3.1971
	21	23	24	24	23	23	21	17	20	24	22	20	19	21	22	21	22	25	59000	6.1971
	31	34	35	36	34	33	30	27	32	37	33	30	29	32	34	30	32	36	80000	10.1971
	35	40	41	41	39	38	35	33	38	42	39	35	34	38	40	35	38	42	106000	2.1972
	39	42	44	45	42	—	39	37	41	44	41	37	36	41	42	37	41	46	115000	6.1972
	39	43	46	46	—	—	—	—.	42	46	42	38	37	42	44	37	41	46	120000	9.1972
	40	43	45	45	—	—	—	46	42	45	41	37	36	41	43	37	41	46	120000	2.1973

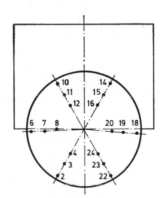

Fig. 2. Settlements, Biblis reactor

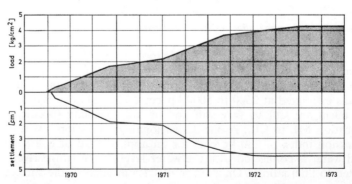

Fig. 3. Average settlement and soil pressure, Biblis reactor

12

roughly 55 mm. Construction work was nearly completed in 1973. The results of the settlement measurements are compiled in *Fig. 2* while *Fig 3* shows the average settlements as a function of the average soil pressure. As the settlement syncline (*Fig. 4*) indicates the floor plate did not behave rigidly but deflected by roughly 10 mm.

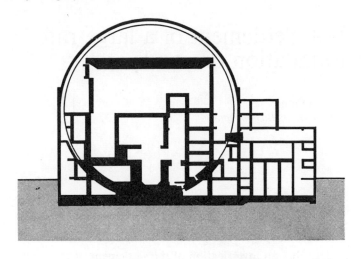

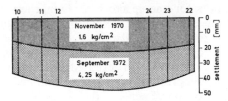

Fig. 4. Cross section and settlement depression, Biblis reactor

I/3. Settlement of a large raft foundation on sand

C. S. Dunn
University of Birmingham

Summary

The paper describes an investigation of the settlement of the sand below the large raft foundation for the reactors of Dungeness 'B' Nuclear Power Station. The raft was founded 9 m below existing ground level on 30 m of beach sand overlying silty clay mudstone.

Prior to construction, metal plates were buried at various depths in the ground below proposed formation level. Movements of these plates were monitored and the compression of the strata immediately under the raft determined. The relative movements of the plates buried under the centre of the raft correspond almost to those which would occur in an elastic homogeneous layer overlying a rigid base at the level of the mudstone.

The load-settlement graph (*Fig. 6*) for the centre of the raft illustrates that there were two phases of settlement; due to the recompression of the ground by pressures up to the initial overburden and the compression due to a net increase in pressure. It is shown that an estimate of settlement due to net load may be made using the results of static cone tests.

Observations were also carried out on settlement studs located at numerous positions on the surface of the raft so that a contour plan of settlement could be drawn and compared with that predicted by a numerical analysis which assumes an elastic raft structure on a Winkler subgrade. There is a significant difference between the predicted deflected form of the raft and that measured.

Geology

Dungeness 'B' Power Station is situated adjacent to and west of the existing 'A' Station on the south facing shore of a very large triangular shaped flint shingle promontory which juts out into the English Channel near Lydd in Kent. The geology of the area has been described by Lewis (1932).

14

Site investigations showed that the shingle has a depth of 8 to 10 m below which lies gravelly sand becoming increasingly sandy with depth until at about 15 m there is fine to medium sand which extends down to the Hastings Beds at 40 m.

The Hastings Beds generally consist of interbedded layers of dense silt and stiff to hard silty clays with siltstone layers at greater depth.

Details of the geological profile below raft formation level which is 9 m below ground level are recorded in *Table 1* along with corresponding average

Table 1. GEOLOGICAL PROFILE UNDERLYING STRUCTURE

Depth below foundation formation level m	Soil description	Average N from S.P.T.	Average C_{kd} from static cone	C_{kd}/N
0–3	Sandy gravel—(grey) (medium to coarse, rounded) progressively changing to	21	—	—
3–5.5	Gravelly sand (grey) (coarse gravel and fine sand) progressively changing to	32	—	—
5.5–8.5	Fine sand (grey)	41	185	4.5
8.5–12.0	Fine sand (grey)	63	247	3.9
12.0–15.0	with occasional shells	>70	253	—
15.0–17.5	Fine sand (dark grey) with traces of gravel and occasional shells	56	265	4.7
17.5–21.5	Fine sand (grey) with some coarse sand, some shells and traces of black silt	51	273	5.3
21.5–24.0	Fine sand (grey)	47	258	5.5
24.0–31.0	Silty fine sand (grey)	>70	247	—
>31	Hastings Beds—stiff silty clays and dense silt	—	—	—

values of N and C_{kd}. The ratio C_{kd}/N ranges from 3.9 to 5.5 and appears to increase slightly with depth.

The water table was found to vary with precipitation between 2.5 and 5.5 m below ground level during a period of three weeks in January 1966.

Stress changes due to construction sequence

A record of the principal changes of stress which took place at raft formation level is illustrated by *Fig. 1*. These show that effective stress changes resulted from (1) temporary ground water lowering, (2) excavation, (3) raft self weight, and (4) structural loading of the raft, (5) re-establishment of natural ground water level.

The ultimate gross dead load on the subgrade is expected to be 1730 MN equivalent to an average pressure of 311 kN/m² resulting in a net change in

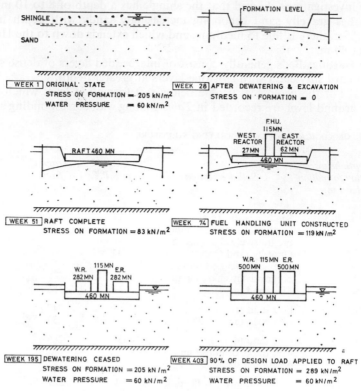

SHINGLE
SAND

WEEK 1 ORIGINAL STATE
STRESS ON FORMATION = 205 kN/m²
WATER PRESSURE = 60 kN/m²

FORMATION LEVEL

WEEK 28 AFTER DEWATERING & EXCAVATION
STRESS ON FORMATION = 0

RAFT 460 MN

WEEK 51 RAFT COMPLETE
STRESS ON FORMATION = 83 kN/m²

F.H.U.
115MN
WEST   EAST
REACTOR   REACTOR
27MN   62MN
460 MN

WEEK 74 FUEL HANDLING UNIT CONSTRUCTED
STRESS ON FORMATION = 119 kN/m²

W.R. 115MN E.R.
282 MN   282 MN
460 MN

WEEK 195 DEWATERING CEASED
STRESS ON FORMATION = 205 kN/m²
WATER PRESSURE = 60 kN/m²

W.R. 115MN E.R.
500 MN   500 MN
460 MN

WEEK 403 90% OF DESIGN LOAD APPLIED TO RAFT
STRESS ON FORMATION = 289 kN/m²
WATER PRESSURE = 60 kN/m²

Fig. 1. Loading sequence

stress of 106 kN/m². About 93% of this had been applied up to the time of writing (June 1973).

Prediction of settlement of the reactor raft

The simple rectangular shape of the raft foundation (101 m × 55 m × 3.4 m thick) and the near symmetry of loading on both the north-south and east-west centre lines made theoretical analysis relatively simple. The main loads were concentrated under the reactors which seat on the annular ring walls shown in *Fig. 2*, and the centrally placed fuel handling unit.

The stress and displacement analysis of the raft was based on the method of Allen and Severn (1960). The raft was divided into a convenient grid and solutions were obtained at node points spaced at 6.2 m. A modulus of sub-grade reaction of 20 ton/ft³ (7.05 MN/m³) was adopted. This was obtained by dividing the gross subgrade pressure applied by an 'A' Station reactor raft by its total observed settlement.

Account was taken of the effect of the surrounding ground surcharge (160 kN/m²). *Fig. 3* shows a contour plan of computed theoretical settlement. Profiles of theoretical settlement are also drawn along the east–west centre line in *Fig. 4*.

16

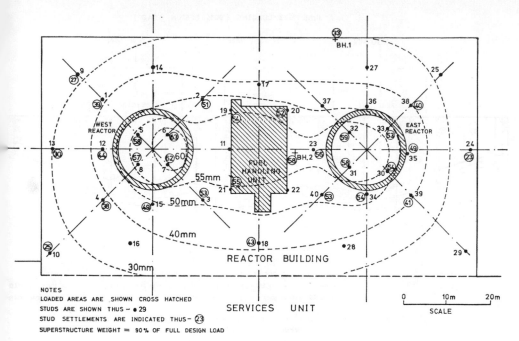

Fig. 2. Plan showing location of settlement studs on reactor raft, borehole positions and contours of settlement due to weight of superstructure on raft. Up to June 1973 (week 403)

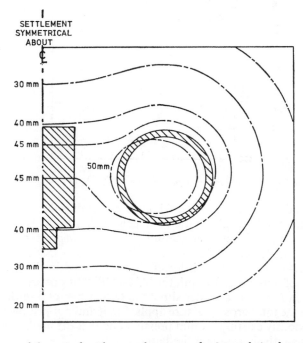

Fig. 3. Contours of theoretical settlement of reactor raft—interpolation between node points

17

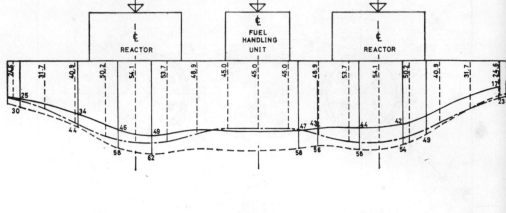

JUNE 1973 LOADING (90% DESIGN LOAD)

Fig. 4. Comparison of theoretical and measured settlements along the east–west centre line through the reactors

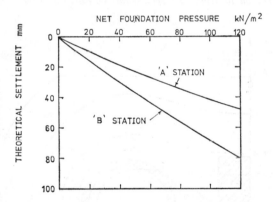

Fig. 5. Relation between net pressure and settlement by De Beer and Marten's method

De Beer and Marten's (1957) method of estimating the maximum settlement using the results of the static cone penetration tests (*Table 1*) was also used. It was calculated that the maximum net increase in pressure of $106\,kN/m^2$ would cause a settlement of 72 mm (corresponding to a modulus of subgrade reaction of $1.47\,MN/m^3$ (much lower than 7.05 used in the design). Theoretical values of settlement for both the 'A' and 'B' Stations are plotted against net pressure in *Fig. 5*. The reader should note that the total settlement of the raft is due to the total gross pressure applied to it and not just the net increase over the previously existing ground pressure and that the method may only be used to conservatively estimate the latter.

Settlement measurements

Prior to construction, metal plates were buried at various depths below formation level in boreholes 1 and 2 at the positions shown in *Fig. 2*. Movements of
these plates were monitored during construction and have enabled the
compression of the strata immediately under the raft to be determined.
Measurements from the ground surface down to each buried plate were taken
by lowering an electrical probe down the inside of P.V.C. tubing. The depth
measurements were made by a steel tape suspended from a clamp mounted on
a rachet and pinion which permitted fine adjustment. The cable and tape
were separated and prevented from twisting by laying them along grooves cut
in the opposite faces of aluminium blocks connected by light flexible steel rods.
Although the arrangement was clumsy to use, readings were reproducible to
± 1 mm.

All levels were measured relative to a datum pile 200 m from the raft and
founded at a depth of 12 m. Levelling was carried out using a Watt's Precise

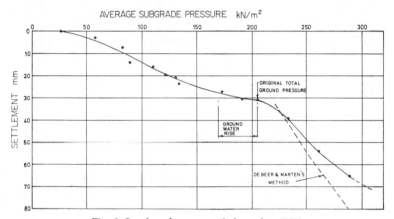

Fig. 6. Load-settlement graph for raft at BH2

Level and if the closing error using a twin scale staff exceeded 0.0035 ft when
levelling between borehole and the datum, the survey was repeated.

Observations on the plates commenced in January 1966 before excavation
started. The second set of observations were taken on 13 June when excavations
had been completed and the raft construction was under way. The June
observations have been taken as zero for the purposes of plotting settlement
with time. The settlement of plate 1 in borehole 2 representing the settlement
of the near centre of the raft is plotted against average applied gross subgrade
pressure in *Fig. 6*.

The most striking feature of this is the point of inflection at the preconsolidation pressure of 205 kN/m² (effective stress 145 kN/m²). At pressures
above this the compressibility has increased confirming that the sand is
behaving as a normally consolidated deposit. It is noteworthy that the
average slope of this part of the graph is nearly equal to that of the predicted
graph in *Fig. 5* indicating that De Beer and Marten's method can give a good
prediction of settlement of normally consolidated sand.

Fig. 7 shows the settlement with time of the individual plates in borehole 2.

19

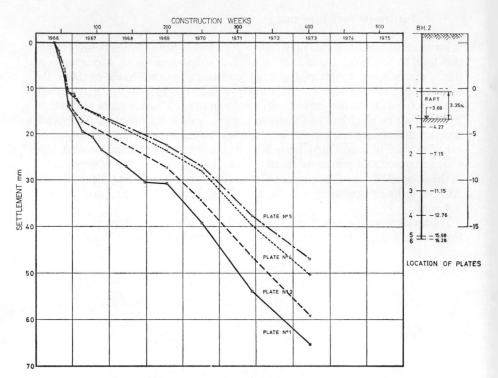

Fig. 7. Settlement of plates buried in BH2 near centre of raft

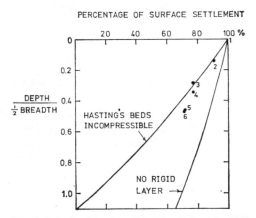

Fig. 8. Distribution of movements of plates in BH2

These movements have been expressed as a percentage of the total surface settlement and plotted against depth in *Fig. 8*. For comparison, theoretical curves representing movement of points at various depths in an elastic layer either semi-infinite or overlying a rigid base, are superimposed. The graph indicates that the Hastings beds are behaving as an almost rigid base.

20

Differential settlement of raft

After construction of the raft, brass studs were set into the surface of the concrete in the positions shown in *Fig. 3*. The settlements of the studs resulting from the weight of the superstructure on the raft (not self weight) are also indicated in *Fig. 2*.

Contours of equal settlement (up to 15 June 1973) have been drawn on *Fig. 2*. In addition, the profile of actual settlement on the east–west centre line have been added to *Fig. 4* for comparison with the predicted ultimate profile.

Although it is evident that the maximum settlement of 63 mm has occurred under the west reactor, there have been almost symmetrical movements about the east–west and north–south centre lines and there has been tilting of both reactors towards the Fuel Handling Unit (F.H.U.). The differential settlement between the reactors and the edge of the raft on the east–west centre line was of the order of 33 mm and the maximum differential settlement gradient was 1 in 550.

Fig. 4 shows that the settlement of the F.H.U. in June 1973 was already in excess of the predicted settlement under full load. The theoretical analysis predicted that the raft would hog upwards in the area of the F.H.U. This has not happened and the differential settlement in the area between the two heavily loaded reactors is small.

The discrepancy between theoretical and actual deformed shape is attributed to the erroneous assumption that subgrade contact pressure is proportional to deflection. The theoretical analysis could be improved if it were to include influence factors to take account of deflection caused by pressure applied at adjacent points as well as by direct pressure.

The settlement of the reactor raft is in excess of the theoretical settlement calculated assuming a modulus of subgrade reaction of 20 tons–ft². This is not surprising, because the value of 20 was obtained from the load-settlement graph for the adjacent 'A' Station, yet De Beer and Marten's method predicted greater settlements under the 'B' Station than under the 'A' Station for the same magnitude of stress.

Further information on the settlement observations may be obtained from Dunn (1972).

REFERENCES

Allen, D. N. and Severn, R. T. (1960). 'The Stresses in Foundation Rafts', *Proc. I.C.E.* Vol. 15, Jan., 35–48
De Beer, E. and Martens, A. (1957). 'Method of Computation of an Upper Limit for the Influence of the Heterogeneity of Sand Layers on the Settlement of Bridges', *Proc. 4th Int. Conf. Soil Mech. & Found. Eng.*, London, Vol. 1, 269–274
Dunn, C. S. (1972). 'An Interim Report on the Settlement Behaviour of the Reactor Raft at Dungeness "B" Nuclear Power Station', *Report RP/97 Construction Industry Research and Information Association*
Lewis, W. V. (1932). 'The Formation of Dungeness Foreland', *Geogr. J.*, Vol. 80, 309–324

I/4. An investigation on settlements of direct foundations on sand

V. K. Garga
Senior Engineer, Geotécnica S/A,

J. T. Quin
Engineer, Geotécnica S/A, Consulting Engineers, Rio de Janeiro, Brazil

Summary

The Paper briefly describes the investigations for a large steel mill complex, founded mainly on micaceous sand, in Brazil. The complete site was well explored by a large number of S.P.T. and Standard Dutch Cone penetration soundings. In view of the lack of settlement observations on existing structures, twelve field load tests on 2.5 meters square plate loaded to 200 ton (1.96 MN) were carried out. A large number of load tests on smaller sized plates, 45 cm square and 1 m square, were also executed to study the size effect on settlement of direct foundations.

A correlation between the estimated 'undisturbed' value of the modulus of elasticity, E, and the Dutch Cone penetration resistance, Ckd, was obtained from the results of a number of *in situ* load tests on 25 cm and 30 cm diameter screw plates. The settlements from all plate load tests have been compared with those obtained from various analytical methods.

Finally, a brief literature review is undertaken on settlements of direct foundations of large widths. This suggests that the settlements are grossly overestimated because the effective depth of strain influence is probably much smaller than the customarily assumed depth equal to twice the width of the foundation. Incomplete settlement observations on two slab foundations with widths equal to 22 m are presented which tend to corroborate this finding. A tentative semi-empirical method based on Gibson (1967) and Som (1968) to evaluate the effective depth of strain influence is suggested.

Introduction

During the twelve years previous to the work presented in this Paper, extensive use had been made of both direct foundations (footing and rafts) and piled foundations for the various facilities of a large steel mill complex at Ipatinga in Brazil. Although no formal settlement records were kept, it is evident that both the direct foundations as well as the significantly more expensive piled foundations behaved satisfactorily. The investigations for large-scale expansion of this mill necessitated a more rational study of the settlement behaviour of foundations as a controlling feature of design. In the absence of settlement records at Ipatinga, recourse was made to load tests on plates of different dimensions. It was recognised that load tests on small sized plates are very susceptible to any local inhomogeneities in the soil, for example, a thin bed of loose sand would not exert any appreciable influence on the behaviour of a large foundation. Also, the settlement of a foundation in natural sands is a non-linear function of its width, Bjerrum and Eggestad (1963). Ideally, if the sand underneath the foundation were to have the same grading and constant relative density (and hence modulus of elasticity) with depth, then the settlements of a foundation of any size could be estimated reasonably well from the results of loading tests on small plates. However, such a condition is rarely encountered in practice and the results of large plate loading tests have been used to investigate the reliability of the various available methods for prediction of settlements on sands.

The site

The subsoil in the area of interest, approximately 3.5 km² in plan, consists of deep, lightly overconsolidated fluvial deposits of fine to medium, micaceous sand with some fine quartzitic gravels, overlain by a 2–7 m thick layer of soft to medium silty or sandy clay. The subsoil profile for the complete site is well known through over 1100 exploratory borings to depths from 20 to 50 m. A typical soil profile is shown in *Fig. 1* where the numbers along the borings refer to the S.P.T. blowcounts (N values). The results of numerous Standard Dutch Cone resistance soundings (C_{kd} values) are also available, and a typical result obtained from a location adjacent to boring No. 6 is also shown in *Fig. 1*. The particle size does not vary significantly with depth, as shown in *Fig. 2*. Occasionally thin seams and pockets of soft clay with some organic matter are also present in sand. The ground water table is generally encountered around El. 230 m with mean ground surface at El. 245 m.

Results of over 100 adjacent percussion borings and Standard Dutch Cone soundings have enabled a correlation to be established between S.P.T. blowcounts (N) and cone penetration resistance (C_{kd} in kg/cm²). The average correlations of C_{kd}/N equal to values of 6.1, 7.1, 7.3, 6.7, 5.6 and 4.7 were obtained for arbitrary effective stress ranges of 0–5, 5–10, 10–15, 15–20, 20–25 and 25–30 t/m² respectively. The ratio was expected to show a significant variation with effective stress in view of the known dependance of N values on overburden pressure, Gibbs and Holtz (1957). These correlations for granular materials are quoted in the literature as varying from 4 for sands, Meyerhof (1956) to 16 for well-graded sandy gravels, Meigh and Nixon (1957).

23

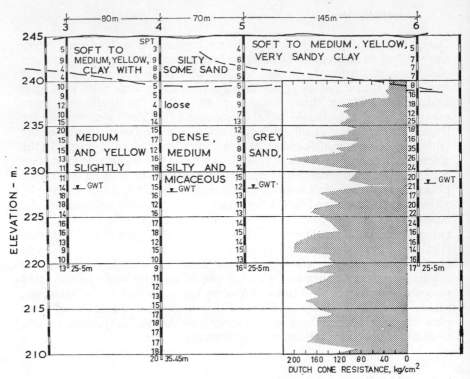

Fig. 1. Typical soil profile of the site

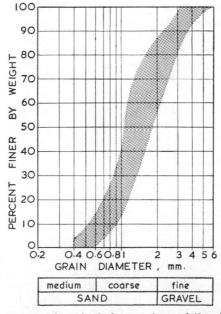

Fig. 2. Grain size variation from depths between 3 m and 40 m for a typical boring

24

Field investigations

Tests on 2.5 m *square plates*

These tests were always done in large open excavations, generally at elevation around 240 m, when all upper clay deposits were removed and sand was exposed. The minimum width of the excavated floor was specified to be 5 m. A Standard Dutch Cone penetration test, every 30 cm, was carried out to a depth of at least 10 m below the excavated surface. The test surface was then lightly levelled and the test plate, 2.5 m × 2.5 m × 0.07 m, was gently positioned. The plate was loaded in stages using a Link-Belt lifting crane to position the large, thick steel plates. After each loading, settlements were measured at the four corners, until they became sensibly stabilised. After the maximum load of approximately 200 tonf (1.96 MN) had been applied, unloading was carried out in stages. In most tests, settlements under a second loading and unloading cycle were also measured.

Tests on 1 m *and 45* cm *square plates*

Loading tests on small sized plates on sand were also carried out in wide excavations. The settlements were monitored at only two sides of the plate. The normal load was applied by a hydraulic jack bearing against a kentledge. Settlements for only the first cycle of loading and unloading were measured.

Screw plate tests

A number of analytical methods require the knowledge of the modulus of elasticity E of sand, for example, Schmertmann (1970) and if possible its variation with depth. The modulus of elasticity may be conveniently estimated from the results of a plate loading test and elastic theory, Timoshenko and Goodier (1951) since for a circular flexible plate

$$S = q(1-\mu^2)2 \, B/E \tag{1}$$

where S denotes settlement, B denotes radius of the plate, μ denotes Poisson's ratio, q denotes loading pressure, and E denotes the modulus of elasticity.

The 25 cm and 30 cm diameter screw plates used for the *in situ* determination of the modulus of elasticity E, were designed after Kummeneje and Eide (1961). The tests were invariably conducted in the same excavations in which load tests on large plate were carried out. The screw plate was gradually screwed into the soil manually and was periodically withdrawn from the hole to remove the excavated sand. The hole was finally cleaned when the depth of excavation had reached 10 cm above the desired depth at which the test was intended to be carried out. At this stage the screw plate was rotated twice into the undisturbed sand, and was now ready for load testing. Hence at each testing stage there was only a 10 cm high column of sand above the plate. Note that the above procedure is different from that suggested by Schmertmann (1970) in which the excavated material along the length of the rod is not removed. Loading was carried out in stages up to a maximum of 2 tonf

25

1.96 MN and the settlements were monitored on the two extensometers. Two cycles of loading and unloading were executed at each test depth. At the end of each test, sand from the upper 10 cm column was recovered for grain size analysis. At each site, measurement of Standard Dutch Cone resistance was taken at every 30 cm depth, adjacent to the location of the screw plate test, to enable a correlation to be made between C_{kd} and E values. Such a relationship has been suggested by various authors. Buisman suggested $E = 1.5C_{kd}$ for normally consolidated sands from Holland, while Schmertmann (1970) has proposed $E = 2C_{kd}$. Schultze and Melzer (1965), on the basis of investigations in an experimental shaft, 3 m diameter, 5.5 m deep,

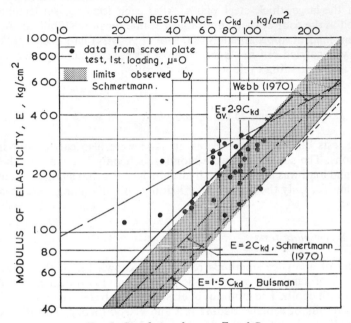

Fig. 3. Correlations between E and C_{kd}

found that the cone resistance was a function of overburden pressure and suggested the following relationship:

$$E = k\sigma^{0.522}$$

where $k = 301.1 \log C_{kd} - 382.3\gamma t + 60.3 + 50.3$, and $\sigma = \gamma t$ denotes the total overburden pressure in kgf/cm². Webb (1970) observed the relationship

$$E = 2(C_{kd} + 25)$$

or the alluvial sandy sediments in Durban, South Africa.

Fig. 3 shows the values of E at Ipatinga derived from Equation 1 for the case of $\mu = 0$, from results of 34 tests, when an average relationship of $E = 2.9C_{kd}$ is found.

As expected, the modulus of elasticity obtained from the second cycle of loading is substantially higher than for the first cycle, and was found to be $E = 4.9C_{kd}$.

26

Note that in all subsequent analysis, the use has been made of the relationship, $E = 2.9C_{kd}$. It is believed that the accuracy of the screw plate test and the natural variations of the subsoil along the same horizon do not warrant the use of Poisson's ratio other than zero. The natural variation in cone resistance of undisturbed sand deposit was investigated by carrying out three

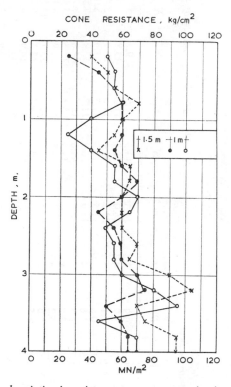

Fig. 4. Natural variation in resistance to cone penetration in uniform sand

Dutch Cone soundings in an open excavation, at a spacing of 1.5 m and 1 m, as shown in *Fig. 4*. The Dutch Cone apparatus was anchored on beams lying across and over the excavation so as not to disturb the *in situ* density of sand as a result of anchoring installation of the equipment. It is evident from *Fig. 4* that the natural variation of cone penetration resistance values is large enough to render any accuracy introduced by the inclusion of Poisson's ratio other than zero meaningless.

Analysis of results

The settlement of load tests have been compared with the predictions from the methods mentioned below. Since detailed reviews of the various methods have already been presented, Schmertmann (1970); D'Appolonia (1967); Webb (1969) only a brief mention is made here.

27

Table 1. MEASURED AND ESTIMATED SETTLEMENTS FOR LOAD TESTS ON PLATES

Load test	Size m	Applied pressure		S_m mm	Calculated settlement mm											
		t/m²	kN/m²		Modified Buisman–De Beer			Schmert-mann	Terzaghi formula	Terzaghi and Peck		Meyerhof		Elastic theory
					K	W	B			N	N'	N	N'	
AF 1	1 × 1	10	98	4.0	0.8	1.0	0.9	2.8	11.5*	13.3	7.5	9.2	5.0	—
		15	147	5.4	1.1	1.3	1.7	4.3	16.0	20.7	11.2	13.8	7.5	—
		25	245	8.0	1.2	1.7	2.1	7.1	22.5	34.4	18.6	22.9	12.4	—
AF 4	1 × 1	10	98	4.0	0.9	1.1	0.8	3.0	11.5	17.9	7.5	11.9	5.0	—
		15	147	5.0	1.2	1.5	1.6	4.6	16.0	26.9	11.2	17.9	7.5	—
		25	245	10.4	1.7	1.9	2.3	7.6	22.5	44.8	18.7	44.8	18.7	—
CG 4	1 × 1	10	98	2.5	2.8	3.4	3.0	6.1	—	19.9	8.1	13.3	5.4	—
		15	147	4.0	3.6	4.2	5.2	9.1	—	29.8	12.2	19.9	8.1	—
		25	245	7.0	3.4	5.5	6.6	15.2	—	49.7	20.3	33.1	13.5	—
A 4	1 × 1	10	98	4.0	1.5	1.8	1.5	5.1	—	19.9	9.4	13.3	6.3	—
		15	147	7.0	1.8	2.2	2.6	7.7	—	29.8	41.1	19.9	9.4	—
		25	245	14.0	2.4	2.7	3.3	12.8	—	49.7	23.6	33.1	15.7	—
AF 6	2.5 × 2.5	10	98	8.0	4.5	5.3	6.7	9.0	17.5*	40.4	17.3	26.9	11.5	12.3
		15	147	10.8	5.6	6.7	8.4	13.5	23.5	60.6	26.0	40.4	17.3	18.5
		25	245	20.5	7.6	8.9	11.0	22.6	41.0	100.9	43.3	67.3	28.9	30.8
AF 7	2.5 × 2.5	10	98	4.2	2.5	3.2	4.1	6.2	17.5	16.2	7.2	10.8	4.8	7.8
		15	147	6.3	3.3	4.1	5.3	9.2	23.5	24.2	10.7	16.1	7.1	11.8
		25	245	10.8	4.6	5.6	7.2	15.4	41.0	40.4	17.9	26.9	11.9	19.6
AF 8	2.5 × 2.5	10	98	3.0	2.6	3.3	4.4	6.3	17.5	15.1	6.7	10.1	4.5	9.6
		15	147	3.9	3.5	4.3	5.7	9.5	23.5	22.7	10.1	15.1	6.7	14.4
		25	245	6.6	4.9	6.0	7.7	15.8	41.0	37.9	16.9	25.3	11.3	24.0

Test	Size								N'					
CG 6	2.5 × 2.5	10	98	4.6	4.2	5.5	7.2	5.2	8.0+	34.6	16.2	23.1	10.8	14.4
		15	147	8.0	5.5	7.2	9.8	7.8	17.5	51.9	24.2	34.6	16.2	21.6
		25	245	17.8	7.7	9.9	12.7	13.0	29.0	86.5	40.4	57.7	26.9	35.9
TM 1	2.5 × 2.5	10	98	2.8	1.9	2.4	3.3	4.6	—	8.7	5.8	5.8	3.9	6.9
		15	147	3.4	2.5	3.2	4.3	6.9	—	13.0	8.7	8.7	5.8	10.3
		25	245	4.4	3.7	4.6	6.0	11.5	—	21.6	14.4	14.4	9.6	17.2
TM 2	2.5 × 2.5	10	98	4.0	3.4	4.2	5.5	7.9	—	18.6	12.1	12.4	8.1	13.2
		15	147	5.4	4.5	5.5	7.1	11.8	—	28.0	18.2	18.7	12.2	19.8
		25	245	9.4	6.2	7.6	9.8	19.7	—	46.6	30.3	31.1	20.3	33.1
TM 3	2.5 × 2.5	10	98	5.3	4.4	5.2	6.2	7.9	—	22.0	16.1	14.7	10.8	9.6
		15	147	6.4	5.5	6.4	7.7	11.8	—	33.0	24.2	22.0	16.1	14.4
		25	245	8.4	7.1	8.2	9.8	19.7	—	55.1	40.4	36.7	28.9	24.0
CA 1	2.5 × 2.5	10	98	3.7	4.7	5.7	7.2	8.9	—	16.2	6.2	6.8	4.2	13.3
		15	147	6.1	6.0	7.3	10.6	13.4	—	24.2	9.3	16.1	6.2	20.0
		25	245	12.5	8.2	9.8	12.3	22.3	—	40.4	15.6	26.9	10.4	33.2
A 5	2.5 × 2.5	10	98	4.4	3.6	4.4	5.6	8.6	5.5+	24.2	11.0	16.1	7.3	10.2
		15	147	6.2	4.6	5.6	7.1	12.8	9.0	36.3	16.5	24.2	11.0	15.2
		25	245	10.4	6.3	7.5	9.3	21.4	14.5	60.6	27.6	40.4	18.4	25.4
A 6	2.5 × 2.5	10	98	3.2	2.3	2.9	3.8	5.8	5.5	12.8	6.4	8.5	4.3	8.2
		15	147	4.7	3.1	3.8	5.0	8.6	9.0	19.1	9.6	12.7	6.4	12.3
		25	245	7.6	4.3	5.3	6.8	14.4	14.5	31.9	16.0	21.3	10.7	20.6
C 3	2.5 × 2.5	10	98	4.8	10.9	13.0	16.0	21.9	—	26.9	13.4	17.9	9.0	19.2
		15	147	7.4	13.7	16.2	20.2	32.8	—	40.4	20.2	26.9	13.5	28.9
		25	245	18.4	17.9	21.0	25.3	54.7	—	67.3	33.7	44.9	22.5	48.1
C 4	2.5 × 2.5	10	98	3.0	1.3	1.8	2.3	3.1	—	6.4	3.6	4.3	2.4	4.8
		15	147	4.2	1.9	2.3	3.1	4.6	—	9.6	5.4	6.4	3.6	7.2
		25	245	6.8	2.6	3.2	4.0	7.6	—	15.9	8.9	10.6	5.9	12.0

S_m—measured settlement
K—Kögler stress distribution
W—Westergaard stress distribution
B—Boussinesq stress distribution
*—Values calculated from test on 43.5 cm plate

N—S.P.T. blow-count as measured
N'—S.P.T. blow-count corrected for overburden pressure
$E = 2.9\, C_M$ was adopted for Modified Buismann–De Beer
+—Values calculated from test on 1 m plate

Methods using S.P.T. results

Methods by Terzaghi and Peck (1948) and Meyerhof (1965) in the form presented by Peck and Bazaraa (1968) were used. Predictions were made using both the measured N values as well as those corrected for the effect of over-burden pressure, Gibbs and Holtz (1957).

Methods using elastic theory

The results of 2.5 m square plates, only, were calculated using Equation 1 and the E value deduced from an average value of cone resistance in the 5 m depth below test level. The loading plate was regarded as rigid and an influence factor of 0.82, Bowles (1968) was used.

Buisman–De Beer method

According to this method, the compressibility index, C, is given by the empirical relationship, $C = 1.5C_{kd}/p'$, where p' is the effective overburden pressure, and C_{kd} is Standard Dutch Cone penetration resistance. The settlement for any layer thickness H, may be obtained from:

$$\frac{\Delta H}{H} = \frac{2.3}{C} \log_{10} \frac{(p' + \Delta p)}{p'}$$

where Δp denotes the applied effective stress increment in the layer due to applied surface load.

Three methods of stress distribution, namely Boussinesq–Newmark, Westergaard and Kögler, and Taylor (1948) have been used to obtain Δp. Also since $C = 2.9C_{kd}/p'$ was used, this method is referred to as the modified Buisman-De Beer method.

Schmertmann's method (1970)

Based on an analysis of theoretical and experimental distributions of vertical strain below the centre of loaded area, leading to a simplified triangular distribution of strain factor, and $E = 2.9C_{kd}$ was used.

Terzaghi and Peck method (1947)

Based on observations on a limited number of tests proposes the evaluation of settlement of a foundation from the observations of plate loading test, by the use of the expression

$$S_f/S_p = [2B_f/(B_f + B_p)]^2$$

where S_f denotes the settlement of the foundations, S; denotes the settlement of the test plate, and B_f and B_p denote the least width of the foundations and the plate respectively.

30

For convenience the test plate generally adopted is a 30 cm diameter or square plate. However, estimates of settlements for the 2.5 m square plates have been obtained on the basis of tests on 0.43 m square and 1 m square plates.

Comments on results

Table 1 presents a comparison of observed and calculated settlements from which the following observations may be made.

In the case of the 1 m square plate, the modified Buisman–De Beer method underestimates settlements, while the Schmertmann method underestimates

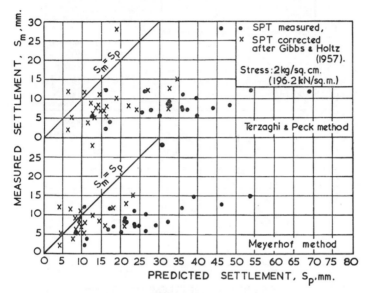

Fig. 5. Comparison of predicted settlements using S.P.T. results with measured settlements

in two cases and overestimates in the other two; the ratio between predicted and observed settlements being 0.8 and 1.6 respectively.

Both the Terzaghi and Peck, and Meyerhof methods using S.P.T. values overestimate settlements for both plate sizes, as shown in *Fig. 5*. However, the Meyerhof method with N corrected for the effect of overburden pressure predicts settlements close to the observed values. *Table 2* shows the average

Table 2. RATIOS OF PREDICTED TO MEASURED SETTLEMENTS

Terzaghi and Peck		Meyerhof	
N	N'	N	N'
4.7	2.3	3.2	1.6

N—S.P.T. blowcount as measured.
N'—N corrected for overburden pressure, Gibbs and Holtz (1957)

31

ratios of predicted to observed settlements for all load tests using these two methods. A similar conclusion has also been reached by D'Appolonia *et al.* (1968) from observations on foundations varying from 3 m to 7 m in width.

The results of the modified Buisman–De Beer and Schmertmann methods for 2.5 m square plates have been plotted in *Fig. 6*, which shows that for a bearing pressure of 10 tonf/m² (98.1 kN/m²), the first method generally predicts the settlements accurately. However, at a pressure of 20 tonf/m² (196.2 kN/m²) this method underestimated in 5 out of 12 cases, and Schmertmann's method would appear to be more appropriate with an average over-estimation by a ratio of 1.8.

Estimates using elastic theory (Equation 1) as well as the Terzaghi and

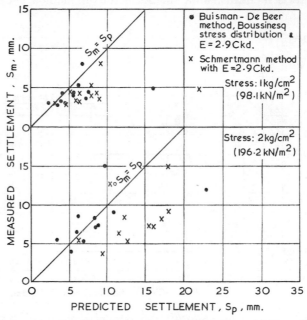

Fig. 6. *Comparison of predicted settlements using modified Buisman–De Beer and Schmertmann methods*

Peck load-test method greatly overestimate the settlements, probably as a result of the non-homogeneous stiffness of the soil. It would appear therefore that for foundation widths of about 2.5 m the settlements can be predicted by either using Meyerhof with corrected N values or by the modified Buisman–De Beer with Boussinesq stress distribution. However, the question still remains whether the experience from loading tests on small plates can also be applied to the prediction of settlements of very large foundations. If the recommendations for small to medium size foundations were to be used for those of large widths, then the effective depth of soil that would contribute to settlements (i.e. the depth corresponding to '2B'), would also be consequently very large. A survey of the literature on this subject, however, reveals that the effective depth is significantly less than twice the width of the foundation, and would appear to be the principal reason for the overestimation of settlements of large foundations. Also the restraining effect of the horizontal stress is likely to be more significant in reducing the compressibility of sand at large depths.

Egorov and Popova (1971) reported measurements of effective compressible depth in both sandy and clay soils for 120–125 m high concrete chimneys founded on rigid circular plates varying in diameter from 23 to 38 m. Their first conclusion states: 'The comparison of actual tower foundation settlements with calculated values proves that in order to calculate soil deformation it is possible to use the linear-deformed medium theory (the elastic theory), considering a finite layer of thickness H. For a bigger foundation area ($R > 10$ m), the H value can be taken equal to the radius ($H = R$) in the case of clayey soils and to two-thirds of the radius ($H = 2R/3$) in the case of sandy soils. It is assumed here that the soils deformation properties do not get worse with depth.'

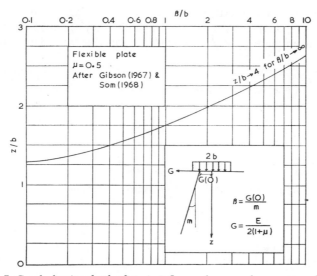

Fig. 7. Graph showing depth of strain influence for a non-homogeneous layer

Some theoretical confirmation of the validity of the above observations is provided by Gibson (1967) and its further extension by Som (1968). Gibson considered the case of a uniformly loaded flexible plate resting over an elastic medium whose elasticity increases with depth (*Fig. 7*). Som compared the settlements for the case of homogeneous elastic medium (Boussinesq case) with those for the Gibson case, from which *Fig. 7* is derived. The effective depth for the case of a non-homogeneous soil layer which would result in the same magnitude of settlement as for the homogeneous case may be determined from *Fig. 7* assuming that for the Boussinesq case and homogeneous elastic layer, all settlements occur in a depth equal to four times the radius as is customarily the case. Similar results for the case of uniformly distributed load over rigid plate resting on non-uniform deposit have been obtained by Shirokov *et al.* (1971).

To summarise, there is considerable evidence of both an observational and analytical type that for large foundations ($R > 10$ m) the effective depth is much less than twice the width—the value being probably closer to the width of the foundation. Hence knowing the variation of the modulus of elasticity (for example from screw-plate tests), the elastic settlement may easily be computed by dividing the estimated effective depth of strain influence (from *Fig. 7*) in layers and by applying Boussinesq stress distribution and the corresponding value of E for each layer. Note that although *Fig. 7* has been prepared for the case of a circular foundation, and $\mu = 0.5$, it is suggested that a rectangular foundation may still be considered by using

$$2b \simeq \sqrt{(cd)}$$

where c and d are the dimensions of the rectangular foundation.

Partial analysis of settlement observations at Ipatinga

Precise, but incomplete settlement records for two large slab foundations, a 22 m wide octagonal slab for a 120 m high chimney and a 75 m $\times$ 22 m slab for a coke oven plant (*Fig. 8*) are available together with the results of cone

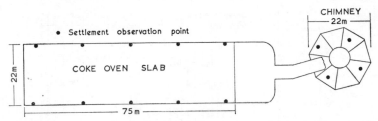

Fig. 8. General layout of settlement points for the cone oven plant and chimney at Ipatinga

penetration tests in the immediate vicinity to a depth of 50 m. Regrettably, the settlements were not measured from the beginning of construction and, therefore, only a comparison between estimates of differences in settlements between two known loading intensities can be made, as shown in *Table 3*. In

Table 3. PARTIAL ANALYSIS OF TWO LARGE FOUNDATIONS AT IPATINGA

				Settlements in mm				
Structure	Foundation size × m	Loading intensities		Observed average	Schmert-mann	Buisman-De Beer		
		t/m²	kN/m²			$d = 2B$	$d = B$	$d = 1.3B/2$
Coke oven plant	22 × 75	17.8	175	13.8	64.4	87.3	67.1	46.5
		9.5	93	6.1	34.4	51.6	41.0	29.5
Differences in settlement				7.7	30.0	35.7	26.1	17.0
Chimney	22 m	25.2	247	10.5	91.1	89.3	75.2	54.5
	(octagonal)	17.1	168	0	61.8	67.7	56.9	42.5
Differences in settlement				10.5	29.3	21.6	18.3	12.0

d—depth of strain influence; B—diameter of equivalent circular foundation (. $2b$ in Fig. 7

34

both cases the effect of the reduced depth of strain influence according to *Fig. 7* is remarkable. It is interesting to note the accuracy of the prediction for the case of the chimney, which is probably accounted for by the fact that the loads are known to the nearest ton and that the shape of the octagonal foundation approximates closely to the circular foundation considered in *Fig. 7*.

Concluding remarks

The settlements of relatively small foundations can be predicted with reasonable accuracy by the use of Meyerhof's method with corrected S.P.T. values, or by the use of the Buisman–De Beer approach using the appropriate variation of elasticity with depth and Boussinesq stress distribution to a depth equal to twice the width of the foundation. However, for large foundations (>10 m) the effect of non-uniform elasticity restricting strains to a more shallow depth should be taken into account and a semi-empirical method has been suggested. The need for settlement observations on large foundations, as well as information regarding the variation of soil stiffness with depth cannot be overemphasised.

Acknowledgements

The Authors are indebted to, USIMINAS, and in particular to Engs. I. Yuasa and A. Augusto for their willing co-operation throughout the investigations, and to Eng. S. P. Velloso and other colleagues at Geotécnica S/A for their assistance with the field work.

REFERENCES

Bjerrum, L. and Eggestad, A. (1963). 'Interpretation of Loading Test on Sand'. *Proc. European Conf. Soil Mech. & Found. Eng.* Vol. 1, 199–204
Bowles, J. E. (1968). *Foundation Analysis and Design.* McGraw-Hill
D'Appolonia, D. J., D'Appolonia, E. and Brisette, R. F. (1968). 'Settlement of Spread Footing on Sand', *J. Soil Mech. & Found. Div., Proc. A.S.C.E.* Vol. 94, 735–759
Egorov, K. E. and Popova, O. V. (1971). 'Comparison of Computed and Factual Settlements of High Smoke Stacks Foundations', *Proc. 4th Asian Conf. Soil Mech. & Found. Eng.* Vol. 1, 9–15
Gibbs, H. J. and Holtz, W. G. (1957). 'Research on Determining the Density by Spoon Penetrating Testing', *Proc. 4th Int. Conf. Soil Mech. & Found. Eng.* Vol. 1, 35–39
Gibson, R. E. (1967). 'Some Results Concerning Displacements and Stresses in Non-homogeneous Elastic Half Space', *Géotechnique.* Vol. 17, No. 1, 58–67
Kummeneje, O. and Eide, O. (1961). 'Investigation of Loose Sand Deposits by Blasting', *Proc. 5th Int. Conf. Soil Mech. & Found. Eng.* Vol. 1, 493
Meigh, A. C. and Nixon, I. K. (1961). 'Comparison of *in situ* Tests for Granular Soils', *Proc. 5th Int. Conf. Soil Mech. & Found. Eng.* Vol. 1, 499
Meyerhof, G. G. (1956). 'Penetration Tests and Bearing Capacity of Cohesionless Soils', *J. Soil Mech. & Found. Div., A.S.C.E.* Vol. 82, No. SM1, Proc. Paper 866
Peck, R. B. & Bazaraa, R. S. (1969). 'Discussion of the Paper by D'Appolonia *et al.*', *J. Soil Mech. & Found. Div., ASCE.* Vol. 95, No. SM3, 905–909
Schmertmann, J. H. (1970). 'Static Cone to Compute Static Elastic Settlement over Sand', *J. Soil Mech. & Found. Div., A.S.C.E.* Vol. 96, No. SM3, 1011–1043

Schmertmann, J. H. (1970). 'Suggested Method for Screw-plate Load Test. Special Procedures for Testing Soil and Rock for Engineering Purposes'. *A.S.T.M.*, *S.T.P.* Vol. 479, 81–85

Schultze, E. and Melzer, K. J. (1965). 'The Determination of the Density and the Modulus of Compressibility of Non-cohesive Soils by Soundings', *Proc. 6th Int. Conf. Soil Mech. & Found. Eng.* Vol. 1, 354–358

Shirokov, N. V., Solomin, V. I., Chereminik, V. A., Malyshev, M. V. and Zaretsky, Y. K. (1970.) 'A Circular Rigid Plate on a Non-linearly Deforming Base', *Proc. 4th Budapest Conf. Soil Mech. & Found. Eng.* 754–764

Som, N. N. (1968). *The Effect of Stress Path on the Deformation and Consolidation of London Clay*. PhD Thesis, University of London

Taylor, D. W. (1947). *Fundamentals of Soil Mechanics*, Wiley.

Terzaghi, K. and Peck, R. B. (1948). *Soil Mechanics in Engineering Practice*, Wiley.

Timoshenko, S. and Goodier, J. N. (1951). *Theory of Elasticity*. McGraw-Hill

Webb, D. L. (1970). 'Settlement of Structures on Deep Alluvial Sandy Sediments in Durban, South Africa', *Proc. Conf. on in situ Investigations in Soils and Rocks*. London, pp. 133–140

I/5. Settlements of two buildings supported on rafts: comparison with predicted settlement calculated from static cone penetrometer data

J. Gielly
Professor of Soil Mechanics, Institut de Technologie de Lyon

P. Lareal
Professor of Soil Mechanics, Institut National des Sciences Appliquées de Lyon

G. Sanglerat
Chief Engineer, Socotec—Professor of Soil Mechanics, Ecole Centrale de Lyon

Summary

The behaviour of two buildings supported on raft foundations are discussed:

(1) A twenty-two-storey building in Colombia with a rigid raft 16×22 m supported on recent sandy and silty marine deposits.
(2) A fourteen-storey-structure on loess, in Lyon.

The settlement predictions for the two buildings were based on cone penetration test data obtained with Gouda apparatus (Delft cone) and the Andina device. The measured settlements indicate the merits of these methods.

Introduction

The authors give the results of two settlement predictions based on static cone penetrometer tests. In each case the in situ tests and geotechnical data used were those made initially for the projects.

The static penetrometer tests were analysed by the methods suggested by the authors, Gielly (1969) and Sanglerat (1972) and by Schmertmann (1970).

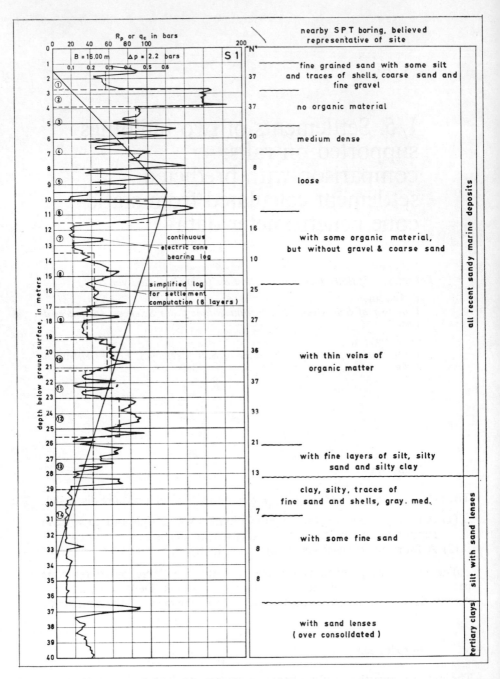

Fig. 1. Soil profile

The results were compared with estimates based on the results of oedometer tests (Lyon). In every case it was possible to make a comparison with the observed building settlements.

Twenty-two-Storey Building, Cartagena, Colombia

Description

The site is located on a peninsula between Cartagena Bay and the Caribbean. The building is supported on a mat founded at a depth of 1.5 m and 16 × 20·5 m in area.

The subsurface soil conditions from the surface and to a depth of 28–29 m consists of recent marine sands over an 8–9 m thick layer of normally consolidated marine silty clay and a basement soil, several hundreds of metres

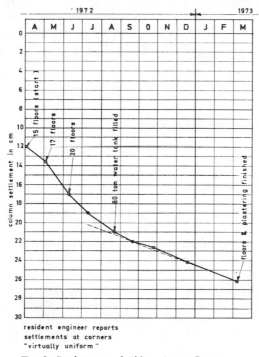

Fig. 2. Settlements—building site at Cartagena

thick of highly overconsolidated, stiff clay, similar to a London Clay. The soil profile is given in *Fig. 1*.

Reference settlement points were installed on the four corners of the mat as soon as its construction was completed. Settlement readings were made systematically after the structure reached fifteen storeys. The settlement curves are shown in *Fig. 2* and give the data for a period of eight months after the end of construction. Settlement magnitudes on the four corners are almost identical.

39

The average foundation pressure was 2.2 bar. S.P.T. and static penetrometer tests (Delft cone) were performed.

Method of analysis

Schmertmann's method uses the various layers for which the cone resistance is constant. These are shown in *Fig. 1*. The values of E_s, the vertical modulus of elasticity, were taken to be equal to two times the static cone resistance for sands. Below a depth of 28 m, the soil is clayey and this correlation between E_s and cone resistance is no longer applicable. However the relationship was assumed to apply in view of the fact that at this depth, the increase in pressure was less than 1/10 of the surface load, therefore causing small settlements.

The calculated settlement at the end of construction was 21.7 cm, and taking into consideration a certain lateral creep of the sand in accordance with the method proposed by Nonveiller (1963) the settlement after one year was calculated to be 26.1 cm. The values are in very good agreement with the observed settlements: 21 cm at the end of construction and 26.2 cm six months thereafter. This example indicates the validity of the method which is the only applicable one for this type of soil in which undisturbed sampling is extremely difficult.

Champvert buildings in Lyon

Site and structure

Two identical fourteen-storey apartment buildings of reinforced concrete construction, founded on a mat 17.15 $\times$ 25.25 m in plan dimension (0.50 m thick), were erected on a hillside, in 1970. The hill consisted of a glacial moraine covered by loess of variable thickness.

The two towers, 50 m apart were constructed with the mat of tower A 5 m lower than that of tower B. The average pressure on the foundation was 1.5 bar.

No soil investigation was performed prior to construction. A few years before, the developer had constructed several structures of six- to eight-stories in the vicinity. These had been built on mat foundations on loess and had performed satisfactorily. He concluded that the same approach would be reasonable for the new construction.

Work started at the level of the footing for tower A. An excavation made in the hillside for the construction of the mat revealed the presence of moraine in the south portion, after having removed the loess to a depth of 6 m. In the north part only loess was exposed.

The contractor became aware of the possibility of differential settlements that such conditions could cause and the authors were called in for advice. Penetrometer tests were made and undisturbed samples were taken for laboratory tests. This investigation showed that the loess in the north part was 12 m thick.

The problem to be solved was to determine if the mat foundation could be

Table 1

Depth	Soil properties					Atterberg's limits			Compressibility		
	W %	γ t/m³	γd t/m³	S %	$<100\mu$ %	W_L	W_p	I_p	e_o	C_c	c_r cm²/s
2.20	20.25	1.88	1.56	77	88	25.2	17.2	8	0.742	0.233	3.76×10^{-3}
3.70	18.65	2.11	1.78	100	77	27.6	19.1	8.5	0.552	0.143	5.65×10^{-3}
5.50	22.55	2.07	1.69	100	88	34.1	21.1	13	0.645	0.136	2.62×10^{-3}
1.80	19.15	2.04	1.71	92.5	84				0.611	0.139	1.03×10^{-2}
4.90	23.65	1.99	1.61	97.5	90	34.0	20.35	13.65	0.704	0.184	1.59×10^{-2}
7.40	20.45	2.00	1.66	91.5	77				0.613	0.170	1.32×10^{-2}

With $\gamma_s = 2.65$ t/m³.

constructed as designed or if a reduction in the building height should be recommended or if construction should be stopped altogether.

The soil investigation took in the location of tower B. The cover of loess was found to be uniform (12 m throughout the site. Some of the properties of the samples of loess recovered are shown in *Table 1*.

Calculated v observed settlements

Settlements were calculated by the authors for the two towers A and B, from the results of the penetrometer tests, Gielly (1969) and Sanglcrat (1972), and from the results of oedometer tests. Another check using Schmertmann's method was applied using the α coefficient proposed by the authors. Settlements of Tower B calculated by three methods are shown in *Table 2*.

Table 2

Method	Oedometer	Authors	Schmertmann	Observed
Settlement cm	27.5	11.5 to 28.5	11.5 to 29.5	~15

It should be noted that the static cone resistances R_p recorded for the loess were between 13 and 20 bar, across the area of tower B. At the location of tower A, the cone resistances of the static penetrometer were higher ($25 < R_p < 40$ bar) and the total settlement of the north side was estimated to vary from 6 to 12 cm, based on data from the static penetrometer.

Although settlement measurements were requested to be made from the start of construction, they were not initiated, as is often the case, until construction was almost completed.

Assuming that the moraine, being only slightly compressible, would however settle 1 cm, led to the conclusion that differential settlements of the order of 5 to 11 cm could be expected. These are rather large but not excessive for a rigid reinforced concrete building frame. The construction of the fourteen-storey building was allowed to continue. The settlement measurements were made too late to be of value. However the measurements made

41

on the vertical faces of the structure in June 1973, indicate that the walls are out of plumb by 15 to 16 cm.

This gives a differential settlement along the north-south axis of 8 cm. It can also be deduced that the total uniform settlement of tower B was of the order of 15 cm.

Conclusion

The two cases presented confirm the validity of the method of estimating settlements, based on the results of the static cone penetrometer.

Acknowledgement

The authors thanks are due to M. Liems for having provided the data for the building in Cartagena (Colombia) and for permission to include it in this Paper.

REFERENCES

Gielly, J., Laréal, P., and Sanglerat, G. (1969). 'Correlation Between *in situ* Penetrometer Tests and the Compressibility Characteristics of Soils', *In situ Invest. in Soils and Rocks*, British Geotechnical Soc. 167, 172, 189, 191. London

Nonveiller, E. (1963). 'Settlement of a Grain Silo on Fine Sand', *Proc. European Conf. Soils Mech. & Found. Eng.* Vol. 1, 285, 299, Wiesbaden

Sanglerat, G., Laréal, P. and Gielly, J. (1972). 'Le Pénétromètre Statique et la Compressibilité des Sols', *Annales I.T.B.T.P.* No. 298, October. Paris

Schmertmann, J. H. (1970. 'Static Cone to Compute Static Settlement over Sand', *J. Soil Mech. & Found. Div., A.S.C.E.* May, No. SM3. 1011–1043

I/6. Loading tests and settlement observations on granular soils

J. F. Levy
Greater London Council

K. Morton
Ove Arup and Partners, London

Summary

This paper describes the results of loading tests on plates and footings carried out on granular soils on 12 sites in the London area. It continues with settlement observations made on two structures supported on pad foundations on granular soils.

The paper concludes with the suggestion that, in general, allowable bearing pressures tend to be over-cautious. It also suggests that the results obtained may lead to a basis for predicting, within limits, the settlement of structures on normal spread footings and hence, by varying design pressures according to the width of footing, to confine differential settlements in a structure within acceptable limits.

Introduction

The loading tests described in this paper were undertaken either by the Structural Engineering Division of the G.L.C.'s Department of Architecture or by Ove Arup and Partners.

In almost every case these tests were carried out to prove that the bearing pressures proposed on the basis of the results of normal site investigations were justified. Settlement readings on actual structures are available from only one site where plate bearing tests were undertaken. However, detailed settlement observations were made on two other structures.

43

Loading tests

Details relating to the loading tests—soil data, plate size and depth below original ground level—are given in *Table 1*. The load was applied either to steel plates or to concrete blocks cast in the ground by means of hydraulic jacks acting against kentledge or tension piles. Tests carried out on different

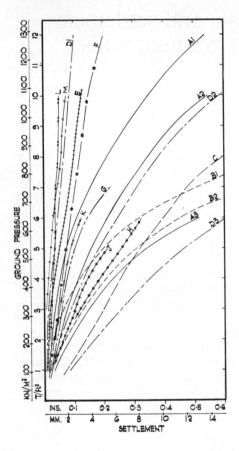

Fig. 1. Ground pressure/settlement curves

sized plates on any one site were all located within a small area of the site in order to obtain, as far as possible, identical soil conditions.

The *N* values given in the table are the average values most closely related to the location and depth of the test. Ground pressure/settlement curves for each of the tests are shown in *Fig. 1*.

Table 1

Ref.	Location	Site conditions (below plate level)	Depth of plate below G.L. (m)	N value (mean)	Plate size (square) (mm)	Comments
A1 A2 A3	Bermondsey	6.10 m sand and gravel over W.R.B.	1.22 1.22 1.22	30 30 30	1 ft (305) 2 ft (610) 3 ft (914)	Overburden not replaced S.W.L. 3.66 m below plate level
B1 B2	Poplar	1.52 m gravel and sand over London Clay	3.96 3.96	40	1 ft (305) 1 ft (305)	Overburden not replaced S.W.L. 0.30 m above plate
C	Finsbury	4.11 m gravel and sand over London Clay	0.46	50	4 ft (1219)	Overburden not replaced Borehole dry
D1 D2 D3	Deptford	6.10 m gravel and sand	3.05 3.05 3.05	20 20 20	1 ft (305) 2 ft (610) 3 ft (914)	3.05 m overburden replaced. S.W.L. 0.91 m below plate
E	Woolwich	4.88 m fine sand over 1.52 m fine silty sand over chalk with flints	1.22	80	1 ft (305)	Overburden not replaced Borehole dry
F	Stepney	3.2 m sand and gravel over London Clay	2.90	80	2 ft (610)	Overburden not replaced Borehole dry
G	Stepney	2.13 m sand and gravel over London Clay	2.44	20	1 ft (305)	Overburden not replaced S.W.L. 1.52 m below plate test site submerged by rain on previous day
H	Southwark	1.52 m sand with some gravel over W.R.B.	3.66	50	4.5 ft (1372)	Concrete footing overburden not replaced S.W.L. 1.52 m below plate
J	Southwark	3.35 m sand and gravel over W.R.B.	1.22	20	3 ft (914)	Concrete footing Overburden not replaced S.W.L. 1.83 m below plate
K	Poplar	4 m sand and gravel over London Clay	2.44	50	1 ft (305)	Overburden not replaced Borehole dry
L	Southwark	7.62 m gravel and sand over London Clay	1.52	20	1 ft (305)	1.52 m overburden replaced S.W.L. 3.35 m below plate
M	Tower Hamlets	2 m sand and gravel over London Clay	4.27	25	1 ft (305)	Overburden not replaced Boreholes dry

Settlement observations on structures

Arts & Commerce Building, University of Birmingham

This building is located on the campus of the University of Birmingham. It consists of two twelve-storey towers, rising through a two-storey podium, connected by a services tower.

Soil conditions.—The Geological Survey for the area shows the succession beneath the site to be Upper Mottled Sandstone, Keuper Sandstone (Trias) and Glacial Drift. A site investigation confirmed the Glacial Drift to be predominantly sands and gravels to a depth of about 12 m overlying Lower Keuper Sandstone. Cone penetration tests in the gravel gave values of N ranging from 22 to 46. Details of one of the boreholes are given in *Fig. 2.*

Structure.—The twelve-storey towers each 10×28 m in plan were designed to produce maximum flexibility of the use of each floor area.

To achieve this the structure was designed around two-storey-height concrete girder frames at every third floor-level. The floor units span the top and bottom flanges of the girder and a third floor is supported on columns rising from the girder. The structure is stiffened against wind by X-wall frames. The towers are each supported by four pairs of columns. The service tower is an independent concrete wall structure laterally restrained against wind. Details of a typical floor and the foundation plan are shown in *Fig. 3.*

Foundations.—Each pair of columns stands in a 7×4 m pad foundation on dense gravel about 5 m below original ground level. The total load on each column is 14 000 kN and produces a gross pressure of 500 kN/m² on the foundations. In view of the extremely stiff superstructure it was considered that only limited differential settlements could be tolerated without producing unacceptable stresses. Allowance was therefore made for jacking by leaving recesses in the starter blocks of the columns to take 0.5 m Freyssinet flat jacks if necessary.

Settlement predictions.—Estimates of probable settlement were made at design stage. These were based on a pro-rata approach to the Terzaghi and Peck (1964) Design Chart. However, consideration was given to the modification of the S.P.T. values by the method suggested by Gibbs and Holst (1957). These predictions gave a maximum settlement beneath the pad foundations of 20 mm.

Settlement observation.—It was decided that jacking procedure would be adopted if the angular distortion reached 1 : 2000. In order to check this reference, points were set up on the columns of each block. The reference points were Ordnance Survey flush-fitting benchmarks. The points were related to a benchmark on an existing structure about 100 m from the towers. Readings were taken with a Wild N3 level and an Invar staff capable of an accuracy of 0.3 mm.

Readings were commenced in October 1967 when the structure was at

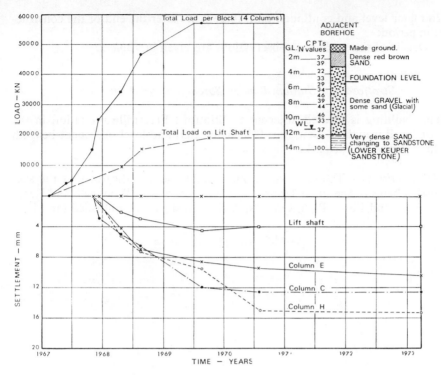

Fig. 2. Load/time settlement record, Birmingham

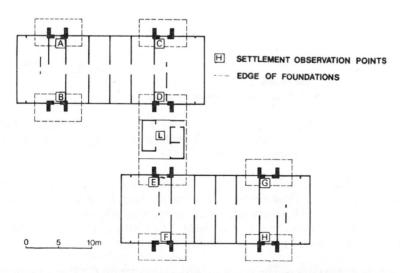

Fig. 3. Arts and Commerce Building—plan of a typical floor and foundations

47

2nd floor level and continued until three years after the end of the construction period.

Details of the settlement observations are given in *Table 2.*

Stratford Bus Station, London Borough of Newham

This building is located adjacent to Stratford East railway station in the London Borough of Newham. It includes a ground-level bus depot beneath three storeys of car park.

Soil conditions.—The Geological Survey of London indicates that the succession beneath the site is Upper Chalk, Thanet Sands, Woolwich and Reading Beds and Flood Plain Gravels. A site investigation confirmed the Flood

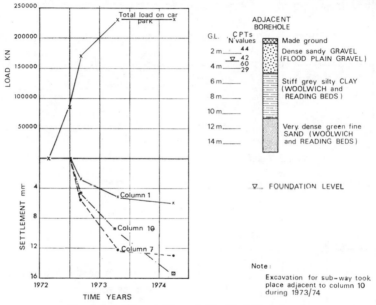

Fig. 4. Load/time settlement, Stratford East, London

Plain Gravel to a depth of 5 m overlying the Woolwich and Reading Beds. Cone penetration tests on the gravel gave values of N ranging from 13 to 69. Details of one of the boreholes is given in *Fig. 4.*

Structure.—The bus station is a simple *in situ* concrete slab, beam and column structure. There are three lines of columns carrying loads of between 1100 and 2700 kN each. These are supported on isolated pad footings at a depth of about 3 m below original ground level. The width of footings is 3 m and 2 m for the centre and perimeter columns respectively.

Settlement predictions.—The proposed gross pressure on the foundations was 300 kN/m² and an estimate of the probable settlements of the footings was made. Estimate of the settlement of the terrace gravel was made by the method

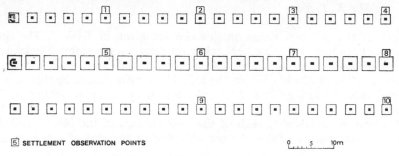

Fig. 5. Stratford Bus Station—foundation plan

suggested by D'Appolonia *et al.* (1970), using an S.P.T. value of 40. Predictions of the elastic and consolidation settlements of the underlying Woolwich and Reading clays were also added.

It was estimated that the final settlement of the interior columns would be 18 mm and for the exterior 10 mm.

Settlement observations.—Settlement observation points were set up by casting brass sockets into a number of columns at ground floor level. The position of these points is shown on the plan of foundations (*Fig. 5*). Readings were taken with a Wild N3 level and invar staff capable of an accuracy of 0.3 mm and were related to three independent benchmarks on structures approximately 50 m from the car park. Observations were commenced in June 1972 when the structure was at 1st floor level and the last reading was taken when about 95% of the dead load of the structure had been imposed. Details of the readings are given in *Table 2*.

Table 2. SETTLEMENT OBSERVATIONS

Arts and Commerce, Birmingham University

Date of reading	Settlement point numbers (settlements in millimetres)									Mis-closure	Stage of construction
	A	B	C	D	E	F	G	H	L		
4 Oct. 1967	0	0	0	0	0	0	0	0	0	0.22	2nd floor
9 Nov. 1967	0.3	0	3.0	0.9	1.2	0.3	0	0.9	0	0.18	3rd/4th floor
13 Mar. 1968	2.7	4.3	4.9	4.0	4.2	1.8	6.4	5.2	2.2	1.80	6th/7th floor
7 July 1968	5.5	6.1	6.4	5.5	7.0	7.3	8.8	7.3	3.0	0.62	9th/10th floor
17 July 1969	7.6	7.6	11.9	7.9	8.5	10.7	9.5	9.5	4.6	1.50	Complete
8 July 1970	11.0	11.3	12.5	9.4	9.4	11.9	12.5	15.0	4.0	N.C.	Occupied
28 Feb. 1973	12.3	12.0	12.6	11.1	10.4	12.0	12.9	15.3	4.0	N.C.	Occupied

Stratford Bus Station

	1	2	3	4	5	6	7	8	9	10		
7 June 1972	0	0	0	0	0	0	0	0	0	0	2.2	1st floor
11 Aug. 1972	2.7	4.0	3.9	4.2	3.1	5.1	5.6	4.9	4.4	4.6	0.7	2½ floor
15 Mar. 1973	5.1	7.1	8.6	7.5	8.6	11.6	12.2	11.8	9.0	9.4	0.2	95% D.L.

N.C.—Not closed

49

Site K, Poplar

Details of the soil conditions on this site are given in *Table 1*. The only settlement readings available were taken on three twenty-storey blocks with raft foundations, approximately 20 m square and applying an average pressure of about 220 kN/m² on the soil. These rafts cannot be regarded as normal spread footings and the results are given, without comment, for information.

Readings were taken at each of the four corners of the blocks and the following settlements (*Table 3*) were observed at the end of construction:

Table 3.

	Block 1 mm	Block 2 mm	Block 3 mm
A	22	32	32
B	22	32	32
C	22	32	29
D	16	32	32

Levels were read in hundredths of a foot and have been converted to the nearest millimetre

Discussion

Loading tests

The results obtained from the loading tests lead to the following general observations:
 (1) As might be expected, on any one site the magnitude of the settlements was clearly related to the size of plate, i.e. the larger the plate the greater the settlement for a given value of ground pressure.
 (2) For similar sized plates, there was an appreciable scatter of results from site to site.
 (3) The relationship between settlement and N value was quite random.
 (4) Despite the three preceding points, settlements were relatively low, even in the sandier soils.

Settlement of Structures

Arts & Commerce Building, Birmingham.—The settlement observations on this structure indicate the following main features:
 (1) The maximum observed settlement of large pad footings subject to pressure of 500 kN/m² is only 15 mm.
 (2) About 80% of the settlement had occurred by the end of construction.
 (3) In the case of a more lightly loaded footing (lift shaft) the equivalent settlement is proportional to the applied load.
 (4) The maximum differential settlement (5 mm) is about 30% of the maximum observed.

(5) The distortion to the structure, which is extremely stiff, is negligible for one tower and takes the form of a small tilt to an angle of 50 seconds on the other.

(6) From extrapolation it seems that about 3 mm of settlement took place before observations were begun. As settlement now appears to be complete the largest actual total settlement is thus 18 mm. This shows very good agreement with the predicted value of total settlement (20 mm).

Stratford Bus Station.—In this case construction has only recently been completed but bearing this in mind the following features seem apparent:

(1) The maximum observed settlement at the end of construction of a 3 m square footing subject to a pressure of 300 kN/m² is 12 mm.

(2) The larger central (3 m wide) footings have settled approx. 25% more than the 2 m wide edge footings.

(3) The maximum differential settlement (7 mm) is about 60% of the maximum observed.

(4) The differential settlement appears to be fairly random and the maximum angular distortion is of the order of 1 : 4000.

(5) In order to make a comparison between the actual and predicted settlements it is necessary to make allowances for the initial settlement of the structure due to the 1st floor loading and the likely subsequent consolidation settlement that will continue. It is estimated that these could amount to 6 and 12 mm for the exterior and interior columns respectively. If these assumptions are correct the largest actual settlements would be 14 mm for the exterior and 22 mm for the interior columns. This suggests that the method of prediction under-estimated the settlement by between 20% and 40%. However, it is considered that the estimate of the order of magnitude of settlement was sufficiently accurate for design purposes.

General

In an attempt to correlate the behaviour of different sizes of plates and footings, values of settlement/width of footing expressed as a percentage were plotted against ground pressure. The result is shown in *Fig. 6*, on which are shown also the equivalent values for the results obtained from the two structures under observation. With the upper and lower limit lines as drawn and up to a ground pressure of 5 tonf/ft² (535 kN/m²) the only values to fall outside the lower limit line are those derived from curves B1 and B2 of *Fig. 1*. At 4 tonf/ft² (428 kN/m²) only those related to B2 fall outside the lower limit line.

Conclusion

From the results obtained it seems evident that ground pressures used in the design of spread footings are often very conservative. *Fig. 6* has been based on the results described in this paper only, but nevertheless seems to suggest

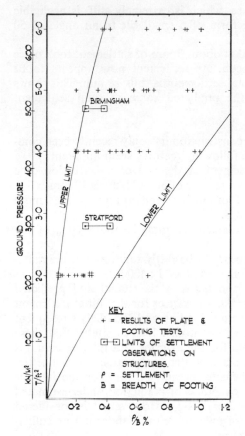

Fig. 6. Relation between pressure and settlement/width of foundation

a possible line of approach to the design of spread footings based on settlement criteria. It would be interesting if other results which may be available were found generally to lie within these limits.

Acknowledgements

The authors are indebted to the G.L.C.'s Department of Architecture for permission to publish the results of the plate tests. Thanks are also due to the Burser of the University of Birmingham and to the London Borough of Newham for permission to publish the results of settlement observations. The authors are grateful to colleagues in the G.L.C.'s Structural Engineering Division and Ove Arup and Partners for their help and in particular to Mr. J. Mitchell and Mr. J. Hopkins of Ove Arup who initiated the surveys of the buildings.

REFERENCES

Terzaghi, K. and Peck, R. B. (1964). *Soil Mechanics in Engineering Practice.* Wiley
Gibbs and Holtz (1957). 'Research on Determining the Density of Sands by Spoon Penetration Testing', *Proc. 4th Int. Conf. Soil Mech. & Found. Eng.* London, Vol. 1, 35
D'Appolonia, D. J., D'Appolonia, E. and Brisette R. F. (1970). 'Discussion of "Settlement of Spread Foundations on Sand" ', *J. Soil Mech. & Found. Div., A.S.C.E.* March

I/7. Settlement of experimental houses on land left by opencast mining at Corby

A. D. M. Penman
Building Research Station, Garston, Watford, Herts

E. W. Godwin
Corby Development Corporation, Corby, Northants

Summary

Ironstone workings at Corby, mine a seam 15 to 30 m below the surface by opencast methods. The seam is overlain by weak oolitic rock and boulder clay in layers of about equal thickness. This overburden is stripped by large walking draglines and left as it is dumped from their buckets. The irregular piles have been levelled by bulldozer and covered with topsoil for agricultural use.

Some of this land is required for buildings, and as an experiment a group of 24 two-storey semi-detached houses was built in 1963. Four different types of foundation were used. Provision was made to measure settlements near foundation level at four positions on each pair of houses and measurements were made during construction and subsequently. Settlement records to date and the types of damage caused to the houses are given in the paper.

Two causes of settlement in the loose fill are creep, proportional to log time, and partial inundation. These effects have been studied by laboratory tests in a large oedometer.

Introduction

The use of opencast mining to win minerals leaves areas of loose, dumped backfill. At Corby iron ore has been won by opencast methods for about 30 years. Large walking draglines were developed for this work and when one was built with a 92 m (300 ft) long boom supporting a 23 m³ (30 yd³) bucket, it was the largest in Britain. These machines strip the overburden from the seam of ironstone and dump it in an area where the economic thickness of the ironstone has been removed. This process tends to invert the natural sequence of the overlying strata and leaves the surface as heaps and hollows which used to be planted with trees. This aforestation hid the unsightly appearance of the ground, but in later years it was not easy to recover the trees as a source of timber.

Regulations, under the Town and Country Planning Act, required the land to be re-instated as agricultural land and since about 1946, this has been done by moving the hills into the hollows with earth-moving machinery and spreading a layer of topsoil on to the smooth surface. To judge by subsequent crops, this produced excellent agricultural land. As a considerable area of such restored land however is included within the designated area of Corby New Town it will be necessary, in due course, to use it for building purposes. It is commonly thought that if loose backfill is left for a sufficient period of time, it will complete its consolidation and become able to support two-storey dwellings without excessive settlement. To check on the behaviour of an area of loose backfill at Corby after a limited time since reinstatement, a group of 24 houses was built as an experiment and the purpose of this note is to describe this experiment.

Loose dumped backfill

Settlement measurements on loose dumped rockfill, which at one time was used for embankment dams, have shown that this material continues to settle for a very long time. Dix River Dam, 84 m high, built in 1925 from dumped limestone, and Salt Springs, 100 m high, built from dumped granite in 1931 are still settling at a rate of about 0.01% of their height a year. These rates have decreased with time, but when settlements are plotted against log time, as in *Fig. 1*, they suggest that settlement may continue for a considerable period.

At one time it was thought that settlement was caused by fines initially holding apart the larger pieces of rockfill and sluicing was introduced as a method of washing the fines from between the points of contact of the larger rocks. Terzaghi (1960) pointed out the error of this assumption and showed that settlement was caused by crushing at the points of contact due to the extremely high stresses developed when only large rocks were used for the fill. Wetting reduced the strength of the rock and increased contact crushing, so increasing constructional settlements and reducing post construction settlement. Clearly the use of a well-graded material can reduce the contact stress by increasing the number of points of contact and so reduce settlement. This principle is used for modern rockfill which is watered to make it workable, and compacted in layers by heavy vibrating machinery to produce a material of low compressibility.

At Corby the overburden consisted of boulder clay overlying weak oolitic rock in layers of about equal thickness. In the area for the experimental houses, the stripping was done in two stages, so that the oolitic rock was dumped at the bottom, with boulder clay on top to retain the natural sequence. The hills and hollows left by the dragline were levelled and topsoil

54

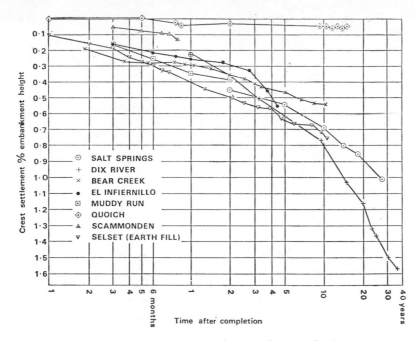

Fig. 1. Post-construction crest settlements of some embankments

spread nine years before the experimental houses were built. The average depth of backfill under the houses was 18 m (60 ft) and the water level was kept down by drainage to the working zone until seven years after the houses were built when pumping ceased.

Design of the experimental houses

Except for the foundations, the two-storey houses were designed as traditional semi-detached pairs in cavity brickwork with ridge tiled roofs and gable ends. Four of the pairs, as shown by the plan *Fig. 2*, were built as two units of four houses, but with a special joint between the pairs which kept them completely separate. The foundations were standard strip footings or one of three types of reinforced concrete raft as indicated by *Fig. 3*. The concrete strip footings (type A) were cast in trenches 0.61 m wide and 0.84 m deep for the outer walls and 0.46 m wide × 0.78 m deep for internal walls. Cavity brickwork was brought up to dpc and a concrete floor slab cast on 0.15 m infilling. A polythene damp-proof membrane was laid under the slab and connected to the dpc in the walls.

The reinforced concrete raft type B was cast on to fill which had been stripped of topsoil. Trenches 0.30 m wide × 0.91 m deep filled with mass concrete formed an edge beam to the raft which was 0.15 m thick, reinforced by two layers of fabric reinforcement to B.S. No. 125. Type C raft was the same except that the edge beam was made only 0.30 m wide × 0.61 m deep and was reinforced. Type D had reinforced edge beams 0.30 m wide × 0.98 m

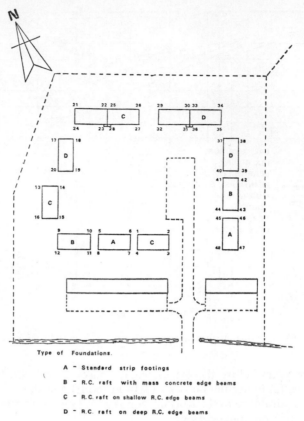

Type of Foundations.

A − Standard strip footings

B − R.C. raft with mass concrete edge beams

C − R.C. raft on shallow R.C. edge beams

D − R.C. raft on deep R.C. edge beams

Fig. 2. Corby-experimental house foundations

deep. In both cases the edge beam had a projecting toe carrying the outer brick skin.

Brass reference points were built into the brickwork below dpc at the four corners of each pair of houses, as shown in *Fig. 2*. There were 48 points and their level was related to a bench mark on an old stone farm building away from the mining area by precise surveying.

Settlement and damage records

Settlement measurements were begun as soon as the brass reference points had been installed and the first readings were taken on 9 July 1963. The houses were virtually complete in November and by 15 January 1964 all building work was completed and block 5, 6, 7, 8 was occupied. First cracks, associated with the special joint, were seen in block 29–36 on Christmas Eve 1963. These houses had not been occupied and once the houses were occupied internal cracks could only be noticed when reported by the householders. A few plaster cracks were reported in 1965 but it was difficult to separate those which might have been caused by differential movements from those due to humidity and thermal strains.

56

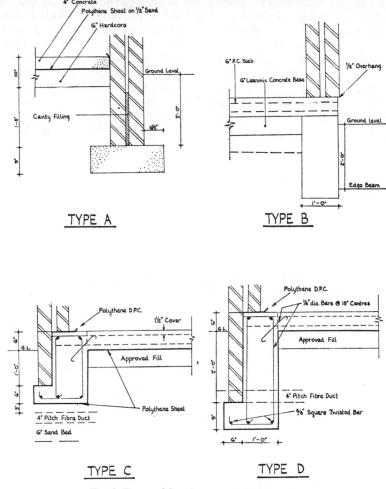

Fig. 3. Types of foundation used for the houses

Block 29–36, with a reinforced concrete raft foundation type D, has had the third largest settlement of all the houses. The maximum settlement occurred at building 37–40 on the same type of foundation and its settlement curves are shown in *Fig. 4*. No special cracking was reported and there were no visible external cracks in the brickwork of this building.

The rate of settlement established by the end of construction appears to have continued until mid 1964 and then decreased until 1968: subsequently the movements have been small. There has been a slight increase in the rate of settlement since July 1970 when pumping was stopped. This general pattern has been shown by all the settlement curves.

Maximum differential settlement occurred at houses 45–48 on traditional strip foundations, type A (see *Fig. 5*) and the houses forming block 29–36 suffered the second largest differential movement. Detailed surveys were made of these houses in November 1967, April 1968 and September 1971. During

57

the first survey there were some diagonal stepped cracks following the brick-work from door and window openings: the widest was less than 2 mm. In addition there were some general areas of fine hair cracks. The gable wall between points 47 and 48 suffered a differential settlement of 0.13 m (0.44 ft) in a distance of 8.45 m (23 ft 7 in) (1 in 54 tilt). It remained completely intact: the movement caused the roof gutters to discharge over point 48 rather than through the downpipe at the other end of the building.

During the second survey the vertical tilt of the wall was found to be 64 mm (2½ in) in a height of 4.9 m (16 ft) (1 in 77 tilt). There were some small additional cracks and the greatest width was about 2 mm. During the third

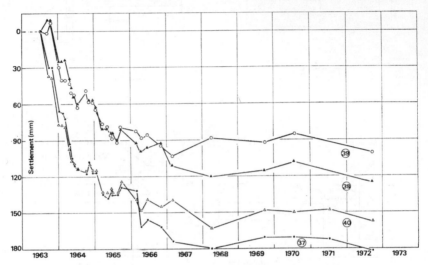

Fig. 4. Settlement curves for house 37–40 (*maximum observed*

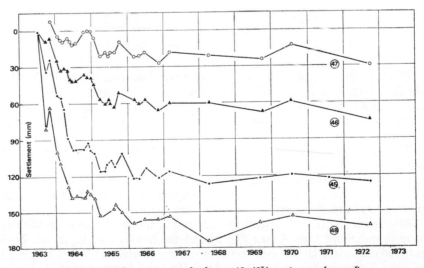

Fig. 5. Settlement curves for house 45–48 (*maximum observed*)

Table 1. EXPERIMENTAL HOUSES AT CORBY

Foundation type	Traditional strip A		R.C. raft with mass conc. edge beam B		R.C. raft with shallow R.C. edge beam C				R.C. raft with deep R.C. edge beam D			
Ref. points, Nos	5–8	45–48	9–12	41–44	1–4	13–16	21–24	25–28	17–20	29–32	33–36	37–40
Max settlement m (ft)	0.135 (0.44)	0.18 (0.58)	0.08 (0.27)	0.13 (0.43)	0.12 (0.39)	0.035 (0.11)	0.045 (0.14)	0.10 (0.33)	0.055 (0.18)	0.155 (0.50)	0.135 (0.44)	0.185 (0.61)
Current differential settlement m (ft)	0.05 (0.17)	0.135 (0.44)	0.065 (0.21)	0.02 (0.07)	0.045 (0.15)	0.015 (0.05)	0.015 (0.05)	0.08 (0.26)	0.02 (0.06)	0.055 (0.18)	0.075 (0.24)	0.08 (0.27)

survey conditions were almost the same: the offset from vertical was still 64 mm and the widest crack 2 mm.

The minimum total and differential settlements occurred at houses 13–16 and the settlement curves are shown in *Fig. 6*. The foundation was a reinforced

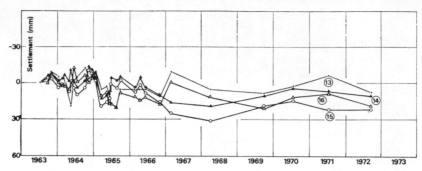

Fig. 6. Settlement curves for house 13–16 (minimum observed)

concrete raft type C. There were some hair cracks in the plaster internal finish but it was difficult to associate them with differential settlements.

Values of the maximum settlement and the current differential settlement measured at all 12 pairs of houses are shown in *Table 1*. There appears to be no direct relationship between the amount of total or differential settlement and foundation type. It is probable that settlements reflect the condition of the dumped backfill rather than the loading or foundation type.

Tests on backfill material

To get some idea of the compressibility, creep rate and effect of inundation samples of the oolitic and boulder clay backfills were tested in a 1 m dia. oedometer under vertical stresses covering the range of overburden pressure. The boulder clay was compacted lightly at its natural water content of 15%, to form a sample with a dry density of 1.43 tonnes/m³. The load was increased in two stages to a vertical stress of 120 kN/m² and left under this load for three days. A total compression of 9.5% occurred. A small head of water was applied to the base of the sample and water emerged from the top within a minute. The supply was turned off, but water continued to emerge as the sample rapidly compressed. After four days the additional compression after adding the water was 10.9%.

The oolitic rockfill was made into a loose sample, using material smaller than 125 mm. The dry density was 1.57 tonnes/m³ and the load was increased in three stages to a vertical stress of 350 kN/m². After three days under this load, the total compression was 9.8%. Water was then added, as before, and after a further three days the compression had increased by 3.7%.

Creep rate can be expressed (Sowers *et al.*, 1965) as the slope α of the settlement v logarithm of elapsed time curve. With the settlement expressed as a percentage of the fill thickness the value for α for the rockfill was 0.12. This was obtained during the three-day period when the load was kept constant before water was admitted to the sample.

60

Discussion

Several of the settlement curves (e.g. *Figs. 4, 5* and *6*) showed some upward movement during the autumns of 1963, 1964 and 1965 and during the spring of 1967. Later measurements, made less frequently, could not show this apparent annual cycle which probably reflects seasonal movements in the boulder clay rather than simply inaccuracies in levelling. There are no large trees near the houses.

In general the maximum rates of settlement occurred immediately after construction and decreased to small rates after about four years. These movements could be due to near surface compaction of the loose dumped boulder clay under the weight of the houses. From mid 1970, when drainage was stopped, the rate has begun to increase. This may reflect a rising water surface in the fill, although, unfortunately there are no observation wells or piezometers near the site.

The structural collapse caused by wetting has been clearly shown by the oedometer tests and it has been suggested that areas of loose backfill to be used for houses could be pre-settled by inundation prior to construction in order to minimise post-construction settlements.

The fill was placed 12 years before the houses were built and although it will still be settling under its own weight, the measured creep rate can account for only a small part of the observed settlements during the 9 years since the houses were built.

The wide range of measured settlements (*Table 1*) do not reflect foundation type but indicate a general change across the site. Least settlement (block 13–16) occurred on the west side of the site and the greatest (37–40) was on the east side, with a slightly irregular distribution in between. Damage to the houses reflected differential settlement and was not connected with foundation type. The amount of damage was small compared with damage which can occur in this type of structure with shallow foundations on shrinkable soils, particularly in the presence of trees, or where there is mining subsidence. The relative magnitude of the damage at Corby may be judged by the fact that the tenants of ten of the houses have bought them recently. These ten include the two houses 45–48 which suffered the largest differential settlements and two of the houses in block 29–36 which suffered the second largest differential settlements.

Acknowledgements

This technical note is published by permission of the Director of the Building Research Establishment and the General Manager of Corby Development Corporation.

REFERENCES

Sowers, G. F., Williams, R. C. and Wallace, T. S. (1965) 'Compressibility of Broken Rock and Settlement of Rockfills', *Proc. 6th Int. Conf. Soil mech. & Found. Eng.*, Montreal. Vol 2, 561–565

Terzaghi, K. (1960). 'Discussion of Salt Springs and Lower Bear River Dams', *Trans. A.S.C.E.* Vol 125, Pt 2, 139–148

I/8. Settlement performance of a raft foundation on sand

J. G. Stuart
Senior Lecturer, Queen's University, Belfast, N. Ireland

J. Graham
Assistant Professor, Royal Military College, Kingston, Ontario, Canada

Summary

Regular settlement measurements have been taken for twelve years on the foundation raft of a thirteen-storey building belonging to Queen's University, Belfast. Site conditions consist of about 25 m of mixed glacio-lacustrine sand and gravel overlying hard boulder clay. Measured settlements have been significantly smaller than those predicted from Standard Penetration tests. Settlement has continued for nine years since completion of the structure.

Introduction

In 1959, Queen's University, Belfast, began construction of the thirteen-storey Ashby Institute building on a site consisting of fine silty glacial sand underlain by compact gravel and hard boulder clay. Most full-size foundations on sand are stable against shear failure, and settlement provides the limiting factor for design. Few published case records were available at the beginning of construction, and it was decided to monitor settlement of the building over a period of years.

The original foundation design was based on the empirical relationship given by Terzaghi and Peck (1948) between Standard Penetration resistance and allowable bearing pressure. Subsequent work by Thorburn (1963), Alpan (1964), Meyerhof (1965) and particularly by D'Appolonia *et al.* (1968), has shown that these recommendations are conservative for spread footings. Similar results have been reported for raft foundations by Meyerhof (1965) and by Macleod and Spence (1972), but the total number of available records is regrettably small. This paper describes the settlement performance of the Ashby Institute foundation raft in the nine years following construction.

Site conditions

The site is 1.5 km south of the centre of Belfast in an area which was formerly a sand pit. The prevailing ground level is 20–25 m above the river Lagan, and the site occupies a shallow pit about 3 m deep. Borings were carried out during 1959 using standard 203 mm (8 in) and 152 mm (6 in) diameter shell and auger equipment. *Fig. 1* shows typical borehole logs, one at each end of the building. The site is covered by sands and gravels which vary considerably in thickness and density, and which overlie stiff boulder clay. Occasional thin lenses of silty clay or silt also occur. The deposit is part of extensive glacio-lacustrine sands and gravels laid down in a late interglacial period.

Standard Penetration tests were taken at 1.5 m intervals using a solid S.P.T. probe (Palmer and Stuart, 1957). Results varied from 3 blows/ft to virtual refusal (*Fig. 1*). The boreholes showed no clearly defined transition from loose to dense material, but rather a general increase in density and particle size with depth. Pockets of loose sand were found below the proposed formation level in Borehole 6 (*Fig. 1*). The water table was at 12.2 m O.D. in all boreholes and was confirmed at this level in the following year.

Construction details

Preliminary designs showed that a raft offered worthwhile savings in cost and construction time over a comparable piled foundation. The average bearing pressures from a raft 55.0 m × 19.8 m in plan were 148 kN/m² dead load, plus 43 kN/m² live load. The average overburden stress at formation level was estimated to be 30 kN/m², giving a stress increase across the raft

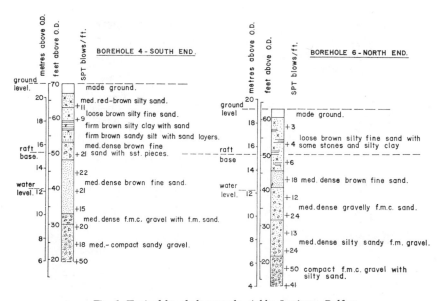

Fig. 1. Typical borehole records, Ashby Institute, Belfast

base of 161 kN/m². On the basis of Terzaghi and Peck's (1948) recommendations, it was estimated that a blow count of at least 15 blows/ft was needed if maximum settlements were to be limited to 5 cm. Because of loose sand in Borehole 6 (*Fig. 1*) further Standard Penetration Tests were proposed for the period of excavation to allow more detailed study of the density of the sand under the raft.

Nine further boreholes were driven during excavation, and 28 penetration tests were carried out. With little extra expense, these tests furnished important information on the extent of the loose sand beneath the north end of the building where readings of 6–7 blows/ft were recorded at depths of 1–2 m below formation level. The sand was excavated and replaced with weak concrete when the blow-count was less than 15. In all, some 1750 m³ of additional excavation was carried out to a maximum depth of 2.3 m below raft base under the north end of the building (*Fig. 2*). In contrast, the south end produced satisfactory values of about 20 blows/ft at this level. An inverted beam and slab was used to reduce the dead weight of the structure. The depth of the raft was 2.1 m, with main beams spanning 7.3 m and secondary beams spanning 5.8 m (*Fig. 2*). Construction of the raft and basement began in April 1960 and finished in March 1961 (*Fig. 3*). The reinforced concrete structure was completed in February 1964.

Settlement performance

Settlements were initally taken on stainless steel pegs cast into the basement floor. Later, when access became difficult, readings were taken on six pegs at ground level (*Fig. 2*) using a parallel plate micrometer precision level and invar staff. Observations are summarized in *Fig. 3*. The work was arranged

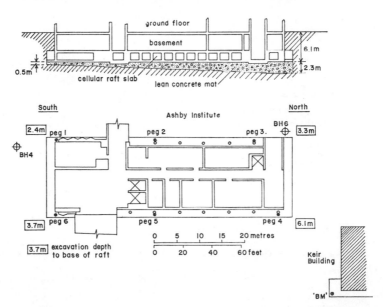

Fig. 2. Foundation details and levelling pegs, Ashby Institute, Belfast

64

in the form of closed circuits, and miscloses of more than 1.5 mm were rejected or re-examined. Readings were referred to a bench-mark on a nearby building constructed about four years earlier on piled foundations (*Fig. 2*). This mark was checked regularly against Ordnance Survey marks and was found to be rising gradually by about 1.5 mm/year. Readings were taken by university students, initially at two-week intervals, and subsequently after longer intervals as the movements slowed down.

Fig. 3 shows average settlements from Pegs 1, 6 at the south end of the structure, and from Pegs 3, 4 at the north end. All pegs rose following excavation and remained above grade during basement construction. Net settlements were observed shortly after commencement of the superstructure, and followed the loading pattern closely. Settlements increased rapidly once the structural loading exceeded the weight of excavated soil. (Later work (*Fig. 3*) suggests that the overburden stress should be larger than the assumed

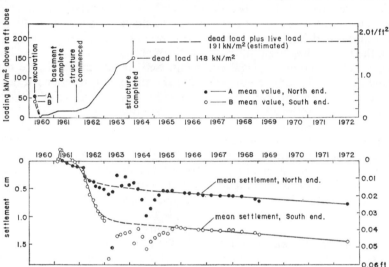

Fig. 3. Load/time settlement records, Ashley Institute, Belfast

design value of 30 kN/m^2). The rate of settlement slowed off slightly before completion of the structure. This may be due to uncertainty in the loading rate in *Fig. 3*, which was prepared from monthly progress photographs. By 1972, maximum settlements were 0.8 cm at the north end, and 1.5 cm at the south end. Both are much smaller than the predicted 5 cm. Major movement ceased at about the date of completion, although smaller movements have continued in the subsequent nine years. There has been no record of damage to interior finishes due to differential settlement of the raft itself, although some minor cracking has occurred at a flexible connecting walkway to a neighbouring pile-supported building.

Discussion

These results draw attention to the usefulness of the Standard Penetration Test as a convenient approximate method of assessing cohesionless deposits.

As reported recently by Macleod and Spence (1972), the test was used during construction to identify loose sand which could cause unacceptable settlements.

It is interesting that smaller settlements under the north end are associated with the replacement of loose sand by weak concrete (*Fig. 2*).

Despite careful planning, difficulties arose in carrying out the complete levelling programme over the full life of the structure. Much of this was during construction, and included problems of access; inadvertent displacement of pegs, and interconnection of basement and ground-floor levels. Other difficulties involved responsibility for continuity of records over a relatively long period. In this connection it is unfortunate that the period of major interest in 1963–64 coincided with a change of responsibility for the observations. Settlements in *Fig. 3* have been checked where possible, and appear accurate. The quality of the basic observation procedures is shown in the periods 1961–62 and 1966–68. Efforts were made to relate apparent oscillations in level with systematic construction sequences, movements of machinery, rainfall and high winds. No convincing correlation was found, and a smoothed curve has been interpolated through these readings.

One feature requires special comment. Settlements on sands are usually thought to be fully complete shortly after the end of construction. In this case however where observations have continued over a longer period than usual, settlement has continued at a slow but measurable rate. Although there are some thin seams of silt and clay in the profile, the drainage times are so short that consolidation settlement is an unlikely explanation. It is possible that continued settlement may result from creep movements in the clay matrix of clay–sand layers, or from strains due to cyclic loading. The design live load is 29% of the dead load of the structure.

The observed settlements confirm the trend noted for example by Meyerhof (1965) and by D'Appolonia *et al.* (1968), that the Terzaghi and Peck correlation over-estimates the measured settlements considerably. The maximum observed settlement was only 1.5 cm, and the differential settlement 0.75 cm.

A complete basement was not originally specified for this structure. In view of Meyerhof's (1965) suggestion that bearing pressures can be increased by 50% above the Terzaghi and Peck value, individual spread footings would probably have provided a more economical solution for this structure.

Conclusions

The Standard Penetration Test remains an important method for investigating foundations on sands, both during design, and later as quality control during excavation. The maximum settlement of the Ashby Institute raft after nine years was only 30% of the value predicted by Terzaghi and Peck (1948). Small settlements have continued since completion of the structure.

Acknowledgements

The authors acknowledge the co-operation of Messrs. F. B. McKee and Co. Ltd., Belfast, Messrs. Cruickshank and Sewards, Manchester, and Messrs.

Kirk, McClure and Morton, Belfast. Settlement observations were supervised by D. R. Crone, R. B. McVilly and J. C. McIlwrath, Department of Civil Engineering, Queen's university, Belfast. Loadings and construction details are from a paper by G. S. Millington and R. S. Beckett to the N.I. Association of the Institution of Civil Engineers.

REFERENCES

Alpan, I. (1964). 'Estimating the Settlements of Foundations on Sand', *Civil Eng. & Public Works Rev.* Vol. 59, Nov. 1415

D'Appolonia, D. J., D'Appolonia, E. and Brissette, R. F. (1968). 'Settlement of spread footings on sand', *J. Soil Mech. & Found. Div., A.S.C.E.* Vol. 94, No. SM3, Proc. Paper 5959, May, 735–760

Macleod, G. and Spence, R. A. (1972). 'Eighteen-Storey Hotel Supported on Granular Fill, *Proc. 25th Canadian Geotech. Conf.*, Ottawa. Dec

Meyerhof, G. G. (1965). 'Shallow Foundations', *J. Soil Mech. & Found. Div., A.S.C.E.* Vol. 91, No. SM2, Proc. Paper 4271, March, 21–31

Palmer, O. J., and Stuart J. G. (1957). 'Some Observations on the Standard Penetration Test and a Correlation of the Test with a New Penetrometer'. *Proc. 4th Int. Conf. Soil Mech. & Found. Eng.* Vol. 1, 231–236, London

Thorburn, S. (1963). 'Tentative Correction Chart for the S.P.T. in Non-Cohesive Soils', *Civil Eng. & Public Works Rev.* Vol. 58, June 1963, 752–753

SESSION II

Normally consolidated and lightly over-consolidated cohesive materials

II/1. A fifty-year settlement record of a heavy building on compressible clay

Knut H. Andersen
Civil Engineer, Norwegian Geotechnical Institute

Carl-Jacob Frimann Clausen
Civil Engineer, Norwegian Geotechnical Institute

Introduction

Oslo Jernbanetollsted is a fifty-year-old storage building situated on a deposit of normally consolidated clay in the south-eastern part of Oslo. The building is divided in two parts by a contraction joint, and the eastern part of the building is heavier than the western part. During the whole life of the building continuous settlement observations have been made.

In this paper the settlement record will be given together with a description of the soil conditions beneath the building. Results from a conventional settlement calculation are also included.

About the building

The building is six storeys high, and it covers an area of approximately 140 by 21 m. A plan and a section of the building are shown in *Fig. 1*. The foundation of the building consists of some 5000 wooden piles with a 1–2 m thick reinforced concrete slab on their tops. Each pile is 9 m long and has a top diameter of 15 cm. The bottom of the concrete slab was originally placed at a level corresponding to 0.3 m below the average sea level. Accordingly, the tip of the piles must have been situated at a level of -9.3 m.

The excavation for the building was started late in the autumn of the year 1919, and the actual construction began in the summer of 1920. In 1919 the surrounding natural ground surface was at a level of 2.2 m. The 2.5 m that was excavated was an old fill. The exact age of this fill is unknown, but it seems reasonable to assume that it was placed there some twenty years

71

earlier. During the period of construction, another 0.5 m of the fill was placed on the surrounding ground surface.

Fig. 2 shows the change in load caused by the excavation and the ensuing construction. Early in 1924 the construction was finished, and two years later the live load had reached its average operational value. As can be seen from *Fig.* 2, the live road was higher in the eastern part of the building than in the

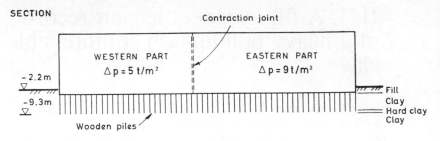

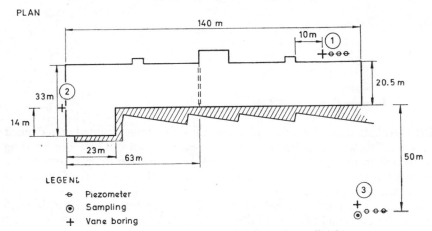

Fig. 1. Plan and section of Oslo Jernbanetollsted

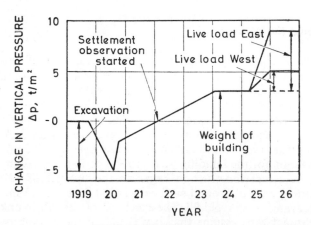

Fig. 2. Construction sequence

western part. These two parts of the building were divided by means of a contraction joint, and they can be regarded as two separate neighbouring buildings.

Because of the high values of the live loads, there can be some doubt about the best values for the net additional loads. An investigation of what kind of goods were stored in the different parts of the building gave the result that the net additional loads of 9 t/m² for the eastern part and 5 t/m² for the western part seemed to be reasonable.

Soil conditions

During the summer of 1968 the Norwegian Geotechnical Institute carried out a soil investigation for the site at Oslo Jernbanetollsted. One sampling and three vane borings were carried out, and six piezometers were installed. The locations of the different borings are shown in *Fig. 1*. Site 1 and site 2 are situated close to the building, while site 3 is located 50 m away from the southern wall. Site 3 can therefore be assumed to represent the soil conditions for natural ground. However, due to the additional fill, these soil conditions are not quite identical to the original soil conditions prior to the construction of the building.

The sampling was performed at site 3, using the NGI 95 mm fixed piston sampler. Ten 1 m long samples were obtained from a depth of 9 to 31 m. Vane borings down to approximately 30 m depth were performed at all the three sites. At sites 1 and 3, open tube piezometers were installed at 4, 12 and 20 m depth.

The boring profile which was obtained, is shown in *Fig. 3*. The soil conditions can be described in the following way:

Depth 0–3 m. Old fill consisting of gravel and stones. As mentioned previously, the lower 2.5 m of this fill was placed there prior to the construction of the building, whereas the upper 0.5 m of fill was placed during the construction period. The ground water table is situated in this layer at level 0.0 m. The unit weight of the fill is assumed to be 2.0 t/m³.
Depth 3–10 m. Sandy and silty clay with weathering in the upper 2 m. Below the weathered zone the vane shear strength is 3/4 tf/m². The unit weight is assumed to be 1.85 t/m³.
Depth 10–11.5 m. Very stiff clay of unknown origin. The undrained shear strength exceeds 6 tf/m². The water content is 27–30%, the liquid limit 40–45%, and the plastic index is thus 15–20%. The content of particles less than 0.002 mm was found to be 30%. The unit weight is 2.0 t/m³. This firm layer has been found at the same elevation in a number of other borings in the area. One hypothesis is that it is formed by resedimented slide masses of quick clay.
Depth 11.5 m. Below the 1.5 m thick crust there is found a fairly homogeneous deposit of marine clay. This deposit can be divided into two layers, the upper one being slightly more plastic than the lower one. The same phenomenon is well known from other sites within the Oslo area, Bjerrum (1967).

In the upper plastic layer, at a depth of 11.5–19 m, the vane shear

73

strength increases from 4 tf/m² to 5 tf/m². The water content is 35–40%, the liquid limit 45–55%, and the plastic limit 25–30%. The plasticity index varies from 20% to 27%.The clay content is 30–40% and the unit weight is 1.92 t/m³.

From 19 m depth down to the end of the boring at 31 m depth, clay with somewhat lower plasticity is found. The vane shear strength increases from 3 tf/m² at 19 m depth to 5 tf/m² at 35 m depth. The water content varies from 30% to 40%, the liquid limit is 38–42% and the plastic limit is 20–25%. The plasticity index is 15–20%. The clay content is about the same as in the layer above. The unit weight is 1.96 t/m³.

The depth to rock is about 80 m.

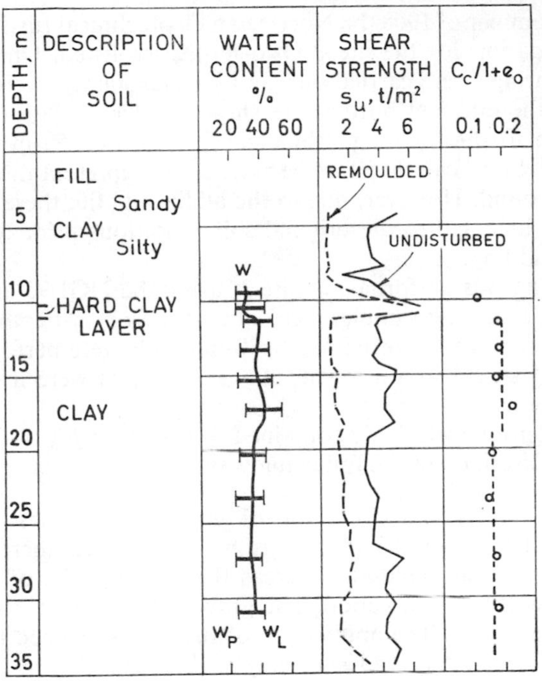

Fig. 3. Boring profile

An investigation of the salinity in the profile gave a fairly constant value of about 27.5 g/l. Comparing this salinity to the 35 g/l for sea water, it can be concluded that no significant leaching has taken place. This is also reflected in the fact that the sensitivity is 5 or less. The organic content is of the order of 1%.

The pore water pressure increases linearly with depth from 0–22 tf/m² at 20 m below sea level. The pore water pressure measurements at sites 1 and 3 indicate that today there is no excess pore water pressure beneath the building when compared with natural ground.

In order to measure the increase in the shear strength of the clay caused by the weight of the building, two vane borings were performed close to the walls as indicated in Fig. 1. Fig. 4 shows the result of these borings together with the results obtained in the boring 50 m south of the building. It can be

74

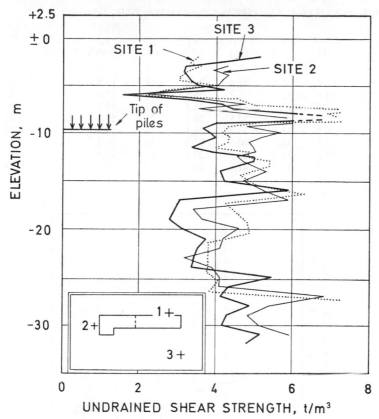

Fig. 4. *Measurement of strength increase beneath the building*

seen from this figure that the strength of the clay beneath the building has increased, and it is interesting to note that the softest clay layers get the highest increase in shear strength.

Results from consolidation tests

A total number of 32 consolidation tests were run. The samples had a cross-sectional area of 50 cm^2 and a height of 20 mm. Results from a typical consolidation test are shown on *Fig. 5*. Data from this test and some other typical tests are given in *Table 1*.

For a deposit of clay with a plasticity as found at this site, one would expect to find an apparent preconsolidation pressure, p_c, which, due to secondary settlements, has become larger than the effective overburden pressure, p'_o, Bjerrum (1967). By comparing the values of p_c and p_o for the tests listed in *Table 1*, the p_c-value is found to be surprisingly low. This is partly due to the fact that the value of p_o' listed in this table does not correspond to the value of p_o' which has been acting for a long time. It corresponds to the p_o'-value for natural ground today, including the weight of the 3 m thick layer of fill,

75

Table 1. RESULTS FROM SOME TYPICAL CONSOLIDATION TESTS.

Test no.	Depth (m)	w (%)	w_L (%)	w_p (%)	I_p (%)	p_o' (tf/m²)	p_c (tf/m²)	$\dfrac{C_c}{1 + eo}$
392	11.4	39.1	48.8	29.4	19	11.5	10	0.19
493	13.3	38.1	45.7	26.5	19	13.0	14	0.20
592	15.4	37.1	47.3	24.1	23	14.7	18	0.18
691	17.2	42.1	52.5	29.2	23	16.3	17	0.22
791	20.2	33.8	42.2	26.1	16	18.8	14.5	0.16
892	23.5	33.3	40.3	22.0	18	21.1	17.5	0.15
992	27.6	34.2	40.7	23.1	18	25.6	17.5	0.17
1091	30.7	36.4	41.6	24.1	18	28.3	22.5	0.18

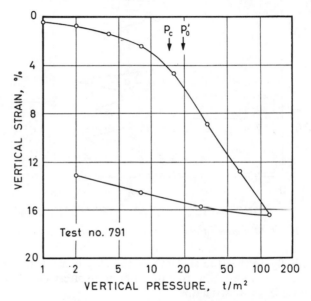

Fig. 5. Results from a typical consolidation test

The compressibility of the clay, expressed by means of $C_c(1 + e_o)$ is found to be 0.20 in average for the plastic clay layer 11.5–19.0 m in depth. For the less plastic clay below 19 m depth, the average compressibility is found to be 0.17.

The average value of the coefficient of consolidation for a pressure in the same range as the p_c-value is found to be 8.5 m²/a for the plastic clay layer and 7.3 m²/a for the underlying clay layer of lower plasticity.

Settlement observations

The settlements of the buildings have been recorded at several points since March 1922. The measurements were started when the net additional load was approximately zero, *Fig. 2. Fig. 6* shows the plotted time–settlement

76

curves for two typical observation points, one for the heavy part of the building (point B) and one for the light part (point A). In the plan of the building, the total settlements observed in June 1973 are given for the different observation points. It can be seen that up to the present the light part of the building has settled approximately 50 cm, whereas the heavy part of the building has settled approximately 70 cm. It can also be seen from *Fig. 6* that during the last forty years the time–settlement curves have been straight lines in a logarithmic plot. They still do not show any tendency to bend upwards.

The surrounding ground, which is not influenced by the weight of the

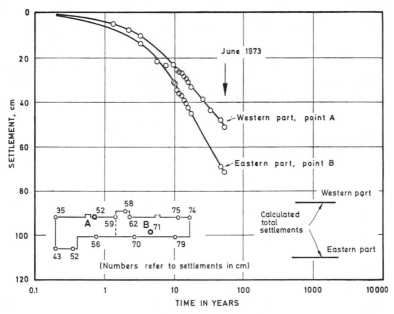

Fig. 6. Settlement observations

building, is also settling. This settlement is estimated to be about 2 mm/a at present.

No damage to the building due to the large settlements has been reported.

Calculation of total settlements

In the settlement calculations the additional loads are assumed to go down through the piles to the top of the plastic clay layer at about 11.5 m depth. The pressure distribution with depth below the tip of the piles is evaluated according to elasticity theory.

The total settlements, δ, are calculated in the conventional way by means of the equation

$$\delta = \sum \frac{C_c}{1 + e_o} \log \frac{p_o' + \Delta p}{p_o'}$$

77

In this equation, Bjerrum, 1967, the values of the compressibility $C_c/1 + e_o$) as found in the consolidation tests are used. Calculating the settlement in this way means that shear deformations are neglected. Δp is the additional vertical pressure, and p_o' is the effective overburden pressure before the additional load is applied.

The old 2.5 m thick fill which had been placed on the ground some twenty years prior to the construction of the building, complicates the settlement calculations. By using a p_o' corresponding to the conditions before the placement of the fill, the total settlement due to both the fill and the building is found. The calculated settlements amounts to 135 cm for the heavy part of the building and 110 cm for the light part of the building. However, some of the calculated settlements are caused by the weight of the fill in the time before the building was constructed, and before the settlement observations were started. In order to calculate settlements that are comparable to the observed settlements of the building, these early settlements have to be subtracted. A reasonable estimate of these settlements is believed to be 25 cm. This means that the calculated total settlements of the building will be 110 cm for the heavy part and 85 cm for the light part.

Summary

The settlement observations from Oslo Jernbanetollsted show that the heavy part of the building has settled approximately 70 cm and the light part of the building 50 cm. For natural ground the rate of settlement is today of the order of 2 mm/a. In spite of the large settlements of the building, no damage has been reported.

Investigations showed that the soil consists of an 80 m thick deposit of normally consolidated clay with a 3 m thick fill of coarser material on the top.

The fill was placed on the ground, partly before and partly during construction of the building, and this complicated the settlement calculations. The total settlements of the building are estimated to be 110 cm for the heavy part and 85 cm for the light part.

Acknowledgement

The settlement observations presented in this paper were collected by the Geotechnical Department of the Norwegian State Railways and the consulting engineers Bonde & Co. The writers are grateful to Messrs. S. Skaven-Haug, H. Hartmark and O. Folkestad for their help and the permission to publish these observations.

REFERENCES

Bjerrum, L. (1967). 'Engineering Geology of Norwegian Normally Consolidated Marine Clays as Related to Settlements of Buildings'. 7th Rankine Lecture. *Géotechnique*, Vol. 17, No. 2, 81–118
(Also published as *Norwegian Geotechnical Institute*, Publication, 71)

II/2. A comparison between the observed and estimated settlements of three large cold stores in Grimsby

B. E. Cowley
Partner, Jenkins and Potter, Consulting Engineers, London

E. G. Haggar
Partner, Jenkins and Potter, Consulting Engineers, London

W. J. Larnach
Senior Lecturer in Civil Engineering, University of Bristol

Summary

The site consists of up to 2.0 m of made ground over a considerable thickness of soft river alluvial deposits overlying boulder clay with chalk at a considerable depth. Experience and laboratory test results indicated that considerable long-term settlement was to be expected. For a number of reasons, largely economic, shallow foundations were adopted for the structures, and the floor slab was ground-supported, although the majority of larger buildings in the vicinity are supported on piled foundations. A deliberate design decision was made to make the single-storey superstructure statically determinate. Other design details ensured no distress to the floor slabs and the maintenance of the vapour tightness of the buildings even though maximum settlements of the order of 280 mm have been observed at the periphery.

The settlements have been monitored over a period of some seven years. Comparisons are made of the observed rate and amount of settlement with those predicted using the results of conventional site investigation and calculation techniques. The results indicate that whereas final settlements were estimated with tolerable accuracy, the rates of settlement increase were less well predicted.

Introduction

By their very nature cold stores are not labour intensive. On the contrary, the capital cost of construction is a significant factor in the economics of running a cold store, and there is a constant search for economy of construction. An essential requirement is that stacking of goods is not restricted and these buildings are therefore generally single-storey having large plan areas and spanned by roofs having the minimum of internal support.

Between February 1966 and September 1970 three large stores were built

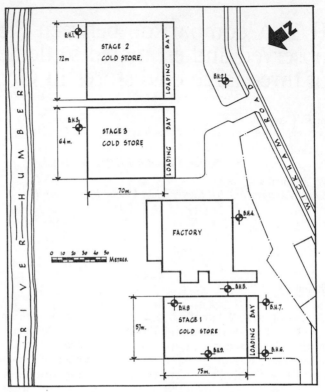

Fig. 1. Site plan

for one client on sites adjacent to the river Humber at Grimsby, as shown in *Fig. 1*. Stage 1 became operational (i.e. received goods for stacking) in July 1966, Stage 2 in May 1967 and Stage 3 in September 1970.

Site investigation

Before planning commenced a commercial site investigation, in two phases, was commissioned. Nine boreholes of 150 mm diameter were made using conventional shell and auger methods to depths up to 21 m. The approximate positions of boreholes are shown in *Fig. 1*, and the borehole logs are recorded in *Fig. 2*; the water levels recorded being those existing at the time of completion of drilling. Standard 100 mm diameter undisturbed core samples were taken at intervals, to provide samples for undrained triaxial tests and for laboratory consolidation tests on 38 mm and 76 mm diameter specimens respectively.

80

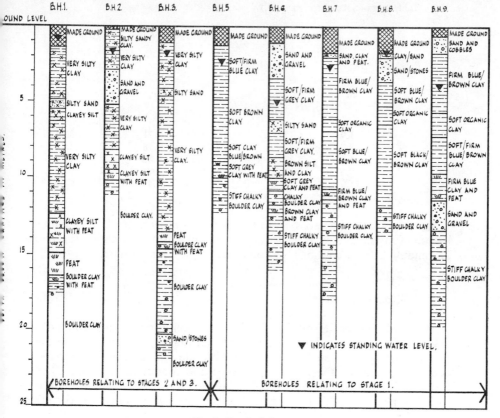

Fig. 2. Borehole logs

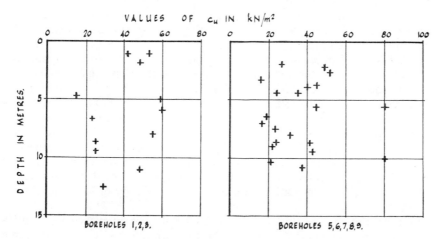

Fig. 3. Undrained shear strengths

81

Table 1. CONSOLIDATION TEST DATA

Borehole	Depth	Pressure range (kN/m²)					
		21–54		54–107		107–214	
	m	m_v	c_v	m_v	c_v	m_v	c_v
		$m^2/kN \times 10^4$	(m^2/a)				
5	1.9	6.15	1.81	3.54	2.71	2.33	3.28
9	2.2	12.71	0.85	10.90	0.78	6.62	1.11
	5.6	8.29	0.75	7.55	0.88	4.28	0.85
	9.5	7.27	0.99	6.99	1.05	5.41	0.75
	11.0	7.55	0.51	7.18	0.49	5.50	0.43

Fig. 3 indicates the variation in the values of c_u with depth in the Stage 1 location (boreholes 5, 6, 7, 8, 9) and in the Stage 2/Stage 3 location (boreholes 1, 2 and 3). *Table 1* contains the consolidation test results data.

Structural concept

A study of the results of the site investigation for the area and general engineering experience locally suggested that due to the considerable thickness of highly compressible soils large absolute and differential settlements could occur due to the proposed storage and foundation loadings. It was first considered that piling of both the cold store floor and foundations would be necessary. Bearing in mind the economic importance of low initial cost of the stores further studies of the structural possibilities were made, and it was concluded that it was possible to construct buildings without piled foundations and having sufficient flexibility to accept the anticipated total and differential settlements.

Structural flexibility was introduced into the superstructure by the use of simple but effective modifications of the constructional details usually employed for such buildings.

Fig. 4 is a typical cross section through one of the cold stores showing the general principles of the construction. Roof construction is asbestos cement sheeting on purlins carried by steel roof trusses spaced at approximately 4.6 m centres. These trusses span between 21.3 m and 29.0 m and are supported internally on *in situ* concrete columns; the perimeter columns being precast concrete. The external wall is formed of precast concrete panels spanning between and fixed to the perimeter columns. The insulation envelope is contained within this shell and is continuous across the floor, where it is covered with a concrete wearing surface. Each store is provided with a heater mat below the floor insulation to prevent the possibility of ice formation in the subsoil.

Foundations to the external columns and walls are, in every case, continuous reinforced concrete inverted tee-beams with the upstand section, some 1.2 m high, forming part of the external wall. Interior columns are supported on a continuous reinforced concrete spread foundation strip.

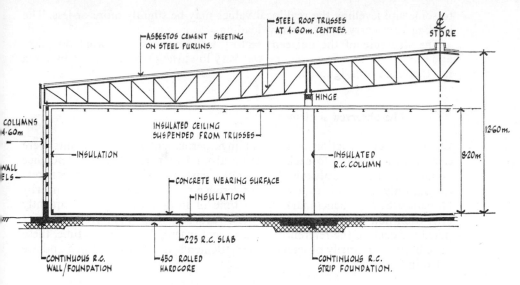

Fig. 4. Typical cross-section of cold stores

Both interior and exterior column and wall foundations are shallow, the thickness at no point exceeding 0.6 m, and are integral with the 225 mm thick reinforced concrete floor slab. The size of foundations is selected to produce a bearing pressure below the foundations similar to that anticipated below the floor of the cold store.

All three stores have movement joints in the external walls, provided by twin columns located at the third points along the length of the store. Stages 2 and 3 have, in addition, a similar joint in the gable and walls.

Whilst the external wall units are clamped to the columns with bolts a flexible packing material is provided between column and wall units, all joints being pointed up in lime mortar. All external precast columns are pinjointed to the upstand wall of the perimeter foundation and all roof trusses are simply supported spans with simple hinges at the supporting columns.

Floor loadings

The superimposed loading on the floors of the cold stores varies with the type and quantity of produce stored. The design superimposed loading for the stores was 36 kN/m², which, after allowing for dead loads, would result in a bearing pressure of 46 kN/m² on the ground. It has not proved possible to monitor accurately the actual superimposed loads arising from stored produce due to rapidly changing storage conditions. but investigations show that the bearing pressure in Stages 2 and 3 is in the region of 43 kN/m². The comparable figure for Stage 1 is 38 kN/m², due to its smaller storage height.

In the subsequent settlement analyses, these values have been taken as the net ground loadings everywhere in each corresponding stage, although due

83

to some site levelling the real local values may be slightly more or less. The loading bays carry no permanent loading.

The structure of the adjacent factory building was supported on piles, and the net ground-floor loading was very low; the presence of this building has therefore been ignored.

The observed settlements

During the construction of the buildings permanent reference points were cast into or attached to each structure around its perimeter. The reference points consisted of galvanised steel bolts cast or drilled into the concrete leaving only the heads exposed. All reference points were related to a nearby Ordnance Survey benchmark at the time of installation and subsequently after each new level survey. From these records settlements have been computed for each of the reference points except where, from time to time, these have been temporarily obscured. The settlements are shown in *Fig. 5* and are listed in *Table 3*.

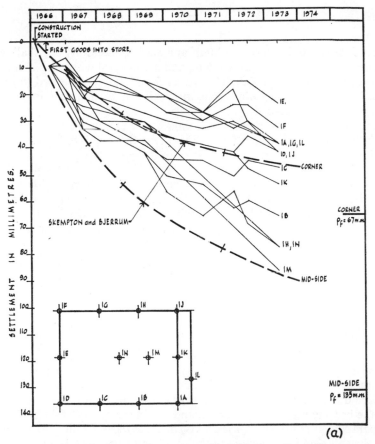

Fig. 5. Time/settlement curve (a) Cold Store, Stage 1

84

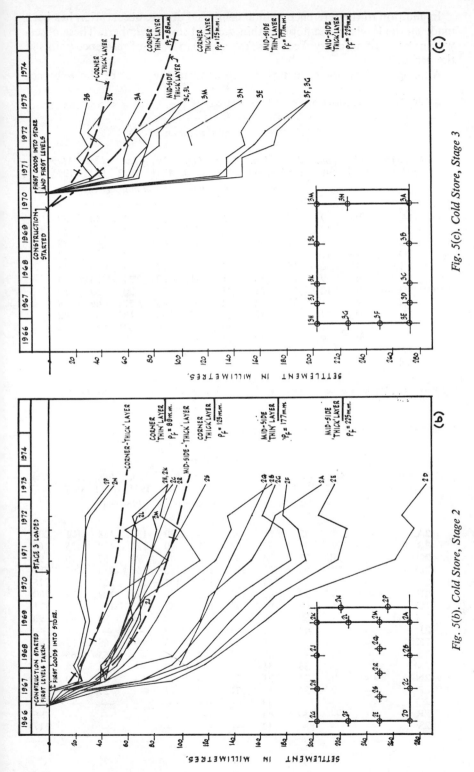

Fig. 5(c). Cold Store, Stage 3

Fig. 5(b). Cold Store, Stage 2

In addition to the reference points mentioned above, observation has been made on the levels of the floors within the cold store chambers. These observations have, of necessity, been less frequent than those of reference points at the perimeter.

Although it has not been possible to confirm the level of the reference

Table 3(a). SETTLEMENT OF STAGE 1 COLD STORE*

					Date of observation					
Reference point	July 1966	Jan. 1967	July 1967	Jan. 1968	May 1969	Jan. 1970	Feb. 1971	Jan. 1972	June 1972	May 1973
					(Settlement at reference points in millimetres)					
1A	9	9	15	15	21	27	27	33	30	39
1B	9	18	30	33	42	51	54	63	60	66
1C	9	18	27	30	39	45	45	51	45	48
1D	9	12	21	24	33	36	39	42	36	42
1E	9	6	15	12	15	18	27	15	15	24
1F	9	9	18	18	21	21	30	24	24	33
1G	9	12	18	21	27	30	33	30	33	39
1H	9	21	33	33	42	57	66	57	69	78
1J	9	12	21	12	21	27	30	18	30	42
1K	9	21	24	—	30	36	39	42	48	54
1L	9	9	15	12	15	21	—	—	—	39
1M	9	12	33	42	42	—	—	—	—	87
1N	9	9	—	30	30	—	—	—	—	78

* The measured settlements have been increased by 9 mm to allow for initial settlement prior to start of recording levels.

Table 3(b). SETTLEMENT OF STAGE 2 COLD STORE

					Date of observations					
Reference point	Oct. 1966	Apr. 1967	July 1967	Jan. 1968	Apr. 1968	May 1969	Jan. 1970	Feb. 1971	Jan. 1972	June 1972	May 1973
					(Settlement at reference points in millimetres)					
2A	—	9	42	66	93	114	120	153	168	171	207
2B	—	9	42	57	63	81	90	126	138	135	168
2C	—	Nil	21	24	27	39	48	90	57	75	93
2D	—	39	93	117	126	165	186	261	276	267	285
2E	—	30	81	111	123	153	171	219	225	204	216
2F	—	30	69	105	114	138	156	195	189	177	180
2G	—	24	69	93	99	123	141	183	177	162	171
2H	—	3	30	45	51	66	72	114	99	84	90
2J	—	12	42	39	45	60	72	—	—	—	—
2K	—	24	42	45	48	57	63	—	66	63	90
2L	—	18	39	39	48	63	60	—	63	66	—
2M	—	9	36	51	54	69	69	—	81	78	—
2N	—	9	24	21	24	30	27	—	27	30	48
2P	—	9	24	18	21	27	21	—	27	27	45
2Q	—	—	—	99	—	—	—	—	—	—	165
2R	—	—	—	45	—	—	—	—	—	—	96
2S	—	—	—	42	—	—	—	—	—	—	117

Table 3 (c). SETTLEMENT OF STAGE 3 COLD STORE

Reference point	Sept. 1970	Feb. 1971	Sept. 1971	Jan. 1972	June 1972	May 1973
			Date of observations			
			(*Settlement at reference points in millimetres*)			
3A	—	57	57	60	57	69
3B	—	33	21	30	24	30
3C	—	69	72	84	84	102
3D	—	—	—	—	—	—
3E	—	135	135	150	147	162
3F	—	147	147	171	174	198
3G	—	129	144	156	165	198
3H	—	—	—	—	—	—
3J	—	—	—	—	—	—
3K	—	42	30	42	30	48
3L		60	66	75	66	102
3M		63	72	75	78	120
3N		—	—	108	105	147

bench mark throughout the recording period, its stability is presumed to be very high due to its location on a massive quay wall of considerable age.

The estimated settlements

Settlements have been estimated using the method proposed by Skempton and Bjerrum (1957) i.e.

$$\rho_f = \rho_i + \mu\rho_{oed} \tag{1}$$

where ρ_f = final settlement
ρ_i = initial elastic settlement
ρ_{oed} = the settlement computed using coefficients of volume decrease (m_v) obtained in the oedometer
$= \Sigma \, m_v(\Delta\sigma_z) \times \Delta z$
μ = a coefficient depending on the geometry of the problem and on the pore pressure parameter A

For normally consolidated soils such as are present on this site it has been assumed that $\mu = 1$. Hence Equation 1 is simplified and the settlement at intermediate times is given by

$$\rho_t = \rho_i + U\rho_{oed} \tag{2}$$

where U is the degree of consolidation as evaluated from the theory of one-dimensional consolidation.

The initial elastic settlement is evaluated using the usual formula

$$\rho_i = q\frac{B}{E}I_\rho \tag{3}$$

in which q = loading intensity
B = foundation width
E = elastic modulus of soil
I_ρ = influence factor which depends on Poisson's ratio v of the soil, and the geometry of the problem.

For immediate (undrained) loading $v = 0.5$, and $E/\bar{c}_u$ has been assumed equal to 100 (Skempton, 1951); I_ρ for a two-layer rigid base situation is available in numerous sources, e.g. Terzaghi (1943).

The idealized soil profiles used on the estimates are given in *Fig. 6*, together

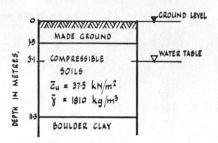

(a) STAGE 1 - FROM BOREHOLES 5,6,7,8,9.

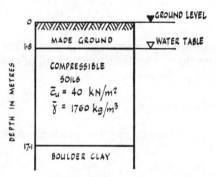

(b) STAGES 2 AND 3 - "THICK LAYER" FROM BOREHOLES 1,3.

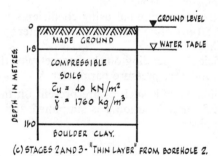

(c) STAGES 2 AND 3 - "THIN LAYER" FROM BOREHOLE 2.

Fig. 6. Assumed ground profiles

with relevant average soils data. Values of m_v appropriate to the initial effective stresses and loading increases were adopted using for any pressure range the mean value of those presented in *Table 1*. The vertical stress increases were evaluated using the two-layer rigid-base solution of Burmister (1956). For the computation of intermediate settlements two-way drainage was assumed, using an overall mean value of $c_v = 12.1$ m²/a.

The results of the settlement computations are presented in Table 2, and some of the values have been added to the observed results in Fig. 5.

Comparison of computed and measured settlements

Bearing in mind the difficulties involved in making an idealisation of the actual soil profiles, it is concluded that the estimates of final settlement for Stages 2 and 3 give a reasonable guide to the order of magnitudes to be expected of settlements at the periphery of the loaded areas. It can be seen from *Tables 2* and *3* that the measured settlements near to the centre of Stage 2 (points 2Q, 2R, 2S) are currently about one third of the estimated final settlements. This discrepancy is large and can be explained mainly by the fact that the central areas of the store are frequently unloaded. Thus the mean long-term loading in the central area is likely to be much smaller than the value assumed. At the periphery however, the storage is more permanent and actual and assumed loadings are more nearly equal. However, the rate of settlement is considerably underestimated, and this is in accord with general experience, e.g. Murray (1971) when the soil structure is such as to provide many secondary drainage paths; the sand layers and silty layers indicated on the borehole logs fulfil this role on this site.

It is interesting to note from *Table 2* that the immediate settlements are relatively small compared with the consolidation component. This is mainly due to the geometry of the problem, in which the loaded areas are very large compared with the thickness of compressible soils, even though these may extend to the surface. In such a situation the possibility of large lateral strains is reduced, and the one-dimensional consolidation theory becomes a reasonable approximation. Also, because of the large lateral extent of the loaded area the addition of an adjacent similar area appears to make little difference to the computed peripheral settlements. However, along the side of Stage 2 which becomes contiguous to Stage 3 (reference points 2A–2D) it was to be expected that settlements would increase when Stage 3 was loaded. *Fig. 5(b)* shows some evidence of this increase, although it is not confined to the expected points.

The correspondence between measured and estimated settlements for Stage 1 is fair during the period of measurement. The rate of increase of measured settlement is less than that estimated and the behaviour at the centre of the store is similar to that in Stage 2.

It will be noted that the measured settlements do not always conform with the expectation, for a fully flexible structure, that the corner settlements are

Table 2. ESTIMATED SETTLEMENTS

	Settlements (mm)								
Stage	*Centre*			*Corner*			*Mid-side*		
	ρ_i	ρ_{oed}	ρ_f	ρ_i	ρ_{oed}	ρ_f	ρ_i	ρ_{oed}	ρ_f
1	25	241	266	6	61	67	13	122	135
2 or 3 separately									
(a) 'Thick layer'	45	404	449	11	114	125	22	203	225
(b) 'Thin layer'	28	323	351	7	81	88	14	163	177
2 and 3 together									
(a) 'Thick layer'	45	407	452	11	114	125	22	203	225
(b) 'Thin layer'	28	353	381	7	88	95	14	176	190

about 50% of those at the middle of the sides. Variations in soil properties clearly account for some of this divergence, but also the inherent stiffness of the upstand beams tends to reduce differential settlements along the periphery.

No satisfactory explanation can be offered of the measured decreases in settlement particularly in the records of Stage 2. The movements are too large to be attributable to datum variations.

Conclusions

It is tempting, when making predictions after the event, i.e. predictions of type C1 according to the classification of Lambe (1973), to make from the available data judicious choices of variables, such as the soil profiles and soil compressibilities, so as to improve the accuracy of the so-called prediction. This temptation has been resisted here: an honest assessment of a very variable site has been made, based on the usual, rather limited, site investigation data provided in a run-of-the-mill commercial investigation. This has resulted in estimates of settlement which are in tolerable agreement with those measured, although the rates of settlement were less well estimated.

Certainly there appears to be little point in using a more refined settlement analysis in situations of this kind. Sufficient accuracy is available in the conventional methods adopted here to provide settlement estimates on which structural design concept decisions can be soundly based. The flexible structure approach to these cold store buildings has been fully justified. In spite of the considerable differential settlements measured and estimated no distress of any kind has been suffered by the structures or floors. The only maintenance which has been necessary because of settlement has been the re-pointing of external wall joints, some of which have opened from 3 mm to about 10 mm.

REFERENCES

Burmister, D. M. (1956). Stress and Displacement Characteristics of a Two-Layer Rigid Base Soil System; Influence Diagrams and Practical Applications, *Proc. Highway Res. Board*. Vol. 35, 773–814

Lambe, T. W. (1973). 'Predictions in Soil Engineering', *Géotechnique*, London. Vol. 23, No. 2, 149–202

Murray, R. T. (1971). 'Embankments Constructed on Soft Foundations: Settlement Studies at Avonmouth, *Dept. Environment, Road Res. Lab.* Report LR 419

Skempton, A. W. (1951). 'The Bearing Capacity of Clays', *Building Res. Cong.* London, Div. 1, Pt 3, 180–189

Skempton, A. W. and Bjerrum, L. (1957). 'A Contribution to the Settlement Analysis of Foundations on Clay', *Géotechnique*, London. Vol. 7, 168–178

Terzaghi, K. (1943). *Theoretical Soil Mechanics*. Wiley

II/3. On the behaviour of a partially floating foundation on normally consolidated silty clays

B. D'Elia
Lecturer in Soil Mechanics, Architecture Faculty, University of Rome
M. Grisolia
Research Assistant, Ist. di Geologia Applicata, Engineering Faculty, University of Rome

Summary

Settlements of a satellite antenna tower during and after construction have been measured. The subsoil consists of homogeneous n.c. soft lacustrine silty clays. The foundation consists of a rigid reinforced concrete cellular structure. The average net foundation loading is 10 kN/m².

Angular distortions smaller than those permissible (1/6000) have been observed. The measured immediate settlements are much smaller than those calculated while the measured consolidated settlements are in good agreement with those calculated.

Introduction

The Telespazio SpA owns a satellite communication station located in the Fucino Plains, about 90 km east of Rome.

One of the antennas is a paraboloid-shaped structure 29.56 m in diameter (*Fig. 1*) placed on a reinforced concrete cellular structure. Settlement measurements of the whole structure were carried out during and after its construction.

Geotechnical properties of the soils

The Fucino basin is a depression, 15 km in diameter, occupied since late Pleistocene by a lake where more than 100 m of fine-grained soils deposited. This lake was fully drained in 1874.

91

Fig. 1. Side view of the antenna structure (courtesy of Telespazio S.p.A.)

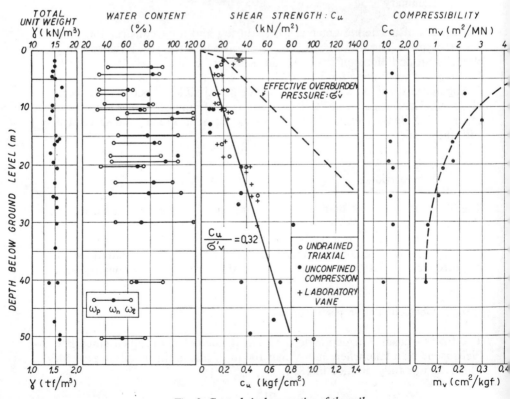

Fig. 2. Geotechnical properties of the soil

Consecutive samples in mud-filled boreholes, some of them more than 70 m deep, were obtained using a double tube rotary core barrel. Undisturbed samples 87 mm in diameter were taken using a Denison-type sampler, available at the time to the contractor.

The subsoil consists of grey silty slightly organic clays. The soil macro-fabric is uniform. The water table is 1.3 m below ground level. The properties of the subsoil are given in *Fig. 2*. The grain unit weight is 25.5–26.0 kN/m³. The oedometric compressibility parameters shown in *Fig. 2* refer to the virgin curve obtained using the Schmertmann construction. The average coefficient of consolidation is $c_v = 10^{-8} \text{m}^2/\text{s}$.

Characteristics of the antenna structure

The geometry of the structure which is highly rigid is shown in *Fig. 3*. With no winds, all the loads are centred. Allowable angular distortions are of the order of 1/6000.

The total weight of the structure (antenna, tower, foundation) is 38.5 MN. The average gross foundation loading is 62 kN/m². The average net foundation pressure is 10 kN/m² for an average foundation depth of 3.50 m.

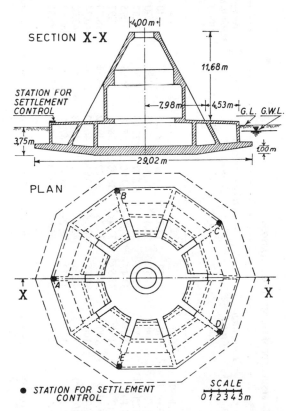

Fig. 3. Geometry of the antenna tower and its foundation

93

Table 1. RESULTS OF SETTLEMENT OBSERVATIONS

Date	Percentage of total load	Settlement (mm) Station					Average settlement (mm)	Average settlement after 100% total load (mm)	Differential settlement* (1)	(2)	Angular distortion
		A	B	C	D	E					
28.7.1969	57	—	—	—	—	—	—	—	—	—	—
22.9.1969	90	20.30	18.77	19.51	21.55	21.89	20.40	—	1.63	B	1 : 7,700
5.11.1969	100	27.51	26.44	27.09	28.66	28.80	27.70	—	1.26	B	1 : 10 000
19.12.1969	100	32.12	31.71	32.27	33.05	32.71	32.37	4.67	0.68	D	1 : 18 400
28.1.1970	100	34.47	34.19	34.90	35.34	34.86	34.75	7.05	0.59	D	1 : 21 200
4.4.1970	100	37.59	37.42	38.08	38.68	38.10	37.97	10.27	0.71	D	1 : 17 600
13.5.1970	100	38.13	38.23	39.00	39.12	39.33	38.76	11.06	0.63	A	1 : 19 800
2.7.1970	100	39.74	40.18	40.91	40.90	39.96	40.34	12.64	0.60	A	1 : 20 800
10.10.1970	100	42.64	42.98	43.62	43.28	42.49	43.00	15.30	0.62	C	1 : 20 200
18.6.1971	100	47.73	48.29	48.45	47.87	47.24	47.92	20.22	0.68	E	1 : 18 400
21.1.1973	100	56.30	56.65	56.78	56.24	55.81	56.36	28.66	0.55	E	1 : 22 700

* Column (1): Difference between the settlement measured at station in column (2) and at the foundation centre.

Settlement observations and discussion

Five stations for settlement control were placed on the foundation upper floor (*Fig. 3*); the settlement values were measured by high precision

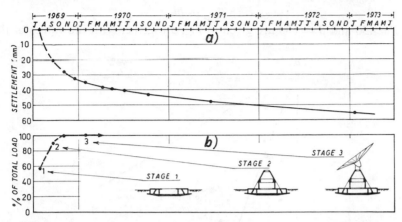

Fig. 4. (a) average settlement v time; (b) percentage of total load and construction sequence

(± 0.05 mm) geometric levelling (*Table 1* and *Fig. 4*). The closing error of the levellings was 0.1 mm.

Measurements of the antenna foundation settlements began when the load transmitted to the soil was equal to 57% total load.

Immediate settlements

An investigation of the measured immediate settlements was carried out to determine the actual values of the subsoil modulus through the commonly accepted relation $E = \beta c_u$. The values of c_u plotted in *Fig. 2* show a linear increase of c_u with depth expressed by

$$c_u = c_{u_o} + k_c z$$

where in the present case $c_{u_o} = 10$ kN/m² at the foundation level, $k_c = 1.5$ kN/m²/m, and z is the depth below the foundation level. It was then assumed that E also increases linearly with depth

$$E = E_o + kz$$

where $E_o = \beta c_{u_o}$ and $k = \beta k_c$. If this is the case, the solution by Carrier and Christian (1973) for a rigid circular plate on a vertically non-homogeneous, linear, elastic half-space with $E = E_o + kz$ and constant Poisson's ratio can be used to determine k and therefore β.

In such solution the settlement is given by

$$\Delta\rho = \Delta q \mathrm{I}'/k$$

where $\Delta\rho$ is the settlement corresponding to a load increment Δq, I' is a non-dimensional influence factor depending upon the ratio E_o/kD and Poisson's ratio v, and D is the diameter of the rigid plate. For $\Delta\rho = 20.4$ mm, corresponding to a gross pressure increment of $\Delta q = 21$ kN/m² (when the load

95

increased from 57% to 90%), $D = 28$ m, $E_o/kD = c_{u_o}/k_c D = 0.24$ and $v = 0.5$, then $I' = 0.65$ and $k = 0.67$ MN/m² per metre. This value of k gives $\beta = 450$ which is a value considerably higher than those (50–100) given by Broms (1972). The graphs of

$$ E = \beta c_{u_o} + \beta k_c z $$

for $\beta = 450$ and $\beta = 100$ are given in *Fig. 5* together with the values of the

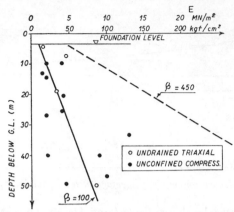

Fig. 5. *Soil moduli based on laboratory undrained compression tests. Solid line and dashed line describe the relation $E = \beta c_{u_o} + \beta k_c z$ for $\beta = 100$ and for $\beta = 450$ respectively*

initial tangent modulus based on undrained compression tests. It can be seen that the line for $\beta = 100$ fits fairly well with laboratory values.

Consolidation settlements

The average values of the settlement measured after 100% of the load was applied were plotted in *Fig. 6* as a function of the square root of time. The

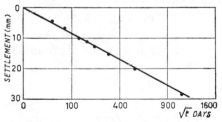

Fig. 6. *Average settlement v square root of time after total load was applied*

settlement v square root of time curve is linear, therefore settlements are mostly due to primary consolidation. The final settlement computed on the basis of the primary consolidation theory was compared with the measured value. The final settlement was calculated as a first approximation in the one-dimensional case using the best-fitting curve of the m_v-values (*Fig. 2*) and a uniform loading of 10 kN/m² on a circular area of diameter $D = 28$ m.

The final computed settlement which is the arithmetic mean between the settlement computed at the centre and that computed at the edge of the loaded area is found to be equal to 250 mm.

The degree of consolidation settlement, U_s, has been computed using Biot's theory of three-dimensional consolidation in the case of a uniformly loaded circle on a semi-infinite mass with permeable upper surface. The time factor T_v has been defined in terms of c_v, based on oedometric tests as suggested by Davis and Poulos (1972). The soil permeability has been assumed as isotropic which is consistent with the clay fabric and is also justified by the low scattering of the c_u values (*Fig. 2*) (Rowe, 1972).

If Biot's theory is used, for $t = 3$ yr and $c_v = 10^{-8}$ m^2/s, U_s has been found to be 0.10 for a Poisson's ratio $v' = 1/3$ of the soil skeleton the resulting settlement is about 25 mm, in good agreement with the settlement measured at the same time (about 28 mm).

Differential settlements

The differential settlements (*Table 1*) are very small and the horizontal rotation axes of the structure have no preferential direction. The angular distortions are about 1/20 000, i.e. three times smaller than those allowable for the antenna.

Conclusions

It has been confirmed that the values of the soil deformability modulus based on the measured immediate settlements are considerably higher than those based on laboratory undrained compression tests.

In the present case β in the equation $E = \beta c_u$ is about 450 against values from 50–100 given by several Authors. The settlements, measured three years after the end of construction, are in good agreement with those calculated on the basis of the consolidation theory.

The foundation angular distortions are very small and largely within the limits of the correct functioning of the antenna. This confirms the assumption of uniform subsoil conditions, and shows that in the case of a centrally loaded axial-symmetric rigid foundation the differential settlements can be very small even if the total settlement is fairly large.

Acknowledgements

The authors wish to thank the Telespazio SpA for giving permission to publish these results.

Thanks are due to Professor F. Esu and Professor S. Olivero for the useful suggestions given during the course of this study.

This study has been carried out with the financial support of the National Council of Researches of Italy (C.N.R.).

REFERENCES

Broms, B. B. (1972). 'Stability of Flexible Structures', *General Report 5th European Conf. Soil Mech. & Found. Eng.*, Madrid. Vol. 2, 239–269

Carrier, W. D., III and Christian, S. T. (1973). 'Rigid Circular Plate Resting on a Non-homogenous Elastic Half-space', *Géotechnique*, London. Vol. 23, No. 1, 67–84

Davis, E. H. and Poulos, H. G. (1972). 'Rate of Settlement under Two- and Three-dimensional Conditions', *Géotechnique*, London. Vol. 22, No. 1, 95–114

Rowe, P. W. (1972). 'The Relevance of Soil Fabric to Site Investigation Practice', *Géotechnique*, London. Vol. 22, No. 2, 195–300

98

II/4. A settlement study within a geotechnical investigation of the Grangemouth area

P. M. Jarrett
University of Glasgow

W. G. Stark
University of Glasgow

J. Green
Imperial Chemical Industries Ltd., Grangemouth

Summary

A brief description of the methods and aims of a geotechnical study of the Grangemouth area of Scotland is presented. As a part of this study, settlement observations taken of fifteen buildings at the Imperial Chemical Industries works in Grangemouth have been examined. These settlement results and the corresponding building loadings have been combined to indicate the form of the compressibility characteristics of the soil profile over a range of stress increases.

Introduction to regional analysis

Grangemouth is a rapidly expanding industrial area in the Forth Valley of Scotland. Foundation conditions in the area are difficult, since the principal drift deposits forming the flat alluvial plain on which the town is located consist of soft, normally consolidated, clayey silts of considerable depth. A widespread study of the foundation problems encountered in the area has been undertaken at Glasgow University, in an attempt to both correlate presently available information into a coherent pattern and to develop new studies as necessary.

The stratigraphy of the region, and a geomorphological explanation of its formation, have been dealt with in considerable detail by Sissons (1969, 1971). Sissons studied over 1500 borehole logs from the district, which were gathered together from the Institute of Geological Sciences, local authorities, consulting engineers, owners, etc. A detailed picture of the drift stratigraphy was developed and presented as maps. The importance of this work is that information that normally would have been tucked away and forgotten was gathered together into a form which could significantly reduce the fieldwork (and expense) necessary to prove the stratigraphy in areas of proposed development.

The objective of the Glasgow study has been to graft geotechnical engineering information on to this basic stratigraphic information providing geotechnical maps or data banks related to foundation problems and their possible solutions. The fact that the stratigraphic units are widespread and reasonably well defined increases the usefulness of such an exercise. The first stage of the procedure was to collect and study the available site investigation reports and abstract values of moisture content, liquid and plastic limits, shear strength and consolidation parameters. Most of this information came from standard site investigations using U4 sampling devices. For the soft soils involved it was realised that these results would not be entirely satisfactory but they nonetheless provided a basis for initial classification.

In an attempt to define a 'standard' for the area block samples were obtained, down to depths of 13 m, in the excavation for a sewage pumping station. Basic tests have, and are, being conducted on these blocks to assess the degree of difference between soil parameters produced from this controlled sampling and testing, and those from standard site investigations. In addition research oriented testing is being performed on the blocks to investigate the fundamental behaviour of these soils.

The importance of field studies in providing both soil parameters from back analysis and details of the behaviour of various forms of foundation was also realised. Approaches were made to authorities, owners and engineers operating in the area for records of structural behaviour. A number of case histories were uncovered, including settlements of buildings, oil tanks and motorway embankments. Not all were the research engineer's ideal of a case history, but, by obtaining an overall view of the parts, it is believed that a picture will emerge.

One of the most complete sets of case histories under study came from the Imperial Chemical Industries works located on the southern boundary of Grangemouth. In this complex (*Fig. 1*) covering some 22 ha, a large number of buildings, representing a variety of loading conditions, have been monitored, for settlement for over 20 years. Results obtained from the study of these buildings are presented in the next section of this note, the emphasis of the presentation being on the overall view rather than on the individual building details.

I.C.I. settlement studies

Soil profile

The soil profile at the works, which is typical of the Grangemouth area, has been determined from the logs of the 14 boreholes located in *Fig. 1*. *Fig 2* presents the average profile of the various soil units together with moisture contents, liquid and plastic limits, and undrained shear strength results, all abstracted from the borehole logs. Whilst an obvious superfluity of results is present the trends are reasonable. Values of compression index from the same source are shown in *Fig. 3*.

Units 1 and 2, known as the 'Carse clay', were formed as post-glacial mudflats. Unit 1 is the desiccated crust of the main deposit, Unit 2, which is a dark grey to black clayey silt, often containing shells and with sensitivity

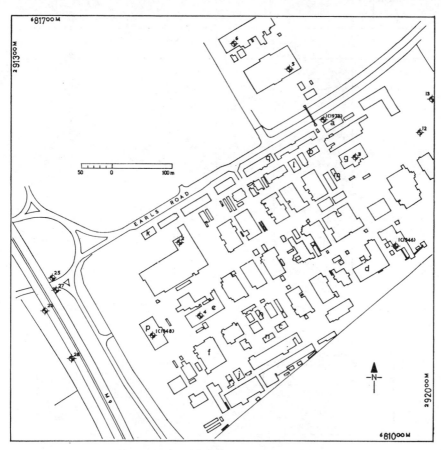

Fig. 1. Layout plan—ICI Works, Grangemouth

values between 4 and 8 being common at its lower levels. A degree of layering exists without being marked. Unit 4 is a soft laminated late-glacial clayey silt. It is usually brown or grey in colour and has a crust formed by previous desiccation. This desiccated crust merges upward into Unit 3, which is a sand and gravel layer with shells. The granular layer was associated with a period of marine erosion which planated the late-glacial deposits in certain areas, and left a layer of coarse deposits to be covered subsequently by the carse deposits.

The variation in soil profile from the average shown in *Fig. 2* is small, strata boundaries over the site being generally within 1 m of the average elevations. Unit 4 is underlain by glacial till above rockhead. The surface of

101

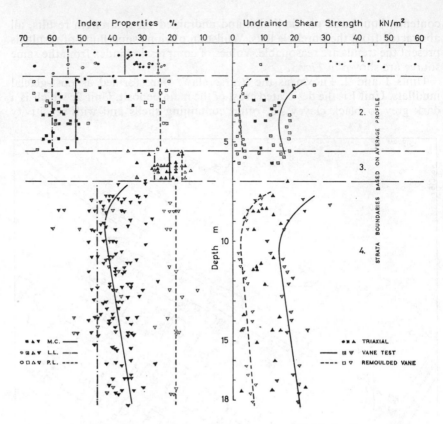

Fig. 2. Index properties and shear strengths results—ICI Works, Grangemouth

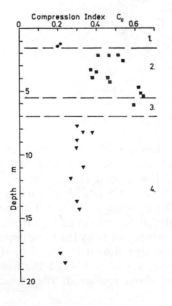

Fig. 3. Compression index values

the glacial till is uneven, its average depth over the site being 18 m. The thickness of the compressible soil of Unit 4 is therefore also variable, but at no location does it appear to be less than 6 m.

Building descriptions

During the period 1947 to 1952 a large number of buildings were constructed at the I.C.I. works. Recognising the importance of settlement, approximately 50 buildings, both old and under construction, had settlement points installed in their outer walls at that time. These were surveyed regularly until 1961, and again in 1972 when this study was undertaken. Fifteen of these records have been included in this note. Of those not included most were the older buildings from the 1930s and 1940s, with incomplete records. Other records with major uncertainties in interpretation were also rejected. The 15 buildings studied are listed in Table 1, together with letters of identification referring to

Table 1. I.C.I. WORKS GRANGEMOUTH, BUILDING DETAILS AND SETTLEMENTS

Identifica- tion	Buildings	Plan dimensions (m)	Foundation type	Nett loadings Δq kN/m^2	Average settlement δ_{ave} mm
a	Analytical Lab. East	12.8 × 33.0	Strip footings	21.9	107
b	East Area Lab.	11.6 × 18.2	Hollow raft	20.2	128
c	West Area Lab.	11.6 × 18.2	Hollow raft	20.2	49
d	G2 Shed	24.3 × 54.8	Hollow raft	31.0	237
e	Q2 Shed	22.8 × 63.3	Hollow raft	24.5	171
f	H2 Shed	24.3 × 45.6	Hollow raft	26.0	152
g	L3 Shed	22.8 × 42.6	Hollow raft	25.0	70
h	West Area Workshop	6.1 × 16.7	Strip footings	13.0	11
j	Welding Shop	12.2 × 18.2	Strip footings	14.7	11
k	Instrument Workshop	11.2 × 21.2	Strip footings	15.6	12
l	New Laundry Block	12.2 × 20.0	Strip footings	13.6	15
m	East Area Workshop	6.1 × 19.8	Strip footings	12.6	8
n	Engineering Lab.	8.2 × 18.2	Strip footings	12.8	12
p	Amenities Block West	24.3 × 56.2	Hollow raft	12.4	18
q	Main Office Extension	12.8 × 17.3	Hollow raft	15.1	40

their plan location in *Fig. 1*. On general description the buildings fall into three main groups:
(1) Buildings a, b, c, p and q are two storeys high, of brick infilled frame construction, with flat reinforced concrete floors and roofs.
(2) Buildings d, e, f and g are large open production sheds of 3 or more storeys, with hollow raft foundations placed at depths of between 1.1 m and 1.3 m. These foundations provide load compensation for the structures to the extent that the installed plant is the main factor influencing settlements.
(3) Buildings h, j, k, l, m and n are light 1 or 1½ storey brick infilled frame structures with pitched roofs of light construction. The walls and columns are founded on continuous strip footings 0.65 m to 1.00 m deep, with 150 mm reinforced concrete slab floors at grade level.

103

By considering the form of construction and building use, net loadings as transmitted from the foundation base were calculated and are shown in *Table 1*. These stress increases are based on the overall area of each building. Whilst this is obviously correct for the buildings on hollow rafts it must be regarded as an interpretation for the strip and slab buildings where the degree of interaction between the walls and the floor slab is uncertain. The fact that in most cases with strip foundations over 40% of the total load is represented by the floor slab and installed plant makes this interpretation more reasonable.

Analysis of Settlements

Settlements were measured at the main corners of all buildings, with a few intermediate points on the large sheds. Differential settlements have occurred, but can in most cases be reconciled with the tilt of a rigid building rather than flexible movements. This is supported by the appearance of the buildings, where, for instance successive window sills appear as a straight line and not a curve. Differential settlements have not therefore caused structural distress in these buildings.

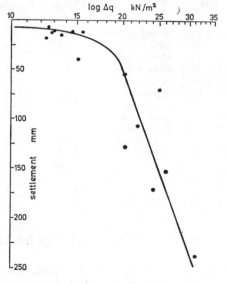

Fig. 4. Net loadings v settlement

To obtain an indication of the soil compressibility under the varying loading conditions created by these buildings, the average settlements, presented in *Table 1*, are plotted against the logarithms of the corresponding stress increases in *Fig. 4*. For the sake of simplicity, all buildings are treated alike no matter what size or type they are. It would be possible to use corrections for load distributions on the basis of width, but this would involve numbers of assumptions leading one away from the original simplicity of a number of differently loaded areas found on similar subsoil. As may be expected with this correlation there is considerable scatter of results, but a reasonable and useful curve may still be fitted to the points.

104

A threshold value of stress increase of about 15 kN/m² is indicated, below which only minor settlements occur. After this threshold a linear relationship between settlement and logarithm of pressure increase is suggested by *Fig. 4*.

Conclusions

The use of rigid hollow rafts to ameliorate the settlements and prevent damage due to flexural differential settlement appears to have been successful at this site underlain by such soft soils. The regional geotechnical study of Grangemouth is a continuing one and it is hoped that as more information is obtained relationships such as *Fig. 4* may be given more significance with possible statistical correlations of soil variability. Further settlement studies are envisaged on proposed road embankments and structures in the area, together with more detailed analysis of some of the individual buildings mentioned in this note. These studies will be supported by good quality soil-sampling.

Acknowledgements

The information used in this paper was released by the Imperial Chemical Industries Ltd. and was collected over the years by their engineers at Grangemouth. The authors wish to thank them all, in particular Mr. P. Whitehead for his recent assistance. Thanks are also due to Professor H. B. Sutherland and the University of Glasgow for continuing help and encouragement with the overall project.

REFERENCES

Sissons, J. B. (1969). 'Drift Stratigraphy and Buried Morphological Features in the Grangemouth–Falkirk–Airth Area, Central Scotland', *Trans. Inst. Brit. Geogr.* Vol. 48, 19–50
Sissons, J. B. (1971). 'Geomorphology and Foundation Conditions Around Grangemouth', *Quart. J. Eng. Geol.* Vol. 3, 183–191

II/5. Predicted and measured settlement of the colonnade building, The First Church of Christ, Scientist, in Boston, Massachusetts

Thomas K. Liu, Harl P. Aldrich, Jr. and John P. Dugan, Jr.
Principal, Senior Principal and Assistant Project Engineer, respectively,
Haley & Aldrich, Inc., Cambridge, Mass.

Summary

As part of the First Church of Christ, Scientist, development programme to expand the Mother Church in Boston, the Colonnade Building was constructed from 1968 to 1972. The structure is supported on pressure injected footings (Franki piles) based in a sand and gravel stratum overlying a nearly 100 ft (30.5 m) thick deposit of over-consolidated to nearly normally consolidated clay. During project design, up to 3 in. (7.6 cm) of long term settlement was predicted for the central portion of the Colonnade Building. This Paper presents the results, to date, of the programme to monitor long-term settlement of the building.

Site conditions

The Colonnade Building is located in Boston's Back Bay where land was created by filling a shallow tidal bay during the nineteenth century. Site grade is essentially level, at about El. +17.5 ft (+5.33 m) (Boston City Base). The land was previously occupied by two- and three-storey tenement buildings.

Two adjacent structures influence stress conditions in the clay beneath the Colonnade Building. One structure, a one-level, underground parking garage located along the east side of the Colonnade Building, imposes a net stress decrease. The existing three-storey Publishing House adjoins the southerly end of the Colonnade Building. A portion of the publishing house was demolished to accommodate the facade for the Colonnade Building. Due to

106

the presence of this existing granite block structure, the increase in effective vertical stress resulting from the Colonnade Building was less in this area.

Subsurface conditions

An extensive programme of test borings was undertaken at the site. A typical generalised soil profile is shown in *Fig. 1*. Beneath about 14 ft (4.27 m) of fill exists a 6 ft (1.83 m) thick stratum of soft, organic silt over 23 ft (7.0 m) of dense sand and gravel. This is underlain by 98 ft (29.87 m) of medium stiff, Boston blue clay over bedrock. Groundwater is separated into two zones by the relatively impervious organic silt stratum. Normally, the water table

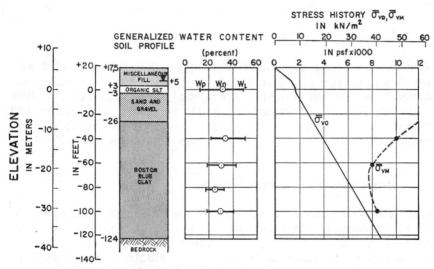

Fig. 1. Generalised soil profile, Atterberg limits and stress history

perched above the organic silt is at about El. +6.5 ft (+1.97 m). The piezometric level in the sand and gravel stratum is generally at about El. +5 ft (+1.52 m).

The stress history of the Boston blue clay from one of the test borings is presented in *Fig. 1*, and is considered typical. The upper, about 50 ft (15.2 m) thick portion of the clay stratum is over-consolidated, with values of over-consolidation ratio ranging up to 4. At lower depths, the clay is only slightly over-consolidated and for design purposes, was considered normally consolidated. Settlement estimates employed a compression ratio, $C_c/(1+e_o)$, of 0.20 and a recompression ratio, $C_r/(1+e_o)$, of 0.03 for the clay.

The structure

The Colonnade Building, a five-storey, reinforced concrete structure with one basement at El. +4 ft (+1.22 m) measures about 120 ft (36.6 m) by 565 ft (172.5 m) in plan dimensions. The major portion of the structure is supported

107

on 120 tonf (1068 kN) design capacity pressure injected footings, based in the upper portion of the sand and gravel stratum overlying the Boston blue clay. *Fig. 2* shows a plan outline of the Colonnade Building, and indicates the three column lines which transmit the building's load to its foundations.

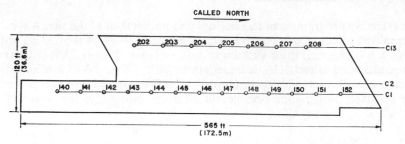

CALLED NORTH

Fig. 2. Settlement reference marker locations on Colonnade building plan

Construction of the Colonnade Building began in October 1968 with excavation of existing fill materials to El. +3 ft (+0.9 m). Excavation was followed by installation of the pressure injected footings from October 1968 to February 1969. It is estimated that by December 1971, construction had proceeded to the point where the foundations carried about 95% of the anticipated dead load. The building was occupied during the first half of 1972.

Predicted and measured settlement

Design settlement estimates were made for the Colonnade Building on the basis of one-dimensional consolidation theory. The ICES-SEPOL computer programme was used to facilitate computations. In addition, a detailed history of settlement observations on the adjoining publishing house, bearing on spread foundations in the sand and gravel stratum, was available for comparison. Colonnade Building settlement estimates for design ranged from essentially no settlement at the southern end to 3 in. (7.6 cm) throughout the central portion and 2 in. (5.1 cm) at the northern end. Due to the over-consolidated nature of the clay, most of the settlement was due to recompression. Relative to time, it was estimated that 60% of the settlement would occur within two years after the start of construction and that at least 90% of the settlement would occur within 10 years after the start of construction. On the basis of this prediction, a decision was made to build the structure's facade with a 3 in. (7.6 cm) camber in the centre, to prevent the possibility of a 'sagging' appearance as a result of long-term settlement.

Precise level survey was used to monitor the progress of settlement of the Colonnade Building, as well as other new structures under construction. A series of 20 settlement reference markers were affixed to selected columns, as shown in *Fig. 2*. Brass reference markers, consisting of a $1\frac{1}{4}$ in. (3.2 cm) diameter base with a protruding portion upon which the level rod is set, were glued with epoxy to the reinforced concrete columns. The elevation measurements were referenced to a benchmark drilled into bedrock through the slab of the adjoining parking garage. The reference markers were installed from June through November 1970, about one and half to two years after the start of construction.

Significant difficulty has been encountered in obtaining continuous un-interrupted elevation measurements on the reference markers. Markers were continually broken off columns by accidental construction or maintenance activities. Of the original 20 reference markers, five have remained undisturbed since their installation and therefore have furnished complete settlement records.

Fig. 3 presents the measured settlements to date on six reference markers, all except no. 149 having no missing data. Settlements from mid-1970 until May 1973 range from 0.33 to 0.65 in. (0.84 to 1.65 cm). It is noted that the

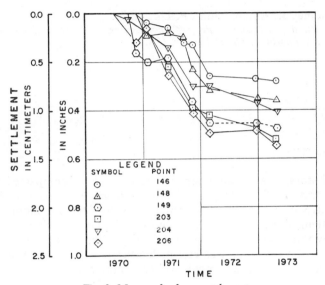

Fig. 3. Measured column settlement

rate of measured settlement ranged from about ¼ to ½ in. (0.6 to 1.3 cm) per year until 1972, after which time settlement rates have dropped sharply.

A comparison of measured settlement to date with predicted settlement must consider the one and a half to two-year delay between the start of con-struction and the initiation of the settlement observations. By extrapolating the measured settlement during 1971 back to the start of construction in early 1969, with consideration of excavation heave and recompression, settlements to date of the order of 1 to 1½ in. (2.5 to 3.8 cm) are obtained, which compares favourably with design estimates.

A significant component of the Colonnade Building settlement is considered to be that due to recompression of the clay stratum following heave resulting from stress release caused by the basement excavation. However, neither the amount of excavation heave nor the resulting recompression were measured.

109

II/6. Elastic settlement behaviour of three steel storage tanks

Thomas K. Liu and John P. Dugan, Jr.
Principal and Assistant Project Engineer, respectively, Haley and Aldrich, Inc., Cambridge, Massachusetts, U.S.A.

Summary

Three steel storage tanks were constructed on pads of compacted granular fill over a stratum of normally to slightly overconsolidated soft and moderately sensitive silty clay. The tanks were instrumented for monitoring full profile settlement of the tank bottom. The initial undrained settlements obtained by the hose settlement gauge during water tests indicate that the behaviour of the materials was essentially elastic in nature. Monitoring of long-term consolidation settlements will be continued.

Introduction

The Texaco sales terminal in South Portland, Maine, is located over tidal flats of the Fore river on land created by hydraulic filling with sands, silts and clays dredged from the river bottom. Underlying the hydraulic fill is from 20 to 60 ft (6.10 to 18.29 m) of soft grey silty clay above glacial till. Over a portion of the site, up to 10 ft (3.05 m) of soft peaty organic silt overlies the clay. The major structures of the terminal consisted of eleven steel storage tanks of different sizes, holding various petroleum products. In the terminal's earlier history, pile foundations had been used for support of the tanks. Several foundation failures of earth-supported tanks had occurred in attempts to find a more economical foundation design.

To expand the storage capacity of the terminal, Texaco recently constructed three additional tanks at the locations shown on the partial site plan, *Fig. 1*. A brief description of the subsurface conditions, foundation design, instrumentation and results of the settlement monitoring programme to date, is presented below.

Site and subsurface conditions

The existing site grade varied from approximately El. +10 ft to El. +20 ft (+3.05 to 6.10 m). A total of 14 test borings were made at the three tank locations. The hydraulic fill and peaty organic silt were encountered down to approximately El. 0, underlain by a stratum of soft, moderately sensitive grey silty clay with occasional streaks of black organic materials. The thickness of this silty clay varied from 25 to 45 ft (7.6 to 13.7 m). The clay stratum overlies very dense glacial till consisting of grey silty fine to coarse sand with a little gravel. The groundwater table was found to be at about El. +7 ft (+2.13 m).

An extensive laboratory testing programme was conducted on samples of the silty clay obtained in test borings. The ranges of test results are summarised as follows:

liquid limit	33 to 47%
plasticity index	15 to 27%
natural water content	30 to 45%
undrained shear strength	500 to 1000 lb/ft²
(increasing with depth)	(23.9 to 47.9 kN/m²)

compression ratio $\dfrac{C_c}{1 + e_o}$ 0.21

recompression ratio $\dfrac{C_r}{1 + e_o}$ 0.024

overconsolidation ratio 1.0 to 2.0

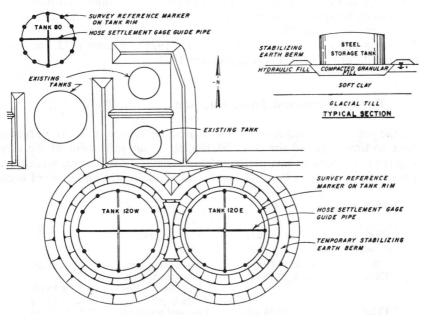

Fig. 1 Site plan and section

111

Tank foundation design

The recently constructed 48 ft (14.6 m) high tanks consisted of one 80 ft (24.4 m) diameter and two 120 ft (36.6 m) diameter tanks. The hydraulic fill and peaty organic silt were too soft and compressible to support the tanks. Therefore, each tank was designed to bear on a pad of compacted sand and gravel. The compacted granular fill thickness, which varies from 16 ft (4.88 m) under tank 80 to 17 ft (5.18 m) under tanks 120E and 120W, was designed to provide complete removal of the unsuitable hydraulic fill and organic silt soils, as well as to provide an adequate safety factor with respect to bearing capacity of the soft clay under critical tank loading conditions. A typical section of a storage tank and its foundation is sketched in *Fig. 1*. The final grade at each tank centre was established at El. + 12.5 ft (+3.81 m). To compensate for long-term differential settlement, the tank pads were designed with a 2% camber.

Water test programme

Each tank was water tested to check the construction and operation of the tank and to test the behaviour of its foundation under a load in excess of its maximum service load. The controlled test consisted of filling the tank in six stages to a full water height of 48ft (14.6 m) with waiting periods of up to two days between stages. In order to improve stability under the 3000 lbf/ft² (144 kN/m²) foundation bearing pressure, each 120 ft (36.6 m) diameter tank was surrounded by an earth berm 7 to 9 ft (2.13 to 2.74 m) high, as shown in the site plan, *Fig. 1*. These berms were to be removed upon completion of water testing. Temporary berms were not required for tank 80 due to a somewhat higher shear strength in the clay at that location.

The tanks were constructed from September 1971 to August 1972, and were water tested during August and September 1972.

Predicted settlement during water test

Since the water test condition represents loading of a very short duration relative to time required for consolidation to occur, settlement of the clay during the test should be essentially undrained or elastic in nature. Settlement under the full water test load was predicted at the centre and edge of each

Table 1. GROUND CONDITIONS

Tank	Compacted granular fill	Soft clay	
80	16 ft. (4.88 m)	Centre and edges	22 ft. (6.71 m)
120E	17 ft. (5.18 m)	Centre, east, west and north edges	20 ft. (6.10 m)
		South edge	28 ft. (8.53 m)
120W	17 ft. (5.18 m)	East and north edges	27 ft. (8.23 m)
		Centre, south and west edges	40 ft. (12.19 m)

Table 2. SOIL PROPERTIES

Materials	Young's modulus	Poisson's ratio
Compacted granular fill	10×10^5 lbf/ft² (4.8×10^4 kN/m²)	0.15
Soft clay	3×10^5 lbf/ft² (1.4×10^4 kN/m²)	0.50

tank on the basis of elastic theory. It was assumed that the dense glacial till did not contribute to the settlement of the tanks. The thickness of the fill and clay under each tank are as shown in *Table 1*.

Soil properties which are required to calculate the theoretical elastic settlement are shown in *Table 2*.

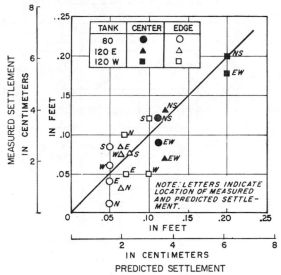

Fig. 2. Comparison of elastic settlement

Young's modulus for the soft clay was based on an analysis of extensive laboratory test data on undrained shear strength from the site and the value for the granular fill was assumed based upon published data (Lambe and Whitman, 1969).

Predicted elastic settlements during water testing at the centre and edges of the tanks are indicated in *Fig. 2*.

Instrumentation

Twelve equally-spaced reference points were established on each tank rim as shown in *Fig. 1*. During water test, elevations on these points were generally surveyed immediately after each filling and at the end of each waiting period.

To monitor the settlement of the tank bottom, two P.V.C. guide pipes $1\frac{1}{2}$ in. (3.81 cm) i.d. were installed under each tank as shown in *Fig. 1*. The hose settlement gauge (Dunnicliff, 1971), was used to measure the full profile tank bottom settlement.

113

Measured settlement during water test

The tanks performed satisfactorily during the water tests, from the aspects of stability and settlement. Measured settlements ranged from about 1 to 2½ in. (2.5 to 6.4 cm) at the tank centres and from about ¼ to 1½ in. (0.6 to 3.8 cm) beneath the tank edges. Tank bottom settlement profiles measured in the east–west direction during the water tests with the hose settlement gauge are presented in *Fig. 3*. Settlements of the reference points on the tank rims

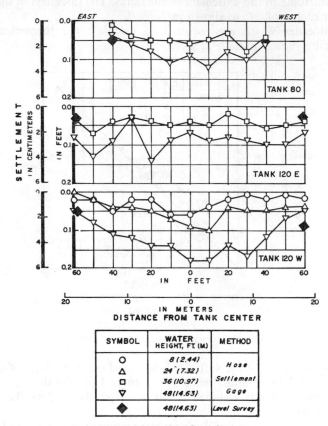

SYMBOL	WATER HEIGHT, FT (M)	METHOD
O	8 (2.44)	Hose
△	24 (7.32)	
◻	36 (10.97)	Settlement
▽	48 (14.63)	Gage
◆	48 (14.63)	Level Survey

Fig. 3. Measured tank settlement

located above the end points of the guide pipes are also plotted on *Fig. 2* for comparison.

The influences of the thickness of clay layer and radial distance from tank centre on the magnitude of elastic settlement were confirmed by the measured results.

Predicted *v* measured settlement

Fig. 2 shows a comparison of measured and predicted elastic settlements during water tests. Points falling on the line drawn at a 45° angle from the

114

origin indicate predicted and measured settlements are equal. Points above this line indicate the estimated settlement was too low and vice versa. This comparison shows that predicted and measured settlements are in fairly good agreement, which indicates that the undrained settlement of the soft clay was essentially elastic.

To date, the tanks have not been loaded sufficiently to produce significant consolidation settlement. The long-term tank settlements will be monitored for the next several years.

REFERENCES

Dunnicliff, C. J. (1971). 'Equipment for Field Deformation Measurements', *Proc. 4th Pan-American Conf. Soil Mech. & Found. Engg*, Puerto Rico. Vol. II, 319–332
Lambe, T. W. and Whitman, R. V. (1969). *Soil Mechanics*. John Wiley & Sons, Inc., U.S.A.

II/7. Settlements of buildings founded on lightly overconsolidated clay in western Sweden

Per-Olof Nordin
Civil Engineer, J. & W. Grundkonsult Sweden

Per Lennart Svensson
Civil Engineer, HSB:s Ricksförbund, Sweden

Summary

This investigation, which is financed by the Swedish Council of Building Research, will continue during a period of 10 years. Its purpose is to increase the knowledge of the properties of lightly overconsolidated clay. Up till June 1973 the measurements have been going on for about one year and both the premises for and the results of the measurements are presented below. The investigations are made in two areas, one situated in Lidköping (the area of Margretelund) and the other in Vänersborg (the area of Lilla Torpa).

Regional geology

Lidköping and Vänersborg are situated in western Sweden near the great lake Vänern. The clays in this region were deposited in saline sea water in front of the receding glacier border during the end of the latest glacial period. The glacier border passed the locality of Vänersborg about 9300B.C. and the locality of Lidköping about 400 years later. The silty and sandy sediments, which cover the clay in Lidköping, were probably deposited in connection with draining of the Baltic Ice Lake.

Soil investigations

The properties of the clay in both areas were carefully analysed. In Lidköping, Margretelund, 14 borings were made. In total, about 400 samples were taken

116

in Margretelund with the Swedish standard piston sampler which has a diameter of 50 mm.

In Vänersborg 14 borings were made with the piston sampler providing about 170 samples. All these were analysed and the preconsolidation load was determined on a large number of samples by means of oedometer tests and by the empirical formula (Hansbo, 1957),

$$\sigma_c = \frac{\tau_f}{0.45 w_F}$$

where σ_c = preconsolidation pressure

τ_f = undrained shear strength (according to the fall-cone test)

w_F = fineness number (according to the fall-cone test) $\approx w_L$ (the liquid limit)

This formula is frequently and successfully used in Sweden.

Measuring equipment

In order to measure the settlements steel bars were driven to firm bottom and the movements of the buildings measured by a gauge which gives the difference between the level of the base slab and the level of the top of the steel bar. Most of the observation points are placed in the staircase-wells of the buildings. The gauge is accurate to one-hundredth of a millimetre.

The pore pressuremeters used are of type NGI with open measuring system. The pore pressure is measured at different levels in the clay layer.

Structure and foundation

In the northern area of Margretelund 17 buildings are under construction and in the area of Lilla Torpa 14 buildings. Most of them are without basements. The houses are founded on a continuous slab of concrete directly on the clay or silt. The frames of the buildings are composed of concrete crosswalls and only short longitudinal concrete walls adjoin the gables. The joists are made of concrete and the facades are made of bricks and are not part of the buildings. Thus the frame of the building is rather flexible in the longitudinal direction and differential settlements might easily occur. As the structure of the building is sensitive to settlements even small differential settlements may cause damage. The load from the buildings is 42 kN/m² in Lilla Torpa and 32.5 kN/m² in Margretelund.

Ground conditions

The ground surface is rather flat and horizontal in Margretelund, Lidköping (*Fig. 1a*). The underlying soil consists of a layer of silty sand with a thickness of 3 m covering a silty clay with very thin seams of silt. The clay deposit has a depth of between 15 and 24 m below the ground surface with the greater depth in the southern part. The clay is underlain by gravel and moraine on bedrock.

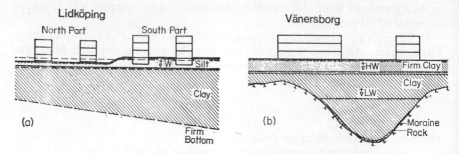

Fig. 1. Cross-section through the two areas

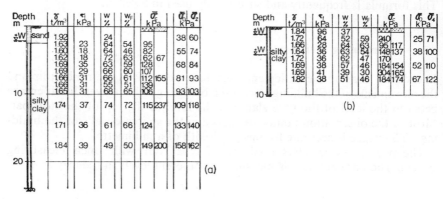

Fig. 2. Typical soil properties

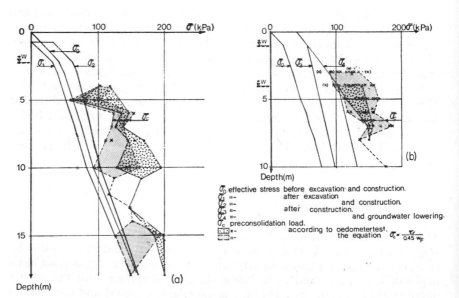

$\bar{\sigma}_0$ effective stress before excavation and construction.
$\bar{\sigma}_1$ "— after excavation and construction.
$\bar{\sigma}_2$ "— after construction.
$\bar{\sigma}_3$ "— and groundwater lowering.
$\bar{\sigma}_c$ preconsolidation load.
"— according to oedometertest.
"— the equation $\bar{\sigma}_c = \dfrac{\tau_f}{0.45\, w_F}$

Fig. 3. Preconsolidation loads and existing overburden pressure

The ground water level in this area is very strongly influenced by the water level in lake Vänern. In the northern part of Margretelund the average ground water level is 2.4 m below the ground surface. *Fig. 2a* gives the results from one of the borings in the area. *Fig. 3a* shows the preconsolidation loads obtained by oedometer tests made on the samples taken in the 15 boreholes and by the previously given empirical formula. It can be concluded from *Fig. 3a* that the clay will remain preconsolidated up to a minimum load of 40 kN/m² above the existing overburden load.

After excavation to 0.7 m depth the additional load to reach the preconsolidation pressure is 52.5 kN/m². The load from the buildings is 32.5 kN/m² corresponding to a lightly loaded building without basement. The remaining 20 kN/m² will be regarded as a factor of safety because of the local differences and the uncertainty of the value. (Because buildings without basements have no stiffness and thus can easily be damaged by settlements.)

The natural ground surface in Lilla Torpa, Vänersborg, is flat (*Fig. 1b*). The soil mostly consists of silty clay covering a thin moraine sheet or bedrock. The thickness of the clay layer is 0–12 m. The local variations of thickness of the layer are large. The upper two metres of the clay layer consist of firm clay. Typical properties of the clay are shown in *Fig. 2b*. Rootfibres perforate the clay layer from top to bottom. The preconsolidation loads of all investigated samples are given in *Fig. 3b*. The scatter of preconsolidation loads is probably caused partly by differences in the property of the clay, partly by differences in the technique of sampling and laboratory investigations. The average ground water level is +48.0, which is 3 to 3.5 m above the water level in lake Vänern. The variations of the ground water level are large. Longer periods without precipitation may lower the ground water level to the level of lake Vänern. As can be seen from *Fig. 3b* the overburden load of natural soil and building is below the preconsolidation load. The total overburden load of natural soil, building and expected ground water lowering is near the preconsolidation stress.

Results

Up till now observations have been made for about one year and the tendency of the results is obvious. Some of the observed results are given in *Table 1* overleaf.

Fig. 4a and *4b* show typical time/settlement curves representative of one building from each area. The results may be summed up as follows:

(1) The settlements were almost finished when the construction activities finished. No increase of pore pressure because of loading has been noticed. Thus the settlements are initial settlements and not consolidation settlements.
(2) The total settlements in Margretelund, Lidköping are 5–9 mm and in Lilla Torpa, Vänersborg 5–15 mm. The settlements are smaller than those calculated from the compression curves. Typical results of settlement calculations are given in *Table 2* overleaf.
(3) The result shows that when the pore pressure increases (due to natural fluctuations) the rate of settlement during the construction period

119

Table 1.

Building	Point	Founda-tion	Joists 1	Joists 2	Joists 3	Roof facade and equipment	Moving in	Last measure-ment
			(settlements in mm)					
Lidköping								
2	1	0.66	1.87	4.94	6.45	8.00	8.42	8.89
	2	0.60	1.35	3.80	4.97	5.89	6.19	6.47
3	3	0.64	1.04	1.87	4.39	6.93	7.24	7.94
	4	0.66	1.34	2.30	4.49	6.12	6.32	6.89
9	1	0.36	0.98	2.40	4.76	5.51	—	—
	2	0.73	1.76	3.51	5.46	6.34	—	—
	3	0.40	1.55	—	5.27	6.29	—	—
Vänersborg								
C	1	0.45	—	1.26	2.32	4.40	4.85	5.43
	2	1.36	—	4.43	5.05	12.99	13.62	15.88
	3	2.29	—	5.46	7.48	10.82	11.40	13.23
	4	2.59	—	6.68	8.76	12.54	12.91	14.38
F	1	—	0.11	1.86	2.55	5.09	—	—
	2	—	3.23	6.64	8.81	10.47	—	—
	3	—	2.99	6.08	80.2	10.47	—	—
	4	—	3.17	6.04	8.02	10.76	—	—
	5	—	2.99	5.38	7.34	9.53	—	—
H	1	0.43	2.47	3.34	4.51	7.16	7.44	8.81
	2	1.85	4.72	5.08	7.59	10.17	10.58	11.61
	3	1.29	4.00	4.24	5.65	8.65	9.33	9.67
	4	0.09	3.63	4.49	5.24	8.07	8.59	9.42

Table 2.

	Calculated settlements	
	elastic undrained (mm)	consolidation (mm)
Lidköping Building 2	30	43
Vänersborg Building C	30	21

decreases and vice versa. Thus the fluctuating pore pressure is important for the development of settlements also in overconsolidated clay.

(4) The results are preliminary. When observations have been made for three years a final report will be prepared which will give a careful analysis of the results.

Conclusions

In Lidköping and Vänersborg, in western Sweden, measurements of settlements of buildings founded on lightly overconsolidated clay have been made for about a year and will continue for a period of 10 years. The measurements up till now indicate that the settlements observed can be regarded as initial settlements and that they will cease after the construction period.

Lidköping (building 2)

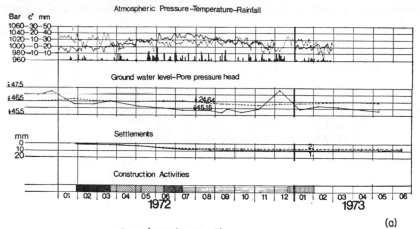

Atmospheric Pressure–Temperature–Rainfall

Bar c° mm
1060–30–50
1040–20–40
1020–10–30
1000—0–20
980–10–10
960

Ground water level–Pore pressure head

‡47.5
‡46.5   24.64
‡45.5   45.15

Settlements

mm
0
10
20

Construction Activities

| 01 | 02 | 03 | 04 | 05 | 06 | 07 | 08 | 09 | 10 | 11 | 12 | 01 | 02 | 03 | 04 | 05 | 06 |

1972   1973

(a)

Vänersborg (building C)

Atmospheric Pressure–Temperature–Rainfall

Bar c° mm
1040–20–40
1020–10–30
1000—0–20
980–10–10
960

Ground water level–Pore pressure head

‡49
‡48   43.30
      50.05 45.66
‡47   42.85

Settlements

mm
0
10
20

Construction Activities

| 01 | 02 | 03 | 04 | 05 | 06 | 07 | 08 | 09 | 10 | 11 | 12 | 01 | 02 | 03 | 04 | 05 | 06 |

1972   1973

(b)

Signs and symbols

■ Foundation	■ Joists 3	----- Atmospheric Pressure (Bar)
▨ Joists 1	▤ Roof, facade and equipment	—— Temperature (c°)
▨ Joists 2	▥ Moving in	-·- Rainfall (mm)

Fig. 4. Typical time–settlement curves

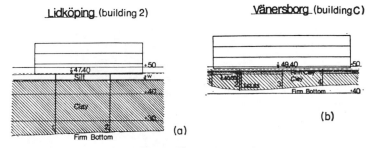

Lidköping (building 2)

‡47.40
+50
Sill
Clay
Firm Bottom
(a)

Vänersborg (building C)

‡49.40   50
Clay
Firm Bottom   40
(b)

Fig. 5. Observation points

121

The measurements also show that in spite of overconsolidation the development of settlements is strongly dependent on the variation of the pore pressure.

REFERENCES

Bjerrum, L. (1972). 'Embankments on Soft Ground'. *Specialty Conf. Perf. Earth & Earth-supported Struct*. Purdue University, Lafayette. Proc., Vol 2, Indiana
Hansbo, S. (1957). 'A New Approach to the Determination of the Shear Strength of Clay by the Fall-Cone Test', *Swed. Geot. Inst. Proc*. No. 14
Parry, R. H. G. (1970). 'Overconsolidation in Soft Clay Deposits'. *Géotechnique*, Vol. 20, No. 4, 442–446

II/8. Unusual settlements of a building at Nantua (France)

G. Sanglerat
Chief Engineer, Socotec-Professor of Soil Mechanics, Ecole Centrale de Lyon
L. Girousse
Engineer, E.E.G.—Soil Mechanics Engineer, Department of Socotec, Lyon
J. Gielly
Engineer, E.N.S.A.M.—Professor of Soil Mechanics at Institut de Technologie de Lyon

Summary

A low residential building was constructed in Nantua during the period 1964–66. During construction settlements and differential settlements of an important magnitude occurred which extended over a long period after construction.

This Paper presents details of the building design, the initial soil investigation (performed with a Static Penetrometer) and the results of the complementary geotechnical study, including sampling and laboratory testing made after construction of the building was completed.

Site and structure

The residential building whose plan dimensions are shown in *Fig. 1* was constructed between 1964 and 1966 in a swampy area in the Alps located near a lake fed by a small river. The building is 130.3 m in total length with three expansion joints. Neither the width nor the number of storeys is constant throughout its length. The ground cellular slab placed directly on the soil is partly embedded. At certain locations the load is G.F. + 1 (G.F. being the load due to the ground floor) or G.F. + 3 (ground floor plus three storeys) (see *Fig. 2*).

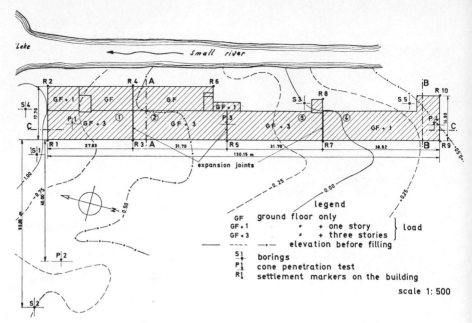

Fig. 1. Location plan

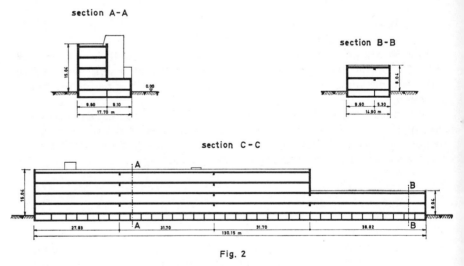

Fig. 2

Fig. 2. Residential building, Nantua

Foundation design and grading

The initial soil investigation consisted of making four static cone penetration tests (Delft cone) whose results are shown in *Fig. 3*. A cellular mat footing was adopted consisting of two slabs separated vertically by 0.9 m and 0.3 m and 0.2 m in thickness, reinforced by longitudinal and transversal beams (*Fig. 4*).

124

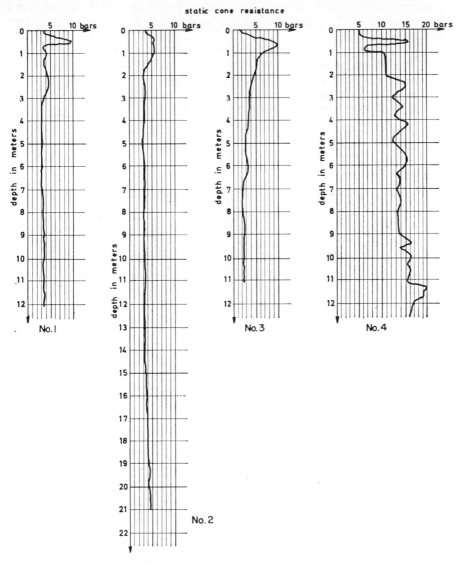

Fig. 3. Static cone penetrometer tests

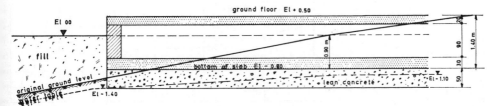

Fig. 4. Longitudinal cross-section of foundation

125

The lower of the two slabs exceeds the building plan dimensions by 0.4 m at the sides and by 2 m at the two front sections (R_1R_5) in order to better distribute the loads.

The embedment of the foundation slab varies from 0.9 to 1.4 m. The slab bottom is placed on lean concrete, 0.5 m thick (see *Fig. 4*).

The site was filled to an El. of 0.00. Any portion of the building site with original ground surface higher than El. 0.00 was left intact, that is to say, the south 1/5 approximately was not graded (see *Fig. 1* and *Fig. 5*).

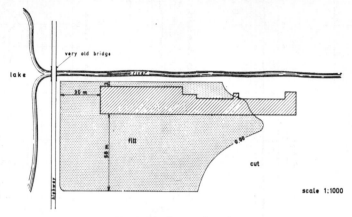

Fig. 5. Plan view of grading

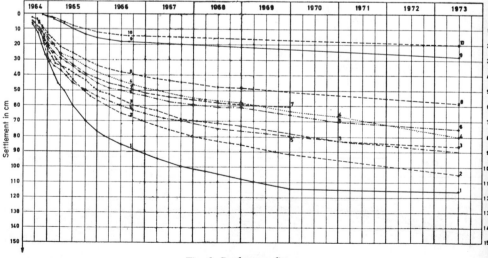

Fig. 6. Settlement diagram

126

Settlement observations

Settlement observations were started when the first floor slab of the first unit was finished. Construction was advanced towards the south, with one floor difference remaining between each unit (see *Fig. 1*).

In view of the unusually large settlements observed, five borings (S1 to S5) were made with recovery of samples and subsequent laboratory testing. Partial results of this second investigation are shown in *Tables 1* and *2* for S1 and S3 and *Fig. 7* which gives results of 4 of the 12 oedometer tests performed. The consolidation test results given correspond to S1, S3 and S4.

Settlements were measured once a week during the first 15 months and

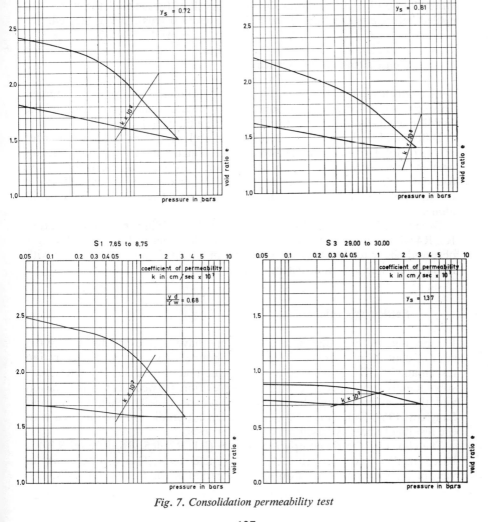

Fig. 7. Consolidation permeability test

127

once a month for a subsequent period of two years and every three months until 1969. The benchmark was located 35 m from the north side of the structure, on an old bridge foundation. In 1970 settlement measurements were interrupted in part for financial reasons but mostly because the engineer in charge left the company. At the special request of the authors four additional measurements were made on 29 Nov. 1972, 6 Feb. 1973, 3 May 1973 and 6 May 1973 and they showed that equilibrium was not quite reached since some differences could be found from measurements made three years ago. The measurements are plotted in *Fig. 6*.

The north-east corner (R1) of the structure settled about 4 ft in 4 years in relation to the design level and nominal differential settlements occurred between the building and the immediately surrounding terrain. The consolidation occurred over a fairly large area.

Numerous single-storey, one-family residential buildings erected on other sites in the neighbourhood showed severe differential settlements. One two-storey house tilted 18 cm from the vertical. All of these small structures had settled before construction of the building under discussion was started.

Consequences of the settlements

Settlements are partially due to the fill, and although there were no appreciable settlements between the building and adjacent fill, settlements of the building were slightly larger than those of the fill near R1.

Differential settlements between the east and west sides were of the order of 11, 6, 15, 14 and 8 cm. Differential settlements along the long axis are greater from one reference point to another. This is due to the fact that the foundation level varies with respect to the original soil level, as well as the thickness of the fill. Near R1, almost 50% of the settlement is due to the fill alone.

Expansion joints opened near the top of the building, mostly at location R3, R4 where, at the top, the separation is of the order of 10 cm. No damages occurred either to the structure or utility lines, which were all provided with movable connections at structural joints.

Settlements calculations

It is possible to show how settlements could have been estimated before the start of construction, based on the data obtained from the cone penetration tests made in 1964. This, of course, is the method of analysis which has been subsequently developed showing the correlation between cone penetration and the compressibility characteristics of clays (Gielly 1969, Costet and Sanglerat 1969, Sanglerat 1972). This settlement prediction is compared with the results of analysis based on oedometer tests.

The choice of the six comparison points (R1, R2, R5, R6, R9 and R10) is based on their proximity to cone penetrometer tests. This was necessary in view of the important variation of stratification of the foundation soils.

Method of analyses

Based on oedometer tests. Settlements were calculated from the results of consolidation tests. Some of the indices of compressibility had to be estimated from the nature of the soils compared with some known values of identical soils in the area.

Table 1. NANTUA— SOUNDING 1

Depth (m)	Soil Description	Undisturbed samples	$\frac{\gamma_h}{\gamma_w}$	$w\%$	$\frac{\gamma_d}{\gamma_w}$	Confining pressure (bar)	$\frac{\gamma_d}{\gamma_w}$	E (bar)	Oedo-meter	W_L %	W_p %	I_p %
1.60	Fill											
2.20	Peaty clay	1.90 H*	1.30	135	0.55	0.250	0.65	35				
			1.26	165	0.48	0.500		38				
	Grey silty clay with decayed	3.00 B*	1.65	80	0.92	0.750		40				
						1.000		47				
	vegetation	3.00 H*	1.14	190	0.29	0.250		22				
4.30			1.17	250	0.33	0.500	0.62	25				
	Grey silty clay with thin layers	4.40 B*	1.34	84.5	0.73	0.750		25				
						1.000		25				
	of decayed	H*	1.40	60	0.87	0.250		17				
	vegetation	7.65 B*	1.43	64.5	0.87	0.500	0.78	17	x	91	36.5	54.5
		B*	1.42	91	0.74	0.750		22				
		8.75				1.000		26				
		10.35 H*	1.45	77.5	0.82	0.250		17				
		11.35	1.50	73	0.87	0.500	0.83	22				
		B*	1.36	86.5	0.73	0.750		22				
						1.000		22				
		12.50 H*	1.56	53.8	1.01	0.250		10				
						0.500	1.03	14		47.5	23.5	24
		13.55 B*	1.59	49.8	1.06	0.750		22				
						1.000		22				
		15.35 H*	1.69	43	1.18	0.250	1.22	10				
			1.70	37	1.24	0.500		19				
		16.40 B*	1.67	35.4	1.23	0.750		20				
						1.000		22				
		17.80 H*	1.66	45	1.14	0.250	1.18	30				
			1.69	44	1.17	0.500		35		50.5	24.5	2.60
		18.85 B*	1.73	42.3	1.21	0.750		35				
						1.000		45				
		20.30 H*	1.73	31.2	1.32	0.250		17	x			
			1.73	34.2	1.32	0.500	1.32	18				
		21.35 B*	1.74	26.4	1.37	0.750		22				
						1.000		25				
		23.50 H*	1.79	36.5	1.32	0.250		13				
			1.72	36.3	1.26	0.500	1.30	17		44.5	25	19.5
24.55						0.750		22				
		24.55 B*	1.73	32.4	1.30	1.000		24				
	Plastic grey clay	26.00 H*	1.78	27.8	1.39	0.250		17				
			1.80	26.1	1.43	0.500	1.40	11				
		27.10 B*	1.81	24.6	1.45	0.750		23				
						1.000		19				
31.00												

* H = Top. B = Bottom of sample

129

Settlement calculations were made for the six reference points R1, R2, R5, R6, R9 and R10 considering the mat foundation to be flexible.

Foundation pressures are 0.11 bar for the mat and 0.07 bar for each floor. At locations R1, R2, R5, the weight of the fill was taken into account. The calculated settlements are in relation to the soil profile to a depth of 32 m. The results are summarised in *Table 3*.

It should be noted that for R1 the influence of the fill on settlement is very pronounced. It accounts for almost half (47%) of the total settlement. For point R5 it accounts for only 18% and is practically nil for points R6, R9 and R10.

Table 2. NANTUA—SOUNDING 3

Depth (m)	Soil description	Undisturbed samples	Physical properties			Triaxial			
			$\dfrac{\gamma_h}{\gamma_w}$	w %	$\dfrac{\gamma_s}{\gamma_w}$	Confining pressure (bar)	$\dfrac{\gamma_d}{\gamma_w}$	E (bar)	Oedometer
1.10	Fill								
	Peaty clay	1.35 H*	1.52	78	0.86	0.250	1.07	23	
			1.67	57	1.07	0.500		45	x
		2.40 B*	1.42	107	0.69	0.750		50	
2.95						1.000		50	
	Grey plastic clay with thin layers of decayed vegetation	4.15							
		5.20							
		7.00 H*	1.40	98.5	0.70	0.250		35	
			1.43	88.5	0.76	0.500	0.65	25	
		B*	1.47	80.5	0.81	0.750		35	
		8.05				1.000		41	
		9.50							
		10.55							
11.50									
	Grey-black clay with vegetation								
13.45									
	Grey plastic clay	14.65 H*	1.59	69	0.94	0.250		45	
			1.69	55.8	1.06	0.500	0.99	48	x
		B	1.60	53.3	1.04	0.750		51	
		15.70				1.000		88	
		19.80							
		20.85							
		25.00 H*	1.79	24.5	1.44	0.250		22	
			1.82	24.5	1.46	0.500	1.46	64	
		B	1.76	23.5	1.43	0.750		64	
		26.05				1.000		77	
		28.95 H*	1.81	32	1.37	0.250		13	
			1.88	33	1.41	0.500	1.41	45	x
		B	1.84	33.2	1.38	0.750		57	
						1.000		64	
30.00									

* H = Top. B = Bottom of sample

Authors' method.—The insufficient depth of penetration of the cone penetrometer tests have lead to an extrapolation of the most unfavourable assumption. Penetration graphs only extend to a depth of 32 m.

For each layer considered, the values of the coefficient α are taken from the preceding tables presented by the authors (Sanglerat, 1972). The results are summarised in *Table 3* below, in which the observed settlements have been

Table 3. SETTLEMENTS IN CENTIMETERS

Method	R_1	R_2	R_5	R_6	R_9	R_{10}
Oedometer	152	126	74	45	8	7
Authors	25–66	20–51	40–103	25–64	5–12	4–10
Settlements observed	116	105	90	75	28	20

included. It appears that the data from the cone penetrometer test P4 is erroneous. No account was made of this particular test.

The two methods give only approximate estimates of settlements in view of the heterogeneity of the soil, the variation in elevation of the original ground surface, the water table and the surcharge imposed by the building and the fill thickness.

Taking into account these uncertainties, the calculated settlements remain, by and large, in agreement with the observed settlements except for points R1 and R2. The authors' method for these two points greatly underestimates the settlements. This is due to the fact that the values of σ_o' are exceptionally low immediately under the foundation ($\gamma' = 0.2$ t/m^3). This influences directly the ratio R_p/σ'_o or q_c/σ'_o which was the basis for the determination of α.

In spite of this difficulty, the authors' method, had it been used for the original study of the site, would have indicated that important settlements were to be expected and that in accordance with the authors' recommendations, for values of R_p or $q_c < 12$ bar it would have been necessary to recover undisturbed samples in order to determine the water content of the soft soils. These values between 80 and 200% would lead to an estimated value of $C_c > 1$ indicating the necessity of additional laboratory testing.

Acknowledgement

The authors are greatful to Mr. Justamon of Pellegrini, Contractors, who collected all settlement data.

REFERENCES

Costet, J. and Sanglerat, G. (1969). *Cours pratique de Mécanique des Sols, Dunod,* Paris
Gielly, J., Laréal, P. and Sanglerat, G. (1969). 'Correlation Between In Situ Penetrometer Tests and the Compressibility Characteristics of Soils', *In situ Invest. Soils, Rock.* 167–172, 189 and 191, London
Sanglerat, G., Laréal, P. and Gielly, J. (1972). 'Le Pénétrometre Statique et la Compressibilité des Sols', *Annales I.T.B.T.P.,* No. 298, October, Paris

II/9. Settlements of three heavy sugar silos in Italy

G. Vefling
Consulting Engineer, Cowiconsult (formerly Chr. Ostenfeld & W. Jonson), Consulting Engineers, Denmark

Summary

It is the purpose of the present Paper to communicate some observations of settlement of three pile-supported sugar silos in the lower Po Valley area in Italy: a group of two sugar silos (capacity 150 and 200 MN) and a single silo (capacity 270 MN). All three silos are supported on very long piles.

The settlements amount to 15–30 cm and the rate of settlement seems to be constant. The effect of the time lag between pile-driving and load application is discussed, as well as the effect of the annular load variation.

For several reasons neither the geotechnical information nor the settlement observations are complete, but it is felt that even the incomplete data might be of interest for further study in view of the unusual size of the loads.

Silo structure

The three silos have a cylindrical shell of prestressed concrete. The silo substructure can be considered rigid as it consists of a lower 1 m thick reinforced concrete raft connected with an upper 0.6 m thick silo floor by a ring wall and numerous columns in the basement, which serves as a sugar recovery area.

Two silos, Russi No. 2 and Molinella, are provided with a central tower. See *Fig. 1*.

Two sugar silos in Russi

In Russi, not far from Ravenna, on the Italian coast of the Adriatic Sea, a prestressed concrete silo of 150 MN capacity was constructed in 1962. In 1968 a new silo of 200 MN capacity was constructed close to the first one. The

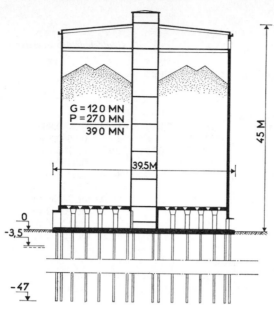

G = 12 0 MN
P = 27 0 MN
390 MN

39.5 M

45 M

0
-3,5

-47

Fig. 1. Molinella, 1970

Table 1. RUSSI

	Silo 1	Silo 2
Year of construction	1962	1968
Dead load (MN)	76	95
+ Sugar load (MN)	+150	+200
−Excavated soil (MN)	− 10	− 15
Total load (MN)	216	280
Height × diameter (m)	42 × 28	42 × 32
Length of piles (m)	33 − 35	34 − 36
Total No. of piles (32 × 32 cm)	226	378
Working load per pile, max. kN	1000	800

main data of the silos are as shown in *Table 1*. Geotechnical borings showed the subsoil to consist of deltaic deposits to considerable depth. Most of the soil belonged to the silt fraction and there were few marked boundaries between different layers. For this reason it is preferred to give a simplified synthesis of the soil conditions in *Table 2*, together with the relevant test results.

Both silos are supported on precast, driven reinforced concrete piles, 33–36 m long and with a square section 32 × 32 cm. *Fig. 2* shows the load–settlement–time relation. It should be noted that the time–load curves are essentially correct (based upon indication from the factory) up to 1969. After that time the time–load curves are symbolically drawn. The following developments took place:

Silo 1 was completed in 1962 and remained unused for more than a year

133

Table 2. RUSSI

Depth (m)	Type of soil	w %	w_P %	w_L %	e	c_v (kN/m²)	m_v (m²/kN)
					Geotechnical properties		
0–8	Clayey silt, traces of peat	28–33	23–27	41–50	0.69–0.85		2.3 × 10–4
8–23	Silt	23–31			0.66–0.86		1.4 × 10–4
23–30	Clay and silt	36	23	39	0.91–0.96		1.6 × 10–4
30–32	Sand	17–24			0.63–0.73		
32–34	Clay and silt	23–33	20–27	38–50	0.62–0.90	150–200	
34–36	Sand				As for 30–32 m		
36–38	Clay and silt				As for 32–34 m		
38–39	Sand				As for 30–32 m		
39–42	Clay and silt				As for 32–34 m		
42–47	Sand				As for 30–32 m		

m_v = 'coefficient of volume compressibility'
w = natural water content
$w_P w_L$ = plastic limit and liquid limit
c_v = shear strength measured by vane tests
e = void ratio

until the factory was ready to start production. During this period the settlement was about 17 mm.

The silo was filled for the first time in the autumn of 1963. The sugar load of 127 MN caused an additional settlement of 30 mm to develop within one month. Only a very slight heave occurred during the following period of emptying.

In March 1968 Silo 1 had settled minimum 110 mm and maximum 120 mm. Unfortunately only a few levellings were taken between 1964 and 1968, but assuming a constant rate of settlement (which is supported by later observations), the yearly increase was approximately 20 mm.

At about that time Silo 2 was constructed next to Silo 1, the free distance between the silo walls being only 5 m. The pile-driving for Silo 2 caused a small tilting of Silo 1 towards Silo 2, corresponding to a differential settlement of about 8 mm. Furthermore, the pile-driving and/or the filling of Silo 1 (which happened to take place at the same time), caused an increase of the overall settlement of Silo 1. The dead load of Silo 2 did not affect the settlement of Silo 1.

At its first filling Silo 2 settled about 100 mm, i.e. almost as much as was reached by Silo 1 *after 5 years*. While Silo 2 remained vertical, Silo 1 underwent a tilting, its settlement increase being minimum 10 mm and maximum 60 mm. Thus, the differential settlement of Silo 1 had reached approximately 80 mm, corresponding to a tilt of between 1:350 and 1:400.

During the years 1968–73 observations were again scarce, but there is reason to believe that the rate of settlement has remained constant for both silos, approximately 20 mm per year. Silo 2 is still almost vertical, while the tilt of Silo 1 has reached 1:300.

Although the observation material is far from perfect, it seems to present two interesting factors:

(1) With due allowance for the disturbance caused by Silo 2, the rate of

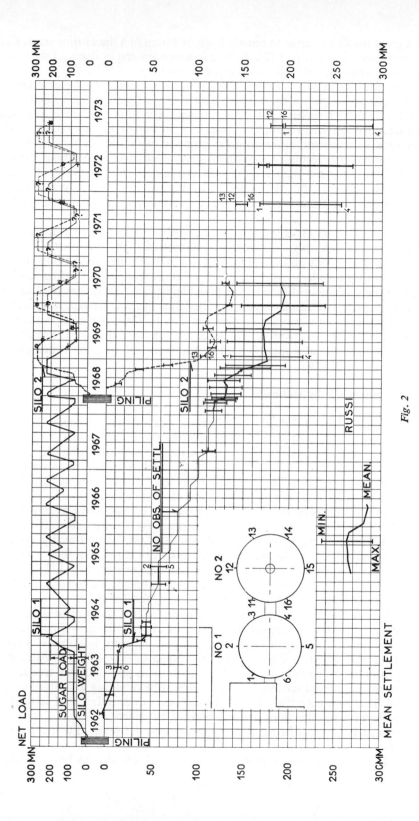

Fig. 2

settlement for Silo 1 seems to remain fairly constant in a linear time scale. For the settlement–log time (see *Fig. 4*) there is no convincing evidence of the usual flattening which ought to have occurred by now.

As mentioned in another contribution to this Conference by the author, a similar phenomenon has been demonstrated by L. Bjerrum for a number of directly founded structures, all of which have a very considerable live load compared with the dead load. The Russi silos, although they also show a constant rate of settlement, differ from this group in two respects: firstly, they are pile-supported and secondly, the annual increase of settlements, about 20 mm, is substantially larger than for the other structures in this group. It seems reasonable, all the same, to ascribe at least a part of the yearly increase of settlements of the Russi silos to the regular variations of a substantial part of their load.

(2) It may seem surprising that Silo 2 settled about 100 mm at the first filling, while Silo 1 only settled 30 mm. Admittedly, the live load for Silo 2 was 30% greater, but on the other hand it was distributed over a 30% larger area (67% more piles). A comparison of the exploratory borings cannot explain the settlement difference. A possible explanation is that the full live load was applied to Silo 2 only six months after the pile-driving, whereas the corresponding period was 20 months for Silo 1. Thus, the silty and clayey layers around the pile group of Silo 2 had less than one-third of the regeneration time of Silo 1 until the full load had to be transferred from the piles.

Sugar silo in Molinella

In Molinella, between Bologna and Ferrara, a prestressed concrete sugar silo was constructed in 1970. The main data of the silo are as shown in *Fig. 1*. It will be seen that, with its 270 MN capacity, the silo is considerably larger than the above-mentioned silos in Russi. Dead load + live load amount to $120 + 270 = 390$ MN. The silo is supported by 392 precast reinforced concrete piles, 35×35 cm, driven to a depth of 47 m.

The subsoil consisted of alternating deposits of fine silt and clay to a depth of more than 70 m. Even at great depths peat layers were encountered and some thin sand layers were also found, although they were too thin to give partial support for a pile foundation as in the case of Russi. The shear strength of the soil as measured by vane tests increased regularly from 40 kN/m^2 near ground level to 100 kN/m^2 at 50 m depth. An outline of the soil conditions is shown in *Table 3*, the symbols being the same as for *Table 2*.

During the first two years, after the start of the concreting, levellings were taken every one or two weeks. *Fig. 3* shows the load–settlement–time relation. A conventional settlement calculation based on the consolidation tests leads to a total settlement of the order of magnitude of 300–400 mm. After two fillings the mean settlement was 150 mm, still with an increasing trend. The maximum differential settlement was 35 mm, corresponding to a tilt of nearly 1:1000. It can be seen that the rate of increase of the differential settlement is considerably smaller than that of the total settlement.

The settlement must be expected to increase in the years to come, for two reasons:

(1) The primarly consolidation can hardly be regarded as having ended.

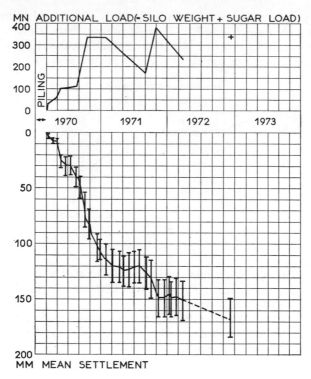

Fig. 3. 270 MN silo, Molinella

Table 3. MOLINELLA

Depth (m)	Type of soil	w %	w_P %	w_L %	e	c_v (kN/m²)	m_{vr} (m²/kN)
				Geotechnical properties			
0–8	Silt, partly organic	32–40	19–35	35–55	0.72–1.31	40	1.8×10^{-4}
8–13	Sand and silt	25–33	18–26	31–67	0.69–0.91	55	0.5×10^{-4}
13–32	Silt, clayey, with organic layers	23–47	17–29	32–89	0.64–1.20	60	0.9×10^{-4}
32–34	Sand, silty	22–25	17–20	40–53	0.72–0.82	75	—
34–45	Silt, clayey, with lenses of chalk	25–35	19–29	30–98	0.66–1.02	86	0.7×10^{-4}
45–48	Silt, clayey	22–30	17–23	32–71	0.62–0.70	90	0.6×10^{-4}
48–49	Silt, sandy	22	17	32	0.62	90	—
49–56	Silt, clayey, with organic content and lenses of chalk	20–43	19–27	29–82	0.56–0.88	140	0.9×10^{-4}
56–57	Silt, sandy	38	25	72	1.12	180	0.8×10^{-4}
57–71	Silt with clay and some organic matter	19–53	22–30	40–97	0.70–1.42	180	0.8×10^{-4}

137

According to a rough estimate, about two-thirds of the total primary settlement should have been reached by now. However, there is every reason to question this because a proper evaluation of the three-dimensional effect of the consolidation is rather difficult under the present conditions.

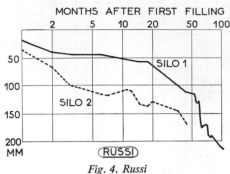

Fig. 4. Russi

(2) The 'secondary settlements', which may be caused by large annual load variations, and which were tentatively indicated for the Russi silos, might easily amount to 10–20 mm per year.

It is hoped that regular levellings of the silos can be resumed, at least to some degree. If so, a note on the further progress of the settlements in Russi and Molinella will be published in due course.

Acknowledgements

The author is indebted to the late Dr. L. Bjerrum for his valuable inspiration and advice during the preparation of the present paper. The author would also like to acknowledge the assistance given by Mr. K. Bennick, M.Sc., Cowiconsult, and Mrs. Ses Inan Kiilerich, M.Sc.

The levellings were carried out by the sugar factory 'Eridania', Succherifici Nazionali, Genua (Russi), and the contractor Ferrobeton, Venice (Molinella). The sugar factory has kindly permitted publication of the results.

The silos mentioned in the Paper were designed by the consulting engineers Cowiconsult (formerly Chr. Ostenfeld and W. Jonson), Copenhagen.

SESSION III

Heavily overconsolidated cohesive materials

III/1. Time-settlement and settlement distribution with depth in Frankfurt Clay

H. Breth
Professor of Soil Mechanics

P. Amann
Institute of Soil Mechanics and Foundation Engineering, Technische Hochschule, Darmstadt, West Germany

Summary

The Frankfurt am Main subsoil consists of tertiary heavily overconsolidated clay down to great depths. For the purpose of studying the influence of the rigidity of the superstructure on the settlement behaviour of the soil, the settlement and the deflection of the foundations of eight multi-storey buildings were measured during and after construction. The report includes a table (*Table 1*) covering the compiled results of these measurements and outlines the settlement behaviour of the buildings with the time. Under one of these tall buildings, which is 116 m high, five extensometers measured the compression of clay down to a depth of 25 m underneath the foundation level. The extensometers were installed in locations where the rigidity of the superstructure does not influence the compression of the clay. The most essential characteristics of construction and installation of the extensometers are described. The measured results are represented graphically as time-settlement curves and depth-settlement-distribution curves. According to these measurements the profile of settlement is substantially higher than for the elastically isotropic half-space. Almost half of the entire settlement occurred in the 5 m directly below the foundation level. The causes leading to such a high settlement focus and the large proportion of initial settlement in total settlement are discussed. Extensometer readings were taken in a 24 m deep excavation during the excavating activities with a view to examining the effect of swelling. These readings represent the effect of swelling on the heave of the foundation level and the underlying soil during excavating. Finally, the report discusses the practical effects of these studies in engineering practice.

Introduction

For several years, the Institute für Bodenmechanik und Grundbau of the Technische Hochschule in Darmstadt has been conducting investigations on deformations and bending stresses of foundations. The influence of the rigidity of the building frame of a multi-storey construction on the differential settlement, distribution of contact pressure and bending stresses of the foundation slab had to be studied. In addition, these efforts had to cover the influence of time such as the creeping of concrete and the time–settlement behaviour of the clay. For these purposes, Heil (1969–71) developed a method of calculation. For the purpose of verifying the theoretical results, the settlement and deflection of the foundation slabs of a number of tall buildings in Frankfurt am Main were measured; eight of these multi-storey buildings are founded on the local tertiary clay. Heil investigated buildings (1) and (2) in *Table 2*. These evaluations indicate that the multi-storey buildings behave much more flexibly than the calculations would suggest. On the basis of the computed stresses damage which might have been expected was not observed.

The calculations assumed a linearly elastic behaviour of concrete and soil. The difference between practical measurements and computed values indicate a deviation of the deformation characteristics of both materials from the theoretical assumptions used in the calculations. As far as the clay in question is concerned, the properties of which will be described later on, the determination of its stress–strain behaviour gives rise to major difficulties in the laboratory. This fact is largely attributable to the susceptibility of the clay to disturbance during sampling and the scatter of the experimental values. Furthermore, the previous history and thus the present stress situation of the clay are unknown. It was therefore decided to perform settlement measurements in the clay at various depths underneath the building of the Division for Educational Sciences of the University of Frankfurt (AfE) in order to obtain more reliable information on the deformation behaviour of the clay. This investigation was not limited to the subject under discussion. It was extended to cover the stress–strain behaviour of the soil in general and thus lends itself to most construction activities in the field of foundation engineering. The knowledge obtained by way of settlement measurements in the subsoil facilitates an improved understanding of the settlement behaviour of clay. Together with the other settlement readings they also helped to review the theoretical assumptions on which the computed settlement values are based.

This paper contains a survey of the settlements of the eight multi-storey buildings. The settlement observed on the AfE-building is described in greatest detail. The measurement of the settlement distribution with depth is the most important aspect of the study. Further measurements of the heaving during excavation are described. Apart from attempting a short explanation, our report is limited to indicating the measured results. A more far-reaching evaluation of the extensometer readings with special consideration of the material behaviour of the clay was carried out by the second mentioned author within the framework of his doctor's thesis; additional details of construction and installation of the measuring devices are appended to this thesis (Amann, 1974).

142

Subsoil

'Frankfurt Clay' is defined as intermittent layers of tertiary clay interbedded with banks of limestone and layers of lime-, hydrobic- and shell-sand which are found in the Frankfurt area under layers of quaternary sand and gravel in depths of more than 7 to 10 m (Breth *et al.* 1970). The proportion of clay in the log differs locally as a function of the depth. In the upper 40 m it averages 85%. The index properties of clay are known from a great number of samples. They are indicated in *Table 1*.

Table 1. INDEX PROPERTIES OF FRANKFURT CLAY

		Average	*Maximum*	*Minimum*
Wet Unit Weight	g/cm³	1.85	2.00	1.70
Void Ratio	%	50	58	40
Water Content	%	35	45	22
Liquid Limit	%	68	80	40
Plastic Limit	%	40	48	18
Consistency	—	0.82	1.1	0.4
Saturation	%	94	100	80
Grains Smaller than 0.002 mm	%	38	50	20
Activity	—	1.0	1.3	0.7
Unconfined Compressive Strength	kg/cm²	3.0	5.5	1.3
Friction Angle (cd)	0	20	25	16
Cohesion (cd)	kg/cm²	0.2	0.6	0.1
Compression Index (0–5 kg/cm²)	kg/cm²	140	250	80

This clay is geologically prestressed, laminated and fissured like most overconsolidated clays. It is calcareous and is therefore frequently referred to as clayey marl. The interbedded layers are water-bearing and often not continuous. They 'float' in the clay. *Fig. 7* which will be discussed in more detail at a later stage shows the log of a 100 m deep bore hole drilled near the AfE building; the upper section of this bore profile may be considered typical for the Frankfurt area.

Settlements of multi-storey buildings on Frankfurt Clay

Table 2 gives a survey of the size of the settlements and deflections measured in Frankfurt and outlines the different dimensions, foundation depths, and weights of the buildings all of which have foundation slabs. With the exception of the AfE building the settlements were only measured after completion of the foundation. The previous settlements were estimated and are included in the settlements indicated in *Table 2*.

In view of the present modes of construction and the short time required for carcase work, the time–settlement of multi-storey buildings becomes particularly important. *Fig. 1* illustrates the scatter of time–settlement curves of the buildings mentioned in *Table 2*.

The time of reference was understood to be the date when the building concerned was finished in the rough. The curves show relatively little scatter

143

Table 2. SETTLEMENTS OF MULTI-STOREY BUILDINGS FOUNDED ON FRANKFURT CLAY

Characteristic data	(1) Hotel (IHC)	(2) University (Juridicum)	(3) University (AfE)	Building (4) Bureau (Zürich I)	(5) Bureau (Zürich II)	(6) Bureau (BHG)	(7) Apartment-house (Biegwald)	(8) Library (Uni)
Height of the building (m)	68	48	116	66	55	82	47	26
Foundation area $b \times a$ (m)	19 × 73	14 × 96	42.5 × 43.5	22 × 22	23 × 35	22 × 48	13 × 22	32 × 52
Foundation depth (m)	6	7	13	7	8	12	4	8
Average contact pressure from weight (kg/cm²)	2.2	1.8	3.3	2.3	2.3	2.3	2.3	0.4
Average contact pressure from live load (kg/cm²)	0.6	0.4	1.2	0.5	0.5	0.6	0.2	0.07
Settlement before finished in the rough (cm)	5.0	4.2	15.0	6.5	4.2	7.7	2.4	1.2
Final settlement (cm)	9.6	6.7	(23.0)	10.0	9.3	11.0	4.2	2.0
Maximum deflection of the foundation f_a/f_b (cm)	3.0/0.7	4.0/− 0.4	(1.0/0.9)	—	−0.9/0.7	2.7/0	0.4/—	1.1/ ± 0.5

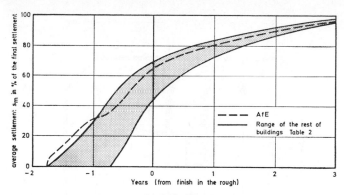

Fig. 1. Range of time-settlement curves of multi-storey buildings founded on Frankfurt Clay (Schwab and Wolff, 1971)

which is due to differences in the sequence of construction procedures. All buildings manifested a similar behaviour. Roughly three years after the buildings had been finished in the rough, the settlement had almost completely stopped. 45 to 70% of the final settlement was reached after the end of carcase work.

Settlements of the building AfE

As compact and heavy as possible a building was selected for settlement measurements with depth. The influence of the adjacent area was to be restricted to a minimum. The building should be clearly structured in order that load stages might be accurately determined. These requirements which correspond to a large-scale loading test were fulfilled by the selected building AfE.

Details of the Building

In its final state, the building AfE was to consist of a multi-storey building and a flat part connected with the side of axis 7, *Fig. 2*. Until now only the multi-storey building and the basement of the building connecting the tall building and the flat building have been completed. The settlement distribution with depth was measured underneath the multi-storey building, which is 116 m high and has a foundation of 13 m depth (*Fig. 3*). The foundation consists of a square slab of 3.6 m thickness and a side length of 43 m on average. The details of the structure have been described by Beck, Schneider (1972). After deducting the uplift of 0.7 kgf/cm² the surcharge due to the structure is 3, 3 kgf/cm² on average. The value of live load used in the static calculations was 1.2 kgf/cm². Apart from the wind load, the load on the foundation is nearly centric.

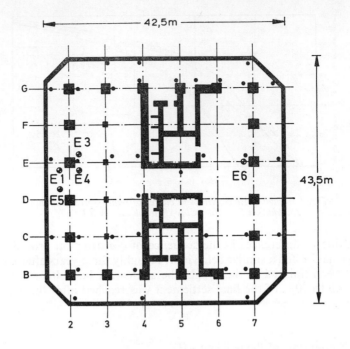

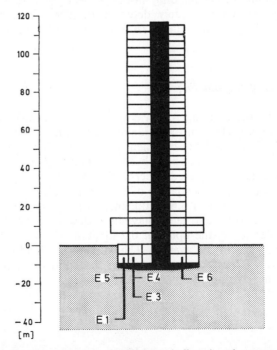

Fig. 2. Plan elevation of the foundation (AFE) with grouping of the extensometers and settlement marks

Fig. 3. Cross-section of the AFE building (tall part) with extensometers

Special extensometers were developed and constructed by the above mentioned institute. *Fig. 4* shows a schematic representation of the extensometer and the measuring device. The inner rod is firmly anchored in the soil (point A) whereas the outer pipe is attached to the foundation slab (point B), (*Fig. 4a*). In order that no forces be transmitted to the inner rod, the outer pipe is furnished with telescoping joints which are 2 to 6 m apart depending on the length of the extensometer. In the vicinity of the seal the protective pipe forms a flanged tube. At the upper edge of the outer tube, the settlement S_B of the building is measured whereas the settlement S_A of the anchor is measured at the upper edge of the inner rod by way of high-precision levelling, *Fig. 4b*. As the levelling procedures are very expensive, a scale

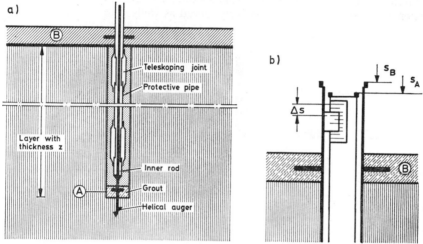

Fig. 4. Schematic drawing of the extensometer and measuring device

with nonious was also installed. It indicates the difference between the mentioned settlements, i.e. the compression ΔS of the soil layer with the thickness z between the foundation level and the anchor.

Grouping and installation of the extensometers

As the rigidity of the structure also influences the settlement distribution with depth, the extensometers were installed wherever this influence is restricted to a minimum. For the purpose of determining these locations considerable computation by difference equations was required for the determination of the settlement distribution under various points of the foundation slab as well as for differing system rigidities (Breth and Heil, 1965). These calculations showed that two points, each on the symmetric axes E and D, fulfil the above condition. In order to keep the influence of the flat building on the measurements to a minimum, the main extensometers were installed on the side of the building opposite the flat section (*Fig. 2*). A total of five extensometers were installed. The extensometers of the main group E1, E3 and E4 are anchored 25.3, 13.8 and 3.8 m below the bottom of the foundation slab (*Fig. 3*). The extensometers E5 and E6 at 3.5 and 4.3 m depth are additional monitors of the compression of the uppermost layers. E6 was furthermore intended to monitor the forthcoming influence of the flat building. In addition, the deformation of the foundation slab was measured at 35 points. As excavating might have been made difficult and the extensometers could have been damaged they were installed from the floor of the foundation. After completion of the drill-hole, the pipes and telescopic joints were screwed together and the rod was lowered with the drilling rope, inserted in the clay with rotational movements and thereafter the anchor-point was grouted. After filling the drill-hole, the outer tube was anchored in the subconcrete, so that settlements could be measured while the foundation slab was concreted. The excavation was kept dry by means of open drainage which required dewatering of the banks of limestone. The drill-holes of the group of extensometers therefore alternatively served as wells.

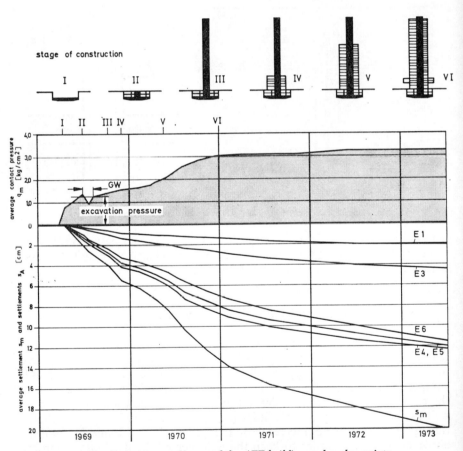

Fig. 5. Average settlement of the AFE building and anchor points

The bottom part of *Fig. 5* indicates the measured average settlement of the building and the related settlements of the extensometers plotted against time. The upper part of the same figure discloses the increases in weight of the building with the most important stages of construction. The concreting of the foundation slab took three weeks. The settlements were plotted from this date. The weight of the slab amounting to 0.8 kg/cm^2 caused a settlement of 1.4 cm. As the plotted curves indicate the settlement ratio at all depths depends on the corresponding stage of construction. After the building has been finished in the rough an average settlement of 13.6 cm had occurred without the initial settlement caused by the foundation slab. The settlement of the AfE building measured during construction may be assumed to account

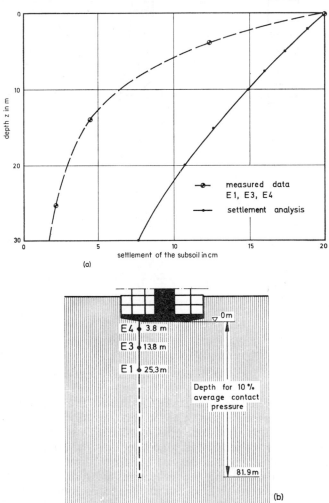

Fig. 6. (a) *Settlement distribution with depth* (b) *Position of anchor points in the extensometer group*

for 65% of the final settlement in accordance with the settlement values obtained for the other buildings. The time–settlement curve of the AfE building shown in *Fig. 1* is based on this assumption. This curve is well in the scatter obtained for the other buildings. Until now, an average settlement of 20.0 cm was measured after concreting of the foundation. *Fig. 5* indicates that the settlement values decrease rapidly with increasing depth.

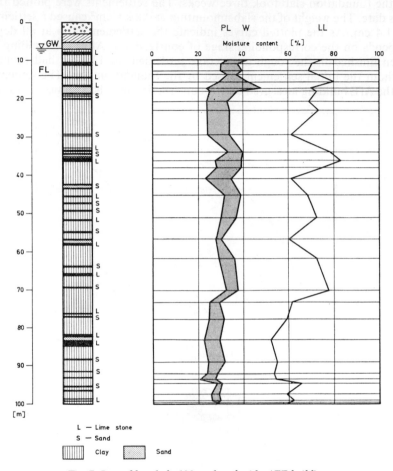

Fig. 7. Log of borehole 100 m deep beside AFE building

The differences revealed by the flat extensometers are insignificant. Extensometer E6 which is the deepest of all, nevertheless discloses too little settlement as compared with extensometers E4 and E5. This is due to the fact that the subsoil is relieved of some of its load after the excavation for the building connecting the multistorey and the flat building. For a better evaluation of the depth effect of the building, the settlements measured in the extensometer group are plotted according to depth in *Fig. 6a*. This diagram furthermore covers the settlement–distribution based on a settlement calculation which also reveals the same average settlement of the building. The settlement calculation is based on the assumption of a homogenous half-space with a

150

constant modulus of rigidity. In this connection, the stresses were considered for a depth where 10% of the contact pressure still occurs. This depth amounts to roughly twice the width of the foundation . *Fig. 6b* is a comparison of the installation depths of the extensometers with the depth of the influence of the building assumed for the purposes of calculation. *Fig. 6a* shows that approximately half of the entire settlement occurred in the upper 5 m. Below extensometer E1, 2 cm settlement or 10% of total settlement were ascertained. According to the settlement calculation 9 cm or about 45% of the settlement of the building should be expected.

To prove that the measuring results were not attributable to a change of the subsoil a 100 m core hole was drilled subsequently, the result of which is shown in *Fig. 7*. Clay was found over the entire depth. The share of sand and lime banks in the clay amounts to 14%, with single layers never exceeding a thickness of 1 m. It is very unlikely that these layers would essentially influence the deformability of the clay. In addition to the drill profile (*Fig. 7*) the plasticity and consistency of the clay are indicated. In accordance with these findings no essential changes in the physical properties of the clay have been detected with increasing depth.

Discussion

The stress state due to the weight of the soil and its non-linear stress-strain-relationship are considered to be the causes of the low involvement of the deep-laying soil strata in total settlement. The dependence of the moduli of deformation on the state of stress has not so far been investigated for a stiff clay. Such studies have, however, been produced for sands, Breth and Schuster (1971) and Domaschuk and Wade (1969). In accordance with their results, the shear and bulk modulus increase with a growing overall state of stress. The shear modulus decreases with an increasing share of the deviatoric stress in the total state of stress. In accordance with triaxial tests, similar tendencies are to be expected for Frankfurt Clay. The weight of the soil causes an increase of the stress state with depth. The shear stresses caused in the soil by the building are greatest in the vicinity of the foundation and decrease rapidly with increasing depth. In agreement with the measured results, the above-mentioned dependence leads to an upward shift of the settlement focus. Such a connection had already been expected by Breth and Back (1963) after comparing the settlements of two buildings with very different foundation areas. Settlement results mainly from shear strain, which also explains the substantial share of immediate settlement in total settlement. This fact is furthermore supported by the fissures in the clay and the water permeable intermediate layers. A more detailed explanation of the connections and interrelated mechanisms was given by Amann and Breth (1973).

The measurements indicated in the present paper were evaluated further by means of a finite-element computation under the assumption of a non-linear stress–strain relationship (Amann, 1974). One approach was chosen in connection with the routine triaxial tests of clay samples 3.6 cm diameter following a suggestion by Duncan and Chang (1970). The results thus obtained coincide well with the measured values.

Measurements of the heaving during excavation

Excavating leads to an unloading of the soil underneath the future foundation level. This relief causes a heaving action. Underneath the excavation there also occurs shear deformation of the soil in the opposite direction, however, from the one later on observed under the finished building. On the other hand, swelling of the highly plastic clay may also occur underneath the subgrade surface due to water penetrating into this area; this offers a further explanation of the high location of the settlement focus.

In order to examine this question, heaving measurements were taken during the digging of a 24 m deep 70 m square strutted excavation in Frankfurt clay. The total heave of the planned subgrade surface and the later foundation

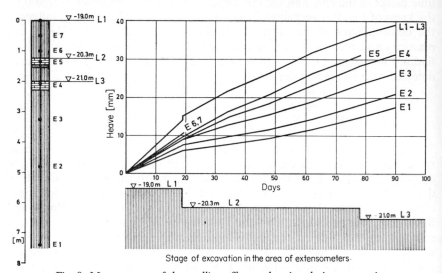

Fig. 8. Measurement of the swelling effect on heaving during excavation

surface amounted to 14 cm. Seven extensometers were installed for the purpose of studying additional possible swelling actions. As these extensometers had to be removed before sealing they were of a much simpler construction than the ones chosen for the AfE building. The outer protective tube was a smooth P.V.C. tube with glued joints. Interaction between the outer tube and inner rod could only happen at the anchor. The depth of anchor points of the extensometers E1 to E7 is indicated in *Fig. 8*. The extensometers were put in place as soon as the excavation had reached a depth of 19 m (L1). The measuring points in the respective excavation level were moved to L2 at 20.3 m and L3 at 21 m as excavation work progressed. During the first and third measuring period (L1, L3) the floor of the excavation was relatively dry in the area surrounding the extensometers whereas the floor of the excavation was wet during the second measuring period. *Fig. 8* shows that subgrade surfaces and anchors were lifted equally with time. Swelling action such as that arising in a consolidometer in case of minimum water penetration could not be observed in the excavation.

152

Summary and conclusions

The settlement measurements taken under multi-storey buildings in Frankfurt show that the greatest proportion of all settlement is immediate settlement. According to extensometer readings under the AfE-building 90% of the entire settlement occurred at a depth not much exceeding half of the overall width of the building. Both of these observations are of practical importance. It is a consequence of the large proportion of immediate settlement that even short-term loads arising during construction may lead to permanent inclinations of parts of the building (Leonhardt, 1973). Owing to the upward displacement of the settlement profile the settlement depression becomes narrower. Thus the effect on the adjacent area is reduced. However, major settlement differentials occur in the immediate vicinity of the building (Sommer, 1973). In continuation of the work by Heil, an examination will have to be carried out to show how the measured depth effect influences the bending moments in the foundation. Its influence will also depend on the location and position of the load. After due account of the measured settlement-distribution the designed bending moments were reduced by as much as 20% for the foundation slab of the AfE building (Hedberg and Schmick, 1972).

Acknowledgements

The measurements referred to in the report were financially supported by the Deutsche Forschungsgemeinschaft. The authors wish to thank the owners of the buildings as well as the contractors for their help in obtaining the measurements.

REFERENCES

Amann, P. (1974). 'Über den Einfluß des Verformungsverhaltens eines steifplastischen Tons auf die Verteilung von Spannungen und Verformungen im Baugrund infolge Bauwerkslasten', *Mitteilungen der Versuchsanstalt für Bodenmechanik und Grundbau,* Heft 13. 1, Darmstadt

Amann, P. and Breth, H. (1973). 'Das Setzungsverhalten der Böden nach Messungen unter einem Frankfurter Hochhaus, Vorträge der Baugrundtagung 1972 in Stuttgart'. Deutsche Gesellschaft für Erd- und Grundbau e.V., Essen

Beck, H. and Schneider, K. H. (1972). 'Tragwerk des Hochhauses AfE der Universität Frankfurt/M.', *Beton und Stahlbetonbau,* Berlin, Heft 1, 1–9

Breth, H. and Back, K. (1963). 'Über die Setzungen von Bauwerken auf Ton', *Proc. 2nd European Conf. Soil Mech.,* Wiesbaden, Vol. 1, 101–106

Breth, H. and Heil, H. (1965). 'Der Einfluß der Steifigkeiten von Hochbauten auf die Verformung der Grundkörper der Bauwerke und die Verteilung des Sohldruckes', *Research Report to Deutsche Forschungsgemeinschaft (DfG),* Bad Godesberg

Breth, H. et al. (1970). 'Das Tragverhalten des Frankfurter Tons bei im Tiefbau auftretenden Beanspruchungen', *Mitteilungen der Versuchsanstalt für Bodenmechanik und Grundbau,* Heft 4, Darmstadt

Breth, H. and Schuster, E. (1971). 'Das Verformungsverhalten von Sand unter anisotroper Belastung', *Research Report to Deutsche Forschungsgemeinschaft (DfG),* Bad Godesberg

Domaschuk, L. and Wade, N. H. (1969). 'A Study of Bulk and Shear Moduli of a Sand', *Proc. A.S.C.E.* Vol. 95, SM2, 561–581

Duncan and Chang (1970). 'Nonlinear Analysis of Stress and Strain in Soils', *Proc. A.S.C.E.* Vol. 96, SM 5, 1629–1653

Hedberg, J. and Schmick, P. (1972). 'Diplomthesis', *Institut für Bodenmechanik und Grundbau der Technischen Hochschule, Darmstadt*

Heil, H. (1969). 'Studies on the Structural Rigidity of Reinforced Concrete Building Frames on Clay', *Proc. 7th Int. Conf. Soil. Mech. & Found. Eng.*, Mexico City. Vol. II, 115–121

Heil, H. (1971). Der Einfluß der Steifigkeit von Stahlbetonskelettbauten auf die Verformung und die Beanspruchung von Gründungsplatten auf Ton', *Mitteilungen der Versuchsanstalt für Bodenmechanik und Grundbau, Heft 8, Darmstadt,*

Leonhardt, G. (1973). 'Setzungskorrekturen an einem im Frankfurter Ton gegrünpeten Hochhaus', *Vorträge der Baugrundtagung 1972 in Stuttgart*. Deutsche Gesellschaft für Erd- und Grundbau, e.V. Essen

Schwab, H. and Wolff, R. (1971). 'Diplomthesis,' *Institut für Bodenmechanik und Grundbau der Technischen Hochschule, Darmstadt*

Sommer, H. (1973). 'Discussion on Amann and Breth (1973), *Vorträge der Baugrundtagung 1972 in Stuttgart*. Deutsche Gesellschaft für Erd- und Grundbau e.V., Essen

III/2. Settlement behaviour of a nuclear reactor

H. Breth
Professor of Soil Mechanics

G. Chambosse
Institut of Soil Mechanics and Foundation Engineering, Technische Hochschule, Darmstadt, West Germany

Summary

A nuclear reactor was built in Gundremmingen, Germany, during 1963–66. The reactor load is 33 150 Mp (325 000 kN). The structure was founded 7.5 m below the ground surface on a plate of 31.5 m diameter. The subsoil consists of Flinz (classification ML). Settlements of 153 mm were measured. The slope angle of the reactor was evaluated at 1 : 1200. The direction of the deflection changed with the eccentricity of the load.

Subsoil

The construction site of the reactor at Gundremmingen is about 1 km from the River Danube. The subsoil was explored by means of a series of borings approximately 40 m deep and one boring of 80 m depth. The tertiary fresh-water deposit 'Flinz' was found (classification ML) under an approximately 7 m thick layer of Danube-gravel. The Flinz was stiff owing to considerable geologic prestressing. The standard penetration resistance was $N \geq 30$. The compressibility of the Flinz was determined by means of one-dimensional consolidation tests. The samples were loaded in steps up to 8 kg/cm². The compression indices fluctuated between 150 and 350 kg/cm² for initial loading and 230 to 470 kg/cm² for reloading. Roughly 50% of the compression occurred shortly after loading. The characteristics of Flinz are compiled in *Figs. 1 and 2*.

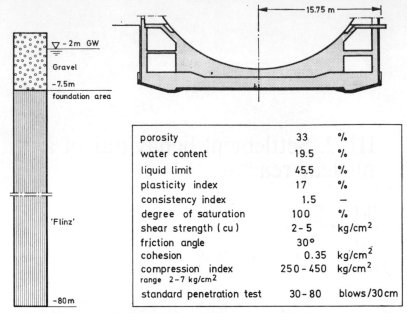

porosity	33	%
water content	19.5	%
liquid limit	45.5	%
plasticity index	17	%
consistency index	1.5	—
degree of saturation	100	%
shear strength (cu)	2 – 5	kg/cm²
friction angle	30°	
cohesion	0.35	kg/cm²
compression index	250 – 450	kg/cm²
range 2–7 kg/cm²		
standard penetration test	30 – 80	blows/30 cm

Fig. 1. Borehole, soil characteristics and cross-section of Gundremmingen reactor

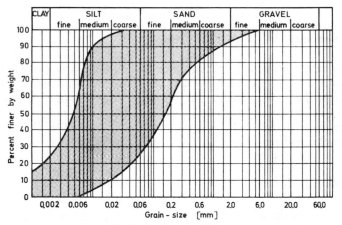

Fig. 2. Grain size of Flinz

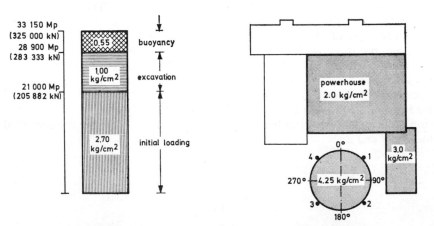

Fig. 3. Reactor load   *Fig. 4. Layout*

Settlement measurements

The circularly cylindric reactor is founded 7.5 m below ground surface on a plate of 31.5 m diameter. The layout, the load factors, and the average stresses in the foundation level are indicated in *Figs. 3 and 4*. Beginning at the start of construction (1963) the settlement of the reactor was measured at 4 points. The settlement curves and the load of the building are shown in *Fig. 5*. In 1970, i.e. roughly four years after completion of the building, the settlement had nearly reached its final value. On average the reactor had settled about 153 mm. Upon termination of construction (duration of construction three and a half years) the settlement volume had already reached 90% of the final value.

Of the entire settlement 25 mm are attributable to disturbance of the Flinz at the foundation level. 20 mm resulted from reloading (building load = excavation), A mean compression index of 360 kg/cm² may be derived from

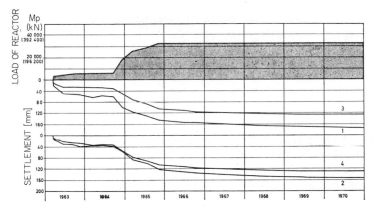

Fig. 5. Settlement and load—Gundremmingen reactor

the measured results. It is applicable for the load of the building minus excavation. This calculated compression index is about 25% higher than the compression index of 290 kg/cm² obtained by way of compression tests. If this load is broken down into reloading and initial loading a compression index $C_R = 850$ kg/cm² for reloading results, whereas a compression index of $C_i = 425$ kg/cm² is obtained for innitial loading. According to the theory of Boussinesq this is applicable up to an effective depth of 40 m.

Inclination of the reactor

According to the measured settlement the reactor only started to behave like a rigid body at the beginning of June 1963. *Fig. 6* gives the magnitude and direction of the inclination of the reactor. It is easy to see how the magnitude and direction of the inclination of the reactor changed during the period of construction. The inclination to be expected in accordance with the load figures given by the contractors are also shown in *Fig. 6*. The measured final inclination was 0.85 mm/m which corresponds to an inclination angle of

157

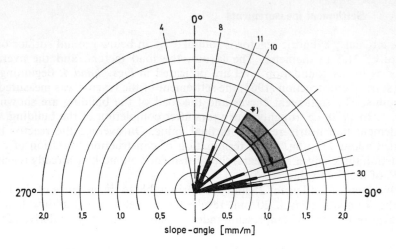

*) range, value and direction of the calculated slope-angle

NR 4 : September 1963
    8 : July   1964
   10 : October   1964
   11 : December   1964
   30 : June   1970

Fig. 6. Slope angle, Gundremmingen

1 : 1200. The direction of the inclination was 83°. The slope of the reactor was less attributable to the neighbouring buildings than to size and direction of its eccentricity. The good agreement between forecast and measured inclination direction suggest a homogeneous subsoil.

III/3. The settlement behaviour of three tall buildings in London

P. A. Green and J. E. Cocksedge
Scott Wilson Kirkpatrick & Partners, London

Summary

This paper presents case histories of three tall buildings in London:
(1) The Shell Centre, built 1957–1962
(2) New Zealand House, built 1959–1962
(3) Commercial Union Building, built 1964–1968
In each case history the factual data concerning the type of construction, loading, settlement behaviour and general performance are given.

The Shell Centre

The Shell Centre development occupies part of the 1951 Festival of Britain site on the South Bank of the River Thames. Prior to the exhibition the site was mainly derelict, being an area of Victorian houses extensively damaged during the 1939–45 war. The houses themselves were constructed on about 3 m of fill which had been placed on the original marsh deposits during the second half of the nineteenth century. Many probably had basements down to original ground level. The site was levelled for the exhibition, generally by using the brick rubble from the houses to fill basements and to produce a flat area with a general level of +3.7 m Newlyn Datum.

The Shell Centre is a complex of office buildings having a total floor area, including basements, of about 175,000 m². The main part of the complex consists of two sections, namely, Upstream and Downstream, separated by the railway viaduct approach to Hungerford Bridge. The Upstream section consists of a tower block of 28 storeys, 107 m height above ground level, and three 12-storey wings forming another block. The Downstream section consists of one main 12-storey block of two wings, and low-level podia.

This paper deals only with the settlement of the tower block, but loadings

159

of the whole Upstream section are considered as they could influence the settlement behaviour of the tower.

Construction details

A section through the tower block, a plan of the cylinder layout under the foundation raft and typical ground conditions are given in *Fig. 1*. The tower is founded on bored, cast *in situ*, concrete cylinders capped with a 0.9 m thick basement raft. The latter is connected to a 14 m high prestressed basement retaining wall. The basement founding level is about 7 m below the top surface of the London Clay. As indicated in *Fig. 1*, most of the cylinders have bells. The toe level of the majority of the cylinders is between 16 and 21 m below the underside of the basement raft, although some cylinders, particularly along grid lines T, and TA, are somewhat shorter.

The basement construction, excluding the retaining wall, is of conventional cast *in situ*, reinforced concrete. For the first two storeys above ground level the main structural members of the block consist of welded steel frames, whilst from the third storey upwards the frame is constructed from braced steelwork with high strength, moment connection bolted joints.

Further details of the ground conditions, design and construction of the Shell Centre have been given elsewhere by Measor and Williams (1962) and Williams (1957).

Loading

The overall gross and net loadings for the Upstream section of the project are shown in the area loading diagram in *Fig. 1*. The percentage increase of load on the tower block foundation as construction proceeded is also shown in *Fig. 1*. The final loading for the tower block is made up as follows:

	kN/m^2
Structural steelwork	20
Concrete (including basement raft)	140
Stone cladding	30
Internal walls	30
Finishes, services, etc.	60
Superimposed loading	30
Gross final loading	310
Unloading due to 12 m of excavation	180
Net final loading	130

For settlement calculations it was considered that in practice only about 20% of the design superimposed loading per floor of 4.8 kN/m^2 would occur and this reduced superimposed loading has been used in the list above. The validity of using such a reduction has been confirmed by recent work by Mitchell and Woodgate (1970). In assessing the loading of the site, consideration must also be given to the loading from previous construction activities as described earlier.

160

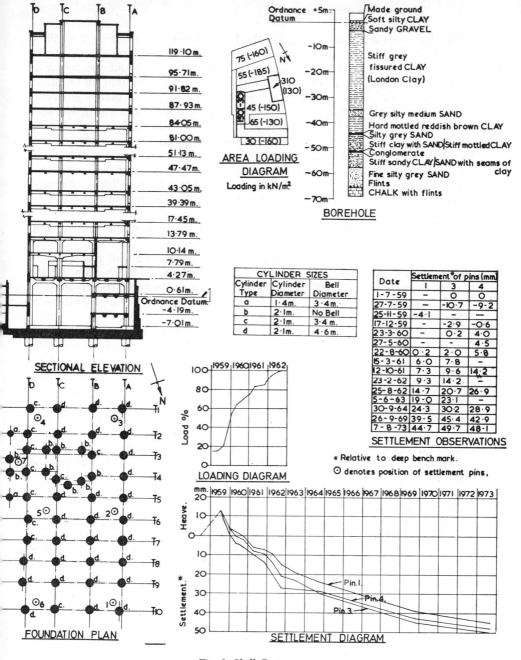

Fig. 1. Shell Centre

Settlement measurements

The first measurement of levels of the tower block was made in July 1959 when virtually all the below-ground construction was completed, the steelwork of the superstructure had reached the tenth floor and the concrete floors had been cased to the second floor. The monuments used for levelling surveys were steel pins cast into the ground-floor slab immediately adjacent to columns. Originally seven pins were installed, as shown in Fig. 1, but one (No. 6) was lost almost immediately and a further three (Nos. 2, 5 and 7) were lost in 1961 towards the end of construction. Surveys were carried out frequently during construction and less frequently thereafter. It will be seen from this figure that there was about 12 mm of heave during the excavation phase of the work, but by early 1960 the heave had been nullified by settlement. At about this time the total net load applied to the foundations was about zero.

The application of load was somewhat irregular through the construction period and the settlements reflect these irregularities.

The levelling was carried out using a precise level and a staff graduated in 0.01 ft divisions. The basic reference level of the surveys was a deep benchmark founded in the Chalk and located adjacent to the Downstream section. This benchmark has now been incorporated into the primary Ordnance Survey network for London.

It will be appreciated that, as the settlement of the tower block was measured against a deep benchmark located in a relatively incompressible stratum well below the level of the tower's foundations, such measurement will give the 'true' settlement. Observations have shown that the ground surface adjacent to the deep benchmark settled 28 mm relative to the deep benchmark during the six years between 1963 and 1969. Other observations of surface benchmarks in the area sited on structures supported by shallow foundations suggest that the regional surface settlement, relative to the deep benchmark, is between 30 and 45 mm in a period of 10 to 15 years (i.e. about 3 mm/a). Thus the settlement of the tower relative to the adjacent ground surface is probably about zero, provided that there is no significant and continuing heave locally adjacent to the tower due to the net unloading of the areas shown in *Fig. 1*.

The measured regional settlement in the vicinity of the Shell Centre is about twice the rate suggested by Wilson and Grace (1942) for the period from 1931 to 2000 due to under-drainage of the London Clay. It is very probable that local effects, such as long-term consolidation of the marsh deposits due to site filling, are significant.

No noticeable damage of the tower block has occurred to date which can be attributed to settlement.

New Zealand House

New Zealand House, which is a 21-storey tower block surrounded by six- and seven-storey podia, stands at the south end of the Haymarket at its junction with Pall Mall, in the heart of London's theatre district. Prior to the new construction, the site was occupied by the nine-storey Charlton Hotel built in 1898. This hotel had basements and vaults extending down to about +7 m Newlyn Datum, namely 3.5 m below pavement level. Demolition of the Charlton Hotel took place in 1958 and a contract for the construction of New Zealand House was awarded in 1959.

Construction details

New Zealand House was constructed using conventional cast *in situ* reinforced concrete. Details of the construction are shown in the elevation and plan in *Fig. 2*. This figure also shows a soil profile which is typical of the site except that in the north-east corner the upper surface of the London Clay is higher being at a level of about +1.5 m Newlyn Datum.

In addition to the boreholes and laboratory tests done for the ground investigation, 1.22 and 2.44 m diameter plate bearing tests were carried out within the upper 3 m of the London Clay to determine its Young's Modulus. These tests showed that the average value was about 86×10^3 kN/m².

The 27 m × 19 m tower block occupies the area bounded by grid lines B, 4, E, and 2. Within this area there is a small pile-cap raft supported by eight cylinders and carrying the main services and lifts. Reinforced concrete ground beams tie together seven of the cylinders in the central area and the small pile-cap raft, whilst three other cylinders have caps to assist in spreading some load to the surface of the London Clay.

The basement floor for the structure is of double-skin construction to permit swelling of the London Clay to occur without inducing upward pressures on the upper part of the basement floor. The floor is isolated from the cylinders, retaining wall foundations and pile caps.

The foundation of the retaining wall, which is of conventional design, is also supported on cylinders. The wall acts as a deep beam, spreading loads from the columns it supports to the cylinders.

Loading

Consideration of the data for the loading (*Table 1*) shows that, had a basement raft foundation been used, the structure would have exerted less

Table 1

	Area (m²)	Gross Loading (kN/m²)	Net Loading (kN/m²)
Previous building	1880	+130	+58
Tower block	530	+180	+5
Six-storey podium	690	+53	−10
Seven-storey podium	1040	+63	−110

net stress on the ground than the previous hotel building and, presumably, there would have been little settlement. However, with the large column spacing of 8 m such a raft would need to have been of considerable thickness,

163

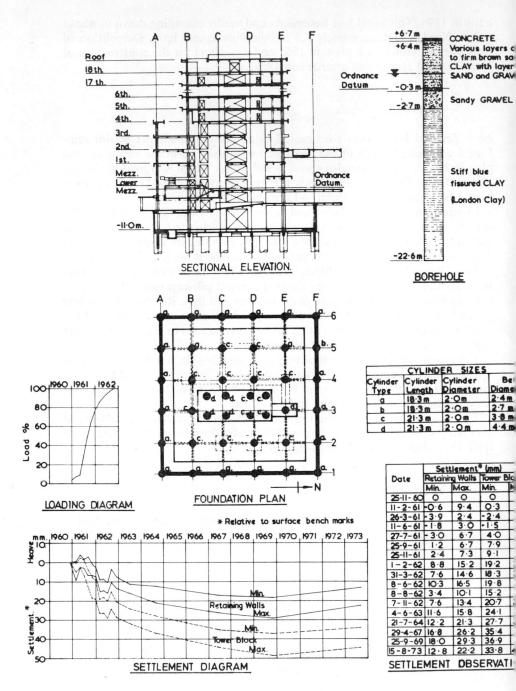

SECTIONAL ELEVATION.

Roof, 18th, 17th, 6th., 5th., 4th., 3rd., 2nd., 1st., Mezz., Lower Mezz., −11·0m.

Ordnance Datum.

BOREHOLE

+6·7 m
+6·4 m
CONCRETE
Various layers c
to firm brown sa
CLAY with layer
−0·3 m — SAND and GRAV
−2·7 m — Sandy GRAVEL
Stiff blue fissured CLAY
(London Clay)
−22·6 m

FOUNDATION PLAN

A B C D E F
6 5 4 3 2 1

├— N

LOADING DIAGRAM

Load %: 100, 80, 60, 40, 20, 0
1960 1961 1962

* Relative to surface bench marks

SETTLEMENT DIAGRAM

mm. Heave: 10, 0; Settlement: 10, 20, 30, 40, 50
1960 1961 1962 1963 1964 1965 1966 1967 1968 1969 1970 1971 1972 1973
Min. / Retaining Walls / Max.
Min. / Tower Block / Max.

CYLINDER SIZES

Cylinder Type	Cylinder Length	Cylinder Diameter	Be Diam
a	18·3 m	2·0 m	2·4 m
b	18·3 m	2·0 m	2·7 m
c	21·3 m	2·0 m	3·8 m
d	21·3 m	2·0 m	4·4 m

SETTLEMENT OBSERVATI

Date	Settlement* (mm) Retaining Walls Min.	Max.	Tower Blo Min.	
25-11-60	0	0	0	
11-2-61	-0·6	9·4	0·3	
26-3-61	-3·9	2·4	-2·4	
11-6-61	-1·8	3·0	-1·5	
27-7-61	-3·0	6·7	4·0	
25-9-61	1·2	6·7	7·9	
25-11-61	2·4	7·3	9·1	
1-2-62	8·8	15·2	19·2	
31-3-62	7·6	14·6	18·3	
8-6-62	10·3	16·5	19·8	
8-8-62	3·4	10·1	15·2	
7-11-62	7·6	13·4	20·7	
4-6-63	11·6	15·8	24·1	
21-7-64	12·2	21·3	27·7	
29-4-67	16·8	26·2	35·4	
25-9-69	18·0	29·3	36·9	
15-8-73	12·8	22·2	33·8	

Fig. 2. New Zealand House

164

or of cellular construction. With the majority of the load structure being supported directly on the cylinders, the basement area between the cylinders is in a state of unloading and, as has been mentioned earlier, possible swelling of the clay under the basement raft due to this effect necessitated a special detail of design.

Settlement measurements

The first measurement of levels was made on 25 November 1960. At that time virtually all the below-ground work and about two-thirds of the first floor was complete. The levels were taken on steel pins fixed into the columns, generally at first floor level, but some pins were also situated at basement and sub-basement level. Subsequent level surveys were carried out on the dates given in *Fig. 2*.

Generally the surveys were carried out using a precise level instrument although in the later surveys a water level was used for measurement in the basement. Several nearby Ordnance Survey benchmarks were taken as basic reference levels, and these were all situated on or adjacent to old buildings, it may be assumed that any general regional settlement (as discussed in the section on the Shell Centre) might affect the benchmarks to a slightly greater extent than New Zealand House because of the latter's deep foundation. This means that the settlements measured for the structure are over-and-above any regional settlement and might be less than would have been obtained had a deep benchmark been used. Taking into account the various factors affecting the accuracy of the survey, it has been assessed that the error for any individual level reading could be ± 3 mm.

A summary of the results of the surveys is given graphically in *Fig. 2*. For the purpose of this summary the results have been divided into two parts, namely, settlements of the tower block (bounded by and including grid lines B, 4, E and 2) and settlements of the retaining walls along grid lines A, 6, F and 1. The settlements at B5, C5, D5 and E5 have not been shown, but these generally lie between those of the retaining walls and of the tower block.

It is worth noting that the settlement readings taken in 1973 suggest that the building may now be rising in relation to the benchmarks, although the amounts are small and not much greater than the accuracy of the surveys. Future observations will be made to determine if this trend is continuing.

No noticeable damage to the structure or finishes has occurred to date which can be attributed to settlement.

Commercial Union Building

The Commercial Union Building is situated on the north side of Leadenhall Street, E.C.3 on a site previously occupied by three-storey shops of brick construction, an eight-storey steel-framed stone-clad Post Office about 70 years old and a 10- to 13-storey office block built in the early 1930s. Site demolition took place in 1963–64. The main part of the new development consists of a 28-storey tower block 118 m high and 1400 m² in plan area, constructed between 1964 and 1968.

Construction details

A plan and section through the tower block and its foundation and typical ground conditions are given in *Fig. 3*. Initially the block was supported by 12 bored cylinders each of 2 m diameter, 34 m long and having a 4.7 m diameter bell founded in the top of the Woolwich and Reading Beds. The sub-structure up to and including the ground-level slab (see *Fig. 3*) consists of a perimeter diaphragm wall, four *in-situ* reinforced concrete floors and a reinforced concrete raft foundation. The tower block consists of a 23 m × 15 m concrete core containing all the services and supporting two canti-lever steel frames at plant-floor levels which, in turn, support trusses and girders around the perimeter of the building. The cladding consists of con-tinuous curtain walls. Fuller details of the construction and ground conditions have been given elsewhere by Williams and Coleman (1965), Williams and Rutter (1967) and Green (1971).

The construction of the cylinders requires more detailed description. It was initially expected these would be constructed without permanent lining, as is normal for large cylinders in London Clay, but water-bearing silty layers were encountered and the resulting sloughing of soil into the boring prohibited the hand digging of the cylinder bells. To overcome the problem, the cylin-ders were constructed in two stages of 2.1 m and 2 m diameter. Before con-struction of the next stage, each earlier stage was lined with a light-weight corrugated steel liner slightly smaller than the bored hole. The cavity between the liner and hole was filled with cement grout containing fly ash and an expanding agent.

Loading

The percentage loading from the structure to the cylinders and, subsequently, to the cylinder/raft system is shown in *Fig. 3*. The overall average gross loading, including the design superimposed loading of 4.8 kN/m² (with an appropriate reduction allowance for the number of floors), is 370 kN/m². The overall average net loading is 90 kN/m². If a realistic value of super-imposed loading resulting from occupation is considered, then the overall average net loading becomes approximately zero.

Settlement measurements

Settlement measurements of the Commercial Union Building were made using two methods. One method was a precise water-gauge system to mea-sure the differential movement of the tower to an accuracy of ±0.25 mm. The other method consisted of geodetic levelling to measure the total settle-ment of the structure and, at the same time, to check the results obtained from the water-gauge system. The levelling used a precise level and staff. The 'basic' benchmark was a pin fixed into a ramp well outside the influence of the building. The level of this pin was itself checked from time to time by com-parison with a number of nearby Ordnance Survey benchmarks mainly situated on buildings. The levelling points in the structure consist of phosphor-

166

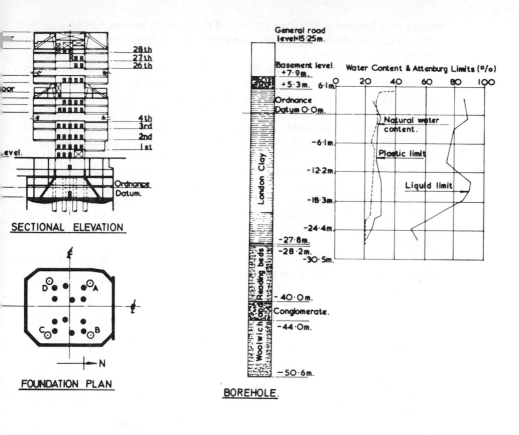

BOREHOLE.

* Relative to surface bench marks

⊙ denotes position of settlement pins

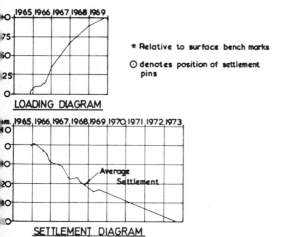

Date	Settlement* (mm)				
	A	B	C	D	Average
25-10-65	0	0	0	0	0
26-10-65	-0·51	0	0	0·25	0
18-11-65	-0·25	0·76	1·27	1·27	0·76
3-12-65	-0·51	—	-0·25	-0·51	-0·25
5-1-66	-2·03	—	1·01	0	-0·25
6-4-66	—	—	4·57	3·30	
13-5-66	1·01	2·80	4·57′	3·30	2·80
17-6-66	—	3·30	5·84	3·05	—
4-8-66	5·59	5·59	7·37	4·83	5·84
9-11-66	5·08	9·14	10·16	8·64	8·13
2-1-67	7·37	9·40	10·92	9·14	9·14
19-7-67	10·92	11·18	11·68	10·41	11·18
16-1-68	17·78	17·02	18·80	18·03	17·78
14-5-68	17·27	16·51	18·54	17·27	17·27
19-7-68	18·80	18·03	20·07	19·56	19·05
6-3-69	24·04	22·54	25·40	25·40	24·29
3-7-69	23·04	22·54	23·79	23·04	23·04
18-7-73	—	—	—	—	40·30

SETTLEMENT OBSERVATIONS

Fig. 3. Commercial Union building

bronze pins located at the four corners of the core. It has been estimated that, with repeat determinations, the accuracy of an individual geodetic level reading is ± 1 mm.

The first determination was made in October 1965. The measurements determined since then are shown in *Fig. 3*. The amount of settlement shown on this figure is that measured in the geodetic levelling and it therefore represents the movement of the tower relative to an external surface benchmark; the latter is probably subject to larger regional settlement than the tower with its deep foundation. During the construction period the settlement readings do not give a smooth curve. The fluctuations appear to be a function of the time of year and it has been calculated that they are probably the result of expansion and contraction in the building between the measurement pins and the basement raft. The expansion and contraction itself is due to seasonal temperature changes; these changes became less pronounced once the building was fully occupied and kept at a fairly constant temperature.

No noticeable damage to the structure or finishes has occurred to date which can be attributed to settlement.

REFERENCES

Green, P. A. (1971). 'Some Aspects of the Foundation Design for the Commercial Union Building', *The Midland Soil Mechanics and Foundation Engineering Society, Proc. Symp. Interaction of Structure and Foundation*, Birmingham. 118–130 (discussion 162–165)

Measor, E. O. and Williams, G. M. J. (1962). 'Features in the Design and Construction of the Shell Centre, London', *Proc. I.C.E.*, Vol. 21, 475–502

Mitchell, G. R. and Woodgate, R. W. (1970). 'A survey of Floor Loadings in Office Buildings', *C.I.R.I.A. Report*, No. 25

Williams, G. M. J. (1957). 'Design of the Foundations of the Shell Building, London', *Proc. 4th Int. Conf. Soil Mech. & Found. Eng.* London. Vol. 1, 457–461

Williams, G. M. J. and Coleman, R. B. (1965). 'The Design of Piles and Cylinder Foundations in Stiff, Fissured Clay', *Proc. 6th Int. Conf. Soil Mech. & Found. Eng.* Montreal. Vol. 2, 347–351

Williams, G. M. J. and Rutter, P.A. (1967). 'The Design of Two Buildings with Suspended Structures in High Yield Steel', *Structural Engineer*, Vol. 45, No. 4, 143–151

Wilson, G. and Grace, H. (1942). 'The Settlement of London due to Underdrainage of the London Clay', *J.I.C.E.*, Vol. 19. 100–127 (also in *A Century of Soil Mechanics*, I.C.E. 1969)

III/4. Settlement at Didcot Power Station

R. B. Hyde and B. A. Leach
Allott and Lomax, Consulting Engineers, Sale, Cheshire

Summary

The settlement of major structures founded on Gault Clay at Didcot Power Station has been recorded for the past seven years and is continuing. The authors, who were invited to submit a short technical paper to illustrate the behaviour experienced over this period, present selected settlement records with brief comments.

Introduction

In 1963 the Central Electricity Generating Board decided to proceed with the construction of a 2000 MW power station at Didcot in Berkshire, a site at which significant movements of heavy structures were likely to occur due to consolidation of the underlying clay strata.

After consideration of the movements to be expected, it was decided to dispense with expensive methods of reducing loading intensity and to design the various heavy structures to allow for the settlements which would arise from bearing pressures derived from the shear strength of the underlying material using normal factors of safety.

Continuous monitoring of the movements of these structures has been carried out from their construction during the period 1965–67, until the present time and selected records are given in this Paper.

The site

The site at Didcot is situated in the flood plain of the River Thames on the northern limits of the London Basin and consequently has river alluvium immediately subsurface. The alluvial deposits vary generally from 1.5 m to

4.5 m in thickness and consist in the main of firm to stiff silty clays, although in some areas the clays were underlain by 0.6 m to 2.1 m of dense fine to medium sandy gravel.

The Gault Clay which underlies the alluvium is weathered in places to a mottled grey and brown clay to a maximum depth of 3.65 m. It is described as

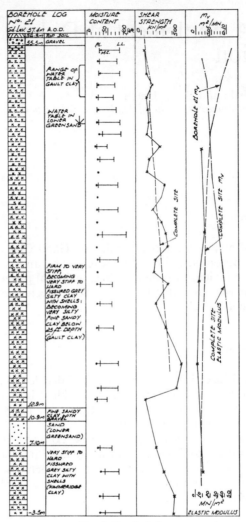

Fig. 1. Borehole log

a firm to stiff—becoming very stiff to hard with depth—fissured grey silty clay with shells, the latter becoming more frequent with depth. Thicknesses of Gault Clay between 29 m and 42.5 m were proved on the site overlying a layer of grey very silty sandy clay with some fine gravel which is believed to be a transitional layer passing into the underlaying Lower Greensand. The Lower Greensand proper, although not present over all the site, was encountered as a dense grey and white fine to coarse sand in thicknesses varying

170

between 0.9 m and 7.6 m. In all cases, however, the material underlying the transitional zone or the Lower Greensand is Kimmeridge Clay, which is a very stiff to hard fissured, becoming shaly in places, grey silty clay with shells, and which, from local evidence, is 26 m thick overlying Corallian Limestone. The total depth of strata proved by boring was 60 m below ground level.

Fig. 1 shows a typical 60 m borehole and gives index properties from tests carried out on driven samples, together with plots of shear strength and coefficient of volume compressibility obtained from conventional laboratory testing. Average values of these parameters for the whole site are also shown.

The structures

The main structures which have been monitored for settlement are the chimney, the cooling towers and the main boiler/turbine buildings.

Main chimney

This structure, 200 m in height, consists of four 6 m diameter concrete flues enveloped by a concrete windshield tapering from 25 m diameter above the flue openings to 20.5 m diameter at the top. The structure is supported on 79 bored piles each of 1.2 m diameter, 18 m in length and set in five concentric rings, the diameter of the outer ring being 36 m. The net imposed dead load on the piles is 25 800 tonne.

Cooling towers

Six cooling towers, 115 m high with a 91 m base diameter were arranged in two groups each of three towers. These structures are each supported on a single ring of 80 No. 1.2 m diameter bored piles 21 m long raked at 1 in 4 (approximating to the line of thrust of the tower shell). The dead load of the structure imposed upon the piles is 14 650 tonne giving a net dead load per pile of 183 tonne; a low value adopted to reduce displacements at the base of the tower shell.

Main buildings

The main buildings have eight column rows running from east to west and 35 column lines running from north to south (see *Fig. 2*). The main columns apply dead loads to the foundations varying from 183 tonne at gable column E2 to 4115 tonne at the main boiler support column F9. Net bearing pressures of up to 245 kN/m² were used in the design. The net dead load bearing pressures are fairly constant under all the bases at 180 kN/m². The need for a double tier cable basement required that in many locations within the building the minimum practical foundation level was some 6/7.5 m below the finished ground floor level and the resulting foundation layout showed that it

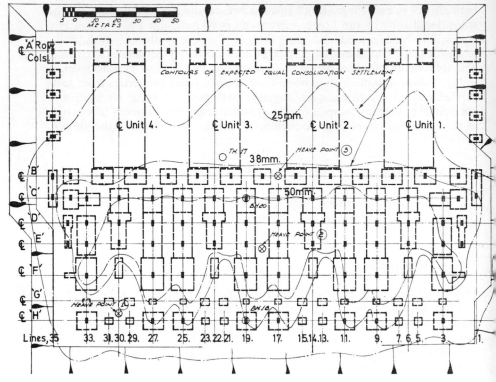

Fig. 2. Main boiler/Turbine buildings

was more economic to excavate the whole area than to excavate separate pits for each foundation.

Heave measurements

The decision to excavate the whole area of the main foundations led to an excavation 205 m × 170 m extending to a depth varying between 6 m and 7.5 m below the finished floor level (see *Fig. 2*). The excavation was carried out to below the ground water table and dewatered by pumping from sumps within the excavation. Difficulties were experienced along the A row embankment due to high ground water flows from extensive overlying gravel deposits and a cut-off drain was installed outside the excavation to prevent water entering this face.

Heaves were recorded at the three locations shown in *Fig. 2* using the measurement device shown in *Fig. 3*. A datum pin was located below the level of the base of the excavation and initial readings taken on it prior to excavation using an extension tube. The casing was removed before excavation commenced and the holes filled with sand of a distinctive colour. This permitted unhindered excavation and also easy relocation of the points after the excavation was complete when further readings were taken on the datum pins. The excavation took two months to complete and the initial and subsequent

172

heaves are as shown in *Table 1*. The heave pins were levelled to remote bench-marks using 'dumpy' levels and it should be noted that even though the closing errors were normally in the order of up to 5 mm it can be observed from the records that the heave at the centre of the excavation was less than at the periphery and that the initial heaves of all points decreased substantially

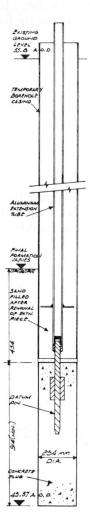

Fig. 3. Heave measurement installation

Table 1

	Heave pin 1	Heave pin 2	Heave pin 3
	Heave in mm (accuracy ±5 mm)		
6 May 1965	0	0	0
10 July 1965	32	30	21
26 July 1965	14	10	15
27 July 1965	5	30	1
2 Aug. 1965	15	15	7
25 Jan. 1966	8	6	1

within the following six months. As the pattern of heave observed does not conform to that predicted by the Steinbrenner formula it is difficult to make a comparison of the laboratory/site modulus of elasticity values. However, heave measurements carried out under similar conditions in excavations below the water table and reported by Bozozuk (1963) and Serota and Jennings (1959) led to somewhat similar anomalies which were attributed to ground water control. This may be the explanation for the behaviour here.

Settlement

Settlement records were taken by the Resident Engineer's staff using 'dumpy' (not precise) levels closing on deep and remote benchmarks with an error normally in the order of 3 mm.

Main chimney

Records of readings taken over five years on the chimney base are given in *Fig. 4.* Four measurement points were set on the chimney base at 90° to each other and the points plotted in *Fig. 4* are averaged readings from these four points.

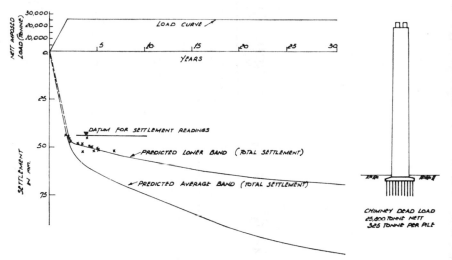

Fig. 4. Main chimney settlement and loading curves

Tilt as measured across the 40 m diameter chimney base has been found to be minimal and within the accuracy of the readings.

Readings were started towards the end of the construction period and the settlement datum is assumed to lie on the predicted total settlement curve at the date of taking the initial reading. The average and lower bound predictions for the expected consolidation movement were based on 'equivalent pier' methods described by Tomlinson and conventional one-dimensional consolidation theory. Average and lowest values of the coefficient of volume

174

compressibility obtained from the site investigation were used. However, in the case of the coefficient of consolidation it was considered that the laboratory values which ranged from about 0.2 to 1.4 m²/a in the Gault and from about 0.4 to 2.7 m²/a in the Kimmeridge Clays were unrealistically low and values of 2.8 m²/a in the Gault and 3.7 m²/a in the Kimmeridge Clays were adopted. The results appear to be following the lower bound prediction of settlement.

Cooling towers

Six years' records taken on the northern group of cooling towers are shown in *Fig. 5*. These again are averaged for each tower from readings taken on four points set at 90° intervals around the perimeter. The ground has been surcharged by fill prior to construction of these towers to depths varying between 1m and 2.5 m and the effect of this has been taken into account in the

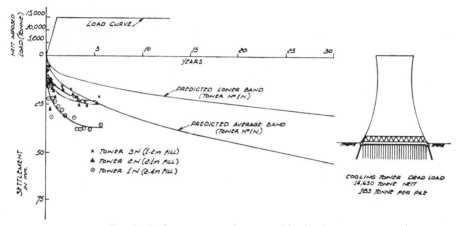

Fig. 5. Cooling towers settlement and loading curves

settlement predictions. A maximum tilt of approximately 20 mm has been recorded across tower 3N in the direction of deepening fill. Readings cover both the elastic and consolidation phases and predictions were calculated as for the chimney. The total settlement appears in the long term to lie between the average and lower bound values, but in this case the rate of settlement occurring appears to be even faster than one dimensional consolidation theory would predict using the increased coefficient of consolidation values adopted.

Main building

In comparison with the chimney and cooling towers, our efforts in recording and analysing the relative movements of the main building foundations have, up to the present time, proved disappointing and have often been frustrated. Difficulty was experienced in obtaining records in the congested main build-

175

ings during construction and plant erection. Some measurement points were destroyed, others became inaccessible, levelling was difficult due to the numerous change-points required and the records are, therefore, not considered to be as reliable as those given previously for the other structures. In addition, owing to station commissioning being extended over the past years, plant and coal bunker loadings have fluctuated abnormally.

Fig. 6 shows interpreted records for three points giving a section across unit 3 of the main building. Predictions were again calculated using conventional one dimensional theory with the effects of adjacent foundations being taken into account but without soil/structure interaction. These predictions indicated the contours of consolidation settlement show in *Fig. 2*, and a recent check of the deflected shape of the structural steel frame with respect to vertical movement has revealed a broadly similar pattern. Future monitoring under expected constant load conditions is expected to yield information of greater value.

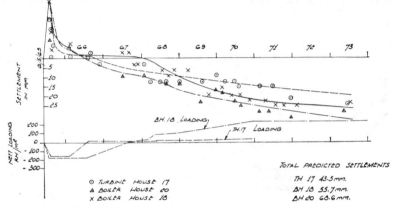

Fig. 6. Main Building settlement and loading curves

The records shown are only representative of a great number taken and in no case has there been recorded to date settlement in excess of the maximum predicted at the design stage.

Acknowledgements

The authors are greatly indebted to Professor H. B. Sutherland, for his advice and assistance during the design and construction of the Works.

The work was undertaken for the C.E.G.B. to whom the authors gratefully acknowledge permission to publish the results.

REFERENCES

Bozozuk, M. (1963). 'The Modulus of Elasticity of Leda Clay from Field Measurements', *Canadian Geotechnical Journal*, Vol. 1 No. 1, September
Serota, S. and Jennings, R. A. J. (1959). 'The Elastic Heave of the Bottom of Excavations', *Geotechnique*, London, Vol. 9, 62–72
Tomlinson, M. J. (1963). *Foundation Design and Construction*, London, Pitman, 389–390

III/5 Heave on a deep basement in the London Clay

J. May
Department of the Environment, Croydon

Summary

This Paper presents facts relating to the heave of a deep basement in the London Clay over a period of more than five years. Measurements are still continuing. The pattern of heave so far follows the theoretical predictions based on one dimensional consolidation theory.

Introduction

This Paper is concerned with the heave taking place on a large deep basement in the London Clay. A basement that is probably unique in that although the building was originally designed for a heavy superstructure this has not yet been built and the basement, after being completed up to ground level, has been left completely empty whilst at the same time detailed and accurate survey measurements have been made of the movements of the base slab for over five years. Thus we have a structure where the dead load can be reasonably accurately assessed and where there has been no indeterminate live load to obscure the results. The recorded movements are considered very relevant to the prediction of heave and the analysis of settlements of buildings in the London Clay with similar deep basements.

When it first became obvious in August 1970 that there would be a long delay in building the superstructure the Foundation and Ground Engineering Section of the Directorate of Civil Engineering Development were asked to give advice on the probable heave of the basement over a period of years and to estimate the effect of this heave on the integrity of the structure itself. It was not a research project and consequently some information that would now be desirable, in view of the history of the basement, may be lacking. The facts presented should therefore be reviewed in this light.

Soil profile

Geological records for the district show that the site is on made ground over alluvium overlying the London Clay. Four wells are recorded on the geological survey map of London near the boundaries of the site (see *Fig. 1*) and these show the London Clay stratum to be approximately 30 m thick.

For the initial geotechnical investigation six boreholes were sunk at the positions shown in *Fig. 2* and the simplified borehole sections of the strata are given in *Fig. 3*.

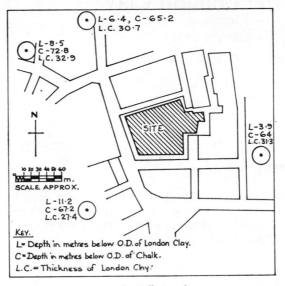

Fig. 1. Well record

Fig. 2

178

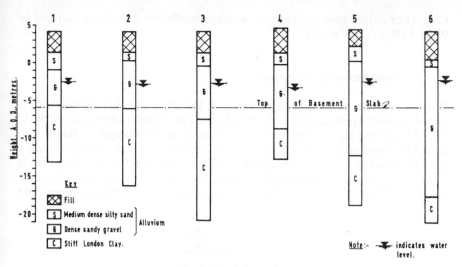

Fig. 3. Borehole sections

Ground water

Ground water was found in the gravel and is shown in the borehole section (*Fig. 3*). In analysing the heave on the basement a water level at −2.5 AOD was assumed, i.e. 3.4 m above the top of the basement slab (*Fig. 3*). This depth of water outside the basement can also be confirmed by the slight seepage of water visible at many of the joints between panels in the diaphragm wall.

A check was made in July 1973 to see if there was any measurable water pressure beneath the floor slab of the basement. At the two probe positions, P1 and P2 in *Fig. 2*, two low pressure Bourdon gauges were grouted in. In the clay no water seepage could be seen but after one week a head of 1.4 m of water above the top of the ground slab was measured. At the position P2, in the gravel, water under pressure was encountered which rose initially about 1 m above the top of the ground slab but the flow rapidly decreased. After the hole was capped and allowed to rest the pressure again built up. The Bourdon gauge after a week showed a pressure equivalent to 2 m of water above the basement slab. Further observations are continuing. These excess pressures will have had no measurable effect on the heave of the basement slab.

Soil properties

The Standard Penetration Tests show the alluvium to be generally dense sandy gravel of high permeability.

No consolidation test results were quoted in the original report and certainly no swelling tests were made. However, it was considered that in estimating the heave these properties could be inferred with sufficient accuracy from other fairly similar London Clay sites and collected data for London

179

Clay. (For calculation purposes an average coefficient of volume increase of 0.035 MN/m² was used.)

Construction of basement

A diagrammatic section of the basement is shown in *Fig. 4*. The basement is approximately 11 m deep over the whole site and forms a two-level car park. The sides of the excavation were retained by using diaphragm walls of reinforced concrete approximately 0.5 m thick cast in 1.5 m widths and extending into the London Clay to provide a cut-off for the ground water. These walls were incorporated in the final construction. A reinforced concrete raft, generally 1.2 m thick but thickened to 1.8 m under the central columns and

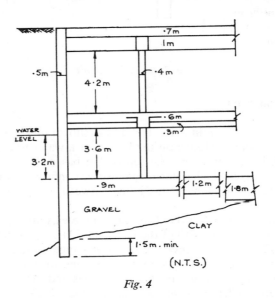

Fig. 4

lift shaft areas and reduced to 0.9 m thick near the perimeter, provides the basement floor. Intermediate and ground floors are of beam and slab construction. The plan size of the basement can be obtained from *Fig. 2* and the dead load of the basement structure is 48 kN/m². (In estimating the heave the raft was approximated to a rectangle 64.2 m × 88.8 m.)

The excavation was begun in June 1966 and finally completed in November 1967. The basement concrete was finished up to ground floor level in May 1968.

Observation of heave

The measured changes in height of the various survey points (*Fig. 2*) are recorded in *Table 1* and have been plotted graphically against the square root of elapsed time in months, in *Fig. 5*. The graphs show a straight line relationship up to the last readings but show a zero error with what seems to be

180

Table 1

Date of reading (initial reading Sept. 1967)	Survey marks										
	1	2	3	4	5	6	7	8	9	10	11
(Accumulative change in height in millimetres—upward changes positive)											
March 1968	0.8	0.7	0.5	0.4	0.1	0	0.5	1.1	0.6	−0.6	0.5
July 1968	7.4	7.7	8.5	9.0	9.2	8.1	5.0	8.6	6.7	6.6	7.3
Sept. 1968	8.8	9.6	10.9	11.4	11.7	10.1	6.0	11.0	8.1	8.0	8.4
July 1969	16.8	18.1	21.1	22.5	22.2	18.7	11.4	20.2	15.7	15.4	15.0
Sept. 1969	19.4	20.5	24.0	25.8	25.5	21.2	12.9	23.3	18.0	17.3	17.0
Nov. 1969	19.7	20.6	24.1	26.2	25.7	20.7	12.6	23.5	18.0	16.8	17.0
Feb. 1970	22.2	23.7	27.4	29.6	28.4	23.9	14.7	26.7	20.3	19.7	19.8
May 1970	27.3	29.5	33.9	36.1	35.6	30.2	19.6	32.3	26.0	25.7	24.7
Aug. 1970	31.5	33.4	38.4	40.9	40.2	34.1	22.2	36.6	29.5	28.8	27.5
Dec. 1970	29.3	31.5	36.7	39.6	31.6	19.2	35.2	27.2	25.9	25.0	
Mar. 1971	32.1	34.7	40.1	43.4	42.3	34.8	21.8	38.1	29.9	28.9	27.4
Aug. 1971	35.0	38.0	44.0	47.1	46.1	38.5	24.1	41.4	32.5	31.7	29.6
Oct. 1971	36.7	39.5	45.8	41.2	48.1	39.3	24.7	43.3	33.8	32.6	30.3
Feb. 1972	39.6	42.7	48.8	52.3	51.1	41.8	27.4	46.1	36.2	35.1	32.7
May 1972	41.7	45.3	51.8	55.3	53.9	44.6	29.4	48.6	38.4	37.7	34.7
Oct. 1972	44.9	48.3	55.1	58.9	57.2	47.4	31.3	51.8	40.8	39.6	36.6
Feb. 1973	46.7	50.1	57.4	60.2	59.5	49.0	32.8	53.0	42.3	41.2	38.3

measurements accurate to 0.1 mm

Points W, X, Y, Z

Initial reading July 1968	W	X	Y	Z
Change to Oct. 1971	12.2	9.1	12.2	15.2

Measurements taken on concrete slab accurate to ±5 mm only

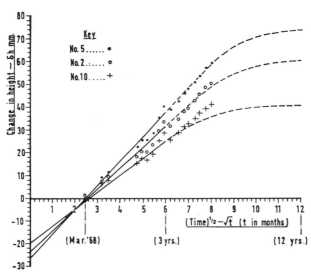

Fig. 5. Recorded heave

virtually no movement between September 1967 and March 1968. The zero error can be explained in two ways:

(1) The change in height that took place up to September 1967 is represented by the negative change in height on the graph. This 'heave', including the immediate heave, of the soil will have been removed during excavation. Between September 1967 and March 1968 the heave taking place has been offset by the settlement due to the placing of the concrete.

(2) There is possibly an error in the original datum level.

Either way there is some doubt as to the amount of heave that has taken place up to the start of levelling.

Calculations indicate that the maximum heave with the basement in its present condition will be approximately 100 mm at the centre of the raft. This in turn indicates a heave of 50 mm at the points of intersection of the axes of symmetry with the actual basement walls. In extending the graph of 'change in height with square root of time' it has been assumed that approximately 27 mm of this swelling occurred during construction, i.e. that at the centre of the raft a further 73 mm of swelling would continue. This is equivalent to a deflection of 1 in 1860 of the span and there should be no structural damage.

Conclusions

The heave is following very much the pattern that would be expected from one dimensional consolidation theory. The rate of heave even after 5 years does not show the expected slowing down and it is possible that the maximum heave will exceed that predicted in 1970 and shown in *Fig. 5.*

III/6 Settlement observations on eight structures in London

K. Morton and E. Au
Geotechnics Division, Ove Arup & Partners

Summary

This Paper describes the results of several years observations of settlement on eight structures in London. Comprehensive records of the structure, foundations, soil conditions and settlement are given. Some general conclusions on the comparative performance of the structures are given.

Introduction

This Paper describes the measurement of settlement of eight structures in London.

Six of the buildings were constructed by the Wates system of precast walls, slabs and cladding using *in situ* stitches, although in each case the ground and first floor were *in situ* concrete. A typical view of these structures is given in *Fig. 1*. Excavation at these sites did not normally exceed 2 m.

The Royal Free Hospital (see *Fig. 3*) was constructed in traditional *in situ* concrete. It has a large plan area and was built into a sloping site so that depth of excavation varied between 3 and 12 m.

The office block at Addiscombe Road, Croydon is a circular structure with floors suspended between an *in situ* concrete core and perimeter columns. It has a basement car park which required excavation to a depth of 11 m. A view of the superstructure is given in *Fig. 2*.

Method of measurement

The observation points on the structures were placed in public access areas at ground or basement level and have generally remained accessible throughout the period of observation.

Fig. 2. Addiscombe Road

Fig. 1. Grantham Road

Fig. 3. Royal Free Hospital

Various types of observation points have been used. Initially, dimensioned pencil marks were made on fair faced concrete ceilings of the ground floor. These proved to be entirely satisfactory and have only occasionally required renewal due to decoration. An attempt to use Hilti pins as observation points proved unsatisfactory as these were lost during the construction period.

In 1968 a brass threaded plug and socket was manufactured to specification and became the standard observation point. These resembled the BRS type and initially attempts were made to fix the plug to the wall shutter to be cast *in situ*. However, it was found that the threads of the plugs were damaged during concreting and a more satisfactory method was to drill and grout in after walls had been cast. Occasionally these sockets have subsequently been covered by services or cupboards but generally have remained accessible.

From the outset, settlement readings were taken using a Wild N3 level and invar staff capable of direct reading to 0.03 mm.

Observations were related to three benchmarks situated between 50 and 100 m from the structure. Usually kerbstones and walls unlikely to be disturbed were chosen and surveyed for re-location. Over the period of observation benchmarks were sometimes lost but the precaution of using three prevented loss of continuity at any site.

No attempt has been made to relate structure settlements to general changes of ground level in the areas concerned.

Surveys were usually commenced when the structures were between ground

185

Table 2. DETAILS OF SOIL CONDITIONS

Site	Location	Geological succession	Soil conditions		In situ test N values	Soil properties		
			Thickness (m)	Soil description		C_u (kN/m²)	C_v (m²/a)	M_v (m²/MN)
Clapham Road, L.B. of Lambeth	South Island Place/ Clapham Road, Nat. Grid Ref. TQ 3177	Made Ground	2	Clay and brick rubble				
		Flood Plain Gravel	5.5	Dense gravel with some coarse sand and flints	25–66			
		London Clay	25 (11 m proven)	Stiff fissured dark grey silty clay with occasional layers of fine sand and claystone		80 — 140	0.3 — 2.8	0.116 — 0.066
		Woolwich and Reading Beds	15	Not proven				
Grantham Road, L.B. of Lambeth	Grantham Road/ Rhodesia Road, Nat. Grid Ref. TQ 3076	Made Ground	1	Clay and brick rubble				
		Flood Plain Gravel	1	Dense gravel with some clayey sand	62/136			
		London Clay	32 (16 m proven)	Stiff fissured grey brown silty clay with pockets of fine sand and layers of claystone		60 180	—	
		Woolwich and Reading Beds	15	Not proven				
Hurley Road Blocks I and II, L.B. of Lambeth	Kennington Lane/ Fairford Grove, Nat. Grid Ref. TQ 3178	Made ground	2	Gravel, clay and brick rubble				
		Flood Plain Gravel	4.5	Very dense sandy gravel with pockets of soft clays	(I) 17–139 (II) 37–215			
		London Clay (17 m proven)	20	Stiff fissured grey-brown silty clay with pockets of grey and brown sand shells and pyrites		90 — 300	0.3 — 1.2	0.140 — 0.049

Location	Site	Stratum	No.	Description				
King Edward's Road, L.B. of Hackney	Balcorne Street/King Edward's Road, Nat. Grid Ref. TQ 3584	Woolwich and Reading Made ground	14	Not proven				
		Flood plain gravel	2	Clay and brick rubble	74–102			
			3	Very dense gravel with some sand				
		London clay	4	Stiff fissured dark grey silty clay		150	2.0	0.018
		Woolwich and Reading Beds (11 m proven)	15	Alternating layers of mottled clays and dense sands with shells and gravel		500	95.0	0.083
Cambridge Road, L.B. of Waltham Forest	Junction of Cambridge Road/Collingwood Road, Nat. Grid Ref. TQ 3688	Made Ground	1	Clay, ashes and brick rubble				
		Taplow Gravel	3	Dense sandy gravel with some flints	23–94	70	0.2	0.18
		London Clay (19 m proven)	22	Stiff fissured grey-brown silty clay with some pockets of sand and claystones		400	1.4	0.05
Royal Free Hospital, L.B. of Camden	Pond Street/Fleet Road, Nat. Grid Ref. TQ 2785	Woolwich and Reading Made Ground	25	Not proven				
		London Clay (weathered)	1	Loamy Clay			0.5	0.11
			9	Stiff fissured brown silty clay with pockets of fine sand and gypsum			2.0	0.08
		London Clay (unweathered)	50	Very stiff fissured dark grey silty clay with some gypsum and pockets of blue-grey silt			7.5	0.04
Addiscombe Road, L.B. of Croydon	Addiscombe Road/Cherry Garden Road, Nat. Grid Ref. TQ 3265	Made ground	2	Clay with some gravel				
		Woolwich and Reading Beds	16	Alternating layers of stiff fissured mottled clays and dense fine sand and Gravels	57–80		60.0	0.03
		Thanet sands (9 m proven)	20	Very dense green-grey silty fine sand with shells	>100			

Table 1. DETAILS OF STRUCTURES AND FOUNDATIONS

Site	Structure	Foundation	Gross applied pressure (kN/m²)	Remarks
Clapham Road, L.B. of Lambeth	22-storey residential block. Ground and 1st floor of *in situ* concrete then Wates system pre-cast units with *in situ* stitches (*Fig. 4*)	0.76 m thick raft	209	2 m of fill encountered over part of site removed and replaced with lean concrete
Grantham Road, L.B. of Lambeth	22-storey residential block. Ground and 1st floor of *in situ* concrete then Wates system pre-cast units with *in situ* stitches (*Fig. 6*)	48 No. underreamed piles. Lengths between 16.2 and 18.7 m. Shaft 0.76 and 0.91 m dia. Base 1.52 to 2.74 m dia.	209	
Hurley Road (Block 1), L.B. of Lambeth	22-storey residential block. Ground and 1st floor of *in situ* concrete then Wates system pre-cast units with *in situ* stitches (*Fig. 7*)	48 No. underreamed piles. Lengths between 17.8 and 19.4 m. Shaft 0.76 and 0.91 m dia. Base 1.22 to 1.98 m dia.	209	
Hurley Road, (Block 2), L.B. of Lambeth	22-storey residential block. Ground and 1st floor of *in situ* concrete then Wates system pre-cast units with *in situ* stitches (*Fig. 7*)	0.91 m thick raft	209	
King Edward's Road, L.B. of Hackney	23-storey residential block. Ground and 1st floor of *in situ* concrete then Wates system pre-cast units with *in situ* stitches (*Fig. 9*)	1.37 m thick raft	244	Soil disturbed by exploded bomb over part of raft. Replaced to depth of 4 m with lean concrete
Cambridge Road, L.B. of Waltham Forest	22-storey residential block. Ground and 1st floor of *in situ* concrete then Wates system pre-cast units with *in situ* stitches (*Fig. 11*)	116 No. straight shafted bored piles. Shaft 0.62 m dia. Length 15.3 m	214	
Royal Free Hospital, L.B. of Camden	17-storey hospital. *In situ* beam, column and slab construction (*Fig. 14*)	2.16 m thick raft	270	Sloping site. Depth of excavation varies between 3.0 and 12.5 m
Addiscombe Road, L.B. of Croydon	23-storey office block, over 3 basement floors of car park. Lightweight concrete floors supported on *in situ* concrete central core and perimeter columns (*Fig. 17*)	2.6 m thick helical raft. 0.5 m thick diaphragm wall at perimeter of car park.	446	Depth of basement about 11m. Diaphragm wall to depth of 18.3 m

and 4th floor level stages of construction. The closing error on any survey ranged between zero and 1.5 mm and was usually a function of the difficulty of the traverse.

Case histories

The structures and their foundations, soil conditions and construction details are described individually below. Most of this information is also summarised in *Tables 1* and *2* and details of settlement observations at each structure are given in *Table 3*.

Table 3a. CUMULATIVE SETTLEMENT READINGS — CLAPHAM ROAD

Date of reading	1	2	3	5	6	7	Closing error (mm)	Stage of construction
			Cumulative settlement readings (mm)					
26 Oct. 1966	0	0	0	0	0	0	0.03	2 storeys
20 Feb. 1967	11.9	8.9	10.3	8.4	9.9	7.7	0.03	10 storeys
26 July 1967	51.5	42.8	40.0	35.4	42.5	35.2	0.28	22 storeys
26 Oct. 1967	58.2	48.8	47.6	41.6	47.8	42.2	0.05	roof complete
1 Feb. 1968	66.0	54.5	52.5	45.0	—	44.5	0.03	occupied
29 May 1968	71.5	61.5	60.6	52.7	61.7	46.8	0.46	
29 Sept. 1969	82.3	72.5	71.0	63.3	73.2	59.7	0.00	
3 Dec. 1970	89.0	74.5	72.0	64.2	74.7	63.5	0.30	
1 Nov. 1972	91.0	81.0	78.5	71.5	81.9	69.0	0.05	

Table 3b. CUMULATIVE SETTLEMENT READINGS—GRANTHAM ROAD

Date of reading	A	B	C	D	E	G	L	M	Closing error (mm)	Stage of construction
				Cumulative settlement readings (mm)						
11 May 1967	0	0	0	0	0	0	0	0	0.53	3 storeys
26 July 1967	3.9	4.2	4.2	4.4	2.4	2.6	4.1	1.0	0.28	10 storeys
26 Sept. 1967	8.5	9.7	9.9	9.8	8.6	—	8.0	—	0.80	16 storeys
20 Nov. 1967	13.1	14.8	15.0	15.0	13.6	—	13.1	—	0.51	20 storeys
26 April 1968	19.0	20.1	20.6	20.8	19.8	15.6	18.3	18.1	1.00	roof complete
27 Jan. 1969	23.8	25.1	25.4	27.9	21.6	22.1	—	22.6	0.61	occupied
25 Sept. 1970	27.2	29.6	29.8	29.4	25.9	23.1	25.3	25.4	0.02	
16 April 1973	31.6	33.7	33.6	33.3	30.3	28.3	27.9	28.7	1.1	

Table 3c. CUMULATIVE SETTLEMENT READINGS—HURLEY ROAD–BLOCK 1

Date of reading	2	4	6	7	8	Closing error (mm)	Stage of construction
			Cumulative settlement readings (mm)				
14 Aug. 1968	0	0	0	0	0	1.50	Gd. + 4 storeys
23 Jan. 1969	8.7	9.1	—	7.7	7.9	0.60	Complete
23 Jan. 1970	—	—	18.4	15.0	16.1		Complete
13 April 1973	22.7	22.1	22.1	19.6	21.3	0.03	Occupied

Table 3d. CUMULATIVE SETTLEMENT READINGS—HURLEY ROAD–BLOCK 2

Date of readings	1	2	3	4	5	6	7	8	Closing error (mm)	Stage of construction
			Settlement point numbers							
			Cumulative settlement readings (mm)							
1 Feb. 1968	0	0	0	0	0	0	0	0	0.41	Ground and 1st Floor
14 Aug. 1968	47.6	51.1	49.5	47.7	48.0	49.8	49.0	53.0	0.95	22 storeys
23 Jan. 1969	63.6	67.3	64.8	63.5	62.0	65.0	66.5	68.3	0.59	Complete
23 Jan. 1970	79.6	86.5	82.3	82.3	82.3	85.3	79.0	83.5	1.25	Complete
31 Aug. 1970	94.8	101.5	100.3	—	97.5	100.1	94.0	98.0	0.56	Occupied
13 April 1973	101.2	108.9	106.8	—	104.1	107.3	98.9	105.0	0.43	Occupied

Table 3e. CUMULATIVE SETTLEMENT READINGS—KING EDWARD'S ROAD

Date of reading	1	3	5	6	7	8	9	10	Closing error (mm)	Stage of construction
			Settlement point numbers							
			Cumulative settlement readings (mm)							
5 Dec. 1967	0	0	0	0	0	0	0	0	0.05	3 storeys
7 April 1968	13.1	13.8	18.3	18.2	19.7	17.0	20.1	18.5	1.50	20 storeys
4 Sept. 1968	25.4	21.9	29.3	30.4	29.6	32.1	30.8	31.4	1.20	Roof complete
3 Oct. 1969	33.1	30.9	—	40.0	—	42.5	—	39.2	0.20	Occupied
9 Sept. 1970	34.5	31.8	—	41.6	—	44.6	—	40.1	1.60	
3 Aug. 1971	40.5	37.5	—	47.6	—	50.0	—	46.5	0.05	
6 April 1973	41.7	39.6	—	50.8	—	52.0	—	48.6	0.20	

Table 3f. CUMULATIVE SETTLEMENT READINGS—CAMBRIDGE ROAD

Date of reading	A	B	C	D	E	F	G	H	Closing error (mm)	Stage of construction
			Settlement point numbers							
			Cumulative settlement readings (mm)							
15 Mar. 1966	0	0	0	0	0	0	0	0	1.5	2 storeys
22 July 1966	5.3	8.6	4.9	8.5	—	—	7.5	8.9	1.80	17 storeys
4 Oct. 1966	7.8	14.9	10.4	16.9	14.0	13.1	13.4	14.7	0.60	21 storeys
15 Feb. 1967	14.3	—	14.0	19.8	19.6	18.5	18.7	20.2	0.20	22 storeys
26 Sept. 1967	16.6	21.6	16.2	20.9	20.6	19.0	20.1	21.8	1.80	Occupied
2 Oct. 1968	18.7	25.4	18.1	26.6	26.0	24.4	25.2	26.9	0.05	
11 May 1973	21.5	27.8	19.4	29.1	29.1	26.6	26.4	29.1	2.20	

Table 3h. CUMULATIVE SETTLEMENT READINGS—ADDISCOMBE ROAD

Date of reading	1	2	3	4	5	6	7	8	9	10	Closing error (mm)	Stage of construction
				Settlement point numbers								
				Cumulative settlement readings (mm)								
29 Oct. 1968	0	0	0	0	0	0	0	0	0	0	1.08	Ground level
21 May 1969	6.9	7.1	5.7	6.4	8.0	8.7	8.0	9.1	6.9	8.1	N.C.	7th floor
7 Jan. 1970	29.2	—	23.8	23.0	28.3	27.3	29.5	29.8	29.2	29.1	1.60	17th floor
1 July 1970	44.0	—	37.2	33.3	41.6	44.7	44.5	46.8	46.1	48.2	N.C	Complete
29 June 1971	53.0	—	43.7	33.6	—	—	53.0	54.7	55.7	—	0.26	Partly occupied
19 April 1973	53.7	48.5	42.4	46.7	—	—	54.7	55.9	58.6	—	0.23	Occupied

N. C. Not closed — Points obstructed

Cumulative settlement readings (mm)

Date of reading	Settlement point numbers																		Closing error (mm)	Stage of construction
	1	2	3	4	5	6	7	8	9	10	11	12	13	17	18	19	20	21		
11 Dec. 1969	0	0	0	0	0	0	0	0	0	0	0	0	0	0	0	0	0	0	0.41	2 storeys
21 June 1970	6.1	5.8	6.5	4.9	2.6	8.7	5.1	6.3	1.5	5.9	5.4	7.1	5.6	2.8	6.5	6.2	6.3	6.6	0.41	7 storeys
5 Nov. 1970	7.9	7.5	6.7	8.5	—	10.5	—	10.3	4.7	8.5	8.0	10.9	9.4	7.3	8.3	9.8	7.8	11.0	1.40	9 storeys
22 Jan. 1971	11.0	11.4	11.5	13.2	5.9	12.9	9.3	13.3	7.7	11.3	10.7	14.2	13.4	9.6	12.7	12.9	10.5	15.1	1.30	11 storeys
26 May 1971	14.8	—	13.2	16.9	7.5	19.9	13.4	16.3	9.7	14.0	14.2	18.1	15.4	10.6	14.8	14.6	14.0	17.5	0.90	16 storeys
17 Nov. 1971	18.5	20.8	16.3	22.9	9.4	21.4	—	21.7	13.5	19.5	21.5	25.0	22.4	17.1	21.6	21.6	19.6	22.8	1.98	17 storeys
30 Mar. 1972	20.0	22.4	17.7	25.0	11.4	22.8	—	23.6	16.3	21.2	22.7	26.7	25.0	18.0	23.2	21.7	20.9	28.0	0.31	Roof
28 Sept. 1972	21.7	26.5	20.5	28.4	13.2	27.6	21.9	27.6	17.2	25.3	27.4	30.8	29.0	21.7	28.0	26.6	—	31.3	0.48	Complete

Table 4. SUMMARY OF OBSERVATIONS

Site	Type of foundation	Max. observed settlement (mm)		Max. observed differential settlement (mm)		ρ_a/ρ_m %	ρ_{md}/ρ_m %	ρ_{ad}/ρ_a %	Type of distortion	Angle of tilt	Max. angular distortion
		During period of observation ρ_m	At the end of construction ρ_a	During period of observation ρ_{md}	At the end of construction ρ_{ad}						
Clapham Road	Raft	91.0	58.2	22.0	16.6	64	24	29	Tilt	6′ 20″	1/2600
Grantham Road	Underreamed piles	33.7	20.8	8.0	5.2	62	24	25	Dish/tilt	0′ 30″	1/1300
Hurley Road (1)	Underreamed piles	22.7	9.1	3.1	1.4	40	14	15	—	—	—
Hurley Road (2)	Raft	108.9	68.3	10.0	6.3	63	9	9	Dish/tilt	0′ 50″	1/650
King Edward Road	Raft	52.0	32.1	12.4	10.2	62	24	32	Tilt	1′ 10″	1/1300
Cambridge Road	Straight shafted piles	29.1	20.2	9.7	6.2	70	33	31	Dish	—	1/800
Royal Free Hospital	Raft	—	31.3	—	10.8	—	—	35	Dish	—	1/1000
Addiscombe Road	Raft	58.6	48.2	16.2	14.9	82	28	31	Tilt	1′ 20″	1/5500

Clapham Road

This is one of two 22-storey Wates systems precast structures constructed at the site for the London Borough of Lambeth. A site investigation whilst the site was occupied by the previous housing showed the soil sequence to be 5 m of Terrace Gravels overlying London Clay.

It was proposed to found this block on a raft just below the surface of the Terrace Gravel. However, during excavation it was discovered that in part of the area of the raft gravel had been worked and replaced with fill, at an earlier time. The fill was removed and 12:1 mix concrete was placed in layers and compacted with a 0.1 tonne vibrating roller.

Construction was fairly rapid and observations were commenced when the structure was at 2nd floor level.

Details of the ground floor plan and position of the observation points are

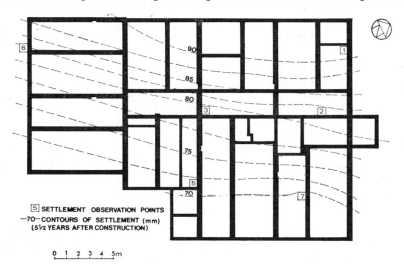

SETTLEMENT OBSERVATION POINTS
—70— CONTOURS OF SETTLEMENT (mm)
(5½ YEARS AFTER CONSTRUCTION)

0 1 2 3 4 5m

Fig. 4. Clapham Road foundation plan

shown in *Fig. 4*. Contours of the total settlements observed to date are also shown. The load–time–settlement relationship for three points is shown in *Fig. 5*.

Grantham Road

This is a 22-storey structure identical to that at Clapham Road but constructed at a later date at another site for the London Borough of Lambeth. A site investigation showed 2 m Flood Plain Gravels overlying London Clay. It was decided that large diameter underreamed piles taken to depths of 19 m should be constructed to support the building.

The disposition and pile sizes are shown together with the ground floor plan and location of observation points in *Fig. 6*. Contours of total settlement are also shown in *Fig. 6* and the load–time–settlement relationship for three points is given in *Fig. 5*

192

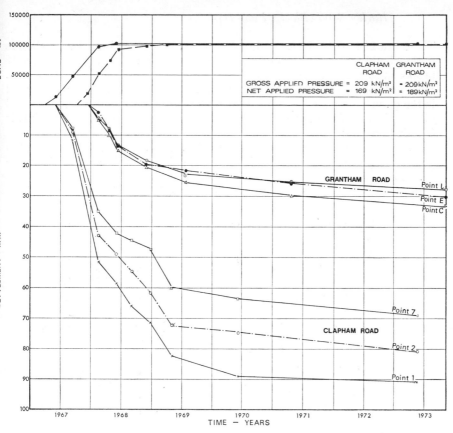

Fig. 5. Clapham Road and Grantham Road load–time–settlement relationship

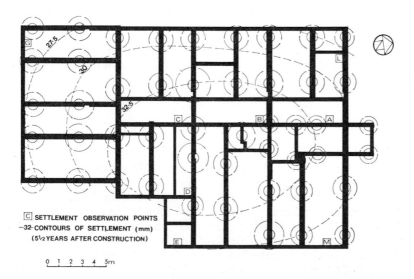

Fig. 6. Grantham Road foundation plan

193

These are 22-storey blocks within 100 m of each other built by the Wates precast building system for the London Borough of Lambeth. It was originally proposed to found both blocks on raft foundations on the Terrace Gravel. However, during site investigation extensive lenses of soft clay were discovered in the gravel at the site of Block I and therefore large diameter underreamed piles to depths of 19 m were constructed to support this structure. Block II is founded on a raft.

Details of the ground floor plan, foundations and settlement observation points are given in *Fig. 7* and load–time–settlement relationship for three points on each block are shown in *Fig. 8*.

In the case of Block I a number of points soon became inaccessible and the

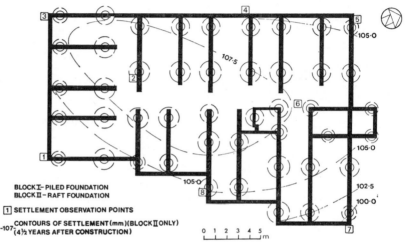

BLOCK I – PILED FOUNDATION
BLOCK II – RAFT FOUNDATION

[1] SETTLEMENT OBSERVATION POINTS

-107 CONTOURS OF SETTLEMENT (mm)(BLOCK II ONLY)
(4½ YEARS AFTER CONSTRUCTION)

0 1 2 3 4 5 m

Fig. 7. Hurley Road Blocks I and II foundation plan

results are somewhat limited. It has not therefore been possible to draw meaningful contours of settlement. There are more comprehensive records for Block II (raft foundation) and the contours of settlement observed to date are given in *Fig. 7*.

King Edward's Road

This is a 23-storey structure built for the London Borough of Hackney. It is also constructed by the Wates precast system but is of different plan layout to the blocks in Lambeth. The site investigation revealed 3 m of Flood Plain Gravel overlying water bearing Woolwich and Reading Beds.

It was designed and constructed on a raft foundation bearing on the Flood Plain gravel. However, during excavation an area of loose contaminated gravel was located. Further investigation showed this to be the site of an exploded bomb with disturbed ground including shrapnel to a depth of 4 m at one end of the raft. The area of disturbed ground was removed and backfilled with lean concrete to the underside of the raft.

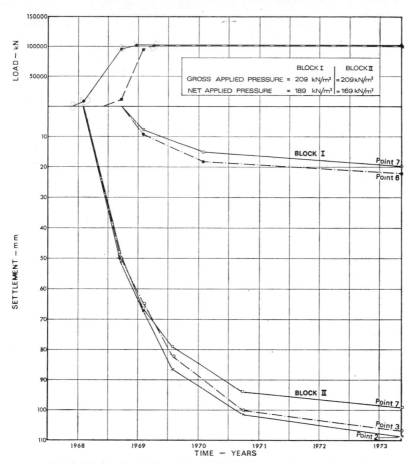

Fig. 8. *Hurley Road Blocks I and II load–time–settlement relationship*

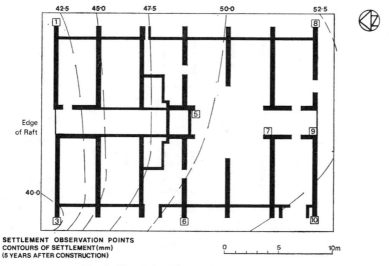

☐3 SETTLEMENT OBSERVATION POINTS
--30- CONTOURS OF SETTLEMENT (mm)
(5 YEARS AFTER CONSTRUCTION)

Fig. 9. *King Edward Road foundation plan*

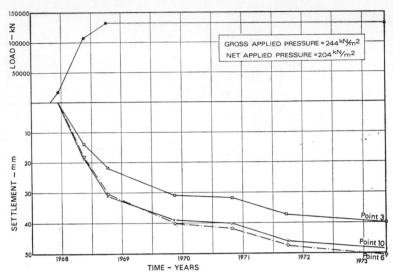

Fig. 10. King Edward Road load–time–settlement relationship

Details of the ground floor plan and settlement observation points are given with superimposed settlement contours in *Fig. 9*. The load–time–settlement relationship for three points is shown in *Fig. 10*.

Cambridge Road

This is one of three 23-storey blocks of maisonettes constructed for the G.L.C. at a site in the London Borough of Waltham Forest. One side of the structure is connected to a semi-basement car park.

The site investigation showed 3 m of Taplow Gravel overlying London

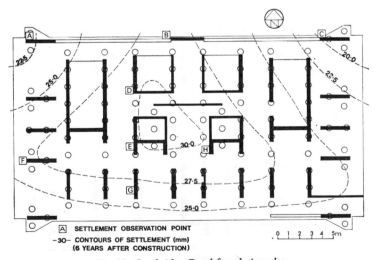

A SETTLEMENT OBSERVATION POINT
—30— CONTOURS OF SETTLEMENT (mm)
(6 YEARS AFTER CONSTRUCTION)

Fig. 11. Cambridge Road foundation plan

196

Clay. The structure was founded on 0.62 m diameter straight shafted piles taken to depths of 15 m.

The ground floor plan, disposition of piles, settlement observation points and contours of settlement are given in *Fig. 11*. The load–time–settlement relationship for three points is given in *Fig. 12*.

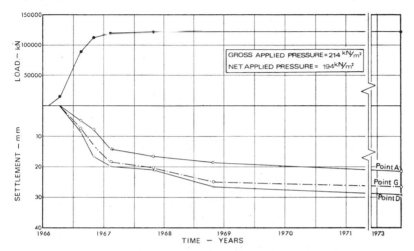

Fig. 12. Cambridge Road load–time–settlement relationship

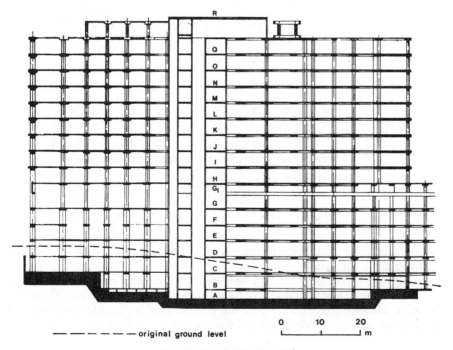

Fig. 13. Royal Free Hospital

Royal Free Hospital

This is a 15-storey *in situ* concrete structure built for the Royal Free Hospital in the London Borough of Camden. The site has a slope of about 6° and the site investigation showed that London Clay is present from the surface to a depth of about 50 m. The structure is founded on a T-shaped raft which was

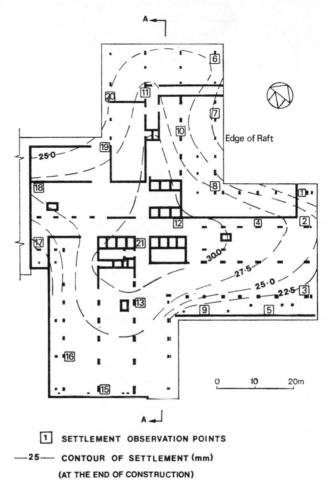

SETTLEMENT OBSERVATION POINTS

—25— CONTOUR OF SETTLEMENT (mm)

(AT THE END OF CONSTRUCTION)

Fig. 14. Royal Free Hospital foundation plan

constructed after excavation to depths of up to 12 m. *Fig. 13* shows a section through the structure.

In this case construction has only recently finished and the settlement record is therefore incomplete. *Fig. 14* shows a plan of the structure giving the location of settlement points and the contours of settlement to date. The load–time–settlement relationship for three points is given in *Fig. 15*.

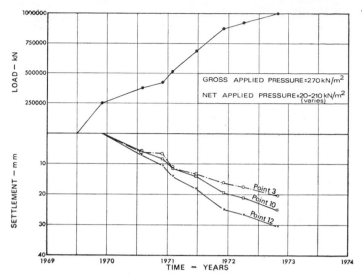

Fig. 15. Royal Free Hospital load–time–settlement relationship

Addiscombe Road

This structure is a 24-storey office development in the centre of the London Borough of Croydon. The structure incorporates a central concrete core and perimeter columns supporting the floors. It has a deep basement car park constructed within diaphragm walls.

The site investigation showed that beneath the basement level the soil sequence was 7 m of Woolwich and Reading Beds over Thanet Sands.

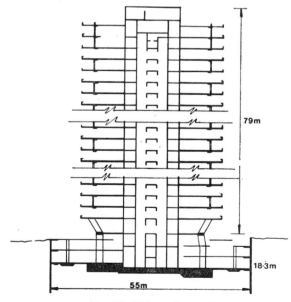

Fig. 16. Addiscombe Road

199

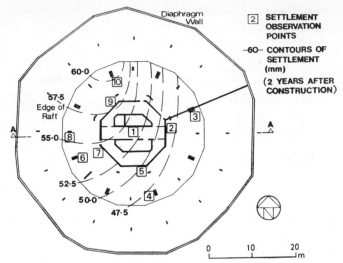

Fig. 17. Addiscombe Road foundation plan

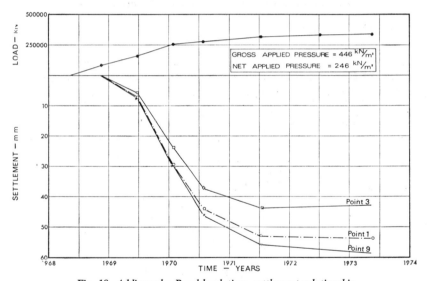

Fig. 18. Addiscombe Road load–time–settlement relationship

This block is founded on a helical circular raft 11 m below ground level on Woolwich and Reading Beds and a section through the structure is shown in *Fig. 16*. A plan of the basement, showing settlement observation points and contours of settlement is given in *Fig. 17*. The load–time–deflection relationship for 3 points is given in *Fig. 18*.

Discussion

Settlement and distortion data for the various structures have been collected in *Table 4*. In this table only actual recorded settlement measurements have been given. No attempt has been made to correct for initial settlements or to estimate the total final settlements. Within these limitations it is possible to draw some general conclusions regarding the settlement performance of these structures.

Maximum settlements

Five of the structures with essentially similar dimensions and net loadings were founded in the London Clay. Those on rafts, namely Clapham Road and Hurley Road Block II, settled about 100 mm whilst those on piles, namely Grantham Road, Hurley Road Block I and Cambridge Road settled about 30 mm.

The structure at King Edward's Road of similar dimensions but with slightly heavier net loading founded on a raft near the top of the Woolwich and Reading Beds settled about 50 mm.

The considerably larger and more heavily loaded structure at Addiscombe Road which was founded near the base of the Woolwich and Reading Beds also settled about 50 mm.

Settlements at the end of construction

The ratio of settlement at the end of construction over maximum settlement observed, expressed as a percentage, has been given in *Table 4*. It can be seen that for the structures on London Clay this ratio is about 60% irrespective of whether the structure is founded on piles or a raft. Hurley Road Block I is an exception with a ratio of only 40%, however, the data from this site is rather sparse.

The results for the structure on a raft over Woolwich and Reading Beds at King Edward's Road also gives a ratio of about 60%. At Addiscombe Road where the raft is near the base of the Woolwich and Reading Beds the ratio is up to 80%, which reflects the thin consolidating layer.

These results confirm those in previous work on overconsolidated clays by Simons and Som (1970) who quoted a ratio of *end of construction settlement* over *50-year settlement* of 57.5%.

Differential Settlements

The largest differential settlements observed varied from 3 to 22 mm. This represents between 9% and 35% of the respective maximum settlement in each case with an average of about 25%. This figure seems to be similar for the London Clay and Woolwich and Reading sites and does not vary significantly with either piled or raft foundations.

The ratio of differential settlement to maximum settlement expressed as a percentage has been shown in *Table 4* for both the end of construction and total period of observation. The ratio shows no significant change with time. This means that the increase of differential settlement is directly proportional to the maximum settlement taking place and that the pattern of settlement established during the construction stages will continue with time.

Distortion and tilting

Figures are given in *Table 4* for angle of tilt and angular distortion. These were deduced by plotting sections through the contours of settlement given in the plans. The tangent of the angle of tilt was calculated by dividing the differential settlement across the section by the width of the section. The angular distortion has been obtained from the maximum radius of curvature of the section considered.

It can be seen that for the raft foundations at Clapham Road, King Edward's Road and Addiscombe Road slight tilting appears to be the most significant feature of the settlement pattern. These are all relatively stiff structures and in the case of Clapham and King Edward's Road it is thought likely that the variation of ground conditions immediately below the raft (replacement of fill by mass concrete) probably resulted in differential settlement over the width of the blocks. At Addiscombe Road the varying depth of the raft probably influenced the pattern of differential settlement.

Hurley Road Block II, a stiff structure on a raft foundation, had a dishlike pattern of settlement combined with slight tilting. At the Royal Free Hospital the settlement pattern at the end of construction is clearly dishlike, the contours closely following the plan shape of the structure. In this case it is considered that the structure and raft would form a relatively flexible unit.

The settlement pattern of the two piled structures shows a dishing with maximum settlements beneath the centre of the structure. At Grantham Road it is combined with a very slight tilting whilst at Cambridge Road the adjacent connected single storey car park has influenced the settlements at the edge.

General observations

It seems from the foregoing observations that the pattern of settlement is likely to be a function of the stiffness, structure and foundation but that it is very sensitive to variations of ground conditions. There is considerable evidence on the distress caused to buildings by differential settlements, Skempton and Macdonald (1956) and Bjerrum (1963). This indicates that damage is rarely caused by angular distortion in excess of 1:300 and a safe limit for design to prevent the cracking of finishes is 1:500. It can be seen that the structures observed are within this limit, and it is confirmed by the fact that no signs of distress are evident in any of the structures studied.

It may be noted that at Hurley Road Block II where a maximum settlement of 110 mm has been observed the angular distortion does not exceed 1:650. It also appears that while the use of piled foundations reduces the maximum settlements compared to rafts they do not significantly reduce the angular distortion.

Conclusions

From the study of the records of settlement of structures founded on over-consolidated soils in London the following general conclusions may be drawn:

(1) For structures with similar dimensions and loadings maximum settlements are of the ratio of 3 to 1 for raft and piled foundations respectively.

(2) Maximum settlements of structures founded on Woolwich and Reading Beds are about half that of similar structures on London Clay.

(3) The settlement at the end of construction is about 60% of the maximum settlement observed irrespective of the type of foundation.

(4) Differential settlements are about 25% of maximum settlements observed for all cases.

(5) Distortion to structures is well within safe limits for all the buildings observed.

(6) The pattern of settlement is very sensitive to minor changes in ground conditions.

Acknowledgements

The authors are grateful to the following for permission to publish the results of the settlement observations: The London Borough of Hackney, The London Borough of Hounslow, The London Borough of Lambeth, The Greater London Council, Livius Investments Limited, The Royal Free Hospital, Wates Limited.

They also wish to thank their colleagues in Ove Arup & Partners, in particular Mr. F. G. Butler and Mr. K. W. Cole, who initiated a large number of the observations.

REFERENCES

Simons, N. and Som,, N. N. (1970). 'Settlement of Structures on Clay with Particular Emphasis on London Clay', *CIRIA Report 22*, July
Skempton, A. W. and Macdonald D. H. (1956). 'The Allowable Settlement of Buildings', *Proc. I.C.E. 5*, No. 3, Part 3, 727–784. Dec.
Bjerrum L. (1963). 'Discussion on the Compressibility of Soils', *Proc. Europ. Conf. Soil Mech. & Found. Engg.*, Wiesbaden, Vol. 2, 16–17

III/7. Guy's Hospital Tower Block, London—settlement records

G. Mould
R. Travers Morgan & Partners, London

Records of settlement—Guy's Hospital Tower Block, London

These notes describe the results of an attempt to record actual settlements of a large structure built on the London Clay, starting from an early stage in the construction. Levelling points have been installed in positions where it is hoped that they will remain accessible for periodic readings after completion of the project. Levelling has been done with reference to a datum established on a nearby but quite separate building, and it has also been possible to correlate that datum with a 'deep datum' extending down to the chalk stratum, situated at the southern end of London Bridge, approximately 500 m away from the site.

The structure

The building itself has been fully described in a paper presented to the Institution of Structural Engineers (Mould, 1971): briefly it consists of two linked towers rising approximately 135 m and 122 m above ground, with a low-rise area (the boiler house, etc.) abutting them. The main towers are founded on a piled raft 3.05 m thick, the underside of which (8.45 m below ground) is in typical London ballast. The piling is distributed uniformly over the area of the rafts, and consists of 762 mm diameter bored piles penetrating some 16 m into the London Clay.

The average gross loading intensity over the area of the piled rafts is of the order of 42.7 tonnef/m^2 (4 tonf/ft^2).

The lighter boiler house structure is founded on individual under-reamed piles placed under each column, the pile diameter and under-ream varying with the load. The average gross loading intensity over this area is 7.1 tonnef/m^2 (0.65 tonf/ft^2).

Construction of the project started in April 1968, and the basic structure

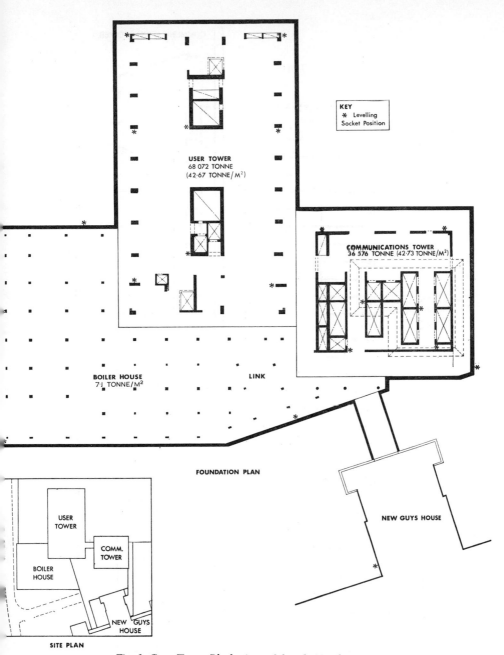

KEY
* Levelling
Socket Position

USER TOWER
68 072 TONNE
(42·67 TONNE/M²)

COMMUNICATIONS TOWER
36 576 TONNE (42·73 TONNE/M²)

BOILER HOUSE
7·1 TONNE/M²

LINK

FOUNDATION PLAN

NEW GUYS HOUSE

USER
TOWER

COMM.
TOWER

BOILER
HOUSE

NEW GUYS
HOUSE

SITE PLAN

Fig. 1. Guys Tower Block–site and foundation plan

was essentially completed by July 1971. Internal partitioning, installation of services and plant, and general fitting out has been proceeding since then and full commissioning is expected by May 1974. The levelling points were installed and the first records obtained in April 1969, when approximately 20% of the structure had been completed. The levelling points consist of

205

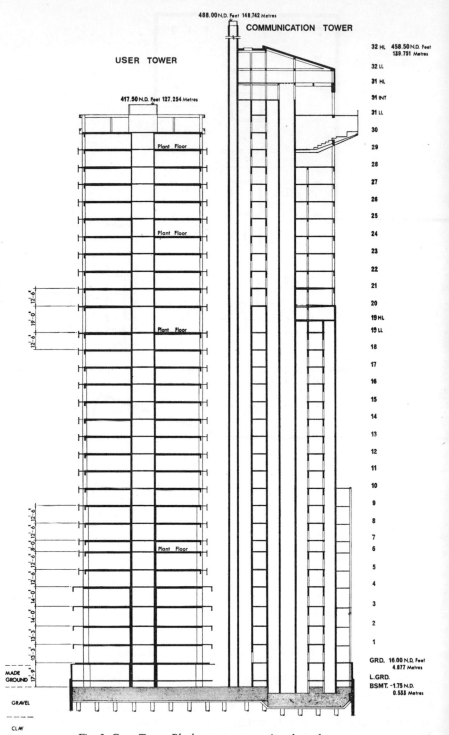

Fig. 2. Guys Tower Block—east-west section through towers

206

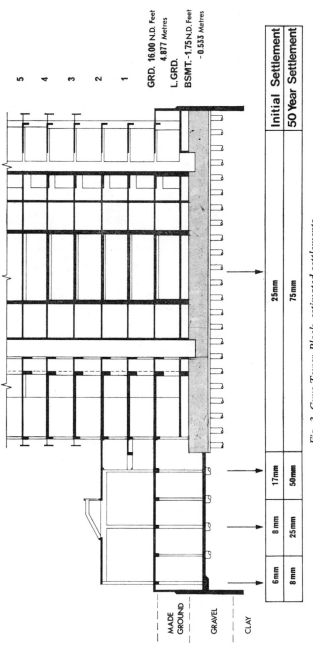

5
4
3
2
1

GRD. 16.00 N.D. Feet
4.877 Metres
L. GRD.
BSMT. -1.75 N.D. Feet
-0.533 Metres

MADE GROUND

GRAVEL

CLAY

Initial Settlement			
50 Year Settlement			
6 mm	8 mm	17 mm	25 mm
8 mm	25 mm	50 mm	75 mm

Fig. 3. Guys Tower Block–estimated settlements

207

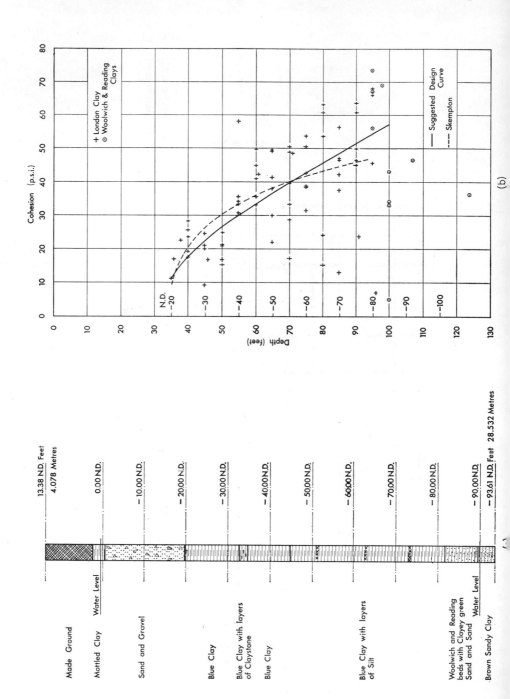

(b)

sockets cast horizontally into columns or walls about 2 m above ground floor level, into which a precision-made peg can be fitted when readings are required.

Fig. 1 shows the layout of the project at foundation level together with a block plan relating the Tower to adjoining and nearby structures. *Fig. 2* is a cross-section through the two towers giving an overall picture of the general proportions, while *Fig. 3* shows the relationship of the low-rise section to the main structure.

The soil

The ground conditions, as revealed by an extensive pre-contract soil investigation are fairly typical of Central London, and reasonably uniform over

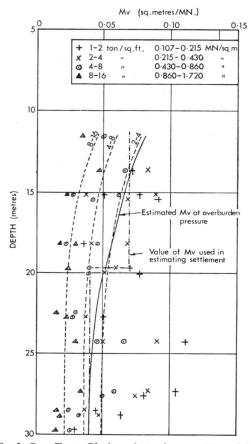

Fig. 5. Guys Tower Block—relation between M_v and depth

the area of the site: approximately 3 m of made ground overlies a gravel bed which is generally 6 m thick. Beneath this the London Clay bed is not so deep as in many other areas of London: it gives place to the Woolwich and Reading beds at some 29 m below ground level.

In *Fig. 4* a typical bore hole plot is reproduced, together with an assessment of the average variation of shear strength with depth in the clay (plotted by averaging triaxial test results from 10 boreholes). Included for comparison is the classic 'Skempton' curve (Skempton, 1959).

Oedometer tests to determine coefficients of consolidation and of volume decrease were carried out, and their results used to produce estimated values of long-term settlements. For this purpose the values of C_v used were:

London Clay, 0.37 m²/a (4.0 ft²/a)

Woolwich and Reading Beds, 0.74 m²/a (8.0 ft²/a)

The average values of M_v obtained, and those used in the calculations, are shown plotted against depth in *Fig. 5*, and a very broad indication of the resulting estimated settlements has been included in *Fig. 3*.

The soil investigation, testing and settlement calculations, were carried out by Terresearch Ltd.

The Measurements

In *Fig. 1* the locations of the levelling points used are shown: readings have been taken reasonably regularly at intervals of two months, and have been plotted individually. The instrument used was a Kern GK1-A, with a 'Mudlark' steel staff. In the interests of simplicity the information obtained is presented in the form of three settlement/time curves (*Fig. 6*) representing respectively the 'User Tower', 'Communications Tower' and Boiler House/Link area. On each curve the ordinates represent the average of the results or that section, i.e. theoretically for eight levelling points on the User

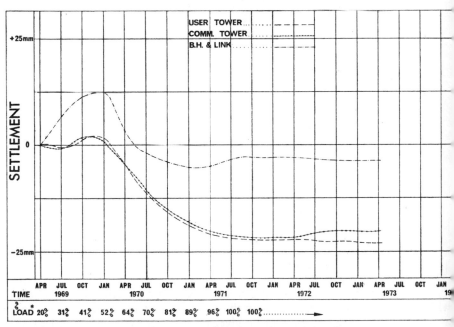

Fig. 6. Guys Tower Block—actual recorded settlement

Tower, six on the Communications Tower, and five on the Boiler House. In practice, as building work has proceeded, accessibility of the various points has varied; in particular the later readings have been made more difficult as finishing trades have proceeded, and the number of points available has been seriously reduced. However, it is believed that the curves plotted can be taken as a fair representation of the variations found over the four-year period under consideration. A definite, though slight, heave effect on the lightly loaded section was observed in the earlier stages, as might have been expected.

The closing error for the level survey has varied from a maximum of 0.010 ft to a minimum of 0.002 ft, the average being 0.005 ft (1.52 mm).

The tying in of the local datum with the London Bridge deep datum has been done with the utmost care three times since the beginning of the operations, the most recent in April, 1973. No record of these surveys is presented, as to date the maximum difference that has been recorded is 0.002 ft (0.61 mm).

Conclusions

The results to date show very reasonable correlation with the calculated 'initial' settlements, but do not at present suggest that movement is continuing as anticipated. Also, somewhat surprisingly, the loading imposed by the new building has not yet had any apparent effect on the nearby existing building.

REFERENCES

Mould, G. (1971). 'Tower Block Development, Guy's Hospital, London', *The Structural Engineer*. Vol. 49, No. 1, 31–53
Skempton, A. W. (1959). 'Cast In Situ Bored Piles in London Clay', *Geotechnique*. Vol. IX, December

III/8. Long-term heave of a building on clay due to tree removal

Stanley G. Samuels
Senior Scientific Officer

John E. Cheney
Senior Scientific Officer

Both at the Building Research Station, Garston, Watford, Herts

Summary

Precision level observations taken over a period of thirteen years on a terrace of cottages at Windsor founded on London Clay are reported. The observations show that the building has been subjected to heave of a long-term nature due to gradual swelling of the clay following the removal of trees during site clearance. The levelling results are presented as graphs to indicate which parts of the building have been affected by ground movement and are discussed in terms of angular distortion.

Methods are described for assessing from simple laboratory tests the amount and extent of clay swell from a knowledge of the ground conditions before land is cleared. Predictions of heave are compared with the lift recorded on the cottages.

Introduction

A single-storey terrace comprising four cottages, about 30 m long and 7.5 m wide, was constructed in 1952 on London Clay at Windsor, Berkshire. The building is of 280 mm cavity brick construction with a steep tiled roof and solid floors. The foundations are concrete strips 0.46 m wide and about 1.2 m deep on the outer walls and about 0.92 m elsewhere.

Six years after construction the Building Research Station was asked to investigate an assumed settlement of the west end of the building; cracks had developed here in the south elevation two years after completion of the property and the damage had worsened over the succeeding years.

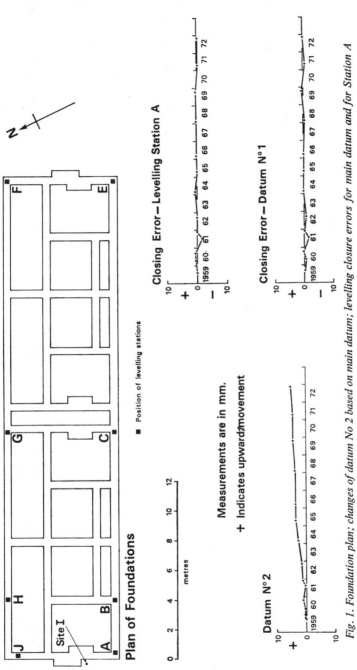

Fig. 1. Foundation plan; changes of datum No 2 based on main datum; levelling closure errors for main datum and for Station A

Preliminary survey

The Station examined the building in October 1958. The west gable wall had tilted outwards with a gap of about 40 mm occurring at roof ridge level; the bearing of the purlins on the gable had reduced by about 25 mm.

The ground conditions adjacent to the chimney in the west gable wall, see *Fig. 1*, hereafter referred to as Site I, were explored in November by means of a trial pit and two borings, up to 3 m deep. There was no evidence of any weakness of the ground below the exposed foundations. The clay was typically fissured and abnormally dry except for a surface layer about 1 m deep which had been affected by recent rain. Water contents ranged from about

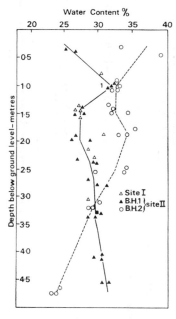

Fig. 2. Variation of water content
with depth

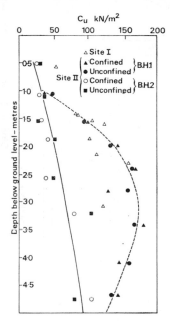

Fig. 3. Variation of undrained
shear strength with depth

26 to 31%, see *Fig. 2*, and comparatively high shear strengths between 80 and 160 kN/m² were measured, see *Fig. 3*. Dead tree roots were in evidence for the whole depth of exploration.

Subsequently, three additional borings were put down which substantiated the particularly dry and strong nature of the clay close to the damaged end of the building whereas it appeared in its normal moist condition near the undamaged end.

Old aerial photographs revealed that a cluster of large elm trees had existed on and close to the west end of the building and it was later confirmed that trees had been cut down in the immediate vicinity of the damaged part of the building only a few months before construction. Two of the trees had been within the confines of the building near to the point H on the plan, see *Fig. 1*, or immediately behind the building in this vicinity.

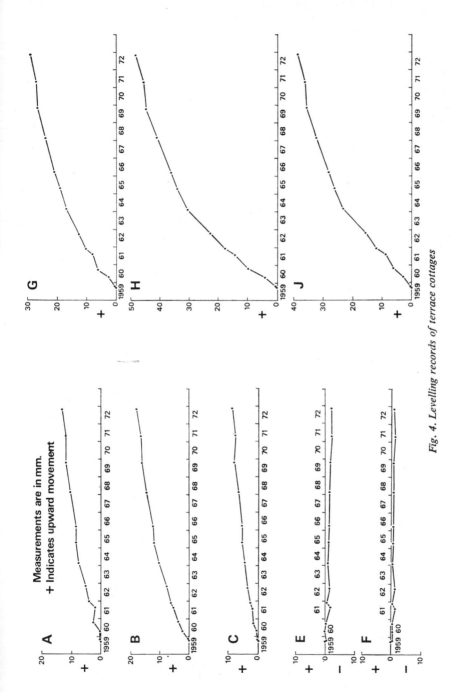

Measurements are in mm.
+ Indicates upward movement

Fig. 4. Levelling records of terrace cottages

215

Level observations

That trees had been cut down on the site suggested that the movements of the building were due to the swelling of the London Clay beneath the foundations. In order to confirm this, and obtain quantitative data, eight levelling stations were installed in the positions A–J, see *Fig. 1*, and a series of measurements of their vertical displacement relative to two levelling datums, 6 m deep and about 15 m from the building, were commenced in August 1959. The techniques for installation and the precautions necessary to accurately record the changes in level are described by Cheney (1973).

The initial levelling showed that the brickwork courses at the west end of the building were some 100 mm higher than the east end, and this observation coupled with the knowledge about the trees in the vicinity suggested that this end of the building had lifted.

Fig. 4 shows the vertical movements which have subsequently occurred. The east end of the building has remained virtually stationary whereas the remainder has continued to lift by varying amounts, the greatest movement being at point H, nearest the site of the trees, where the building has lifted a further 50 mm. Assuming that the brickwork was built level, a maximum upward movement of about 150 mm has taken place at the west end of the building in the twenty years since construction. Heave is still continuing although at only a very slow rate and the west end will presumably continue to lift until the moisture deficiency developed in the ground is eliminated at the site of the trees.

Assuming that the trees were very close to station H it would appear that the ground about 30 m away at station E has suffered insignificant movement, whereas at half this distance, close to stations C and G, the heave since construction has varied between 50 and 100 mm. At a distance of about 6 m, see further investigations described later, the heave is as much as 120–130 mm.

It has been possible by extrapolation to deduce the maximum angular distortion through stations A, B and C on the south elevation about two years after construction where at that time cracking was first noted. The estimate of approximately 1 : 400 is less than the angular distortion of 1 : 300 often associated with the onset of cracking in normal brickwork and plaster. At the time of the initial observations in 1959 the angular distortion in this part of the building was about 1 : 170 and during the two-year period between March 1958 and April 1960 there was a substantial increase in crack width from 3 mm to 14 mm close to station B. Since 1959 the angular distortion has increased to about 1 : 130.

The distortion through stations J, H and G on the north elevation does not appear to have been initially so severe. In 1959 the value was a little less than 1 : 400 and no serious cracking appeared until late 1960 when the distortion had increased to about 1 : 250 and, finally, to 1 : 170 in 1972.

Further field and laboratory investigations

An opportunity to make a further investigation arose in July 1963 from a proposal, subsequently abandoned, to build on land, *Fig. 2*, Site II, about 120 m away from the building under observation.

216

The area was mainly grass-covered with occasional elm trees growing singly or in groups; two with girths of 4 m were standing about 9 m apart alongside one of the proposed buildings.

A boring (BH 1) was made about 5 m deep in the London Clay in a position about 6 m from each tree. During boring the clay appeared very dry and brittle and contained many gypsum crystals to the full depth investigated. Roots were found to penetrate at least 2 m below ground surface. No ground water was encountered. Almost continuous undisturbed sampling was carried out using standard 102 mm diameter open-drive sampling equipment.

A second boring was put down to a similar depth in completely open grassland and frequent undisturbed samples again taken. The clay was noticeably softer and very few gypsum crystals were found. At a depth just below 4 m a thin water bearing pebble stratum was encountered, water entered the boring and rose eventually to 1 m below ground surface.

The principal soil properties investigated were:
(1) natural water content, liquid and plastic limit, and particle size distribution;
(2) undrained (quick) strength; and
(3) swelling characteristics.

The swelling test involved the determination of the swelling pressure employing the method described by Ward, Samuels and Butler (1959); each specimen was then unloaded in stages and the amount and rate of swelling determined. Usually at least 2 to 3 days were allowed under each pressure which was sufficient time for all swelling of a primary nature to take place.

Fig. 2 shows the variation in water content with depth and clearly suggests that the ground in the vicinity of borehole 1 and at Site I is deficient in water to a depth of about 3 m below ground level.

The liquid limits ranged generally between 71 and 93 with the majority of values close to 80; it was apparent, however, that some variation existed in the composition of the clay which was reflected in the water content profiles and mention should be made of a very low value of 50 for the liquid limit of the bottom sample from borehole 2.

Below the zone subjected to normal seasonal variations the water contents close to the elms were generally at the plastic limit whereas in the open grassland they were typically some 4% above the plastic limit at all depths.

Some 60 to 70% clay fraction was found in most samples with the activity ranging generally between 0.75 and 0.95.

The undrained shear strengths are plotted against depth in *Fig. 3* and it is significant that there is a large difference between the strength of the clay in the two situations. The strength data from Site I, also plotted in *Fig. 3*, suggests that the ground conditions close to the west gable wall of the cottages, although not as severe, were more typical of the conditions close to the elm trees on Site II than of the open grassland.

The swelling pressure results are presented in *Fig. 5*. The relationships are similar in form to the undrained strength plots. The difference between the two curves suggests that close to the trees the whole profile down to more than 5 m is deficient in water compared to the open grassland condition, the zone of maximum water deficiency occurring where the highest values of swelling pressure were obtained.

When unloading individual specimens some variation was noted in the

217

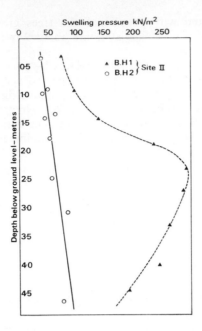

Fig. 5. *Variation of swelling pressure*
with depth

slopes of the pressure-void ratio (rebound) curves, values of the swelling index ranging between 0.05 and 0.07. This variation may be attributed primarily to variations in composition of the London Clay.

Individual values for the coefficient of swelling are not reported here but the coefficient was found to decrease markedly as the pressure on the specimen was reduced; this decrease is completely unlike the small variations in the coefficient of consolidation with applied pressure normally found in this clay.

Prediction of heave from laboratory tests

The likely heave of the ground on removal of the trees can be predicted from the soil data in two ways:
 (1) by assuming that the water content profile in *Fig. 2* close to the trees takes up water to the values under grass and the ground expands in proportion to the water uptake;
 (2) by assuming that the high swelling pressures in *Fig. 5* near the trees decrease to those under grass and estimating the corresponding expansion with full lateral restraint.

In the first case the results from *Fig. 2* give a gain of the equivalent of 200 mm depth of water. The associated heave depends on the nature of the cracking in the ground, the degree of saturation and many other factors. Ward (1953) studied the problem of drying shrinkage under grass and found that the ratio between equivalent depth of water loss and vertical shrinkage was usually greater than 3. Accepting this figure gives a predicted ground surface heave of about 70 mm with the depth of influence extending to at least 3 m. If differences in soil type and moisture conditions are taken into account a heave of 100 mm at the site of the cottages is estimated with the depth of influence extending to about 6 m. If the depth of founding of the cottages is taken into account the above figures become 40 mm and 65 mm respectively.

Adopting the second method the predicted heave at foundation level amounts to about 70 mm with the depth of influence extending to about 6 m. No account has been taken of the load of the building which is relatively small. Use of the oedometer swelling curves involves the assumption of no lateral swelling so that the predicted heave is likely to be overestimated by this method.

Any comparisons between observed heave and that predicted from the geotechnical data are, unfortunately in this case, of an indirect nature and must be treated with some reserve. For example, the amount of heave which the cottages have been subjected to since 1959 adjacent to Site I, obtained by interpolation of the observed levels at stations A and J, is only about 25 mm, which is less than the laboratory estimates of heave based on the ground conditions on Site II, 6 m from the elms; however, it appears that the ground conditions on Site I at the time of the initial survey in 1958, see *Fig. 3*, were less severe than those on Site II. In contrast, the movements of the building 6 m from station H, mentioned previously, varying between 120 and 130 mm are greater than the laboratory estimates of heave. It is evident that comparison between observed and predicted movements are not easy, but the predictions are of the right order.

The rate of heave of the ground on removal of the trees is much more difficult to predict. In particular, it is not easy to select an appropriate value for the swelling coefficient. Even more important is to decide on the source of water which is necessary for heave to take place. If it is assumed that water is readily available to the affected zone from above and below and an average value for the swelling coefficient is used, the predicted rate of heave is appreciably slower than that observed on the cottages.

Conclusions

In recent years the Station has given advice concerning an increasing number of buildings which have become distorted and cracked a year or two after construction and found to be due to ground heave following the removal of trees during site clearance. Differential movements of 50 to 75 mm are not uncommon with heave continuing more than five years after construction.

Apart from initially determining the direction of movement the level observations at Windsor clearly demonstrate the long-term nature of the heave mechanism and indicate that in a heavy clay area, severely affected by tree root action, the amount of heave may be considerable. Therefore, it would seem important on sites where building development is anticipated on or close to areas previously occupied by trees that the order of magnitude of subsequent clay swell is determined geotechnically by methods such as those described in this paper and the type of foundation is chosen which is most able to withstand the expected ground movement, e.g. specially designed raft or deep strip, or pile and beam foundation with suspended floor. In any case, if trees are to be removed with development in mind they should be cut down well in advance of the building operations.

Alternatively, if normal strip foundations are proposed it would be necessary, if the Windsor site is typical, to delay construction for many years to allow a major part of the ground heave to take place, as illustrated by the level observations which have been presented.

Acknowledgements

The authors wish to acknowledge the kindness and interest of the consultant architect for the Windsor site, Mr. R. Tatchell, and of the Crown Estate Office in allowing the installation and site work.

The authors are also indebted to their colleagues for the interest shown and encouragement given and also wish to acknowledge the invaluable help given both with the observations and laboratory work, particularly Mr. D. Burford, Mrs. J. Butcher and Mrs. V. Coleman.

The Paper is published by permission of the Director of the Building Research Establishment.

REFERENCES

Cheney, J. E. (1973). 'Techniques and Equipment Using the Surveyors Level for Accurate Measurement of Building Movement', *Proc. Symp. on Field Instrumentation, Brit. Geotech. Soc., May 30–June 1*. Butterworths, London
Ward, W. H. (1953). 'Soil Movement and Weather', *Proc. 3rd Int. Conf. Soil Mech. and Found. Engg*. Switzerland. Vol. 1, 477–482
Ward, W. H., Samuels, S. G. and Butler, M. E. (1959). 'Further Studies of the Properties of London Clay', *Geotechnique*. Vol. 9, No. 2, 33–58

III/9. St Paul's Cathedral— measurements of settlements and movements of the structure

J. B. Thomas and A. W. Fisher
Freeman Fox & Partners, London

Summary

St. Paul's Cathedral has been settling and moving ever since building started 300 years ago. Combined with deterioration of the stonework, these movements have made extensive renovations necessary at various times to maintain the fabric in good order.

Since 1923 the movements have been meticulously observed and recorded at six-monthly intervals by levelling at fixed points at five levels of the cathedral to an accuracy of 0.0001 ft; linear measurements and plumbings have been checked to an accuracy of 0.001 ft and movements of cracks in the fabric to 0.0005 mm; and continuous records have been maintained of the variations in subsoil water level.

Though all the movements have been very small, plots have been carefully maintained so that variations from the general trend could be quickly recognised, the cause ascertained and, where necessary, remedial action taken.

A modern soil analysis of the ground below the foundations has also been made and movements of the structure have been related to ground behaviour with time.

It is believed that the duration, accuracy and range of these measurements are unique amongst building surveys.

Introduction

St. Paul's Cathedral, designed by Sir Christopher Wren, was built between 1675 and 1710 on the site of an earlier, even larger cathedral destroyed in the Great Fire of 1666. It is founded on a thin stratum of Potter's (brown sandy)

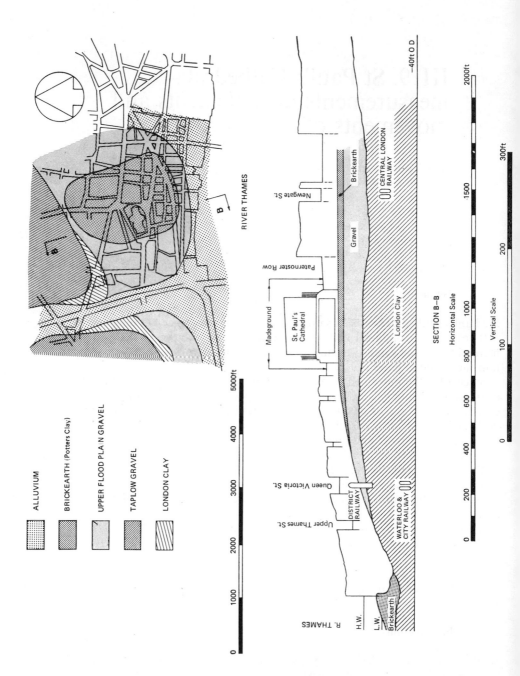

ALLUVIUM

BRICKEARTH (Potters Clay)

UPPER FLOOD PLAIN GRAVEL

TAPLOW GRAVEL

LONDON CLAY

RIVER THAMES

0   1000   2000   3000   4000   5000ft

SECTION B–B

Horizontal Scale

0   200   400   600   800   1000   1500   2000ft

Vertical Scale

0   100   200   300ft

Newgate St.

Paternoster Row

Madeground

St. Paul's Cathedral

Queen Victoria St.

Upper Thames St.

DISTRICT RAILWAY

WATERLOO & CITY RAILWAY

CENTRAL LONDON RAILWAY

Brickearth

Gravel

London Clay

–40ft O D

R. THAMES

H.W.

L.W.

Brickearth

222

Clay, 4 ft to 6½ ft thick, that overlays deposits of water-bearing sand and gravel, 10 ft to 20 ft thick. Still deeper is a 100 ft bed of London Clay and below that the Woolwich and Reading beds and the Upper Chalk. Outside the building, the Potter's Clay is covered by made ground, 13 ft to 16 ft thick. The geological section in the vicinity of the cathedral is shown in *Fig. 1*.

The cathedral is 500 ft long from east to west and 250 ft north to south. The plan at crypt level is shown in *Fig. 2*. Above where the nave and choir meet stands the Great Dome with the lantern, the ball and the cross. From the nave floor to the top of the cross is 365 ft. The weight of the lantern, ball and cross is carried by a brickwork cone hidden between an inner, decorative dome and an outer protective dome. To help retain the cone and prevent it spreading, Wren fitted his 'Great Chain' of iron links around its base. The cone and the two domes are carried on two concentric drum walls, joined by buttresses at the level of the whispering gallery.

The whole of the dome structure is carried on eight piers, joined at their heads by the Great Arches, and by four large bastions. The weight of the dome, piers, arches and bastions produce a total load at foundation level of about 70 000 tonf and a maximum load on the foundations of 5 tonf/ft².

Differential settlements occurred during building, requiring localised 'levelling'. Settlement, with cracking of the structure, continued after completion and spasmodic repair and patching followed until 1780 when extensive renovations were made. Throughout the nineteenth century repairs to the fabric continued to be required.

During the second half of the nineteenth century, when areas north of the cathedral were built upon, and drainage systems laid, the subterranean flows towards the Thames would have been markedly reduced. Such a reduction would have lowered the ground-water level and might well have altered the consolidation of the sand and gravel stratum below the cathedral. It is certain that damage to the cathedral fabric continued so that, in 1913, repairs were required to the dome and main piers below. However, movements and cracking continued to occur and, in 1921, a major programme of extensive repairs was prepared including new chains around the drum walls, grouting of the main piers and installation of numerous stainless steel ties. This work was completed in autumn 1930.

During the restoration (1925–30) a comprehensive programme of measurements was instituted. The programme continued checks already being made on 47 cracks in the stonework and required more precise observations to be made of linear measurements between selected stations, levels in the crypt, the church floor and at other heights, and of plumbings of vital points. In 1932, as part of an investigation by Sir Alexander Gibb and Mr. (later Sir) Ralph Freeman (1933) into the foundations of the soil conditions below and around the cathedral, eight automatic water level recorders were installed. Another borehole in the crypt is also used, but this is not fitted with a recorder.

Gibb and Freeman were of the opinion that stability had been reached but they were concerned that any interference whatsoever with either the subsoil or the water levels in the vicinity might 'upset the equilibrium' and cause further settlement. On their recommendation, The City of London (St. Paul's Cathedral Preservation) act 1935 was obtained which gives the Dean and Chapter of St. Paul's control over any building work within a defined area that might endanger the cathedral. Now, because of the increase

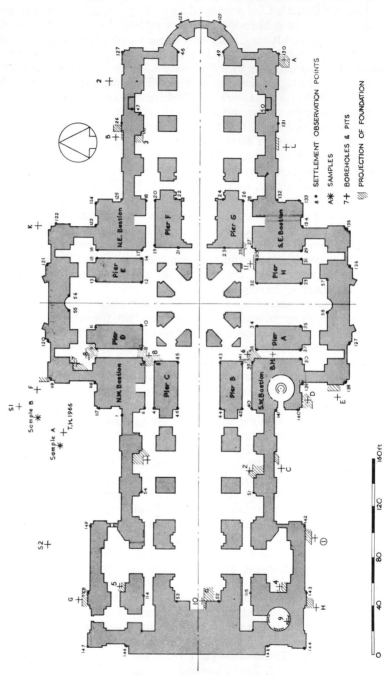

Fig. 2. Plan at crypt level

224

since the 1930s in the size of buildings, thought is being given to asking Parliament to extend the area covered by the 1935 Act.

The measurements in the fabric surveys of the past 50 years have been made with a high degree of precision: levels to 0.0001 ft, linear measurements to 0.001 ft, and crack movements to 0.0005 mm. Every measurement has been recorded in duplicate and plots on stable cloths have enabled accurate diagrams of the movements to be drawn over a period of 50 years. About 75 sets of measurements have been made over a long period with a degree of care that is probably unequalled in any building survey. That this very real achievement has been possible can be attributed to the fact that, ever since these measurements began, one surveyor has been continuously engaged on the work.

The movements have been found to be exceedingly small and this must be largely attributed to the restoration work of the late 1920s when the new chains were fitted round the drum wall and the main piers were strengthened by grouting. That the movements have been small has been a matter for satisfaction but the accuracy of the surveying makes it possible to assess changes in the pattern of movements with complete confidence so that remedial action can be considered and undertaken in advance of any noticeable damage to the fabric occurring.

Measurements and records

Crack measurements

To check on movements at the cracks, a pair of ground carbon steel pins, $\frac{5}{8}$ in. in diameter, were set in sulphur cement, protruding 3 in. from the stone face and each protected by a grease impregnated felt-lined brass cap. Three different instruments, each bearing on machined faces on the pins and measuring to 0.0005 mm, are used to measure changes in the vertical and horizontal distance between the pins, and of relative movement (projection) at rightangles to the walls. A pair of 'control' pins set in uncracked stone enable the instruments to be checked for consistent readings. The air temperature is recorded each time a crack is measured.

Levelling

Five main groups of levels are regularly checked: crypt, church floor, cornice and whispering gallery, and externally just above ground level. Each round of levels is related to a datum point near to the cathedral (SPDP) which, in turn, is related to a datum embedded in the 10 ft thick concrete plug of an old cast-iron lined access shaft about 70 ft deep located 1200 ft from the cathedral.

Standard sequences of readings, in established rounds or traverses, have been evolved, each group containing one point that is related via other traverses to the SPDP. Each round starts and finishes on the same point and a closure error of 0.0001 ft per reading is accepted except that, if, even within this limit, a significant change in level between two points is observed, then the reading is checked. At each point, a bronze Morse taper socket has been

225

set into the stonework. When levelling, a ground steel Morse taper plug with a spherical head is inserted in the socket. Staves, 3 ft, 8 ft and 10 ft long, are used, each consisting of a mahogany case with a ground steel shoe and an invar steel strip insert fixed at its lower end only. The case is graduated in tenths of a foot while the invar steel strip is graduated in tenths and fiftieths of a foot. The staves are also fitted with spirit bubbles to ensure their uprightness in use. A 14 in precision reversible level fitted with a parallel plate micrometer permits readings to 0.0001 ft.

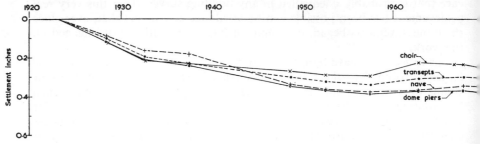

Fig. 3. Average settlements 1923–71

To connect the levels at church floor level with the cornice and whispering gallery levels, an invar steel tape, graduated at each end and at the centre in tenths and fiftieths of a foot, is suspended from a fixed bracket and tensioned by a 25 lb weight.

The changes in level have been plotted as developed sections along each traverse. Initially, it was assumed that all points were at the same level and therefore could be represented by a full horizontal line. Subsequent changes at each point above or below this base line have been drawn. From these detailed plots, average settlements at dates roughly five years apart have been deduced and are shown in *Fig. 3*.

Plumbings

The out-of-plumbness of the walls is checked by suspending a 25 lb plumb-bob fitted with vanes immersed in dashpots filled with water or oil. Measurements are taken from the suspension wire (24 gauge—0.022 in. diameter) to a bronze dome-headed pip, or pips, set in the adjacent masonry. The wire is suspended from purpose-made brackets fixed into the stonework.

The distance from the pips to the wire is measured with a steel T-section bar having one end machined to form an anvil that is presented to the pip. To facilitate use, the bar is fitted with a spirit level and is supported at the far end from the pip by a steel rod with a holding nut.

The bar is graduated in tenths and hundredths of a foot and the readings can be interpolated to 0.001 ft.

In the case of the dome, the fixing point for the wire is 265 ft above the floor, and a bob weighing 45 lb is used with a steel centre point just clearing the ground.

An existing brass grille set in the floor immediately under the dome has been scribed with the north–south and east–west axes. The centre of oscillation of the bob is determined and measured in the two directions of the grille axes using a graduated steel set square and straight edge, readings being interpolated to 0.001 ft. The polar distance from the intersection of the grille axes to the centre of the bob oscillation is also measured.

Linear measurements

Linear measurements are taken to determine variations in distance from plumb lines set up over the grille centre point and hung from the keystones of the four main arches or to bronze pips set in the bastions; and also of several diameters of the whispering gallery and across the corridor behind it. All the linear measurements are interpolated to 0.001 ft.

Base-line apparatus is used to make these measurements: an invar steel tape under a tension of 25 lb is suspended over adjustable pulleys mounted on tripods. The tripods are set up so that the tape is on the alignment to be measured and, where appropriate, sets of readings are taken and averaged to give the distance between the plumb wires. For the distance to each pip on the bastions, a steel scale is used to measure the distance from the tape to the pip.

To measure the diameter of the whispering gallery two steel scales are used to measure the distances from a graduation mark at each end of the tape to protruding lips on the bronze sockets used for levelling, ten separate readings being taken and averaged to determine the distance.

For the corridor behind the whispering gallery, measurements are made as extensions of those for the gallery diameters. An extending measuring rod is used to measure between pips or sockets set in the stonework. It has an anvil at one end and a scribed line at the other which registers on a sliding extension of the rod which is also fitted with an anvil.

Water levels

Fluctuations in ground water levels have been obtained from recorders in which a pen makes a continuous trace on a chart fixed to a clockwork driven cylinder that rotates once a month.

In addition, the depth of the water below the Cathedral is measured by lowering a weighted steel tape into a borehole driven from the crypt floor.

Observed movements

Cracks

The movements at the cracks have generally been small. They show a clear correlation with temperature fluctuations (*Fig. 4*).

Of the cracks, 40 have generally widened, three have shown no movement and four have narrowed. The movements have always been converted into inches and the largest observed are: width, 0.029 in.; level, 0.048 in.; and projection, 0.025 in.

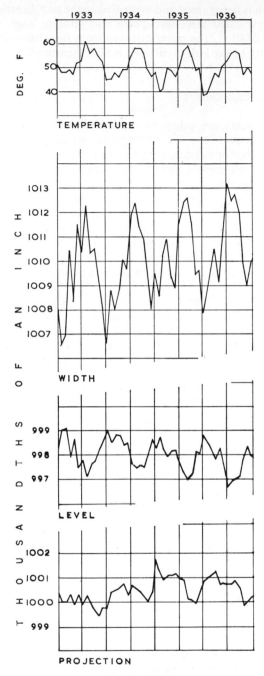

Fig. 4. Typical movements at a crack in the fabric are seen to correlate with changes in ambient temperature

Levels

The plots of the levels show that, during the period 1923–71, the Cathedral has settled more under the dome (0.625 in.) than at the ends of the transepts (0.375 in.).

Levelling has been precise enough to observe small trends. During con-struction of the Choir School (1940) and the Undercroft (1968) the adjacent parts of the Cathedral tended to rise, possibly due to expansion of the Potter's Clay caused by increasing moisture content.

Between 1923 and 1938 the whispering gallery and the crypt settled equally but, between 1939 and 1947 the gallery settled by 0.197 in. more than the crypt. This is accounted for by the lack of heating during the later period which would have caused the piers to contract. After the war, and until 1956, the gallery remained at a constant level while the crypt continued to settle. It is suggested that this was due to the reinstatement of heating which caused the piers to expand at roughly the rate of settlement.

Overall, the gallery has settled more than the crypt by 0.063 in. in the SE. quadrant and 0.125 in. in the NW., causing the dome to tilt. No expla-nation is offered for overall greater settlement of the gallery relative to the crypt but the differential settlement of the gallery is possibly attributed to the bomb damage sustained by the north transept, and the subsequent delay to repairs, which may have weakened the stonework of the piers on the north side.

Plumbings

Continuous plumbings of the dome showed that, until March 1940, the dome plumbing oscillated within a circular area of 0.1 in. diameter. After the war the plumbing centre appeared to have shifted to the NW., possibly due to bomb damage to the north transept or to the reference point at floor level having been disturbed. Since 1940, however, oscillations have occurred within an elliptical area of about 0.125 in. by 0.062 in. with the major axis in the NW. direction (*Fig. 5*).

Both the towers have also leaned outwards, the northern one by 0.125 in. in a northerly direction between 1925 and 1965 and since then by 0.06 in. in a westerly direction. Since 1923 the southern tower has tilted by about 0.15 in. in a southerly direction.

The movement of these towers serves as a good example of the usefulness of precise long-term measurements. At one time it was thought that movement was being caused by the settlement of the foundations; this has now been discounted by Skempton (1973). Similarly, the theory that it was due to the thrust from the intervening arches has been shown by R. J. Ashby and L. Chitty (see Skempton, 1973) to be false, since they exert a horizontal thrust of only 70 tonf compared with the weight of the towers of 12 000 tonf each.

It is now suggested, though not confirmed, that these tilts have been caused by the rising temperature of the stonework of the gallery at the west end (due to improved central heating of the cathedral) providing an outward thrust on the towers. Coarser, less frequent measurements might have led to a wrong conclusion, resulting in remedial action which would not only have incurred

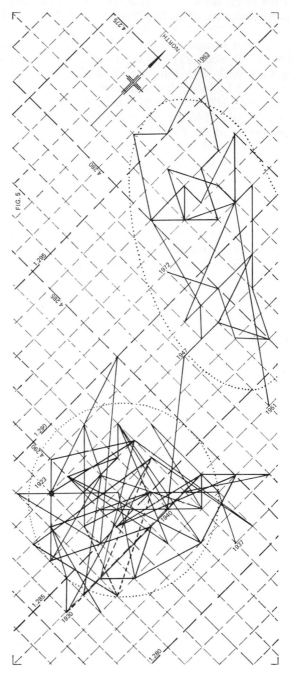

Fig. 5. Cone plumbing movements

230

unnecessary expenditure but may, in fact, have created worse conditions than those it was intended to rectify.

Linear measurement

The range of movements of the main arches (0.156 in.) and of the bastions (0.125 in.) has also been small.

Originally, in 1923, the twelve selected whispering gallery diameters varied from 110.221 ft to 110.722 ft. By 1947 the changes in diameters ranged from +0.141 in. to −0.078 in., with increases on the N–S axis and decreases on the E–W axis. Since 1928 (when the new chains were installed around the drums) the diameters have remained sensibly constant.

There have been virtually no changes in the width of the corridor behind the gallery.

Water levels

Since the recorders were first installed, the water level has fallen by an average of 1 ft 8 in. Below the crypt, however, the reduction is only 6 in.

Abrupt rises in level (18 in. in 24 h) have followed breaks in the local water mains but after repairs the water has returned to its previous level. A typical plot is illustrated in *Fig. 6*.

Whilst further continual reduction in water level below the cathedral would give some cause for concern, the effect is not yet thought to be detrimental.

Ground conditions

A detailed analysis of the soil conditions below the cathedral, leading to an interpretation of the settlements and changes in plumbings, has been prepared by Professor A. W. Skempton.

Calculations (using Terzaghi's equation) of the bearing capacity of the Potter's Clay (which must have been fully consolidated ever since the earliest period of construction) show that there is a factor of safety of the order of 3 against shear failure of the clay under the foundations of the dome piers and the western towers. Under the more lightly loaded walls of the nave and choir, the factor of safety is correspondingly greater.

On the assumption that the foundations below the walls of the nave and the choir and below the piers supporting the dome were built at the same level a differential settlement from when construction began to 1900 of between 2 in. and 3 in. was observed for the east, north and west piers and of between 4½ and 6 in. for the south piers. This greater settlement of the south piers is attributed to a 30 ft deep excavation made close by in 1831 for a proposed sewer that was abandoned because of risk to the safety of the cathedral.

From his measurement of the characteristics of the Potter's Clay, Skempton estimated that the 'natural' settlement under the piers for the first 200 years

231

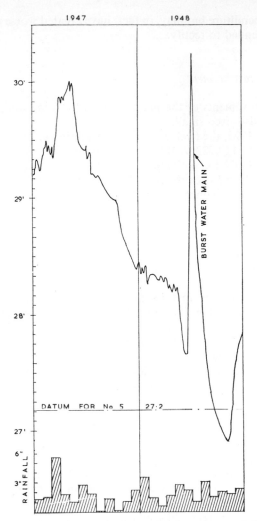

Fig. 6. *Typical water level record and concurrent rainfall*

(up to 1900) would be of the order of $2\frac{1}{2}$ in. or much the same as mentioned above. He also found a close agreement between his estimates of the expected differential settlements since the 1920s (when the piers were grouted and the loads on their foundations consequently increased) and settlements actually measured. The possible contribution of consolidation within the London Clay was found to be very small.

By considering the possible 'rigid body' movement of the structure below the dome, a good correlation was found between the out-of-plumb movement of the dome since 1923 and the difference in settlements measured at the north and south piers. Since, however, both these movements (of the dome tilt and its foundations) have practically ceased for the last 20 years, the present conditions of the pier foundations can be considered satisfactory, as was believed by Gibb and Freeman (1933).

232

A rough calculation (all that is possible) of likely settlement due to consolidation of the London Clay below the piers during the period 1923–71 is about 0.1 in., whereas the surveys put the figure at about 0.4 in. This could be accounted for if the datum point, 1200 ft away, was less affected by regional subsidence than the cathedral. Such a difference would have little or no influence on the differential settlements. It is concluded, however, that the rate of real settlement due to continued consolidation in the London Clay, and due to the present slow lowering of ground water level in the chalk, is very small and could have very little bearing on considerations of safety of the structure.

Conclusion

For 50 years, the structure of the cathedral has been surveyed with a precision that has made it possible, because of the smallness of the movements, to recognise the slightest changes in their pattern. Only thus has it been possible to forestall damage to the fabric by taking early remedial action. The future preservation of the cathedral requires that the surveys should be continued with similar precision that will provide the data essential to understanding and monitoring the structural behaviour of the cathedral and its foundations in the event of changes in the external conditions.

Acknowledgement

The authors are indebted to the Dean and Chapter of St. Paul's Cathedral for their kind permission to present this paper.

Gibb, Sir Alexander and Freeman, R. (1933). 'St Paul's Cathedral—Investigation of Foundations and Subsoil'. Report to the Dean and Chapter (unpublished)

Longfield, T. E. (1932). 'The Subsidence of London'. Paper read before the British Association. *Ordnance Survey Professional Paper—New Series*, No. 14. H.M.S.O.

Peach, C. and Allen, G. (1930). 'The Preservation of St Paul's Cathedral', *R.I.B.A. Journal.* 9 August

Skempton, A. W. (1973). 'St Paul's Foundations'. Unpublished report prepared for Freeman Fox & Partners

Ceosthwaite, C. D. and Crawford, A. S. (1973). 'A Study of the Internal Environment of Cathedrals, with Particular Reference to St Paul's'. Paper presented to August 1973 meeting of the British Association for the Advancement of Science

III/10. Displacement of over-consolidated clay/sand soil during construction of a building

K. Ueshita
Professor of Civil Engineering, Nagoya University, Japan

K. Matsui
Structural Engineer, Nikken Sekkei, Ltd.

T. Ohoka
Structural Engineer, Nikken Sekkei, Ltd.

S. Nagase
Nagoya University

Summary

The central part of the city of Nagoya in Japan is underlain by alluvial deposits which are known to be suitable for supporting buildings. The authors recently had an opportunity to predict and to measure the vertical displacements occurring during the extension of an existing building founded on these deposits. The base of the extension had to be placed on a layer of over-consolidated clay at G.L. -14.75 m.

Whereas the old part of the building had been built on piers, preliminary investigations showed that piers would not be necessary for the extension because the underlying layers of clay were heavily over-consolidated and the weight of the building was nearly equal to the weight of the excavated soil.

However, no measurements were available of the displacements occurring during the construction of buildings on these alluvial deposits of clay and sand. The displacements were therefore estimated using the theory of elasticity with the results of preliminary soil surveys. Observations made during construction of the new part of the building showed reasonable agreement with these predictions.

Introduction

The central part of the city of Nagoya in Japan is underlain by alluvial deposits. An extension to a fifteen-storey building in this area was planned in 1970.

The cross-section of the building and subsoil is shown in *Fig. 1*. The old part of the building was supported on piers and piles because the engineers at that time were not confident of the load carrying capacity of the underlying layers of clay. It was reported that the piers could not be constructed through the lower layer of clay because of the pressure of water in the underlying layer of gravelly sand. Consequently piles had to be used so that the piers

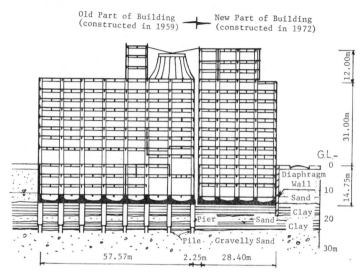

Fig. 1. Extension work of Building on overconsolidated clay/sand ground (section through south side)

could be supported by the gravelly sand layer without cutting the second layer of clay.

In order to construct the extension, it was planned to support the new part of the building directly on the upper layer of clay using a raft foundation, avoiding excessive foundation works, because it was known from the soil investigation that the underlying clay layers were heavily over-consolidated and the weights of the new building and of the excavated soil were nearly equal. However, no information was available on the displacements occurring during construction of buildings on these alluvial clay and sand deposits.

The displacements were estimated using the theory of elasticity with the results of preliminary soil surveys and these estimations were confirmed by means of observations made during the operations of excavation and construction.

235

The new building and foundation

The new part of the building on a raft foundation was linked to the old part but was kept structurally separate as shown in *Fig. 1.* This new part of the building is about 30 × 30 m in plan. The overburden pressure at foundation level was 2.43 kgf/cm², and the average pressure exerted by the building was 2.65 kgf/cm². The foundation was made as rigid as possible in order to minimise differential settlement. Before excavation, the area to be excavated was surrounded by diaphragm walls of reinforced concrete in order to protect adjacent ground and to prevent inflow of ground water.

The subsoil was investigated by testing samples from boreholes. The investigation was terminated at the depth of G.L. −30 m because the layers

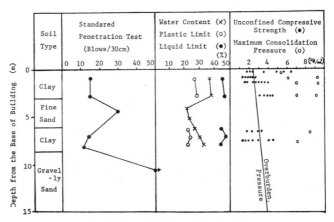

Fig. 2. Typical soil investigation data

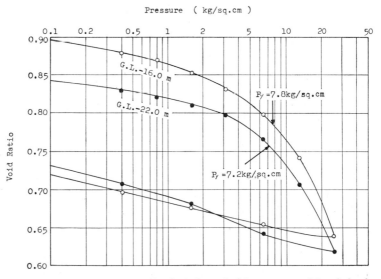

Fig. 3. Consolidation curves of samples from the underlying overconsolidated clay layers

236

of soil below this depth were considered to be very dense or hard. Consequently it is regretted that the conditions of the ground below G.L. -30 m cannot be shown here.

Fig. 1 shows the types of soil and Fig. 2 shows typical soil properties that were found in the investigation. Typical consolidation curves are shown in Fig. 3. The average value of the maximum previous consolidation pressure is 7.8 kgf/cm^2 which is about three times the overburden pressure or the pressure exerted by the completed building.

Therefore, the authors considered that it would be safe to omit piers and to place the new building directly on the layers of clay. However, an estimate was required of how much the new extension would settle during its construction.

Estimation of displacements during excavation and construction

It was considered that these layers of sand and over-consolidated clay would behave elastically during the construction. Therefore, moduli of elasticity were measured in boreholes using lateral loading tests similar to the pressure-meter. Measured elastic moduli are shown in Table 1.

Table 1. MODULI OF ELASTICITY (kg/cm^2) MEASURED BY LATERAL LOADING TESTS IN BORED HOLES

| Test process | Tested depth | | |
	G.L. -16.0 m (Clay)	G.L. -18.3 m (Fine sand)	G.L. -28.2 m (Gravelly sand)
Rebound	334	923	4612
Recompression	325	890	3064

Calculations of the rebound and recompression of the soil were made using Boussinesq's equations for three layers under the excavation together with Holl's expression (1940, mentioned by Harr, 1966) to allow for the gravelly sand and underlying layers as shown in Fig. 4. The elastic moduli assumed for rebound and recompression are shown in Table 2.

Although the condition of the soil below G.L. -30 m was not known exactly, Holl's expression proved to be convenient for taking account of

Table 2. ASSUMED MODULI OF ELASTICITY BASED ON LATERAL LOADING TESTS IN BORED HOLES

| Layer | Modulus of elasticity (kg/cm^2) | | Poisson's ratio |
	Rebound	Recompression	
1st clay layer	334	325	0.5
Fine sand layer	1130	1090	0.2
2nd clay layer	334	325	0.5
Gravelly sand and underlying layers	$923\left(\dfrac{z}{18}\right)^3$	$890\left(\dfrac{z}{18}\right)^3$	0.2

z = depth (m) below the ground surface

237

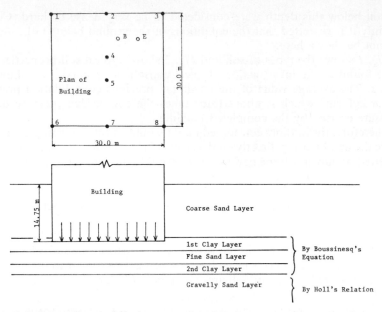

Fig. 4. Calculation of vertical displacement—B and E were measurement points during excavation and 1–8 were measured points during construction of building

these deeper layers. According to this expression, the modulus of elasticity increases with depth so that below a certain depth the effect on the surface displacement is negligible. The authors considered that the total deformation of the deeper layers estimated by this method might be close to the true value for this case.

The results of the calculations are shown in *Table 3*.

Table 3. CALCULATED AND MEASURED VALUES

	Layer	Rebound (cm)	Recompression (cm)
Calculated values	1st clay	0.35	0.56
	Fine sand	0.33	0.54
	2nd clay	0.98	1.03
	Gravelly sand and underlying layers	0.55	0.58
	Total	2.21	2.71
Measured values	Total	2.2	2.8

Measurement of displacements during excavation and construction

In order to measure the rebound of the base of the excavation five measuring rods of the type shown in *Fig. 5* were installed in the excavated area, and a benchmark was established about 60 m away from the construction site.

238

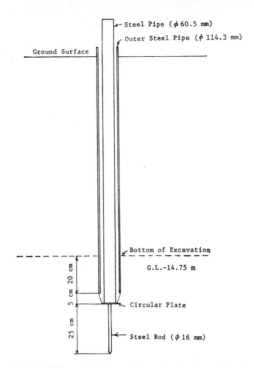

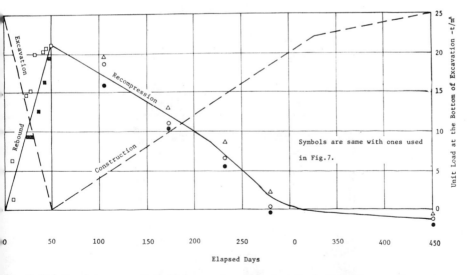

Fig. 5. *Method for measuring rebound at bottom of excavation*

Fig. 6. *Relation between vertical displacement, unit load at the bottom of excavation and elapsed time*

239

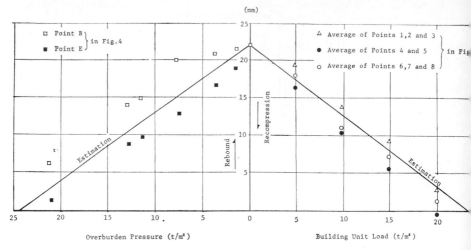

(mm)

□ Point B ⎱
■ Point E ⎰ in Fig.4

△ Average of Points 1,2 and 3 ⎱
● Average of Points 4 and 5 ⎰ in Fig
○ Average of Points 6,7 and 8 ⎰

Overburden Pressure (t/m²)

Building Unit Load (t/m²)

Fig. 7. Estimated and measured displacement during construction work

Fig. 8. The completed new part of Building (left) and the old part (right) photographed from the north side

240

The levels of these rods were measured using a Carl Zeiss N1–2 level (with an accuracy of 0.01 mm) and these results were corrected for the effects of changes in temperature on the steel rods.

Although five measuring rods were installed, only the two marked in *Fig. 4* were successfully monitored until the completion of excavation.

The settlement of the building was measured using the eight measuring points shown in *Fig. 4*. These were set on the lowest floor when construction of the building was started.

The variation of vertical displacement and pressure at the base of the excavation with elapsed time is shown in *Fig. 6*. Comparisons of estimated and measured displacements during excavation and construction showed good agreement as indicated in *Table 3* and *Fig. 7*.

The completed building is shown in *Fig. 8*.

Conclusion

In extending a fifteen-storey building on layers of sand and over-consolidated clay, the authors proposed to dispense with the unnecessary foundation works used under the old part of the building, and to support the new part directly on an over-consolidated layer of clay using a raft foundation.

The authors estimated the vertical displacements of the base of the excavation (G.L. −14.75 m) during the excavation of the soil and the construction of the building, using the theory of elasticity. The ratio of the previous maximum consolidation pressure of the clay to the overburden pressure was about three, and the pressure imposed by the completed building was nearly equal to the overburden pressure.

In order to evaluate the displacement's Boussinesq's equations were used for the upper three layers of the soil—two layers of over-consolidated clay, and a sandwiched layer of sand; with Holl's expression being used for the lower parts consisting of a layer of gravelly sand and unknown underlying layers. The values of elastic moduli used were based on values measured in lateral loading tests in boreholes.

Comparisons between the estimated displacements and those measured during excavation and construction showed good agreement, and the extension was constructed without any difficulties.

REFERENCES

Harr, M. E. (1966). *Foundation of Theoretical Soil Mechanics*, McGraw-Hill, New York, 109–110
Holl, D. L. (1940). 'Stress Transmission in Earth', *Proc. Highway Res. Board*, Vol. 20

III/11. Settlement of the sugar silos on moraine clay in Gørlev, Denmark

G. Vefling
Cowiconsult (formerly Chr. Ostenfeld & W. Jonson), Consulting Engineers, Denmark

Summary

Settlement observations have been taken over a period of about 15 years for two sugar silos on heavily preconsolidated moraine clay. The live load is applied and removed once a year and amounts to 90% of the net total additional load on the ground.

Settlements are small but still increasing with time. It is shown that this increase may be caused by the load variation.

Introduction

Settlement observations on sugar silos are particularly interesting for three reasons:
(1) Sugar loads are considerable.
(2) The magnitude of the load is well known.
(3) The sugar load represents a substantial part of the total load with a large and rather regular yearly variation.
In the present paper and in another paper submitted to the present Conference (Vefling, 1974) some typical silo plants have been selected.

Previous records of a similar kind can be found, for instance, in a comprehensive paper by Parker and Bayliss (1971) and in a discussion contributed by the author (Vefling, 1971).

Constructional data

The present paper deals with a sugar silo group in Gørlev, Denmark, 90 km west of Copenhagen. The group comprises two silos with a capacity of

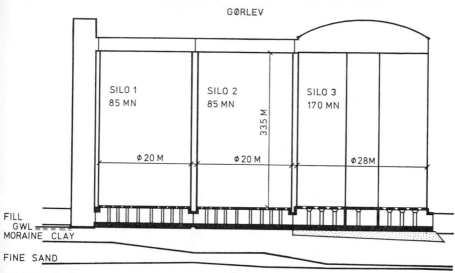

SILO 1
85 MN

SILO 2
85 MN

SILO 3
170 MN

33.5 M

Φ20 M

Φ20 M

Φ28M

FILL
GWL
MORAINE CLAY

FINE SAND

MORAINE CLAY

Fig. 1

85 MN each and one silo with a capacity of 170 MN. The two smaller silos were constructed in 1956 and the big silo in 1970–71.

The settlement behaviour of these silos is typical of a number of other Danish silos with virtually the same construction and soil conditions.

All three silos in Gørlev are directly supported. The substructure of each silo can be considered rigid, as it consists of a lower 1 m thick reinforced concrete raft connected with an upper 0.5 to 0.6 m thick silo floor by a ring wall and numerous columns in the basement recovery area. The main cylindrical silo shells are of prestressed concrete. For further details see *Fig. 1* and *Table 1*.

Table 1

	Silo 1	*Silo 2*	*Silo 3*
Structural weight (MN)	38	38	70
Sugar load capacity (MN)	+85	+85	+170
Excavated soil (MN)	−28	−28	−60
Net load increase (max) (MN)	95	95	180
Slab area (m²)	380	380	760
Net pressure increase (max) (kN/m²)	250	250	237

Subsoil conditions

The silos are located on a glacial plain 11 m above sea level. A simplified stratification of the somewhat erratic glacial formation is shown on *Fig. 2* together with the position of the borings.

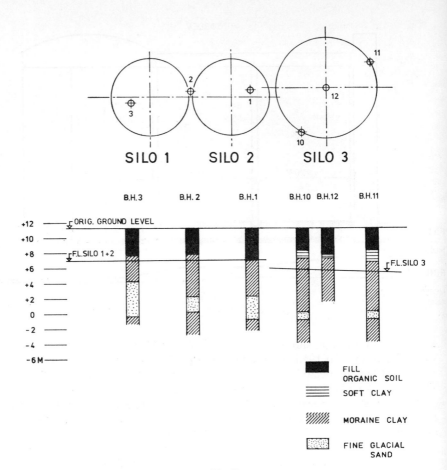

SILO 1   SILO 2   SILO 3

B.H.3   B.H. 2   B.H.1   B.H.10 B.H.12   B.H.11

FILL
ORGANIC SOIL
SOFT CLAY
MORAINE CLAY
FINE GLACIAL SAND

Fig. 2

Table 2

Depth below ground level(m)	Subsoil
4	Foundation level
4–7(11)	Lean, stiff moraine clay
	$w = 13\%$
	$e = 0.3$–0.4
	c_v increasing from 100 to 250 kN/m²
7(11)– 12	Water-bearing, glacial sand
12–	Lean, hard moraine clay
	$w = 13\%$
	$e = 0.3$–0.4
	c_v increasing from 250 to 700 kN/m²

w denotes mean value of the natural water content.
e denotes the void ratio.
C_v denotes shear strength as measured by vane tests.

244

The main features of the subsoil conditions are given in *Table 2*.

The moraine clay is a heavily consolidated, poorly graded material containing clay, silt and sand in varying proportions. Small deposits of sand occurred in the lower stratum, but no real sand layer was found. No boring was carried to greater depth than 15 m.

Consolidation tests were performed in 1956 prior to the construction of Silos 1 and 2. A typical value of the coefficient of volume compressibility was $m_v = 16 \times 10^{-3}$ m²/MN, measured on the recompression curve.

However, since the performance of these tests a new oedometer has been developed in which the error due to apparatus deformation has been virtually eliminated even when testing heavily precompressed clay (Moust Jacobsen, 1970). This new oedometer would probably have produced more accurate m_v values, which according to experience were likely to be 2–4 times smaller than the one quoted above (that is, leading to smaller settlements).

Development of settlements, 1956–60

The settlements were measured with a standard instrument and the possible error might be a few millimetres.

The load–settlement–time curves for the years 1956–60 are shown in *Fig. 3*. In addition to the mean settlement curve for each silo two additional curves are given representing the two points showing the smallest and the greatest settlement. Further details on the vertical movements during this period are provided in *Table 3*.

Due to their rigid construction (*Fig. 1*) silos of the type discussed here are supposed to settle as stiff bodies. This seems to be confirmed by the settlement observations when the error in measurement is taken into account. The angular distortion can be estimated from the curves in *Fig. 3*. After the first

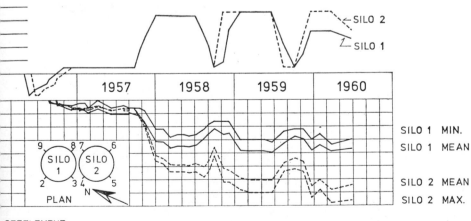

Fig. 3

245

Table 3. SETTLEMENTS 1957–60 CAUSED BY SUGAR LOAD (SETTLEMENTS FROM WEIGHT OF CONSTRUCTION OMITTED)

| | Silo 1 | | | | | Silo 2 | | | | |
| | \multicolumn{5}{c}{Point no. (Fig. 3)} | | | | | | |
	2	9	3	8	Mean	4	7	5	6	Mean
Settlement increase for first filling (mm)	8	11	13	18	12.5	17	19	25	25	21.5
Heave after first filling (mm)	6	6	8	8	7	8	7	7	6	7
Settlement increase for second filling (mm)	7	7	10	10	8.5	11	11	14	14	12.5
Heave after second filling (mm)	5	4	6	7	5.5	7	8	9	9	8.2
Settlement increase for third filling (mm)	5	6	6	8	6.2	10	12	13	13	12

The load variations are shown in *Fig. 3*.

load cycle both silos already showed a slight southward inclination (i.e. at right-angles to their common axis, see *Fig. 3*), which has been maintained and has even increased since then.

After three fillings the mean total settlements were 20 mm and 34 mm, and the southward tilts were roughly 1 : 2000 and 1 : 1700 for Silos 1 and 2, respectively.

Development of Settlements, 1960–70

During these years the settlements showed quite a regular course. Only a few yearly levellings have been performed and in some cases they probably did not coincide with the max. or min. peak values of the settlements. The logarithmic time-settlements curve in *Fig. 4* may, however, give an impression of the development.

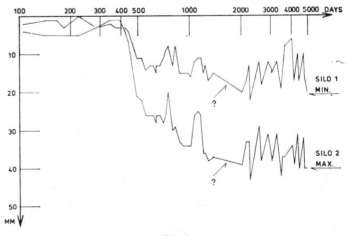

Fig. 4

246

After thirteen load cycles the mean total settlements were 26 mm and 38 mm for Silos 1 and 2 respectively, and the southward tilt was still 1 : 1700 for Silo 2, but had increased to 1 : 1000 for Silo 1.

Development of settlements, 1970–72

In 1970–71, Silo 3 was constructed immediately south of Silo 2. For dimensions etc., see *Table 1*.

The direction of the tilt of Silos 1 and 2 suggested that the shear strength of the clay might be lower at the new site. This proved to be the case and 1–2 m of comparatively weak soil under Silo 3 had to be replaced by compacted sand fill (*Fig. 1*).

When first loaded, Silo 3 underwent an immediate mean settlement of 24 mm, increasing to 30 mm after five months of full load. The angular distortion was westward and amounted to roughly 1 : 3000. The sugar load in Silo 3 was 169 MN, corresponding to 222 kN/m^2.

Conclusions

The following conclusions deal mainly with the settlements of Silos 1 and 2 during the first few years after their construction in 1956. It is intended to discuss the settlements of Silo 3 in a later publication.

Theoretically the stress distribution under each silo should be influenced by the neighbouring silo and the two silos should tend to lean towards each other. As, however, the angular distortion has retained roughly the same direction (southwards) for both silos, this mutual influence must have been small in the present case.

Although the settlements increase from north to south and a similar variation can be found in the thickness of the upper layer of moraine clay, it is not possible to explain the differential settlements by the stratification alone, not even if the wedge-shaped sand layer is assumed to be incompressible. This leads to assume a reduction of the deformation properties of the moraine clay towards south.

For several reasons it may seem surprising that the settlements continue during the years:

(1) A settlement calculation has been performed, based on the m_v-values which according to Danish experience are valid for moraine clay of the Gørlev-type. This calculation results in total settlements, equal to (or even smaller than) those which had occurred after the first year. At any rate, no further settlement would have been expected after the first loading.

(2) Another estimate based on current permeability values (k) for Danish moraine clays (Moust Jacobsen, 1970) shows that the full settlement of the upper clay layer will occur within a few days of loading. Furthermore, the estimate shows that there is every probability that the total degree of settlement at the end of one load cycle will be 70% (or more).

(3) This agrees well with concurrent experience from several countries (e.g. De Jong and Harris, 1971 and Moust Jacobsen, 1970), showing

that, even for large structures, a predominant part of the settlement of heavily preconsolidated moraine clay will occur during the construction period.

From these considerations, therefore, it may be concluded that no appreciable settlement increase should be expected after the first few months of loading.

However, this has proved not to be the case. *Table 3* shows that the settlement increases due to the second and third loading more than compensate the preceding heave. (The same may be the case even for the subsequent load cycles (*Fig. 4*), but the observation material does not permit of definite conclusions). The net settlement increase per load cycle is given in *Table 4*.

Table 4

	Silo 1	Silo 2
Net increase for second filling (mm)	1–2	5–6
Net increase for third filling (mm)	1	4
Net increase for second through eighth filling (mean value) (mm)	1	2

A similar phenomenon has been pointed out by Bjerrum (1968) for a number of directly founded structures with the common feature that the live load accounts for a large percentage of the total load. Bjerrum mentions a grain silo in Moss, Norway, with a live load of 80–85% of the total load and a yearly settlement increase of approximately 1 mm. The Gørlev silos belong to the same category of structures, since the live load amounts to roughly 90% of the net total additional load on the ground.

Acknowledgements

The author is indebted to the late Dr. L. Bjerrum for his valuable inspiration and advice during the preparation of the present paper. The author would also like to acknowledge the assistance given by Mr. K. Bennick, M.Sc., Cowiconsult, and Mrs. Ses Inan Kiilerich, M.Sc.

The levellings have mainly been executed by the sugar factory 'De Danske Sukkerfabrikker', Copenhagen, through its staff in Gørlev, to whom the author offers his sincere thanks. The sugar factory has kindly permitted publication of the results.

The silos mentioned in the paper were designed by the consulting engineers Cowiconsult (formerly Chr. Ostenfeld and W. Jonson), Copenhagen.

REFERENCES

Bjerrum, L. (1968). 'Secondary Settlements of Structures subjected to Large Variations in Live Load', *Norwegian Geotech. Inst. Publ.* No. 73 (originally published 1964)

De Jong, J. and Harris, M. C. (1971). 'Settlements of Two Multistory Buildings in Edmonton', *Canadian Geotech. J.* Vol. 8, No. 2, 217–235

Moust Jacobsen (1970). 'New Oedometer and New Triaxial Apparatus for Firm Soils', 'Strength and Deformation Properties of Preconsolidated Moraine Clay', 'Time Dependence of Settlements', *Danish Geotech. Inst. Bull.* No. 27

Parker, A. S. and Bayliss, F. V. (1971). 'The Settlement Behaviour of a Group of Large Silos on Piled Foundations', *Behaviour of Piles.* I.C.E., London, 59–69

Vefling, G. (1971). Discussion to Parker and Bayliss, ibid., 96–100

Vefling, G. (1974). 'Settlements of Three Heavy Sugar Silos in Italy', *Conf. on Settlement of Structures*, Cambridge

III/12. Settlements of a brick dwelling house on heavy clay (1951–73)

W. H. Ward

Building Research Station, Garston, Watford, Herts

Summary

A 20-year record of the settlement of the narrow concrete strip foundations of a typical house at Hemel Hempstead demonstrates the effects of forming the foundations in wet trenches. Rather surprisingly the settlement continues fairly steadily for the whole period at net rates between 0.12 and 0.39 mm/a. The angular distortion of the foundations continues to increase, and since erection of the superstructure has increased to 0.10% without cracking the walls.

Introduction

In 1949 the Building Research Station recommended the Hemel Hempstead Development Corporation to use, for reasons of economy, the narrow concrete strip foundation proposed by Ward (1947) in the construction of houses on heavy clay, provided the houses were kept clear of trees and large bushes either immediately prior to construction or subsequently.

This foundation was adopted throughout the many thousands of houses in the new town. Only two cases of cracking due to foundation movement have been reported. In one, a pair of houses rose up and cracked during construction about the line of a large blackthorn hedge removed just prior to construction. In the other, a detached house, built within 3 m of a row of Lombardy poplars, cracked so badly four years later that it was pulled down.

Since this was the first large-scale use of this house foundation it was decided, in co-operation with the Development Corporation, to measure the settlement performance of a typical pair of semi-detached houses to be built well away from trees at Adeyfield.

250

Site conditions and house construction

The site is a plateau of glacial drift overlying the drained Upper Chalk. The drift consists of a stiff, heavy, mottled clay with variable amounts of flint. The clay appears to be derived from the strongly mottled clays of the Reading Beds; it contains some concentrations of manganese dioxide. The irregular

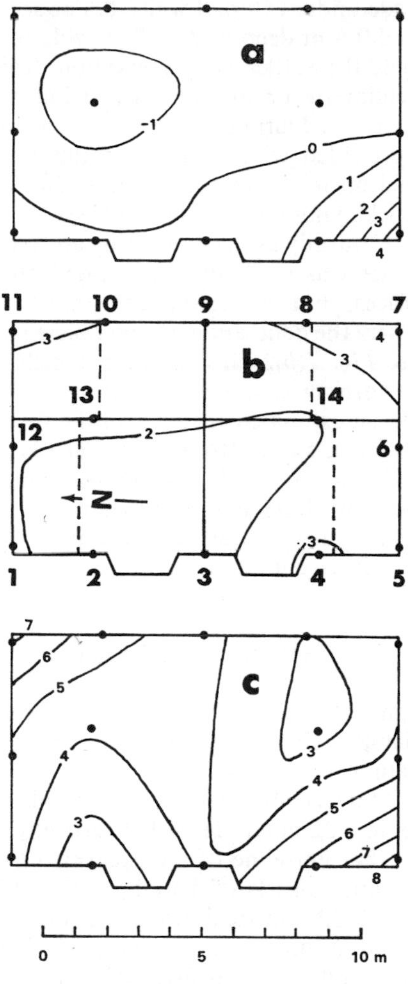

Fig. 1. Settlement contours in millimetres (a) for the period October 1951–March 1953 when only the foundations were in position; (b) for the period March 1953–Sept. 1953 during erection of the super-structure; (c) for the period Sept. 1953 to June 1973

erosion surface of the Chalk showed up in the bottom of some deep trenches in the area at 4.5 m below ground level and it is deeper than this under the houses. Undrained tests showed that the clay had an ultimate bearing capacity of at least 350 kN/m^2 at a depth of 0.6 m and increased substantially lower down. There is a perched water table in the drift which rises to the surface in wet winters and falls sufficiently in most summers to allow the grass roots to extend to a depth of 1.5 m.

The house has two-storeys with gable ends and a tiled roof. All walls are of Fletton brick in a sand–lime–cement mortar, external and party walls are of 0.33 m cavity construction and internal walls of single brick. Internal finishes are gypsum plaster, the ground floor is concrete resting on flint hardcore, the first floor and roof are of timber with ceilings of plasterboard with a skim coat of plaster. The bay windows are single storey.

A plan of the centre lines of the foundations is shown in *Fig. 1(b)* by the solid lines, the dotted lines showing lower storey walls resting on the floor slab. The foundation trenches were excavated by hand 0.9 m deep and 0.46 m wide under external walls, 0.6 m deep and 0.46 m wide under the party wall, and 0.6 m deep and 0.38 m wide under the main internal wall. During this time the settlement observation devices were installed. Rain fell and left a centimetre or two of water lying in the slightly deeper parts of the outer trenches. During the next two days with the trenches in this condition they were filled with concrete to the ground surface and two courses of bricks laid up to damp course level. The settlement zeros were established during bricklaying (end October 1951).

Naturally as the concreting advanced around the outer trenches the muddy water was gradually impounded, thicker mud being trapped underneath in places, but it accumulated most and the greatest entrapment occurred where the concreting was completed in the vicinity of observation point No. 5, see *Fig. 1(b)*. This practice, which had an important effect on settlement performance, was witnessed in silence to keep the performance typical.

The estimated loading on the foundations, assuming the structure has no stiffness, varies from about 60 kN/m^2 under the internal, front and back walls, to about 80 kN/m^2 under the gable ends and 100 kN/m^2 under the party wall. Thus the foundations had an ample factor of safety against the minimum ultimate bearing capacity of 350 kN/m^2 of the natural ground in its winter condition.

Settlement installation

A novel method of settlement measurement is used, which is considered to be simple, reliable and quick to read. Each observation point has a separate datum.

At every observation point numbered 1 to 14 in *Fig. 1(b)* a sleeved rod was set in a 32 mm hole drilled vertically to a depth of 2.45 m below ground level on the centreline of the excavated foundation trench. The rod, which is the datum, consists of a well-greased, 9.5 mm bore, galvanised steel pipe fitted with a 32 mm steel disc at the lower end. The sleeve, which moves with the foundation, is a similar pipe of 19 mm bore. After inserting a sleeved rod in a drillhole it was driven down firmly and the sleeve pulled up well clear of the disc on the rod. The sleeve was then cast into the foundation concrete. Each sleeved rod terminates above the damp course in the wall opposite a soot box. Here, changes in the protrusion of the inner pipe above the top of the sleeve are measured with a caliper gauge divided into millimetres and estimated to tenths. The changes represent the vertical motion of the foundations relative to points 2.45 m below ground level.

252

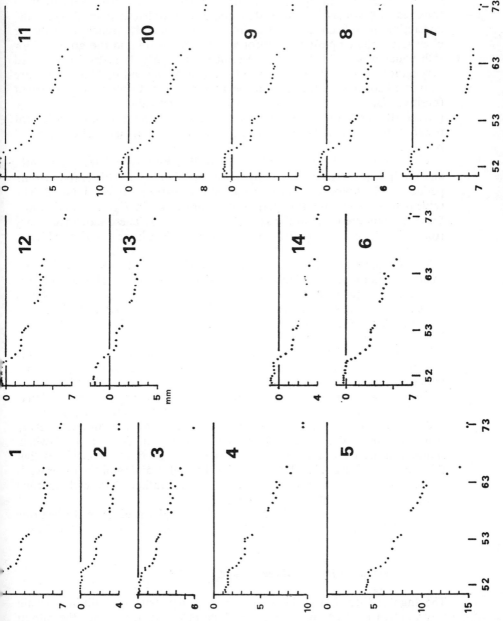

Fig. 2. Settlement–time records in millimetres: time scale from beginning of 1953 is one-eighth of former time scale.

Performance

The settlement at each point is plotted against time in *Fig. 2*. In this figure the time scale for the period 1953 to 1973 is one-eighth of the earlier period.

A fortnight after the foundations had been formed no further construction took place, but a settlement of 0.37 mm was recorded at point 5 where the concreting had been completed in the most muddy trench conditions; elsewhere, minor uplifts were recorded. The hardcore and the ground floor slab which were added in mid-December and early January 1953 caused very minor movements and the displacements to the end of March 1953 are associated solely with foundations being in position. Contours of settlement (positive) for this period are given in *Fig. 1(a)* and values for each point are given in the first row of figures in *Table 1*. The concentration of settlement towards the very muddy corner at point 5 and a minor uplift elsewhere will be seen.

Wall construction commenced in mid-April and was completed in a month, and the houses were completed by mid-September 1953. This is the loading period which shows up clearly in the settlement-time curves of *Fig. 2*. The settlement contours for this period alone are given in *Fig. 1(b)* and values for each point are given in the second row of *Table 1*. These settlements vary from 1.5 to 4.0 mm and follow no special pattern. The largest settlement is not at point 5.

The outstanding feature of the movements in the next 20 years is the very small but continuing settlement which shows no sign of diminishing, see *Fig. 2*. The contours of settlement from September 1953 to June 1973 shown in *Fig. 1(c)* have features in common with both the *Fig. 1(a)* and *(b)* diagrams, in particular the concentration of settlement at point 5, noticeable in *(a)* but not *(b)*, has reappeared.

The long-term observations were made at irregular intervals but attempts were made from time to time to elucidate the minor irregular movements that appeared rather consistently at every point. Seasonal thermal and moisture movements in both the superstructure and the ground are possible explanations. The house was subjected in 1963 to the coldest winter of the century and observations 10 days before the thaw came show a small rise at every point. But more dramatic is the next observation in October 1964 which shows a rather large settlement at points 3, 4, 5 and 6. It is considered that this movement is associated with a major thermal straining of the outer walls causing a redistribution of foundation loading rather than with thermal or moisture movements of the ground.

The net long-term rates of settlement of each point is given in the last row in *Table 1*. These vary from 0.12 to 0.39 mm per year with a mean of 0.22 mm.

Angular distortion of foundations

The angular distortion of the walls of a particular structure built in a certain way is often used as an index of cracking of the structure, but the rate at which the distortion occurs and the frequency and magnitude of thermal and moisture oscillations are probably important. The house superstructure has shown no signs of cracking which might be attributed to major structural

254

Table 1. SETTLEMENT AND LONG-TERM RATES AT EACH POINT

	11	10	9	8	7
Settlement (+ ive) of foundations alone (mm)	−0.5	−0.5	−0.7	−0.7	−0.2
Settlement during erection of superstructure (mm)	3.2	3.0	2.8	3.2	4.0
Settlement since end of construction to 1973 (mm)	7.0	5.6	4.3	3.0	3.3
Average rate of settlement since construction mm/a	0.34	0.27	0.21	0.14	0.16

	12	13		14	6
Settlement (+ ive) of foundations alone (mm)	−0.5	−1.5		−0.6	0
Settlement during erection of superstructure mm	2.1	2.1		2.0	2.5
Settlement since end of construction to 1973 mm	4.7	4.1		2.6	4.2
Average rate of settlement since construction mm/a	0.23	0.20		0.13	0.20

	1	2	3	4	5
Settlement (+ ive) of foundations alone (mm)	0.4	0.1	0.3	1.5	4.5
Settlement during erection of superstructure (mm)	2.1	1.5	1.5	3.3	2.3
Settlement since end of construction of 1973 (mm)	4.3	2.4	4.1	6.2	8.1
Average rate of settlement since construction mm/a	0.21	0.12	0.20	0.30	0.39

distortion and it is of interest to know the maximum angular distortion of the foundations in the vertical plane associated with this satisfactory performance. For this purpose the distortions between the observation points since construction was complete have been examined. They are steadily increasing with time and the maximum value has always lain between points 5, 6 and 7 across a gable end wall where the most mud was originally trapped at one end. The current value is 0.10%; elsewhere the value in the plane of a wall does not exceed 0.05%.

Acknowledgements

The co-operation and interest of the Hemel Hempstead Development Corporation and George Wimpey and Co. Ltd., and in latter years the owners of the houses, in providing the facilities for making the measurements is gratefully acknowledged. The author is also indebted to his colleagues, Mr. J. E. Cheney and Mr. D. Burford, for invaluable help with the installation and the observations. The paper is published by permission of the Director of the Building Research Establishment.

REFERENCE

Ward, W. H. (1947). 'House Foundations', *R.I.B.A. Journal*. Vol. 54, 226–235

SESSION IV

Rocks

IV/1. Short-term settlement of a five-storey building on soft chalk

J. B. Burland,
Head, Geotechnics Division, Building Research Station, Garston, Watford Herts

R. Kee,
Senior Lecturer, Hatfield Polytechnic, College Lane, Hatfield, Herts

D. Burford,
Higher Scientific Officer, Building Research Station

Summary

The settlements during construction of a five-storey building founded in soft low grade Chalk at Reading are presented and analysed. The settlements are very small and give values of equivalent elastic modulus E which are in good agreement with values obtained previously at Mundford. The results confirm that even for poor quality Chalk with a high water table the settlement of spread foundations will be small for moderate bearing pressures.

Introduction

In this paper the settlements during construction of a five-storey building founded in soft Chalk at Reading are presented and analysed. The building is of reinforced concrete frame construction with brick cladding.

The site investigation indicated that the Chalk was exceptionally soft to a depth of about 14 m. A simple analysis based on an empirical correlation between the Standard Penetration Test value and an equivalent elastic modulus E (Kee and Clapham, 1971) had indicated that the settlements might exceed 50 mm. Settlements of this magnitude were felt to be unacceptable and the use of a piled raft foundation was considered.

At this stage the consulting engineers, A. E. Butler and Associates, sought the advice of the Building Research Station. The work carried out by BRS on the Chalk at Mundford (Ward *et al.*, 1968; Burland and Lord, 1969) had indicated that at moderate foundation pressures (400–500 kN/m²) values of

259

E could be many times higher than values obtained from pile tests and small diameter plate tests. On the basis of the experience at Mundford it was estimated that the settlement of the building was likely to be of the order of 10 mm and was unlikely to exceed 20 mm. The decision was therefore taken to found the building on a raft without piles. A settlement study was undertaken to check the design assumptions and to obtain further information about the compressibility of very soft Chalk.

Ground conditions

Fig. 1 shows an outline plan of the building at 25/31 London Street, Reading. The site is situated on Upper Chalk close to the point at which the chalk outcrops from beneath the more recent Eocene deposits of London Clay and Woolwich and Reading beds.

The site investigation was carried out by sinking three 150 mm diameter boreholes by means of a shell and auger rig. The positions of the boreholes

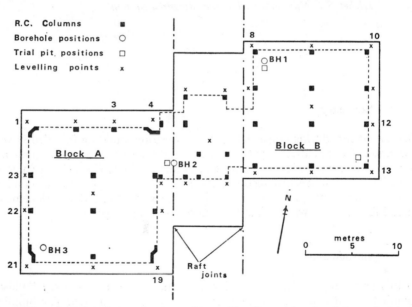

Fig. 1. Plan of raft foundations showing positions of levelling stations boreholes and trial pits

are shown in *Fig. 1.* Standard Penetration Tests (S.P.T.) were carried out in the boreholes at various depths. Three trial pits were dug at the points indicated in *Fig. 1.*

The site investigation showed that the surface of the Chalk is very soft and weathered and generally lies between one and two metres below ground surface. The water table is at a depth of 2.5 m. The S.P.T. results are given in *Fig. 2.* It can be seen that for boreholes 2 and 3 the *N*-values increase from 8 near the surface of the chalk to about 17 at a depth of 13 m. In borehole 1 the *N*-values remain constant over this depth at between 8 and 9. At a depth of about 14 m in all the boreholes the N-values increase very sharply to greater than 50, some being as high as 80.

On the basis of the S.P.T. values it appears that the Chalk is of low quality to a depth of 14 m. It was not possible to grade the Chalk by *in situ* visual inspection (Ward *et al.*, 1968) due to the high water table. However, following the approximate correlation between grade and S.P.T. value given by Wakeling (1969) it appears that the grade varies from VI at the surface to between V

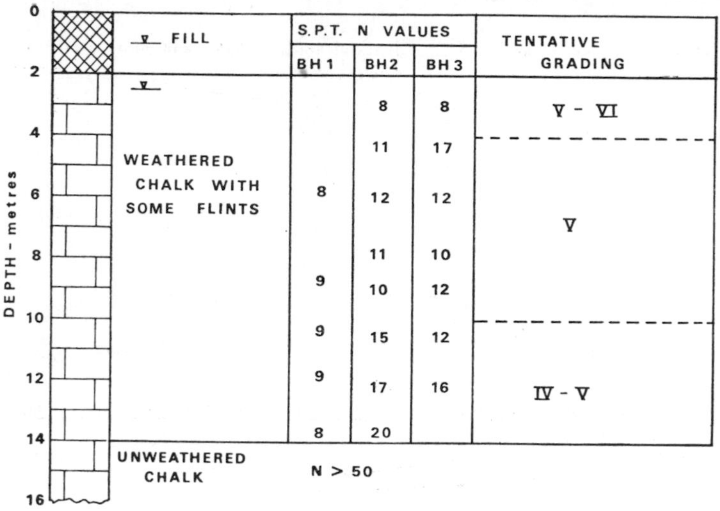

Fig. 2. Ground profile showing results of S.P.T. tests

and IV at 14 m. As pointed out by Wakeling unusually low S.P.T. values are sometimes obtained for weathered Chalk beneath the water table. The above grading may therefore be conservative.

Foundations

Fig. 1 shows a plan view of the foundations which consist of three discrete rafts with joints between them. The rafts are 1 m thick and were cast directly onto the surface of the Chalk, occasional soft spots having first been excavated and filled with lean concrete. 1 m of hardcore was then placed over the rafts and the ground floor slab was cast on this.

Settlement observations and results

Levelling stations for settlement observations were positioned as shown in *Fig. 1*. These were installed in the upper surface of the rafts outside the cladding line immediately after the rafts had been constructed. The positions of services and details of ground surface finishes had not been fixed at the start of the work and a large number of points were installed in the hope that a reasonable number could be preserved.

Two datums were located on nearby buildings. The main one (D1), located some 15 m south of point S21, is located in the massive front wall of a chapel which had been built in 1866. A subsidiary one (D2) is located some 25 m south-west of S21 in the wall of an old brick building on the opposite side of London Street. No measurable differential settlement has taken place between D1 and D2.

Levelling was carried out using a Wild N3 precision level and invar staff following the procedures given by Cheney (1973). Closing errors seldom

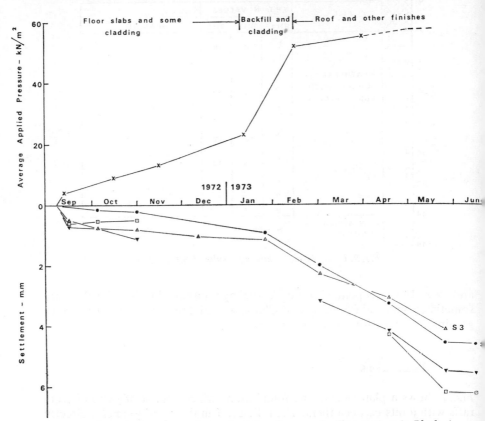

Fig. 3. Relationship between settlement and time for some levelling stations in Block A

exceeded 0.3 mm. Three initial rounds of levels were run to establish a reliable set of zero readings. Thereafter levelling was carried out after each floor was cast. Inevitably a number of the levelling stations were inaccessible during construction although the majority have been preserved.

The foundations for the building were constructed in August 1972 and by May 1973 the building was substantially complete. *Table 1* lists the observed settlements of a number of the points by the middle of June when the average pressure, exclusive of the weight of the raft, had reached 57 kN/m² which is very close to the maximum design pressure of approximately 60 kN/m². The average settlements for blocks A and B are 4.09 mm and 2.82 mm respectively.

Fig. 3 shows some typical curves of settlement of various points against

262

time for block A (see *Fig. 1*). Also shown is the relationship between average net foundation pressure and time. Most of the points show between 0.5 mm and 0.1 mm settlement at very low load, possibly due to the bedding of the raft. During the initial stages of construction the rate of settlement with time is small and is probably elastic. Towards the end of construction, after the

Table 1. OBSERVED SETTLEMENTS ON COMPLETION OF MAIN STRUCTURE

Block	Station	Settlement (mm)	Average Settlement (mm)
A	1	4.63	4.09*
	3	4.15	
	4	5.61	
	19	1.13	
	21	3.56	
	22	6.26	
	23	8.87	
B	8	2.51	2.82
	10	4.13	
	12	2.28	
	13	2.35	

* Calculated from the mean settlement of each side.

placing of the backfill, the rate of settlement is significantly higher. It is not yet clear whether this increased rate of settlement is due to a reduction in E at higher pressures or to the onset of creep. Observations over a long period of time are required before any conclusions can be drawn.

Analysis

For the purpose of this analysis, and to remain consistent with previous work, it is assumed that the ground can be represented by an elastic material having an equivalent elastic modulus E. Each foundation raft is represented by a uniform square load on the surface of an isotropic homogeneous elastic layer with a rigid stratum at depth. The value of E is given by the following expression:

$$E = \frac{qB}{\rho} I \qquad (1)$$

where ρ is the settlement
q is the average pressure ($= 57$ kN/m²)
B is the breadth (16.5 m and 14.5 m for blocks A and B respectively)
I is a factor dependent on the geometry of the foundation, the depth of the rigid stratum ($= 12$ m) and Poisson's ratio (taken as 0.24, following Burland and Lord, 1969). Using Steinbrenner's method average values of I for blocks A and B are 0.43 and 0.46 respectively.

Using the settlement values quoted previously and subtracting 0.5 mm to allow for the apparent bedding of the raft the values for E for blocks A and B are 113 MN/m² and 164 MN/m² respectively. Although homogeneity has been assumed for simplicity the stiffness probably increases with depth as indicated by the S.P.T. results. The softer material near the surface makes the major contribution to the settlement and the values of E are likely to be representative of the Chalk at a depth of about 6 m which is grade V–VI.

Discussion and conclusions

As mentioned in the introduction the work at Mundford indicates that for pressures in excess of 200 kN/m^2 and possibly as high as 400–500 kN/m^2 the values of E, even for poor grades of Chalk, are many times higher than indicated by the results of pile tests and small diameter plate tests. The decision to found the present building on a raft rather than on piles was based largely on the assumption that the Chalk at this site has broadly the same stiffness as the Chalk at Mundford. One of the objectives of the settlement study was to check this assumption.

The results of the full-scale tank loading test at Mundford gave an immediate value of E for grade V chalk of 360 MN/m^2. The period of loading was only two days. The load was left on for a year during which time settlement continued at a decreasing rate corresponding to a progressive reduction in equivalent elastic modulus. At the termination of the experiment the values of E in the top few metres of the Chalk (grades V and IV) had reduced to about a third of the initial values. Thus for the conditions at Mundford the value of E for grade V Chalk after a period of a year was about 120 MN/m^2.

At Reading the major part of the loading was applied over a period of about six months. In the previous section average values of E of 113 MN/m^2 and 164 MN/m^2 were obtained for the two rafts. These values are in remarkable agreement with the results obtained from Mundford and confirm that even for poor quality Chalk with a high water table settlements will be very small at moderate bearing pressures. Settlement observations at Reading will be continued to investigate the long-term behaviour of the Chalk.

Acknowledgements

The authors gratefully acknowledge the co-operation of the owners, Galliford Estates and the consulting engineers, A. E. Butler and Associates. The Paper is published by permission of the Director of the Building Research Establishment.

REFERENCES

Burland, J. B. and Lord, J. A. (1969). 'The Load-deformation Behaviour of Middle Chalk at Mundford, Norfolk: A comparison between full-scale performance and in situ laboratory measurements', *Proc. Conf. on In Situ Investigations in Soils and Rock.* British Geotechnical Society, I.C.E., 13–15 May

Cheney, J. E. (1973). 'Techniques and Equipment Using the Surveyors Level for Accurate Measurement of Building Movement', *Symp. on Field Instrumentation*, British Geotechnical Society, 30 May–1 June. Butterworths, London

Kee, R. and Clapham, H. G. (1971). 'An Empirical Method of Foundation Design in Chalk', *Civ. Eng. and Pub. Wks Rev.*, September

Wakeling, T. R. M. (1969). 'A Comparison of the Results of Standard Site Investigation Methods Against the Results of a Detailed Geotechnical Investigation in Middle Chalk at Mundford, Norfolk', *Proc. Conf. on In Situ Investigations in Soils and Rock*, British Geotechnical Society, London, 17–22

Ward, W. H., Burland, J. B. and Gallois, R. W. (1968). 'Geotechnical Assessment of a Site at Mundford, Norfolk for a Large Proton Accelerator', *Géotechnique*. Vol. 18, No. 4, 388–431

IV/2. Comparison between measured and estimated settlements at two Spanish aqueducts on gypsum rock

J. L. Justo and L. Zapico

Summary

Laboratory tests and plate loading tests were carried out on gypsum rock and stiff clays for two long aqueducts. Observations were made on the piers and the results compared with the test parameters.

Introduction

Part of the water of the River Tagus is being transferred to the south-east of Spain through 241 km of channel with a capacity of 33 m³/s. This channel has two aqueducts (*Fig. 1*), 2.9 and 6.2 km long respectively, which constitute unique structures of its type.

Both aqueducts are divided into frames with two, three or four piers each, separated by joints. Due to the hyperstatic character of the frames, the differential settlement between adjacent piers at the conduit level once these piers were tied by prestressing of the dowels, was limited to 1 cm in the design stage. Special means were considered to enable the structures to support greater settlement, but in any case there were other limitations to settlement: first the slope of 0.001 of the channel means a difference of level of only 4 cm between adjacent piers, and, secondly, although the joint between dowels at the middle of the span might admit some movement, this movement ought to be small. In the elastic calculations a modulus of elasticity of concrete of 14 700 MN/m² was introduced to find the stresses produced by a differential settlement of 1 cm. Thus, time-dependent deformations of concrete (with an instantaneous modulus of elasticity of the order of 25 000 MN/m²), were considered.

266

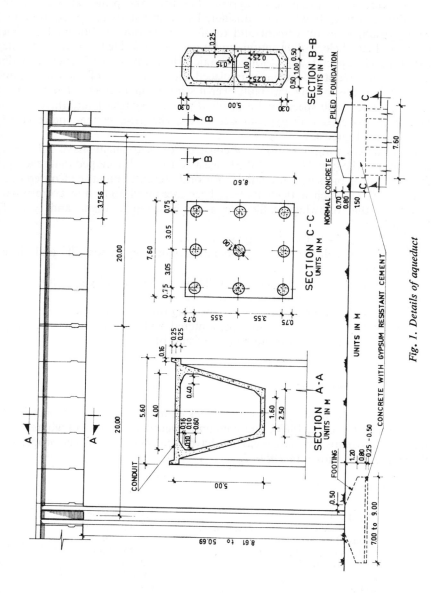

Fig. 1. Details of aqueduct

Owing to the settlement limitation, all foundations were taken to firm ground: overconsolidated clays with or without gypsum, and marls, gypsum rock with caves, soft calcarenites, limestone, and clay with boulders (sometimes the boulders prevail).

The widespread presence of gypsum has been the principal problem in foundation work. It was not considered convenient to make the footings with gypsum resistant cement and the piers with normal Portland cement owing to fear of electrolyte action, and it was not possible to make the whole structure with special cement for economic reasons. A layer of concrete made with gypsum resistant-cement was placed on the bottom of the excavation (*Fig. 1*), and the footing was built over it with normal cement. The footings were protected with special paint (asphalt plus epoxy resin), and the excavations were covered with gypsum free earth. The arrangement indicated in *Fig. 1* was adopted for pile caps.

Up to now, owing to the state of construction, settlements have only been measured in the first 84 piers of the longest aqueduct. It is hoped that reading. for most of the piers will be available for oral presentation at the Conference

The piers in which settlements have been measured up to now correspond to one of the following profiles, or a combination of them, and type of foundation:

(1) Footings 7 × 10 m over gypsum rock.
(2) Piles through very soft to medium organic, gypseous silts and clays (bearing on gypsum rock.
(3) Footings 9 × 13 m over firm to hard clay followed by gypsum.
(4) Piles through firm to hard clay.
(5) Piles through firm to hard clays bearing on gypsum rock.

The different types of ground encountered will now be reviewed. They have been fully described by Justo (1971a and b).

The gypsum rock

The gypsum which comprises the bedrock along the entire aqueduct is of Miocene age (M.O.P., 1969), and contains caverns up to 80 cm high as deep as 25 m; above this level larger caverns have been encountered. The properties of this rock are shown in *Table 1*.

Table 1

Dry density	2.1–2.4 g/cm^3
Unconfined compression strength, q_u	4.5–42 MN/m^2
Uniaxial deformation modulus, E_u	300–35 000 MN/m^2
Plate loading modulus, E_p	$\geq 510 \ MN/m^2$
Poisson's ratio	0.2–0.3
$SO_4.Ca.2H_2O$	61–100%
$CO_3.Ca$	0–6%

Joints were not common in the gypsum rock, although there were frequent firm clay pockets which had to be cleaned during the excavations for footings. These pockets were afterwards filled with gypsum-resistant concrete. The

268

area occupied by each footing was investigated with a waggon-drill to detect caves which were subsequently grouted.

Fig. 2 shows the stress-strain curve in an unconfined compression test with measurement of longitudinal and transversal strains in gypsum rock. The recoverable longitudinal deformations range from 60% of the total strain for a stress of 25% the failure stress, and 100% for 7% the failure stress. The recoverable transversal strains reached 45% for axial stresses ranging from 12 to 25% the failure stress. Up to stresses ranging from 15 to about 70% of

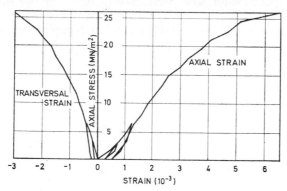

Fig. 2. Uniaxial compression test on gypsum rock

Fig. 3. Plate loading test on gypsum rock. 30 × 30 cm plate at 4·9 m depth

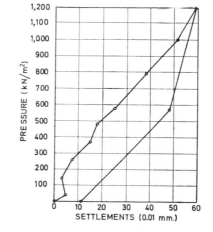

the failure stress in the different cores tested, there was a linear relationship between axial stresses and longitudinal or transversal strains. The moduli of deformation and Poisson ratios listed above correspond to this interval. Time dependent deformations are important in gypsum rock, especially when saturated, although perhaps they correspond to the proximity of failure (Justo, 1971c). The lower strengths correspond to samples with important proportions of carbonates or clay.

Two plate loading tests over gypsum rock were made. In the first, the settlement once the elastic deformation of the steel column was deducted was practically nil. The second is represented in *Fig. 3*. The recoverable deformation is 80% for a maximum load of 1.2 MN/m² and much smaller for large loads. During the excavations water appeared in 27% of the footings.

Properties of the firm to hard foundation clays

The range of these properties is shown in *Table 2*.

The clays were often mixed or interlayered with gypsum, sand or gravel. The standard penetration tests were mostly made in the more gravelly or gypseous zones.

The average depth of the water table was 6.5 m. Sampling was made mostly by a composite sampler with adequate area ratio.

Table 2. CLAY PROPERTIES

Clay properties	Average	Minimum	Maximum
Water content			
w (%)	23.6	5.9	41.9
Dry density			
ρ_d (g/cm³)	1.58	1.28	1.93
Unconfined compression strength			
q_u (KN/m²)	280	44	850
I_L	0.24		
$SO_4Ca.2H_2O$ (%)	13.9	0.11	86.3
CO_3Ca (%)		0	94
Compression index			
C_c	0.160	0.095	0.300
Swelling index			
Ci_s	0.048	0.015	0.084
Uniaxial deformation modulus			
E_u (MN/m²)	9.6	2.1	58
Pedometric modulus			
E_m between 0 and 200 KN/m² in saturated samples (MN/m²)	13	8.2	23
E_{mp} in partly saturated samples (MN/m²)	58	12	100
Preconsolidation pressure			
σ' in partly saturated samples (KN/m²)	340	100	490
Plate loading modulus			
E_p (MN/m²)	90	11	160
Standard penetration test			
N (blows/30 cm)	107	3	450

A detailed examination of the oedometer curves showed that below the water table the clays had preconsolidation pressures by Casagrande's method of the same order as that of the effective weight of soil. Notwithstanding that the strength indicated that the clays were overconsolidated.

The partly saturated clays above the water table were tested in special oedometers in which they suffered no desiccation during the test. *Fig. 4* shows one of these compressibility curves. *Fig. 5* shows the result of a plate loading test.

The moduli of *Table 2* for partly saturated soils and plate loading tests correspond to the stage of loading from the overburden pressure to the pressure finally reached below the footing.

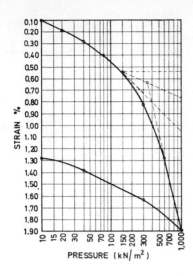

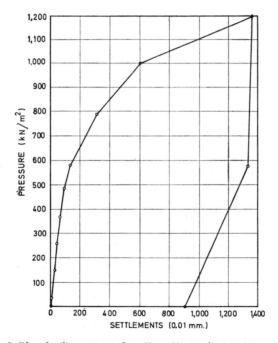

Fig. 4. Compressibility test in partially saturated clay; $w_i = 23\cdot2\%$ $w_f = 23\cdot3\%$

Fig. 5. Plate loading test on clay. 30 × 30 cm plate at 4·1 m depth

271

Pile loading tests

Two loading tests were carried out on two 30 m long piles which were opposite members of a pile group.

The profile of the ground was:

0–21.8 m: Firm to hard clay with boulders and gypsum, $q_u = 406$ KN/m², $E_u = 12$ MN/m²

>21.8 m: Very hard clay interlayered with gypsum, $q_u = 900$ KN/m², $E_u = 36$ MN/m²

Fig. 6 shows the results of the loading test.

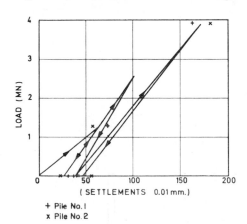

Fig. 6. Pile loading tests

+ Pile No. I
x Pile No. 2

The deformation of concrete working as a column would have been 6.1 mm (assuming $E = 24\,500$ MN/m²).

An elastic analysis was carried out (Morgan and Poulos, 1968) and the modulus of linear deformation of the clay, E, was found to be 350 MN/m².

Settlement observations

In 37 piers (footings over gypsum rock) the references were placed once the piers were constructed. In the remaining 47 piers the references were placed after the footings were constructed. An M-3 level was used (precision 1/10 mm). Metal rods with 2 mm divisions were employed. The closing error was always less than ± 2 mm.

The vertical loads on the footings are given below in MN,

Footing (7 × 10 m)	2.53	Average pier (32 m)	3.64
Footing (9 × 13 m)	3.95	Conduit	4.64
Pile cap	4.21	Water	4.91

There are also non-permanent additional loads due to wind and traffic. During service life, the maximum load over any pier, below the footing or pile cap will be:

$$N = 21 \text{ MN}$$
$$Q_y = 0.5 \text{ MN}$$
$$M_y = 19 \text{ MN m} \left.\right\} \text{wind loads}$$

272

Up to now, the conduit is mounted on 51 piers, and a loading test has been made on 13 piers applying a load of 6.13 MN. The maximum measured settlement is 5 mm. Owing to the small deformations. the settlements have been evaluated statistically, considering average values for the same type of foundation and ground. Levellings have been repeated so as to have good averages. The E values obtained are given in *Table 3*. The calculations of

Table 3. MODULI OF LINEAR DEFORMATION WHICH FIT THE SETTLEMENT OBSERVATIONS

	Minimum (MN/m²)	Average (MN/m²)	Maximum (MN/m²)
Footings over gypsum rock	390	570	670
Point bearing piles over gypsum rock		710	770
Floating piles in very stiff clay (q_u . 278 KN/m² $E_u = 9.6$ MN/m²)		100	
Footing over 2 m of firm clay followed by gypsum rock (average E)		220	

Poulos (1968), Poulos and Mattes (1969) and Butterfield and Bannerjee (1970, 1971 a and b) have been used for pile groups. Some results of observations on the over-consolidated clay are included.

It seems that there was an important increase in the settlements of footings and point-bearing piles over gypsum rock in a period of one to three months after loads were applied. Two possible reasons for this are (1) the slow deformation of the rock or (2) plastic deformation under the action of surcharges. The E values in *Table 3* are based upon the final settlement and loads applied when measurements were made. On the other hand, it seems that the settlement of firm to hard clays are immediate. The presence of water in the footings over gypsum rock produced no increase in the settlements.

Conclusions

A comparison of the test results and the observations leads to the following conclusions:

(1) The field modulus of deformation of gypsum rock is of the same order as the smallest value found in unconfined compression.

(2) The field modulus of deformation of firm to hard Spanish clays (in part non-saturated) is from 10 to 25 times larger than the unconfined modulus, and from 4 to 10 times the oedometric modulus. Partly saturated samples must be tested in special oedometers. Casagrande's method is not adequate to find the preconsolidation pressure in these clays, but even using the unloading curve throughout, the calculated settlements may be perhaps from three to six times the measured ones.

(3) The modulus of deformation found in plate loading tests is of the same order as the field modulus in either gypsum rock or firm to hard clays.

Butterfield, R. and Bannerjee, P. K. (1970). 'A Note on the Problem of a Pile Reinforced Half-space', *Géotechnique*. Vol. 20, 100–103

Butterfield, R. and Bannerjee, P. K. (1971a). 'The Elastic Analysis of Compressible Piles and Pile Groups', *Géotechnique*. Vol. 21, 43–60.

Butterfield, R. and Bannerjee, P. K. (1971b). 'The Problem of Pile Group–Pile Cap Interaction', *Géotechnique*. Vol. 21, 135–142

Justo, J. L. (1971a). 'Informs Sobre el Acueducto del Cigüela', *Lab. Transporte*. Madrid, unpublished

Justo, J. L. (1971b). 'Informe Geotécnico Sobre la Cimentación del Acueducto del Riansares', *Lab. Transporte*. Madrid, unpublished

Justo, J. L. (1971c). 'Propiedades Geotécnicas de los Terrenos Yesîferos', *I. Cong. Hispano-luso-Americano Geol. Económica*. Madrid, Sección 5,75–93

M.O.P. (1969). 'Paso de los rîos Riansares y Cigüela'. Madrid, unpublished

Morgan, J. R., and Poulos, H. G. (1968). 'Stability and Settlement of Deep Foundations', *Soil Mechanics, Selected Topics*. Butterworths, London, 528–609

Poulos, H. G. (1968). 'Analysis of the Settlement of Pile Groups', *Géotechnique*. Vol. 18, 449–471

Poulos, H. G. and Mattes, N. S. (1969). 'Analysis of End-bearing and Floating Pile Groups', *Univ. Sydney Research Report*. 115

274

IV/3. Settlement of a 12-storey building on piled foundations in chalk at Basingstoke

R. Kee
Senior Lecturer, Civil Eng. Dept., Hatfield Polytechnic
A. S. Parker
Associate Engineer, John Laing Design Associates Ltd., London
J. E. C. Wehrle
Principal Engineer, John Laing Design Associates Ltd., London

Summary

Using deep datum rods and precise levelling an attempt has been made to measure settlements of a reinforced concrete building founded in the Upper Chalk during the period of construction. The results have been plotted and the load/settlement characteristic investigated.

Geology and soils data

Basingstoke lies near the northern edge of the Chalk Downs, the geological sequence of which is given in *Table 1*. The site is at approximately 81 m Ordnance Datum some 5 m above the River Loddon. No soils investigation was carried out on the specific site of the building, but three boreholes were sunk on a site at similar level approximately 100 m away, and piling characteristics indicated similar conditions. *Fig. 1* gives typical Standard Penetration Values recorded and indicates the classification of the chalk (Wakeling, 1969).

Building and foundations

Situated on the side of a hill of approximately 1 in 14 slope, Belgrave House is an office block of reinforced concrete construction (*Figs. 2 and 3*). The

Table 1. BELGRAVE HOUSE BASINGSTOKE SETTLEMENT MEASUREMENTS

Levelling points	8 Aug. 1972 DL (kN)	Level readings relative to datum	7 Dec. 1972 DL (kN)	Settlement (m)	1 Mar. 1973 DL (kN)	Settlement (m)	26 Apr. 1973 DL (kN)	Settlement (m)	14 June 1973 DL (kN)	Settlement (m)	No. of piles in group
2	1750	96.6473	2830	—	6069	0.0024	6652	0.0026	6652	0.0029	12
4	792	99.6677	1764	0.0019	2431	0.0033	2897	0.0034	2897	0.0035	4
5	2568	99.6708	6280	0.0026	7970	0.0038	9684	0.0033	9684	0.0041	20
6	1149	99.6729	3387	0.0021	4642	0.0033	6093	0.0041	6093	0.0041	10

GEOLOGY
Period   Cretaceous
Classification Upper Chalk (approx. thickness 60 m)
                Upper Chalk (60 m)
                Middle and Lower Chalk (60 m)
Stratification Greensand
                Gault Clays

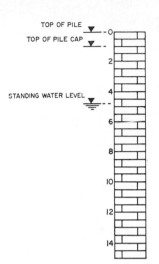

	AVERAGE S.P.T. "N"	TENTATIVE GRADING
	N=6 (4 to 10)	V / VI
	N=19	IV
	N=30 to 40	II - I
	N > 50	I

Fig. 1. Simplified strata profile showing average S.P.T. N and tentative grading of chalk

floors consist of precast units supported on *in situ* concrete beams and columns, stability being provided by *in situ* concrete walls forming the service and access cores. The loads are transferred to pile groups by pile caps immediately below a non-suspended basement slab. There are no ground beams between pile caps, the piles being in stable groups of varying size.

The piles are Franki driven cast *in situ* heavy duty (nominal 530 mm) of average length 11.50 m (maximum 15 m, minimum 10 m) the final driving characteristic prior to 'bulbing' being an average 20 mm set for 10 blows of a 3.4 tonne hammer dropping 1.25 m (maximum set 40 mm minimum 10 mm).

Piling was carried out during March and April 1972 and the main period of superstructure construction was from August 1972 until April 1973. At the time of writing the building is still being finished and is not occupied.

Datum and levelling

Realising that settlement would be likely to be of small order it was determined to establish stable datum benchmarks and to use high quality instrumentation.

Three datum points were sunk to a depth of some 12 m into the chalk at a distance of some 20 m from the building. These were drilled using a 50 mm fish-tailed bit on 28 mm diameter coupled grout rods which were then left in place to form the datum rods. Cement grout was introduced through the rods to fix the bottom 2 m or so and the remainder of the hole was filled with bentonite. The tops of the rods were fitted with hemispherical heads which are kept greased and protected as the datum points.

Six levelling points were cast into structural columns and walls at basement level. These are B.R.S. type A373/52. The level being used is a 'Wild NA2 automatic engineer's level with parallel plate micrometer enabling a reading

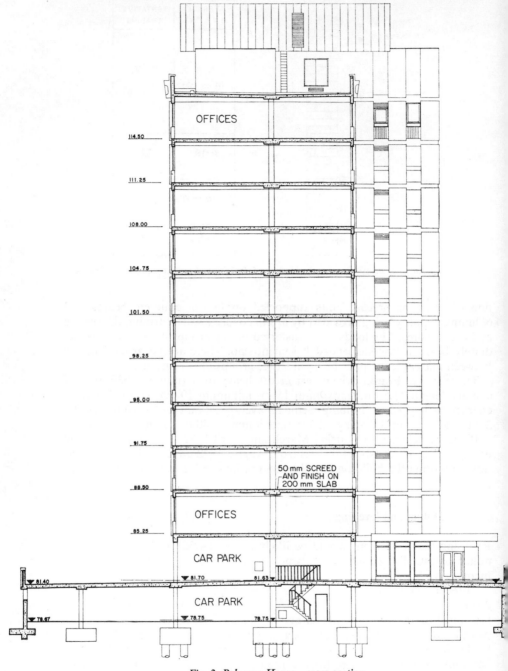

OFFICES

114.50

111.25

108.00

104.75

101.50

98.25

95.00

91.75

50 mm SCREED
AND FINISH ON
200 mm SLAB

88.50

OFFICES

85.25

CAR PARK

81.40   81.70   81.63

CAR PARK

78.67   78.75   78.75

Fig. 2. Belgrave House—cross-section

278

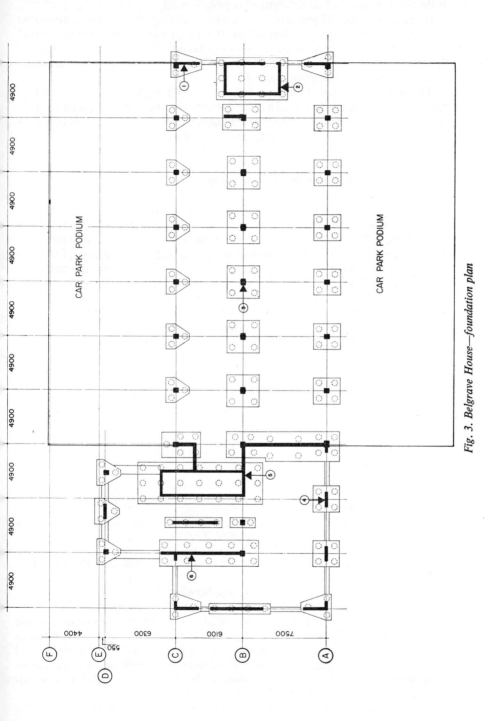

Fig. 3. Belgrave House—foundation plan

of 0.0001 m to be taken and the fifth place of decimal to be estimated. A Wild GWL Invar staff with 'vertical' bubble is used.

In practice factors of weather, accessibility, sighting, etc., make it impossible to achieve the accuracy potential of the instrument, and certain 'rogue' readings have had to be rejected. Closing errors have been of the order of 0.0005 to 0.001 m and it is reckoned that errors in the settlement readings adopted should be less than ± 0.0005 m.

The first 'datum' set of readings was taken on 8 August 1972 and four further sets have been taken during the construction period (see *Table 1*). It is intended to take regular settlement readings over a period of years to determine the long term behaviour and characteristics.

Results of settlement measurements

In view of the complex stress conditions associated with the load carrying characteristics of pile groups, straightforward interpretation of the results is not practical. Therefore in an attempt to analyse the results it is assumed that the load carried by a group is uniformly distributed over the area of the pile cap and that the load will be transferred to a lower level due to interaction of the foundation system. It is further assumed that each pile group under

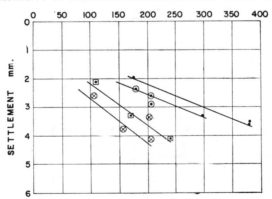

Fig. 4. Relationship between pressure and settlement of pile group (pressure assumed to be uniform over area of a group)

observation acts as a discrete unit in that it is not significantly influenced by adjacent groups.

Fig. 4 shows the typical relationship between foundation pressures and settlements for four pile groups. The consistency in the pattern of behaviour is apparent, in each case there is an initial settlement due to bedding of the foundation unit followed by a linear portion which has been interpreted as being 'elastic'. The long term behaviour is not known and this will be studied by further observations over a period of time. The bedding of foundations placed in soft chalk has also been observed by Burland *et al.* (1974) at Reading (Burland, Kee and Burford, 1974).

For the purpose of this analysis, the linear portion of the foundation pressure/settlement relationship has been used to derive an 'equivalent' elastic modulus E. Although there is evidence that the stiffness of the chalk increases with depth, homogeneity has been assumed for simplicity. The elastic equation for the vertical deformation of a loaded area may then be applied, i.e.

$$E = \frac{qB(1 - \mu^2)I}{\rho}$$

where E = Elastic modulus. In this analysis it represents an equivalent value as it may include the effect of some consolidations. Furthermore, the deformation in this range may not be fully recoverable.

q = Uniform foundation pressure

B = Width of loaded area

μ = Poisson's ratio, taken as 0.24 for chalk (Burland and Lord, 1969)

I_ρ = Influence factor depending on the geometry of the problem. In this case it is assumed that the surface loadings are uniformly transferred to a depth of 6 m above the base of the piles.

ρ = Vertical deformation of the loaded area

Results of analysis

Assuming that the loads are transferred to an area acting at about 6 m above the pile bases, a depth correction of 0.7 (Fox, 1948) has been applied in deriving the E values given in *Table 2* below.

Table 2

Settlement points	Derived equivalent elastic modulus E MN/m^2
2	160
4	135
5	106
6	89

Discussion

Fig. 1 shows a simplified strata profile together with average Standard Penetration Test N values and a tentative grading, Wakeling (1969). It can be seen that the piles are embedded in soft chalk (grade V changing to grade IV with depth) with the bases in the harder variety (grade II–I).

The derived E values ranging from 89 MN/m^2 to 160 MN/m^2 suggest that all the loads are virtually carried by the soft chalk occurring above the ends of the piles. This was borne out by the analysis of a pile test at the Basingstoke site (Kee and Clapham, 1971) to derive a relationship between working shaft friction and S.P.T. results in chalk. According to the E values given by Ward, Burland and Gallois (1968) for different grades of chalk at Mundford, the derived E values are pertinent to grade V chalk which occurs

281

to a depth of some 6 m below the top of the piles. Furthermore, the derived E values are in close agreement to those found by Burland, Kee and Burford (1974) for the settlement studies in soft chalk at Reading where E values were 113 MN/m^2 and 164 MN/m^2.

Conclusions

This study has shown that, within the range of foundation pressure imposed (less than 400 kN/m^2), the 'end of construction' settlements of foundations placed in soft chalk comprise a portion due to bedding followed by 'elastic' deformation, see *Fig. 4*. Although the range of pressure differs, the derived E values compare well with those obtained for similar grade of chalk at Mundford, Ward *et al.* (1968) and at Reading, Burland *et al.* (1974).

Acknowledgements

The Authors wish to thank Grosvenor Estate Commercial Developments Ltd., the building owners, Wiggins Teape Ltd., the tenants, and the Laing Group of Companies who designed and constructed the building, for allowing the investigation to be carried out and for their kind co-operation and assistance.

REFERENCES

Burland, J. B. and Lord, J. A. (1969). 'The Load-deformation Behaviour of Middle Chalk at Mundford, Norfolk: A Comparison Between Full-scale Performance and *In Situ* and Laboratory Measurements', *Proc. Conf. in In Situ Investigations in Soils and Rock*, British Geotechnical Society, I.C.E., London

Burland, J. B., Kee, R. and Burford, D. (1974). Presented for publication at the *Conference on Settlement of Structures*, Cambridge University

Fox, E. N. (1948). 'The Mean Elastic Settlement of a Uniformly Loaded Area at a Depth Below the Ground Surface', *Proc. 2nd Int. Conf. Soil Mech.* Vol. 1, Rotterdam

Kee, R. and Clapham, H. G. (1971). 'An Empirical Method of Foundation Design in Chalk', *Civ. Eng. and Pub. Wks Rev.* September

Wakeling, T.R.M. (1969) 'A Comparison of the Results of Standard Site Investigation Methods Against the Results of a Detailed Geotechnical Investigation in Middle Chalk at Mundford, Norfolk', *Proc. Conf. on In Situ Investigating in soils and Rocks*, British Geotechnical Society, London

Ward, W. H., Burland, J. B. and Gallois, R. W. (1968). 'Geotechnical Assessment of a Site at Mundford, Norfolk for a Large Proton Accelerator, *Géotechnique.* Vol. 18 No. 4

IV/4. Some observations on the settlement of a four-storey building founded in chalk at Basingstoke Hampshire

L. M. Lake
Senior Geotechnical Engineer, Mott, Hay & Anderson, Croydon, Surrey
N. E. Simons
Reader in Soil Mechanics, University of Surrey

Summary

Settlement observations have been carried out on a four-storey reinforced concrete framed structure supported on spread foundations in chalk, designed to a bearing pressure of 430 kN/m². The observed settlements are compared with predictions made on the basis of standard laboratory test data, tests on intact specimens cut from a block sample, *in situ* dynamic penetration tests and plate bearing tests using plates of varying diameter. Tests with plates larger than 0.6 m appear to provide the most reliable basis for settlement prediction, all other methods considered leading to gross errors.

Introduction

Prior to the construction of roads, sewers and subways for the proposed Eastrop Business Area, Basingstoke, an investigation was carried out in 1967 by means of trial pits to a maximum depth of 3.6 m. In 1969 one of the first buildings to be erected in the Business Area was a four-storey office and laboratory building. From his earlier knowledge of the site, the first author advised a design bearing pressure of 430 kN/m² for spread foundations placed in the clean jointed chalk, which was confirmed by visual inspection as each base was excavated. With the co-operation of the contractor, brass levelling points were set into selected stub columns shortly after the shuttering was struck and were surveyed periodically.

In 1970 an investigation using borings, trial pits and plate bearing tests

283

was carried out on an adjacent site for a building complex incorporating a 20-storey tower block. This paper compares the observed settlements on the four-storey building with those predicted using data from the latter investigation.

Site and geology

The Business Area is situated to the east of Basingstoke and the site is located adjacent to the northern distributor road. The natural ground surface falls southwards at a gradient of about 5° towards the River Loddon. Beneath 1 m of topsoil and stiff chalk with traces of brown clay, very weak to weak jointed chalk with flints occurs; the bedding is sensibly horizontal and the formation is several hundred feet thick. The chalk forms part of the Upper Chalk, probably the Micraster coranguinum zone. Ground water was encountered around 73 m, A.O.D., about 18 m below site level.

Properties of the chalk

Description and classification

The intact chalk was very weak to weak. At shallow depths, joints and fractures were at 50 to 100 mm spacing, generally closed, but sometimes stained brown. Below 2 to 3 m depth the spacing increased up to 200 mm and the fractures were generally tight. As boring proceeded, Standard Penetration Tests were carried out in the base of undisturbed sample holes giving N values typically around 20 at 3 m increasing to 30 at 30 m depth. The fracture spacing and penetration values correspond to chalk grades III to II in Wakeling's (1970) classification, which is consistent with a visual assessment based on an examination of numerous excavations within the business Area.

Properties from Laboratory Tests

A limited programme of laboratory testing was carried out, mainly on standard open-drive 100 mm samples. These tests were supplemented by tests on intact specimens hand cut from a block sample. The results are summarised in *Table 1*. Of special interest are the c_v values obtained from dissipation tests on the intact specimens which are more than a thousand times faster than values obtained from oedometer tests. Very similar values for c_v have since been measured on 100 mm specimens taken from another site with comparable chalk.

Consolidation tests were carried out on specimens prepared from open drive samples and a block sample. Standard 24 h loading stages were used for the former and m_v values between 0.02 and 0.03 m²/MN were determined. With the intact specimens the correction for the apparatus deflection accounted for all the initial compression and the remaining settlement for periods up to 80 h produced a straight line when plotted against the logarithm

284

Table 1. PROPERTIES OF CHALK FROM LABORATORY TESTS

Property	Open drive Samples	Block Sample
Water content %	25–28	26
Bulk density (t/m³)	1.87–2.02	1.80–1.87
Dry density (t/m³)	1.44–1.60	1.34–1.43
Coefficient of vol. decrease (m_v) (m²/MN)	0.019–0.028	≃ 0.005*
Coefficient of consolidation (c_v)		
Oedometer (m²/yr)	2.8	—
Triaxial dissipation (m²/yr)	—	$5–12 \times 10^3$*
Shear strength		
Total stress (C_u) (kN/m²)	200–500	—
Effective stress (c') (kN/m²)	—	250
ϕ'	—	36°
Tangent modules E (triaxial specimen) MN/m²	<21.5	31–52

* See discussion.

time. The compressibility from the intact specimens was about one tenth of that quoted above. Burland (1970) reported a similar relationship from field tests at Mundford.

Tangent modulus values were calculated from the tri-axial stress strain curves. From the 100 mm open drive specimens very few values exceeding 20 MN/m² were obtained whilst values from the intact specimens were in the range 31 to 52 MN/m². These are all much smaller than would be expected.

In situ plate bearing tests

Plate bearing tests were carried out in a pit 3.1 m square and 1.5 m deep. A reaction of 600 kN was provided by kentledge and the loads were applied through a ball joint by means of a calibrated hydraulic jack. Settlement was measured by four 0.001 in. micrometer gauges set equidistant around the perimeter of the test plate. Loads were applied incrementally, each being maintained until the movement of all four guages became less than 0.001 in. in two minutes. At least one unloading and reloading cycle was included in each test. Ten tests were carried out at essentially the same level, two using 0.91 m diameter plates, two with 0.61 m diameter plates and six with 0.30 m diameter plates, two of which were located at the centre of the areas already used for the 0.91 m tests.

Typical pressure settlement curves for a 0.30 m plate and a 0.91 m plate followed by a 0.30 m plate are plotted in *Fig. 1*. The results of the tests are summarised in *Fig. 2* where the average pressure/settlement curves for the plates of each size are plotted. For this purpose the second tests on the 0.91 m plate areas have been excluded.

Interpretation of plate bearing tests

The pressure/settlement curves for all tests exhibit an initial low deformation phase followed by a phase where the settlement is larger and increases

285

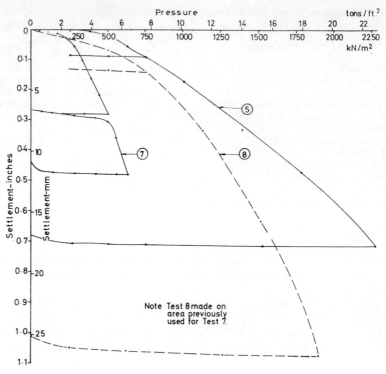

Fig. 1. *Results of plate bearing tests, No. 5 (0.30 m diameter), No. 7 (0·91 m diameter) and No. 8 (0·30 m diameter)*

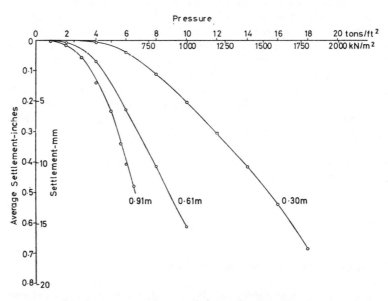

Fig. 2. *Average settlement of plates of various diameters*

286

roughly proportional to the applied pressure. Tests on the larger plates were necessarily terminated in this phase but in the 0.30 m plate tests, pressures up to 2.46 MN/m² were reached and there is some evidence of a change of slope beyond about 1.93 MN/m², possibly denoting the onset of general shear failure.

The plate tests were required primarily for settlement prediction. Therefore, the results were interpreted in terms of a deformation modulus E_D for the chalk from the expression:

$$E_D = \frac{kqB}{S}(1 - \mu^2) \tag{1}$$

where k = influence value depending upon the geometry of the situation
q = applied pressure causing settlement S
B = diameter of plate
μ = Poisson's ratio
It was assumed that stresses below the plate were only significant to a depth equal to 1.5B, and based on the work of Burland (1970), Poisson's ratio was taken to be 0.25.

Three values of E_D were determined from each test, (1) a tangent modulus E_{Dt} from the initial part of the pressure/settlement curve, (2) a reloading modulus E_{Dr} using a best fit through the unloading/reloading cycle and (3) a secant modulus, E_{Ds} taking the settlement at a pressure of 365 kN/m². The results are given in *Table 2*.

Table 2. RESULTS OF PLATE BEARING TESTS

Test no.	Plate diameter (m)	Deformation modulus E_D (MN/m²)		
		(i) Tangent E_{Dt}	(ii) Reloading E_{Dr}	(iii) Secant E_{Ds}
5	0.30	910	1365	548
6		391	683	273
9*		910	1365	65
10		547	912	548
Average		690	814	458
1	0.61	322	2730	195
2		303	911	103
Average		312	1820	149
3	0.91	512	747	103
7		373	455	68.5
Average		443	602	85.5
Reloading tests				
4	0.30	91	391	78
8		68.5	391	68.5

* *Note:* Void found adjacent to plate, result unreliable.

The most notable feature is the large differential between the E_{Dt} and E_{Ds} values obtained from the 0.30 m plate tests and the remainder. This is also shown by the curves plotted in *Fig. 2*. It is inferred that the 0.30 m plate is too small relative to the size of the chalk blocks and that the 0.61 m plate is the minimum size appropriate.

Foundation settlement

The settlement observations were carried out on a four-storey reinforced concrete framed structure, having the columns in the middle and around the outside at 9.8 and 6.7 m centres respectively. A simplified foundations plan is shown in *Fig. 3.*

Observations were continued for 36 weeks commencing within a few days of the stub column shuttering being struck. Levels were taken on brass angle plates grouted into pockets formed low down on the columns. Initially a quickset level was used but subsequently this was changed for a more accurate geodetic precise level with a parallel plate micrometer by Vickers Instruments

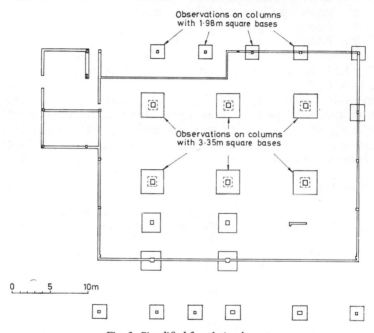

Observations on columns
with 1·98m square bases

Observations on columns
with 3·35m square bases

0   5   10m

Fig. 3. Simplified foundation layout

Limited. Closure on the datum was generally better than 2 mm but traverses within the building were often tortuous and became more so as construction advanced. Observations were made on six internal columns with bases 3.35 m square and four external columns with bases 1.98 m square. The larger bases were located 1.5 m to 3.0 m below the ground floor slab level and the smaller bases 1 m below this level. The design bearing pressure was 430 kN/m², and that due to structural loading was calculated to be 365 kN/m², which is the value adopted for all the settlement calculations.

The measured settlement for the 3.35 m and 1.98 m square bases and a mean settlement curve for each are plotted in *Fig. 4.* For the former, a total average settlement of about 6.5 mm and for the latter 4 mm were recorded, about half taking place within six weeks, before the first floor slab had been cast. Thus, only part of the initial movement can be attributed to recovery of elastic heave. It is inferred that the remainder of this early movement results

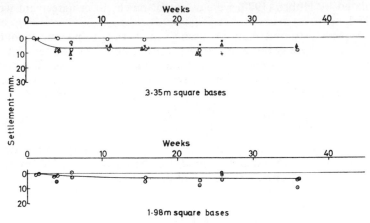

Fig. 4. Average measured settlement

from the bedding down of chalk disturbed by excavation for the foundations. Very little settlement was recorded after ten weeks. From the total observed settlements, deformation moduli of 105 and 95 MN/m² can be deduced for the 1.98 m and 3.35 m bases respectively.

Prediction of foundation settlement

It is known that the settlement of real foundations in chalk are consistently over-estimated from laboratory test data. This has been confirmed by calculations based on tangent modulus values obtained from triaxial tests, consolidation tests on open drive samples and the linear log time/settlement

Table 3 SUMMARY OF SETTLEMENT PREDICTIONS

| Method of prediction | Settlement (mm) | | | |
| | 1.98 m *base* | | 3.35 m *base* | |
Measured settlement	4.0	*% difference*	6.5	*% difference*
Standard consolidation tests	7.4	+85	10.8	+66
Block specimen consolidation test	6.1	+52	10.3	+59
Tangent modulus from triaxial test	11.2	+180	16.3	+150
Standard penetration test	7.8	+95	11.4	+75
Terzaghi and Peck plate relationship for sands	1.6	−60	4.4	−32
Plate bearing tests	3.5	−12	5.9	−9

curves obtained from intact 'block' specimens (*Table 3* shows that predicted settlements are over-estimated by 52% to 180% compared to the observed values.

Wakeling (1970) produced a correlation of deformation modulus and S,P.T. values. Using his line A, applicable to general foundations, settlement is still over-estimated by 75% to 95%. If the upward trend in Line A towards Line

B postulated by Hobbs (1970) were adopted, much closer agreement would be reached.

Plate bearing tests are commonly used to predict the behaviour of foundations, although there is no accepted way for extrapolating the test data for this purpose. In a homogeneous elastic medium, it follows from Equation 1 that the elastic settlements S_1 and S_2 of two footings of width B_1 and B_2 loaded to the same intensity are simply related by the expression

$$\frac{S_2}{S_1} = \frac{B_2}{B_1} \tag{2}$$

Applied to sands, it is known that Equation 2 can be very inaccurate and Terzaghi and Peck (1967) therefore modified the relationship. However, this can still lead to great errors and a more general relationship has been suggested of the form,

$$\frac{S_2}{S_1} = \left(\frac{B_2}{B_1}\right)^{\alpha} \tag{3}$$

where α varies between 0.3 and 1.0 and is a function of several variables.

The modified expression of Terzaghi and Peck (1967) corresponds to an α value of approximately 0.4 and applied to chalk, *Table 3* shows that it underestimates settlement by 32% to 60%. Lake and Simons (1970) suggested that for chalk it would be prudent to take $\alpha = 1.0$ and from *Fig. 5*, it will be seen

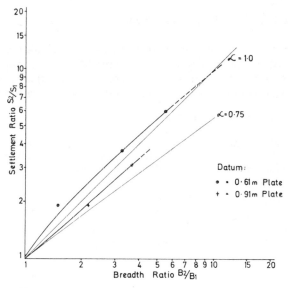

Fig. 5. Settlement ratio versus breadth ratio for spread foundations in chalk

that there is fair agreement between the observed data and the simple elastic behaviour implied by this assumption for foundations up to about 5 m wide. For the reasons given earlier, the 0.30 m plate was not thought to be appropriate and the 0.61 m plate was therefore used as the datum. Even the latter plate could be too small however, although *Fig. 5* shows that with the 0.91 m plate as datum, the result is still similar.

Calculations show that this relationship is essentially independent of stress level, at least to about 550 kN/m². It should be noted that the measured settlement is due in part to excavation disturbance which would become less significant as the size of the loaded area increases and the stress influence deepens. Further, a general improvement in the quality of chalk with depth and hence an increase in modulus would normally be anticipated. Both these factors suggest that for very large foundations producing the same pressure intensity, the settlement should not be proportionately large and there is an indication from *Fig. 5* that for the larger sizes, settlements would be less than that predicted with $\alpha = 1.0$.

Conclusions

(1) The tests with small plates have shown the chalk to have an ultimate bearing capacity in excess of 2 MN/m².
(2) Laboratory test results lead to gross over-estimates of the settlement of foundations in chalk.
(3) Of the techniques considered, plate bearing tests provide the most reliable method of predicting settlement of shallow spread foundations up to 5 m wide in chalk. The use, however, of data from small plates under-estimates settlement and a minimum plate size of 0.6 m appears necessary for the size of blocks encountered.

Acknowledgements

The authors are indebted to Mr. A. H. Penny, Chief Civil Engineer, Unilever Limited for permission to make the settlement observations and also to the agent Mr. L. Blows of Hawkins Contractors (Gosport) Limited for his help and co-operation. Thanks are also due to the Directors of Mott, Hay and Anderson for permission to publish this Paper.

REFERENCES

Burland, J. B. and Lord, J. A. (1970). 'The load Deformation Behaviour of Middle Chalk at Mundford, Norfolk', *Proc. Conf. on In Situ Investigations in Soils and Rocks*. British Geotechnical Society, London
Hobbs, N. B. (1970). Discussions. *Proc. Conf. on In Situ Investigations in Soils and Rocks* British Geotechnical Society, London, 47–50
Lake, L. M. and Simons, N. E. (1970). 'Investigations into the Engineering Properties of Chalk at Welford Theale, Berkshire', *Proc. Conf. on In Situ Investigations in Soils and Rocks*. British Geotechnical Society, London
Terzaghi, K. and Peck, R. B. (1967). *Soil Mechanics in Engineering Practice*, 2nd Ed. Wiley
Wakeling, T. R. M. (1970). 'A Comparison of the Results of Standard Investigation Methods Against the Results of Detailed Geotechnical Investigation in Middle Chalk at Mundford, Norfolk', *Proc. Conf. on In Situ Investigations in Soils and Rocks*, British Geotechnical Society, London

IV/5. Settlement studies of two structures on Keuper Marl

J. A. Lord
Engineer, Geotechnics Division, Ove Arup & Partners, London
D. F. T. Nash
Engineer, Geotechnics Division, Ove Arup & Partners, London

Summary

Settlement records for two structures founded on partially weathered Keuper Marl are presented. One structure is supported by pad footings bearing directly on the marl, whilst the other is supported by groups of end-bearing piles founded in a thin gravel stratum just above the surface of the marl.

Plate loading tests and pile tests carried out at the two sites have been compared with the subsequent performance of the structures. The load–settlement behaviour of both the structure and the preliminary tests have been used to determine the apparent moduli of elasticity for the Keuper Marl at the two sites. Comparison is made with previously published results for Keuper Marl; the zoning is based on visual and site investigation techniques. The settlement records indicate that the majority of settlement occurred during construction and that there has been very little creep of the marl.

Introduction

Published work about the performance and settlement of structures founded on Keuper Marl is very sparse. The difficulty of sampling marl by either rotary coring or soft ground boring techniques, on account of its alternating hard and soft bands, has long been recognised. In consequence laboratory testing has been restricted and the behaviour of the material 'en masse' has been in doubt. Greenland (1964) and Meigh and Greenland (1965) described pressuremeter tests and a plate loading test carried out in boreholes in the Keuper Marl. However it was not until the two structures to be described had virtually reached the construction stage that two major contributions on Keuper Marl appeared: Chandler (1969) tabulated Skempton and Davis's descriptions of the zones of weathering of Keuper Marl and discussed their effects, and Chandler, Birch and Davis (1968) dealt with its engineering properties. In the latter, settlement studies of bridge abutments on Keuper Marl were presented. Foley and Davis (1971) described pile tests in Keuper Marl and concluded that the majority of the load is transmitted as skin friction.

In the mid 1960s Arup Associates undertook the design of two structures founded on Keuper Marl. In the absence of published information loading tests were carried out to study the short term behaviour of the marl; settlement measurements both during and after construction were made to assess the performance of the structure and to investigate possible creep of the marl. The findings of these studies are presented below.

Loughborough University chemical engineering building

The Chemical Engineering building was amongst the first to be built on a new 35 ha site at the University of Loughborough. The surface of the site is underlain by a considerable depth of weathered Keuper Marl, and dips at an

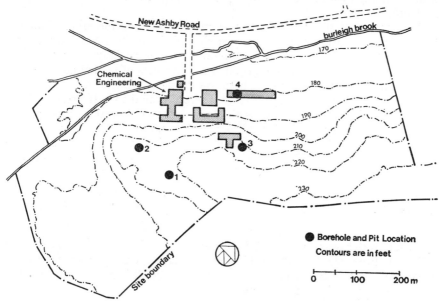

Fig. 1. Site plan showing development up to 1971—Loughborough University

average of 4° towards Burleigh brook (see *Fig. 1*). At the time when the buildings were being planned (1964), little was known about the field behaviour of Keuper Marl, and a site investigation consisting of boreholes and plate loading tests was carried out. This has been followed up by a settlement survey during construction of one of the blocks.

293

Site investigation

Boreholes

An initial survey of the site using a hand auger revealed that the Keuper Marl was up to 1.0 m below ground level beneath a thin cover of topsoil and glacial drift. The surface of the marl was often desiccated beneath which it was weathered to the consistency of a firm to stiff red silty clay. Subsequently four boreholes were put down by Soil Mechanics Ltd. to a maximum depth of 30 m using a 131 mm diameter rotary core and soft ground equipment. Details of these boreholes are shown in Fig. 2 together with the results of the Standard Penetration Tests which were carried out in Boreholes 1 and 2. It was found that beneath a variable depth of the highly weathered marl, the consistency changed to that of a mudstone (with the S.P.T. blowcount ranging from 60 to 200 blows for 300 mm penetration). This red silty mudstone, which sometimes had the consistency of a stiff clay even at depth, was interbedded with layers of medium hard grey silty sandstone. Hard Keuper Marl laminated with thin veins of gypsum was found in Borehole 4 below 22 m depth.

Recently the examination of a series of trial pits and the excavations for footings has yielded a better understanding of the profile. The harder marl at the crest of the hill is capped by several metres of glacially reworked marl, a very stiff silty clay containing gravel size lithorelicts of mudstone and some medium rounded gravel. This is absent on the side of the hill at the location of the Chemical Engineering building where the mudstone is present about 2 m below the original ground level. However at the lowest part of the site the depth of heavier weathering of the marl is greater.

Chandler (1969) has classified the Keuper Marl according to the degree of weathering. It seems probable that the complete weathering profile is present at the site of the Chemical Engineering building and that the profile is:

0–2 m	Zones IV and III	fully and partially weathered marl
beneath 2 m	Zones II and I	mostly unweathered marl but containing bands of Zone III marl.

Groundwater was encountered during the site investigation (December 1964) in Borehole 2 at about 10 m depth. However, subsequent measurements have shown that except at the top of the site it is present 1.0 to 2.0 m below the ground level during the winter months.

Plate loading tests

In addition to the boreholes, a series of plate loading tests were carried out on the marl in order to assess the likely settlements of individual pad footings. These tests which were carried out during the summer of 1965 by Soil Mechanics Ltd. were in pits located adjacent to Boreholes 1 and 2. Two pits measuring a minimum of 1.8 m by 2.8 m were dug to depths of 1.5 and 3.05 m. No groundwater was encountered in these pits. 305 mm and 610 mm square plates were then carefully positioned at the base of each pit (see *Fig. 3*). These were loaded to a maximum of 2680 kN by a hydraulic jack

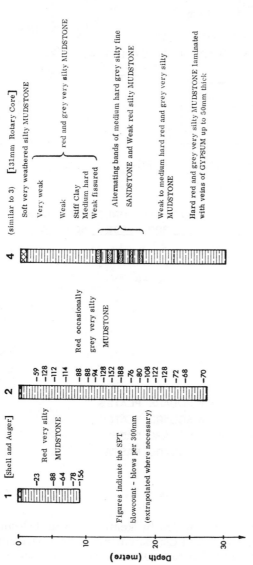

Fig. 2. Borehole records—Loughborough University

295

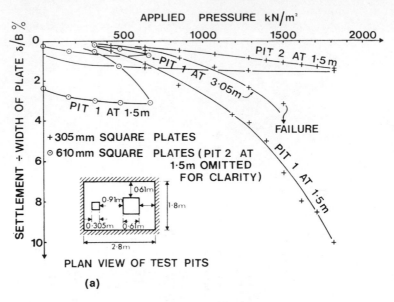

(a)

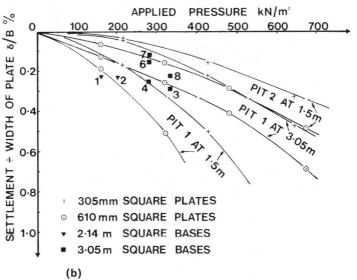

(b)

Fig. 3. (a) Observations on plate loading tests (b) enlargement of Fig. 3(a)

with the reaction being provided by kentledge. The deflections were measured by four dial gauges and the increments of load were applied when the rate of settlement had fallen to 0.001 mm/min, or two hours after the previous increment. In general, increments were applied at approximately hourly intervals.

Full details of the load-deflection characteristics are shown in *Figs. 3(a)* and *3(b)* in the form of applied pressure *v* settlement divided by the breadth of the plate. Only one plate test reached the ultimate load and that plate

296

suffered a premature rotational failure. The rebound characteristics were measured for four of the tests and showed that for settlements up to 2% of the plate width one half was recoverable. Beyond that the deformations were largely irrecoverable. It can be seen that the larger plates settled proportionately more than the adjacent small ones. This point will be discussed further in a later section.

At the base of each pit, Standard Penetration Tests and Mackintosh Probe Tests were carried out in an attempt to relate the properties of the marl from one part of the site to another. (The Mackintosh probe is a hand dynamic penetration test in which a 27 mm cone is hammered into the ground by a 4.16 kg weight dropping freely 330 mm.) Details of these tests are given in Table 1 together with the results of some moisture content

Table 1. SUPPLEMENTARY TESTS IN PITS

Pit No.	Mackintosh tests in base of pit Blows for penetration 150–225 mm 225–300 mm		S.P.T. in base of pit 150–450 mm	S.P.T. in adjacent borehole 150–450 mm	Moisture content
1 (at 1.5 m)	42 ± 2	43 ± 2 (3 tests)	12 ± 2 (4 tests)	—	—
1 (at 3.05 m)	45 ± 20	68 ± 40 (8 tests)	10 ± 3 (4 tests)	23	$20 \pm 8\%$ (6 tests)
2 (at 1.5 m)	130 ± 50	310 ± 120 (5 tests)	27 ± 2 (4 tests)	—	$16 \pm 3\%$ (5 tests 1–3 m depth)

determinations for the marl. These show that the marl tested at both levels in Pit 1 was much weaker than that tested in Pit. 2. Pit 1 was probably located in Zone III marl or even the glacial deposits, whilst the harder marl in Pit 2 was probably Zone II. It is possible that the harder marl was close below the base of Pit 1 at 3.05 m depth, since the settlements of the plates was somewhat less than at the shallower depth. It should be noted that the blowcount from the Standard Penetration tests at the base of the pits was considerably lower than that measured in the adjacent boreholes at the same level. This was probably due more to the differing conditions under which the tests were carried out than to variation of the marl.

Chemical Engineering building

The Chemical Engineering building is a three-storey structure which is stepped down the hillside (see *Fig. 4*). The building is made up of a number of units, 16.23 m square, each supported at its corners by individual pad footings 2.14 or 3.05 m square bearing on the Keuper marl. From the results of the plate loading tests, the footings were designed to exert a maximum bearing pressure of 400 kN/m². The structure is formed from 1.2 by 1.2 m cruciform columns cast *in situ* with precast girders spanning between them. The members around the perimeter of each square support the intermediate girders, and also support the external walls or alternatively the link strip joining two adjacent units.

297

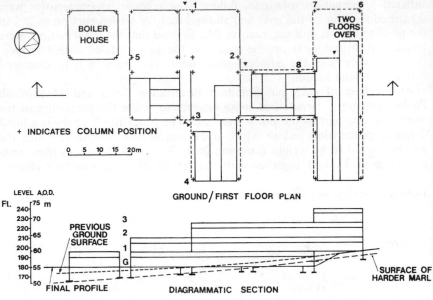

BOILER
HOUSE

TWO
FLOORS
OVER

+ INDICATES COLUMN POSITION

0   5   10   15   20m

GROUND / FIRST FLOOR PLAN

LEVEL A.O.D.
Ft.
240
230
220
210
200
190
180
170

75 m
70
65
60
55
50

PREVIOUS
GROUND
SURFACE

3
2
1
G

SURFACE OF
HARDER MARL

FINAL PROFILE

DIAGRAMMATIC   SECTION

Fig. 4. Loughborough University Chemical Engineering Building

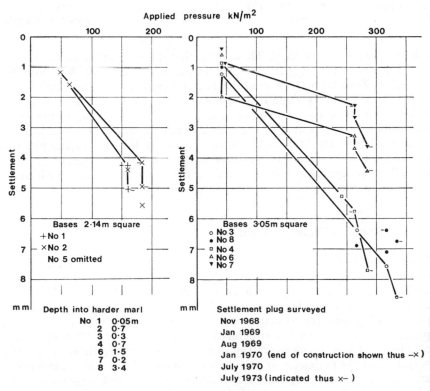

Applied   pressure   kN/m²

Settlement

Bases 2·14m square
+No 1
×No 2
No 5 omitted

Settlement

Bases 3·05m square
o No 3
● No 8
□ No 4
△ No 6
▼ No 7

mm   Depth into harder marl
No 1   0·05m
2   0·7
3   0·3
4   0·7
6   1·5
7   0·2
8   3·4

mm   Settlement plug surveyed
Nov 1968
Jan 1969
Aug 1969
Jan 1970 (end of construction shown thus →x)
July 1970
July 1973 (indicated thus x–)

Fig. 5. Settlement of footings of Chemical Engineering Building

298

The bases were founded on the harder marl between 2.0 and 5.0 m below previous ground level as shown in the section in *Fig. 4*. Before excavating for the footings Mackintosh probe tests were carried out at each base position to establish a suitable founding level. A criterion of a blowcount of 100 blows for 75 mm penetration was used, similar to the condition found at the base of Pit 2, in order to define the surface of the harder marl. At many locations tested the probe met refusal at this level. Where groundwater was encountered above this depth (as was found at the lowest side of the building) the allowable bearing pressure was reduced to 250 kN/m².

Construction was started in the autumn of 1968, the bases were installed by the following January and the full dead load was applied by the next year. During excavation any surface water was intercepted, but it was still necessary to pump some seepage water from the deeper bases. However no serious problems were experienced due to the deterioration of the marl. The majority of the bases were founded at the anticipated levels on weak mudstone. However at the lower part of the site, where there is generally a greater thickness of weathered marl, it is possible that the bases were founded on marl with the consistency of a stiff clay.

Early in construction settlement plugs were installed in eight of the columns Using a precise level, settlement readings were taken at approximately six-monthly intervals during construction, and another survey was carried out recently, three and a half years after the end of construction. Details of these settlements are shown in *Fig. 5*.

Discussion

Settlement records

The settlement records indicate that the columns had settled up to 7 mm by the end of construction, and that since then the creep settlement has been very small. It was necessary to extrapolate a zero point since the sockets were installed in columns when they had been constructed above final ground level. There is inevitably a certain amount of scatter in the readings which is partially due to inaccuracies in the survey (typical closing error 0.5 mm) and also to uncertainties over the precise load on a base at any time. (The measurements on Column No. 5 have not been included; the xero point is in doubt and was not surveyed until after the full dead load was on the base. However, it is likely that there has been approximately 2 mm creep since then.)

The final settlements have also been included with the results of the plate loading tests in *Fig. 3(b)*. It can be seen that two of the 3.05 m bases (Nos. 6 and 7) have settled only half as much as the others and also that the 2.14 m bases have settled the most in proportion to their width. Bases Nos. 6 and 7 support columns which are isolated from the structure and support two floors overhead. They are thus completely unrestrained. Founded approximately 2 m below ground level they are at the highest end of the building. In contrast, the 2.14 m bases (Nos. 1 and 2) are at the lower side of the building where the bearing pressure was reduced on account of the proximity of the water table.

These differences in behaviour are probably due both to considerable variation of the marl and to the effect of water. Whilst the same criterion was used to determine the foundation level of each base, there is a general tendency that the surface marl is more weathered at the lowest part of the site. This could account for the similar trend in the settlements of the footings. However, comparison of base No. 8 with Nos. 6 and 7 reveals that whilst it was founded 3 m further into harder marl it settled twice as much. This difference can be more readily ascribed to the relative levels of the water table, with the water both causing deterioration of the marl and also affecting its settlement characteristics.

Behaviour of footings of different widths

The plate loading tests indicated that larger footings might settle proportionately more than smaller ones founded on similar marl. They also showed that contrary to bearing capacity theory the ultimate pressure of the 610 mm square plates appeared to be smaller than those of 305 mm. Terzaghi and Peck (1967) recommended the use of the relationships

$$\delta = \delta_0 \times B \quad \text{and} \quad \delta = \delta_0 \times \left(\frac{2B}{B+1}\right)^2$$

for predicting the settlement of a footing width B feet on clay and sand respectively from the results of a 1 ft plate loading test. Bjerrum and Eggestad

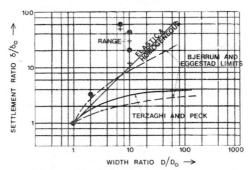

δ_0 IS THE SETTLEMENT OF 305mm PLATE AT 1·5m DEPTH IN PIT 2

\+ INDICATES END OF CONSTRUCTION SETTLEMENT

● INDICATES FINAL SETTLEMENT

Fig. 6. Settlement of footings of different widths

1963) have plotted out the results of various observations of footings on sand and concluded that generally they fell within the envelope shown in *Fig. 6*.

The settlements of the footings at Loughborough have been related to the settlement of the 305 mm plate tested in Pit 2 and these are also shown in *Fig. 6*. Assuming that the marl beneath the footings was similar to the material tested in Pit 2, then the behaviour of the footings confirm the trend observed for the two sizes of plates, that the larger the footing the greater is its settlement in proportion to its width. This is in contrast to the observed behaviour of footings on sand and clay and even allowing for variation of the marl and the presence of water indicates the need for caution in extrapolating the results of other plate bearing tests on Keuper Marl.

Apparent elastic modulus

From the observations of these plates and footings, calculations have been made of the apparent modulus of elasticity of the marl. These are shown in *Table 2* and indicate that it ranges from 0.5 to 3×10^5 kN/m² for all the

Table 2

Pit no.	Plate tests		Footings			
	0.305 m *width*	0.610 m *width*	*No.*	2.14 m *width*	*No.*	3.05 m *width*
1 (at 1.5 m)	0.86	0.53	1	0.5	3	0.9
1 (at 3.05 m)	2.2	1.0	2	0.6	4	0.9
2 (at 1.5 m)	3.4	1.7			6	1.5
					7	1.9
					8	1.2

Apparent modulus E calculated from

$$\delta = 0.88 \left[\frac{qB(1 - \mu^2)}{E} \right]$$

where δ is derived from the settlement of a rigid circular plate of the same area as the footing. No correction is made for depth as the footings are mostly at the surface of the harder marl.
q is the applied pressure and μ is Poisson's ratio which has been taken here as 0.3.
   Values of E in 10^5 kN/m², determined at 300 kN/m² for plates, and calculated from final settlements of footings.

results. If the results of settlements of footings on hard marl somewhat above the water table are separated from those placed close to the water table there are then two ranges 1.5 to 3×10^5 kN/m² and 0.5 to 1.2×10^5 kN/m². These may be compared with the values computed from the settlements of three motorway bridges on Keuper Marl in Leicestershire given in CIRIA Report No. 13. There the modulus was found to range from 1×10^5 to 1.4×10^5 kN/m².

Factory for John Player & Sons, Nottingham

The new factory for John Player & Sons (known as the Horizon Project) has been built on an 11 ha site on the Lenton Industrial Estate, located to the south-west of Nottingham. The site is generally level with a steep bank down to the Beeston Canal which bounds the west side. Originally the area was part of the flood plain of the River Trent, crossed by dykes and used only occasionally for farming on account of periodic flooding. During the past twenty years, it has been infilled with general fill and tipped council refuse to a depth of approximately 6 m to bring it above flood level.

301

Site investigation

A site investigation consisting of 34 boreholes to a maximum depth of 15 m was carried out by Structural Soils Ltd. over the whole site. This was supplemented by a further 10 boreholes put down by Nuttall Geotechnical Services primarily to study the chemical nature of the fill immediately beneath the structure. The location of these boreholes is shown on the site plan (*Fig. 7*).

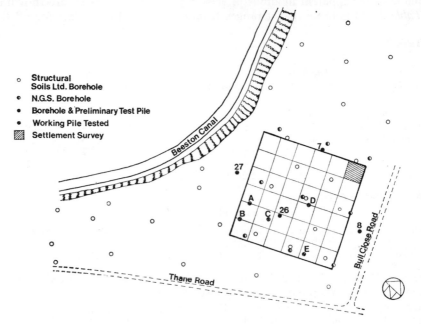

Fig. 7. Site plan showing location of boreholes and test piles—Nottingham

The general succession encountered in the boreholes beneath the structure was,

0 – 6.4 m	General fill and tipped refuse
6.4– 7.5 m	Soft to firm brown sandy clay
7.5–10.4 m	Very dense coarse sand and gravel with some cobbles
10.4–14.2 m	Red marl or laminations of red and grey⎤Keuper
(proved)	sandstones and red marl ⎦Marl

The fill consisted of an entirely random mixture of silty clay, sand and gravel, intermixed with pieces of wood, glass, metal, clinker, brick, concrete, paper and rags. The thickness of the sand and gravel layer was found to be variable across the site. Beneath the structure, apart from one or two local depressions, the top of the gravel is generally at a level of +21.0 to +21.3 m O.D.N. The level of the surface of the marl as observed in the boreholes was found to be irregular and it has not proved possible to contour it.

Standard and cone penetration tests were carried out in the sand and gravel; the recorded blowcount was in general between 18 and 25 blows/300 mm. Penetration tests made in the marl seldom penetrated more than

302

50–75 mm; the extrapolated blowcount increases from about 40 at the top of the marl to between 100 and 150 blows/300 mm at a depth of 0.7 m.

Moisture content measurements made on samples of the Keuper Marl showed a decrease from 18% at the top to between 11 and 14% about 0.9 to 1.5 m below the surface. These values, together with penetration test results, indicate that the top of the Keuper Marl is comparatively unweathered and that it might be generally classified as Zone II on the basis of Chandler (1969).

Design of foundations

The main structure is 180 m square, based on a 30 m primary grid, and consists of three floors (ground, 'void' and production) supported on columns at 7.5 m centres. The roof consists of a steel-framed deep truss spanning 30 m. The exterior of the structure is clad in precast concrete units. On account

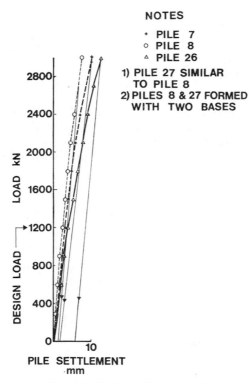

Fig. 8. Preliminary pile test results

of the thickness and variability of the fill underlying the structure, it was decided to support the structure on piles. As the water table was about 5 m below ground level and the fill contained lumps of concrete, driven Franki piles designed to carry 1200 kN were employed with the base enlarged in the thin sand and gravel stratum.

The columns on the primary grid were designed to carry loads of up to

303

8800 kN and these are supported on groups of eight piles. The smaller intermediate columns are each supported by a pair of piles.

Four preliminary piles were installed immediately adjacent to boreholes so that the strata in which they were founded could be known with reasonable certainty. These were tested to a maximum load of 3000 kN. In addition,

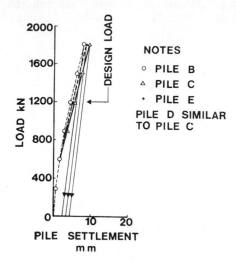

Fig. 9. Working pile test results

loading tests were carried out on working piles. The location of these piles is given in *Fig. 7* and their load settlement characteristics in *Figs. 8* and *9* respectively. The levels of the base of each pile and the top of the gravel and marl strata are given in *Table 3*. The working piles settled slightly more than

Table 3

Pile	7	26	B	C	D	E
Ground level	+29.1	+28.5	+28.4	+28.0	+28.2	+27.7
Top of gravel	+22.4	+21.0	+21.0	+21.0	+20.7	+21.0
Base level	+19.2	+19.8	+19.1	+19.0	+19.5	+19.0
Top of marl	+17.5	+18.7	+18.5	+18.3	+17.8	+18.1
δ_1 (900 kN) mm	2.9	2.7	2.7	3.3	3.1	3.0
δ_2 (1800 kN) mm	7.2	6.5	7.8	9.7	8.4	8.4
$E_{LB} \times 10^5$ kN/m²	1.2	1.4	1.3	0.9	0.7	1.0
$E_{UB} \times 10^5$ kN/m²	3.1	3.3	2.3	1.8	2.2	2.1
Stress increment on marl for load 900–1800 kN (kN/m²)	180	320	630	490	180	400

All levels in m above Ordnance Datum.
Apparent modulus E calculated from

$$\delta = \frac{\pi}{4} \times \frac{\Delta q D}{E} \quad (1 - \mu^2)$$

for a rigid circular plate of diameter D
To eliminate the contribution of shaft friction

$$\delta = (\delta_2 - \delta_1) - \text{(Shaft shortening)}$$
$$\Delta q = (1800 - 900) \text{ kN}$$

except for Pile 7 where δ_1 measured at KN 1200 and δ_2 measured at 2400 kN
$E_{LB} \equiv$ Lower Bound modulus based on assumption that gravel stratum incompressible
$E_{UB} \equiv$ Upper Bound modulus based on assumption that elastic moduli for the gravel and marl are the same
Preliminary Piles 8 and 27 not analysed, as formed with two bases.
Working Pile A disregarded on account of movement of reference beam supports.

the similarly formed preliminary piles at the same load. This can possibly be ascribed to the effects of slight uplift (Cole, 1972) caused by driving groups of piles. However, the difference in performance of the piles was sufficiently small as to be neglected and confirmed the design assumption of a maximum settlement at working load of between 5 and 15 mm with differential settlement substantially less.

Settlement measurements

Four settlement plugs were installed on columns as shown in *Fig. 10*, located in the north-east corner of the building (see general layout *Fig. 7*). This area was chosen as it combined ease of access with the least incidence of live

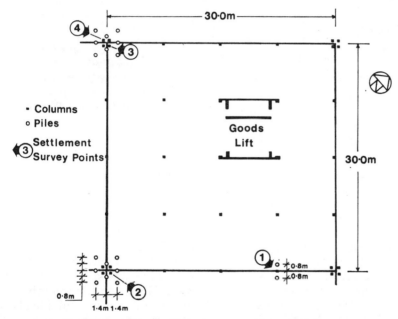

Fig. 10. Location of settlement survey points and pile layout

loading whose magnitude cannot be always accurately assessed. The following situations were chosen:

Point 1 —intermediate column supported on 2-pile group
Point 2 —internal columns on primary grid supported on 8-pile group
Points 3 and 4—external columns on primary grid supported on 8-pile group.

Two points were installed on the external columns in order to measure the tilt that was anticipated as a result of eccentric loading.

The settlements have been related to a datum established on the preliminary pile adjacent to Borehole 8. Settlements were measured six times over the period July 1970 to June 1971 during construction and three times sub-

305

sequently. The load/settlement observations for the three columns are presented in *Fig. 11*. (The load in this instance has been taken as the load/pile assuming a uniform distribution of load). Subsequent to the completion of the structure the columns settled slightly more than 2.5 mm up to November 1972, with little or no further movement in the following eight months.

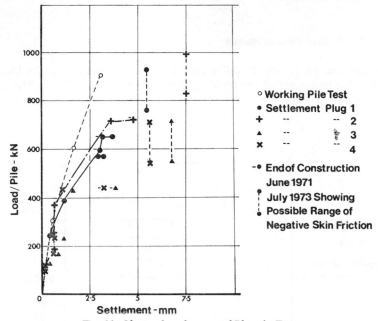

Fig. 11. Observed settlements of Player's Factory

This additional movement may be attributed both to the effects of creep of the marl and, as the ground surface has settled 150 mm relative to the pile cap due to consolidation of the fill, to negative skin friction on the shafts of the piles. As the fill is so variable upper and lower bound estimates of this friction have been made on the basis of 25 and 10 kN/m^2 respectively acting over a 6 m length. These loads have been added to the structural loads and included in *Fig. 11*. It is inferred from these results that little creep of the marl has occurred.

Discussion

Assessment of the foundation performance

Examination of the observed settlements in *Fig. 11* indicates that the settlement of the groups of piles is significantly greater than that of the individual piles tested. However, in making such a comparison it is necessary to consider the geology of the strata underlying the piles. Typical situations are illustrated in *Fig. 12*. (The thickness of the gravel beneath the piles has been inferred from adjacent boreholes.) Comparison of the pile test records indicates that the settlement characteristics appear to be independent of the

306

thickness of gravel beneath the base of the pile (see *Fig. 9* and *Table 3*). It is therefore concluded that the gravel has a stiffness similar to or possibly greater than that of the marl.

The settlement of the 8-pile groups may be seen in *Fig. 11* to be between $2\frac{1}{2}$ and 3 times greater than that of a single pile at the same load. This is consistent with a greater volume of marl being stressed under an 8-pile group. However, the settlement of the 2-pile group under settlement Plug 1 was somewhat greater than might be expected. Comparison of the driving records for piles in this area with those elsewhere suggest that the gravel might be slightly less dense.

The performance of the groups of piles clearly demonstrates that the marl is very hard. The observed settlement under 8000 kN column loads is only 7.5 mm.

The effects of creep in the marl have been masked by the incidence of negative skin friction on the shafts of the piles, but estimates of the magnitude of such friction indicate that any creep is very slight.

Modulus of elasticity of the marl

An assessment of the modulus of elasticity of the marl may be made from the behaviour of those piles founded closest to it. Examination of the results of tests carried out by Beer and Walays (1972) indicates that the formation of the enlarged base of Franki piles artificially creates high modulus values in granular strata as a result of 'tightening up'. Consequently it is possible to consider two extremes using simple elastic theory:

(1) The gravel layer is incompressible and the load is distributed through it from the base of the pile in a 2:1 spread, as shown in *Fig. 12*. The modulus of elasticity for the marl derived in this way can be regarded as a lower bound value.

(2) The modulus of elasticity of the gravel layer is assumed to be the same as that of the marl. This will probably be an upper bound value for the marl.

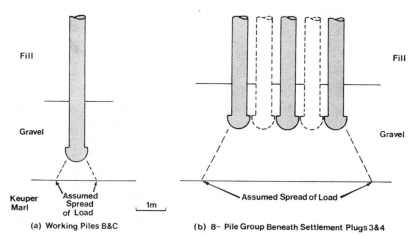

(a) Working Piles B&C

(b) 8- Pile Group Beneath Settlement Plugs 3&4

Fig. 12. Founding level of piles

307

The upper and lower bound modulus values for the marl based on the pile test results are presented in *Table 3*. It is to be expected that the piles founded closest to the top of the marl (Piles 26, B, C and E) would give the most consistent results. The shape of the pile bulb has been taken as a semi-ellipsoid (from observations by Beer and Walays, 1972) with a diameter of 0.75 m, based on the piling records. As the shafts of the piles penetrated up to 2.0 m into the gravel stratum, some of the load would be carried by friction in the gravel. It is anticipated that shaft friction will have been fully mobilised after a settlement of 2.5 mm. Consequently the increment of load from 900 to 1800 kN which would be entirely carried by the base of the pile, has been used to derive the modulus of the marl.

Table 4

Settlement plug	1	2	3/4
Column	Intermediate	Internal primary	External primary
Foundation	2-piles	8-piles	8-piles
Ground level	+27.5	+27.6	+27.9
Top of gravel	+19.8	+19.8	+20.1
Base level	+18.1	+19.2	+18.6
Top of marl	+17.1	+17.0	+16.9
Shaft friction/pile (kN)	315	100	225
Load on base/pile (kN)	335	620	210
$E_{LB} \times 10^5$ kN/m²	1.0	1.6	0.8
$E_{UB} \times 10^5$ kN/m²	1.6	2.6	1.2
Stress on marl (kN/m²)	115	140	60

All levels in m above Ordnance Datum.
Apparent modulus of the marl E calculated for the 8-pile groups from

$$\delta = \frac{\pi}{4} \frac{qD}{E} \ (1 - \mu^2)$$

for a rigid circular plate of diameter D, whose area is the same as that enclosed by the perimeter of the 8-pile group.
Apparent modulus of the marl E calculated for the 2-pile group from

$$\delta = q \frac{2B}{E} \ (1 - \mu^2) \ I_\varrho$$

for a flexible rectangular footing $L \times B$.
where δ is the settlement at the centre of the footing
  q is the load intensity
  I_p is Terzaghi's (1943) influence coefficient given by L/B.
  $E_{LB} \equiv$ Lower Bound modulus based on assumption that gravel stratum incompressible.
  $E_{UB} \equiv$ Upper Bound modulus based on assumption that elastic moduli for the gravel and marl are the same.

The primary columns for the structure are supported on groups of eight piles arranged as shown in *Fig. 10*. It is assumed that the densely compacted gravel will arch between the enlarged bases of the piles so that they effectively form a rectangular footing 3.94 m × 3.52 m. The intermediate columns are supported on pairs of piles at 1.6 m centres, so forming a rectangular footing 2.35 m × 0.75 m. It has not been possible from the settlement records to isolate the influence of friction in the gravel on the shafts of the piles. An allowance has been made for this on the basis of $K_0 = 1$ and $\delta = 40°$, leading to an average value of 80 kN/m². Upper and lower bound modulus values of the marl have been calculated as previously and are presented in *Table 4*.

The lower bound modulus values for the marl range between 0.8 and 1.6 × 10⁵ kN/m² which is in agreement with those quoted in CIRIA Report

No. 13 and the range of values quoted by Hobbs and Dixon (1969) for 'medium strong' Devonian marl. The upper bound modulus values exhibit greater scatter, between 1.2 and 3.3 $\times$ 10^5 kN/m^2, although the majority of results lie between 1.6 and 2.6 $\times$ 10^5 kN/m^2.

Conclusions

The site investigations showed that at the Loughborough and Nottingham sites the structures were founded on predominantly Zone II marl. The foundations of both structures performed satisfactorily. The maximum observed settlement in both cases was 8 mm for a bearing pressure of approximately 300 kN/m^2 on foundations up to 3.5 m square. The creep of the Keuper Marl at a bearing pressure of 300 kN/m^2 has been very small measured over a period of three years.

It is reasonable to treat Keuper Marl as an elastic media. Moduli of elasticity of between 0.5 and 2.0 $\times$ 10^5 kN/m^2 have been derived for Zone II marl at both sites. The range of values reflects local variations in the marl and are in broad agreement with previously published results for Keuper Marl.

The settlement of the footings was affected both by the variations in the marl and by the presence of water. However, the behaviour of footings of different sizes on marl was not consistent with previously published data for other materials. Care must therefore be exercised in extrapolating the results of plate tests to predict the behaviour of footings on marl.

Acknowledgements

The authors would like to thank their colleagues in Arup Associates and the Geotechnics Division of Ove Arup and Partners for their ever-willing assistance in assembling data and providing helpful advice, and in particular to Mr. E. Au who undertook much of the precise levelling.

The authors are indebted to Loughborough University and John Player & Sons for allowing the results of the investigations to be used in this Paper.

REFERENCES

Beer, E. E. de and Walays, M. (1972). 'Franki Piles with Overexpanded Bases', *La Technique des Travaux*. No. 333, Jan.–Feb., 1–48

Bjerrum, L. and Eggestad, A. (1963). 'Interpretation of Loading Tests on Sand', *Proc. European Conf. Soil Mech. & Found. Eng.*, Wiesbaden. Vol. 1, 119

Chandler, R. J. (1969). 'The Effect of Weathering on the Shear Strength Properties of Keuper Marl', *Geotechnique*. Vol. 19, 3, 321–334

Chandler, R. J., Birch, N. and Davis, A. G. (1968). 'Engineering Properties of Keuper Marl', *CIRIA Research Report* No. 13

Cole, K. W. (1972). 'Uplift of Piles due to Driving Displacement', *Civ. Eng. and Pub. Wks Rev.*, March, 263–269

Foley, G. P. and Davis, A. G. (1971). 'Piling in Keuper Marl at Leicester', *Civ. Eng. and Pub. Wks Rev.*, Sept., 987–991

Greenland, S. W. (1964). 'Economic Loading Tests with the Menard Pressuremeter',

Proc. Symp. on Economic Use of Soil Testing in Site Investigation. Birmingham, 3–7 and 3–8

Hobbs, N. B. and Dixon, J. C. (1969). 'In situ Testing for Bridge Foundations in the Devonian Marl', *Proc. Conf. on In situ Investigations in Soil and Rock*, 13–15 May. British Geotechnical Society, London

Meigh, A. C. and Greenland, S. W. (1965). 'In situ Testing of Soft Rocks', *Proc. 6th Int. Conf. Soil Mech. & Found. Eng.*, Montreal. Vol. 1, 73–76

Terzaghi, K. (1943). *Theoretical Soil Mechanics*, 383. Wiley, New York

Terzaghi, K. and Peck, R. B. (1967). *Soil Mechanics in Engineering Practice*, 2nd Ed., 489 Wiley, New York

IV/6 In situ deformation of Bunter Sandstone

J. F. A. Moore
Principal Scientific Officer, Building Research Station, Garston, Watford, Herts

C. W. Jones
Principal Professional and Technology Officer, Daresbury Nuclear Physics Laboratory, nr. Warrington, Lancs

Summary

Structural movements adjacent to a synchroton due to applied stresses of 100 kN/m² were measured by precise levelling to an accuracy better than 0.025 mm. The greatest settlement was less than 2 mm. Strain measurements beneath a current plate test suggest a Young's modulus higher than previously assumed and also an increase in its value with depth. Finite element predictions using these new parameters accord with measured settlements. There is no evidence of settlement at depth nor of creep due to these stresses, thus suggesting the use of simple foundations.

Introduction

The 5 GeV synchrotron, Nina, at the Daresbury Nuclear Physics Laboratory was commissioned in 1966. The satisfactory operation of this research facility required the provision of exceptionally stable foundations for the synchrotron magnets. In particular, problems were expected from differential settlements caused by changes in the arrangement of the radiation shielding. Some influence was also expected on the floor of the adjacent experimental hall and regular very precise levelling was performed to monitor the behaviour of the magnets and the floor in service. Comparisons are made between the actual behaviour, the predicted behaviour and more recent assessments of the behaviour. The conclusions are interpreted in terms of more simple foundation design.

311

Geology and ground properties

The site was investigated in 1962 and 1963. It comprises Bunter Sandstone to a depth of at least 80 m. The sandstone is generally fine to medium grained, and well-cemented below about 5 m from the surface. But in the weathered zone above this depth there is considerable lateral and vertical variation in strength. The water table has fallen considerably in the period 1962–72 and now rests over 25 m below ground level. The main floor level of the synchrotron is at 30.8 m above Newlyn datum (ND).

Laboratory compression tests were made on samples from borehole cores at 25, 56 and 101 mm diameter. Some pressuremeter and plate loading tests were also preformed and the results have been discussed by Meigh and Greenwood (1965). For calculation purposes a mean value for Young's modulus of 300 MN/m² was adopted, although test results ranged from 100 to 600 MN/m². Measurements of Poisson's ratio gave many low results but a value of 0.25 was taken as being more realistic. Further laboratory tests on 50 mm diameter specimens in 1969 reinforced this choice.

A long-term plate test was started in 1970 on this site by the Building Research Station. Preliminary results of settlement measurements below the plate suggest that a modulus value nearer 1000 MN/m² would be more realistic. Although the variability of the strata may be significant both the earlier and the current plate tests were performed below the strongly weathered zone and at similar elevations some 150 m apart. *In situ* inspection of the ground by closed circuit television shows that it is far less jointed than conventional cores indicate, thus correlating with the higher *in situ* modulus.

Structure and loading

The main feature of the synchrotron is a ring of 40 magnets, each weighing about 15 000 kg, set in a circular path of 35 m radius as shown in *Fig. 1*. A typical cross-section applicable to most of the ring is also shown at A–A in *Fig. 1*. The foundations for the magnets comprise 60 piles founded 11.6 m deep at 19.2 m N.D. The piles are concrete cast in 0.91 m diameter bitumen-lined pipes set in holes augered in the sandstone. The gap outside the pipes is filled with pea gravel in an attempt to reduce lateral instability without introducing vertical restraint. Each pile shaft was examined to ensure sound sandstone at its base. At basement level the piles are connected by a continuous capping beam. Concrete columns above each pile (the magnet 'upstands') are connected in threes at ground level, each structure of three columns supporting two magnets. A completely separate concrete tunnel surrounds the magnets at ground level and the upstands and capping beam at basement level.

Where the magnet ring emerges into the experimental hall removable shielding is required as shown in section B–B in *Fig. 1*. The basement tunnel continues as in section A–A but the permanent structure is replaced by precast concrete blocks and beams. The total removable load is about 35×10^5 kg distributed over an overall area of 33.5 m × 10.7 m. The average load intensity on the ground is therefore 98 kN/m² although local stresses on the floors up to 300 kN/m² may occur. Most of this stress is distributed to the

312

ground through the tunnel structure. The complete arrangement of block shielding around the magnet ring in the experimental hall was built initially before installing the magnets. It was then dismantled at the end of 1965 and subsequently reassembled in August 1966 before commissioning.

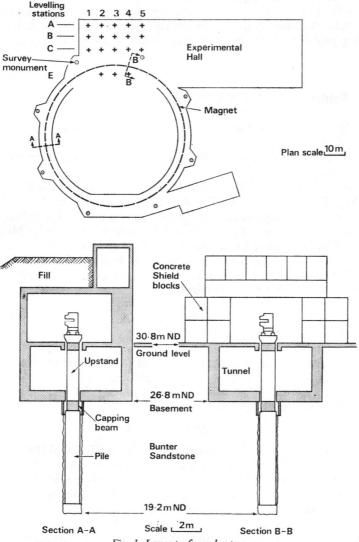

Fig. 1. Layout of synchroton

The floor of the experimental area is continuous with the main level of the magnet ring and is 200 mm thick reinforced concrete cast on a blinding surface on the intact but weathered Bunter sandstone excavated to a level surface. Equipment set up on this floor may apply stresses up to 100 kN/m².

The grid of floor levelling stations installed in the experimental hall is shown in *Fig. 1*. These stations consist of a brass conical seating located within a 400 mm steel box cast in the floor slab.

Observations were made with a Watts precise level on an invar staff fitted with a spherical base. A reference monument 105 m from the magnet ring was taken as a stable datum. This and other survey monuments were 4 m long freestanding reinforced concrete columns set at 26 m N.D. on cast *in situ* bored piles 7 m deep. Accuracies of much better than 0.025 mm were achieved for the elevations of the levelling stations, reading being direct to 0.0012 mm. A major contribution to this accuracy was the reduction in refraction problems due to levelling indoors and to the use of very short equal sites.

Settlements during initial loading

The design concept of magnet upstands, ring beam and piles founded at depth and separated from other structures had two objects:
 (1) To achieve long-term stability in more uniform well-cemented sandstone.
 (2) To avoid differential settlements caused by moving shielding blocks. Any immediate settlements due to construction could be ignored because final alignment of magnets could be undertaken subsequently.
The original predictions of foundation behaviour were made by Meigh and Skipp (1964). They have been reworked using the actual loads imposed by the concrete shielding rather than the assumed loading. The behaviour may be considered in two parts, the movement at depth as affects the piles and the surface movement as affects experiments.

Subsurface settlement

Early predictions.—The Boussinesq equations may be used to give the settlement ρ of the surface of a homogeneous semi-infinite isotropic solid due to a flexible vertical load of intensity q. For a rectangular area, length L and breadth B, the settlement at a corner is

$$\rho = qB(1 - v^2)I_p/E \tag{1}$$

where the influence value I_p is a function of L/B, see Terzaghi (1943). Assuming a modulus of 300 MN/m^2 and a uniform vertical stress of 100 kN/m^2 at basement level (26.8 m N.D.), a settlement of 5.1 mm is predicted at that level. Using Steinbrenner's method given by Terzaghi (ibid.) the settlement 7.6 m below at 19.2 m N.D. will be 3.5 mm. These values could be reduced by up to a third if the higher modulus were assumed. As the maximum allowable differential settlements were about 0.05 mm only local rearrangement of the shield blocks could be permitted.

Measurements.—The overall load of the initial shielding was built up at a fairly uniform rate between 2 February and 30 July 1965, although locally the loading was irregular. From the end of March the tops of all the magnet upstands settled between 0.75 and 1.0 mm although the load was changed adjacent to only a few of them. The tops of the survey monuments also settled between 0.5 and 1.0 mm during the same period but settled a further

314

0.25 mm in the following six months without load change. Although a large part of these settlements occurred during the trial loading their uniformity suggests no correlation with the loading. Continued drying shrinkage due to space heating in the tunnel is considered the most likely reason, and it is concluded that, contrary to original expectations, removal of shielding at ground level did not cause significant movement at a depth of 12 m.

Plate test results.—The current plate tests at 0.91 m diameter show that, even at a stress of 1500 kN/m²—that is much higher than the shielding applies—some 90% of the total 1.0 mm measured settlement occurs within one diameter below the loaded plate. Using this result as a guide, the basement

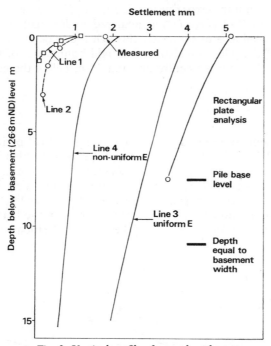

Fig. 2. Vertical profile of central settlement

settlement of 5.1 mm given by Equation 1 might be expected to reduce to 0.5–1.0 mm at the base of the piles.

The settlement profile with depth below the loaded plate is plotted as line 1 in *Fig. 2* for a stress of 1500 kN/m². The corresponding Boussinesq distribution for modulus of 1000 MN/m² is plotted as line 2. The modulus was deduced from the measured surface settlement using the relationship

$$E = \frac{q}{\rho} \times \frac{\pi}{2} R (1 - v^2)$$

where R is the radius of a rigid punch, e.g. Timoshenko and Goodier (1951). Similar differences between measured and predicted settlement at depth have been found by Burland and Cole (1972) in the London Clay and Burland, Sills and Gibson (1973) in the Chalk. In both cases analyses based on the assumption of a modulus increasing uniformly with depth removed the discrepancy.

Finite element predictions.—Such analyses are readily performed using finite element techniques. In *Fig. 2* a comparison may be drawn between the early settlement predictions and line 3 which shows finite element predictions based on a uniform modulus of 300 MN/m^2. The effect of a modulus increasing from zero at the surface to 1000 MN/m^2 at a depth of 13.2 m, being constant thereafter is seen from line 4 to radically alter the settlement profile.

In composing the finite element mesh the magnet tunnel and basement were assumed to be entirely concrete of modulus 1000 MN/m^2. It was not practicable to model the very thin (and therefore flexible) concrete floor of the experimental hall but the error due to this omission is thought to be negligible. The cross-section B–B (*Fig. 1*) was analysed under plan strain conditions assuming a vertical axis of symmetry through the magnet ring and ignoring the presence of the piles and shafts. The rigid boundaries of the mesh were 38 m horizontally from the centre line and 48 m below the ground surface.

Surface

Surface settlements were measured during the initial load period at the floor datum points B1, 3, 5, C1, 3, 5, E2, 3 and 4 shown in plan in *Fig. 1*. The composite settlement profile in *Fig. 3* shows settlements of these points plotted against the distance of these points from the magnet ring. The idealised loaded area is assumed to be symmetrical about the centre line of the ring but in fact there were a number of minor irregularities in the load arrangement which are one reason for the scatter of the measured settlements about line 1 in *Fig. 3*.

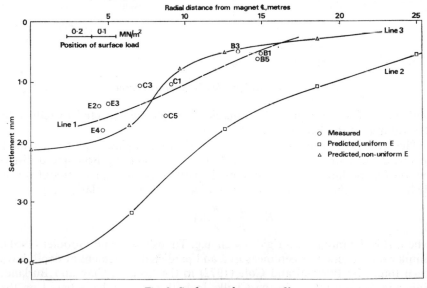

Fig. 3. Surface settlement profile

316

Line 2 in *Fig. 3* shows the settlement predicted by a finite element analysis assuming a uniform modulus. Much greater settlements are predicted extending much further away from the load. By contrast the modulus increasing with depth (line 3) produces very satisfactory agreement.

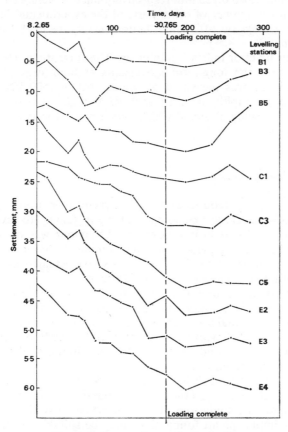

Fig. 4. Surface settlement during loading

The behaviour of the nine floor datums is plotted in *Fig. 4* on a base of time. The effect of local rearrangements of load are evident on a number of points which actually show uplift after earlier settlement. By contrast a magnet stored on the floor near C5 caused rather greater settlement at that point.

Long-term behaviour

During the subsequent final reassembly of the shield blocks in August 1966 and erection of experiments many floor datums became inaccessible. When levelled in September 1966 points B1, 3, C1, 3 were within an average of 0.075 mm of their level before the initial load was dismantled. C5 had risen so that its net overall settlement was much nearer that of C1 and C3.

317

Although the applied stresses are quite low the possibility of time–dependent settlement or creep could not be completely discounted, although it was unlikely to have a significant effect on floor or magnet stability. Of the original points only B1 and C1 remained accessible in April 1973, having then moved up 0.030 mm and down 0.125 mm respectively since September 1966. Records are available for a number of other parts of the experimental hall and show movements in a range of ±0.75 mm. Unfortunately the load dispositions are imperfectly known.

There is thus no clear indication of creep. The current plate test was established to examine the long-term movements at high stresses. Very slight creep of 0.075 mm was measured at stresses below 1500 kN/m^2, but the movement decayed rapidly over a few days. At higher stresses creep lasting several months is being observed which may account for as much as 20% of the total settlement.

Concluding remarks

The exacting stability requirements of a synchrotron were monitored by extremely precise levelling extending over many years. However, the most useful information for interpreting ground behaviour was obtained from a period of initial loading prior to commissioning. It has been shown that earlier concern about excessive settlements, particularly at depth, were unfounded although small movements did occur.

In the light of more recent assessment of the *in situ* properties of the Bunter sandstone at Daresbury and with the use of new analytical techniques it proved possible to make predictions close to the lower settlements which were measured. The behaviour is consistent with a modulus which increases with depth and corresponds to the behaviour found under experimental long-term plate test conditions.

It is concluded that at fairly low stresses the Bunter Sandstone, even though weathered near the surface, constitutes a sound founding material not requiring complicated foundations. The rapid decay of settlement with depth suggests that simple spread foundations may be suitable, even for sensitive structures. The effect of higher stresses, in particular creep behaviour, is being investigated by a large *in situ* load test.

Acknowledgements

The design of the foundation system was undertaken by the Works and Building Department of the U.K.A.E.A. Risley who acted as consultants for all building civil works (structural engineer, G. E. A. Haden). The authors would like to thank H. Stott at Daresbury who was responsible for all the survey observations and H. D. St John at B.R.S. who performed the finite element analyses. Numerous other colleagues have greatly contributed to the work described. This paper is published with the permission of both the Director of Daresbury Nuclear Physics Laboratory and the Director of the Building Research Establishment.

REFERENCES

Burland, J. B., Sills, G. C. and Gibson, R. E. (1973). 'A Field and Theoretical Study of the Influence of Non-homogeneity on Settlement', *Proc. 8th Int. Conf. Soil Mech. & Found. Eng.*, Moscow

Cole, K. W. and Burland, J. B. (1972). 'Observations of Retaining Wall Movements Associated with a Large Excavation', *Proc. European Conf. Soil Mech. & Found. Eng.* Madrid

Meigh, A. C. and Greenwood, S. W. (1965). '*In situ* Testing of Soft Rocks', *Proc. 6th Int. Conf. Soil Mech. & Found. Eng.* Montreal

Meigh, A. C. and Skipp, B.O. (1964). Private communication

Terzaghi, K. (1943). *Theoretical Soil Mechanics.* Wiley, New York

Timoshenko, S. and Goodier, J. N. (1951). *Theory of Elasticity.* McGraw-Hill, New York

319

Allowable and differential settlements,
including damage to structures and
soil-structure interaction

V/1 Differential settlements of petroleum steel (tanks)

L. Belloni, A. Garassino and M. Jamiolkowski
Studio Geotecnico Italiano

Summary

An extensive study of petroleum tank foundations was started by the writers about two years ago. The performance of large floating roof tank foundations during and after controlled water tests has been documented.

The scope of this investigation is to establish:

(1) procedures and standard instrumentation for water tests of large floating roof tanks laying on soft soils and

(2) distortional settlements of the tank periphery to determine acceptable values for the type of tanks under consideration.

In this Paper due to space limitation, a short account of the measured settlements for five tanks is given. The full results of this study will be published elsewhere.

Table 1. CHARACTERISTICS OF THE TESTED TANKS (FLOATING ROOF)

No.	Location	Diameter (m)	Height (m)	Volume (m³)
1	Ravenna	67.0	14.7	50,000
2	Quiliano	69.8	17.0	65,000
3	San Nazzaro	91.4	18.3	120,000
4	Ancona*	96.2	22.0	160,000
5	Venezia	30.6	10.5	7,700

* Tank still under water test—maximum load reached.

Soil conditions for five of the considered sites are shown schematically in *Figs. 1* to *3*. The adopted foundation schemes are shown in *Figs. 4* and *5*.

In *Table 2* the measured total and differential settlements are briefly summarized.

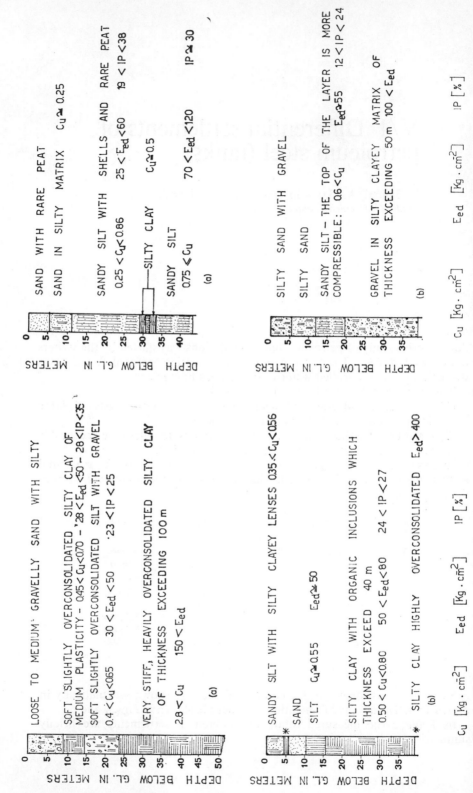

Fig. 1. Soil profile under (a) tank No. 4, Ancona (b) tank No. 5, Venezia

Fig. 2. Soil profile under (a) tank No. 1, Ravenna (b) tank No. 3, San Nazzaro

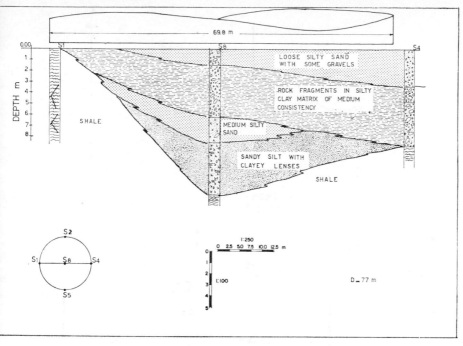

Fig. 3. Soil profile under tank No. 3, Quiliano

Table 2. MEASURED SETTLEMENTS OF FIVE TANKS

Tank No.	S^c (cm)	S^p_{max} (cm)	S^p_{min} (cm)	δ_{max} (%)	δ'_{max} (%)	ϕ_R
1	130	73.8	59.6	0.241	0.148	0.126
2	*	6.3	0.4	0.140	0.104	0.066
3	12.5	6.85	4.65	0.025	0.016	0.014
4	†	21.0	15.00	0.233	0.098	0.018
5	*	14.70	12.00	0.208	0.122	0.043

* Not measured.
† Measured not yet elaborated.
S^c = settlement at the centre of the tank
S^p_{max} = maximum settlement along the perimeter of the tank
S^p_{min} = minimum settlement allong the perimeter of the tank
δ_{max} = maximum distortional settlement along the perimeter of the tank

$$\delta_{max}{}^{jk} \frac{S^j - S^k}{l} \times 1000$$

where l = distance between points j and k
δ'_{max} = maximum differential settlement referred to the plane of rigid tilting according to De Beer (1968)

$$\delta'_{max} = \left[S'_i - \left(\frac{S_{i-1} + S_{i+1}}{2} \right) \right] \frac{1}{l}$$

S_i' = displacement with respect to the plane of rigid tilting
ϕ_R = angle of rigid tilt according to De Beer (1969)

325

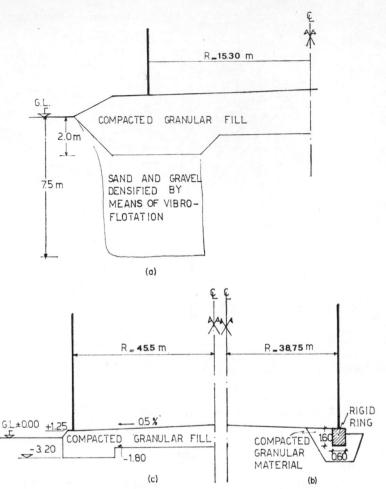

Fig. 4. Foundation schemes for (a) tank No 1, Ravenna, (c) tank No. 3, San Nazzaro, (b) tank No. 2 Quiliano

From the analysis of the data in *Table 2* (and of other data not reported in this note) the following preliminary observations can be drawn:

(1) The maximum differential settlement that a large floating roof tank may withstand is relatively high. At this stage the following values may be taken as a working hypothesis:

$$\delta_{max} < 0.3 \text{ to } 0.35\%$$
$$\delta'_{max} < 0.2 \text{ to } 0.22\%$$

(2) The above mentioned limits for δ_{max} mean that engineers may adopt simple and cheap foundations for large floating roof tanks, even in the most unfavourable soil conditions, provided a carefully designed and controlled water test is carried out. The control of the water test can be performed by means of high precision levelling of at least 16 points along the tank periphery and the pore pressure, horizontal and vertical displacement measured.

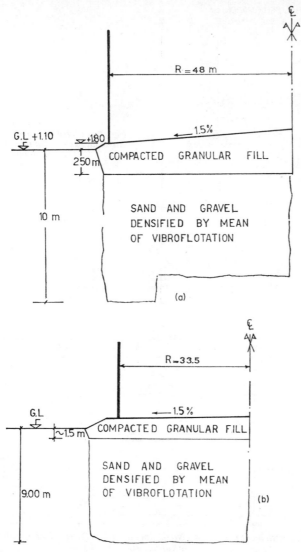

R = 48 m

G.L +1.10   ▽ +180

250 m   COMPACTED GRANULAR FILL

1.5%

10 m

SAND AND GRAVEL
DENSIFIED BY MEAN
OF VIBROFLOTATION

(a)

R = 33.5

G.L

~1.5 m   COMPACTED GRANULAR FILL

1.5%

9.00 m   SAND AND GRAVEL
DENSIFIED BY MEAN
OF VIBROFLOTATION

(b)

Fig. 5. Foundation schemes for (a) tank No. 4, Ancona, (b) tank No. 5, Venezia

An example of measurement of soil deformations in both directions with a high precision inclinometer is shown in *Figs. 6* and *7*.

The Writers think that a rational approach to the determination of the allowable distortional settlements for floating roof tanks will require in addition to the above-mentioned measurements also the evaluation of the stresses induced in the shell. Such an analysis will be carried out in the further stage of the present research programme.

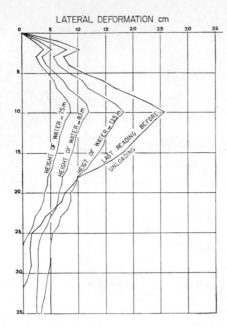

Fig. 6. *Lateral movement of tank No. 1, Ravenna using Galileo Vibrating Wire Torpedo Inclinometer*

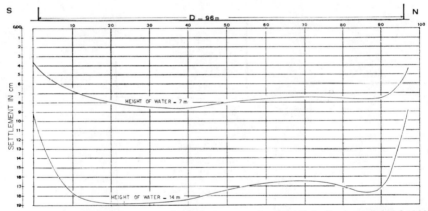

Fig. 7. *Vertical settlement under tank No 4, Ancona using Digitilt inclinometer (model 50301 plastic casing, 2.75 in o.d.)*

REFERENCE

De Beer, E. E. (1969). 'Foundation Problems of Petroleum Tanks,' *Annales de L'Institute Belge du Petrole*. Vol. 6, 25–40

V/2. Settlement behaviour of buildings above subway tunnels in Frankfurt Clay

H. Breth
Professor of Soil Mechanics

G. Chambosse
Institute of Soil Mechanics and Foundation Engineering, Technische Hochschule, Darmstadt, West Germany

Summary

In some tunnel sections of the Frankfurt/Germany subway system the settlements of overlying buildings were measured. The tunnels were located in the Frankfurt Clay (classification CH). The settlement depressions of the ground surface were also measured and vary from 3 cm to 15 cm. At the same time the deformation of the buildings were observed. The examinations showed that the outside walls of many buildings behaved like rigid plates. Most resistant to differential settlements proved to be a barrel vault, which experienced a rotation of 1 : 130 without damage. Furthermore the measurements show that the slope of the settlement depression of the soil is always steeper than the rotation of the buildings.

Subsoil

Frankfurt Clay which is a tertiary clayey marl, occurs beneath a 7 m thick layer of quaternary gravel and sand. The clay is irregularly interbedded with bands of limestone, silts and water-bearing sands (*Fig. 1*). The clayey marl is heavily over-consolidated and laminated (classification CH). The jointed limestone bands interbedded in the clay contain artesian water. During the course of construction the ground-water table was lowered by 20 m. The diameter of the tunnels is 6.7 m and the overhead cover is 10–12 m. The settlement attributable to the driving of the tunnel was measured directly on the buildings above the tunnel as well as on the free ground surface.

329

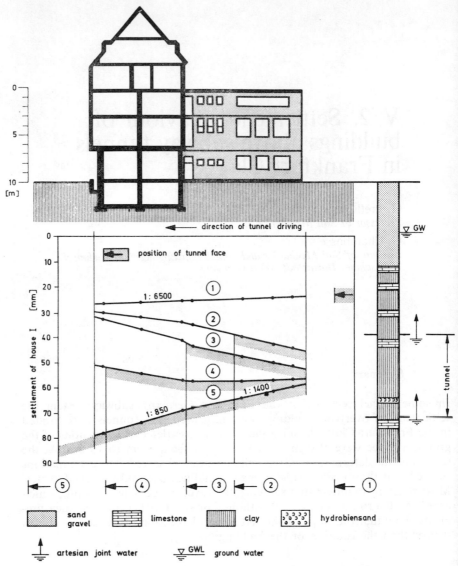

position of tunnel face

settlement of house I [mm]

1 : 6500

1 : 850

1 : 1400

5	4	3	2	1

sand gravel	limestone	clay	hydrobiensand

artesian joint water   ▽ GWL ground water

Fig. 1. Settlement as function of tunnel driving, House 1

Settlement measurements on Building 1

Two tunnels were driven simultaneously under Building 1 (New Austrian Tunnelling method). Building 1 (*Fig. 2*) had been constructed around the year 1955 with a reinforced concrete frame structure. The massive cellar walls consist of low-grade concrete. The western outside walls of both houses were panelled with sandstone. The settlement of the buildings were measured by means of 41 measuring points on the cellar walls. *Fig. 1* represents the settlement of the houses during driving of the tunnel in section AA (*Fig. 2*).

Initially a 25 mm settlement had occurred due to the lowering of the ground water. As the tunnels were excavated the buildings moved towards the excavated space and tilted, the walls of the buildings behaving relatively rigidly. Distinguishable differential settlement could only be observed on the partition wall separating the main building and its annexe. The layout of the buildings shows that the cellar wall is interrupted at this junction and the overlying building has large windows and doors. Apart from two insignificant 0.3 mm wide cracks (*Fig. 2*) in the cellar walls, the settlement did not cause any damage. After the tunnels had been driven underneath the buildings, their maximum inclination to the west was 1 : 900. No distortion could be observed in the main buildings. As the length-to-height ratio was 1 : 1.6 the

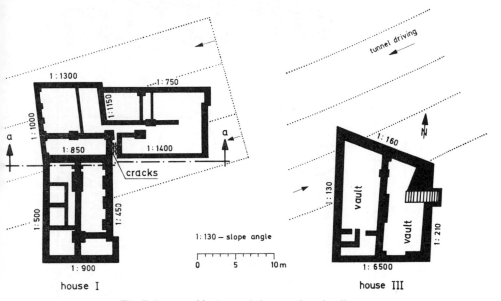

Fig. 2. Layout of houses and slope angles of walls

narrow side of the building was sufficiently rigid to prevent distortion. In the annexe, on the other hand, a minor distortion of 1 mm in the longitudinal direction could be observed. In the annexe the length-to-height ratio was 1 : 0.9. *Fig. 3* allows a comparison to be made between the settlement of the houses along the western side and the settlement syncline in the surface of the ground. The observed settlements of the buildings and of the ground are not identical. The walls of the buildings behaved rigidly also in this context. In the settlement curve of the two buildings there was a clear bend in the area of the fire-proof wall. The torsion of the two building fronts towards each other was roughly 1 : 1100. At roof level the two house fronts should thus have moved about 15 mm towards each other. In spite of careful examination no cracks or other damage could be observed in the structure. The two house fronts were obviously able to accept the necessary shearing deformation. A great number of jammed windows and doors support this suspicion.

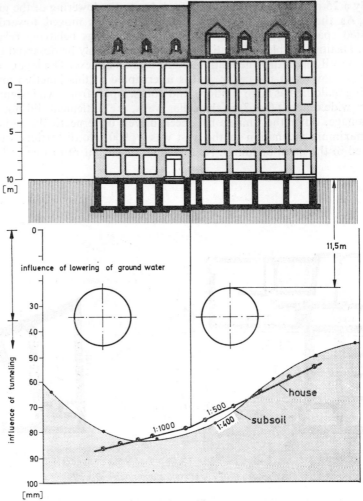

Fig. 3. Settlement of building and subsoil, House 1, facing west

Settlement measurements on Building 2

Building 2, a reinforced concrete frame structure, is approximately 20 years old and connected with a larger building of similar construction. Maximum settlement of the open ground surface amounted to 13 mm and that of the building reached 10 mm (*Fig. 4*). The building deflected by about 3 mm and separated from the neighbouring house at the fire-proof wall by about 3 mm. Furthermore, a settlement crack of 1 mm occurred in the fire-proof wall. No other damage arose in addition to these cracks. An archway for motor vehicles did not have any detectable effect on the settlement behaviour. A comparison between the settlement curves of the front and rear sides of the house revealed that there had been torsion. This is attributed to the fact that the tunnel tubes were not at right angles to the building.

332

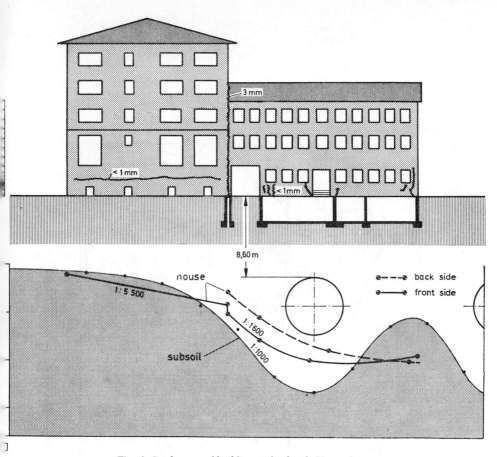

Fig. 4. Settlement of building and subsoil, House 2

Settlement measurements on Building 3

In contrast to the buildings discussed so far, the house referred to in this section was exposed to two settlement actions as the two tunnel tubes were driven successively and in the opposite direction (shield driving method). The former monastery cellar of Building 3 (eighteenth century) consists of two mighty barrel vaults made of sandstone, which reach a height of roughly 2.5 m. In the area of the abutments, the outside walls are approximately 1 m thick. The storeys above ground level are more recent and were built in brickwork. Extending south from the building there is a storage building made of brickwork without a cellar.

Fig. 5 shows the northern side of the house, the settlement syncline of the free ground and the inclination of the building. The biggest settlement (192 mm) was measured at the north-western corner (influence of both tunnels). The smallest settlement (46 mm) occurred at the south-eastern corner. Even during the construction of the first tube, very substantial inclinations were detected. After driving of one of the tunnels underneath the northern

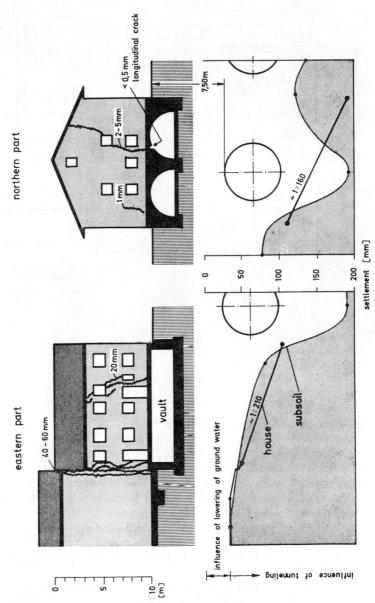

Fig. 5. Settlement of building and subsoil, House 3

334

front, a differential settlement of 64 mm, equalling an inclination of 1:210 had arisen. In the western barrel vault an almost perfectly horizontal crack (width 0.5 mm) could be seen at a height of roughly 2 m, over the entire length of the vault. As *Fig. 5* shows, this crack continues vertically across the northern front of the building. No additional cracks occurred when the second tube was built, only the inclination increased to 1:160 (*Fig. 5*). The location of the cracks and of the settlement syncline suggest that cracks have to be expected at those points where the settlement curve of the building intersects with the settlement syncline of the soil. *Fig. 6* shows the western side of the house with the corresponding settlement curves. The entire

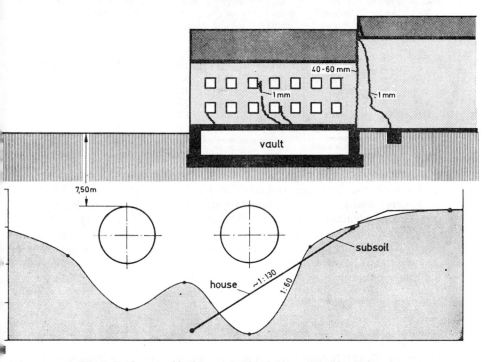

Fig. 6. Settlement of building and subsoil, House 3, facing West

building moved 40–60 mm away from the attached storage building. Owing to the very high longitudinal rigidity of the barrel vault the western side of the building did not flex. After the building of the first tunnel the inclination of this wall amounted to 1:150. After completion of the second tunnel tube (*Fig. 6*) it increased to 1:130. Here also the statement applies that 'cracks occur in those areas where the settlement synclines of the soil and building cross'. The settlement behaviour of the eastern side of the building is shown in *Fig. 5*. The final inclination on that side was 1:210. The layout shown in *Fig. 2* includes the final inclination for all outside walls. The building was exposed to torsion from south to north. The massive barrel vault led to a relatively uniform settlement and manifested no major damage in spite of the extreme inclinations and torsions. The stability of the vault was not jeopardized. The above-ground building had, however, suffered considerably.

Not only did cracks occur on the outside walls but even on inside walls. The measured values suggest that the building would have hardly survived the imposed stresses had it not been for the rigid vaults.

In conclusion, it can be stated that the walls of buildings above tunnels behave like rigid plates as long as the length-height-ratio is not less than one. Window openings do not reduce this plate effect. If the tunnels are not at right

Table 1. SETTLEMENTS AND SLOPE ANGLES

House No.	Type of house	Settlement of subsoil (mm)	Settlement of houses (mm)	Slope angle of subsoil	Slope angle of houses	Houses length/ height	Houses flexion (mm)	Cracks (mm)	Note
II	a	13	10	1 : 1000	1 : 1600	1 : 0.7	3	3	1
I	a	84	87	1 : 400	1 : 450	1 : 1.4	0	<0.5	2
IV*	b	120	90	1 : 120	1 : 500	1 : 0.4	4	<1	2
V*	b	120	87	1 : 120	1 : 400	1 : 0.4	2.5	30	1,3
VI*	b	190	116	1 : 60	1 : 770	1 : 1.3	0	0	2
III	c	190	190	1 : 60	1 : 130	1 : 0.8	—	60	1

* not described in this Paper
1—house added to another building
2—detached house
3—very different in construction

a—concrete frame
b—brick work
c—vault

angles with the building, uniaxial or biaxial torsion occurs. Differential settlements and the resulting inclinations up to 1 : 450 do not damage concrete frame structures. Even in the case of facade panelling with sandstone slabs no damage was caused by this inclination. Buildings with a length-height-ratio greater than one moved without further damage. Major damage must, however, be anticipated when tunnels are driven underneath buildings consisting of various parts with extreme rigidity differentials. *Table 1* shows the results of the settlement measurements.

REFERENCES

Chambosse, G. (1972). 'Das Verformungsverhalten des Frankfurter Tons beim Tunnelvortrieb', *Mitteilungen der Versuchanstalt für Bodenmechanik und Grundbau*, No. 10, Technische Hochschule, Darmstadt

V/3. Damaging uplift to a three-storey office block constructed on a clay soil following the removal of trees

John E. Cheney
Senior Scientific Officer, Building Research Station, Garston, Watford, Herts

D. Burford
Higher Scientific Officer, Building Research Station

Summary

Precise levelling records of the movement of a three-storey office block for a period of 14 years since construction are presented. The measurements were taken because the building was located on a clay site from which trees had been removed two months prior to construction. Uplift was expected and indeed occurred, resulting in cracking of the building. Indications are that cracking commenced at angular distortions of 0.05% and was well established at 0.1%.

Introduction

It was anticipated that damaging uplift might occur to a shallow-founded brick building to be constructed on an afforested clay site. Damage did indeed occur. The building construction, foundations, bearing pressures and soil conditions are described. A precision levelling system was installed at the time of construction. Results of 14 years observations are given. The levelling observations are used to calculate the angular distortions associated with the observed onset of cracking. The angular distortion at a levelling station is defined as the change in the angle subtended in the vertical plane at that station by straight lines joining it to the adjacent stations. It is measured in radians and expressed as a percentage.

337

Site history

The building is situated at Garston, Hertfordshire, on the edge of Bricket Wood. Eight oak trees about 15 m tall and 2.5 m in girth were felled in October 1958 within two months of the start of construction. The position

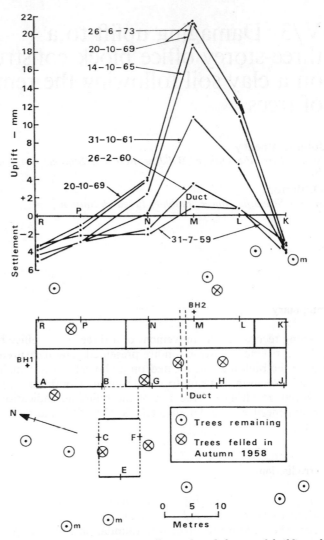

Fig. 1. Vertical movements of the East wall at selected dates and building plan showing position of levelling stations, A–R, and trees

of these trees and of the remaining nearest trees is shown in *Fig. 1*. Three of these trees, which have since been felled, are indicated on the plan by the symbol *m*.

Previously, since about 1938, the area between the trees had been occupied by an assortment of raised timber huts and paved areas.

The Soil

The subsoil is a heavy boulder clay to a depth of at least 6 m overlying chalk. There are known to be pockets of gravel overlying the chalk in the area. Two 6 m deep boreholes (BH1, BH2, *Fig. 1*) were made on the site in August 1964 and revealed the clay to be silty, varying in colour from yellow-brown near the surface to red-brown with some grey. At all depths there was a chalk content, generally in the form of small fragments. Varying quantities of small flints were present in the clay to a depth of about 3.7 m in BH1, but only to about 1.1 m in BH2. In BH2 some fine gravel was found at about 5 m depth. Both boreholes were dry and remained so during the several days they were left open. It is to be noted that this was at the height of summer. Generally the area is very wet at the surface and Bricket Wood is wet under foot in all but the driest summers.

Fine roots were found to a depth of 3.6 m in BH1 and to 2.3 m in BH2. Close examination of disturbed samples showed root fibres in both holes to 5.8 m but it is not known if these came from recent trees; they could be of great antiquity.

Laboratory tests

Water content, liquid limit, plastic limit and swelling pressure tests were carried out on samples taken at about 300 mm intervals of depth from BH1 and BH2, and also from a borehole in open grassland about 50 m from the building. The results of these tests are reported in detail by Samuels (1967). Water contents, which were adjusted to exclude material retained on a No. 7, sieve were consistently between 17–19% below 1.8 m rising to about 25% near the surface. Similarly liquid limits lay between 45–55% below 1.8 m, rising to about 70% nearer the surface. Plastic limits were between 18–21% rising to about 23–26% above 1.8 m depth. Activity of the clay was between 0.5 and 1.1 at all depths.

Swelling pressure tests were carried out by the method described by Ward, Samuels and Butler (1959). They generally indicated higher swelling pressures of the clay near the building than in open grassland, but did not define precisely a zone of drying, which similarly could not be determined from water content tests.

The building

The main building (see *Fig. 1*) is of three storeys and is 42.7 m long by 11.3 m wide at its gable ends. The external walls are of London stock brick cavity construction 395 mm thick on the first two storeys and 280 mm on the third. There are 11 windows each 1.57 m² on each storey on the long walls, nine similar openings on the south gable and three on the north.

The concrete ground floor slab is 125 mm thick on minimum 150 mm hardcore. The slab is reinforced over its full width between a point 600 mm north of the duct and K. The remaining floors are of cast *in situ* clay pot T-beam construction spanning across the building. The eaves height is 8.5 m,

339

the 20° double pitched roof has a ridge height of 10.7 m and is of felt-on-woodwool-board construction.

There is a single storey annexe similarly floored and roofed to one side of the main building and attached to it by a covered lobby. The annexe is 15.2 m in protrusion by 7.6 m. It has an eaves height of 3.2 m and is 4.7 m to the ridge. Its west gable is of 394 mm cavity brick as is also the east wall. The sides are glass and 115 mm brick infill between two brick piers, on each side, which support the roof.

Foundations

The foundations are shown as full lines on the outline plan (see *Fig. 1*). Those for the external walls of the main building are 0.91 m wide and for the inner long wall 1.07 m wide. The width of the few cross-wall foundations varied between 0.69 m and 1.07 m. The depth of the foundations is 1.22 m below finished floor level (F.F.L.). However, the original ground sloped upwards in a north-westerly direction, K being the lowest part of the site, so the foundations at the south end are only 600 mm below original ground level (O.G.L.) and at the north 1.30 m below O.G.L.

A duct for lagged steam pipes passes under the main building (see *Fig. 1*). It is of reinforced cast *in situ* concrete 203 mm thick and is externally 1.32 m wide. It is founded at a depth of 1.73 m (1.41 m below O.G.L.) under the west wall and 1.61 m (1.24 below O.G.L.) under the east wall. Its roof is formed by the building floor slab. Reinforced lintels carrying the main walls rest on the duct sides and span the gap in the wall foundations. The duct is dry with an air temperature of 29 °C.

The annexe east-wall foundation is 1.07 m wide by 2.13 m deep (2.11 m below O.G.L.) to carry below the invert of a 300 mm pipe running between this wall and the nearest roof pillars. This pipe was laid to contain water in a ditch that runs parallel to the main building. The east roof piers are on pads 0.76 m by 1.2 m at 2.13 m depth (2.21 m below O.G.L.). The west pads are 0.61 m by 0.76 m and 1.22 m deep (1.30 m below O.G.L.). The infill walls rest on sub-ground beams of 1.06 m square section (about 1.1 m below O.G.L.). The west gable foundation is 0.69 m wide by 1.22 m deep (1.35 m below O.G.L.).

Foundation bearing pressures

The building load distributed by traditional methods gives pressures of about 100 kN/m² for each of the three long main walls. The few internal cross foundations would locally reduce these figures. The duct, assuming a 45° distribution transverse to long walls as it passes under them, reduces pressures to about 30 kN/m². The actual effect of the duct may well be to reduce this figure even lower. The pressure at the main gables is 66 kN/m².

For the annexe, the west gable pressure is 45 kN/m² and the east wall 39 kN/m². The pier pressures, west 47 kN/m² and east 23 kN/m², are probably greatly reduced by the integral sub-ground beam under the infill sections.

The levelling installation

Fourteen levelling stations were incorporated in the brickwork at a height of 900 mm above damp-proof course as the building was constructed. Their positions are shown A–R in *Fig. 1*. The levelling stations took the form of stainless steel sockets into which a plug could be screwed for support of a precision levelling staff. A Wild N3 level was used to refer levels from two sleeved datums founded at a depth of 6 m situated 13 m north of station C and 35 m east of station M. The system was generally as described by Cheney (1973).

Observations and comments

Fig. 2 shows the levelling records to date. The levelling of the main building commenced on 2 February 1959 when the brickwork was to first floor level and that floor ready to concrete. By 26 March the second floor slab had been cast and by 31 July 1959 most of the load was on the main building. On this date the first levels were taken on the annexe which was then under construction.

Comment in this paper is restricted to the subject of cracking. In July 1959 hair cracks appeared in the external ground floor walls near points H and L, in the ground floor slab on this section, and internally on the second story above the duct where it passes under the long internal foundation. *Table 1* indicates −0.05% angular distortion (hogging) at this time. Levelling point L

Table 1. ANGULAR DISTORTION IN THE PLANE OF THE EXTERNAL BRICKWORK AT SELECTED LEVELLING STATIONS

Date	P %	N %	M %	L %	G %	H %
31 July 1959	−0.01	+0.04	−0.04	−0.04	−0.03	−0.05
26 Feb. 1960	−0.03	+0.08	−0.10	−0.02	−0.04	−0.08
26 June 1973	+0.02	+0.17	−0.34	−0.09	−0.07	−0.27

The negative sign indicates that the levelling station has risen relative to the straight line joining the two adjacent levelling stations (hogging).

was damaged in August 1959 and a replacement was re-zeroed in February 1960. Movement between August and February has been assumed for this point as zero as one of the adjacent points moved up and one down. By January 1960 wall cracks were well apparent. By the end of February the width of the cracks were 2.5 mm at M and 1.3 mm at H at ground floor level and of the same order on the second floor above the duct near M and were sufficient to attract the attention of the casual observer. As indicated by *Table 1*, this stage of well established cracks corresponded to about −0.1% angular distortion.

Over the years cracking has continued and has been repaired at intervals. By December 1962 second-storey room occupants were complaining of draughts through the cracked external walls, one or two windows have cracked and electrical conduit in the walls has failed in tension.

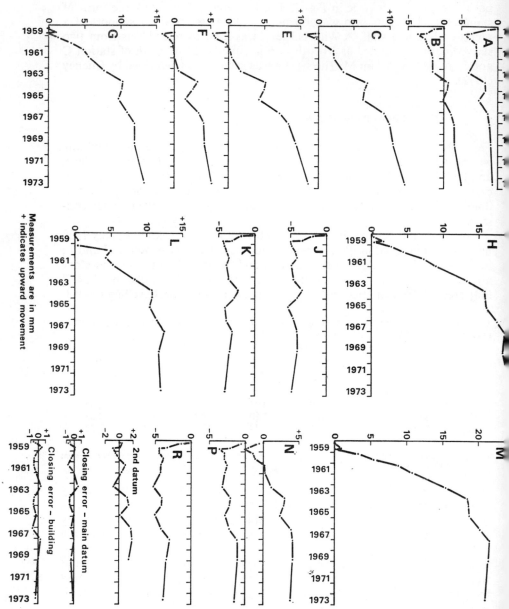

Fig. 2. Vertical movement of the levelling stations, A–R, with time

In this Paper no attempt has been made to give a detailed explanation of the observed damage. Many factors are involved, the major ones probably being the inherent non-homogeneity of the ground due to earlier afforestation and certain structural complications such as the reinforced concrete duct shown in *Fig. 1*. It is intended to continue observations for so long as significant movements continue.

Acknowledgement

This Paper is published by permission of the Director of the Building Research Establishment.

REFERENCES

Cheney, J. E. (1973). 'Techniques and Equipment Using the Surveyors Level for Accurate Measurement of Building Movement', *Proc. Symp. on Field Instrumentation*, British Geotechnical Society, 30 May–1 June. Butterworths, London

Samuels, S. G. (1967). 'The Uplift of Buildings on Swelling Clays', Building Research Station unpublished note

Ward, W. H., Samuels, S. G. and Butler, M. E. (1959). 'Further Studies of the Properties of London Clay', *Geotechnique*. Vol. 9. No. 2, 33–58

V/4. Settlement and contact pressure distribution of a mat-supported silo group on an elastic subgrade

John C. Crowser
Engineer, Shannon and Wilson, Inc., Seattle, Washington

Robert L. Schuster
Professor and Chairman, Department of Civil Engineering, University of Idaho

Ronald L. Sack
Associate Professor, Department of Civil Engineering, University of Idaho

Introduction

The observed settlement behaviour of an existing mat-supported silo group is presented and compared with theoretical behaviour. The foundation soils have been idealised as a non-homogeneous elastic medium, underlain by a rigid base at a depth somewhat less than twice the mat width. Nonlinear stress-strain relations determined from laboratory testing with the associated mathematical expressions have been utilized for individual soil strata.

The finite element method (F.E.M.) has been used to evaluate contact pressure and stress distributions beneath the foundation by imposing the observed mat-deflection shape on the soil subgrade. Comparable settlements were computed by elastic theory using a stress-dependent modulus of elasticity and by one-dimensional consolidation theory using the F.E.M. stress distribution.

The storage silo group consists of eight, 40 ft (12.2 m) diameter silos rising 141 ft (43.0 m) above exterior grade, supported by a 5 ft. (1.52 m) thick concrete mat, 100 ft $\times$ 183 ft (30.5 m $\times$ 55.8 m) in plan. The mat foundation has an average design bearing pressure of about 6000 lbf/ft^2 (287.7 kN/m^2). An isometric view of the silo group is shown in *Fig. 1.*

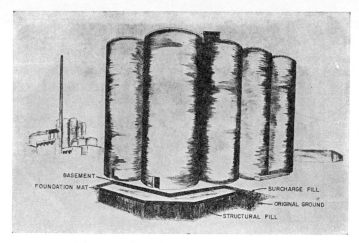

Fig. 1. Isometric view of mat-supported silo group

Geotechnical properties

Geologically, the silo site is situated on recent tidal sediments underlain by dense silt, sand, and gravel (glacial till). The recent sediments, consisting of layered silt and sand, have been deposited in a tidal flat environment. In general, the relative densities and/or consistencies of the sand and silt layers show a gradual increase with depth ranging from loose to dense and soft to very stiff, respectively. The subsurface conditions are summarized by the generalised cross section in *Fig. 2*.

Field explorations consisted of six subsurface borings to depths of 184 ft (56.1 m). The average standard penetration value was 28 blows/ft for the fine sand, and 8 for the sandy silt and silt soils. Water contents of the silt samples ranged from 30 to 40% below a depth of about 10 ft (3.05 m).

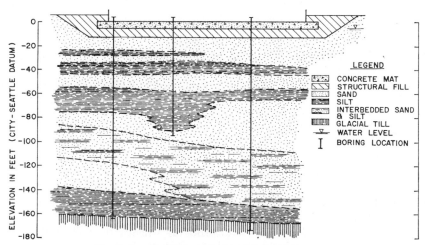

Fig. 2. Longitudinal profile beneath foundation mat

345

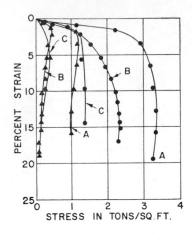

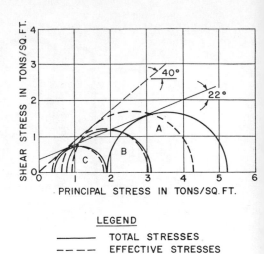

LEGEND

——————		TOTAL STRESSES
— — — —		EFFECTIVE STRESSES
●	$\sigma_1 - \sigma_3$	TOTAL DEVIATOR STRESS
▲	u	PORE PRESSURE

Fig. 3. Representative consolidated undrained triaxial test results on silt soils

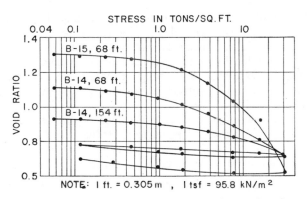

Fig. 4. Representative oedometer test results on silt soils

Atterberg limits showed the silts to be non-plastic or of very low plasticity. Relative density values of the sand ranged from about 55 to 70%; these results appear to agree with the Standard Penetration Resistance at the same locations. Representative results of consolidated-undrained triaxial compression tests are shown in *Fig. 3*. Representative oedometer test results are presented in *Fig. 4*.

Loading and settlement *v* time

Load and settlement histories are shown in *Fig. 5*. Very little settlement took place until the total structural load reached about 50–60 × 10⁶ lbf (22.7 × 10⁶ to 27.2 × 10⁶ kgf). This load probably stressed the soil to about the preconsolidation load, which generally agrees with the laboratory test results

346

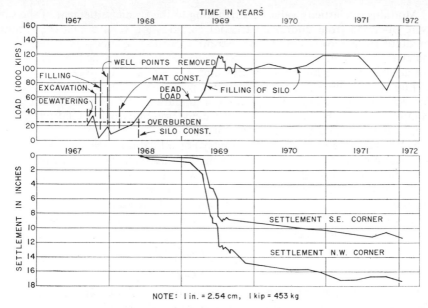

Fig. 5. Load and settlement history of mat-supported silo group

on the silt layers. During the initial filling of the silos, settlement responded very rapidly to loading; therefore it can be concluded that the predominant settlement of the silt layers is a combination of immediate or elastic settlement and primary consolidation. This settlement is referred to here as initial settlement. Settlement that occured after initial filling was at a relatively constant rate (time dependent) and has been considered as secondary consolidation.

Analysis of contact pressures and stress distributions

Contact pressures and stress distributions beneath the mat foundation were analysed for two cross-sections through the mat centrelines using F.E.M. and the assumptions of plane strain. The soil strata were divided into a number of discrete elements connected at their joints or nodal points. Each mesh of elements had a trapezoidal shape incorporating the mat width plus approximately 80 ft at the top and enlarging at the bottom boundary to a width of about four times that of the mat. The bottom boundary (in contact with the more rigid glacial till) was assumed to be fixed, and the side boundaries were constrained laterally. The nodal points beneath the mat were subjected to vertical displacements corresponding to the observed mat deflection shape attributed to initial settlement.

Subsurface properties were idealized using the results of the six borings. Zones containing less than 30% silt were idealized as sand, and those with more than 70% silt were idealized as silt. Soils with estimated percentages between these values were proportioned accordingly. Groundwater level at the site was observed to be immediately beneath the mat foundation.

347

Modulus of elasticity values contributing to the elastic settlement of the silt were obtained from undrained triaxial compression tests taking the secant modulus from zero strain to 0.3% strain. Constrained modulus values were obtained from conventional oedometer test results, taking the tangent modulus at effective overburden pressures and at 2 and 4×10^3 lbf/ft^2 (95.8 and 191.6 kN/m^2) above effective overburden pressures. Values of drained modulus used in conjunction with primary consolidation settlement of the silt were computed from the determined constrained modulus using relations from elastic theory and a Poisson's ratio of 0.3.

Since the value of elastic and primary consolidation settlement of the silt soils cannot be readily separated from the observed initial settlement, an equivalent modulus and Poisson's ratio were used in relating initial settlement to stress distribution. Determination of an approximate equivalent modulus was accomplished by adding the respective strains at various stress levels for the computed elastic and drained modulus values. This resulted in a constitutive relation such that

$$E_s = 4000 p_a \left(\frac{\sigma_3'}{p_a}\right)^{0.41}$$

where σ_3' is the effective minor principal stress in lbf/ft^2 and the term p_a, denoting atmospheric pressure, is introduced such that the constant (4000) is a pure number. An equivalent Poisson's ratio of $\mu = 0.33$ was used.

For the sand strata, a constitutive relation in the form suggested by Hardin and Drnevich (1970),

$$E = K(\sigma_m')^{0.5}$$

was used, where σ_m' is the mean effective stress in lbf/ft^2, and K is a function of void ratio and strain amplitude. K was equal to 15 000 with E being expressed in lbf/ft^2.

The F.E.M. analysis was carried out using a piece-wise linear step procedure in conjunction with an iterative process to establish the elastic properties of each element for each step.

Contact pressures analysed using F.E.M. and preceding constitutive relations are presented in *Figs. 6* and *7*, together with the mat deflection shape imposed on the soil subgrade. Distribution of vertical stress within the soil subgrade beneath the centre and edges of the mat, analysed using F.E.M. and Boussinesq theory, is presented in *Fig. 8*. Note that the calculated pressures at the edges of the mat are considerably greater than the average applied load. This condition should occur in order to allow the relatively uniform settlements which result from the rigid structure. Even though the contact pressure at the edges is considerably greater, the distributions of vertical stress below the centre and edges of the mat are very similar below a depth of about 35 ft (10.7 m) beneath the mat. Further, the vertical stress does not dissipate rapidly below that depth. In comparison, the distributions of vertical stress at depth beneath the edge points determined by Boussinesq theory vary considerably from those at the centre and are much lower than the distributions determined by the F.E.M. This is probably due to the assumptions used in evaluating the Boussinesq pressure distribution, i.e. a flexible area load and a semi-infinite elastic subgrade. These assumptions do not correspond to the behaviour of the mat and the finite subgrade thickness.

348

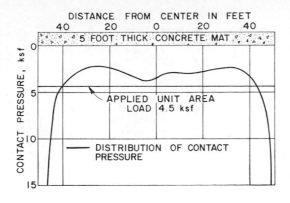

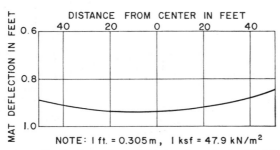

NOTE: I ft. = 0.305m, I ksf = 47.9 kN/m²

Fig. 6. Contact pressure distribution and mat deflection shape for the 100 ft (30·5 m) dimension at centre of mat-supported silo group

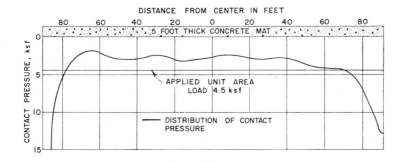

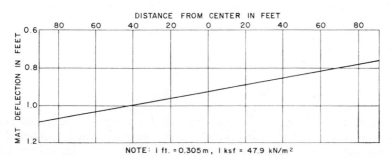

NOTE: I ft. = 0.305m, I ksf = 47.9 kN/m²

Fig. 7. Contact pressure distribution and mat deflection shape for the 183 ft (55·8 m) dimension at centre of mat-supported silo group

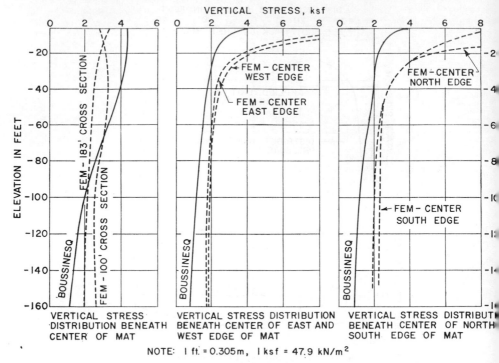

VERTICAL STRESS, ksf

VERTICAL STRESS DISTRIBUTION BENEATH CENTER OF MAT

VERTICAL STRESS DISTRIBUTION BENEATH CENTER OF EAST AND WEST EDGE OF MAT

VERTICAL STRESS DISTRIBUTION BENEATH CENTER OF NORTH SOUTH EDGE OF MAT

NOTE: 1 ft. = 0.305m, 1 ksf = 47.9 kN/m²

Fig. 8. Comparison of vertical stress distributions determined by finite element analysis and Boussinesq theory

Behaviour of silo walls

Diagonal cracks in the silo walls, inclined upward from the lower floor slab toward the centre of gravity of the structure, were observed during initial filling. They were continuous through the wall and ranged from hairline to a few thousandths of an inch in width. The predominant cracks occur in the four corner silos.

It appears that the foundation mat for the silo group essentially acts as a rigid body as illustrated by the very modest dish-shaped settlements shown in *Figs. 6 and 7.* From the computed contact pressures between the rigid mat and the underlying soils, it can be shown that the resulting reaction forces imposed upon the mat are at least three times as great at the extremities as at the centre. The resulting forces applied by the redistribution undoubtedly impose severe stresses in the corner silos and could be the primary cause of the observed cracking in those silos.

Settlement predictions

Settlement predictions of the mat-supported silo group were computed from: (1) elastic analysis using Steinbrenner's equation (Steinbrenner, 1934 and Terzaghi, 1943) in conjunction with the previously defined modulus

350

relationships, (2) conventional consolidation theory and stress distributions obtained from the Boussinesq equation and the F.E.M. and (3) Standard Penetration Test data.

For computation of the immediate elastic settlement of the silt strata, an elastic modulus determined from the undrained triaxial tests and $\mu = 0.5$ was used. For the drained or primary consolidation settlement of the silt, the drained modulus determined from the confined compression test data was used with $\mu = 0.3$. For the sand, E determined from the relationship of Hardin and Drnevich (1970) was used with $\mu = 0.25$. The modulus value used for each of these relationships was evaluated using average confining pressures obtained from F.E.M. results.

Predicted initial settlements of the combined silt and sand strata at various

Table 1. TOTAL PREDICTED AND OBSERVED INITIAL SETTLEMENTS

| | Predicted settlement | | | | |
| | Based on Standard Penetration Test data (in) | Steinbrenner elastic analysis (in) | Consolidation theory* | | Observed settlement (in) |
Location			Finite element stress distribution (in)	Boussinesq stress distribution (in)	
35 ft north of centre	3.1	9.3	13.5	11.0	11.8
South-west corner	2.6	7.5	10.3	8.8	9.6
North-east corner	3.1	9.4	13.2	7.4	11.5
50 ft north of south-east corner	3.0	8.7	10.2	5.7	9.1
Differential settlement	0.5	1.9	3.3	5.3	2.7

* Predicted settlements from consolidation theory include average settlement predictions of sand from standard penetration data and analyses from elastic theory.

boring locations are compared with observed values in *Table 1*. Total settlements computed using Standard Penetration Test data are considerably lower than those calculated using either of the other methods. However, the relationship used to evaluate Standard Penetration Test data has been established for granular soils and is probably not applicable for settlement calculations involving silt.

Settlement values obtained using Steinbrenner's elastic analysis are about 4–22% lower than the observed values. The observed settlements, however, probably include some secondary consolidation which is not taken into account in the elastic analysis. The predicted differential settlement is near that observed.

Settlement values obtained using one-dimensional consolidation theory in conjunction with stress distributions determined from F.E.M. were about 7–15% greater than those observed. These somewhat larger values are probably the result of the conservative nature of consolidation theory. Predicted settlement values near the edge of the mat based on the Boussinesq stress distribution were considerably smaller than observed values.

Conclusions

Contact pressures can be predicted for mats on irregularly layered systems involving nonhomogeneous and nonlinear soils using F.E.M. The F.E.M. provides a mechanism for quantitatively predicting large pressure gradients in the vicinity of the edges of rigid mats. For this study, the calculated contact pressures in the vicinity of the edge of the mat could impose severe boundary stresses on the bases of the silos. These applied stresses could cause cracking of the silos.

From the results discussed in this Paper, it appears that the stress distributions obtained from F.E.M. can be used in conjunction with consolidation theory to provide reliable settlement predictions. Boussinesq stress distributions used in conjunction with consolidation theory would appear to provide excessive differential settlement predictions and unconservative total settlements for rigid mats on finite subgrade thicknesses.

Predicted settlements computed by Steinbrenner elastic analysis are comparable to observed values. The elastic analysis provided a relatively simple and straightforward means of predicting settlement for a stratum of limited thickness overlying a rigid base. Further, the modulus of elasticity obtained from conventional oedometer test results can be used effectively in elastic analysis for prediction of primary consolidation.

Settlement predictions determined using standard penetration test data resulted in considerably lower settlement than observed. It would thus not appear reasonable to use Standard Penetration Test results for prediction of settlement of a structure on a silt subgrade.

REFERENCES

Hardin, B. O., and Drnevich, V. P. (1970). 'Shear Modulus and Damping Soils; 2, Design Equations and Curves', *Tech. Report* 27–70–CE–3: *Soil Mech. Series*, No. 2. Univ. of Kentucky, College of Engineering

Steinbrenner, W. (1934). 'Tafeln Zur Setzungsberechnung', *Die Strasse*. Vol. 1, 121–124. (See also *Proc. 1st Int. Conf. Soil Mech. and Found. Eng.*, Cambridge, Mass., 1936. Vol. 2, 142–143

Terzaghi, K. (1943). *Theoretical Soil Mechanics*. Wiley, New York

V/5. The failure of two oil-storage tanks caused by differential settlement

P. A. Green and D. W. Hight
Scott Wilson Kirkpatrick & Partners, London

Introduction

This Paper describes an investigation into the failure of two oil-storage tanks carried out in 1967 and 1968 by the senior author and colleagues. In addition to describing the site, the failures and the design of the tanks, the Paper also discusses the likely magnitude of differential settlements that caused the failures and the design of oil-storage tanks in relation to their ability to tolerate such settlements.

The site

The site of the tank farm at which the failures occurred is on the north bank of the Thames Estuary about 30 km east of Central London. Originally the site consisted of tidal marshes at a general level within 1 m of Newlyn Datum. Reclamation and drainage works and a flood bank adjacent to the Thames enabled cultivation of the area within the last century. In 1965 about 3 m of silty sand was pumped onto the site to form a more-or-less level area for the construction of the tank farm. In the first phase of development of the farm 33 tanks were built consisting of 10 tanks 7.6 m diameter × 9.1 m high, 10 tanks 14.6 m diameter × 12.8 m high, 9 tanks 17.1 m diameter × 12.8 m high and 4 tanks 24.4 m diameter × 12.8 m high.

Several ground investigations have been carried out at the site (or nearby) since the early 1950s, but generally the detailed records of these investigations are not available. These investigations included one in 1963 for the tank farm development which consisted of 22 Dutch soundings, 8 drive-sample borings and 3 shell-and-auger borings. It is known that samples from these boreholes were subjected to cell tests of the type described by De Beer

(1950), as well as consolidation and permeability tests. The original design of the tank farm was based on the results of these tests.

To investigate the failures and to assist in the design of the remedial works 5 shell-and-auger borings were made in June 1967. One of these boreholes (No. 2) was near the locations of the tanks which collapsed and the log for this hole is given in *Fig. 1*, together with a summary of the laboratory test results. Continued observations of settlement of the site since 1967 have shown that the ground conditions probably vary rapidly from place to place. (It has

N	Undisturbed Sample No.	Depth Below Surface (m)	DESCRIPTION OF STRATA			Sample Number	Natural Water Content(%)	Liquid Limit	Plastic Limit	Cohesion (φ = 0) (kN/m²) *							
											Coefficient of Volume Compressibility m_v (m²/MN) †						
											Coefficient of Consolidation C_v (m²/year)						
											for pressure range in kN/m²						
												13,–27	27–54	54–107	107–214	214–322	322–42
4		1·83	Loose fine brown SAND and BALLAST	Recent Fill													
5		3·20	Loose grey silty SAND														
		3·81	Soft black SILT														
		5·03	Stiff brown mottled CLAY	Old Fill													
	4		Soft PEAT and decaying wood			4	390	370	190	45	m_v	3·2	3·4	3·2	1·9	1·1	0·6
		7·93									C_v	0·28	0·15	0·22	0·15	0·02	0·01
	7		Soft to firm grey-blue very silty peaty CLAY	Estuarine Deposits		7	58	73	27	12	m_v	—	1·6	1·7	0·8	0·6	0·4
											C_v	—	0·31	0·20	0·19	0·09	0·07
	10					10	62	92	32	41	m_v	—	0·8	1·1	0·9	0·5	0·5
			Firm dark brown organic silty CLAY								C_v	—	0·44	0·40	0·23	0·06	0·02
	12	12·50				12	80	125	30	41	m_v	—	—	0·4	0·4	0·5	0·3
		13·41	Firm grey green silty sandy CLAY								C_v	—	—	0·44	0·35	0·55	0·07
		14·82															
13			Medium loose fine silty sandy GRAVEL	Flood Plain Gravel													
17		18·30															
			CHALK	Chalk													

* From UU triaxial tests on 76mm x 25mm diameter specimens

† From oedometer tests on 19mm x 76mm diameter specimens

Fig. 1. Borehole log and summary of laboratory test results

been observed that the amounts of settlement for equal loading can differ by a factor of two or more over distances of less than 100 m). However, it is believed that Borehole No. 2 is reasonably representative of much of the site and, particularly, of the area in which failures occurred.

The undisturbed samples from Borehole No. 2 were taken with a piston sampler. All testing was carried out as far as practicable in accordance with BS 1377: 1967.

As can be seen from *Fig. 1* a number of strata occur at the site, of which the most significant and compressible are the estuarine deposits. These deposits are of post-glacial age and probably similar to those at Tilbury and Shellhaven described in detail by Skempton and Henkel (1953).

354

The failures

Following completion of the first tanks in February 1966, all the 12.8 m high tanks were subjected to full water testing, generally two at a time. This testing probably consisted of pumping water from one tank to the next taking not more than one week for the filling process and not more than one week for emptying. It is likely that no one tank was kept full for more than a few days.

Fig. 2. View of inside of ruptured 24.4 m diameter tank (note rupture occured at junction of sketch and annular plates)

Fig 3. 24.4 m diameter tank after failure

355

In August 1966, at the end of this sequence, the final two tanks one of 14.6 m diameter and one of 24.4 m diameter, were left full of water. Both tanks collapsed within a week or two of each other after they had been under test for about one month. Subsequent investigations of the failures indicated that rupture of the tank floors had taken place, as shown in *Fig. 2*; the resulting exodus of water produced a slight vacuum in the tanks causing an implosion-type failure of their shells and roofs. A view of the mode of failure for the 24.4 m diameter tank is shown in *Fig. 3*. It was not possible to determine conclusively if, at the time of the failure, the tanks' vacuum valves were operating correctly.

Inspection of the other 12.8 m high tanks at the site showed that their floors had all suffered some damage from water testing, some being severely damaged, probably just short of rupture.

Design of tanks

All the tanks at the site had fixed roofs and were designed, fabricated, erected, inspected and tested in compliance with BS 2654 Part 1 (1965) and Part 2

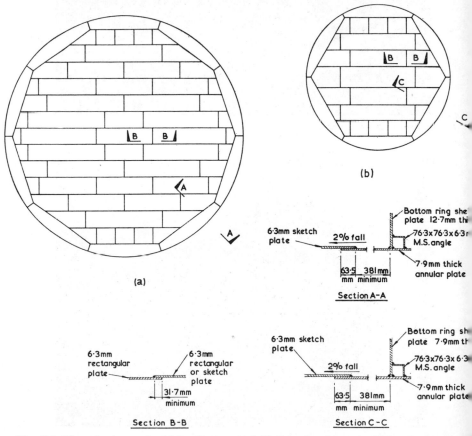

Fig. 4(a). Floor plate layout for 24.4 m diameter tank (b) floor plate layout for 14.6 m diameter tank

(1961) 'Vertical Mild Steel Welded Storage Tanks with Butt Welded Shells for the Petroleum Industry'. The bottom ring of shell plates of such tanks are fillet welded to horizontal annular plates which are in turn welded to sketch and rectangular floor plates. Details of the construction of the floors of the two tanks that collapsed are shown in *Fig. 4*. It should be noted that the design of these tanks is slightly unusual because of the 76.3 mm × 76.3 mm × 6.3 mm mild steel angle on the outside of the tanks to stiffen the joint between the shell and annular plates.

The floors of the tanks, which were constructed before the tank shells, were placed directly on shaped fill and consisted of rectangular plates laid from the centre outwards, each successive plate overlapping the ones previously laid. The floors initially sloped *down* from the periphery to the centre at a grade of 2%. The British Standard recommends that such floors should be sloped *up* at a minimum grade of 0.83% and that on poor soils this slope should be increased to compensate for extra settlement at the tank centre.

Settlements

At the time of the investigation into the failures, the two collapsed tanks had been dismantled and it was not possible to obtain direct measurements of the settlements at the time of failure. However, to assist in planning future development of the site a full-scale tank loading test was carried out at the end of 1967 using a 17.1 m diameter × 12.8 m high tank. The site of this test was

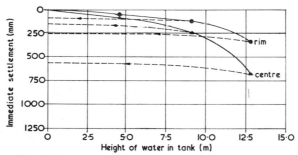

Fig. 5. Immediate settlement of 17·1 m diameter test tank

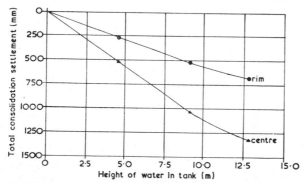

Fig. 6. Total consolidation settlement of a 17.1 m diameter tank calculated from the performance of the test tank

357

on an unloaded area in the general vicinity of the tanks that failed and, therefore, the direct observations of settlement are probably applicable in assessing the likely settlements that occurred under the failed tanks. The length of this Paper does not permit a detailed description of the full-scale test, but the following summarises the apposite results obtained:

(1) The immediate settlement[1] of the tank was large, see *Fig. 5*. Much of this settlement was irreversible; on first unloading from a head of 12.8 m it was estimated that the recoverable (elastic) settlement at the rim was about 85 mm. The non-recoverable settlement was 75% of the total, although as can be seen from *Fig. 5*, this percentage decreases with decreasing stress level.

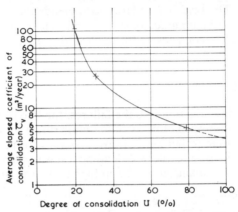

Fig. 7. Variation of average elapsed coefficient of consolidation with degree of consolidation

(2) The maximum pore pressures in the compressible strata were about 50% of the applied load.

(3) The ratio of settlement at the centre of the tank to that at its rim remained more-or-less constant at a value between 1.9 and 2.0 (This compares with the theoretical settlement ratio, using a Boussinesq stress distribution, of about 2.0 for the strata and tank geometry in question.)

(4) The amount of consolidation settlement was calculated to be as shown in *Fig. 6*.

(5) The rate of consolidation decreased markedly with increasing consolidation. An analysis of pore-pressure dissipation indicated that at 20% consolidation the average elapsed coefficient of consolidation[2] was about 100 m²/a reducing to an estimated 4 m²/a at 100% consolidation, as shown on *Fig. 7*. From these rates of consolidation it can

---

[1] It should be noted that during the test it was only possible to fill the tank at a rate of 1 m per day. This meant that some consolidation occurred during filling which made it difficult to separate accurately the immediate settlements from the early consolidation settlements.

[2] If the coefficient of consolidation changes as the degree of consolidation (U) increases, then the coefficient can be considered to have two values for the value of U in question, i.e. an instantaneous value and an 'average' value from $U = 0\%$ to the value of U in question. This latter value is the apposite one for calculating the time to reach any particular degree of consolidation and is referred to in this Paper as the 'average elapsed coefficient of consolidation $\bar{c}_v$'.

be calculated that 30% consolidation occurs in 26 days, 50% in six months and 95% in seven years.

Using the results of the full-scale test it is possible to estimate the immediate and consolidation settlements that occurred under the 14.6 m and 24.4 m diameter tanks which collapsed after being full of water for about one month. These estimated settlements are given in the following table.

Table 1. ESTIMATED SETTLEMENTS OF COLLAPSED TANKS

Tank diameter	Immediate settlement (mm)		30% Consolidation settlement (mm)		Immediate + 30% consolidation settlement as % of tank diameter		
(m)	Centre	Rim	Centre	Rim	Centre	Rim	Differential
14.6	570	300	380	200	6.5	3.4	3.1
24.4	1090	530	460	210	6.0	3.0	3.0

Discussion

An examination was made of the quality of the welds and steel plate used in the construction of the failed and intact tanks; this showed that no significant difference in quality existed between the tanks and that poor workmanship did not cause the failures. The essential difference between the failed and intact tanks was the larger total and differential settlements which occurred below the failed tanks after they had remained full of water for approximately one month. Both the mode of rupture of the tank floor illustrated in *Fig. 2* and the observed damage to the floors of some of the intact tanks strongly suggested that differential settlement was the cause of the failures.

Differential settlement across the initially horizontal or dished floor of a storage tank produces tensile stresses in the plates, and distorts the welds. It is possible to calculate these tensile stresses for a homogeneous tank floor subjected to differential settlement, for example see the method described by Hanna and Dawood (1965); however, the value of such calculations is questionable when, as occurs in practice, it is difficult (if not impossible) to determine the other local and general stresses in the floor caused by, inter alia:

(1) The hoop deformation of the shell under hydrostatic pressures.
(2) Friction between the floor and underlying soil.
(3) Residual stresses due to welding and other construction processes.
(4) The non-homogeneous construction, i.e. the rectangular and sketch plates lapped and fillet welded on orthogonal axes.

Clearly failure of the welds due to bending and shear distortion significantly limits the extent to which the tank floors can deform safely and it is interesting to note that for the 24.4 m diameter tank that failed there is evidence to indicate that failure occurred at the sketch plate/annular plate fillet weld (see *Fig. 2*). In this respect the design of tanks outlined in BS 2654 appears to be less than ideal and, in theory, it would be preferable to construct oil-storage tank floors using a series of petal-shaped plates joined by radial and tangential butt welds.

It can be seen that allowable differential settlements for the tanks and loading in question could not easily be determined since appropriate methods of analysis do not exist and permissible stress and weld distortion levels for this form of construction cannot be quantified. On the evidence of the estimated settlements given in Table 1, it can be concluded that the two tanks probably failed when the differential settlements were about 3% of their diameters. In addition, inspection of the other 12.8 m high tanks after water testing suggested that some permanent damage could occur if the diffe ential settlement exceeds 2% of the tank diameter. It should be stressed that these conclusions do not necessarily apply to tanks greater than 25 m diameter, or to tanks with floors which are initially domed, or to tanks not constructed in the manner described in BS 2654.

There was a significant difference between the rates of consolidation measured in the laboratory (see *Fig. 1*) and those determined from field observations (see *Fig. 7*), although both methods indicate that the rate decreases with increasing effective stress. Generally the rates measured in the laboratory are one to two magnitudes smaller than those observed in the field.

REFERENCES

De Beer, E. E. (1950). 'The Cell Test', *Geotechnique*. Vol. 2, 162–172
Hanna, M. M. and Dawood, R. H. (1965). 'Stresses in Circular Tank Floors and Walls Subject to Settlement on Uniform Soils', *Proc. I.C.E.*, Vol. 30, 167–170
Skempton, A. W. and Henkel, D. J. (1953). 'The Post-Glacial Clays of the Thames Estuary at Tilbury and Shellhaven, *Proc. 3rd Int. Conf. Soil Mech. & Found. Eng.*, Zurich, Vol. 1 302–308

V/6. Differential settlement tolerances of cylindrical steel tanks for bulk liquid storage

David A. Greenwood
Cementation Specialist Holdings Limited, London

Summary

Settlement records are tabulated for water tests on 48 cylindrical steel liquid storage tanks built on soft alluvial silts on 16 different sites. Limits of tolerable differential settlement are proposed: for tilting and flexure of the floor these are likely to be large and hence controlled respectively by aesthetic and operational considerations, but for shell differentials a limiting deformation from the mean tilted basal plane of 40 mm is proposed as an alternative to usual limits on angular distortion.

Introduction

Cylindrical unlined mild steel storage tanks for bulk liquids are large and simple structures. They are often built on soft estuarine or alluvial soils initially too weak to support them without large settlements which would be unacceptable for most other structures. Such tanks have operated satisfactorily after substantial differential settlements, but there is a divergence of opinion in the specification of limiting deformations. Ever larger tanks are being constructed (up to 160,000 m³) and whilst tight specifications are prudent when extrapolating from experience with smaller tanks, much more correlation of operating experience with drawing board forecasts is necessary.

Observations on settlement tolerances

It is important that the shell should not distort causing ovality at the top, so affecting the operation and sealing of floating roofs; nor that secondary shell stresses or tension in the floor plates should affect integrity of the structure or welds.

Tilting

It is clear that plane tilting of the shell, unless very exaggerated, is comparatively unimportant. Simple geometry shows that the extra 'wetted' height of the wall is approximately $D/2x$ and the extra 'wetted diameter' parallel to the maximum tilt is D/x^2 where D is the initial shell diameter and D/x the difference in level on the diameter of maximum tilt. Common specifications limit tilt to $x = 400$ and 300 respectively for the floating and the stiffer coned roof tanks. This seems too severe since even for 100 m diameter, the change in axis length of the liquid surface is less than 1 mm whereas seal tolerances are usually up to 200 mm and the 'as built' diameter at the top may vary by up to 150 mm. Extra water head at the lowest point due to tilt is typically of order 2 to 3%, which should be well covered by designed safety factors; bending of a floating roof when partially supported on the floor when the tank is nearly empty is unimportant as the roof, like the floor, is very flexible.

It seems likely therefore that much larger tilt could be tolerated, with appropriate limits being more likely aesthetic than technical; tilts of more than about 1 : 200 can be detected by eye. Examples of satisfactory performance with large tilt are the tanks at Canvey (Site J) in *Table 1* which sets out tank settlement records during water testing for structural integrity.

Peripheral distortions of shell

Whilst inherent shell stiffness tends to reduce peripheral differential settlements, these remain the major influences on secondary steel stresses and ovality. They are also the most difficult to control since tanks cover large areas within which the properties of surface soils may vary by two or three times over short distances or depths, giving rise to investigation and settlement prediction problems. Tilting associated with peripheral level differentials is critical since it tends to arise from shallow shear deformation of the soil.

Settlements are measured at levelling points near the base of the shell on an arc of length l usually 1/8 to 1/12 of its circumference (Column 12, *Table 1*) the differential s defined by $l/s = y$ (Column 11, *Table 1*) is the level difference between the relevant levelling point and the mean level of two equidistant points. De Beer (1969) considered shell differentials s by reference to an imaginary vertical circle through the chord of the arc $2l$, the differential s being at the mid point; an arc of the vertical circle running through the three levelling points considered (of which there were eight in total). He assumed a minimum radius for this circle R_c of 1500 m for tanks up to 20 m diameter, but found this criterion insufficiently stringent for larger tanks, e.g. it is equivalent to $y = 109$ for 70 m diameter. He therefore arbitrarily chose

$y = 450$ for larger tanks by reference to known behaviour of structural frames. This is equivalent to $R_c = 89D$.

A common British specification is $y = 250$ over arc l, approximately 10 m.

Table 1 suggests that the larger the arc l measured the lesser is the differential $1/y$ at which need for action is signalled.

A fixed criterion y suitable for large arcs may be unnecessarily severe for shorter ones; locally severe deformation does not affect ovality as much as general 'dishing' of the shell, and stress redistribution prevents damage even with moderately severe local deformations.

A better alternative would be an absolute limit on deformation t (Column 13, *Table 1*) of the peripheral base plate from the mean basal plane. From the tabulated results an appropriate criterion for any tank size appears to be $t = 40$ mm measured perpendicular to the mean plane of maximum tilt. However this may be too severe for tanks over 50 m diameter, which might well accommodate up to 60 mm deformation; further records for large tanks are needed to confirm this.

A 40 mm deformation corresponds to values of $y = 125$, 250 and 750 respectively for the cases of $l = 5$, 10 and 30 m; this is equivalent to $R_c = 12.5l^2$. Shell stiffness increasingly inhibits deformation with shorter arcs so the criterion in this form is unlikely to be invoked over lengths less than about 5 m.

It should be noted that new tanks prior to water test may have built in deviations up to $y = 500$ and $t = 40$ mm: data in *Table 1* take pre-test levels as zero datum. Test differentials tend to centre on built-in deviations which become exaggerated during test loading.

Differentials between perimeter and centre-floor

Tank floors are extremely flexible transversely and can tolerate large settlements relative to the shell (Column 14, *Table 1*). It is usual to drain stored liquid to the periphery and so the main problem is accurate settlement estimation to maintain a positive (upward) floor camber. Column 15 of *Table 1* shows pre- and post-water test calculated cambers, with z defined by the camber at the centre ($= D/2z$). The extra diametral plate length to provide camber is only about 10 mm, and in most cases complete loss of camber results in calculated compressive stresses of order 75 MN/m² only. It is probable therefore that the onset of the floor 'buckling' observed at the end of water testing is due to locally uneven deformation of soft surface soils: this probably begins before significant plate compression can occur. However, providing calculated negative camber of greater magnitude than initial positive camber does not cause excess tensions, tanks can function properly, e.g. *Table 1*, Site K–Old Harbour. Acceptance of 'dead' volume or addition of floor drainage at the lowest point may be necessary in such cases.

It is frequently noted (de Beer, 1969; Carlson and Fricano, 1961; Penman and Watson, 1967) that the greatest floor deformation occurs between $\frac{1}{4}$ and $\frac{1}{3}$ radius from the perimeter; this happens when shear deformation of soft soils is a major contributor to settlement. Carlson reported floor settlements near the perimeter about 1 m more than in the centre, whereas de Beer noted floor

363

Table 1. TANK SETTLEMENT RECORDS

Location (1)	Tank No. (2)	Tank data — Dia. (m) (3)	Ht. (m) (4)	Roof (5)	Mean depth of soft soils (m) (6)	Soil data (kN and m² units) (7)	Time for water test (days) (8)	Peripheral Settlement — Mean total (m) (9)	Max. tilt x (10)	Max. diff. y (11)	Dist. l (m) (12)	dist. t (mm) (13)	Centre settlement — Max. total (m) (14)	Diff. from tilted plane z (15)	Remarks (16)
Stanlow U.K.—lubricating oil plant	4901	44.0	20.1	C	5.0		77	0.090	400	708	17.2	24	NM	—	{ Initial test—tilted on local peat lens
	4902	24.6	18.3	C	9.6		77	0.230	140	1050	9.6	12	NM	—	Second test after tilt correction
	4903	24.6	16.5	C	9.0	$c_u = 20-35$	44	0.060	1333	2100	9.6	5	NM	—	
							42	0.120	726	1260	9.6	6	NM	—	
Shell (U.K.) Ltd.	4904	24.6	16.5	C	10.0	$m_v = 2-10 \times 10^{-4}$	41	0.100	445	630	9.6	11	NM	—	{ Initial test-shell relevelled for tilt
	4905	24.6	16.5	C	9.6		38	0.040	2660	2100	9.6	5	NM	—	Second test
	4906	14.7	20.1	C	7.0		82	0.090	800	1050	9.6	9	NM	—	
	4908	19.6	20.1	C	8.0		58	0.070	480	210	5.7	21	NM	—	
	4909	19.6	20.1	C	6.0		59	0.080	426	315	7.7	17	NM	—	
(A)	4910	19.6	20.1	C	8.5		59	0.100	640	840	7.7	3	NM	—	
							58	0.100	915	1000	7.7	5	NM	—	
Porto Marghera Italy—NE zone I.R.O.M.	726	39.0	12.5	C			92	0.199	950	850	15.3	35	0.319	+49/+70	
	729	39.0	12.5	C	5.5	$c_u = 10-20$	92	0.200	1390	1020	15.3	22	0.311	+53/+75	
	730	39.0	12.5	C		$m_v = 4-8 \times 10^{-4}$	103	0.256	1440	955	15.3	13	0.421	+51/+89	
(B)	731	39.0	12.5	C			103	0.256	355	364	15.3	5	0.431	+52/+99	
Porto Marghera Italy—Isola dei Petroli I.R.O.M.	158	70.0	14.8	F		0-6 m $c_u = 10-20$ $m_v = 8-10 \times 10^{-4}$	61	0.280	11680	1960	27.5	12	0.293	+70/+169	
	159	51.0	14.8	F		6-10 m $c_u = 110-125$ $m_v = 1 \times 10^{-4}$	72	0.369	450	240	13.4	62	0.402	+51/+260	
	160	70.0	14.8	F		10-30m $c_u = 30-50$ $m_v = 1 \times 10^{-4}$ ~30.0 m	143	0.369	1490	1020	18.4	18	NK	—	
	161	70.0	14.8	F			{ 123	0.355	666	334	18.4	55	0.453	+70/+87	{ Initial test—caused ovality
							67	0.071	999	613	18.4	50	0.229	+87/+144	{ Second test after shimming perimeter
	162	70.0	14.8	F			97	0.426	875	558	18.4	30	0.698	+70/+154	
(C)	163	70.0	14.8	F			103	0.499	824	1840	18.4	25	0.841	+70/+222	

Site	Tank														Remarks
L.D.F. tank. B.P. Chemicals Ltd. (D)	LDF	44.0	16.5	F	5.8	$m_v = 5-6 \times 10^{-4}$	41	0.128	2880	705	17.2	12	NM	—	
Teesport U.K.—crude oil. Shell (U.K.) Ltd. (E)	102	67.0	16.5	F	5.0		45	0.046	2440	736	9.6	18	NK	—	Soil replacement by slag to 1.6 m. No V/F
	104	67.0	16.5	F	6.0	$c_u = 9-19$	107	0.076	4400	960	9.6	10	0.127	+147/+188	
	106	67.0	16.5	F	6.4	$m_v = 4-0.6 \times 10^{-4}$	119	0.068	1570	960	9.6	8	0.079	+147/+154	
	109	55.0	16.5	F	5.5		115	0.144	460	565	9.6	37	NK	—	
Ravenna, Italy —Porto Corsini fuel oil tank E.N.E.L. (F)		70.0	14.0	F	26.0	$c_u = 19-40$ $m_v = NK$	483	0.660	585	688	27.5	60	1.300	+58/−875	Bottom plates cut and re-welded to bed down. Centre drainage added
Swansea U.K.—Shell-Mex BP for United Carbon Black (G)	11	24.4	18.5	C	15.2	$c_u = 20-40$	~30	0.171	666	250	7.7	31	NK	—	
	12	19.5	18.5	C		$m_v = 4-7 \times 10^{-4}$	~30	0.132	2130	420	7.7	15	NK	—	
Goa, India—phosphoric acid tanks Zuari-Agro Chemicals (H)	2	25.0	10.0	C	10.0	$c_u = 4-21$	30	0.140	1670	1780	19.6	11	0.320	+50/+1790	
	3	25.0	10.0	C	10.0	$m_v = 3-7 \times 10^{-4}$	20	0.090	835	980	19.6	20	0.240	+50/+1250	
Grangemouth U.K.—Xylene complex B.P. (I)	T121	24.4	14.7	C	42.5	$c_u = 9-34$ $m_v = 4-6 \times 10^{-4}$	322	0.514	320	216	4.8	25	NM	—	
Canvey Island U.K.—Oil tanks London & Coastal Oil Wharves (J)	Nth.	39.0	16.5	C	~27.0	$c_u = 5-13$	168	0.631	166	3140	15.3	8	NM	—	
	Sth.	39.0	16.5	C		$m_v = 6-9 \times 10^{-4}$	141	0.785	135	3350	15.3	6	NM	—	
Old Harbour, Jamaica–Fuel oil tanks (K)	1	18.3	15.3	C	>13.0	NK	~21	0.195	1930	1130	7.2	6	0.441	+120/−600	Dye penetrant tests showed sound welds after water test on both tanks
	2	18.3	15.3	C			~28	0.183	522	373	7.2	19	0.330	+120/−555	

(See footnotes at end of table)

Table 1. TANK SETTLEMENT RECORDS (cont.)

Location (1)	Tank data					Soil data (kN)(m²) (7)	Time for water test (days) (8)	Peripheral Settlement					Centre settlement		Remarks (16)
	Tank No. (2)	Dia. (m) (3)	Ht. (m) (4)	Roof (5)	Mean depth of soft soils (m) (6)			Mean total (m) (9)	Max. tilt x (10)	Max. diff. y (11)	Dist. l (m) (12)	Max. dist. t (mm) (13)	Max. total (m) (14)	Diff. from tilted plane z (15)	
Nottingham U.K. —Fuel depot	7595	24.4	12.8	F	3.0		~18	0.048	850	806	7.65	42	NM	—	0.75 full water load —perimeter re-levelled
	7596	24.4	14.6	C	3.0		74	0.083	700	1200	7.65	17	NM	—	
	7797	24.4	12.8	C	None	$c_u = 4-9$	42	0.041	2560	1600	7.65	11	NM	—	{ Foundation excavated to gravel
	7598	29.3	12.8	F	None	$m_v = $NK	~20	0.022	1535	805	7.65	10	NM	—	{ Foundation excavated to gravel
(L)	7599	24.4	16.5	F	3.0		44	0.041	513	805	7.65	14	NM	—	0.75 full water load —perimeter re-levelled
	7600	24.4	12.8	F	3.0		38	0.039	8400	1200	7.65	21	NM	—	
Avonmouth, U.K. —Fuel depot Shell Mex-B.P. (M)	18	24.4	16.5	C	} 19.4	$c_u = 14$	22	0.026	214	NK	—	—	1.120	+25/−32	Test abandoned at full load-bottom plate tension—pad rebuilt
	20	24.4	16.5	C		$m_v \simeq 6 \times 10^{-4}$	270	0.735	137	NK	—	—	1.320	+25/−20	Negative floor camber—pad rebuilt
Antwerp, Belgium— de Beer (1969)	A	53.7	15.6	C	10.0	$c_u = 25-100$	~198	0.390	383	2846	21.1	7	NK	—	Re-levelled to relieve shell stress/ovality
	B	76.2	16.5	F	10.0		~198	0.398	318	315	29.9	86	NK	—	
	C	51.2	14.7	F	10.0	$m_v = 2-7 \times 10^{-4}$	NK	0.937	160	242	14.6	76	NK	—	Re-levelled to relieve shell stress
(N)	D	44.0	14.6	NK	10.0		150	0.568	232	495	17.3	45	NK	—	Sand drains used in foundation
Teesmouth, U.K.— Penman & Watson (1967) (O)	2402F	13.7	14.7	None	4.0	$c_u = 7-12$ $m_v = 5-16 \times 10^{-4}$	2	0.338	300	NK	—	—	0.640	+45/−46	
U.S.A. Avon Refinery (Darragh 1964) (P)	650	42.6	14.7	F	15.0	$c_u = 10-14$	~147	0.142	NK	<275	33.5	NK	NK	—	Perimeter relevelled to reduce ovality

Symbols: F. floating roof; C = coned roof; NK = not known; MN = not measured.

Notes: (1) All tanks are of mild steel unlined during water tests and all founded on asphalt topped rolled granular pads without concrete ring beams.

(2) All foundation soils were normally consolidated alluvial, estuarine or littoral silts; those tabulated above the double line were strengthened by granular columns formed by

deformations of 1 : 10 over 3.1 m horizontal length for which the equivalent 'calculated' camber was 1 : 138; both tanks were undamaged. Thus subject only to wide bending tolerance, floor settlement controls depend mostly on operating requirements.

Settlements in service

Finally, additional differential settlements in service must not run much beyond water test tolerances. They depend on operating loads and the degree of soil consolidation achieved during water test (usually high). Recovery of total settlement on completion of water test is normally about 2 to 5%. Typically for oil storage (specific gravity = 0.8) total settlements and tilts may be 20 to 25% more after three to five years, whilst associated differentials either incrase only marginally or may even decrease relative to those recorded during water tests. Periodic checks in service are rarely made but generally unlined tanks which meet the suggested water test tolerances also behave satisfactorily in service.

Acknowledgements

Thanks are due to the site owners named in *Table 1* for permission to publish the data contained therein and also to Cementation Ground Engineering Ltd. for free use of their files.

REFERENCES

De Beer, E. (1969). 'Foundation Problems of Petroleum Tanks', *Annales de l'Institut Belge du Petrole*, Vol. 6, 25–40

Carlson, E. D. and Fricano, S. P. (1961). 'Tank Foundations in Eastern Venezuela', *Proc. A.S.C.E.*, Vol. 87, No. SM5, Oct., 69–90

Darragh, R. D. (1964). 'Controlled Water Tests to Pre-Load Tank Foundations', *Proc. A.S.C.E.*, Vol. 90, No. SM5, Sept., 303–329

Penman, A. D. M. and Watson, G. H. (1967). 'Foundations for Storage Tanks on Reclaimed Land at Teesmouth', *Proc. I.C.E.*, 19–42 May

V/7. An assessment of the effects of inter-action between a structure and its foundation

G. J. W. King
Senior Lecturer

V. S. Chandrasekaran
Research Student, Department of Civil Engineering, University of Liverpool

Summary

A method of analysis for the solution of soil–structure interaction problems under plane strain conditions in which the superstructure, foundation and soil medium are treated as parts of a continuous system is presented using finite element modelling. In the method described the superstructure and the foundation are represented by bending elements, the interface between the structure and the soil by friction elements and the soil by rectangular continuum elements.

The procedure is used to study the effects of interaction between superstructure and foundation by analysing a two bay continuous frame supported on a raft resting on an elastic stratum. It is shown that when the raft is relatively flexible the effects of superstructure–raft interaction are significant, but for this type of problem the influence of the roughness of the raft is small.

The immediate and long term response of the same structure founded on a clay stratum possessing inhomogeneity and transverse isotropy is then considered and it is shown that both immediate and long term settlements are affected by transverse isotropy, the effect on immediate settlement being more significant. It is also shown that differential settlements and hence the bending moments in the raft and superstructure are not appreciably affected by transverse isotropy.

Introduction

It is common practice to analyse a structure and calculate loads transferred to the foundation assuming complete restraint at the junctions between superstructure and foundation. Subsequently the behaviour of the foundation is determined under these loads. The resulting designs are considered adequate if the estimated differential settlements of the foundation can be tolerated by the structure.

The procedure outlined above does not allow for the effects of interaction between the superstructure and the foundation. An early approach to account for interaction was made by Meyerhof (1947). He used slope-deflection equations together with empirical bearing pressure-settlement curves to analyse simple two dimensional frames founded on footings. In a later work (1953) he suggested an approximate procedure in which the superstructure was replaced by a beam of equivalent stiffness. Chamecki (1965) proposed an iterative method, using load transfer coefficients for analysing structures on isolated footings. Both Meyerhof's and Chamecki's approaches do not satisfactorily account for moments at the column–foundation junctions. Larnach (1970) and Larnach and Wood (1972) computerised Chamecki's approach and used the familiar expression

$$s = \sum um_v \Delta z$$

for calculating the settlements of footings.

Grasshof (1957) examined the influence of perfectly rigid and perfectly flexible superstructures on the behaviour of combined footings resting on an homogeneous, isotropic, elastic half space. Sommer (1965) extended this approach to account for the flexural rigidity of the superstructure by replacing it with a beam of equivalent stiffness. Haddadin (1971) used the finite element method to analyse frames on foundation mats resting on a Winkler foundation. Lee and Brown (1972) considered both Winkler and elastic half space models for the soil medium in their analyses of two dimensional frames on combined footings. Most of the approaches mentioned above make use to some extent of a linear elastic, isotropic, homogeneous half-space or use the Winkler model to represent the soil medium.

With advances in numerical techniques such as the finite element method and the increase in speed and storage capacity of modern computers, it is now possible to perform a more sophisticated analysis to study problems involving soil–structure interaction. The advantage of the finite element method is that even the most complicated soil foundation properties and geometries can be considered. It is possible to model the soil medium with regard to inhomogeneity and transverse isotropy and also non-linear stress dependent properties can be considered by using an incremental approach.

In this paper an integrated approach to the soil–structure interaction problem is presented using finite element modelling. In the method adopted the superstructure, the foundation and the supporting soil medium are treated as parts of a single continuous system. The superstructure and the foundation are represented by bending elements, the interface between the foundation and the soil by friction elements and the soil by rectangular continuum elements. The procedure is only applicable to plane–strain

369

problems such as those involving strip rafts supporting long continuous frames, tunnels, buried conduits or sheet pile retaining walls.

Method of analysis

The procedure followed here is based on the displacement method. The superstructure and the raft are discretized by a number of bending elements. The bending element has stiffness relationship corresponding to that of a

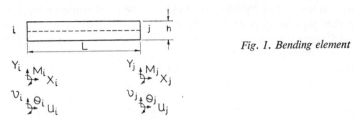

Fig. 1. Bending element

long rectangular plate bending into a cylindrical surface. The equation relating the nodal forces per unit width at the sides of the element to the corresponding displacements (*Fig. 1*) is

$$
\begin{Bmatrix} X_i \\ Y_i \\ M_i \\ X_j \\ Y_j \\ M_j \end{Bmatrix} =
\begin{bmatrix}
\dfrac{A}{L} & 0 & 0 & \dfrac{-A}{L} & 0 & 0 \\[2mm]
 & \dfrac{12D}{L^3} & \dfrac{-6D}{L^2} & 0 & \dfrac{-12D}{L^3} & \dfrac{-6D}{L^2} \\[2mm]
 & & \dfrac{4D}{L} & 0 & \dfrac{6D}{L^2} & \dfrac{2D}{L} \\[2mm]
 & & & \dfrac{A}{L} & 0 & 0 \\[2mm]
 & \text{Symmetric} & & & \dfrac{12D}{L^3} & \dfrac{6D}{L^2} \\[2mm]
 & & & & & \dfrac{4D}{L}
\end{bmatrix}
\begin{Bmatrix} u_i \\ v_i \\ \theta_i \\ u_j \\ v_j \\ \theta_j \end{Bmatrix}
\tag{1}
$$

in which

$$D = \frac{Eh^3}{12(1 - v^2)}$$

$A = Eh/(1 - v^2)$
E = Young's modulus of the material of the element
v = Poisson's ratio of the material of the element
h = thickness of the element, and
L = length of the element.

In order to account for the effects of frictional forces between the raft and the soil surface, friction elements like those developed by Goodman et al. (1968) are introduced at the contact interface to transmit normal and shear stresses.

In formulating the stiffness relationship for these elements it is assumed that both normal and tangential displacements vary linearly along the length of the element. The stiffness of the friction element is dependent on the interface properties k_n and k_s. These are stiffness coefficients which relate the normal and shear stresses with the relative normal and tangential displacements so that

$$\sigma_n = k_n \Delta n$$

and

$$\tau_s = k_s \Delta s \qquad (2)$$

in which Δn = relative normal displacement, Δs = relative tangential displacement, σ_n = normal stress and τ_s = shear stress.

In the present analysis, because the nodal points located on the neutral

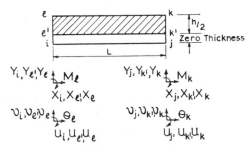

Fig. 2. Modified friction element

axis of the raft do not coincide with the nodal points on the ground surface, the stiffness relationship for the friction element was reformulated. The modified friction element is shown in *Fig. 2*. The relationship governing the nodal forces and displacements with respect to nodes i, j, k', l', is given by

$$
\begin{Bmatrix} X_i \\ Y_i \\ X_j \\ Y_j \\ X_{k'} \\ Y_{k'} \\ X_{l'} \\ Y_{l'} \end{Bmatrix} = \frac{L}{6}
\begin{bmatrix}
2k_s & 0 & k_s & 0 & -k_s & 0 & -2k_s & 0 \\
 & 2k_n & 0 & k_n & 0 & -k_n & 0 & -2k_n \\
 & & 2k_s & 0 & -2k_s & 0 & -k_s & 0 \\
 & & & 2k_n & 0 & -2k_n & 0 & -k_n \\
 & \text{Symmetric} & & & 2k_s & 0 & k_s & 0 \\
 & & & & & 2k_n & 0 & k_n \\
 & & & & & & 2k_s & 0 \\
 & & & & & & & 2k_n
\end{bmatrix}
\begin{Bmatrix} u_i \\ v_i \\ u_j \\ v_j \\ u_{k'} \\ v_{k'} \\ u_{l'} \\ v_{e'} \end{Bmatrix} \qquad (3)
$$

Now we have

$$
\left. \begin{aligned}
u_{k'} &= u_k - \theta_k \frac{h}{2} \\
u_{l'} &= u_l - \theta_l \frac{h}{2}
\end{aligned} \right\} \qquad (4)
$$

The statically equivalent force system for the *ijkl* element is

$$[X_i Y_i X_j Y_j X_k Y_k M_k X_l Y_l M_l]^T$$

$$= \left[X_i Y_i X_j Y_j X_{k'} Y_{k'} -X_{k'} \frac{h}{2} X_{l'} Y_{l'} -X_{l'} \frac{h}{2} \right]^T \qquad (5)$$

371

Using Equations 3, 4 and 5, the stiffness matrix of the modified friction element with nodes i, j, k, l becomes

$$
\begin{Bmatrix} X_i \\ Y_i \\ X_j \\ Y_j \\ X_k \\ Y_k \\ M_k \\ X_l \\ Y_l \\ M_l \end{Bmatrix}
= \frac{L}{6}
\begin{bmatrix}
2k_s & 0 & k_s & 0 & -k_s & 0 & k_s\frac{h}{2} & -2k_s & 0 & k_s h \\
 & 2k_n & 0 & k_n & 0 & -k_n & 0 & 0 & -2k_n & 0 \\
 & & 2k_s & 0 & -2k_s & 0 & k_s h & -k_s & 0 & k_s\frac{h}{2} \\
 & & & 2k_n & 0 & -2k_n & 0 & 0 & -k_n & 0 \\
 & & & & 2k_s & 0 & -k_s h & k_s & 0 & -k_s\frac{h}{2} \\
\text{Symmetric} & & & & & 2k_n & 0 & 0 & k_n & 0 \\
 & & & & & & k_s\frac{h^2}{2} & -k_s\frac{h}{2} & 0 & k_s\frac{h^2}{4} \\
 & & & & & & & 2k_s & 0 & -k_s h \\
 & & & & & & & & 2k_n & 0 \\
 & & & & & & & & & k_s\frac{h^2}{2}
\end{bmatrix}
\begin{Bmatrix} u_i \\ v_i \\ u_j \\ v_j \\ u_k \\ v_k \\ \theta_k \\ u_l \\ v_l \\ \theta_l \end{Bmatrix}
\quad (6)
$$

Here M_k and M_l are the moments at the neutral axis of the raft produced by the shear stresses on its base.

In the analysis described k_n has been made equal to the arbitrary large value 10^9 lb/ft^3 (1.57×10^8 kN/m^3); k_s has been made equal to a small value 10 lb/ft^2 (1.57 kN/m^3) to model a smooth interface and a large value 10^9 lb/ft^3 (1.57×10^8 kN/m^3) to model a perfectly rough interface. Experimentally determined values of k_s could be employed to model real interface behaviour.

The soil medium is divided into a number of rectangular continuum elements. These elements have quadratic internal displacements with linear variation of displacement along their edges. The derivation of the stiffness matrix for this element for a transversely isotropic material can be carried out as indicated by Zienkiewicz (1967).

The overall stiffness matrix $[K]$ of the entire assembly is formed by systematically adding the stiffness matrices of individual elements after transformation from local to global co-ordinates if necessary. The system of equations representing the equilibrium of the nodes can be written simply as

$$[K]\{\delta\} = \{F\} \quad (7)$$

in which $[K]$ = overall stiffness matrix, $\{\delta\}$ nodal point displacement vector and $\{F\}$ = nodal force vector.

These equations are modified to include prescribed boundary displacements and then solved for nodal displacements. In the solution procedure which uses Gaussian elimination for band symmetric matrices a forward pass is first given to triangularise the system and the displacements are evaluated during back substitution. The overall stiffness matrix is stored peripherally in the form of a rectangular array by retaining terms only on and above the leading diagonal and only the part of it corresponding to size 2U.B.W $\times$ U.B.W., where U.B.W. is the upper bandwidth of the overall stiffness matrix is brought into the active core during solution time enabling a large number of equations to be solved.

The bending moments, axial force and transverse shear in the elements of the superstructure and the raft, the normal stress and shear stress in the friction elements and the stresses in the continuum elements are then calculated from the respective element properties and the evaluated displacements.

Assessment of accuracy

The accuracy of the finite element procedure can be examined by comparing solutions for a strip raft with a smooth base resting on a homogeneous elastic solid and loaded with a central line load. The contact pressure distribution depends on the relative rigidity factor k defined as

$$k = \frac{1}{6} \frac{E}{E_s} \frac{(1-v_s^2)}{(1-v^2)} \frac{h^3}{a^3} \qquad (8)$$

in which $E =$ Young's modulus of the material of the raft, $E_s =$ Young's modulus of the underlying medium, $v =$ Poisson's ratio of the material of the raft, $v_s =$ Poisson's ratio of the underlying medium, $h =$ thickness of the raft and $a =$ half the width of the raft. The theoretical contact pressure distribution obtained for an infinitely rigid strip (Boussinesq's solution is compared with the finite element solution corresponding to a highly rigid strip ($k = 100$) in *Fig. 3*. This figure also shows a comparison of the contact

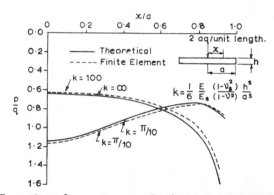

Fig. 3. Comparison of contact pressure distribution for rigid and flexible strip

pressure distribution for the relatively flexible strip ($k = \pi/10$) obtained by Borowicka (1939) with the corresponding finite element solution. The agreement is good for both rigid and relatively flexible rafts. Similar comparisons for a circular raft have been carried out by Smith (1970).

Comparative analyses of two bay frame on elastic medium

Consider the two bay framed structure shown in *Fig. 4*. The structure is continuous in the longitudinal direction and unit length of it is considered. The members of the superstructure have values of $D = 152 \times 10^5$ lb ft (206×10^2 kN m) and $A = 324 \times 10^6$ lb/ft (473×10^4 kN/m), D and A being defined in Equation 1. It is acted upon by uniformly distributed loads of intensity $q/2$ on the top member and $q/2$ on the raft. Since it is symmetrical only one half of the frame is analysed. The portion of the raft below each bay is divided into ten bending elements and there are ten friction elements attached to them which transmit the normal and shear stresses to the soil

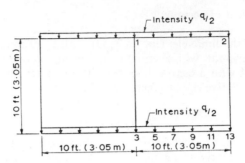

Fig. 4. Two-bay frame

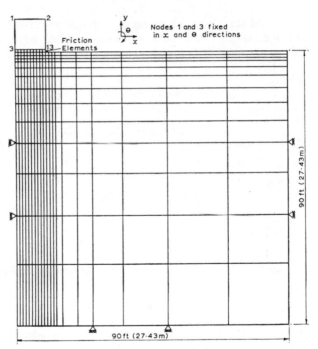

Fig. 5. Finite element idealisation

stratum. The side boundaries of the soil stratum are located sufficiently far away from the raft (4.5 times the breadth of the raft) so that their influence is small. The finite element idealization is shown in *Fig. 5*. The nodes on the side and the centre line are restrained in the lateral direction and are free to move in the vertical direction. The nodes on the bottom boundary are restrained in the vertical direction and are free to move in the lateral direction.

Two methods of analysis were performed. In the first, referred to as inde-

374

pendent analysis, the superstructure and the raft were considered separately and in the second referred to as interactive analysis, the superstructure, raft and soil medium were treated as parts of one continuous system.

The soil stratum was assumed to have constant Young's modulus of 50 000 lb/ft² (2394 kN/m²) and a Poisson's ratio of 0.48. The relative rigidity of the raft was varied by using different raft thicknesses and its base was assumed to be perfectly smooth.

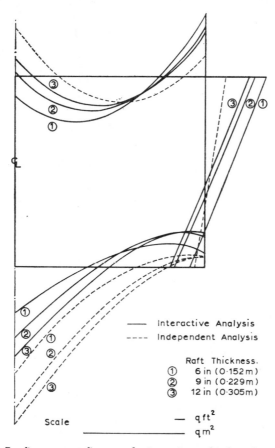

Fig. 6. Bending moment diagrams for interative and independent analyses

The bending moment diagrams from independent and interactive analysis for raft thicknesses 6, 9 and 12 in (0.152, 0.229 and 0.305 m) are shown in *Fig. 6* and for raft thicknesses of 24 and 36 in (0.61 and 0.915 m) in *Fig. 7*. The contact pressure distributions for raft thicknesses of 6, 9 and 24 in (0.152, 0.229 and 0.305 m) obtained from independent analysis are shown in *Fig. 8* and those from interactive analysis in *Fig. 9*.

It is seen that for a relatively flexible raft the differences in the bending moment diagrams obtained from independent and interactive analyses are significant. For a value of q equal to 1000 lb/ft² (47.88 kN/m²) the raft thickness would usually be of the order of 9 in to 12 in (0.229 m to 0.305 m)

375

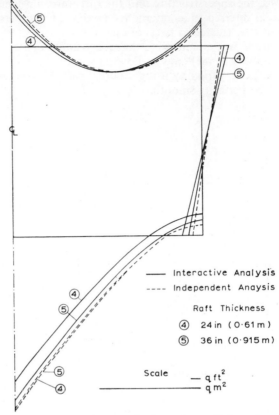

Fig. 7. Bending moment diagrams for interative and independent analyses

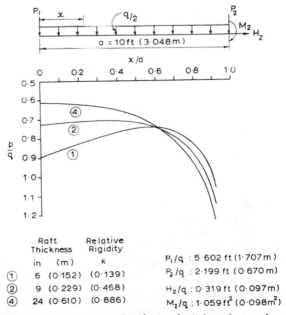

	Raft Thickness		Relative Rigidity
	in	(m)	k
①	6	(0·152)	(0·139)
②	9	(0·229)	(0·468)
④	24	(0·610)	(0·886)

P_1/q : 5·602 ft (1·707 m)

P_2/q : 2·199 ft (0·670 m)

H_2/q : 0·319 ft (0·097 m)

M_2/q : 1·059 ft^2 (0·098 m^2)

Fig. 8. Contact pressure distribution for independent analyses

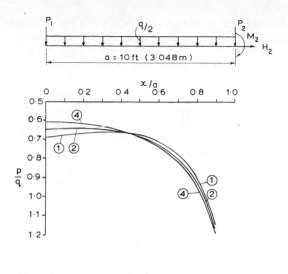

Fig. 9. Contact pressure distribution for interactive analyses

	Raft Thickness in (m)	P_1/q ft (m)	P_2/q ft (m)	H_2/q ft (m)	M_2/q ft^2 (m^2)
①	6 (0·152)	3·270 (0·997)	3·365 (1·026)	0·8 (0·224)	1·458 (0·135)
②	9 (0·468)	4·008 (1·222)	2·996 (0·913)	0·857 (0·261)	3·192 (0·297)
④	24 (0·810)	5·232 (1·595)	2·384 (0·727)	0·516 (0·157)	2·199 (0·204)

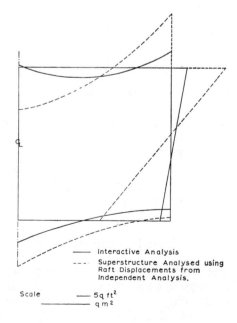

Fig. 10. Comparison of bending moment diagrams for raft thickness 9 in (0.229 m)

Interactive Analysis

Superstructure Analysed using Raft Displacements from Independent Analysis.

Scale — 5q ft^2
        — q m^2

377

and for this thickness it is important to consider the influence of the super-structure in the analysis. The difference in the contact pressure distributions obtained from interactive analysis for a relatively flexible and relatively rigid raft is small. This is due to the superstructure contributing additional stiffness to the raft.

It is of interest to consider the effect of the settlements obtained from inde-pendent analysis on the bending moments in the superstructure. For this purpose one analysis was carried out in which the wall base displacements obtained from independent analysis of the raft were applied to the super-structure. The resulting bending moment diagram is compared with that obtained from interactive analysis for a raft thickness of 9 in (0.229 m) in *Fig. 10*. It is evident that this procedure does not give realistic bending moments in the superstructure and would lead to gross overdesign.

Effects of base friction

To assess the effects of the roughness of the base of the raft on the response of the structure, some of the previous analyses were repeated assuming the raft base to be perfectly rough. The distribution of horizontal contact pressure the raft of thickness 24 in (0.610 m) is shown in *Fig. 11*. The bending moments,

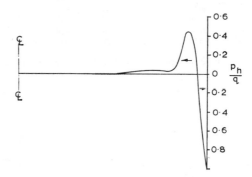

Fig. 11. Horizontal contact pressure distribution for raft thickness 24 in (0·610 m)

M, at different points in the structure for raft thicknesses of 9 in (0.229 m) and 24 in (0.610 m) are compared for perfectly smooth and perfectly rough base in *Table 1*. There is no appreciable difference in the bending moments in the superstructure and the raft in this example because of the small magnitude of the net horizontal force. The effect of base roughness on vertical contact pressure distribution and displacements was also insignificant.

The influence of transverse-isotropy of foundation medium

The most common type of anisotropy exhibited by soil media is transverse isotropy in which the vertical axis is an axis of symmetry. The stress–strain relationship of a transversely isotropic material is governed by five elastic constants, E_1, ν_1, E_2, ν_2 and G_2 where E_1 = modulus of elasticity in the

378

Table 1. VALUES OF M/q FOR SMOOTH AND ROUGH BASES

	Raft thickness			
Position Mij Mji $\overset{\frown}{i}$ $\overset{\frown}{j}$	9 in (0.229 m)		24 in (0.610 m)	
	Smooth (ft²)	Rough (ft²)	Smooth (ft²)	Rough (ft²)
1–2	−0.420	−0.430	−4.122	−4.131
2–1	5.377	5.375	2.963	2.957
13–2	−3.192	−3.209	−2.199	−2.192
3–4	7.478	7.470	15.311	15.231
4–5	5.546	5.538	12.748	12.668
5–6	3.759	3.751	10.296	10.215
6–7	2.116	2.108	7.960	7.881
7–8	0.620	0.615	5.755	5.678
8–9	−0.714	−0.718	3.703	3.631
9–10	−1.864	−1.866	1.836	1.772
10–11	−2.790	−2.785	0.204	0.165
11–12	−3.421	−3.421	−1.117	−1.151
12–13	−3.638	−3.507	−2.001	−1.627
13–12	3.192	3.076	2.199	1.841

Multiplication factor for M/q (m²) is 0.0929.

plane of isotropy, v_1 = Poisson's ratio in the plane of isotropy, E_2 = modulus of elasticity in planes perpendicular to the plane of isotropy, v_2 = Poisson's ratio representing the strain in the plane of isotropy due to unit strain normal to it, and G_2 = shear modulus in planes perpendicular to the plane of isotropy.

If the soil skeleton exhibits behaviour corresponding to that of a linear elastic transversely isotropic body then it can be shown that under conditions of full saturation and no drainage the stress–strain behaviour in terms of total stress is also transverely isotropic. Equations relating the total and effective stress elastic parameters are given by Uriel and Canizo (1971). denoting the effective stress elastic parameters with primes these relationships are

$$
\left.
\begin{aligned}
n &= \frac{2(n' - n' v'_1 - 2n'^2 v'^2_2)}{(2 + n' - 4n' v'_2 - 2v'_1) - (1 - v'_1 - n' v'_2)^2} \\
v_2 &= 1/2 \\
v_1 &= 1 - n/2 \\
G_2 &= G'_2 \\
E_1 &= \frac{(4 - n)E'_1}{2(1 + v'_1)}
\end{aligned}
\right\} \tag{9}
$$

with $\quad n = E_1/E_2$ and $n' = E'_1/E'_2$

The two-bay frame with the same loading considered earlier was assumed to be founded on an inhomogeneous transversely isotropic medium similar to the clay beds at Ashford Common near London. The raft was assumed to have a smooth base and was 9 in (0.229 m) thick. The properties of clay beds at Ashford Common with regard to inhomogeneity and transverse isotropy have been established recently by Wroth (1971) and Henkel (1971). In this paper a similar clay stratum was assumed to extend up to the ground surface and to be fully saturated.

The value of $v_2' = 0.12$ was determined by Wroth (1971) and the values of $n' = 1.6$ and $v_1' = 0.156$ by Henkel (1971). These values were adopted

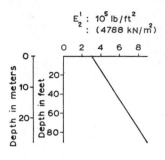

	Isotropic	$G_2 = \dfrac{E_2'}{2(1+v_2')}$	$G_2 = \dfrac{E_1'}{2(1+v_1')}$
n'	1	1·6	1·6
v_1'	0·12	0·156	0·156
v_2'	0·12	0·12	0·12
G_2/E_2	0·446	0·446	0·692
n_1	1	1·219	1·219
v_1	0·48	0·391	0·391
v_2	0·48	0·48	0·48
G_2/E_2	0·333	0·283	0·438
E_2	1·339E_2'	1·579E_2'	1·579E_2'

In the chart at left of the table: $E_2' : 10^5$ lb/ft^2 : (4788 kN/m^2), depth in meters and depth in feet axes.

Fig. 12. Elastic parameters used in the analyses

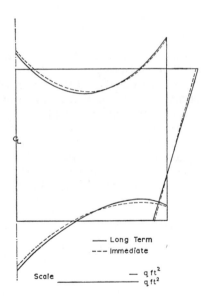

Fig. 13. Bending moment diagrams for immediate and long term behaviour $(G_2 = E_2'/2(1 + v_2'))$

—— Long Term
- - - Immediate

Scale —— $q\,\mathrm{ft}^2$ / $q\,\mathrm{ft}^2$

and assumed to be constant with depth. The assumed variation of E_2' with depth is shown in *Fig. 11*. This straight line variation is the same as given by Wroth (1971) at depths greater than 30 ft (9.15 m) and has been extended up to the ground surface. The value of the independent shear modulus G_2 remains to be prescribed. Although Barden (1963) and Raymond (1970) placed restrictions on the value of G_2, it has been shown by Hearmon (1961), Pickering (1970) and King (1973) that there are no restrictions on this value except that it be positive. Two arbitrary values of G_2 have been adopted here corresponding to the values $E_2'/2(1 + v_2')$ and $E_1'/2(1 + v_1')$ respectively.

380

Table 2. FOUNDATION DISPLACEMENTS

Foundation Displacements $q = 1000$ lb/ft² $= 47.9$ kN/m²		Immediate			Long-term		
		Case (i)	Case (ii)	Case (iii)	Case (i)	(Case (ii))	Case (iii)
Vertical settlement of node 3	ft	0.0316	0.0256	0.0218	0.0719	0.0688	0.0644
	(mm)	(9.6)	(7.8)	(6.6)	(21.9)	(20.9)	(19.6)
Vertical settlement of node 13	ft	0.0276	0.0216	0.0181	0.0672	0.0642	0.0597
	(mm)	(8.4)	(6.6)	(5.5)	(20.5)	(19.6)	(18.2)
Differential settlement between nodes 3 and 13	ft	0.0040	0.0040	0.0037	0.0047	0.0046	0.0047
	(mm)	(1.2)	(1.2)	(1.1)	(1.4)	(1.3)	(1.4)
Rotation at node 13 (rad)		0.3806×10^{-4}	0.3479×10^{-4}	0.4182×10^{-4}	0.4228×10^{-4}	0.4561×10^{-4}	0.3717×10^{-4}
Horizontal displacement at node 13	ft	0.1750×10^{-4}	0.1729×10^{-4}	0.1659×10^{-4}	0.1853×10^{-4}	0.1835×10^{-4}	0.1847×10^{-4}
	(mm)	(53.34×10^{-4})	(52.70×10^{-4})	(50.56×10^{-4})	(56.48×10^{-4})	(55.93×10^{-4})	(56.30×10^{-4})

Case (i)   Soil Medium Isotropic
Case (ii)   Soil Medium Anisotropic $G_2 = E_2'/2(1 + v_2')$
Case (iii) Soil Medium Anisotropic $G_2 = E_1'/2(1 + v_1')$

Because the clay is overconsolidated, if the imposed stresses are small, elastic theory can be used with reasonable accuracy to calculate both immediate and long term settlements. Total stress parameters evaluated using Equations 8 were used for analysing the immediate behaviour of the structure and effective stress parameters for long term behaviour. Interactive analyses were performed using the parameters defined above and also parameters E_2' and v_2' alone and assuming isotropy. The numerical values of these parameters are given in *Fig. 12*.

The immediate and long term displacements of the centre and edge of the raft for a value of $q = 1000 \text{ lb/ft}^2$ (47.88 kN/m²) are given in *Table 2*. Both immediate and long term settlements were less when transverse isotropy was considered in the analysis and the effect on immediate settlement was the more significant. The differential settlement between the centre and the edge of the raft and the rotation at the edge of the raft were not appreciably affected by transverse isotropy. Further it may be seen that these quantities showed little change during consolidation. Because of this, there were only small changes in the bending moments in the superstructure and the raft during consolidation. The bending moment diagrams corresponding to immediate and long term settlements with

$$G_2 = E_2'/2\,(1 + v_2')$$

are shown in *Fig. 13*.

Conclusions

For a relatively flexible raft, as is often so in practice, the bending moments in the superstructure and the raft tend to be significantly different if the superstructure raft interaction is considered in the analyses. This difference will of course diminish as the soil medium becomes stiffer. For vertical loadings the roughness of the base has negligible effect on settlement, vertical contact pressure distribution and bending moments in the superstructure and the raft.

While both the immediate and long-term settlements were affected if transverse isotropy was considered, the effect on immediate settlement was the more significant. Both immediate and long-term differential settlements and hence the bending moments in the superstructure and the raft were not greatly affected by transverse isotropy. Also the bending moments in the superstructure and the raft changed only slightly during consolidation.

REFERENCES

Barden, L. (1963). 'Stresses and Displacements in a Cross-anisotropic Soil', *Géotechnique*, London. Vol. 13, No. 3, 198–210
Borowicka, H. (1939). 'Druckverteilung unter Elastichen Platten', *Ingenieur-Archiv*. Vol. 10, 113–125
Chamecki, S. (1956). 'Structural Rigidity in Calculating Settlements', *J. Soil Mech. & Found. Div.*, *A.S.C.E.*, Vol. 82, No. SM1, 1–9
Goodman, R. E., Taylor, R. L. and Brekke, T. L. (1968). 'A Model for the Mechanics of Jointed Rock', *J. Soil Mech. & Found. Div.*, *A.S.C.E.*, Vol. 94, No. SM3, 637–659
Grasshof, H. (1957). 'Influence of Flexural Rigidity of Superstructure on the Distribution of Contact Pressure and Bending Moments of an Elastic Combined Footing', *Proc. 4th Int. Conf. Soil Mech.* London, Vol. 1, 300–306

Haddadin, M. J. (1971). 'Mats and Combined Footings—Analysis by the Finite Element Method', *Proc. Am. Concrete Inst.* Vol. 68, No. 12, 945–949

Hearmon, R. F. S. (1961). *An Introduction to Applied Anisotropic Elasticity*. Oxford Univ. Press.

Henkel, D. J. (1971). 'The Relevance of Laboratory Measured Parameters in Field Studies', *Proc. Roscoe Memorial Symp.*, Cambridge Univ. 669–675

King, G. J. W. (1973). 'Discussion on Prediction of Undrained Deformations and Pore Pressures in Weak Clay under Two Embankments', *Géotechnique*, London. Vol. 23, No. 1, 133–135

Larnach, W. J. (1970). 'Computation of Settlements on Building Frames', *Civ. Eng. and Pub. Wks Rev.* Vol. 65, No. 770, 1040–1044

Larnach, W. J. and Wood, L.A. (1972). 'The Effect of Soil-Structure Interaction on Settlements', *Int. Symp. on Computer Aided Design*, Univ. of Warwick

Lee, I. K. and Brown, P. T. (1972). 'Structure Foundation Interaction Analyses', *J. Struct. Div., A.S.C.E.* Vol. 98, No. ST11, 2413–2431

Meyerhof, G. G. (1947). 'The Settlement Analysis of Building Frames', *The Structural Engineer*. Vol. 25, 369–409

Meyerhof, G. G. (1953). 'Some Recent Foundation Research and its Application to Design', *The Structural Engineer*. Vol. 31, No. 6, 151–167

Pickering, D. J. (1970). 'Anisotropic Elastic Parameters for Soil', *Géotechnique*, London. Vol. 20, No. 3, 271–276

Raymond, G. P. (1970). 'Discussion on Stresses and Displacements in a Cross-anisotropic Soil', *Géotechnique*, London. Vol. 20, No. 4, 133–135

Smith, I. M. (1970). 'A Finite Element Approach to Soil-Structure Interaction', *Canadian Geotech. J.* Vol. 7, No. 2, 95–105

Sommer, H. (1965). 'A Method for the Calculation of Settlements, Contact Pressures and Bending Moments in a Foundation including the Influence of the Flexural Rigidity of the Superstructure', *Proc. 6th Int. Conf. Soil Mech.*, Montreal. Vol. 2, 197–201

Uriel, A. O. and Canizo, L. (1971). 'On the Elastic Anisotropy of Soil, *Géotechnique*, London. Vol. 21, No. 3, 262–267

Wroth, C. P. (1971). 'Some Aspects of the Elastic Behaviour of Overconsolidated Clay', *Proc. Roscoe Memorial Symp.*, Cambridge Univ. 347–361

Zienkiewicz, O. C. (1967). *The Finite Element Method in Structural and Continuum Mechanics*. McGraw Hill, London

V/8. Observations of brick walls subjected to mining subsidence

G. S. Littlejohn
Geotechnics Laboratory, Department of Engineering, University of Aberdeen

Introduction

This technical note describes a field experiment carried out at Peterlee New Town, County Durham. Three brick walls on unreinforced concrete strip footings and one footing without a wall were built (*Table 1*), and the purpose

Table 1. STRUCTURAL DETAILS

Item	Structure No.			
	1	2	3	4
Foundation				
Length (m)	15.8	31.2	15.8	15.8
Width (mm)	610	610	610	610
Thickness (mm)	150	150	150	150
Brick wall				
Length (m)	15.2	30.4	—	15.2
Height (m)	1.22	1.83	—	1.22
Thickness (mm)	230	230	—	230

Materials—Foundation
  Unreinforced concrete = 1:2:4 O.P.C Mix
  28-day strength = 30 MN/m^2
  Density = 2.37 Mg/m^2

Materials—Brick wall
  Wire cut bricks without frogs (Class B)
  Mean strength (on flat) = 55.9 MN/m^2
  Mortar 9:1 sand:cement
  28-day strength = 1.38 MN/m^2

of the experiment was to investigate in detail the behaviour of these surface structures when subjected to mining subsidence.

A few of the individual observations are included to illustrate the form of monitoring employed on the structure and adjacent ground, and some results relating degree of damage to angular distortion and strain are presented.

Surface and underground development

The layout at the site and the coal extraction programme are shown in *Fig. 1*. The walls were constructed parallel to the direction of mining advance and situated centrally over the face so that the walls might be subjected to maximum dynamic ground strains and minimal twisting effects. The depth of the level working was 256 m, the main coal seam of 1.36 m being extracted

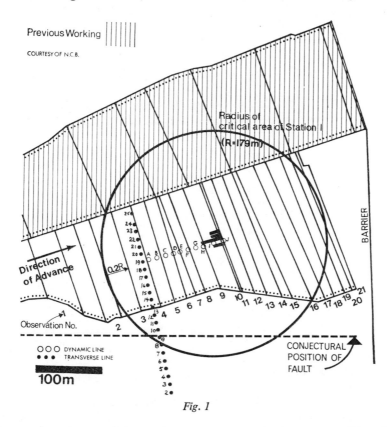

Fig. 1

under total caving conditions. The average rate of advance was 6.1 m/week and *Fig. 2* indicates the general profiles of surface subsidence which were monitored during the experiment. These results were obtained from a transverse and dynamic line of stations, spaced at 12.2 m intervals (*Fig. 1*).

Ground conditions

A conjectural section of strata in the area is shown in *Fig. 3*, and the main properties of the natural sandy loam subgrade beneath the footings tabled.

385

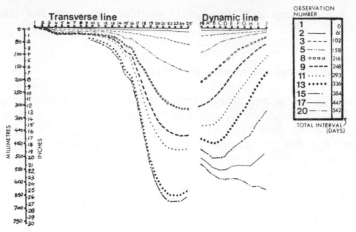

Fig. 2. Verticals subsidence, Oakerside—Sunny Blunts

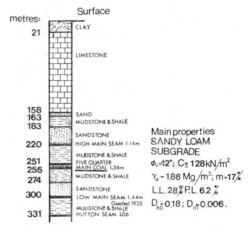

Fig. 3. Conjectural section of strata at Sunny Blunts—Oakerside

Field measurements

To study in detail the behaviour of the structures and the adjacent ground during mining, monitoring stations were inserted at 1.22 m intervals, along the footings, brick walls, and adjacent ground. Using these stations and special instruments (Littlejohn, 1973), horizontal strains, differential settlements and changes of slope were monitored which in turn allowed relative movements at the soil-foundation interface to be estimated.

Whilst the site was being affected by the coal extraction, measurements were taken at monthly intervals but only a few readings have been plotted in an attempt to illustrate the main trends more clearly. In addition, all structural and ground deformations are related to position of the advancing face and not time, since the development of mining subsidence is primarily a function of the area of seam worked, Wardell (1953).

386

In *Fig. 1*, the point 1 begins to subside as soon as the working enters its critical area, and it is commonly observed that about 95% of the ultimate subsidence has taken place, when the extraction moves out of the critical area, i.e. when the advance $>2r$.

Structure No. 1

Total settlement

The foundation tilted as the face advanced, although the structure concerned eventually regained a level position after a total settlement of 635 mm, when the face had passed (2.0r). The overall movements shown in *Fig. 2* are often the only results available for analysis.

Differential settlement

In all cases, differential settlements increased gradually and evenly, reaching maximum values of 52 mm at 1.3–1.4r (*Fig. 4*). Thereafter, the values gradually decreased although erratically in the case of the ground.

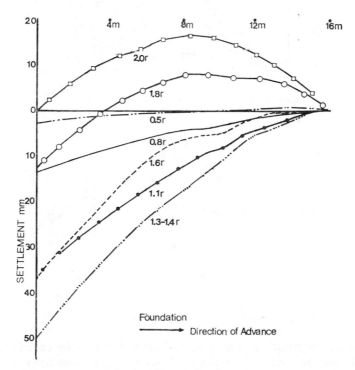

Fig. 4. Differential settlements

Deflection

Deflection diagrams (*Fig. 5*) show the above effect more clearly. Hogging increased gradually on all lines until 1.3*r*. A zone of compressive ground strain then entered the vicinity of the wall and surprisingly two peaks developed in the ground profile. As a result, hogging continued to increase and the hogging deflection of the structure was greater than the corresponding ground deflection.

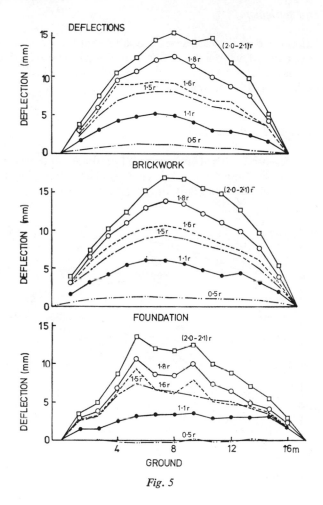

Fig. 5

Relative settlement

The greater magnitude of the structural deflection may be explained using *Fig. 6*. During initial ground movement (0.4–0.5*r*) the foundation penetrated the ground at the ends and separated under the central section of the structure. As the tensile ground strain increased, the entire foundation penetrated the subgrade, although the former pattern remained. Maximum penetration and tensile ground strain were recorded at the same time (1.1r–1.2r). The development of tensile ground strain, therefore, altered significantly the load–settlement properties of the subgrade and the distribution of the supporting forces. As anticipated, the development of compressive ground strain caused a reduction in penetration.

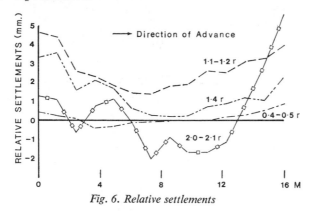

Fig. 6. Relative settlements

Analysis of average figures indicates that a tensile ground strain of -50×10^{-5} produced an end penetration of 4 mm whilst a subsequent compressive ground strain of $+50 \times 10^{-5}$ caused a reduction in penetration of 0.5 mm. The use of pressure cells in the foundation is recommended in future to indicate variation of vertical stress with change in relative settlement.

Horizontal strains

Tensile ground strains (E_g) developed as soon as ground movements entered the vicinity of the wall and maximum values of -200×10^{-5} were recorded at 1.1r (*Fig. 7*). During this time foundation strains (E_f) were only -12×10^{-5}, although brickwork strains (E_b) of -80×10^{-5} were registered at a height of 1.2 m. At 1.5r maximum compressive ground strains of $+380 \times 10^{-5}$ had developed but the structure remained in tension due to curvature ($E_f = -18 \times 10^{-5}$ and $E_b = -135 \times 10^{-5}$). As the compressive ground strains decreased, tensile strains in the structure increased steadily and maximum values of $E_f = -24 \times 10^{-5}$ and $E_b = -267 \times 10^{-5}$ were developed at 2.0r.

Longitudinal slip

Fig. 8 shows the manner in which ground strains induced horizontal forces in the structure. During the development of tensile ground strain, the ground slipped outwards from the centre of the foundation thus inducing tensile forces in the structure. Maximum slips were produced at the ends of the structure during the period of maximum tensile ground strain and values of $+5.6$ and -2.5 mm were recorded at distances 0.61 and 15.2 m respectively. During maximum compressive ground strain the equivalent slips in the

389

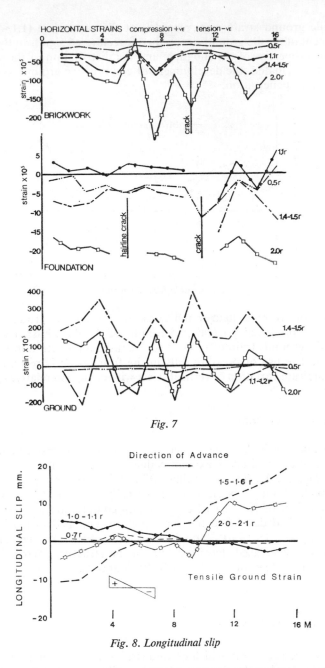

Fig. 7

Fig. 8. Longitudinal slip

opposite direction were -11.9 and 19.6 mm respectively. The maximum rate of relative displacement or slip was recorded at $1.4r$ equal to 0.36 mm per day, as the ground strain changed from tension to compression. The character of the graphs was influenced by the uneven distribution of ground strain, and the formation of a crack at 9.15 m. During the development of compressive ground strain, however, when the crack was virtually closed, the relationship

390

between longitudinal slip and structural length was approximately linear. Analysis of these results indicated that the longitudinal slip at any point l metres from the neutral axis of the footing could be estimated from Equation 1. Longitudinal slip

$$H_l = 1.02lE_g \qquad (1)$$

Since subgrade restraint is a function of slip, a knowledge of this function from a foundation–soil shear test combined with Equation 1 would enable foundation stresses to be calculated along the length of structure.

Change of slope

Changes of slope based on the difference in settlement over a distance of 1.2 m were monitored throughout and maximum values on all stations were obtained at 1.3–1.4r equal to 5.4 mm/m. At the time of maximum change of slope, curvatures of 22×10^{-5} and 30.8×10^{-5} m^{-1} were recorded for the ground and structure respectively.

Structural damage

Table 2 illustrates some of the ground and structural deformations related to damage. The cracks occurred during the development of tensile ground strain and it was observed on each occasion that longitudinal slip was zero in the damage area. Consequently, these sections were subjected to the maximum tensile forces acting at the time of damage.

All crack profiles and movements were monitored to study the influence of crack geometry on subsequent structural behaviour. In Structure 1 vertical forces were transmitted across horizontal sections of the crack at mid height, thus creating a shearing resistance to horizontal movements. Under these circumstances the crack operated to some extent as a hinged joint which enabled the structure to act as a composite unit throughout the investigation. In Structure 2, crack no. 1 acted as a free edge with the result that the damaged structure had to be analysed as two separate units.

Where failure occurred as a result of tensile strains initial cracking invariably resulted from a breakdown of the brick–mortar bond.

Conclusions

Although a simple form of structure has been investigated, the behaviour of a brick wall subjected to subsidence is complex. Detailed measurements are considered essential if the mechanisms governing this complex behaviour are to be understood.

The foundation movements observed in relation to the adjacent ground serve to confirm that a knowledge of soil-structure interaction is required if safe, economical designs are to be produced for new structures to be erected in subsidence areas, or damage, likely to be caused to existing structures by proposed mining operations, is to be assessed.

Table 2. STRUCTURAL DAMAGE

Damage	Foundation			Angular distortion (deflection/span)			Average strain (tension −ve)			Remarks
	Position of mine face	Max differential settlement (Δmm)	Span (m)	Ground	Found	Brick	Ground	Found	Brick	
Structure 1 Foundation crack extending 0.66 m up brick wall width = 0.1–0.26 mm	0.6r	8.6	15.8	1:20800	1:6000	1:6130	-22×10^{-5}	-7×10^{-5}	-22×10^{-5}	Aesthetic damage
Crack width = 0.92 mm	1.1r	36.6	15.8	1:4450	1:2540	1:2740	-55×10^{-5}			Crack operated as a hinge and wall acted as a composite unit. No damage due to excessive compression.
Crack width = 0.46 mm	1.5r	47.2	15.8	1:2300	1:1730	1:2030	$+213 \times 10^{-5}$			Crack extension due to increase in curvature
Extension of crack to top of wall, width = 0.5–2.5 mm	1.6r	37.1	15.8	1:1600	1:1270	1:1390	$+211 \times 10^{-5}$			
Maximum width recorded = 13.2–14 mm	2.0r	5.5	15.8	1:1160	1:930	1:920	$+20 \times 10^{-5}$	-17×10^{-5}	-80×10^{-5}	Severe damage
Structure 2 Hairline cracks in foundation and wall Crack width = 0.1–0.38 mm	0.5r	8.9	31.2	1:120000	1:15000	1:16530	-30×10^{-5}	-7×10^{-5}	-5×10^{-5}	Aesthetic damage
Crack width = 7.1 mm	1.1r	31.8 / 41.9	12.5 / 18.7	1:10670 / 1:3790	1:4360 / 1:3600	1:4360 / 1:3950	-54×10^{-5}			Main crack acted as a free edge, i.e. wall separated into two units. No damage due to excessive compression
Crack width = 0.25 mm	1.5r	31.8 / 54.6	12.5 / 18.7	Existence of 7 cracks in the structure prevented simple tabulation of average angular distortions			$+190 \times 10^{-5}$			
Maximum width recorded = 12.1–19.6 mm	2.0r	1.0 / 2.5	12.5 / 18.7				-11×10^{-5}	-30×10^{-5}	-40×10^{-5}	Severe damage

In mining subsidence the degree of damage observed is often a function of ground strain, which can cause cracking at very small angular distortions (*Table 2*). This fact should be borne in mind when assessing the influence of convex ground curvature on buildings, e.g. ground movements adjacent to excavations.

REFERENCES

Littlejohn, G. S. (1973). 'Monitoring Foundation Movements in Relation to Adjacent Ground', *Ground Engineering*. July, 17–22
Wardell, K. (1953). 'Some Observations on the Relationship Between Time and Mining Subsidence', *Trans. Instn Min. Engrs*. Vol. 113, 471–83

V/9. An effective stress finite element analysis to predict the short- and long-term behaviour of a piled-raft foundation on London Clay

D. J. Naylor
University College of Swansea

J. A. Hooper
Ove Arup & Partners, London

Summary

The construction of a large foundation on London Clay is simulated by finite elements. Elastic parameters relating effective stress to strain are used. The analysis is divided into three stages: (1) Excavation; (2) construction of building; (3) dissipation of excess pore pressure in clay. It is assumed that there is no drainage in stages (1) and (2).

The settlements and distribution of load under the foundation at the end of stage (2) are compared with measured values. In addition, a prediction is made of the changes to be expected when drainage is complete.

The paper also brings out two points of analytic interest: (A) the use of effective stress parameters for undrained analyses, and (B) a means of proceeding from the undrained to the fully drained condition.

Introduction

The building under consideration is the Hyde Park Cavalry Barracks tower block in central London. It is 90 m high and has a 9 m deep basement. Support is partly on 51 under-reamed piles, 25 m long, and partly by direct bearing on the raft which caps the piles and extends the full width of the foundation. London Clay extends from 5 m below ground surface to well below the base of the piles. *Fig. 1* shows a cross-section through the foundation.

Of particular interest is the distribution of load between the piles and the raft. To measure this, load cells were installed in three of the piles, and three pressure cells were installed at the base of the raft. In addition, levelling sockets were built into the substructure. After the building was completed a finite element analysis—distinct from that described here—was carried out. A full description of this work is given by Hooper (1973a). The analysis showed that a linear elastic isotropic model with a modulus of elasticity varying linearly with depth fitted the measurements adequately. Realistic values for the elastic parameters were thus obtained, and these are used in the work described here.

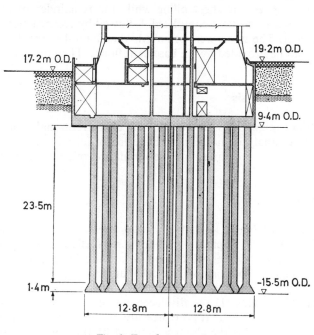

Fig. 1. Foundation section

The present analysis differs from the earlier one in two respects. Firstly, a *total stress* approach was used in the earlier analysis. Secondly, the construction sequence is treated rather differently in the two analyses. In the earlier analysis it was assumed that no uplift forces acted upon the foundation until the raft had been cast; in the present analysis it is assumed that the entire uplift force is mobilised *before* the casting of the raft. On this count alone, different results are to be expected from the two analyses.

Some unconventional analytical methods are featured. The effective stress approach has been utilised to predict excess pore pressures, and has allowed the same material properties to be used for both drained and undrained analyses. In addition, the effects of consolidation have been predicted without invoking the time-dependent consolidation equations.

395

Construction

As the construction sequence doubtless has an important effect on the final distribution of load, a brief description is appropriate.

The 51 piles were installed first, the concrete shafts being poured to the level of the raft base. Sheet piles were then driven to form a continuous wall around the edge of the foundation. They extended below the final excavation level. Excavation started in the centre where it was carried to full depth. A wedge of soil was left around the edge to support the sheet piling. The centre part of the raft was then concreted. It was used as support for temporary struts which propped the sheet piling while the remainder of the soil was excavated. The raft was then completed, followed by construction of the rest of the basement. The struts were removed when lateral support for the sheet pile walls could be provided by the basement floors. Then the tower was built.

Basis of analysis

In this section explanations are given of the unconventional procedures used in the computer analysis; namely the handling of initial stresses, the computation of excess pore pressures, and the simulation of consolidation.

Initial stresses

The program is designed so that an initial stress field can be input. Total stresses are defined by separate specification of effective stresses and pore pressures. The nodal forces which equilibrate the initial total stresses, but which do not cause displacements, are calculated within the programme. Then the *applied* loads are input and the main part of the finite element analysis is carried out. This yields (1) the displacements caused by the applied loads, and (2) the effective stress and pore pressure distributions, which are the sum of the initial values plus the changes induced by the applied loading.

Excess pore pressures

Under truly undrained conditions there is no relative movement between the soil skeleton and pore fluid. By treating the soil as a composite of solid skeleton and pore fluid components, each of which undergo the same strains but which have independently specified stiffness properties, it has been shown (Naylor, 1974) that the total stiffness of the soil can be expressed as the sum of the soil skeleton stiffness and the pore fluid stiffness. The pore fluid stiffness is defined by its apparent bulk modulus, K_a, which is related to the actual bulk modulus of the pore fluid, K_f, by the approximate relation (Bishop, 1966)

$$K_a = \frac{1}{n/K_f + (1 - n)/K_s} \tag{1}$$

where n is the porosity and K_s is the bulk modulus of the soil particles. Once

396

the strains are known, the induced excess pore pressures, u_e, are calculated from

$$u_e = K_a \varepsilon_v \qquad (2)$$

where ε_v is the volumetric strain.

Consolidation

Having predicted the distribution of effective stress and pore pressure resulting from the undrained application of load, the question arises as to what further deformation will occur and what will be the stress distribution when the excess pore pressures have finally dissipated? The procedure is as follows. The initial stresses and pore pressures are taken to be the values reached at the end of the previous (undrained) analysis. In addition, the steady state pore pressures are input. The excess pore pressures, being the difference between the two sets of values, are now converted to equivalent nodal forces. This is done using the standard virtual work procedure for equilibrating internal stresses with nodal forces. These nodal forces are applied to the soil skeleton. (This is achieved by setting $K_a = 0$). The displacement and stress fields which result correspond to the end of consolidation.

This assumes that the soil skeleton is linear elastic. If the soil stiffness varies significantly with effective stress then the method will give incorrect results. To obtain a rigorous solution it would be necessary to solve the time–dependent consolidation equations, a procedure which involves much more computation (Sandhu and Wilson, 1969; Hwang, Morgenstern and Murray, 1971). If, however, the non-linearity is mild, the simple procedure described here could have useful application.

Analysis

The finite element mesh

An axisymmetric mesh was used to represent the foundation. It was the same as that used in the earlier analysis and consists of 180 quadrilateral 8-noded elements (Zienkiewicz, 1971) with a total of 587 nodes. In the real structure the piles are arranged approximately in concentric rings; in the mesh these become annuli. The cross-sectional area of the annuli is greater than that of the piles, and a reduction in the modulus of elasticity of the concrete was made to compensate. The raft stiffness was also adjusted—in this case increased—to take account of the additional stiffness contributed by the basement and the tower.

Material properties

For the concrete, Young's modulus (un-modified) and Poisson's ratio were taken as 13.8 GN/m² and 0.3 respectively. For the clay, Poisson's ratio, v', (in terms of effective stress) was taken as 0.1 and the corresponding Young's modulus, E', was assumed to vary with depth according to the relationship

$$E' = 7.5 + 3.9z \qquad MN/m^2 \qquad (3)$$

where z is the depth below ground surface in metres. These are the same as the *drained* values used in the earlier work.

The value of K_a was taken as 479 MN/m² (i.e. 10^7 lbf/ft²). This was a somewhat arbitrary choice, but provided K_a is large compared with the stiffness of the soil skeleton, its numerical value is not critical. It must not be too large, however, otherwise very poor stress results will be obtained, precisely as occurs if a Poisson's ratio close to 0.5 is used. The value chosen is about one quarter of the actual bulk modulus of water.

Excavation, construction and consolidation

Three computer runs were required, one for each of the three stages. Stress and pore pressure distributions at the end of the first and second runs were stored on magnetic file to become the initial stresses and pore pressures for the subsequent run. To obtain the initial stresses for the excavation analysis, a gravity stress field based on a bulk unit weight of soil of 19.2 kN/m³ was calculated. Horizontal and vertical stresses were assumed equal. The pore pressure was assumed to vary hydrostatically, having zero value at 4.0 m below ground surface. This initial stress distribution was applied to soil and piles alike, the implication being that the piles were installed without disturbance to the adjacent soil.

The raft elements were assigned a negligible Young's modulus for the excavation analysis, thus simulating the non-existence of the raft at this stage; here the presence of the concrete which had actually been placed before completion of the excavation was neglected. For the construction and consolidation analyses, a Young's modulus corresponding to the stiffness of the completed structure was assigned to the raft elements.

Excavation was simulated by applying an inward distributed force to the boundary of the excavation which exactly cancelled the initial total stresses. The analysis was undrained. Construction was simulated by applying the total weight of the tower plus raft (228 MN) to the raft. Following the earlier work, the load was distributed parabolically, its intensity being three times as great at the centre as at the edge. Again undrained conditions were assumed.

The consolidation analysis followed the procedure already described. The steady state pore pressures were assumed to be the same as those existing at the start of excavation.

Results

The radial distribution of raft settlement, raft contact pressure, and pile load are given for the end of construction and end of consolidation cases in *Fig. 2*. Both total and effective soil pressures are shown, the pore pressure at the base of the raft being defined by the separation of the full and broken lines. Settlement is assumed zero at the start of construction. Measurements taken at the end of construction (40 months after the start) are included for comparison with the computed values.

398

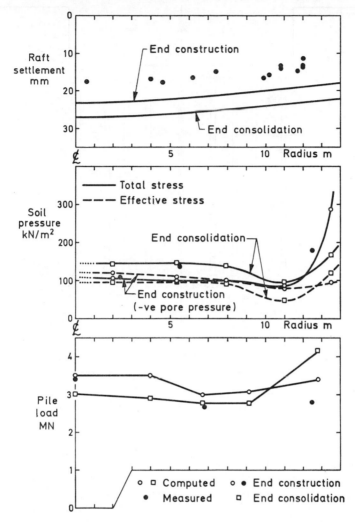

Fig. 2. Radial distribution of settlement, pressure and pile load

Distributions of excess pore pressure obtained from the excavation and construction analyses are shown in *Fig. 3*. Crosses mark the element centres. The contours are based upon spot values obtained for the element centres.

Care was required in the prediction of contact pressures. It had been found previously that with elements of the type used here an averaging process whereby the stress at the element centre was taken as the average of values from four points within the element gave excellent results (Hooper, 1973b, and Naylor, 1974). These points are the Gauss sampling points used in the numerical integration process. Stresses and pore pressures were obtained in this way. Both the pressures on the raft base and the pile loads were obtained by extrapolation of the element average stresses from each column of elements under the raft.

399

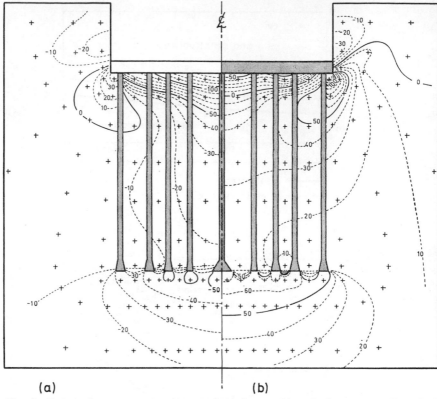

Fig. 3. Computed excess pore pressure in kN/m² units (a) end of excavation (b) end of construction

The distribution of computed total vertical support, obtained by integrating over the base of the raft, is given in *Table 1*.

Table 1. PERCENTAGE DISTRIBUTION OF SUPPORT AT BASE OF RAFT

	End of construction	End of consolidation
Pore water	6	10
Soil skeleton	21	17
Piles	73	73

Discussion

The computed settlements are about one third higher than the measured values, whereas in the earlier work, using the same elastic moduli, good agreement was obtained. This apparent inconsistency is explained by the different datum used for settlement in the two cases. In the present work, the

400

datum was taken at the completion of the excavation whereas previously it was taken at the start of the excavation.

The predicted increase in settlement due to consolidation must be caused by the dissipation of excess pore pressures. But the analyses show an *increase* in pore pressure on the raft base (except near the edge) during consolidation. It appears that this is more than compensated by the dissipation of excess pore pressures beneath the pile bases. In practice, the high negative pore pressures induced just below the bottom of the excavation due to the removal of soil would most likely dissipate before much of the construction had been carried out. The opening of fissures in this region would accelerate the dissipation process. Consequently the pore pressures on the raft base at the end of construction are likely to have been somewhat higher than predicted, i.e. the upper broken line in *Fig. 2* would be depressed. The pore pressure predictions at depth (*Fig. 3*), however, would not be affected.

The capabilities of the finite element method are being somewhat stretched in the present analysis, but confidence is gained by comparing the integrated soil pressures and pile loads with the applied load. Agreement was within 2%, although such close agreement is probably fortuitous. Furthermore there is satisfactory agreement between measured and computed stresses and pile loads at the end of construction (*Fig. 2*).

The transfer of load from the centre to the outer piles as a result of consolidation (*Fig. 2*) is of interest. The trends indicated by recent load cell readings are consistent with this prediction.

Conclusions

The feasibility of using the finite element method to (1) simulate the behaviour of a foundation during the successive stages of excavation, construction, and consolidation, (2) predict deformation and stress fields at the end of these stages, and (3) separate the total stresses into effective and pore pressure components, has been demonstrated. In achieving this, unconventional techniques for handling initial stresses, predicting excess pore pressures, and simulating consolidation have been used.

The pile loads, soil pressures and raft displacements computed for the end of construction have been found to be in reasonable agreement with measured values. The analysis also suggests that the overall distribution of load between the piles and the base of the raft will change very little due to consolidation of the soil. However, some transfer of load from the centre piles to the edge piles is to be expected, accompanied by a transfer of soil pressure from the edge of the raft to the centre.

REFERENCES

Bishop, A. W. (1966). 'Soils and Soft Rocks as Engineering Materials', *Inaug. Lect. Imp. Coll. Sci. Technol. 6*, 289–313
Hooper, J. A. (1973a). 'Observations on the Behaviour of a Piled-Raft Foundation on London Clay', *Proc. I.C.E.* Vol. 55, Pt. 2, December 855–877
Hooper, J. A. (1973b). 'Some Finite Element Results for a Circular Raft on a Thick Elastic Layer' (unpublished work)
Hwang, C. T., Morgenstern, N. R. and Murray, D. W. (1971). 'On Solutions of Plane Strain Consolidation Problems by Finite Element Methods', *Canadian Geotech. J.* Vol. 8, No. 1, 109–118

Naylor, D. J. (1974). 'Stresses in Nearly Incompressible Materials by Finite Elements, with Application to the Calculation of Excess Pore Pressures', *Int. J. for Num. Methods in Eng*, Vol. 8.

Sandhu, R. S. and Wilson, E. L. (1969). 'Finite Element Analysis of Seepage in Elastic Media', *J. Eng. Mech. Div., A.S.C.E.* Vol. 95, No. EM3, 641–652

Zienkiewicz, O. C. (1971). *The Finite Element Method in Engineering Science*, McGraw-Hill, London

V/10. Differential foundation movement of domestic buildings in South-East England
Distribution, investigation, causes and remedies

John F. S. Pryke
Pynford Design Limited, London

Summary

Three major groups of strata outcrop in South-East England, an area which has a population of some twenty million. An important feature is the London Clay basin where many millions live. An analysis of subsidence cases related to surface strata shows that the risk of damage to domestic property by differential subsidence is 10 to 20 times greater in London Clay areas than elsewhere. The Author concludes that the minimum foundation depth in shrinkable clay recommended by CP 2002 should be increased.

The most important symptom of differential foundation subsidence is cracking in brick or blockwork walls. The interpretation of cracking patterns is discussed and the importance of noticing which way cracks taper is made clear.

Causes of subsidence are listed and each is commented upon. Particular attention is paid to the growth and effect of tree roots and it is emphasised that it is the very small roots that mainly absorb moisture from the subsoil and cause shrinkage in dry seasons. Most cases of subsidence are due to instability in the subsoil within 3 m of the surface.

A typical case of subsidence is taken and four different solutions to the problem are suggested, using either contiguous piers, micropiles beams and piers or beams and piles.

To conclude the paper, actual cases that occurred in North London and Sevenoaks are considered.

Geology and population of South-East England

Three principal groups of solid strata outcrop in South-East England. Firstly chalk, greensands and gault clays of the Cretaceous period form the chalk ridges of the North and South Downs, the Weald clays of Kent, and the uplands running through Bedfordshire and north Hertfordshire out on to the Cambridgeshire plain. Secondly, and enclosed by these Cretaceous areas, the Tertiary London Clay and Woolwich and Reading beds form the great wedge of the lower Thames basin including Greater London, Berkshire, Middlesex, south Hertfordshire and Essex; there is also a Tertiary outcrop around Portsmouth, Southampton and Poole. Thirdly, Norfolk and Suffolk are mainly Quaternary.

Drift deposits include sands and gravels, which form river terraces along the banks of the Thames, and alluvial silt and mud on the east coast and lower reaches of the Thames. Glacial deposits of sands and gravels and stony clays, often less than a metre thick, are scattered widely north of the Thames.

About twelve million people live on about 400 000 ha in the Greater London conurbation and another eight million people live on a further 3 650 000 ha in East Anglia, the Home Counties and along the south coast from Kent to Hampshire.

Design and construction of typical domestic properties

Until the mid seventeenth century houses in South-East England were mainly built with timber frames, either clad with a lath and plaster skin or infilled with brick. Brick work had begun to replace this style of construction in the mansions of the nobility in the late sixteenth century and this process was hastened by the great fire of London in 1666. The typical domestic building then became a brick outer shell with internal walls of timber studding, lath and plaster, and with floors supported by timber joists boarded over and often plastered beneath. The bricks were bedded in soft lime mortar and the buildings were relatively flexible. There was usually a basement under at least part of the building often built only partially below ground level. The roofs were tiled on timber framing.

In the nineteenth century, Portland cement was introduced, and this with harder plasters made possible stronger, thinner walls resulting in construction that is more sensitive to smaller differential movement of the foundations.

Materials and methods are continually refined and today the majority of houses are built with 260 mm thick external cavity walls, internal partitions of light-weight block work, upper floors of timber, ground floors of solid construction upon hard filling, and tiled prefabricated timber roof trusses that span the full depth of the house, enclosed between brick gable walls.

Distribution of cases of subsidence

In *Fig. 1* numbers of cases of subsidence are plotted in their geographical location. There is a heavy concentration of cases in those parts of the London area built since 1918. This reflects the fact that large parts of inner London

404

are now commercial or industrial, and that nearer the river the subsoil is mainly gravel.

The high concentration of cases of subsidence spreads south of the river up to the boundary with the chalk, with relatively heavy concentrations in Bromley, Beckenham, Croydon, the Leatherhead area and Guildford.

North of London cases concentrate in an area running through Chingford and Enfield to Harrow and Stanmore and then fade away towards the north. It will be noted that when the Cretaceous zone is reached, north of the river, the numbers do not drop off suddenly as they do to the south. This is explained by the large areas of boulder clay that are found north of the river.

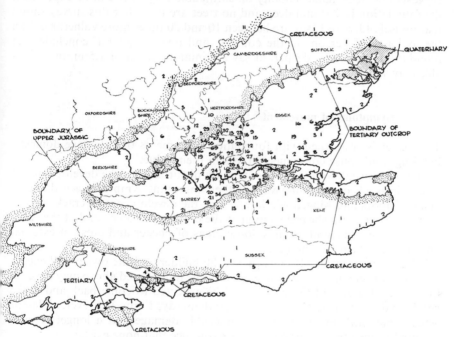

Fig. 1. Distribution of cases of subsidence in South-East England related to the main geological strata

Note in *Fig. 1* a tendency for cases to concentrate along the clay soils that outcrop again just south of the North Downs and north and south of the South Downs; also note the cases reported along the coast and along river valleys, for example along the Thames riverside areas in north Berkshire and Oxfordshire.

It is interesting to follow the A3 road north from Portsmouth. Ten cases of subsidence out of thirty-four reported in Hampshire appear in the housing estates built on the clay lowlands between the chalk ridge behind Portsmouth and the North Downs, only one other case is then found before the road leaves the Hog's Back beside Guildford and runs out on to the Thames Valley plain, where we find twenty-one cases.

Out of 2767 cases analysed 2434 occurred in the counties of Hertfordshire, Essex, Middlesex, Greater London and Surrey. The relationship between numbers of cases of subsidence and population in each county is even more revealing. In Essex the number of cases reported per 100 000 is 42.8, nearly 1 in 500, compared with roughly 1 in 10 000 in Sussex for example.

Note that the analysis may be weighted in a number of ways; firstly, the cases are those reported to the Author's company, which is based in London; secondly, the value of a property may influence the probability that the owner will take action. Despite this the Author believes the results to be reasonably representative of the relative risk of building on shrinkable clay soils compared with other soils.

The new foundation Code of Practice (CP 2004 : 1972, Clause 3.2.7.1) suggests that reasonable stability on shrinkable clay soils can be expected if the foundation is 900 mm deep and no trees are near, but this survey shows that householders in Essex are between 10 and 20 times more vulnerable than the majority of persons in the country, and one cannot but conclude that foundations should be deeper, at say a minimum depth of 1.5 m, in shrinkable clay soils.

Symptoms of movement

This paper is concerned with foundation movements that damage domestic buildings and cause consequential loss in their value.

The most usual symptom of foundation disturbance is the development of cracks in brick or masonry walls. Other important symptoms are the jamming of doors and hinged window sashes, the development of cracks across ceilings, cracks and shearing action at the corners of rooms and at the junction between walls and ceilings, loose areas of plaster and areas of external pointing that begin to fall out.

When houses are newly constructed the builder and the owner will expect an initial settlement period. Some or all of the faults listed above may develop to a lesser degree and the builder will be expected to come back and make good the defects, including cracks larger than, say, 1 mm. However, all latent defects may still not have caused noticeable damage and a longer term is required to give the owner reasonable assurance that none exist.

The Author has noticed that his company is rarely consulted about cracks unless at least some have reached about 3 mm in width. Often a householder will not notice a crack until his attention is drawn to it by a surveyor employed in connection with a proposed sale.

Investigation of a building in which subsidence is suspected

The first stage in any investigation will be a visual survey of the property. A systematic inspection both internally and externally must be made in order to study the construction and layout of the property in detail, to record the pattern of cracking and other defects, to estimate the age of the defects and to relate the property to its surroundings and note important features. Enquiries should also be made about the history and development of the defects, as this may give important clues about the cause of subsidence.

The engineer will bear in mind that the timber framing and soft mortar

406

used in the houses built before about 1900 can accommodate substantial movement without cracking; whereas houses built in the twentieth century, and particularly since 1950, will be more brittle and will crack with less differential foundation movement.

The relative flexibility of the timber based parts of the house, compared

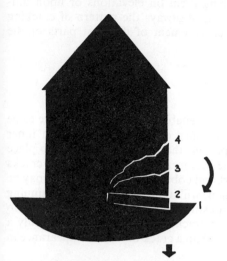

Fig. 2. Settlement of the corner of a house following withdrawal of support from below the foundation

Fig. 3. Cardboard cut-out illustrating crack pattern arising from settlement at the centre of a wall with window and door openings

Fig. 4. Cardboard cut-out illustrating crack pattern arising from settlement at the ends of a wall with window and door openings. Note that the cracks appear in the same positions but that they all taper in a different direction

with the brickwork and blockwork walls, emphasises the importance of cracking in walls as a means of analysing movements caused by subsidence

Figs. 2, 3 and 4 illustrate typical cracking patterns. Note that subsidence cracks are almost invariably tapered and that they are caused by rotation of one part of a building compared with another.

When analysing cracking patterns the Author uses a trial and error process thus; first observe the cracks, then propound a theory of movement, search for further cracks which would confirm the theory propounded and finally either modify the theory or confirm it by further observation.

A building should be inspected systematically, for example, by following through the rooms in a clockwise manner when looking at the plan, starting at each floor level from the same corner of the building.

Cracks are best recorded by sketching them on elevations or upon 'unfolded' diagrams of internal rooms. Almost always the pattern of cracking alone will clearly indicate the relative movement of different parts of the house.

The age of cracks

Careful inspection of a crack will generally enable an estimate of its age to be made. Relatively new cracks will have sharp edges, and clean fresh inner faces; often tiny pieces of mortar or plaster, that would fit neatly into place if the crack were closed, loosely bridge the open gap. Conversely old cracks will often be painted over and have dirt and cobwebs inside, and there may be old making good which, if cracked again, will indicate movement continuing over a long period, and may allow an estimate of the rate of movement to be made. In heavily industrialised areas cracks become very dirty in only a few months and thus appear much older than a crack of similar appearance in a rural area.

Discussion with the occupant is often very helpful, particularly useful information being the age of decorations. Remember that cracks may have occurred before, and possibly long before, they were first observed.

Site layout

It is important to note the layout of the site upon which the building stands. Features to note include the slope of the ground if any; the compass direction of the various elevations; the position and depth of drains; the position, size and type of nearby trees; the position and height of retaining walls, ditches or cuttings; damage, if any, which may be observable in nearby buildings; and clues to the surface strata which may be obtained by observing, for example, banks or ditches in the area. It is a useful habit to begin close observation of the locality as one approaches the house to be surveyed.

Causes of foundation failure

Foundation failures may be due to the following reasons—(1) differential clay shrinkage; (2) removal of lateral support from the ground beneath the foundations; (3) differential consolidation where basements provide deep foundations under part only of a building; (4) building on filled ground; (5) clay swelling; (6) over loading; (7) swallow holes; (8) creep on clay slopes; (9) poor quality concrete; and (10) other causes including sheer negligence. These causes are enlarged upon in following sections.

It is, of course, essential to identify the cause or causes, for remedial work based upon faulty diagnosis can lead to expensive failures. Engineers should be particularly wary of assuming that poor foundation concrete or leaking drains are main causes of subsidence.

Clay shrinkage and trees

Clay shrinkage was analysed in detail by the Building Research Station in the late 1940s. As a consequence the short-bored pile foundation was proposed and has been used widely in clay areas since then (Ward and Green, 1948). The harmful effects of tree root systems in clay soils has been widely publicised and most designers and builders are aware of the dangers of poplars and elms. However, it is not generally realised that all trees can be harmful, and many cases have been investigated where oaks, planes, and even thorn trees have caused serious cracking.

A large deciduous tree transpires tens of thousands of gallons of water each year, which are absorbed from the soil through a network of tiny rootlets mainly less than 0.5 mm in diameter. The surface area of a root system in the soil is enormous. For example, in the 1930s it was found that a rye wheat plant developed a root system 387 miles long with a surface area of 2554 ft^2 in a soil volume of less than 2 ft^3 (Epstein, 1973).

The amount of water absorbed by a particular tree depends upon many factors including the spacing of the fine rootlets. Elms and poplars, which frequently cause trouble, have closely spaced rootlets, whereas oaks and plane trees, that are usually only troublesome in very dry seasons, have a less dense root system.

In dry conditions roots extend towards more moist ground and the shape and size of a tree root system will depend upon many factors, including the type of tree, the nature and fine structure of the subsoil, competition from the root systems of adjoining trees, the water table, sources of water, and sources of nourishment, which may be supplied in generous quantity by leaking soil drains. Over a period of dry summers root systems will advance relatively fast, perhaps 1–1.5 m per season, and in wet seasons they may wither back.

The extent of a root system is for practical purposes impossible to predict in detail, but experience of many cases establishes that there is an unacceptable risk of damage in shrinkable clay soils when single trees are closer than their mature height to shallow foundations, or rows of trees are closer than 1.5 times their mature height. (National House Builders Registration Council Handbook, Clause Ex 2.) For example, the roots from a copse of full grown elm trees *may* penetrate 6 m down and spread 40 m from the edge of the copse, but they will not necessarily do so. Many trees that put houses on shrinkable clay soils at risk do not, in fact, cause damage.

Fig. 2 shows typical cracking caused when the ground shrinks away from beneath the corner of a house. The crack will form initially beneath the foundations and it may be that the brickwork will cantilever out and no sign of damage will appear for some time. Then, usually in late summer, a crack will appear above ground level and develop to perhaps 10 mm wide within a few days. This will first show along the damp proof course (Crack 2) then irregular diagonal cracks will reach up across the wall from the pivot point about which the movement is taking place (Cracks 3 and 4). If the cracks are wedged up the development of further fractures at higher level can often be avoided. The visible cracks develop very quickly but in fact the underlying cause has developed very much more slowly.

The subsoil most susceptible to shrinkage is London Clay. However other susceptible clays exist in South-East England.

Removal of lateral support

Subsidence damage is often caused when excavations for drains and basements are made deeper than, and near to, existing foundations. If proper precautions are taken to strut the face of the excavations nearer the foundation then damage can be avoided, but this requires a higher standard of strutting than is required for the safety of the workmen. It is this 'safety' strutting that is customary.

Reference is often made to the 45° line indicating that excavations outside

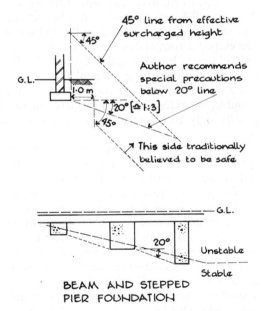

Fig. 5. Recommendations for excavation close to existing foundations

this can cause no harm. This is not so and the line should be drawn from a point above the edge of the foundation (see Fig. 5). The Author recommends 20° as a rule of thumb. Excavation below this line should not proceed unless careful consideration has been given to the strength of the ground, the temporary support and the final construction details of both the foundation and the adjoining work. All excavations below the water table also need particular care. A great deal of unnecessary damage to buildings and consequential disturbance to adjoining owners could be avoided if a proper standard of care was maintained when excavating close to existing or proposed foundations.

Partial basements

Basements under only part of a house often cause small cracks which spring from the edge of the basement and sometimes increase to widths of about 3 mm near the eaves. The movement will be more exaggerated if there are heavy loads, such as a substantial internal chimney breast, just outside the basement area.

Many severe cases of differential foundation subsidence occur when partial basements combine with other causes, for example clay skrinkage or deep drain trenches.

Filled ground

Failure of foundations built on filled ground is to be expected where normal strip foundations are extended over patches of fill, for example, filled-in ditches. The use of heavy earth-moving equipment increases the risk of this. Reinforced concrete raft foundations for houses on filled sites may also give trouble if long term consolidation is continuing beneath the raft infilling which is not uniform in quality and depth. A failed reinforced concrete raft can be very expensive to underpin.

Clay swelling

Where large trees are felled on clay sites, the ground beneath them will have been reduced in moisture content to depths of up to 6 m and it may take years for the natural moisture content to be regained. Moisture deficiencies within 1.5 m of the surface will usually be made up within a few months, but if the drying has penetrated much more deeply, instability of the surface will continue for many years, particularly if the surface stratum is a good puddle clay which will form a very low permeability barrier nearer the surface. A site in Windsor has been swelling for nearly 15 years.

The effects of swelling will begin to appear within a year or so of construction and will continue with decreasing effect for a very long time. Swelling is unlikely to affect an old building unless a large tree has been recently felled nearby and there is evidence that the tree disturbed the foundations during dry seasons.

Overloading

Overloading of virgin subsoils rarely causes damage to one or two storey houses but taller buildings and blocks of flats may be affected. Overloading will cause settlement cracks that start early in the life of the property but will tend to die away within the first few years.

411

Swallow holes

Swallow holes are caused when surface deposits are washed into cracks in deeply fissured rocks. Examples are quite common on the fringes of the chalk outcrops, for example in north Hertfordshire and in the Cray valley in north west Kent. Swallow holes tend to follow lines across the countryside where fissures in the chalk have given rise to local weaknesses and solution holes. Similar failures occur less frequently in other strata. A case that occurred over greensand beds at Sevenoaks is described later.

Surface creep

Creep of clay slopes may occur when they are steeper than about 1 in 10. Creep will be spread over large areas and may affect a number of properties. Inland, damage due to creep is nearly always due to excavation into the slope at lower level, for example, for roads, and sliding retaining walls must be suspected. On the coast, erosion has made large areas behind the foreshore unstable; examples of this occur near Folkestone and at Ventnor.

Poor quality foundation materials

The Author knows of only one or two cases where weak concrete in the foundations was the principal cause of subsidence. If poor quality foundations are the main cause of trouble there will either be clear evidence of compression or shearing of the foundation, or subsidence soon after construction will be coupled with badly honeycombed concrete into which the subgrade has been squeezed.

Leaking drains

Drains may be tested and found to be leaking but be very wary of assuming this to be the cause of subsidence, for drains are often fractured by subsidence. During wet weather most foundations dug into clay soils are permanently surrounded by water, and in dry weather a leaky drain may, in fact, protect the foundation.

Remedies

It is usually very much cheaper to strengthen and deepen an existing foundation by underpinning than it would be to demolish and reconstruct the house on new foundations.

A satisfactory underpinning scheme will be a re-designed foundation devised to solve the problems that have led to the subsidence and damage. There is no universal solution to problems of foundation failure. However, it is generally found that the unstable ground that causes differential movement

412

of the foundations of domestic buildings lies within the top three metres of ground and stable strata are found beneath.

Many hundreds of successful underpinning schemes designed to support the building on a stratum about 3 m below ground level have been completed by the Author's company.

Alternative methods of refounding buildings with swallow foundations that have failed are illustrated in *Figs. 6, 7, 8 and 9*. In each scheme it is

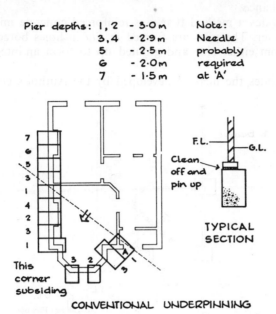

Pier depths: 1, 2 - 3.0 m
              3, 4 - 2.9 m
              5 - 2.5 m
              6 - 2.0 m
              7 - 1.5 m

Note:
Needle
probably
required
at 'A'

CONVENTIONAL UNDERPINNING

Fig. 6. Underpinning by the traditional method

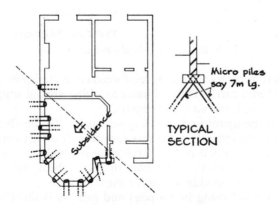

'Micro pile' - positions determined by strength
of existing foundations and need
to balance reactions

Work from both sides of the wall

Fig. 7. Underpinning with micropiles

413

assumed that the affected part of the building is a front external corner and that a stable stratum exists at a depth of 3 m.

Fig. 6 illustrates the traditional method which is to form piers beneath the existing foundation, either contiguously placed, or so close together that the foundation between the supports will bridge the gap. Difficulty may be encountered with isolated piers, for example between the front bay and the front door, and a needle may be required through the pier with consequent internal disturbance.

Fig. 7 illustrates a method that uses reinforced concrete micropiles 100–225 mm diameter. The piles are formed *in situ* in holes bored through the foundation from either side and grouted up to form an integral structure with it.

Fig. 8 illustrates the method developed by the Author's company using

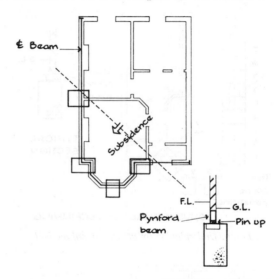

Fig. 8. Underpinning by the Pynford method

Pynford stools to form a continuous reinforced concrete beam above foundation level. The beam carries the wall loads to widely spaced support piers and acts as a tie to resist horizontal movements.

It is useful to compare underpinning schemes with the foundations that would have been designed had the problem been foreseen before the house was constructed. Beam and pier underpinning is usually a close approximation to the 'open site' solution, the main difference being alterations in base shape and position to provide access for excavation.

In *Fig. 9* a method using bored piles and needles is illustrated. A disadvantage of piling is that the boring equipment must be mounted above the position of the piles, which cannot be formed directly below the walls unless large pits are dug beneath them. If piling within the building is to be avoided pairs of piles and a cantilever pile cap will be required. The wall is supported by beams on either side spanning between pile caps and in turn carrying needles through the wall at intervals, which necessitates the removal and

414

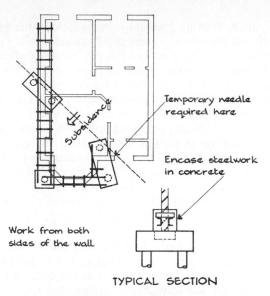

Temporary needle required here

Encase steelwork in concrete

Subsidence

Work from both sides of the wall

TYPICAL SECTION

Fig. 9. Underpinning with bored piles

Fig. 10. Damaged property in Hampstead. Note the plane tree

415

replacement of much flooring internally. Alternatively the beams could be constructed with less disturbance using the Pynford stooling method.

Difficulties of access have a greater influence on piling schemes than upon those obtaining support from underpinning piers, and it is usually found that the least expensive proposals are obtained with fewer pile caps and larger beam spans, Schemes based upon conventional bored piles are most competitive when the unstable ground is 5 m deep or more.

At first sight a partial underpinning scheme, such as those illustrated, will create the same problems as partial basements. If the disturbance extends beyond the area underpinned and the whole foundation is moving, but part is moving more than the remainder giving localised relative subsidence, then partial underpinning will not be successful. Causes (1), (2) and (3), and in some circumstances (4) and (7), all of which cause local subsidence, may be successfully prevented from causing further damage by underpinning the unstable part.

It will be noted in *Figs. 8 and 9* that the foundation beams are extended beyond the support points into the unaffected part of the foundation. This effectively avoids an abrupt change in level at the edge of the last support piers. Any tendency for the foundation immediately adjoining this pier to subside will transfer load on to the adjacent underpinning support and away to the far end of the projecting beam where the ground is stable and where, usually, the foundation has a substantial safety factor. This complies with the '20° rule' given earlier and is found effective in practice

Case histories

A typical example of clay shrinkage aggravated by partial basements is provided by a group of problems that arose in Hampstead, London, N.W.3. Terraced properties about 100 years old, three storeys high with two-storey bays on the front, are built upon a hillside which is about 1 in 10 at the steepest part. Alternate houses are handed and thus the construction is generally of party walls flanked by entrance halls, staircases and semi-basements, alternating with party walls combined with chimney breasts between the living rooms. Thus relative to the ground surface the party walls between adjoining entrances have deeper foundations than the heavier party walls containing the chimneys.

A clear pattern of cracking can be seen in the properties, that has developed over a long period and shows little sign of recent movement. Old cracks spread away from the basement on either side indicating relative subsidence of chimney walls.

Following the very dry summer of 1969 a number of front bays sank away from the main terraces causing severe cracks up to 20 mm wide. Large plane trees grow in the street (*Fig. 10*) and in three cases these were immediately opposite the damaged bays. Inspection pits revealed shrinkable clay containing plane tree roots beneath the shallow bay foundations. The most severe cracks were immediately adjoining the deeply founded walls where the relative settlement was most dramatic. The underpinning scheme adopted is shown in *Fig. 11*.

Fig. 12 shows one bay where, mainly for legal reasons, two complete seasonal cycles passed before remedial works were carried out. By setting in temporary needles beneath the bay, and by tightening the wedges under the needles each summer the brickwork above the needles was kept from cracking, although below the needles the bay deteriorated badly.

Another property in the same street also had severe internal cracks and it was found that a main internal partition had been built approximately

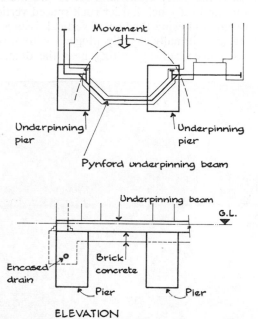

Fig. 11. Underpinning scheme for the bay of a house in Hampstead

Fig. 12. Cracked bay of a house in Hampstead. Note the brickwork dropping away from beneath temporary needles

417

600 mm off line from the internal basement wall causing severe local differential subsidence.

Figs. 13 and 14 illustrate the interesting and unusual example met at Sevenoaks, Kent and referred to earlier. In 1960 a large hole some 5 m across appeared at the back of a house following a severe storm. The outcropping stratum in the area is the lower Greensand series and the subsoil was very soft sandstone with bands of much harder rock spaced vertically at 0.5–1 m intervals; vertical vents following the line of vertical fissures are common in the strata but they are normally densely plugged with washings from the surface. There was a deep railway cutting near the damaged house, and

Fig. 13. Collapse of subsoil beneath foundations of a house at Sevenoaks, Kent

Fig. 14.

between the cutting and the house local surface water drainage was discharged into another fissure. The hole beneath the foundations was clearly a deep swallow hole and the Author proposed a split ring of concrete jacked tight against the side of the hole and pinned in place so that the jacks could be removed. To support the foundations a reinforced concrete beam spanned across the ring which was then filled and sealed with tarmacadam to keep out water. The remedial works have been successful despite further consolidation within the vent.

Acknowledgement

The Author gratefully acknowledges the permission given by the Directors of Pynford Limited to publish the analysis of cases and photographs illustrating this Paper.

REFERENCES

Epstein, E. (1973). 'Roots', *Scientific American*. May, 48
Ward, W. H. and Green, H. (1948). 'House Foundations; The Short Bored Pile', *Final Report Proc. Public Works and Municipal Services Congress and Exhibition*. London

V/11. Differential settlements of cylindrical oil tanks

Richard A. Sullivan
Managing Director, McClelland Engineers, Ltd., Harrow, Middlesex

Joseph F. Nowicki
Manager, Civil Engineering Division, The M. W. Kellogg Company, Houston, Texas

Introduction

A major expansion to a refinery in Colombia, South America, included constructing four cylindrical steel storage tanks in 1966 over weak marsh deposits. Two of the tanks are of the fixed cone roof type and are located about 300 m south of two smaller diameter floating-roof tanks. The tanks are supported at grade and were preloaded by controlled water testing (Darragh, 1964) over periods of several months to improve tank stability by densifying and strengthening the foundation soils, and to force substantial settlements of the tanks to occur prior to placing them in service thereby minimizing later operational settlements.

Soil conditions

Soil conditions were explored by five wash borings advanced using a chopping bit and 6.5 cm casing. The soils were generally sampled with a 5 cm split-spoon sampler in accordance with the standard penetration test procedure, and some selected undisturbed samples of soft clays were obtained with 5 cm thin-walled tubes pushed into the soil.

General soil stratigraphy at the two tank sites is depicted by the profiles in *Fig. 1* showing the soils may be divided into three principal zones. Installation of piezometers and inclinometer wells used for controlling the water load testing of the larger tanks confirmed expectations that the Tertiary surface is extremely irregular.

The surface fill is 1 to 2.5 m thick consisting of well compacted sand, silt, clay and gravel. The groundwater level fluctuates within the fill.

The uppermost stratum of Holocene deposits is soft-to-firm clays having undrained shear strengths ranging from 12 to 45 kN/m². To about 6-m depth the soils are grey clays and silty clays with moisture contents between 30 and 40%; below 6 m the clay becomes highly plastic with organic matter and moisture contents ranging from 50 to 90%. Compressibility characteristics

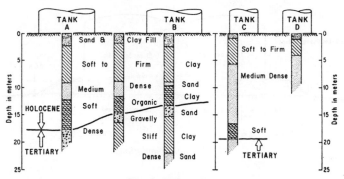

Fig. 1. Soil profiles

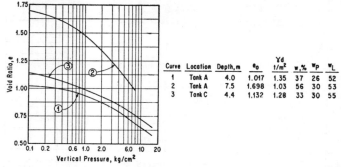

Fig. 2. Compressibility characteristics

of these lightly overconsolidated clays are presented in *Fig. 2* with pertinent classification properties. The middle stratum consisting of silty fine sand is medium-dense in condition based on standard penetration test results. The lowest stratum is soft and highly compressible organic clay containing wood with moisture contents generally greater than 140%.

The Tertiary deposits varied from light-coloured stiff clays and sandy clays with gravel having moisture contents of about 20% to dense fine sands with some gravel. These soils are strong and of low compressibility.

Tank foundations

The four tanks 12-m-high for storing refined product having a specific gravity of about 0.8 are supported on compacted fill pads as illustrated in *Fig. 3*. Tanks A and B are 41 m in diameter with a cone roof supported by interior columns bearing on the tank floor. Tanks C and D have floating roofs and are 20.5 m in diameter. Although large settlements and non-uniform shell settlements were anticipated, it was economically preferable to support the tanks at grade rather than on costly pile foundations and accept the possible need to relevel floors after water testing.

The rate of water loading to each tank is shown by the load diagrams on

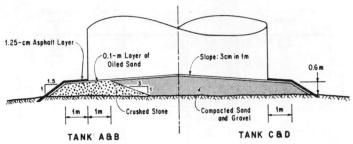

Fig. 3. Tank foundations

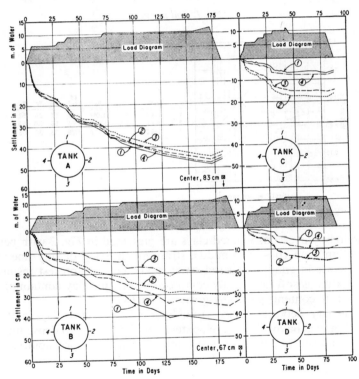

Fig. 4. Tank shell settlements

Fig. 4. The foundation soils beneath the cone-roof tanks were precompressed for a substantial period with a surcharge ratio of 0.165 and for a short time with a ratio of 0.33, where surcharge ratio is defined as the ratio of the excess load to the sustained load. Soils beneath the floating-roof tanks were similarly surcharged.

422

Settlement observations

Tank settlement readings were made on eight equally spaced clip angles welded on each tank shell about 0.5 m above the tank floor using a surveyor's level. These observations were referenced to stable benchmarks bearing on the strong Tertiary soils, and their accuracy was within 0.3 cm.

Observed settlements for the four tanks are presented in *Fig. 4*. The time–settlement curves of Tank A show relatively uniform settlement of the tank shell whereas the spread in the time–settlement curves of Tanks B, C and D reveal non-uniform shell settlements. Tank A settled 46 cm at the shell and 83 cm at its centre, as measured after the tank was empty, giving a shell-to-centre settlement ratio of 0.58 which is within the expected range for lightly overconsolidated clays (Casagrande, 1964). The edge settlement of Tank B ranged from 20 to 43 cm and the tank centre settled 67 cm. The smaller floating roof Tanks C and D experienced less than one-half the settlement of the cone-roof tanks with edge settlements ranging from 7 to 18 cm and 7 to 16 cm, respectively.

Analysis of tank settlements

Initial distress of cone-roof tanks with interior columns supporting the roof is generally buckling of the roof due to distortion of the tank bottom. Floating roof tanks or cone roof tanks without interior columns can tolerate greater differential movements before distress is noticeable by pulling of the shell inward. Tests run on single-fillet welded bottom joints and analysis of membrane stresses produced by settlements (Rinnie, 1963) indicate that the safe permissible angular distortion between the shell and centre of a tank is 1/45. Investigations of structural tank bottom failures (Clarke, 1971) suggest an allowable differential tank bottom settlement of 5 cm in 9 m (1/180) with maximum shell settlements of 30 cm.

The average angular distortion between the centre and shell of Tank A was 1/55 while for Tank B it ranged from 1/44 to 1/85. The roof of Tank A remained stable indicating a bottom distortion of 1/55 was tolerable. After Tank B was emptied, the roof had some dimples in its surface and was considered unstable. Interior columns on the low side of the tank were raised by cutting windows in the tank bottom and ramming fill under the portion of the bottom plate supporting the column.

Tilting of a tank shell can be examined by plotting the observed settlement of equally spaced survey points on the developed tank cylindrical surface to produce a settlement curve that is compared to a sine or cosine curve. Curves of this type are shown in *Fig. 5* for Tanks B and C. The expression for the cosine curve is:

$$y = (\Delta s/2) \cos \phi$$

where Δs is the maximum differential settlement between any two settlement points, and ϕ is the horizontal angle between the radial line through the settlement point and the reference radial line. If the observed settlement readings fall on the cosine curve then the shell has tilted in a plane.

Excessive planar tilting of the four tanks was not significant but it can

423

cause elliptical deformation of a tank shell (Carlson and Fricano, 1961). Non-planar tilting can produce distortion of a shell that may result in a tank being out-of-round which is particularly troublesome with floating-roof tanks because the seal can be broken and the roof may bind against the shell. As shown by the curves in *Fig. 5*, the observed settlement curves for Tanks B and C deviate from the cosine curves, and the vertical distance between two curves is the amount of distortion from planar tilt.

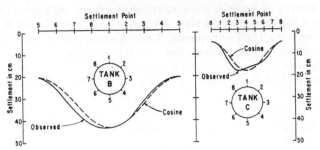

Fig. 5. Differential shell settlements

Study of the performance of 21 floating roof tanks ranging in diameter from 12 m to 110 m and in height from 12 m to 19 m (McClelland Engineers, 1973) indicated a safe tolerable out-of-plane distortion of 3 cm that was essentially independent of tank height and diameter. If the out-of-plane settlement is greater than 4.5 cm, then the probability of difficulties with a floating roof is high. The maximum observed out-of-plane distortion of cone-roof Tank B was almost 4 cm. Floating-roof Tank C experienced 3 cm of out-of-plane distortion and the floating roof continued to operate without difficulty.

Conclusions

Steel cylindrical tanks with crowned bottoms can tolerate considerable total and differential settlements, however, floor movements are significant for cone-roof tanks with interior columns. Tank bottom angular distortion of 1/55 was tolerable for a column-supported cone-roof tank that settled 40 cm more at the centre than at the shell. Out-of-plane shell distortion of 3 cm did not cause difficulty with floating-roof tanks.

REFERENCES

Carlson, E. D. and Fricano, S. P. (1961). 'Tank Foundations in Eastern Venezuela', *J. Soil Mech. & Found. Div. A.S.C.E.* Vol. 87, No. SM5, 69–90
Casagrande, L. (1964). 'Effect of Preconsolidation on Settlements', *J. Soil Mech. & Found. Div. A.S.C.E.* Vol. 90, No. SM5, Part 1, 349–362
Clarke, J. S. (1971). 'How to Handle Tank-bottom and Foundation Problems', *Oil & Gas J.* 5 July, 82–84
Darragh, R. D. (1964). 'Controlled Water Tests to Preload Tank Foundations', *J. Soil Mech. & Found. Div. A.S.C.E.* Vol. 90, No. SM5, Part 1, 303–329
McClelland Engineers, Inc. (1973). Unpublished report
Rinnie, J. E. (1963). 'Tanks on Soft Soils are Economic Challenge', *Petro/Chem Engineer.* September, 56–58

V/12. The performances of buildings founded on river alluvium

S. Thorburn
Thorburn and Partners, Glasgow

R. S. L. McVicar
Chief Engineer, Corporation of Glasgow, Department of Architecture and Civic Design

Introduction

The main objective of the Paper is to record the performances of medium and high rise buildings which have been founded on the alluvial plain of the River Clyde during the last decade.

The Paper does not seek to establish relationships between structural deformations and the measured soil parameters but rather to indicate the general structural performances of buildings for the guidance of Engineers and Planners. It is fortuitous that the case histories concern buildings which are of relatively traditional construction and which have been founded on relatively uniform ground conditions.

Basic and comprehensive information concerning the geological structure at the site of each building together with the field and laboratory tests is provided and observations made if relationships between different forms of testing are evident. A significant financial investment has been made on site investigation work within the alluvial plain of the River Clyde in order to establish the geological structure, the soil properties of each geologically significant stratum and permit future development work to proceed with confidence and economy of decision.

The measurements of the settlement of each building due to consolidation and compaction of the underlying soils were made by means of accurate but ordinary levelling instruments and the survey work was carried out by the Contract Engineers as part of their normal duties. The settlement records are of necessity limited in extent and accuracy since the investigation work cannot be placed in the category of planned research but is rather an attempt to accumulate useful data for guidance in making reliable but economical engineering decisions.

425

Future research into the performance of buildings founded on the alluvial plain of the River Clyde has already been planned in light of experience and all measurements will be made by precision instruments.

The performances of the various buildings are discussed in relative terms and all buildings were examined for evidence of structural distortion which would readily be commented upon by the occupants due to superficial damage to finishes.

Geological history

The ancient landscape within the Glasgow district is essentially preglacial and was formed during a period when the land surface was at an elevation at least 90 m higher than the present elevation. Although the ancient primary drainage system is similar in direction to the existing system, an examination of the configuration of the contours of the ancient land surface within the district, indicates that two large valleys exist along the axis of synclinal folds. These two large valleys in the preglacial landscape had been infilled with alluvium before glaciation and ice-movement sculptured the district. The pre-glacial valleys converge at Clydebank and Renfrew and the river Kelvin flows for a considerable distance along the axis of the deepest buried valley known as the Bearsden or Kelvin channel.

The main drainage channel of the district at the present time is the River Clyde and it is of interest to note that the course of the river is co-incident with the axis of the shallower of the two buried valleys.

The central portion of Glasgow is mainly situated on the relatively flat plain which exists on either side of the River Clyde. On the north side of the river the ground surface slopes steeply upwards from the perimeter of the alluvial plain and the northern portion of the City is underlain by glacial till. The superficial deposits within the alluvial plain of the River Clyde may be divided in accordance with three geological epochs:

(1) The Glacial period.
(2) The period of Marine Deposition.
(3) The Recent period.

The first period is represented by the formation of an extensive mass of glacial till, commonly described as 'boulder clay'. The glacial till contains intercalations of sands, silts and silty clays, as a result of the ice-sheet having taken up and incorporated within the till the pre-glacial deposits which existed within the valleys and depressions in the ancient landscape. An interesting feature of the glacial till is that it has been deposited in the form of lenticular hills or mounds, locally known as drumlins, with their long axes oriented parallel to the direction of the movement of the ice-sheet as shown in *Fig. 1.*

The second period was a time of quiet marine deposition when a vast body of salt and brackish water covered the lowlying land surface within the Glasgow district and sand, silt and silty clay were deposited in that geological sequence. The uppermost layer of silty clay is always laminated and, in many places, near the surface of the deposit, are found decaying roots or stems of water plants.

The third and last period was a time of elevation of the land and the deposition of the recent freshwater alluvium and fluviatile deposits. The

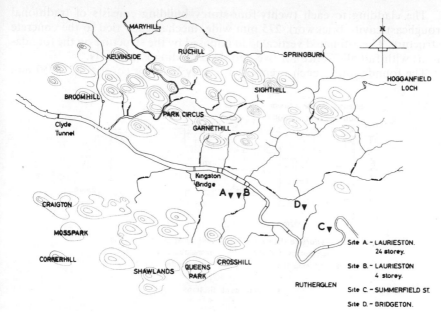

Fig. 1. Plan of Glasgow district

third period is sometimes referred to as the human epoch, since canoes were discovered within the fluviatile deposits at various locations along the present course of the River Clyde.

Case histories

Reference should be made to *Fig. 1* for the locations of each building relative to the present course of the River Clyde.

Laurieston/Gorbals
Twenty-four-storey blocks of flats—Block Reference Nos. 1 and 2

Structural Elements

Each structure consists of a combination of cast-in-place reinforced concrete columns and shear walls having the plan distribution shown in *Fig. 2*. The floors are of cast-in-place reinforced concrete flat slab construction and provide the necessary horizontal diaphragm action to ensure interaction between the vertical support elements.

Three-dimensional structural analyses of the basic combined structure indicate that the overturning moments due to the wind forces may be expected to have the distribution shown in *Table 1*. In order to convey an impression of the stiffness of the unclad basic structure, the theoretical horizontal translations and rotations of the structure at roof level have been super-imposed on *Fig. 2*.

427

The cladding to each twenty-four-storey building consists of traditional roughcast cavity brickwork 275 mm wide, mechanically tied to the concrete structure. The estimated vertical load imposed by the structure on the foundations with full allowance for live and dead loading is 110 MN.

The foundation of each twenty-four-storey building consists of a 750 mm

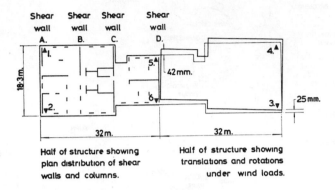

Half of structure showing
plan distribution of shear
walls and columns.

Half of structure showing
translations and rotations
under wind loads.

▾ – Settlement Stations

Fig. 2

Table 1

Shear wall reference	Percentage of total overturning moment
A	26
B	5
C	51
D	18

thick reinforced concrete flat slab supported by groups of driven West's Concrete Shell Piles. The average vertical load per pile neglecting wind forces may be expected to be of the order of 650 kN.

The pile groups were designed on the basis that a sufficient number of piles were provided locally to each column and shear wall such that these individual structural elements could act as stable independent units although interconnected by the slab foundation.

Site investigation work

The site investigation boreholes which were sunk within the solum of each of the twenty-four-storey buildings revealed the geological structure shown in *Fig. 3*. Undrained triaxial compression tests carried out on samples of the laminated silty clay deposit obtained by means of 102 mm standard open-drive tube samplers, gave the normal undrained strength–depth relationships, *Fig. 3*, which may be expected for this deposit using such methods of sampling and testing. The mean line which has been drawn does not indicate the

428

undrained strength profile for the deposit but has been drawn purely as a line of reference.

The Authors have observed over a period of fifteen years of site investigation work that the volume of anomalous test results is such that serious consideration must be given to certain standard methods of sampling and testing since the measured soil parameters appear to be misleading as the

SPECIFIC CONE RESISTANCE

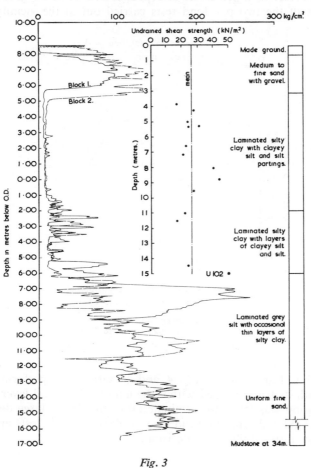

Fig. 3

result of sample or ground disturbance. McGowan, Barden, Wilby and Lee have carried out a limited investigation into the problems presented by sample disturbance.

Fig. 3 indicates the results of quasi-static cone penetration tests using an electrically operated cone. If the static cone resistances, measured as the cone penetrated the laminated silty clay layer, are used in conjunction with the empirical relationship given by Sanglerat (1972) (undrained shear strength $= C_r/15$) then the strength–depth relationship is relatively uniform and linear

429

with increasing depth and the average undrained shear strength is of the order of 45 kN/m².

Precontract pile tests were carried out at the location of each of the twenty-four-storey buildings in order to provide supplementary information to that obtained from the site investigation work and enable a decision to be made concerning the depths of penetration of the piles for the contract. This supplementary investigation work indicated that the bases of the driven large displacement piles could be founded within the layer of uniform fine sand, the upper surface of which exists at depths of 21.5 m below ground surface. The results of the two pile load tests carried out in the vicinity of Block

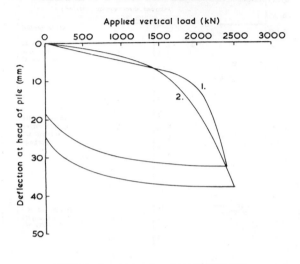

CURVE I. Constant rate of penetration test
Block I. 525 mm dia to 23.6 m

CURVE 2. Maintained load test
Block I 525 mm dia to 35.7 m

Fig. 4

Reference No. 1 are shown in *Fig. 4*, and the satisfactory performance of the test pile with its base founded within the uniform fine sand layer is evident.

The decision was taken to terminate all of the driven large displacement piles within the uniform fine sand layer. Subsequent pile tests and building performance justified this early decision.

Recorded settlements

The levelling stations were located as shown in *Fig. 2* and consisted of levelling plugs screwed into metal sockets inserted into the surface of the raft foundations. It is unfortunate that additional levelling stations could not be installed between the centre and corners of each building but storage space requirements for the construction works did not permit more than six levelling stations.

In view of the scarcity of information concerning the relative building deformations between the gables and the centres of both twenty-four-storey buildings, no attempt has been made to indicate settlement contours since such action could provide misleading information. *Table 2* indicates the ultimate total settlements measured at the six levelling stations installed in both twenty-four-storey buildings.

Table 2

Block No.	Settlement Station No.	Maximum total settlement (mm)
1	1	16.8
	2	10.7
	3	12.2
	4	15.3
	5	24.7
	6	17.4
2	1	16.8
	2	18
	3	21
	4	16.8
	5	30.5
	6	30.2

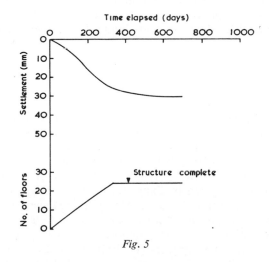

Fig. 5

It will be observed from *Fig. 5* which shows the time–settlement curves that the settlement response was rapid and closely conformed to the rate of application of the construction loads. If one assumes that the rigidity of each half of the building was such that a linear settlement relationship exists between the levelling stations then the recorded maximum relative deflection is 1: 1666.

The settlement records indicate that Block Reference No. 2 has settled more than Block Reference No. 1 and it is also of interest to observe that for both buildings the maximum settlement occurs at Settlement Station 5.

Laurieston/Gorbals
Four-storey blocks of flats—Block Reference C

Structural elements

The basic structure consists of 100 mm thick precast concrete panel construction with cross-walls at about 3.5 m centres. It is a relatively monolithic reinforced concrete cell-type structure after completion of the interconnection of the wall and floor panels. An outer leaf of 100 mm brickwork acts as cladding to the gables and the cellular elevations are clad with roughcast cavity brickwork 275 mm wide, mechanically tied to the precast concrete structure. The foundation of the 100 m long four-storey building elements consists of a 250 mm thick reinforced concrete slab with edge-beams having downstand dimensions of 300 mm wide by 550 mm deep. The slab foundation is about 8 m wide.

The 100 m long block was divided into four unequal lengths by complete separation joints which divide both structure and foundations. In order to avoid the development of high soil stresses at each separation joint the reinforced concrete slab foundation was overlapped and a horizontal sliding joint constructed to provide complete separation. The estimated net loading intensity transmitted by the raft foundation to the ground with full allowance for dead and live load is 40 kN/m².

The 100 m long by 8 m wide discontinuous flexible concrete slab foundation was reinforced to provide resistance to not only local bending moments but also the general bending moments induced by the differential settlement of the building.

Site investigation work

Fig. 6 shows the geological structure revealed by site investigation boreholes sunk in close proximity to the 100 m long four-storey building. *Fig. 6* also indicates the results of a quasi-static cone penetration test carried out within the low rise development site. If the static cone resistances measured as the cone penetrated the laminated silty clay layer are used in conjunction with empirical relationships given by Sanglerat then the strength–depth relationship is relatively uniform and the average undrained shear strength is of the order of 75 kN/m². Undrained triaxial compression tests were carried out on samples of the laminated silty clay deposit obtained by means of 102 mm standard open-drive tube samplers (U. 102) and 152 mm fixed piston type samplers (UP. 152). The undrained strength–depth relationships are given in *Fig. 6*. The mean lines which have been drawn do not indicate the interpreted strength profiles for the laminated cohesive deposit but have been drawn only as lines of reference. It is the opinion of the Authors that it would be incorrect on the basis of current knowledge to deduce the strength–depth relationships for this extensive deposit of laminated silty clay in view of the known sample disturbance and the excessive variations in undrained shear strength measured by numerous investigators.

It is of interest to observe from *Fig.* 6 that the undrained shear strengths obtained from UP. 152 samples have a mean value of 33 kN/m² in direct contrast to a mean shear strength of 20 kN/m² obtained from tests on U.102 samples.

Fig. 7 shows the results of 76 mm standard oedometer tests carried out on samples of the laminated silty clay deposit obtained by means of 102 mm standard open-drive tube samples. The coefficients of consolidation of the

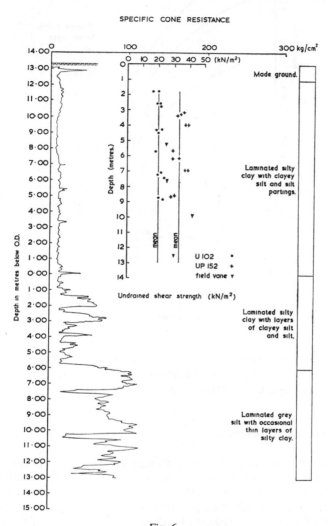

Fig. 6

laminated silty clay deposit for both vertical and horizontal directions have been determined for samples taken from a depth of about 4 m below ground surface. The relationships between the coefficients of consolidation and effective pressures were determined from UP. 152 samples tested in 152 mm Rowe cells and the graphical relationships are given in *Fig.* 7. The test results indicate that there are relatively small variations in the coefficients of consolidation within the range of effective pressure of 700 kN/m² for drainage

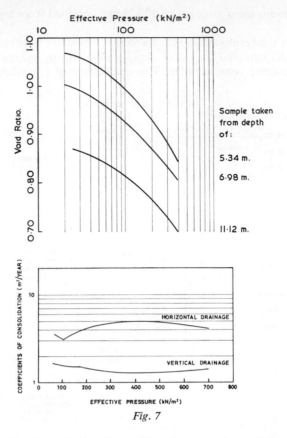

Fig. 7

in vertical and horizontal directions. The average values of the coefficients of consolidation for vertical and horizontal drainage are 1.5 and 4 m²/a respectively giving a C_h/C_v ratio of about 2.67.

Recorded settlements

The levelling stations consisted of levelling plugs screwed into metal sockets inserted in the surface of the raft foundations. The contours of ultimate total settlements have been interpreted from the levelling measurements and incorporated in *Fig. 8*. The configuration of the contours is irregular and not such as would readily be predicted by theory. The influence of the variations of ground support on the performance of the 100 m long four-storey building is evident from the variations in the contours. The maximum relative deflection of the four-storey building is 1 : 500.

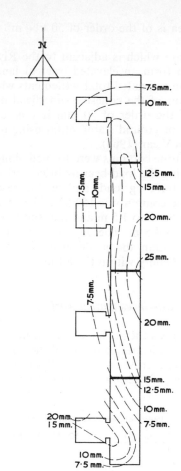

N

—7·5mm.
—10 mm.

12·5mm.
15mm.

20mm.

25 mm.

7·5mm.
10mm.

7·5mm.

20mm.

15mm.
12·5mm.

10mm.

20mm.
15 mm.

7·5mm.

10mm.
7·5 mm.

Fig. 8. Settlement contours,
Laurieston 1B

Summerfield Street—Dalmarnock
Five-storey blocks of flats—Block Reference Nos. 1 to 6

Structural elements

The basic structures consist of load-bearing 275 mm cavity brickwork, partly roughcast, with precast reinforced concrete floor and roof slabs. The plan dimensions of each block of flats is about 15 m square but each block consists primarily of two separate rectangular blocks measuring about 15 m by 6 m interconnected by the reinforced concrete floor slabs. The central core between the rectangular primary blocks of load-bearing cavity brickwork is utilised as the stair access to the dwellings.

The individual primary blocks are relatively rigid units because of their compact shape and the disposition of the internal walls. The load-bearing brickwork walls are supported by reinforced concrete continuous footings founded on stabilised made ground. The average net loading intensity imposed on the made ground by the five-storey buildings over their entire plan area is of the order of 50 kN/m² with full allowance for dead and live loads.

The site which is adjacent to the River Clyde has been infilled in former times to form an elevated tract of land above the flood level of the river. Since the estimated total settlements which would take place due to the consolidation of the cohesive superficial deposits underlying the made ground were of the order of 50 mm it was decided to employ the depth vibrator method of ground stabilisation using the techniques described by Thorburn and MacVicar (1968).

The stone columns were formed along the lines of the continuous footings supporting the load-bearing brickwork at various spacings depending on the applied building loads and on the basis that the maximum allowable load per stone column was 150 kN. The spacings of the stone columns vary from about 0.68 m to 1.5 m and the lengths of the stone columns vary from about 3.66 m to 4.5 m. The depths of penetration of the vibrator were contained mainly within the made ground and the stone columns, therefore, were formed mainly within the fill materials.

Site investigation work

Fig. 9 shows the geological structure revealed by site investigation boreholes sunk at the locations of five of the buildings. Undrained triaxial compression tests were carried out on samples of the laminated silty clay deposit obtained by means of 102 mm standard open-drive tube samplers, and the undrained

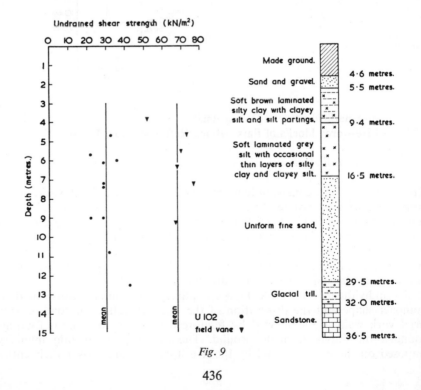

Fig. 9

436

strength-depth relationships are given in *Fig. 9*. The results of field vane tests carried out within the laminated silty deposit have also been incorporated. The mean lines have been drawn for reference purposes only and it will be observed that the mean shear strength obtained from the field vane tests is more than twice the results obtained from the undrained triaxial compression tests. As previously stated, serious consideration must be given to certain standard methods of sampling and testing in order to establish the best methods of investigation. A considerable expenditure can be incurred on site investigation work resulting in the production of anomalous data.

A representative sample of the laminated silty clay deposit was subjected to mechanical analysis by sedimentation and Atterberg limit tests. The results of these classification tests are given in *Table 3*.

Table 3. PARTICLE SIZE DISTRIBUTION

Particle size (mm)	*Percentage passing*
0.02	100
0.006	74
0.002	50
0.0006	33
Liquid limit:	55
Plasticity index:	32

Atterberg limit tests carried out on the thin layers of clayey silt which are intercalated with the laminated silty clay deposit gave the result: Liquid limit, 36, Plasticity index, 17.

A borehole sunk to a depth of 11 m within the area of block reference No. 2 revealed the presence of a 4.3 m thick layer of very soft dark grey very silty clay containing organic matter, immediately beneath the made ground.

The Authors carried out trials involving the construction of stone columns through the very soft highly compressible deposit by means of a depth vibrator within the area of Block Reference No. 2. It was decided on completion of the trials that ground stabilisation was an acceptable solution with the full realisation that relatively much greater settlements would be experienced by this five-storey block due to the inability of the very soft high compressible deposit to prevent abnormal lateral yield of the stone columns when subjected to the building loads. The average undrained shear strength of the very silty clay deposit was 14 kN/m^2.

Recorded Settlements

The levelling stations were located in the brickwork underbuilding at each corner of the six blocks of flats. The measurements were taken over a period of about four years in view of the apparently continuous and significant settlement of Block Reference No. 2. *Fig. 10* shows the time settlement curves for each of the six five-storey buildings and the abnormal behaviour of Block Reference No. 2 is evident.

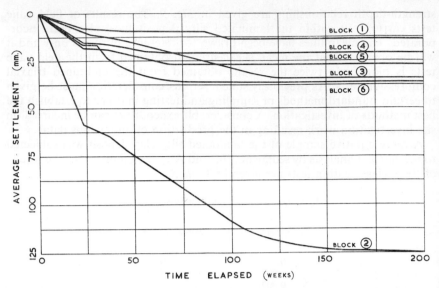

Fig. 10. Time–settlement curves for 5-storey buildings at Summerfield Street, Glasgow

The recorded maximum and average total settlements for the six buildings are given in *Table 4*.

Table 4

Block reference	Maximum total settlement (mm)	Average total settlement (mm)
1	16	14
2	143	129
3	38	35
4	25	21
5	35	28
6	41	37

The settlement records indicate that although the maximum total settlement of Block Reference No. 2 is as great as 143 mm, the maximum differential settlement is only 20% of the maximum total settlement. It would appear that the individual cell construction of the five-storey blocks combined with the more favourable distribution of vertical stresses under stabilised essentially granular made ground has resulted in minimal relative deflections. The records indicate that the maximum differential settlement experienced by the five-storey buildings is 1 : 472.

Contours of ultimate total settlements have not been presented because of the scarcity of information on file concerning any measurements within the central core of each five-storey building.

438

Bridgeton
Fifteen-storey blocks of flats

Structural elements

Each structure consists of a reinforced concrete frame structure with no-fines concrete cladding having a roughcast finish. The floors are of cast-in-place reinforced concrete flat slab construction and provide the necessary horizontal diaphragm action to ensure interaction between the vertical support elements. The no-fines concrete walls act as shear walls and in combination with the reinforced concrete frame resist the wind forces. The resulting structure is relatively monolithic and rigid.

In view of the existence of former coal workings beneath the site it was considered prudent to found the structures on the upper superficial deposits rather than transmit the building loads by means of piles to the surface of the bedrock which had been extensively undermined. Provision was made in the design of the foundations for jacking the combined frame and shear wall structures to correct any tilting or distortion caused by mining subsidence. It was considered advisable to limit the loading intensity on the laminated silty clay deposit, and, therefore, a decision was taken to utilise semi-buoyant reinforced concrete raft foundations.

In order to reduce the total loading intensity of 135 kN/m² on the ground imposed by the 15-storey structures, the semi-buoyant foundation was founded at a a depth of 4.3 m below ground surface. The resulting net loading intensity imposed by the semi-bouyant foundation on the underlying compressible soils was of the order of 54 kN/m².

Site investigation work

The site investigation boreholes which were sunk within the site revealed the geological structure shown in Fig. 11. Samples of the laminated silty clay and silt deposits were obtained by means of 102 mm standard open drive tube samplers. The results of undrained triaxial compression tests carried out on the samples obtained by means of 102 mm standard open-drive tube samplers gave the undrained strength–depth relationship shown in *Fig. 11*. The mean line which has been drawn does not indicate the undrained strength profile for the laminated silty clay deposit but has been drawn purely as a line of reference.

76 mm standard oedometer tests were carried out on samples of the laminated silty clay deposit obtained by means of 102 mm standard open-drive samplers and final consolidation settlements ranging from 50 to 56 mm were predicted, see Somerville and Shelton (1972).

Recorded settlements

The levelling stations, totalling nineteen in number, were located within the solum of each fifteen-storey building and consisted of levelling plugs screwed into metal sockets inserted in the reinforced concrete columns at a height of

439

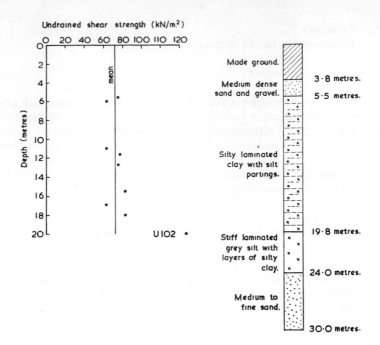

Fig. 11. Bridgeton

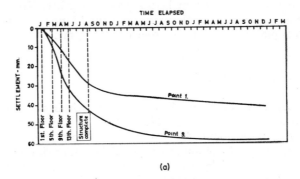

(a)

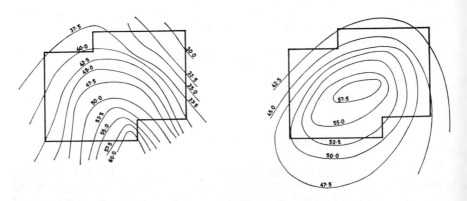

(b) Approx. Maximum Angular Distortion 1 in 550   (c) Approx. Maximum Angular Distortion 1 in 730

Fig. 12. Bridgeton (a) Block C (b) Block B and (c) Block C

about 300 mm above ground floor level. Measurements of the change in level at each station were made at approximately three monthly intervals. *Fig. 12* shows the time-settlement curves for levelling stations References 1 and 9 approximately located at the corner and the centre of the foundations of Block Reference C. The contours of ultimate total settlements have been interpreted from the levelling measurements and incorporated in *Fig. 12*. It will be observed from the time settlement curves that the ultimate settlements were attained after a period of about two years had elapsed.

Performances of buildings

The medium- and high-rise buildings have performed in a most satisfactory manner and their performances have justified the decisions to found these major structures on or within the superficial deposits. No settlement cracks have been observed during the inspections by the District Engineers and the only cracks which have developed may be attributed to elastic compression, concrete shrinkage, and thermal movements.

Conclusions

Current knowledge indicates that traditional medium- and high-rise buildings can be safely founded on or within the superficial deposits of the alluvial plain of the River Clyde provided that the foundations are designed in accordance with soil mechanics principles. The expensive solution of founding the bases of piles on the bedrock which exists at depths in excess of 30 m need not be adopted as a means of ensuring satisfactory building performances.

The comprehensive investigation work has indicated that serious and urgent consideration must be given to certain standard methods of sampling and testing since the measured soil parameters appear to be misleading as the result of sample or ground disturbance.

Acknowledgements

The Authors wish to thank the Corporation of Glasgow for permission to publish the results of the extensive field and laboratory investigations and to the University of Strathclyde for information obtained from their research work using the Rowe cell.

The Authors also wish to record their thanks to their colleagues, P. Wilby, W. M. Reid, J. Barr, S. Sheen and W. Ormond for their assistance in obtaining and preparing the information incorporated in the Paper.

The assistance of the numerous District Engineers is also acknowledged.

REFERENCES

McGowan, A., Barden, L., Wilby, P. and Lee, S. H. 'Sample Disturbance in Soft Alluvial Clyde Estuary Clay'. Private communication

Sanglerat, G. (1972). 'The Penetrometer and Soil Exploration'. Elsevier Publishing Company

Thorburn, S. and McVicar, R. S. L. 'Soil Stabilization Employing Surface and Depth Vibrators', *The Structural Engineer*. October, No. 10, Vol. 46

Somerville, S. H. and Shelton, J. C. 'Observed Settlement of Multi-Storey Buildings on Laminated Clays and Silts in Glasgow', *Géotechnique*. Vol. 22, No. 3, September (1972)

V/13. Observed settlement and cracking of a reinforced concrete structure founded on clay

D. L. Webb
D. L. Webb & Associates, Consulting Civil Engineers, Durban, South Africa

Summary

Level observations on a two-storey reinforced concrete structure founded on reclamation fill underlain by deep normally consolidated silty clay, indicated that certain columns were settling very much more rapidly than others. The ground profile is described and the results of laboratory tests given. Details of the structures are shown in diagram form, and from analysis of accurate time settlement records available it is shown that while a rotational strain of about 1 in 175 was tolerable as a result of slow differential settlement of the new concrete, cracking occurred at this strain where the settlement was rapid.

Introduction

An area of some 400 ha of low-lying partly swampy ground under sugar cane cultivation, 15 km to the south of Durban Harbour, was reclaimed for light industrial purposes in 1968 and 1969 by filling with clayey fine dune sand, known as Berea red sand and described by Webb and Hall (1967). The sand fill, which was traffic compacted, varies in thickness from 1 m to about 3 m over the area, and is underlain by normally consolidated post-Glacial estuarine and alluvial clays and silts to depths of up to 50 m in places.

One of the earliest buildings to be erected in the reclaimed area was a large structural steel warehouse and two-storey reinforced concrete office block. It was decided, for economic reasons, to found the entire building on shallow spread footings after vibroflotation to a sand layer at a depth of about 2.5 m, and to design the building to tolerate the anticipated settlement. A month after completion of the reinforced concrete framework survey levels revealed that two groups of office block columns had settled excessively. The remaining 40 or more columns were settling at a slow rate compatible with that of the regional settlement.

Ground profile

To investigate the reason for the large settlements, six borings were put down close to the columns, to depths of 15 m below top of reclamation fill. Results obtained from each boring were similar and a representative profile

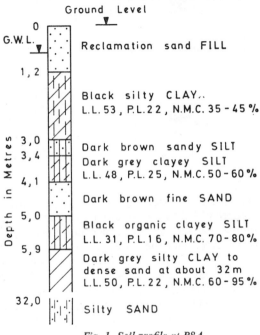

Fig. 1. Soil profile at P8A

is shown in *Fig. 1*. The clay minerals present are of the illitic or hydrous mica type with small proportions of montmorillonite. From previous borings put down in the general area it is known that the dark grey silty clay extends to a depth of about 32 m here and that it is underlain by silty sands 12 m thick resting on fresh Dwyka tillite bedrock.

The borings were advanced by washing through 150 mm diameter casing and samples, 100 mm in diameter, were taken by hydraulically jacking U4 tubes into the clay at intervals of 1.5 m using a 'Wayfarer' drilling rig. Standard Penetration Tests carried out below the level of each U4 sample yielded N values in the range 1 to 15 blows per 300 mm in the clays and sands to 6 m. Between 6 and 32 m the N values range from 0 to 4, but are generally less than 2 and frequently less than 1. Cone resistances in the Dutch Cone penetrometer test at depths of between 6 and 32 m in the general area are commonly 500 to 1000 kPa, and reflect the existence of a sandy layer 1 to 2 m

444

thick within the clay, at a depth of 15 to 17 m, by increasing to as much as 7000 kPa.

Laboratory test results

There is considerable variation in the grain size distribution of the shallow alluvial clayey and silty strata to a depth of 6 m with clay contents, as reflected by representative Curves 1 and 2 in *Fig. 2*, from about 12 to as much as 40%. The underlying dark grey silty clay, Curve 3, is, however, relatively consistent in composition, having a clay-size fraction of about 20% by weight. Results of representative consolidation tests on samples from the general area are

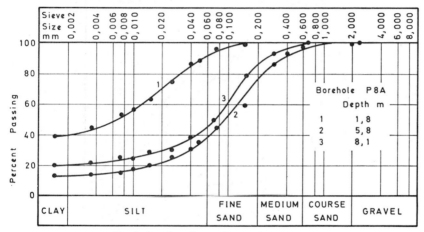

Fig. 2. Results of particle size determination

Table 1. COMPRESSIBILITY AND UNDRAINED SHEAR STRENGTH

Material	m_v kPa^{-1}		c_u kPa	
	Range	Average	Range	Average
Black silty clay	0.0002⎫ 0.0006⎭	0.0004	35–55	40
Dark grey clayey silt	—	—	30–45	35
Black organic clayey silt	0.0006⎫ 0.0012⎭	0.0008	15–25	18
Dark grey silty clay	0.0008⎫ 0.0014⎭	0.0010	12–30	20

plotted as void ratio-pressure curves in *Fig. 3*. Curve 1 reflects the pre-consolidating effect of partial dessication of the shallow materials during their recent geological history. Corresponding curves for the underlying normally consolidated clayey silt and silty clay are numbered 2 and 3 respectively. These curves were corrected for sampling disturbance using the method given by Schmertmann (1953), and on this basis values of m_v the coefficient of volume compressibility based on 3 to 5 tests in each group, are as shown in *Table 1*, for pressure increments of overburden plus 200 kPa.

Undrained shear strengths were determined in the triaxial apparatus using 38 mm diameter × 76 mm high cylindrical specimens cut from U4 tube samples. Some 6 to 8 test results are available for each of these clayey strata in the area and the results are given in *Table 1*. Average values of 4 to 10

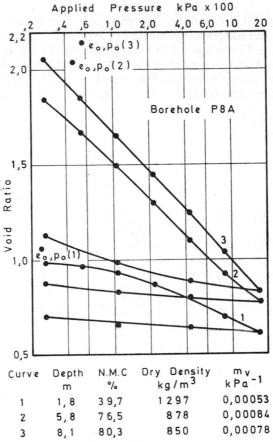

Curve	Depth m	N.M.C %	Dry Density kg/m³	m_v kPa⁻¹
1	1,8	39,7	1 297	0,00053
2	5,8	76,5	878	0,00084
3	8,1	80,3	850	0,00078

Fig. 3. Consolidation test results

Atterberg limits and natural water contents (NMC) determined for samples from each of the shallow clayey strata are given in *Fig. 1*. The liquid and plastic limits of the deeper dark grey silty clay varied in the ranges 36 to 64 and 18 to 31 respectively, with average values as given in *Fig. 1*.

Details of the structure

The office block comprises a two-storey reinforced concrete structure 60 m long and approximately 12 m wide. To minimise detrimental effects of uneven settlement column spacing in the longitudinal direction was made as great as possible and the beam and slab system in that direction as flexible as possible. Columns are bulky to resist severe impact and the cross beams are correspondingly stiff to permit transverse tilting of the structure. The proportion of reinforcing steel employed was 97 kg/m³ which is somewhat greater than usual to permit some redistribution of bending moments without endangering the structure. Additional stirrups were provided in the main beams to resist torsion. The first floor slab was cast 150 mm higher than necessary to permit some subsequent reduction in overhead clearance due to

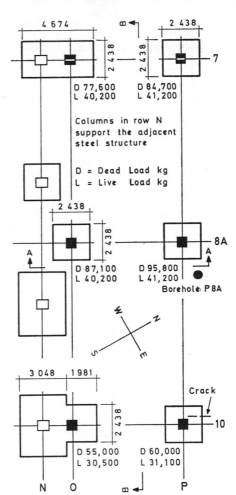

Fig. 4. Foundation layout of eastern end of office block

settlement of the office block in excess of that of the adjacent warehouse. Expansion or movement joints were provided transversely through the structure, including the columns as shown in *Figs. 4. and 5.* A representative cross-section showing the ribbed floor construction is shown in *Fig. 6.* The footings in row *N* along the southern side of the office block support long span steel portals of the warehouse and, although relatively lightly loaded, are larger than those in rows *O* and *P* to provide stability against uplift during wind loading.

Beneath each of the office block columns 4 vibroflot compactions were

447

provided to form compact columns of crushed stone in the clayey soils underlying the sandy reclamation fill, and transmit the foundation load to a layer of sand which generally occurs over the site at a depth of 2 to 2.5 m below ground level and which is 2 to 3 m thick. The vibroflots were located symmetrically beneath the footings at a spacing of 1.70 m and formed stone columns about 800 mm in diameter.

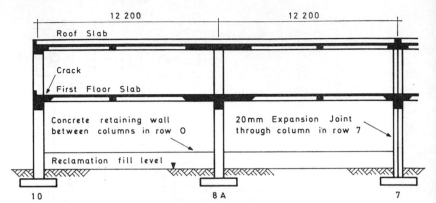

Fig. 5. Longitudinal section B–B

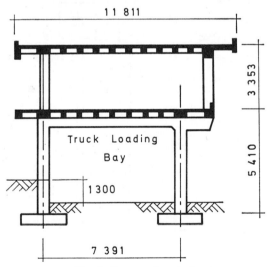

Fig. 6. Cross-section A–A

The construction sequence in 1971 was as follows: End of January—Vibroflotation for office block and warehouse foundations completed, Mid April—Foundations for office block and warehouse completed; End of May—Erection of steelwork, and cladding to warehouse roof completed; Mid June—Concrete to first floor of office block cast; End of July—Concrete to roof of office block cast. Propping was left in position beneath the first floor during construction of the roof and was removed at the beginning of September 1971, thereby transferring the full structural load to the foundations.

448

Settlement observations

Time settlement curves for each of the six columns at the eastern end of the office block, shown in *Fig. 4*, are plotted to a linear scale in *Fig. 7*. After detection of the large settlements in September 1971, level observations were made initially at 2- to 3-day intervals, increasing a month later to weekly and then, in 1972, to monthly intervals.

The site level control beacons were situated some 60 m from the edge of the building, and although unlikely to be affected by settlements due to structural loading and recent earth fill in the warehouse area, it was considered

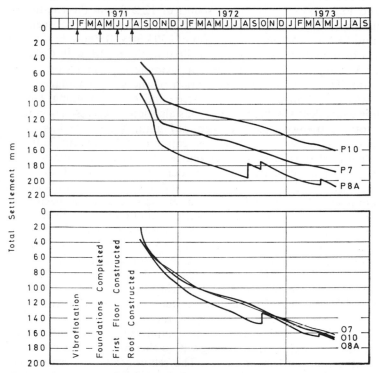

Fig. 7. Time-settlement curves

that they might be affected by regional settlement resulting from the original reclamation fill. For this reason the level of each site beacon, which comprised a short steel peg in a block of concrete cast at ground level in an area remote from building operations and traffic, was related to three stable beacons 300 to 500 m away. Level observations were made on steel pegs in the sides of the columns, and closure errors of the traverses were within 2 mm with respect to the site beacons and within 3 mm with respect to the stable beacons.

Although not very clear, owing to the small scale necessarily employed, the curves in *Fig. 7* reflect an increase in rate of settlement as the propping was removed.

449

Performance of the structure

Despite the relatively large settlements there was no apparent damage to the structure in any form until the end of the third week in September 1971, when a hair-line crack appeared in the top of the north side edge beam of the ribbed floor adjacent to Column P10. The crack extended a distance of 1.5 m from the edge beam parallel to the main cross beam between columns P10 and O10 as shown by the dashed line in *Fig. 4*. At no time was there any sign that the crack had penetrated to the soffit of the floor slab or to the rib beams themselves, but it did penetrate to the neutral axis of the edge beam.

The crack appeared when the differential settlement δ between columns P10 and P8A was 69 mm. For a span l of 12 200 mm the rotational strain δ/l was 1 in 177. At the same time the differential settlement between Columns O8A and P8A was 54 mm over a span of 7393 mm. Because of the transverse stiffness of the structure, however, it tilted to the north, with Columns O8A and P8A being out of plumb by approximately 35 mm over a height of 7620 mm. The effective rotational strain of the stiff undamaged transverse beam was thus 1 in 296. At the western end of the building, Column P4, which is equivalent to P8A in *Fig. 4*, has also settled excessively, and by June 1973 total settlement had reached 187 mm. Corresponding, maximum differential settlement between this column and adjacent column P2A. equivalent to P10, is 71 mm, which represents a rotational strain of 1 in 172. No cracking has occurred during the deformation of this beam which has increased steadily for 26 months. The cladding, partition walls and finishes were commenced in September 1972, after the structure had been preloaded, and thus far, as seen from the time settlement curves, subsequent differential settlements have been small and there has been no damage.

The excessive settlements are attributed to the thinning of the shallow sand stratum in the two localised areas, with the result that here the crushed stone columns formed by vibroflotation terminated in clay some 600 mm above the sand. Softening of clayey sands during vibroflotation, as observed by Webb and Hall (1969), is considered to have contributed to the large initial shear deformations of the clay overlying the sand.

To control differential settlement columns O8A and P8A were raised periodically by jacking, after cutting with a thermic lance, as reflected by the steps in the time settlement curves. The relationship between displacement and increase in jacking load was linear, and for the maximum increase in load of 40% of the initial column load before jacking, vertical displacement was 15 to 20 mm.

REFERENCES

Schmertmann, J. H. (1953). 'Estimating the True Consolidation Behaviour of Clay from Laboratory Test Results', *Proc. A.S.C.E.* Vol. 79, Separate No. 311
Webb, D. L. and Hall, R. I. (1967). 'Characteristics of a Clayey Sand in the Durban Area', *Proc. 4th Reg. Conf. for Africa on Soil Mech. & Found. Eng.*, Cape Town. Vol. 1, 161–167
Webb, D. L. and Hall, R. I. (1969). 'Effects of Vibroflotation on Clayey Sands', *J. Soil Mech. & Found. Eng. Div.*, A.S.C.E., Vol. 95, No. SM6, 1365–1378

V/14. Calculated and measured settlements of a mat foundation in Arlington, Va., U.S.A.

Ernest Winter
Associate, Schnabel Engineering Associates, Bethesda, Md., U.S.A.

Summary

Results of settlement observations are presented for a mat foundation founded on overconsolidated sedimentary geologic formations of clay and sand. Measured values are compared to calculated settlements. Calculations are based on the Winkler subgrade method. The effects of the structural rigidity of the superstructure are evaluated in the calculations. A limited agreement is obtained between measured and calculated deflection curves.

Subsoil conditions

The mat foundation was installed on a site located in the Metropolitan Washington Area. The geology showed surface soils being river terrace deposits of the Pleistocene geologic age. Soils below foundation levels were, however, Cretaceous sedimentary deposits. It was further known that considerable erosion occurred in the upper river terrace deposits and as a result lower layers were significantly preconsolidated. It is estimated that a preconsolidation in excess of 6 tonf/ft^2 (574 kN/m^2) has occurred in the Cretaceous sediments. These soils are locally called the Potomac Formation and are considered good foundation soils in the area.

A typical soil profile is indicated by *Fig. 1*. The cretaceous formation of sand and clay was generally too dense and stiff to recover good quality undisturbed samples. Standard Penetration Test results indicated blowcounts between about 21 to 50 in the sand and clay. Typical grain size distribution for the fine to coarse silty sand is indicated by two grain size curves in *Fig. 2*. A typical consolidation test in the clay is shown in *Fig. 3*. A summary of soil test data is presented in *Table 1*.

451

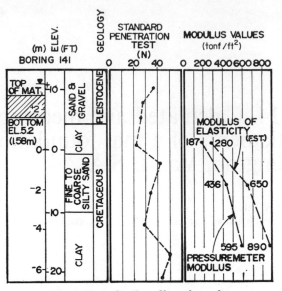

Fig. 1. Typical soil profile and test data

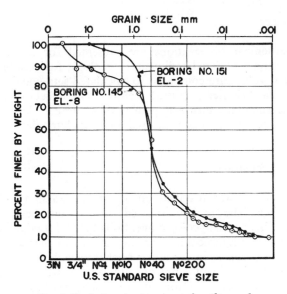

Fig. 2. Typical grain size curves for silty sand

The modulus of elasticity values were evaluated from the considerable test data available from this and previous projects. The primary source of information was the pressure meter as introduced by Menard (1957) which made *in situ* testing possible in the relatively hard subsoils and was also well correlated from many previous tests. A great number of tests were performed in local soils in the Washington area. Typical pressure meter test curves are shown by *Fig. 4*. Menard further recommends, that to obtain the modulus of

452

Table 1. INDEX PROPERTIES, CLAY AND SILTY SAND (1 lb/ft³ = 156.9 N/m³)

Description of material	Index properties			
	Natural moisture (%)	Plasticity index (%)	Natural dry density (lb/ft³)	Compression index
Brown fine to coarse silty sand, trace clay and gravel	18–21	—	98–105	—
Brown and grey clay and silty clay with trace of fine sand and lignite	24–44	25	93–100	0.10

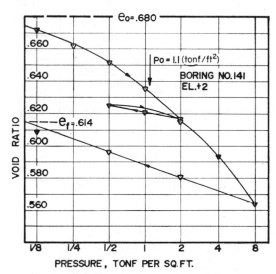

Fig. 3. Typical consolidation test curve for clay

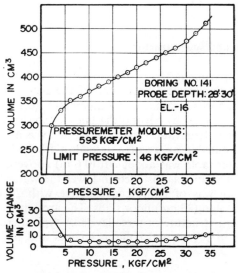

Fig. 4. Typical pressuremeter test curves

elasticity of the soil, pressure meter modulus values be divided by a rheologic coefficient, which in case of hard clays was found to be about 0.66. This ratio is believed to be dependent on material type and degree of preconsolidation among other factors, but is believed to show only minor variation within one geologic formation.

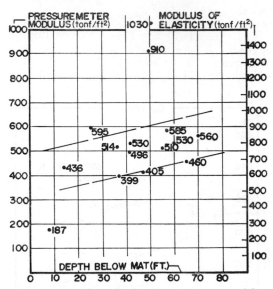

Fig. 5. Pressuremeter modulus v depth and conversion to modulus of elasticity

The subsoil stratification at greater depth was well known. Several borings penetrated this stratum up to 70 ft (21.4 m) below the project grade on adjacent sites and indicated pressure meter modulus values between 580 to 1030 tonf/ft² (55 500 to 98 570 kN/m²), generally increasing with depth. Fig. 5 indicates a plot of modulus values v depth from the Cretaceous sand and clay layers on this site.

Foundation and superstructure

The fifteen-storey apartment tower was founded on a mat foundation of 3 ft 10 in (1.17 m) thickness. Dimensions of the mat were 106 by 246 ft (32.3 by 75.0 m) supporting columns at a 20 by 20 ft (6.1 by 6.1 m) spacing. The mat extended 3 ft (0.9 m) beyond the line of exterior columns to compensate for end effects. Excavation to mat subgrade was an average of about 22 ft (6.7 m) and resulted in an average unload of 1.3 tonf/ft² (124 kN/m²). The hydrostatic water table was about 2 ft (0.6 m) above the top of the mat and a subdrainage system was installed to lower the water level.

The superstructure was of reinforced concrete with 24 by 24 in (61 by 61 cm) columns and 9 in (23 cm) floor slabs. The area surrounding the office building was a garage structure of three levels. A typical cross section of the structure is shown in Fig. 6 and column layouts and loads are shown in Fig. 7.

454

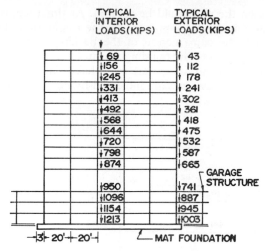

Fig. 6. General structural layout

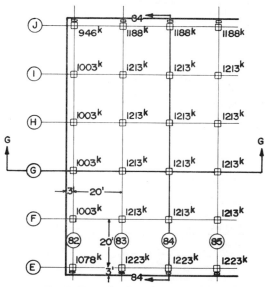

Fig. 7. Typical column layout and loads

Settlement observations

Settlements were measured at 13 column locations. The points were on two sections along the width and length of the mat. Readings were made on the floor and marks established on the columns. Two benchmarks were used. These were located outside the perimeter of the project and at an elevation about 25 ft (7.6 m) higher than the mat level. Measurements were made to an accuracy of 0.001 ft (0.3 mm).

The initial readings were made after concrete in the mat was set and

455

access to points on the mat could be secured. At this time, three additional floor levels were poured and supported by the foundation. The observation continued through the construction of the building and for an additional period of about two years. The building was not occupied during this period and accordingly, only dead loads were considered on the foundation.

Results of settlement readings are indicated by *Fig. 8*. Measurements show

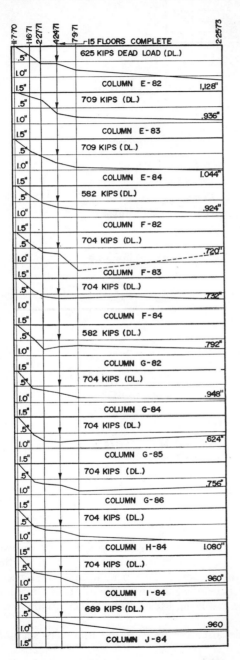

Fig. 8. Typical settlement curves

settlement between about 0.6 to 1.1 in (1.52 to 2.79 cm). Movements were more rapid during construction and settlements generally level off after all floor levels were complete.

Movements along the two typical cross sections are shown in *Fig. 9.*

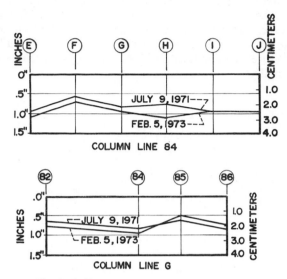

Fig. 9. Deflections at two typical cross-sections

Differential movements shown by the measurements may be due to a combination of subsoil conditions, edge effects, and some loading pattern during construction. Unfortunately, no information was available regarding any sequence of construction and release of superstructure loads at these particular points.

Evaluation of data and calculation of deflections

This foundation represents a mat founded on relatively stiff and dense soils with almost completely symmetrical loading. The design is well suited for correlation with usually simplified theoretical assumptions. The theory using the Winkler subgrade was chosen for this analysis. This widely known theory considers the mat as a homogeneous, isotropic, elastic, thin slab resting on an ideal subgrade which exerts at any point a vertical reactive pressure, p, proportional to the deflection, w, of the slab at the same point. The constant

$$k = \frac{p}{w}$$

is called the coefficient of subgrade reaction. The ELAS75 digital computer programme for the linear equilibrium problems of structures, Utku (1971), adapted to the Winkler subgrade theory was used for calculations. This programme provided the capability of evaluating several loading and subsoil schemes in detail for comparison with the measured values.

The calculated settlement matched relatively well with the general magnitude of measured deflections by using a subgrade modulus $k = 35$ lbf/in³ (9500 kN/m³). It was obvious, however, that to approximate the shape of deflections, the stiffness of the superstructure must be considered. The evaluation of loading from the superstructure imposes several difficulties including the constantly changing stiffness condition during construction and consequent rearrangement of load transfer. While methods are available, Larnach (1970), to take into account structural stiffness in building frames when the structure is completed, stiffness during construction would be difficult to estimate.

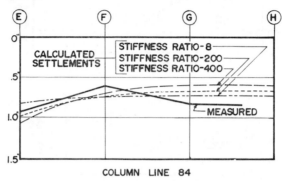

COLUMN LINE 84

Fig. 10. Calculated and measured settlements

It was attempted to study several stiffness conditions with various stiffness ratios. The stiffness ratio for comparative purposes was calculated as follows:

$$S_r = \frac{E_s BL^3}{EI}$$

where $E_s = 400$ tonf/ft² (38 280 kN/M²), modulus of elasticity of soil
$B = 105$ ft (32.0 m), width of the mat
$L = 240$ ft (73.2 m), length of mat
$E =$ modulus of elasticity of concrete (250 000 tonf/ft²,
      23 925 000 kN/m²)
$I =$ modulus of inertia of the mat (structure)

Fig. 10 shows the results of calculations. The flattest curve represents the completed building with a stiffness ratio of 8 and the largest inertia value. Two curves considering the stiffness of the mat only ($S_r = 400$) and an intermediate condition during construction ($S_r = 200$) show increased bending. The particular shape of the measured deflection curve was, however, different from the calculated values due to construction, loading and subsoil patterns.

Total settlements indicate an effective modulus of elasticity of 300 to 400 tonf/ft² (28 710 to 38 280 kN/m²) for the average subsoil. While this value is considerably lower than the soil tests would indicate, it was considered that soils disturbance at subgrade level due to construction and heave effects account for a significant portion of measured settlements.

Mat settlement measurements are badly needed and poorly represented in the literature. It was with the purpose to increase knowledge on mat behaviour from case histories, that this study was presented.

458

Larnach, W. J. (1970). 'Computation of Settlements in Building Frames', *Civ. Eng. and Pub. Wks Rev.* September

Menard, L. (1957). 'Measures in Situ des Proprietes Physiques des Sols', *Annales des Ponts et Chausses.* Vol. 127, 357–377

Utku, S. (1971). 'Concrete Thin Shells', *A.C.I. Publication SP-28.* Detroit, Michigan

V/15. The effects of soil-structure interaction on raft foundations

L. A. Wood
Engineer, Geotechnics Division, Ove Arup, London
W. J. Larnach
Senior Lecturer in Civil Engineering, University of Bristol

Summary

A suite of computer programs designed to evaluate the settlements and ground reactions of a soil-structure system, together with the stress resultants induced within the structure, is described. The structural behaviour is determined from a finite element analysis, whilst the soil is treated in the usual manner as a continuum within which a Boussinesq stress distribution is applicable. Emphasis is placed upon the behaviour of a raft foundation, using both linear and non-linear soil models. The non-linear soil behaviour is analysed using an incremental numerical analysis based upon a tangential deformation modulus defined by a hyperbolic function. It is shown that the critical edge condition associated with a stiff raft foundation can be correctly modelled only if a non-linear soil analysis is used.

Introduction

A suite of computer programs designed to evaluate the ground reactions and settlements of a soil-structure system, together with the stress resultants induced within the structure, has been developed to provide an insight into the reciprocal action of the structure and the soil on which it stands.

The computer programs are very general in nature. Non-homogeneous, non-linear soil properties are allowed for and provision has been made for the use of deformation characteristics and induced pore-pressures determined from suitable experimental data. Furthermore, the behaviour of both skeletal frame structures resting on isolated pad foundations (Larnach and Wood, 1972) and raft foundations may be analysed. In this paper the emphasis is placed upon the immediate and long term (i.e. time-independent) behaviour of raft foundations, so that pore-pressures need not be considered. Time dependence is fully discussed by Wood (1972).

The structure is analysed using standard finite element techniques (Zienkiewicz and Cheung, 1967), and is coupled to the soil, which is treated as a single element, at common points on the interface so as to maintain vertical equilibrium.

Soil-structure interaction model

For a linear, elastic soil model the behaviour of the structure and soil combined may be represented (Cheung and Zienkiewicz 1965, Wood 1972) in the usual finite element nomenclature by:

$$w = (K_p + K_s)^{-1}Q, \tag{1}$$

where K_s is the stiffness matrix of the soil
$\quad K_p$ is the stiffness matrix of the structure
$\quad w$ is the displacement vector
and $\quad Q$ is the corresponding force vector.

For the non-linear case

$$(K_p + K_s)^{-1} = f(Q) \text{ a function of } Q \tag{2}$$

therefore
$$w = f(Q) \cdot Q. \tag{3}$$

Differentiating Equation (3) we obtain

$$\delta w = f'(Q) \cdot \delta Q \cdot Q + f(Q) \cdot \delta Q. \tag{4}$$

Hence, if $f'(Q)$ approaches zero, as δQ decreases, it is possible to say that

$$\delta w = f(Q) \cdot \delta Q. \tag{5}$$

Provided that δQ is small then $f'(Q)$, the change in $f(Q)$, will be negligible in comparison with $f(Q)$ itself. Therefore, replacing the derivatives by finite increments Equation (5) becomes

$$\Delta w = (K_p + K_s)^{-1}\Delta Q, \tag{6}$$

where K_s is the stiffness matrix of the soil appropriate to the total amount of applied load $\Sigma\Delta Q$. (In the present analysis the structure is considered to act as an elastic body and therefore K_p is independent of $\Sigma\Delta Q$.) Hence, the increment in ground reaction, ΔF (assuming contact is maintained between soil and structure), is given by

$$\Delta F = K_s^{-1}\Delta w \tag{7}$$

Wood (1972) fully describes the procedures required if contact is lost.

Thus, the total displacement w is given by,

$$w = \Sigma\Delta w. \tag{8}$$

and the total ground reaction F by

$$F = \Sigma\Delta F \tag{9}$$

Soil model

The soil model employed here is based upon the use of the incremental tangent modulus approach described by Duncan and Chang (1970). During

461

each increment the soil is assumed to act as an isotropic, elastic material, the increments of vertical strain, Δe_z, being related to the stress increments by

$$\Delta e_z = \frac{1}{E_t}(\Delta \sigma_z - \nu(\Delta \sigma_x + \Delta \sigma_y)) \tag{10}$$

where E_t is the tangent modulus, governed by the stress level at the end of the previous load increment;

where $\quad \nu \quad$ is Poisson's ratio, assumed constant

$\Delta \sigma_z$ is the increment in vertical stress

$\Delta \sigma_x$ and $\Delta \sigma_y$ are the increments in horizontal stress.

The value of the tangent modulus, E_t, is defined by the semi-empirical hyperbolic function:

$$E_t = \left| 1 - \frac{R_f(1 - \sin \phi)(\sigma_1 - \sigma_3)}{2c \cos \phi + 2\sigma_3 \sin \phi} \right|^2 K \cdot p_a \left(\frac{\sigma_3}{p_a} \right)^n \tag{11}$$

where c, ϕ, R_f, K and n are soil parameters obtained from conventional triaxial tests and p_a is a reference pressure (usually atmospheric). The derivation of Equation (11) is based upon the Mohr–Coulomb failure criterion (i.e. the effect of the intermediate principal stress, σ_2, is ignored). However, Wood (1972) has derived a similar equation based on the extended Von Mises failure criterion, whence

$$E_t = \left| 1 - \frac{R_f(3 - \sin \phi)\tau_{oct}}{2\sqrt{2}c \cos \phi + 2\sqrt{2}\sigma_{oct} \sin \phi} \right|^2 K \cdot p_a \left(\frac{\sigma_{oct}}{p_a} \right)^n \tag{12}$$

where τ_{oct} is the octahedral shear stress and σ_{oct} is the octahedral normal stress; c, ϕ, R_f, K and n are again soil parameters determined from triaxial tests.

The vertical displacement of a point on the soil surface may be obtained by integration of the vertical strain beneath the point. That is

$$\Delta w = \int_0^z \frac{1}{E_t(z)} \{\Delta \sigma_z - \nu(z)(\Delta \sigma_x + \Delta \sigma_y)\} dz, \tag{13}$$

which is usually approximated by:

$$\Delta w = \sum_{i=1}^n \frac{1}{E_{t,i}} \{\Delta \sigma_{z,i} - \nu_i(\Delta \sigma_{x,i} + \Delta \sigma_{y,i})\} h_i. \tag{14}$$

Equation (14) may be re-written as

$$\Delta w = \sum_{j=1}^m \Delta F_j \sum_{i=1}^n \frac{1}{E_{t,i}} \{I_{z,i,j} - \nu_i(I_{x,i,j} + I_{y,i,j})\} h_i \tag{15}$$

where m is the number of junctions between the soil and structure, and $I_{z,i,j}$, $I_{x,i,j}$ and $I_{y,i,j}$ are the normal stresses due to a unit force at point j. Thus, the term

$$W_{k,j} = \sum_{i=1}^n \frac{1}{E_{t,i}} \{I_{z,i,j} - \nu_i(I_{x,i,j} + I_{y,i,j})\} h_i \tag{16}$$

is the contribution to the displacement w at a point k, due to a unit load at the point j.

462

The total settlement increment, Δw_k, at point k is given by

$$\sum_{j=1}^{m} W_{k,j} \Delta F_j \qquad (17)$$

where ΔF_j is the force increment at point j. Hence, for the whole soil system a series of equations such as Equation (17) can be written, giving

$$\begin{aligned}
\Delta w_1 &= W_{1,1}\Delta F_1 + W_{1,2}\Delta F_2 + \ldots + W_{1,m}\Delta F_m \\
\Delta w_2 &= W_{2,1}\Delta F_1 + W_{2,2}\Delta F_2 + \ldots + W_{2,m}\Delta F_m \\
\Delta w_m &= W_{m,1}\Delta F_1 + W_{m,2}\Delta F_2 + \ldots + W_{m,m}\Delta F_m.
\end{aligned} \qquad (18)$$

Thus
$$\Delta w = f\Delta F \qquad (19)$$

where f is a square matrix composed of $W_{k,j}$ terms. It is the flexibility matrix for the soil. Hence,

$$\Delta F = K_s \Delta w \qquad (20)$$

where $K_s = f^{-1}$ is the stiffness matrix of the soil.

It will be apparent that by forming the stiffness matrix in this manner the method is not confined to the three-dimensional, homogeneous, elastic situation (Cheung and Zienkiewicz, 1965). The $W_{k,j}$ components may be expressed in terms of the Terzaghi one-dimensional soil model (and its refinements due to Skempton and Bjerrum, 1957, and Simons and Som, 1970). Furthermore, realistic variations of soil parameters within the soil mass may be modelled.

In this Paper the stress components are assumed to be given the by Boussinesq stress distribution, and are therefore independent of the elastic modulus of the soil. This assumption, while it imposes limitations on the usefulness of the method, has been made so as to permit the study of the truly three-dimensional situation, together with non-linear and non-homogeneous soil properties, within the bounds of computer time and storage available at present.

Raft foundation: example of analysis

An 18.3 m (60 ft) square raft (with no superstructure) resting on an 18.3 m (60 ft) soil deposit exhibiting non-linear stress-strain characteristics has been analysed. To facilitate the computation and presentation of the results advantage has been taken of the symmetry of the system, and only one-quarter of the raft has been analysed.

For a raft resting on an isotropic, homogeneous, elastic continuum it has been shown (Cheung and Zienkiewicz, 1965; Wood, 1972) that the pressure distribution underneath the raft, the differential settlements, and the stress resultants induced within the raft are functions of the relative rigidity of the raft-soil system. The relative rigidity may be expressed as

$$\gamma = 180\pi \frac{E_s(1 - v_p^2)}{E_p(1 - v_s^2)} \cdot \left(\frac{a}{t}\right)^3 \qquad (21)$$

where E_s is the elastic modulus of the soil
E_p is the elastic modulus of the raft
v_s is the Poisson's ratio of the soil

ν_p is the Poisson's ratio of the raft

a is a characteristic dimension of the raft (in this case one sixth of the sidelength)

and t is the thickness of the raft.

It should be noted that a value of $\log \gamma = 3.0$ represents a very flexible system and a value of $\log \gamma = -1.0$ represents a very rigid system. Furthermore, it has been found that the most rapid change in the performance of the raft occurs within the limits $\log \gamma = 0.0$ and $\log \gamma = 2.0$. That is, increasing the rigidity of the raft below $\log \gamma = 0.0$ (or conversely increasing the flexibility of the raft above $\log \gamma = 2.0$) produces only a very small difference in the over-all performance of the system.

The typical soil parameters used to define E_t in Equation (11) in the present non-linear analysis are listed below:

$$\phi_u = 0.0$$
$$c_u = 47.7 \text{ kN/m}^2 \text{ (0.445 tonf/ft}^2)$$
$$K = 47$$
$$n = 0.0$$

(Thus, the initial tangent modulus is

$$E_i = 4.87 \text{ MN/m}^2 \text{ (45.32 tonf/ft}^2)$$

and is independent of depth);

$$R_f = 0.9$$

K_0, the coefficient of lateral earth pressure at rest, has been taken equal to 1.0 and Poisson's ratio equal to 0.5.

The elastic properties for the raft have been taken as

$$E_p = 13\,800 \text{ MN/m}^2 \text{ (128 600.0 tonf/ft}^2) \quad \text{and} \quad \nu_p = 0.15.$$

Three rafts of different thicknesses have been analysed. The raft thicknesses and corresponding values of the relative rigidity ($\log \gamma$) based upon the nitial tangent modulus are given below in Table 1.

Table 1

Thickness m (ft)	Log γ	Comment
0.305 (1.0)	2.41	Flexible
1.22 (4.0)	0.61	
1.83 (6.0)	0.08	Rigid

Results for the uniformly loaded raft

Figure 1 shows the normalised contact pressure distribution, p/q (i.e. pressure beneath raft divided by the applied pressure) across the centre line of the 0.305 m thick raft. It should be noted that at low loadings the pressure at the

464

edge actually increases above the elastic solution value. This is caused by an increase in the overall rigidity of the system as the initial load is applied and is accompanied by an increase in the intensities of the bending moments, which are also evaluated using the programmes. The system is initially very flexible and local softening of the soil does not occur at the edges of the raft but rather over the whole area of the raft. That is, the elastic modulus of the soil decreases, with a corresponding increase in the rigidity of the system. However, as general failure is approached the contact pressure distribution tends to even out. A corresponding reduction is noted in the intensities of the bending moments.

The normalised contact pressure distributions, p/q, across the centre line of the 1.83 m thick raft are shown in *Fig. 2*. In this case, unlike that of the more flexible 0.305 m thick raft, there is no initial increase in the overall rigidity of the system and the contact pressures at the edge and corner decrease relative to the elastic values, with a corresponding relative increase at the centre, tending to even out as general failure is approached.

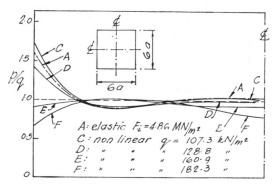

Fig. 1. *Contact pressure distribution along centre line of 0·305 m thick raft subjected to a uniformly distributed load q*

The model is inadequate to predict the growth of the plastic zones in the soil as failure is approached, and the highest loading investigated corresponds to F, where $q = 182.5$ kN/m², which is nevertheless a high proportion of the classical ultimate load.

Once again the moment intensities decrease as general failure is approached, although as with the 0.305 m thick raft there is a small increase in the bending moments at low loadings.

So as to compare the behaviour of the two rafts, together with that of the 1.22 m thick raft, the normalised contact pressure, p/q, has been plotted in *Fig. 3* against the applied pressure q for each raft. As failure is approached, not only do the contact pressures tend to even out, but they tend to the same value for all three rafts. A similar feature was found by Hoeg *et al.* (1968) when analysing a uniformly loaded strip footing on an elastic-plastic soil.

Although the pressure distribution under the rafts at failure would appear to be independent of the rigidity of the system, it has been found (Wood, 1972) that the settlement profiles are radically different. The maximum differential settlement at failure of the 1.83 m thick raft was 0.06%, expressed

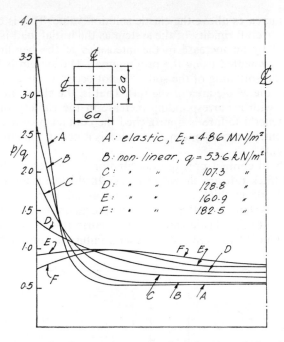

Fig. 2. Contact pressure distributions along centre line of 1·83 m thick raft, subjected to a uniformly distributed load q

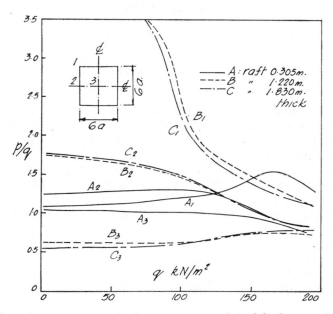

Fig. 3. Variation of normalised contact pressure p/q with loading intensity q

466

as a percentage of the total settlement at the centre. (The comparable figure was 9.1% for the elastic analysis.) For the 0.035 m thick raft the value was 14.2% (74% in the elastic analysis). This leads to the conclusion that the bending moments given by an elastic analysis will in general be an over-estimate of those occurring in all but the most flexible rafts.

Concentrated Point Load at the Centre

The behaviour of the two rafts discussed above, but now subjected to a concentrated applied load at the centre of the raft has been investigated. In this analysis the extended Von Mises failure criterion has been employed to define the tangent modulus, E_t, see Equation 12. It should, however, be noted that Wood (1972) found that the use of the extended Von Mises failure criterion rather than the Mohr–Coulomb failure criterion did not produce any significant variation in the performance of the soil-structure system. So as to allow comparison of these results with those for the uniform loading case, the local contact pressure p under the raft has been normalised with respect to the total load at the centre, P_3, divided by the total area of the raft ($36a^2$); the normalised pressure is therefore given by

$$36a^2 \, p/P_3. \tag{22}$$

Figure 4 shows the pressure distribution under the 0.305 m thick, comparatively flexible raft. For low values of applied load the raft loses contact with

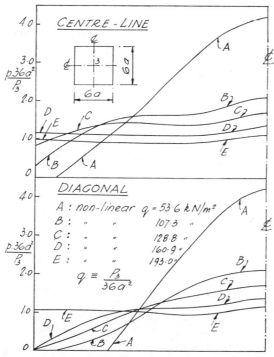

Fig. 4. *Contact pressure distribution along centre line and diagonal of a 0·305 m thick raft subjected to a central point load P_3*

467

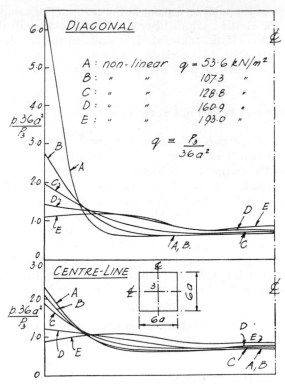

Fig. 5. Contact pressure distribution along centre line and diagonal of a 1·83 m thick raft
subjected to a central point load P_3

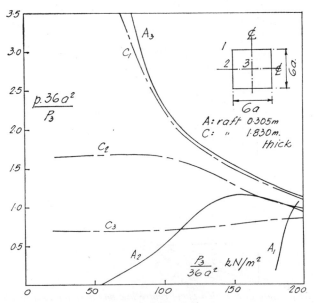

Fig. 6. Variation of normalised pressure $p.36a^2/P_3$ with equivalent loading pressure $P_3/36a^2$

the soil at the edge (and corner). As the applied load is increased the soil and raft come into contact at the edge but contact is regained at the corner only as failure is approached. Again, as failure is approached the contact pressures even out. The 1.83 m thick, comparatively rigid, raft does not lose contact with the soil at any time, *Fig. 5*, and as might be expected the pressure distribution underneath resembles that of a uniformly loaded raft. Again, as failure is approached the pressure distribution tends to even out.

The performance of the two rafts is compared in *Fig. 6* where it can be seen that for the flexible raft the highest pressures occur at the centre, curve A_3, and for the rigid raft at the corner, curve C_1; but the pressures become more uniform as failure is approached.

It has been found that for both rafts the moment intensities increase as failure is approached. However, the settlement profiles are radically different. To illustrate this the maximum differential settlements (expressed as a percentage of the total settlement at the centre) are given below

0.305 m thick raft—100% approximately,
1.83 m thick raft—4.1% approximately.

Conclusions

The following detailed conclusions may be drawn from the work reported.

(1) The performance of a raft foundation is governed by the relative rigidity of the soil-structure system, and the factor of safety which exists against general soil failure. As the factor of safety is decreased (i.e. as the load q is increased) the contact pressure distribution changes from that given by the elastic solution (curves A in *Figs. 1 and 2*). There is a decrease in the edge pressure due to local soil failure and a corresponding increase towards the centre of the raft.

(2) For a cohesive soil ($\phi = 0.0$) the contact pressure distribution becomes more uniform as general soil failure is approached. (Subsequent work on a cohesionless soil, $c = 0.0$, shows that in this case the pressure distribution increases towards the centre as failure is approached.)

(3) The contact pressure distribution as general soil failure is approached is independent of the rigidity of the raft.

(4) The contact pressure distribution under a symmetrically loaded rigid raft is independent of the manner in which the load is applied.

(5) In general, except for a very flexible raft subjected to a low load, the contact pressure distribution under uniformly distributed loadlines given by elastic theory over estimates the contact pressure at the corner and along the edge of the raft, and underestimates that at the centre.

(6) For a uniformly loaded raft the bending moment intensities within the raft are reduced as general failure is approached. (Further work indicates that for a cohesionless soil a reversal of sign may occur and the magnitude of the moments increase.)

(7) For a raft subjected to a concentrated point load at the centre the bending moment intensities increase as general soil failure is approached.

Whilst in this paper greatest emphasis has been laid on contact pressure distributions, it must be stressed that the method provides a design tool in the sense that bending moments, shears and settlements throughout the structure are computed.

The method of combining the soil and structure as a single unit described here has considered the compatibility of only vertical forces and displacements at the common interface. No attempt has been made to take into account compatibility between the horizontal displacements of the structure and soil. However, the major force component governing the settlement of a structure is that due to vertical ground reactions, and the inclusion of horizontal displacements would represent an increase in computer time and storage requirements, which, in the light of the approximate nature of the method is not justifiable at present, since so little field data is available for comparison with predictions.

The general solution presented here is based upon accepted structural and soil mechanics methods, which in the latter case are known to be rather crude approximations to the true behaviour. The major advantage of the present solution over previous approaches is its application, albeit approximate, to realistic variations in soil strata and indeed types of structure.

REFERENCES

Cheung, Y. K. and Zienkiewicz, O. C. (1965). 'Plates and Tanks on Elastic Foundations—an Application of the Finite Element Method', *Int. J. Solid Struct.* Vol. 1, 451–461

Duncan, J. M. and Chang, Chin-Yung (1970). 'Nonlinear Stress and Strain in Soils', *J. Soil Mech. & Found. Div.*, *A.S.C.E.* Vol. 96, No. SM5, 1629–1653

Hoeg, K., Christian, J. T., and Whitman, R. V. (1968). 'Settlement of Strip Load on Elastic-Plastic Soil', *J. Soil Mech. & Found. Div.*, *A.S.C.E.* Vol. 94, SM2, 431–445

Larnach, W. J. and Wood, L. A. (1972). 'The Effect of Soil-Structure Interaction on Settlements', *Int. Symp. Computer-Aided Structural Design.* Dept. Eng., Univ. Warwick

Simons, N. E. and Som, N.N. (1970). 'Settlement of Structures on Clay, with Particular Emphasis on London Clay', *CIRIA Report*, No. 22

Skempton, A. W. and Bjerrum, L. (1957). 'A contribution to the Settlement Analysis of Foundations on Clay', *Géotechnique.* Vol. 7, No. 4, 168–178

Wood, L. A. (1972). 'Some Aspects of Soil–Structure Interaction'. Ph.D. Thesis, University of Bristol

Zienkiewicz, O. C. and Cheung, Y. K. (1967). *The Finite Element Method in Structural and Continuum Mechanics.* London, McGraw-Hill

PART 2

SESSIONS I–V

Review Papers
and
Discussion

Granular materials

General Reporter, Hugh B. Sutherland,
Cormack Professor of Civil Engineering,
University of Glasgow

Introduction

State-of-the-art reviews are being presented to this Conference on the question of settlements of structures founded on granular and cohesive soils. The allowable bearing pressures for both materials are determined from a consideration of an adequate factor of safety being provided against soil rupture and the limitation of settlement to guard against detrimental total and differential settlements.

With cohesive soils, soil rupture considerations more usually predominate in the assessment of allowable bearing pressure. In granular soils the allowable settlement is usually exceeded before soil rupture considerations become significant. In general however the total settlements of footings on granular soils are small, recorded settlements being of the order of 25 mm or less and rarely exceeding 50 mm. The commonly accepted basis of design is that the total settlement of a footing should be restricted to about 25 mm as by so doing the differential settlement between adjacent footings will be confined within limits that can be tolerated by a structure. If the method of assessing settlement for a given pressure is conservative then the footings will be over-designed. Therefore, reliable and accurate methods for estimating footing settlements are essential for rational design.

It is of interest to compare the procedures that are used in the prediction of the settlement of structures founded on cohesive and cohesionless soils. With cohesive soils the approach is fairly standardised, and follows the method developed by Terzaghi whereby the compressibility characteristics of the soil are obtained from oedometer tests and then applied in the Terzaghi theory. This procedure, based on laboratory tests, requires good and representative soil samples and site investigation techniques exist to permit relatively undisturbed samples of cohesive soil to be obtained. With granular soils the situation is different. There is the basic difficulty of obtaining an undisturbed sample apart from the question of whether the sample is representative. As with cohesive soils a knowledge of the compressibility characteristics of the soil is required for the calculation of settlement. Since this property cannot readily be obtained from laboratory tests, methods of settlement calculation have been developed based on *in situ* tests from which the relative density and compressibility of the soil can be assessed. The two main types of field test used are small scale plate load tests and penetrometer tests in which the resistance of the soil to penetration is measured under dynamic or static driving conditions. The choice of field test depends on the settlement calculation method and its origin. The most common dynamic penetration test is the Standard Penetration Test (S.P.T.). It originated in the U.S.A. and the settlement calculation associated with it is widely used in North and South America and the United Kingdom. Static penetrometer methods of estimating settlements based on the Dutch Cone Test were developed in the Low Countries, have been used extensively in Europe and in recent years have been increasingly used in Britain and the U.S.A. Plate bearing test results and their correlation with S.P.T. results were a basic factor in the development of the method of calculating settlements using S.P.T. values. They are expensive to carry out, however, and the methods of estimating settlements using penetrometer test results are now more widely used, although some plate bearing test results are presented in the papers to this session, and this type of test appears to be used in the U.S.S.R. in the prediction of settlement. While *in situ* tests are those most frequently used in the assessment of settlements, laboratory tests using the oedometer or 'stress-path' triaxial tests can also be used.

Before considering the individual papers presented to this session, the various methods of calculating settlement will be reviewed.

Use of small scale loading tests

At first sight a plate loading test on a soil would appear to give the best prospect of assessing the bearing pressure which could be applied to that soil. In fact plate bearing tests have been carried out for centuries but it is only comparatively recently that their limitations have been appreciated and that more rational methods have been developed for their interpretation. For the results of a plate test to be applied directly to the prototype foundation the subsoil conditions in the zones of influence beneath the plate and the prototype must be exactly similar. This is a condition that rarely exists in practice. The results of plate load tests can be substantially affected by minor variations in soil density near the base of the plate, variations which would have much less significance with the prototype. If, as is usual, the subsoil conditions vary with depth, plate tests should ideally be carried out at various depths and with different size plates within the zone of influence of the actual footing and the execution of these tests is further complicated, apart from expense considerations, if they have to be carried out below the groundwater level. A large number of plate tests would therefore be required to give information which is representative. A more recent version of the plate loading test is the screw plate, which can be rotated into the ground, a loading test carried out, and the plate then rotated to a greater depth for another load test. No excavation is required and tests can be carried out below the water table. Schmertmann (1970) used the results from screw plate tests to obtain a correlation between Dutch Cone penetration resistance and sand compressibility.

To obtain the settlement of the actual foundation the results of the plate load test must be extrapolated.

Terzaghi and Peck (1948) proposed the following relationship between the settlement δ_B of a footing of width B ft and the observed settlement, $_1\delta$, of a 1 ft square plate, loaded to the same loading intensity

$$\frac{\delta_B}{\delta_1} = \left[\frac{2B}{B+1}\right]^2 \tag{1}$$

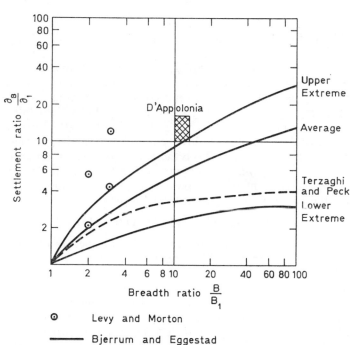

Fig. 1. Settlement ratio versus breadth ratio for footings on sand

The implication of this relationship is that the settlement of a footing, how ever wide, would never exceed four times the settlement of a 1 ft square plate loaded to the same intensity. The validity of this relationship was investigated by Bjerrum and Eggestad (1963). Their study of case records indicated that there can be an appreciable scatter in the correlation between settlement and the width of the loaded area, and more important, that settlement ratios could be very much greater than four. As shown in *Fig. 1*, they plotted the settlement ratio for equal unit pressure against the ratio of the diameter or width of the loaded area to the width of a standard plate. They drew three curves representing the 'average' correlation and the two 'extreme' boundaries. The curve from Equation 1 is also shown for comparison. They could not distinguish between the correlation of settlement and dimension

475

of a footing for dense, medium and loose sand. They detected trends which indicated that the points representing dense sands were located between the average and lower extreme sands, whereas the upper extreme curve is valid for very loose, slightly organic sand.

D'Appolonia *et al* (1968) carried out plate load tests on a fine dune sand compacted to a high relative density. Their results which are also shown in *Fig. 1*, indicate a settlement ratio of 10 for a dense sand, a finding contrary to the trend suggested by Bjerrum and Eggestad that the correlation for dense sand is between the average and lower curve.

Results of plate load tests therefore require careful interpretation. It would appear that at the present time no reliable method exists for extrapolating the settlement of a standard size plate to the settlement of a prototype footing.

Use of Standard Penetration Tests

The Standard Penetration Test (S.P.T.) was developed in the U.S.A. in the 1920's presumably for use in the assessment of the compactness of soils. It is a commonly used empirical test and consists of driving a standard split spoon sampler into the soil using a 140 lb. (63.6 kg) weight falling 30 in. (762 mm). The sampler is first driven 6 in. (152.4 mm) into the soil to penetrate disturbed soil and the number of blows required to drive the sampler a further 12 in. (304.8 mm) is recorded. This is the Standard Penetration Test N value. If the soil consists of very fine or silty sand below the water table, a correction is made to take account of the excess pore water pressure set up during driving. If the measured N is greater than 15, then

$$N(\text{corrected}) = 15 + 0.5\,(N - 15) \tag{2}$$

As experience with the S.P.T. increased, correlations between the blow count N, soil density and compressibility began to develop.

Terzaghi and Peck (1948) were the first to propose a correlation between blow count and the allowable bearing pressure on sands but subsequent observations on actual structures indicated that their method was unduly conservative. In recent years various modifications and refinements have been proposed to the basic Terzaghi and Peck approach in order to give better agreement between predicted and observed settlements on granular soils. However, unless there is a proper appreciation of the basis on which an empirical procedure such as Terzaghi and Peck was developed, there is a danger that sophisticated refinements infer an accuracy that is incompatible with the assumptions implicit in the original treatment. It is of interest therefore to examine the original Terzaghi and Peck method and the subsequent modifications.

Terzaghi and Peck (1948 and 1967). The Terzaghi and Peck correlation between allowable bearing pressure p, blow count N and footing width B is shown in *Fig. 2*. The allowable bearing pressure obtained from this chart corresponds to a settlement δ_1, of 1 in. and it is intended that it should be obtained for the largest foundation on the loosest part of the granular deposit. Bazaraa (1967) has discussed how the correlation was originally developed. According to Bazaraa the correlation was based on limited plate load test data and in fact some of the N values were obtained using a smaller diameter spoon than in the S.P.T. It is doubtful if any settlement measurements on structures were actually available, and the limitation of settlement

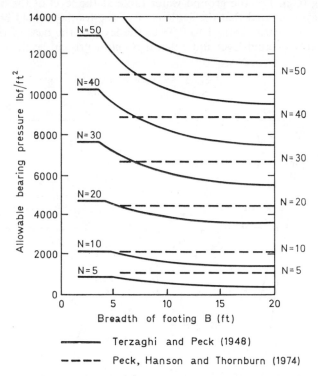

Fig. 2. Terzaghi and Peck (1948) and Peck, Hanson and Thornburn (1974) correlations of allowable bearing pressure for 1 in. settlement on sand with S.P.T. value.

to 1 in. was based on the assumption that this was the maximum settlement in buildings where settlement damage had occurred. The curves in *Fig. 2* can be approximated by the expression

$$\delta_B = \frac{3p}{N}\left[\frac{2B}{B+1}\right]^2 \tag{3}$$

which incorporates the settlement ratio value given in Equation 1.

Substitution in Equations 1 and 3 gives

$$\frac{p}{\delta_1} = \frac{N}{3} \tag{4}$$

Bazaraa plotted a large number of plate load tests in the form $\frac{p}{\delta_1}$ versus N and it can be seen from *Fig. 3* that the relationship $\frac{p}{\delta_1} = \frac{N}{3}$ is a very conservative interpretation of the data. In fact $\frac{p}{\delta_1} = \frac{N}{2}$ more closely represents a lower limiting condition and had this been adopted originally by Terzaghi and Peck would have given a 50% greater allowable pressure than that obtained from *Fig. 2.* Terzaghi and Peck applied correction factors for the effect of the position of the ground water table (C_w) and for the depth of embedment of the footing (C_D). For the ground water table at the level of the base of the footing they estimated that the settlement of the footing would be doubled, i.e. $C_w = 2$. C_D varies from 1 to 0.75 as the level of the base of the footing varies between ground level and a depth B below ground level.

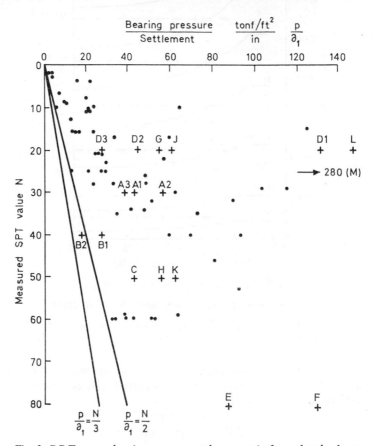

Fig. 3. S.P.T. versus bearing pressure settlement ratio from plate load tests

The use of *Fig. 2* has been, and probably still is, the most widely practised method of estimating the settlements of structures founded on granular materials. However, as observations on actual structures became available, it became clear that the settlements predicted from *Fig. 2* were substantially greater than the actual settlements. It should be emphasised nevertheless that Terzaghi and Peck clearly stated that their correlations did not take into account the geologic origin and environment of the sand deposit and were therefore necessarily a conservative basis for design.

The discrepancies which were found between predicted and observed settlements led to a critical examination of the various factors entering into the

Terzaghi and Peck correlation. Among other things, attention was directed to the S.P.T. procedure by a number of investigators; Fletcher (1965) has summarised their work and listed thirteen important factors which can affect S.P.T. results. It is clear that the greatest care must be exercised when carrying out the Standard Penetration Test otherwise misleading results can be obtained. Tomlinson (1969) further emphasises the precautions that must be observed in obtaining S.P.T. values, particularly when tests are being carried out where water is present. However it was the work of Gibbs and Holtz (1957) which led to the first major modification of the Terzaghi and Peck method.

Gibbs and Holtz. The penetration resistance of a soil is a function of both the *in situ* density and effective stress. Gibbs and Holtz from a laboratory investigation demonstrated the influence of effective overburden pressure on penetration resistance, their results being as shown in *Fig. 4*. Mansur and Kaufman (1958), Philcox (1962) and Zolkov and Wiseman (1965) confirmed from field work that the effective overburden pressure has a substantial influence on the penetration resistance of sands. Several authors, Sutherland

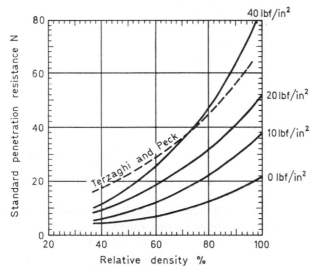

Fig. 4. Gibbs and Holtz correlation between S.P.T., relative density, and effective overburden pressure

(1963), Thorburn (1963) and Alpan (1964) suggested that the N value should be corrected to reflect the influence of effective overburden pressure, the method of correction being based on the approximate coincidence of the 40 lbf/in² effective overburden pressure curve in *Fig. 4* with the relationship between blow count and relative density implied from the Terzaghi and Peck chart. The N values corrected for overburden pressure are then used in *Fig. 2*. Tomlinson (1969) has proposed the simple correction chart for overburden pressure shown in *Fig. 5* from which it can be seen that the correction is greatest at shallow depths. Tomlinson's chart indicates that the measured N

479

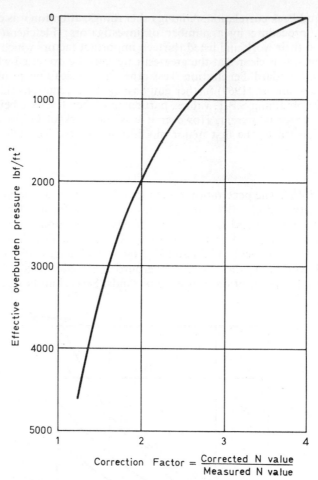

Fig. 5. *Correction factor for influence of effective overburden pressure on S.P.T. (Tomlinson)*

should be increased up to fourfold for shallow depths but a 'correction' of this magnitude should be applied with caution.

The use of corrected blow counts results in a much improved settlement prediction compared with the original Terzaghi and Peck method.

Meyerhof (1965) from a comparison of predicted and observed settlements of eight structures proposed a revision of the original Terzaghi and Peck method, a proposal which ignored the effect of the influence of effective overburden pressure. He recommended that the allowable bearing pressure obtained from *Fig. 2* should be increased by 50%. He also recommended that the presence of the ground water table should be ignored on the basis that its effect is already reflected in the measured blow count. This particular recommendation means that where the ground water is at the level of the base of a foundation, the allowable bearing pressure would be increased by a further 100% over the Terzaghi and Peck value. Meyerhof applied a similar

480

correction to that of Terzaghi and Peck for the depth of embedment of the foundation. Even with the proposed modifications Meyerhof found that for the eight cases he reviewed his modified predicted settlements were 1.2 to 4 times greater than those observed.

D'Appolonia et al (1968) presented a valuable comparison of estimated and observed settlements of over 300 shallow footings constructed on an over-consolidated and vibratory compacted fine dune sand. They compared the ratios of predicted to observed settlement obtained from different methods of calculating settlements including those based on S.P.T. values. For the original Terzaghi and Peck method, the average ratio of predicted to measured settlement was 5.0 (ranging from 2.5 to 18, approximately). Using Meyerhof's recommendations of a 50% increase on Terzaghi and Peck with no account being taken of the ground water table, the overall ratio was 2.4 (range 0.9 to 7.0). When the N values were corrected according to *Fig. 4* and used with Meyerhof's recommendations the average ratio was 1.0 (range 0.5 to 2.1).

In closing the discussion on their paper, D'Appolonia *et al.* (1970) stated that the Terzaghi and Peck correlation and its subsequent modifications had obvious shortcomings. In view of these shortcomings they advocated a different approach when applying S.P.T. data in the estimation of footing settlements on sand. This different approach is based on the theory of elasticity and will be discussed later. However, further modifications have been proposed to the original Terzaghi and Peck method.

Peck and Bazaraa (1969) in discussing D'Appolonia *et al.* (1968) recognised the overconservatism associated with the original Terzaghi and Peck method. They proposed three modifications. First, the allowable bearing pressure given by *Fig. 2* should be increased by 50% as proposed by Meyerhof. Second, they conceded that the influence of the effective overburden pressure on blow count should be taken into account but argued that application of the Gibbs and Holtz values lead to an overcorrection. They proposed arbitrarily that the original Terzaghi and Peck relationship between blow count and relative density corresponded to about the 10 lbf/in² curve instead of the 40 lbf/in² curve as suggested in *Fig. 4* by Gibbs and Holtz. This assumption leads to the following values for N_B, the corrected value of blow count, in terms of p', the effective overburden pressure (Kips/ft²)

$$N_B = \frac{4N}{1 + 2p'}, \quad \text{for} \quad p' \leqslant 1500 \text{ lbf/ft}^2 \ (1.5 \text{ Kips/ft}^2) \quad (5a)$$

$$N_B = \frac{4N}{3.25 + 0.5p'}, \quad \text{for} \quad p' \geqslant 1500 \text{ lbf/ft}^2 \quad (5b)$$

Third, they proposed a slightly different correction for the ground water position. They recommended that when the water table is at a distance D_w below the base of a shallow footing of width B then the settlement δ' may be estimated from $\delta' = k\delta$ where δ is the settlement of the same footing when the sand is dry and k is the ratio of the effective overburden pressure at depth $0.5B$ below the base of the footing when the sand is dry to that at the same depth when the water table is present.

Peck (1973) has further revised his approach to the calculation of settlement of footings on sands in the latest edition of Peck, Hanson and Thornburn (1974). The researches of Bazaraa (1967) have been taken into account but a different detailed procedure has been proposed from that advanced by Peck and Bazaraa in 1969. The influence of the effective overburden pressure on

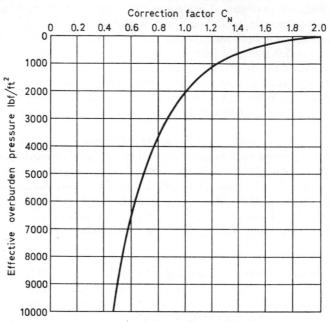

Fig. 6. *Correction factor for influence of effective overburden pressure on S.P.T. (Peck, Hanson and Thornburn, 1974)*

blow count is considered differently. The correction factor C_N to the blow count is given as

$$C_N = 0.77 \log_{10} \frac{20}{p'} \qquad (6)$$

where p' is the effective overburden pressure in $\mathrm{tonf/ft^2}$ at the elevation of the penetration test. The blow count at a depth corresponding to an effective overburden pressure of 1 $\mathrm{tonf/ft^2}$ (2000 $\mathrm{lbf/ft^2}$) is taken as a 'standard'. Values of C_N are given in *Fig. 6* and it can be seen that for $p' < 1$ $\mathrm{tonf/ft^2}$ the blow count is increased, while for $p' > 1$ $\mathrm{tonf/ft^2}$ the blow count is decreased. The corrected N values are used in a new chart shown in *Fig. 7*. This new chart correlates allowable bearing pressure with N and B for 1 in. settlement, and this correlation is superimposed on *Fig. 2* to give a comparison with the original Terzaghi and Peck correlation. A correction for the water table position is given as

$$C_w = 0.5 + \frac{D_w}{D + B}$$

482

where D_w is the depth of the water table from the surface of the surcharge surrounding the footing and D is the depth of the footing. This revised value for C_w means that the effect of the groundwater table on settlement is now considered to be less than originally proposed by Terzaghi and Peck for cases other than a surface foundation with the groundwater level at ground surface. For a specific case of $D_w = D = 0.25B$, the allowable bearing pressure would now be taken as 0.7 of that for dry sand compared with a reduction of 0.5 by the original Terzaghi and Peck procedure.

It is always difficult to generalise from isolated examples on the comparative values of settlement obtained from different methods. Peck, Hanson and Thornburn (1974) however have presented worked examples to demonstrate their method. For the straightforward case of a footing founded on sand at or

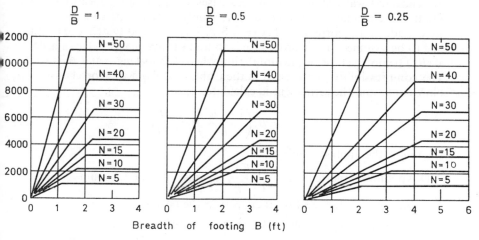

Fig. 7. *Peck, Hanson and Thornburn (1974) correlation of allowable bearing pressure for 1 in. settlement on sand with S.P.T. value*

near the original ground level, their revised method gives an allowable bearing pressure of the order of 50% greater than that from the original Terzaghi–Peck method. However in one other case they consider, where the S.P.T. results were obtained from the existing ground level before the site was finally graded, the relief of overburden pressure was greater than their 'standard' of 1 tonf/ft². This led to the corrected S.P.T. values being less than the original values and the resulting allowable bearing pressures from the revised method were less than those obtained from the original Terzaghi–Peck method.

The foregoing methods of estimating settlements from S.P.T. results have been based on the use of correlations between allowable bearing pressure and blow count. D'Appolonia *et al.* (1970) have expressed dissatisfaction with this approach and have proposed a method in which a correlation between S.P.T. and sand modulus is established from field case studies and then used with elastic theory to estimate footing settlement. Parry (1971) advances a somewhat similar approach in so far as he attempts to establish a relationship between blow count and compressibility which allows a direct calculation of settlement to be made based on conventional elastic theory. These proposals are attractive in that they appear to place the calculation of settlement on a more rational basis, but as with the Terzaghi and Peck method and its modifications, a method is only as good as the empirical correlation on which it is based, however more sophisticated the subsequent operations appear to be.

D'Appolonia et al. (*1970*) have proposed separate correlations between the S.P.T. blow count and sand compressibility for preloaded or compacted sands and normally loaded sands. Their correlation for preloaded sands was obtained from the results reported in their 1968 paper which dealt with settlement observations on structures on fine dune sand which had been compacted by vibration. Their correlation for normally loaded sands was based on less evidence and was obtained from field cases where at some locations S.P.T. results were available but where at others only Dutch Cone penetration tests were made and a correlation between S.P.T. results and Dutch Cone results had to be assumed. Their procedure is to find the average N value in a depth $\sqrt{(BL)}$ below the base of the foundation where L is its length and B its width. The compressibility of the sand is then obtained from the correlation with blow count and the settlement is calculated from

$$\delta = q \cdot B \cdot \left(\frac{1 - \mu^2}{E}\right)I = \frac{qB}{M}I \tag{7}$$

where q is the applied loading.

D'Appolonia plotted $M = \dfrac{E}{1 - \mu^2}$ against the blow count N, and if Poisson's ratio μ for sand is taken as 0.25 the correlations are

$$E(\text{kgf/cm}^2) = 540 + 13.5N \text{ (for a preloaded sand)} \tag{8a}$$
$$E(\text{kgf/cm}^2) = 216 + 10.6N \text{ (for a normally loaded sand)} \tag{8b}$$

The influence factor I takes account of the footing dimensions and the depth of the sand layer. D'Appolonia *et al.* (1970) give charts from which factor I can be obtained, but D'Appolonia (1974) has indicated that these charts are under review and recommends some caution in using them in their present form.

D'Appolonia *et al.* argue that a distinct advantage of their method over the methods based on the Terzaghi and Peck correlations is that the effects of footing embedment, varying foundation dimensions, and sand layer thickness can be taken into account.

Points to be noted are that they do not apply corrections to the measured blow count and that they ignore the presence of the static ground water level on the premise that its effect on the sand modulus is reflected in the measured S.P.T. value.

Parry (1971) dissents from the Terzaghi and Peck correlation since it is based on Equation 1 which Bjerrum and Eggestad (1963) have shown can lead to errors in estimating settlement. He also points out that the original Terzaghi and Peck approach does not allow stress changes in the ground to be taken into account, stress changes which could be due to site grading and foundation excavations. He sets out to obtain a relationship between blow count and sand compressibility and then applies his appropriate compressibility value into Equation 7 and makes corrections to the settlement so calculated. These correction factors take into account stress changes which occur due to site grading or excavation and also allow a specific thickness of compressible sand layer to be considered. No correction is applied for ground water effects where surface footings are used or for footings in back-filled excavations provided the water table does not vary after the site investigation and during the life of the structure. He obtained his correlation between blow count and sand compressibility, not from actual structures but from a limited number of plate bearing tests published by other authors. His equations for calculation of settlement indicate that the modulus of elasticity of the sand is given by

$$E \ (kgf/cm^2) = 50N \ (approximately) \tag{9}$$

A comparison of the E values obtained from Equation 9 with those obtained from the D'Appolonia Equations 8(a) and 8(b) indicate that Parry gives much higher values for the modulus particularly when $N > 20$.

Parry justified his method from a comparison of his predicted settlements with those observed in 24 published case records. He found that the overall ratio of predicted to observed settlements was 1.2, the range being from 0.8 to 2.6. In applying his method to the various cases, he had to make a number of assumptions as to the soil and other properties. He concludes that his method is suitable for feasibility studies and minor structures but recommends that his calculated values of settlement should be increased by 50% for design purposes. Moreover he recommends that for major structures additional calculations should be made by at least one other method such as the stress path method.

The D'Appolonia and Parry methods in general lead to lower estimates of settlement than those obtained from the Terzaghi and Peck approach and its modifications. They point the way to a re-assessment of the traditional methods of assessing settlements from S.P.T. results. However, further work requires to be done on obtaining correlations between blow count and sand compressibility preferably from case records of actual foundations.

Use of Static Penetrometer Test results

It is of interest to note that the more recent work on the application of S.P.T. results in the calculation of settlements of footings on sands has moved away from the original Terzaghi and Peck approach and its modifications. D'Appolonia and Parry have advocated methods of more direct calculation of settlement based on the theory of elasticity and similar to the procedures adopted for the calculation of the immediate settlement of foundations on cohesive soils. Their methods require the blow count to be related to sand compressibility. This concept is one that has been followed from the outset when calculating settlements from cone penetrometer tests where the crux of the method is a correlation between penetration resistance and modulus of elasticity of the sand.

The Dutch Cone test is the most widely used type of static penetrometer test in which a 60° cone with a cross-sectional area of 10 cm² is forced into the ground with a constant rate of penetration, and provision is made to measure independently the point resistance and the resistance due to side friction. The test was first devised to assess the bearing capacity of piles but is now used to predict the settlement of structures on sands. The original prediction methods were developed by de Beer and his co-workers but more recently Schmertmann has proposed a different approach based on cone penetrometer results.

De Beer and Martens (1957) and de Beer (1965) give details of their method. It is based on the semi-empirical Terzaghi–Buisman formula for the calculation of settlement of foundations on soils, i.e.

$$\delta = \frac{2.3}{C} \log_{10} \left[\frac{p' + \Delta p}{p'} \right] H \tag{10}$$

where C is a constant of compressibility, p' is the effective overburden pressure at the depth considered, Δp is the increment of pressure at that depth due to the foundation loading, and H is the thickness of the particular layer considered. The pressure increment Δp is normally determined from the Boussinesq stress distribution theory.

Buisman developed a relationship between the end cone penetration resistance q_c and the constant of compressibility C as

$$C = 1.5 \frac{q_c}{p'} \tag{11}$$

or

$$Cp' = 1.5 q_c = \frac{1}{m_v} = E \tag{12}$$

De Beer recommended that at least three penetration tests should be carried out and from these the maximum and minimum values of C determined. The average settlement and the limits of settlement can then be calculated. In the field cases considered by de Beer and Martens the average ratio of the predicted settlements to the calculated settlements was of the order of 1.9.

De Beer (1965) states that the foregoing method only applies to normally loaded sands. When the soil has already been previously loaded to higher pressures than will be imposed by the foundation, a reduction factor is applied to the settlement calculated as above, the reduction factor being obtained from cyclic loading tests carried out in an oedometer. The obvious difficulty is that in many applications the degree of overconsolidation of a sand is not known and cannot be easily determined.

Meyerhof (1965) proposed a modification to the Buisman-de Beer method on the basis of a comparison of the observed settlements of 17 structures with those predicted by the method. He noted that the actual settlements were overestimated by a factor of about two and recommended increasing the allowable bearing pressure by 50% for the same calculated settlement. Schmertmann (1970), points out that this arbitrary modification by Meyerhof is approximately equivalent to changing the Buisman relationship $E = 1.5q_c$ to $E = 1.9q_c$.

Schmertmann (*1970*) proposed a different approach to the use of cone penetration tests in the calculation of settlement of footings on sands. The Buisman–de Beer method involves the determination of the induced vertical stress under the foundation due to the applied loading. Schmertmann makes the point that the distribution of vertical strain under the centre of a footing based on a uniform sand is not qualitatively similar to the distribution of the increase in vertical stress. He adopts a simplified distribution of vertical strain under a foundation expressed as a strain influence factor I_z and considers the significant vertical strain to extend to a depth of $2B$ below the base of the foundation. He also recommends a less conservative method of estimating E than in the Buisman–de Beer method using $E = 2q_c$ instead of $E = 1.5q_c$. He obtained this relationship from *in situ* screw-plate load tests and claims support for this from the work of Webb (1969) who obtained a similar relationship from similar tests. It should be noted however that Schmertmann's relationship $E = 2q_c$ was obtained from a line drawn between a factor-of-2 band on the plot of E against q_c obtained from his screw-plate field tests.

Schmertmann draws attention to the work of Thomas (1968) who from laboratory tests determined a relationship between q_c and E which can be applied to the elastic theory settlement formula. Thomas found that on average $3q_c < E < 12q_c$ but Schmertmann considers that the E value as obtained from Thomas is overestimated and its use can lead to serious underestimate of settlement.

Schmertmann gives the following equation for calculating settlement

$$\delta = C_1 . C_2 \Delta p \sum_0^{2B} \left(\frac{I_z}{E} \right) . \Delta z \tag{13}$$

C_1 is a depth embedment factor which ranges from 0.5 to 1.0. Schmertmann argues that many published settlement records of structures on sands indicate that settlement continues with time. From Nonveiller's work (1963), Schmertmann introduces a creep factor

$$C_2 = 1 + 0.2 \left(\frac{t \text{ yr.}}{0.1} \right)$$

application of which implies that the settlement increases by a factor of 1.34 five years after construction. Schmertmann's view that long-term movements can occur is not shared by other investigators such as Sanglerat (1970).

Schmertmann claims that not only is his method simple to apply but its use leads to more accurate estimates of settlement than the Buisman–de Beer method. He bases this claim on an examination of 16 sites but in his analysis he necessarily had to make in a number of cases fairly wide assumptions as to the penetration resistance of the soil and the magnitude of the applied loading. Nevertheless, on average, the calculated Buisman–de Beer settlements were about 50% greater than those obtained by the Schmertmann method.

Correlation between S.P.T. and Dutch Cone tests

It can be useful to be able to estimate the equivalent S.P.T. value from the results of Dutch Cone penetration resistance and vice-versa. Meyerhof (1956) attempted to correlate the results of the two tests and suggested the following relationship.

$$q_c(\text{kgf/cm}^2) = 4N \qquad (14)$$

Further work by Meigh and Nixon (1961), Rodin (1961) and Sutherland (1963) showed that this simple relationship did not take into account the effect of grain size which is indicated in *Table 1*.

Table 1

Soil description	q_c/N
Sandy silt	2.5
Fine sand and silty fine sand	4
Fine to medium sand	4.8
Sand with some gravel	8
Medium and coarse sand	8
Fine to medium sand	10
Gravelly sand	8–18
Sandy gravel	12–16

Schmertmann points out that if only S.P.T. values are available at a site they can be converted to cone penetrometer values and so allow the Buisman–de Beer or Schmertmann methods to be applied. He gives a table of q_c/N values for different soil types which differ from the above q_c/N values and which purposely lead to conservative values of q_c. Schmertmann considers a foundation problem which he treats in detail by both his and the Buisman–de Beer methods, and for which he obtains settlements of 40.5 mm and 68.6 mm respectively. His recommended conservative q_c/N value of 3.5 can be used to obtain equivalent N values. If the N values so obtained are applied to the D'Appolonia and Parry methods the predicted settlements would be of the order of 25 mm and 17 mm respectively.

Settlement predictions from laboratory tests

Settlements can be estimated using oedometer tests or from stress path triaxial tests.

Oedometer tests

De Beer (1965) states that the use of oedometer results leads to crude approximations of the settlements, his main objection being the difficulty of simulating in the laboratory the real relationship between horizontal and vertical stress in the field. Nevertheless a number of authors have reported the use of oedometer tests carried out on samples recompacted to the field density.

488

Martins, Furtado and Silva (1963) found that the settlement of a 10-storey building founded on a fine silty sand predicted from oedometer tests was 1.64 times the observed settlement. D'Appolonia *et al.* (1968) found that settlements estimated from the first loading cycle were about 1.8 times those from compressibility coefficients obtained from cyclically loaded specimens. The average predicted settlement based on recycling data for a number of foundations was found to be 1.1 times the average observed settlement. Although de Beer does not favour the use of the oedometer test results for the estimation of settlements, he proposes, as previously discussed, the use of cyclic loading oedometer test data when estimating the settlement of preconsolidated sands. Eggestad (1963) and others have shown in model tests that a significant contribution to total vertical deformation can be due to horizontal deformations in a sand. The oedometer test is carried out under zero horizontal strain conditions and this would suggest that compressibilities measured in the oedometer will not, in general, lead to reliable predictions of settlement.

Stress path triaxial tests

Lambe (1964, 1967) has described the stress path method which requires that a sample must be reconstructed in the laboratory to the same *in situ* density and stress history as possessed by a corresponding element in the field before the foundation load is applied. The increments of stress arising from the foundation load are then applied to the laboratory sample and the strain produced by the change in stress recorded, and assigned to the corresponding soil element in the field. The triaxial tests are repeated for an appropriate number of elements in the field, the distribution of strain with depth is obtained, and then integrated to find the total settlement. Lambe and D'Appolonia *et al.* found good agreement between observed and predicted settlements using the stress path method. Lambe (1967) recognises the difficulties that can arise in applying the stress path method, probably the greatest of which is the obtaining of and the reconstruction of the samples to their original condition, particularly if the sand has been preconsolidated.

The various papers presented to this Session of the Conference can now be reviewed in the light of the review of the various methods for assessing the settlement of foundations on granular material.

Review of papers

Paper 1/1. Bratchell, Leggatt and Simons

The settlements which occurred when two large tanks were hydrotested are presented. The tanks, 79 m diameter, were founded on a 10 m thick layer of sandy gravel fill which had been compacted by vibroflotation, the effect of the vibroflotation being assessed by S.P.T. values and dynamic and static cone penetration tests. The gravel fill was underlain by Barton Clay with shear strengths of the order of 48–96 kN/m^2. The settlements of the compacted fill have been calculated by different methods and compared with the observed values.

The observed settlement of the granular fill was from 1.8 to 8.3 times greater than the values of settlement predicted by the various methods based on S.P.T. values, a situation the reverse of which is normally found. The de Beer method using Dutch Cone penetrometer results estimated the settlement at 110 mm compared with the observed value of 60 mm, i.e. an overestimation of 1.83, while the Schmertmann method underestimated the settlement by a factor of 4.3. Leggatt and Bratchell (1973) gave the predicted settlement of the fill as 50 mm and in the light of the conflicting results obtained from the various methods of calculation it would be of interest to learn how this value was decided upon.

The authors suggest that lateral deformations of the surrounding soft alluvium and the underlying Barton Clay contributed to the settlement of the gravel and this appears to be the most likely explanation of the disparity between predicted and observed settlements of the fill.

A correlation made between the S.P.T. and static cone penetration results obtained both before and after vibroflotation indicates that $q_c/N = 1.7$. This ratio is much lower than any previously published values for granular materials and discussion on this difference would be welcomed.

Paper I/2. Breth and Chambosse

The authors present the records of settlements observed on a nuclear reactor 60 m diameter founded at a depth of 5 m below original ground level. The base of the reactor was underlain by 55 m of what is described as dense fine to medium prestressed granular material. The subsoil conditions were investigated for a further depth of 40 m and found to consist of sandy silt and sandy clays. The centre of the reactor had settled about 55 mm by the end of construction while the perimeter had settled about 38 mm. No predictions of settlement are given in the paper. The site was explored by static penetration tests but incomplete data are presented. Nevertheless using the data given, estimates of settlement arising from the sand layer alone can be made using Schmertmann, Parry and D'Appolonia (for preloaded sands). The estimates are approximately 60 mm, 40 mm and 60 mm respectively compared with the maximum measured value of 55 mm. There are not sufficient data presented to allow assessment of the contribution made by the underlying clay to the total settlement. It would be helpful if the authors could give some additional information regarding their predictions of settlement.

One point of interest is that after six months of unchanged loading after completion there is no evidence of 'creep' movements. It would be interesting to learn if the settlements have increased since the last reported observations of February 1973.

Paper I/3. Dunn

The author gives the settlement records of a reactor raft foundation 101 m long, 55 m wide and 3.4 m thick founded at a depth of 9 m on a fine sand which extends to a depth of 31 m below the base of the raft and is underlain by a material which the author considers to be relatively incompressible.

490

The author had the advantage of the results of a "full scale load test" in predicting the settlement and behaviour of the raft as observations carried out on an adjacent reactor raft foundation enabled a value of 7.05 MN/m³ to be calculated for the modulus of subgrade reaction of the sand.

This value was used in the design of the raft and the average actual settlements across the raft which can be anticipated under the full design pressure appear to be 1.25 times the average predicted settlement.

While there is therefore reasonable agreement between the predicted and observed settlement, the author draws attention to the fact that settlement proceeded at a fairly uniform rate until the overburden pressure was re-established. Beyond this pressure (i.e. for a net increase in pressure), further loading produced settlements at 2.5 times the previous rate of settlement. It would be of interest to learn if the same phenomenon was observed during the settlement of the first reactor raft.

The maximum settlement that occurred during the restoration of the overburden pressure corresponds to a modulus of subgrade reaction of the order of 6.7 MN/m³ as against the design value of 7.05 MN/m³. The modulus appropriate to applied pressures greater than the overburden pressure is about 2.7 MN/m³.

There are a number of points on which further information would be helpful. The author describes the plates that were buried at various depths below the base of the raft. He states that observations were made on the plates before the start of any excavation. However he takes June 1966 as his zero time for plotting settlements, i.e. after excavation had been completed. What movements occurred during the excavation process, and how did they compare with the settlements observed during the restoration of the overburden pressure? The author also presents records of the movements of some of the plates beneath the base of the raft. Has the distribution of vertical strain with depth been investigated and compared with the vertical strain distributions discussed by Schmertmann (1970)?

The restoration of the ground water table seemed to have little effect on the settlements but the author may have further information on this matter.

The settlements were predicted from the derived modulus of subgrade reaction and have previously been commented on. However, the author has also applied the de Beer method to estimate the settlement and obtained a maximum settlement of 72 mm for the maximum net increase in pressure of 106 kN/m² compared with the actual maximum settlement of 65 mm which occurred under the gross average pressure of 289 kN/m². However the actual settlement that occurred under a net increase in average pressure of 83 kN/m² was only 34 mm.

The main conclusion that can be drawn is that there is nothing to compete with the information from a prior full-scale loading test when it comes to predicting settlements in granular materials.

Paper I/4. Garga and Quin

The authors describe an impressive programme of tests carried out as part of the site investigation for a steel complex in Brazil. A large number of S.P.T. and Dutch Cone tests were made along with plate load tests on 45 cm, 1 m

491

and 2.5 m square plates. *In situ* load tests were also carried out using screw plates and from these tests a new correlation of $E = 2.9q_c$ is proposed compared with $E = 1.5q_c$ used by de Beer and $E = 2q_c$ determined by Schmertmann.

The actual settlements of twelve 2.5 m square plates have been compared with those predicted by various methods. The authors conclude that when working from S.P.T. values the Meyerhof recommendations, with N values corrected according to Gibbs and Holtz, can be used to predict settlement. Their conclusions are not quite so specific regarding predictions based on Dutch Cone tests. They used the modified correlations $E = 2.9q_c$ when applying the de Beer and Schmertmann methods and found that De Beer on average slightly underestimated the observed settlements while Schmertmann overestimated settlements by a ratio of 1.8, but the authors appear to recommend both methods.

The settlements of the 2.5 m square plates were also estimated from the settlements of the smaller plates using the Terzaghi and Peck Equation 1. Contrary to Bjerrum and Eggestad the authors found that Equation 1 substantially underestimated the observed settlements. This is not surprising since the actual settlements of the 2.5 m square plates were on average only just over 20% greater than the average actual settlement of the 1 m square plates under the same load intensities. It would appear from the tabulated data that some of the smaller plates settled less under the same load intensity than did larger plates at the same location. It would have been interesting to superimpose the results of the plate tests in *Fig. 3* but the N values at each plate location were not given.

The manner in which the results of this extensive and expensive plate loading programme were applied in the design of the actual foundations is not described and a limited amount of information is presented on the performance of actual foundations. The settlements of two wide slab foundations are discussed but unfortunately settlement records were not taken from the beginning of construction. However from a comparison of the settlements arising from estimated differences of loading it is concluded that for foundations of width greater than 10 m the effective compressible depth of subsoil below the base of the foundation is substantially less than twice the width of the foundation as assumed by de Beer and Schmertmann for 'relatively small' foundations. The authors make a plea for the need of settlement observations on large foundations on sands and it is hoped that more such observations will become available from their investigations.

Paper I/5. Gielly, Lareal and Sanglerat

The authors compare the measured and calculated settlements of two structures. The first structure was a 22-storey building founded on recent marine sand deposits on which S.P.T. and Dutch Cone penetrometer tests had been carried out. The authors used the Schmertmann method to estimate settlement, the modulus of elasticity of the sand being taken as twice the static penetration resistance. The settlements so predicted both at the end of construction and six months after construction were practically identical to the observed values, and this might be regarded as fortunate when it is remembered

492

that the data from which Schmertmann derived his relationship between modulus and cone resistance encompassed a factor of 2. The settlement six months after construction had increased by 25% over that observed at the end of construction as against a 17% increase predicted by Schmertmann. However the borehole log indicates the presence of some organic material in the sand and this may have contributed to the continuing settlement.

Settlements can also be estimated from the S.P.T. results but the predictions by all the S.P.T. methods give at least 50% greater settlements than those observed. A comparison of the Dutch Cone and S.P.T. results indicates that $q_c/N = 2.5$, a low value for the material as described.

At the second site the authors consider two tower blocks, one of which was founded on a uniform thickness of loess and the other on a varying thickness of loess. The data presented show that this material was cohesive rather than cohesionless and the authors predicted settlements from consolidation test data and from zone tests carried out on the loess. Insufficient data are presented to enable the details of settlement prediction to be followed, and since settlement observations were not initiated from the start of construction it is difficult to draw a comparison between predicted and observed settlements. Nevertheless the authors were able from their predictions and deductions to authorise the construction to continue.

Paper I/6. Levy and Morton

This paper consists of two parts in the first of which the results of 17 plate load tests carried out at 12 sites on plates varying from 1 ft (305 mm) square to 4.5 ft (1219 mm) square are presented. At ten of the sites single size plate tests were made. At the remaining two sites tests were carried out on square plates of size 305 mm, 610 mm and 914 mm. S.P.T. values are given for each plate test. The authors found that there was no relationship between settlement and S.P.T. value. In their *Fig. 6* they plotted bearing pressure against the ratio of settlement to breadth of footing expressed as a percentage, but since the data as presented does not relate to sand properties it cannot be used in the prediction of settlement of a particular foundation.

The plate test information given can however be used to supplement that presented by Bazaraa and as shown in *Fig. 3*. The authors data can be plotted on a similar basis and is superimposed on *Fig. 3*. It confirms Bazaraa's contention that the interpretation of plate test data used by Terzaghi and Peck is very much on the conservative side. Points $B1$ and $B2$ are closest to the Terzaghi limit of $\frac{p}{\delta_1} = \frac{N}{3}$ but in both these plate tests the groundwater table was apparently above the base of the plates during the tests. At the two sites where different size plates were tested the results can be examined to see if they conform to the Terzaghi and Peck relationship given in Equation 1. In the first series of tests (site A) the actual $\frac{\delta_B}{\delta_1}$ values were 2 and 4.2 compared with 1.78 and 2.25 predicted by Equation 1. The series of tests at site D gave $\frac{\delta_B}{\delta_1}$ values of 5.4 and 14 as against the Terzaghi–Peck values of 1.78 and 2.25.

493

The results are plotted on *Fig. 1.* The test on the smallest plate appears to be unrepresentative and the position of the groundwater would appear to have affected each plate test to a different extent.

In the second part of the paper the authors present settlement records from two sites for which S.P.T. values are available but at which no plate load tests were made.

At the Arts and Commerce Building at Birmingham eight footings 7 m by 4 m were loaded to 500 kN/m². The authors predicted settlement using the Gibbs and Holtz correction applied to the Terzaghi and Peck method, and were apparently satisfied that the settlement so obtained was in good agreement with the maximum settlement observed in the various footings 3.6 years after construction. The average settlement of the eight footings as tabulated was 9.15 mm at the end of construction. It had increased to 11.63 mm one year after construction and to 12.33 mm after 3.6 years. The authors estimate that a further 3 mm should be added to all settlements to allow for movements which occurred before observations commenced. These observations can be considered in the light of the creep factor C_2 used by Schmertmann. For the revised observed settlements, the values of C_2 at this site are 1.20 after one year and 1.26 at 3.6 years compared with Schmertmann's predictions of 1.2 and 1.3 respectively.

Settlement observations are also presented for the Stratford Bus Station where the foundations consist of 3 m and 2 m square footings founded at a depth of about 3 m below original ground level. The foundations are underlain by about 1.6 m of dense sandy gravel below which lies stiff grey silty clay. The settlement of a footing would be made up of deformation in the granular deposits plus elastic and consolidation settlement of the underlying silty clay. The authors predicted the movement in the granular material using D'Appolonia's correlation of E with N. They do not state the value they obtained but it would probably be less than half the total settlement. It is not possible to separate the components of settlement and so make a comparison of the predicted and observed settlement in the granular material.

Settlement measurements at a third site are given where the bases of 20 m square raft foundations overlie about 4 m of sand and gravel below which lies London Clay. A plate load test 1 ft square was made at this site but again it is not possible to compare predicted and observed settlements for the granular material due to the indeterminate contributions made by the sand and gravel and the clay.

Paper I/7. Penman and Godwin

This paper describes the settlement behaviour of a group of 24 two-storey semi-detached houses constructed on loosely placed backfill which consisted of approximately 9 m of dumped weak oolitic rock overlain by about the same thickness of dumped boulder clay. The houses were constructed as an experiment and the loose backfill had been in position 12 years before the houses were built. Four different types of foundation were used in the design of the houses. Large diameter oedometer tests were carried out in the laboratory on samples of the oolitic and boulder clay backfills to investigate the compressibility of the materials before and after inundation with water.

494

The observed total and differential settlements were irregular between the blocks and could not be related to the type of foundation. The settlements appeared only to reflect the irregularity of the fill material. The authors found that the imposed loading from the houses caused settlements which were substantially greater than any movements that could be attributed to creep but the influence of creep of the backfill would be difficult to evaluate in the absence of observed deformations of the fill in the period between placing and construction of the houses. The authors discuss the question of settlement of loose dumped rockfill and point to the increasing settlements which can be induced by sluicing loosely placed rockfill tips. In their laboratory tests they assessed the effect of inundation on compressibility and found in fact that the boulder clay after inundation had an additional compression of about three times that of the oolitic rock fill.

The paper is an interesting specific case record which does not give guidance on the more general problem of the evaluation of settlements in granular materials.

Paper I/8. Stuart and Graham

The authors present settlement observations made for 12 years on the foundation raft of a 13-storey structure founded on sand and gravel. The original design was based on the 1948 Terzaghi and Peck recommendations from which the authors predicted a settlement of 5 cm. In fact the maximum observed settlement was 1.5 cm at the south end of the building, the differential settlement between the south and north ends being 0.75 cm despite the average S.P.T. values being practically identical.

As the authors indicate the more recent developments in methods of predicting settlements from S.P.T. results would have led to estimates of settlements more in line with those actually measured. The authors state that if these methods had been available a more economical foundation design would have resulted.

One point to which the authors have drawn attention is the continuing nature of the settlements. Both ends of the building have continued to settle at approximately the same rate since completion. This has meant that the settlement at the north end has increased by just over 50% and the south end by 27%. The average creep factor is 1.39. Schmertmann's creep factor C_2 for the same period of time is 1.38. No explanation can be offered by the authors for these continuing movements. There is the possibility however, that the boulder clay which underlies the site might have contributed to the continuing movements but no properties for this clay are given.

Summary and conclusions

The various methods of predicting the settlements of footings on sands using the results of plate bearing tests and penetrometer tests have been reviewed. There is no evidence to suggest that any one of these methods will in all situations give an accurate prediction of settlement although some of the methods consistently give more correct predictions than others. Before a designer becomes entangled in the details of predicting settlement he must clearly satisfy himself whether a real problem actually exists and ascertain what advantages and economies can result from refinements in settlement prediction. Although settlements of footings on sand are generally small, over-conservative methods of design can result in uneconomic foundation design, a point which is brought out by Graham and Stuart in their paper.

A number of conclusions can be drawn from the review of the methods and of the papers presented to this Session.

1. There is no reliable method for extrapolating the settlement of a standard size plate load test to the settlement of an actual footing at the same location. Bjerrum and Eggestad (1963) examined the Terzaghi and Peck extrapolation using Equation 1 and concluded that its use could lead to an underestimate of settlement. The trends they indicated for loose medium and dense sands were not in agreement with the plate test results of D' Appolonia et al. (1968) or those reported by Levy and Morton. The differences probably arise from the difficulty of obtaining reliable results from small-scale plate tests as appears to be demonstrated by some of the plate test results presented by Garga and Quin.

2. Settlement predictions using S.P.T. values originated with the 1948 Terzaghi and Peck correlation. It is now generally agreed that the use of this method leads to over-conservative results, this being mainly due to the conservative nature of the interpretation of the plate tests on which the correlation was based. Various modifications have been proposed. Meyerhof (1963) increased the allowable bearing pressure given by the Terzaghi and Peck correlation by 50% and neglected the influence of the groundwater table position by the argument that its effect is reflected in the measured S.P.T. value, a view shared by D'Appolonia et al. (1970) and Parry (1971). Peck and Bazaraa (1969) and Peck, Hanson and Thornburn (1974) adhere to the view, on the basis of field evidence, that the position of the groundwater level is significant in settlement prediction. However their correction for groundwater effects is now less severe than that originally proposed by Terzaghi and Peck.

It now appears to be generally accepted that some correction should be made for the influence of the effective overburden pressure when using a Terzaghi and Peck type correlation. There is a difference of view on the value of the correction factor, but it becomes pointless to make a direct comparison between the correction factors as the different corrected S.P.T. values are applied subsequently into charts based on differing correlations.

D'Appolonia et al. (1970) and Parry (1971) have advanced direct methods of predicting settlements from S.P.T. values. It is considered that these methods are worthy of further study using a greater number of case records to try and ascertain a relationship between sand compressibility and blow count.

Comparisons of the results obtained when each of the various methods is applied to different problems shows that no one method consistently gives a higher or lower estimate of settlement than another method. The designer must therefore regard the methods as aids to design which cannot replace the critical role of engineering judgment. Nevertheless at the present stage of knowledge it would appear that the Meyerhof (1965) procedure gives a reasonably good estimate of the maximum probable settlement.

3. The de Beer method of prediction based on Dutch Cone penetrometer values and working to $E = 1.5q_c$ leads to conservative results. Meyerhof (1965) recommends that the de Beer allowable bearing pressures should be increased by 50% for the same computed settlement. This proposal corresponds to assuming that $E = 1.9q_c$ and can be compared with the relationship $E = 2q_c$ found by Schmertmann (1970) and used in his strain distribution approach to the evaluation of settlement. Schmertmann concluded that Meyerhof's modification to de Beer results in predictions similar to those obtained by his own method. Settlement predictions so obtained are still conservative on average, but are less conservative than those obtained from the original de Beer procedure. What must not be overlooked is that all correlations between E and q_c are based on 'average' lines and Schmertmann's correlation of $E = 2q_c$ is obtained from a line drawn through fairly widely scattered points. Having this in mind it appears that Schmertmann's approach or Meyerhof's modification of de Beer's method can be recommended to give a reasonable estimate of the maximum probable settlement. Garga and Quin from their screw plate tests derived the correlation $E = 2.9q_c$. When they used this relationship in the de Beer method they found that on average the settlements were accurately estimated, but this means that half the settlements were underestimated. When $E = 2.9q_c$ was used in the Schmertmann method the actual settlements were substantially overestimated. Garga and Quin essentially confined their analysis to the settlements measured in their 1 m and 2.5 m square test plates and since there appear to be anomalies in the results from some of their plate tests it is not considered that their finding of $E = 2.9q_c$ can as yet be generally applied.

4. Settlements can be predicted from laboratory test results. The difficulties of preparing proper specimens and of taking account of the stress history of the *in situ* deposit make the general use of this method unlikely.

5. Most methods, with the exception of Schmertmann's, assume that sands are instantaneously compressible. However some of the settlement records presented to this Session indicate that there have been continuing movements under conditions where there is no clear evidence of the influence of other compressible strata. Further information on this question would be of interest.

6. the plea that has been made many times before that more reliable observations of settlement are required is once more repeated. It is gratifying to receive the records of settlements that have been presented to this Session. While appreciating the difficulties involved in obtaining records, it is disappointing to note that in some cases circumstances were such that observations were not made from the start of construction. It is hoped that the increasing awareness of the need for more information on the behaviour of structures generally will lead to an increase in the number of settlement observations.

7. The methods available for predicting settlements of footings on granular materials are improving but the state-of-the-art is such that the judgment and experience of the designer are still of paramount importance.

REFERENCES

Alpan, I. (1964). 'Estimating the Settlements of Foundations on Sand', *Civ. Eng. and Pub. Works Rev.*, Vol. 59, Nov., 1415–1418

Bazaraa, A. R. S. S. (1967). 'Use of the Standard Penetration Test for Estimating Settlements of Shallow Foundations on Sand', *Ph.D. Thesis*, University of Illinois, Urbana-Illinois

Bjerrum L. and Eggestad, A. (1963). 'Interpretation of Loading Tests on Sand', *Proc. European Conf. Soil Mech. & Found. Engng*, Wiesbaden, Vol. 1, 199–204.

D'Appolonia, D.J., D'Appolonia, E., and Brisette, R. F. (1968). 'Settlement of Spread Footings on Sand', *J. Soil Mech. & Found. Div.*, *A.S.C.E.*, SM3, May, 735–760

D'Appolonia, D. J., D'Appolonia, E. and Brisette, R. F. (1970). 'Discussion on Settlement of Spread Footings on Sand', *J. Soil Mech. & Found. Div.*, *A.S.C.E.*, SM2, March, 754–761

D'Appolonia, E. (1974). Personal communication

De Beer, E. (1965). 'Bearing Capacity and Settlement of Shallow Foundations on Sand'. *Proc. Sym. Bearing Capacity and Settlement of Foundations*, Duke University, 15–33

De Beer, E. and Martens, A. (1956). 'Method of Computation of an Upper Limit for the Influence of Heterogeniety of Sand Layers in the Settlement of Bridges', Proc. 4th *Int. Conf. Soil Mech. & Found. Engng*, London, Vol. 1, 275–281

Eggestad, A. (1963). 'Deformation Measurements Below a Model Footing on the Surface of Dry Sand', *Proc. European Conf. on Soil Mech. & Found. Engng*, Wiesbaden, Vol. 1, 233–239

Fletcher, G. A. (1965). 'Standard Penetration Test—Its Uses and Abuses'. *J. Soil. Mech. & Found. Div.*, *A.S.C.E.*, Vol. 91, SM4, July, 67–75

Gibbs, H. J. and Holtz, W. G. (1957). 'Research on Determining the Density of Sands by Spoon Penetration Testing, *Proc. 4th Int. Conf. Soil Mech. & Found. Engng*, London, Vol. 1, 35–39

Lambe, T. W. (1964). 'Methods of Estimating Settlement', *J. Soil Mech. & Found. Div.*, *A.S.C.E.*, Vol. 90, SM5, Sept., 43–67

Lambe, T. W. (1967). 'The Stress Path Method,' *J. Soil. Mech. & Found. Div.*, *A.S.C.E.*, Vol. 93, SM6, Nov., 309–31

Leggatt, A. J. and Bratchell, G. E. (1973). 'Submerged Foundations for 100 000-Ton Oil Tanks', *Proc. I.C.E.*, Vol. 54, 291–305

Mansur, C. I. and Kaufman, R. I. (1958). 'Pile Tests, Low-sill Structure, Old River, Louisiana', *Trans. A.S.C.E.*, Vol. 123, 715–748

Martins, J. B., Furtado, R. J. and da Silva J. V. (1963). 'Settlements of a Ten-storeyed Building', *Proc. European Conf. Soil Mech. & Found. Engng*, Wiesbaden, Vol. 1, 313–317

Meigh, A. C. and Nixon, I. K. (1961). 'Comparison of In Situ Tests for Granular Soils, *Proc. 5th Int. Conf. Soil Mech. & Found. Engng*, Paris, Vol. 1, 499

Meyerhof, G. G. (1956). 'Penetration Tests and Bearing Capacity of Cohesionless Soils'. *J. Soil Mech. & Found. Div.*, *A.S.C.E.*, Vol. 82, SM1, Jan.

Meyerhof, G. G. (1965). 'Shallow Foundations,' *J. Soil Mech. & Found. Div.*, *A.S.C.E.*, Vol. 91, SM2, March 21–31

Nonveiller, E. (1963). 'Settlement of a Grain Silo on Fine Sand', *Proc. European Conf. Soil Mech. & Found. Engng*, Wiesbaden, Vol. 1, 285–294

Parry, R. H. G. (1971). 'A Direct Method of Estimating Settlements in Sands from S.P.T. Values', *Proc. Symp. Interaction of Structure & Found.*, Midlands Soil Mech. & Found. Engng Soc., Birmingham, pp 29–37

Peck, R. B. and Bazaraa, A. R. S. S. (1969). Discussion of Paper by D'Appolonia *et al.*, *J. Soil Mech. & Found. Div. A.S.C.E.*, Vol. 95, SM3, 305–309

Peck, R. B. (1973). Personal communication

Peck, R. B., Hanson, W. E., and Thornburn, T. H. (1974). *Foundation Engineering*, 2nd edn., Wiley, New York

Philcox, K. T. (1962). 'Some Recent Developments in the Design of High Buildings in Hong Kong', *Struct. Engr*, Vol. 40, No. 3, October

Rodin, S. (1961). 'Experiences with Penetrometers with Particular Reference to the Standard Penetration Test', *Proc. 5th Int. Conf. Soil Mech. & Found. Engng*, Vol. 1, 517

Sanglerat, G. (1972). *The Penetrometer and Soil Exploration*, Elsevier, Amsterdam

Schmertmann, J. H. (1970). 'Static Cone to Compute Static Settlement Over Sand', *J. Soil Mech. & Found. Engng Div.*, *A.S.C.E.*, Vol. 96, SM3, 1011–1043

Sutherland, H. B. (1963) 'The Use of In Situ Tests to Estimate the Allowable Bearing Pressure of Cohesionless Soils', *Struct. Engr.*, Vol. 41, No. 3, March, 85–92

Terzaghi, K. and Peck, R. B. (1948). *Soil Mechanics in Engineering Practice*, 1st edn., Wiley New York

Terzaghi, K. and Peck, R. B. (1967). *Soil Mechanics in Engineering Practice*, 2nd edn., Wiley, New York

Thomas, D. (1968). 'Deep Sounding Test Results and the Settlement of Spread Footings on Normally Consolidated Sands', *Geotechnique*, London, Vol. 18, Dec., 472–488

Thorburn, S. (1963). 'Tentative Correction Chart for the Standard Penetration Test in Non-cohesive Soils', *Civ. Engng and Pub. Works Rev.*, June, 752–753

Tomlinson, M. J. (1969). *Foundation Design and Construction*, 2nd edn., Pitman, London

Webb, D. L. (1969). 'Settlement of Structures on Deep Alluvial Sand Sediments in Durban, South Africa', *Brit. Geotechnical Soc. Conf. on In Situ Investigations in Soils and Rocks*, London, 133–140.

Zolkov, E. and Wiseman, G. (1965). 'Engineering Properties of Dune and Beach Sands and the Influence of Stress History,' *Proc.6th Int. Conf. Soil Mech. & Found. Engng*, Montreal, Vol. 1, 134

499

Normally consolidated and lightly over-consolidated cohesive materials

General Reporter, N. E. Simons,

Department of Civil Engineering,
University of Surrey

Introduction

This conference is concerned with the prediction of the settlement of structures, and Session II is confined to normally and lightly overconsolidated clays. Embankments are not specifically considered in this session, partly because the associated deformation problems have been discussed at recent meetings, for example, the Purdue Conference on Performance of Earth and Earth-Supported Structures in 1972, and the London Symposium on Field Instrumentation in Geotechnical Engineering in 1973.

This general report is not intended to be a state-of-the-art contribution, mainly because of space limitations, and the reporter has attempted to concentrate on factors which affect significantly the accuracy of settlement predictions.

Because of the advances that have been made in recent years, and having available the power of the finite element method of analysis, there may be a temptation to believe that settlement prediction has become an exact science. This is simply not true and the following quotation, Terzaghi (1936), is particularly relevant and valid at the present time:

'Whoever expects from soil mechanics a set of simple, hard-and-fast rules for settlement computation will be deeply disappointed. He might as well expect a simple rule for constructing a geological profile from a single test boring record. The nature of the problem strictly precludes the possibility of establishing such rules. If a supervising or construction engineer wants to enjoy the benefits of recent developments in this field he should first of all study the rules for securing settlement records, and then start to observe the buildings of his district. After he has done this for a certain period he will discover for himself the value of the information which he can obtain from soil mechanics.'

It should be made clear that although initial, primary consolidation and secondary settlements are discussed in this report under separate sections, this does not imply that they are separate components taking place at different times. The settlement at the end of construction is sometimes taken as being equal to the initial settlement but, even if the construction period is short, this will include some part of the primary consolidation settlement. Furthermore, if initial settlement is considered to be settlement associated with lateral strains, it can continue over long periods of time, particularly with low factors of safety. Secondary settlement also, considered to be the settlement occurring after changes in effective stress have taken place, will develop during the primary consolidation period as the porewater pressure dissipates and the effective stresses increase.

No specific comments on stress distribution are made in this report, as this aspect has been covered in the report on Session 4, Hobbs (1974), whose comments on elastic displacements are also directly relevant to Session 2.

Initial settlement

The initial settlement is generally considered to be the settlement which takes place under constant volume (undrained) conditions when the clay deforms to accommodate the imposed shear stresses. It is of interest to note that such settlement is accompanied by a change in effective stress, either horizontally or vertically, or in both directions. If, however, no change of shape is allowed to take place, for example in the oedometer or under global loading conditions in the field, then no changes in effective stress, either horizontally or vertically, can occur; for this condition the increases in vertical stress, horizontal stress and excess porewater pressure are equal.

The initial settlement may be calculated using various procedures and those which seem to be of most use in practice have been tabulated by Lambe (1973a), *Table 1*.

Table 1. METHODS OF CALCULATING INITIAL SETTLEMENT (after Lambe (1973a))

Method	Formula	Reference
Elastic displacement	$\delta_i = \dfrac{q.B.\mu_0.\mu_1}{E_u}$	Janbu, Bjerrum and Kjaernsli (1956)
Elastic strain summation	$\delta_i = \dfrac{\Sigma[\sigma_z - 0.5(\sigma_x + \sigma_y)]\delta h}{E_u}$	Davis and Poulos (1968)
Modified elastic displacement	$\delta_i = \dfrac{S_R\,q.B.\mu_0.\mu_1}{E_u}$	D'Appolonia *et al.* (1970)
Finite element	By computer	

While finite element analysis provides a powerful tool for solving the initial settlement problem and non-linearity, non-homogeneity and anisotropy can

501

be taken into account. Such an analysis requires detailed information concerning the undrained stress–strain relationships for the soil, which in most routine problems is not available, and therefore the calculation of the initial settlement for many problems in practice is based on the elastic displacement method.

A convenient form of this solution has been given by Janbu, Bjerrum and Kjaernsli (1956) and is expressed as follows:

$$\delta_i = \frac{q \cdot B \cdot \mu_0 \cdot \mu_1}{E_u} \tag{1}$$

This equation gives the average initial settlement for a flexible foundation, taking Poisson's ratio equal to $\frac{1}{2}$. The factors μ_0 and μ_1 given in *Fig. 1* are

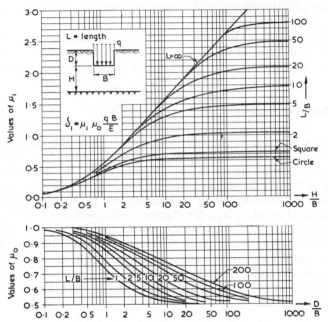

Fig. 1. Calculation of initial settlement (after Janbu, Bjerrum and Kjaernsli (1956))

functions of the geometry of the problem and allow the depth of the footing and the thickness of 'elastic' material, which in practice is often limited, to be taken into account in a simple manner. By using the principle of superposition, it is also possible to carry out calculations for multi-layered soil systems.

It is obvious that this simple approach imposes fairly severe limitations on the accuracy which may be expected in any specific problem.

Although the use of an average value of E_u may, with experience, give a reasonable estimate of the average initial settlement of a structure, if it is necessary to predict the initial deflected shape then non-homogeneity and anisotropy must be taken into account. The effects of non-homogeneity and anisotropy have been considered, for example, by Lambe (1964), Gibson (1967), Davis and Poulos (1968), Gibson and Sills (1971), Burland, Sills and Gibson (1973), Carrier and Christian (1973) and solutions to a limited number of problems are available.

An example of the influence of non-homogeneity is given in *Fig. 2*, taken from Burland, Sills and Gibson (1973), where the theoretical deflected shapes of the ground surface for the Mundford tank, founded on chalk, Ward, Burland and Gallois (1968), are compared with the observed displacements. It can be seen that the observation that the settlement of the ground surface is localised around the loaded area, much more than the simple elastic Boussinesq theory predicts, is accounted for almost entirely by the influence of non-homogeneity.

Foundations for structures are generally designed with a factor of safety against ultimate failure of the order of $2\frac{1}{2}$ to 3. It has been shown (Davis and Poulos (1968)) that for normally consolidated clays yielding and deviation from linear behaviour will first occur when the factor of safety against a bearing capacity failure is between 4 and 8. For slightly overconsolidated clays the corresponding factor of safety at first yield is 2 to 3, and for heavily overconsolidated clays first yield does not occur until the factor of safety is below 2. These factors of safety depend on the initial shear stress ratio

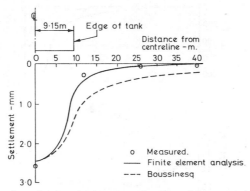

Fig. 2. Effect of non-homogeniety on ground surface deformation (after Burland, Sills and Gibson (1973))

which is the ratio between the initial shear stress before any external load is applied and the undrained shear strength. For truly normally consolidated clays, therefore, use of Equation 1 may well lead to underestimates of the initial settlement. A simple method to correct for the effect of first yielding has been proposed by D'Appolonia, Poulos and Ladd (1971)—it requires a knowledge of the ultimate bearing stress and the initial shear stress ratio.

The main difficulty in the prediction of initial settlement is the determination of the value of the deformation modulus, E_u, which is usually obtained from laboratory compression tests or occasionally, on more important projects, from field loading tests. Sample disturbance is known to affect considerably the value obtained, Simons (1957), Ladd (1969), and Raymond *et al.* (1971). For example, it was found for two structures in Norway (Simons (1957)) that E_u values obtained by carefully conducted unconfined compression tests carried out on samples obtained by a 54 mm thin walled stationary piston sampler, were about one-third of the values which could be estimated from the settlement observations. Ladd (1969) suggested that sample disturbance can be partly overcome by proper reconsolidation in the laboratory.

It has been suggested that more realistic determinations of E_u will be obtained if (1) samples are reconsolidated under a stress system equal to that existing in the field, e.g. Simons (1957) and Berre (1973), or (2) samples are reconsolidated isotropically to a stress equal to $\frac{1}{2}$ to $\frac{2}{3}$ of the in situ vertical stress, Raymond *et al.* (1971).

If samples of sensitive clays in particular are significantly disturbed, however, then reconsolidation may well lead to large changes in moisture content and hence a stiffer structure, with E_u determinations which are on the high side.

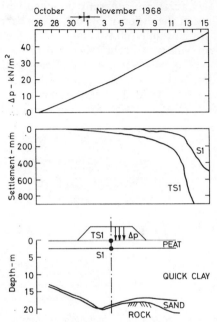

Fig. 3. Settlement observations for a test fill at Mastemyr, Norway
(after Frimann Clausen (1970))

Factors which should be considered when using consolidated undrained tests to estimate the deformation modulus are
(1) type of consolidation, i.e. whether isotropic or anisotropic
(2) stress level
(3) consolidation period
(4) stress path followed
(5) rate of strain
(6) times elapsed between opening up a test pit or drilling a borehole, taking a sample and then testing it
(7) size of the sample
(8) orientation of the sample
It is not possible within the length limitation of this report to discuss these factors in any detail and reference may be made to Marsland (1971), Ward (1971), Berre and Bjerrum (1973), Bjerrum (1973) and Lambe (1973a, b).

One factor of particular importance is the shear stress level. The field loading tests which have been carried out over the past few years in Norway have yielded invaluable information. Two field loading tests on a soft quick clay were carried out at Åsrum, Høeg, Andersland and Rolfsen (1969),

504

another at Mastemyr, Frimann Clausen (1969, 1970) and two at Sundland, Engesgaar (1970). These tests showed clearly that, for surface loads up to approximately one-third to one-half of the failure load, the measured undrained settlements (and the induced porewater pressures) were small; as the loading approached failure much larger settlements were naturally observed. The corresponding E_u values were thus shown to be very sensitive to the shear stress level. The results of the settlement observations taken at Mastemyr are shown in *Fig. 3*. Even though the filling was placed on peat and soft quick clay with a total thickness of some 19 m, very small settlements were recorded during the early stages of the test.

Table 2. VALUES OF E_u/s_u

Site	$\dfrac{E_u}{s_u}$	Reference
Test embankment, Kings Lynn	40	Wilkes (1974)
Oil tanks, Arabian Gulf	50–70 (depending on factor of safety)	Meigh and Corbett (1970)
Skabo Office Building, Oslo	150	Simons (1957)
Turnhallen (Heavy) Drammen	190	Simons (1957)
Tank, Shellhaven	220	Bjerrum (1964)
Northeast Test Embankment, Boston	240	Lambe (1973)[a]
Preload test, Lagunillas	250	Lambe (1973)[a]
Preload test, Amuay	250	Lambe (1973)[a]
Loading test, Skå Edeby	340	Bjerrum (1964)
Storage tanks, South Portland	400	Liu and Dugan (1974)
Satellite antenna tower, Fucino plains	450	D'Elia and Grisolia (1974)
Loading test, Fornebu	500	Bjerrum (1964)
Loading test, Åsrum	1000	Høeg, Andersland and Rolfsen (1969)
Økernbråten, Oslo	1500	Simons (1963)
Loading test, Mastemyr	3000	Frimann Clausen (1969, 1970)

Carefully conducted laboratory tests on a variety of clays, Berre (1973), confirm the conclusion. In addition, Berre found that the undrained stress–strain relationships were somewhat anisotropic, and also time dependent, the smaller the strain rate the smaller the E_u value, by a factor of approximately one-third per log cycle of time. It should, however, be pointed out that the E_u value for the Fornebu Clay was not found to be dependent on the strain rate, Bjerrum, Simons and Torblaa (1958), and Madhloom (1973) found an increase in E_u with increasing time to failure.

Because of the many difficulties involved in selecting a modulus value from the results of laboratory tests it has been suggested that a correlation between the deformation modulus and the undrained shear strength may provide a basis for a settlement calculation. Different authors have quoted different values for the ratio of E_u/s_u, for example Bjerrum (1964), 250 to 500, and Bjerrum (1972a) 500 to 1500.

Table 2 shows values of the E_u/s_u ratio for a number of structures on normally and slightly overconsolidated clay and the ratio ranges from 40 to

3000. As discussed previously, the shear stress level is a factor which has great influence on E_u; low values of E_u/s_u would be expected for highly plastic clays with a high shear stress level, and higher values for lightly loaded clays of low plasticity, Ladd (1964), Bjerrum (1972a).

The use of the Menard pressuremeter has also been advocated for the determination of the undrained modulus of clay, Hobbs (1971) and Calhoon (1972). In principle, the pressuremeter, testing a comparatively large volume of soil *in situ*, should provide a reliable basis for calculating the initial settlement, provided that the soil is not greatly anisotropic with respect to E_u. It is of interest to note that Calhoon (1972) found that initial settlements computed from pressuremeter data are of the same order of magnitude as initial settlements computed for E_u values in the range of 500 to 1000 times the undrained shear strength.

A new development is the Norwegian field compressometer, Janbu and Senneset (1973), which may well provide a reliable method for obtaining the undrained deformation modulus of clays although up to the present it has been used mainly in sands and silts.

An interesting investigation into the undrained deformation properties of a clay crust has been carried out by Bauer, Scott and Shields (1973). They found that *in situ* plate loading tests, using a rigid 46 cm diameter steel plate, gave a good correlation with the results from 3.1 × 3.1 m heavily reinforced rigid footing, 66 cm thick, while values deduced from laboratory tests were generally too low and erratic.

Primary consolidation settlement

Prediction of the magnitude and rate of the primary consolidation settlement of saturated clay strata was first made possible by Terzaghi some 50 years ago, when the theory of one dimensional consolidation was presented, and an oedometer designed which enabled determinations to be made of the parameters, c_v and m_v or C_c, necessary for the calculation procedure.

Since then considerable progress has been made in refining the computation process, and the reviewer has attempted to isolate some of the factors of particular importance which influence the accuracy of settlement predictions.

Net increase in stress

Consolidation settlement calculations are based on the assumption that the settlement is a function of the *net* increase in foundation pressure; if this net increase is zero it is assumed that the consolidation settlement will be zero. The view has been expressed that because the changes in porewater pressure for equal unloading and loading stages may well be different (Bishop and Henkel (1953)), it is possible that residual porewater pressures could be set up due to an unloading and reloading cycle, particularly in cases when the factor of safety against a bottom heave failure during excavation is small. It would appear, however, that there is a considerable body of evidence to show that only very small settlements are experienced with fully floating foundations, Casagrande and Fadum (1944), Aldrich (1952), Bjerrum (1964),

Golder (1965), Bjerrum and Eide (1966), Bjerrum (1967) and D'Appolonia and Lambe (1971). The data given by Bjerrum and Eide (1966) are summarised in *Table 3*.

Table 3. SETTLEMENTS OF COMPENSATED FOUNDATIONS (after Bjerrum and Eide (1966))

Structure	Date completed	Undrained shear strength (kN/m²)	Net foundation pressure (kN/m²)	Heave (mm)	Total settlement (mm)
Håndverk Industri, Drammen	1958	7–15	1	26	34
Werringgården, Drammen	1956	6–15	0	29	30
Norløff, Drammen	1955	6–15	0	—	c. 30
Park Hotel, Drammen	1961	10–15	0	—	40
Idunbygg, Drammen	1963	30	0–5	—	45

It can be seen that the settlement during reloading was approximately equal to the heave. In addition, it was noted that the settlements terminated a short time after the end of construction. It can therefore be concluded that it is satisfactory to perform consolidation settlement calculations based on net increases in foundation pressure only.

Sample disturbance

The importance of sample disturbance has been recognised for many years— e.g. Terzaghi (1941), Rutledge (1944), Schmertmann (1953 and 1955)—and procedures have been suggested to correct laboratory stress–strain relationships for such disturbance. For normally consolidated clays, sample disturbance will lead to measured oedometer compressibilities which are too low, while for overconsolidated clays the measured compressibilities may be too high.

Bjerrum (1967) pointed out that the measurement of the preconsolidation pressure was particularly sensitive to sample disturbance and that only the highest standards of sampling and laboratory technique would result in consistent and reliable determinations.

Berre, Schjetne and Sollie (1969) compared values of $C_c/(1 + e_0)$ and p_c'/p_o' obtained from a new 95 mm piston sampler and from the 54 mm piston sampler (itself a very high quality instrument) which has been in use for many years. They found more scatter in the results from the 54 mm sampler and, in addition, the average value of p_c'/p_o' derived from the 95 mm sampler was some 5% higher than that obtained from the smaller tube. They also noted that chemical changes may take place in a clay if stored for several months in steel sampling tubes and that these changes could well alter the geotechnical properties of the soil.

Bjerrum (1973) noted that several types of soft and sensitive clays at very small strain show a critical shear stress which in many cases governs their

507

behaviour. This small strain behaviour is destroyed even if the clay is subjected to relatively small strains either during the sampling operations or during handling in the laboratory. Furthermore, sample disturbance leading to redistribution of moisture content may lead to significant errors in measured parameters.

There can be no doubt that high standards in sampling and testing must be adopted if reliable settlement predictions are to be obtained.

Induced porewater pressures under a structure

A most important contribution to settlement analysis was made by Skempton and Bjerrum (1957) who pointed out that an element of soil underneath a foundation undergoes lateral deformation as a result of applied loading and that the induced porewater pressure is in general less than the increment in vertical stress on the element, being dependent on the A value. For the special case of the oedometer test, where the sample is laterally confined, then, irrespective of the A value, the porewater pressure set up is exactly equal to the increment in vertical stress.

They proposed that a correction factor should be applied to the settlement calculated on the basis of oedometer tests and showed that the factor was a function of the geometry of the problem and the A value, the smaller the A value the smaller the correction factor.

For heavily overconsolidated clays, A values less than 1 would be expected and the Skempton–Bjerrum correction factor is usually applied in such cases. It should be noted that in the working range of stress for normally consolidated clays, particularly when there is no question of overstressing occurring, A values less than 1 may well be found and the correction factor should then be applied.

Stress path settlement analysis

As discussed previously, Skempton and Bjerrum (1957) recognised that an element of soil underneath a foundation undergoes lateral deformation as a result of applied foundation loading and that the subsequent consolidation would be a function of the excess porewater pressures set up *in situ*. It was assumed, however, that the relationship between axial compressibility and effective stress could be determined in the standard oedometer, i.e. the influence of lateral stresses on the stress deformation characteristics of the soil was not taken into account. A better laboratory procedure to predict settlement would be to test the soil by applying, as closely as possible, the same stress changes as those to which the soil will be subjected in the field. Moreover, soil behaviour being generally non-linear the various deformation properties will vary with stress level and it is also desirable that a soil specimen, after sampling, be first brought back to the stress system initially prevailing in the ground before subjecting it to the stress changes it is likely to undergo on loading.

It has been shown that the use of stress path testing results in smaller (and more reliable) predictions of settlements for structures on heavily overconsolidated clay, e.g. Lambe (1967), Simons and Som (1970). Fewer data are available for normally or lightly overconsolidated clays, although Lambe (1973a) showed for the North East test Embankment that settlements predicted from oedometer tests and stress path triaxial tests did not differ greatly.

Preconsolidation pressure. An excellent example of the importance in practice of determining the preconsolidation pressure is given by Vargas (1955) who showed from settlement observations of a number of buildings in Saõ Paulo that

(1) where the applied foundation stress did not exceed the difference between the preconsolidation pressure and the *in situ* effective pressure, the observed settlements were generally less than 10 mm, and were very much smaller than those computed directly from oedometer curves, on occasions equalling only 10% of the computed values.

(2) where the applied foundation stress exceeded $p_c' - p_o'$, albeit only marginally, much larger settlements were measured. Reasonable agreement between calculated and observed settlement was obtained only if the calculations were based on stress increases in excess of p_c', not p_o'. Detailed studies of the properties of various clays by a number of investigators have shown that there are different factors which can give rise to the preconsolidation pressure being greater than the present effective overburden pressure, for example
  (a) removal of overburden
  (b) fluctuations in the ground water table
  (c) cold-welding of mineral contact points between particles
  (d) exchange of cations
  (e) precipitation of cementing agents
  (f) geochemical processes due to weathering
  (g) delayed consolidation.

Reference can be made to Casagrande (1936), Terzaghi (1941), Schmertmann (1955), Moum and Rosenqvist (1957), Zeevaert (1957 and 1960), Bjerrum and Wu (1960), Hamilton and Crawford (1960), Leonards and Ramiah (1960), Casagrande (1964), Crawford (1964 and 1965), Leonards and Altschaeffl (1964), Bjerrum (1967), Ladd (1971), Moum, Løken and Torrance (1971), Berre (1973).

It is well known that the determination of p_c' is partly a function of the test procedure adopted in the laboratory, for example, the rate of loading adopted, whether or not rest periods have been permitted, and the effects of sample disturbance. The value of p_c' determined in the laboratory may well be different from that which can be relied upon in the field.

The development of preconsolidation pressure due to delayed consolidation and the effects of such preconsolidation on the settlements of structures has been discussed in detail by Bjerrum (1967, 1972a and 1973).

The concept of instant and delayed compression as proposed by Bjerrum is illustrated in *Fig.* 4(a). Instant compression is the settlement which would result if the excess porewater pressures set up by a foundation loading could dissipate instantaneously with load application, and delayed compression is the settlement then developing at constant effective stress.

509

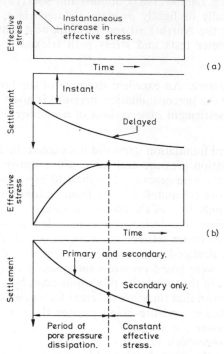

Fig. 4. Instant and delayed compression (after Bjerrum (1967))

A clay which has recently been deposited and come to equilibrium under its own weight but has not undergone significant secondary consolidation may be classified as a 'young' normally consolidated clay. Such a clay is characterised by the fact that it is just capable of carrying the overburden weight of soil and any additional load will result in relatively large settlements. If an undisturbed sample of a 'young' clay is tested in an oedometer, the resulting e–log p curve will show a sharp bend exactly at the effective

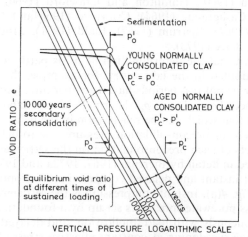

Fig. 5. Voids ratio–log pressure–time relationships (after Bjerrum (1967))

overburden pressure p_o', which the sample carried in the field. A consolidation curve of this type is shown in *Fig. 5*, marked 'young', being characterised by the fact that $p_c' = p_o'$. To this group of clays thus belong only the clays which are recent from a geological point of view. A clay which has just consolidated under an additional load—as for instance a fill—will also be classified with respect to its compressibility as a 'young' clay deposit.

If a young clay is left under constant effective stresses for hundreds or thousands of years, it will continue to settle. The result of this secondary or delayed consolidation is a more stable configuration of the structural arrangement of the particles which means greater strength and reduced compressibility. With time a clay undergoing delayed consolidation will thus develop a reserve resistance against a further compression. It can carry a load in addition to the effective overburden pressure without significant volume change. If an undisturbed sample of such an 'aged' normally consolidated clay is subjected to a consolidation test, the resulting e–$\log p$ curve will follow the curve marked 'aged' on *Fig. 5*. The curve shows an abrupt increase in compressibility at a pressure p_c' which is greater than p_o'.

As pointed out by Bjerrum (1972a) considerable patience is required to determine experimentally in the laboratory the basic information necessary

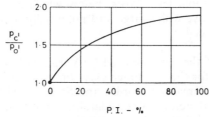

Fig. 6. p_c'/p_o' versus plasticity index (after Bjerrum (1972))

for constructing such a diagram as *Fig. 5*. In fact, the only diagrams existing originate from areas where long-term observations of settlement are available.

A p_c' effect developed as a result of a delayed consolidation is characterised by the fact that the developed value of p_c' increases proportionally with p_o', the effective overburden pressure the clay carried in the period it experienced delayed consolidation. In a homogeneous clay deposit, the ratio p_c'/p_o' is consequently constant with depth and this ratio can conveniently be used to describe the effect. The p_c'/p_o' ratio of clay deposits of the same age will increase with the amount of secondary consolidation which the clay has undergone under the existing overburden pressure. Because secondary consolidation increases with the plasticity of the clay the p_c'/p_o' ratio will increase with the plasticity index. *Figure 6* shows the correlation between p_c'/p_o' ratio and the plasticity index observed in some normally consolidated clays which all have aged over a period of thousands of years.

It is of considerable importance to distinguish between preconsolidation pressure due to (a) overconsolidation as a result of removal of overburden pressure, or ground water level fluctuations or chemical and weathering effects and (b) delayed consolidation. There is some evidence to indicate that for the first category clays can be loaded up very close to the preconsolidation pressure with small resulting settlements, e.g. Vargas (1955), but if p_c' is due

511

to delayed compression then significant settlements, probably too large to be considered acceptable, develop when the applied pressure exceeds 50% of $p_c' - p_o'$, Bjerrum (1967), and *Table 4*.

It can be seen from *Table 4* that settlements after 20 years amounting to about 200 mm may be expected for values of $\Delta p/(p_c' - p_o')$ of 58% (Scheitliesgate 1), increasing to 400 to 460 mm (Konnerudgate 16 and Danvikgate 3) for $\Delta p/(p_c' - p_o')$ about 80%.

A completely different picture emerges from the regional studies carried out in Sweden by Nordin and Svensson (1974), Paper II/7. At Margretelund, Lidköping, the measured settlement to date is less than 9 mm (and appears to be virtually complete) with a $\Delta(p/(p_c' - p_o')$ value of about 0.5, while at Lilla Torpa, Vänersborg, the observed settlements are up to 16 mm, with very little more expected, and here the net increase in stress due to the building loads and anticipated ground water lowering is nearly equal to $p_c' - p_o'$. At the present time, laboratory studies alone will not allow accurate settlement predictions to be made. Long term regional studies are vitally necessary to determine, in particular

(1) whether in the field primary consolidation and/or secondary settlements will develop over a long period of time, and
(2) whether a threshold level exists, below which acceptable settlements develop and above which large and potentially dangerous settlements will be experienced.

Regional studies of the type reported by Vargas (1955) for the Saõ Paulo clay, Bjerrum (1967) for the Drammen clay, Jarrett, Stark and Green (1974) for the Grangemouth area and Nordin and Svensson (1974) for Lidköping and Vänersborg, are therefore of considerable significance.

It should be stressed that at the present time it has been shown that only remarkably few clays exhibit a preconsolidation pressure due to delayed consolidation, Brown (1969). Indeed the Oslo clay, Andersen and Frimann Clausen, Paper II/1, is an interesting and surprising example. The settlement records for the structure indicate marked delayed compression and oedometer tests also show appreciable secondary settlements, yet in spite of high quality undisturbed sampling and extremely careful handling and testing techniques p_c' was found to be approximately equal to p_o' while from *Fig. 6*, a value of 1.2 for the p_c'/p_o' ratio would be indicated. Berre (1973) has quoted a value of 1.25 for the p_c'/p_o' ratio for the very similar Oslo clay at Vaterland, $w_L = 47$, $w_p = 27$, $w_c = 40$ and $s_t = 5$.

Berre and Bjerrum (1973) have shown from a series of triaxial and simple shear laboratory tests that the development of a quasi-preconsolidation pressure may be associated with a critical shear stress. If the applied shear stress together with the initial shear stress is less than this value then only relatively small deformations will be experienced. If the critical stress is exceeded, however, appreciable deformations will occur. A similar result was obtained by Hartlen and Pusch (1973). This concept is confirmed by the test embankments at Åsrum, Mastemyr and Sundland.

Further confirmation is given by the settlement observations of various structures in Drammen and the relevant data are summarised in *Table 4*. It can be seen that there is a correlation between the 20-year observed settlement and the ratio $\Delta p/s_u$, Δp being the net applied foundation pressure on the surface of the plastic clay layer and s_u the undrained shear strength, the greater $\Delta p/s_u$ the greater the observed settlement.

The determination of the preconsolidation pressure can also be affected by temperature, Habibagahi (1973), and it may therefore be necessary to carry out laboratory tests at the same temperature as that existing *in situ*.

Rate of settlement. The observed rate of settlement of structures is almost invariably very much faster than that calculated using one dimensional consolidation theory based on oedometer tests carried out on small samples. Rowe (1968 and 1972) showed that the drainage behaviour of a deposit of clay depends on the fabric of the soil. Thin layers or veins of sand and silt, or rootholes, can result in the overall permeability of the clay *in situ* being many

Table 4. TWENTY-YEAR SETTLEMENTS FOR STRUCTURES IN NORWAY RELATED TO $\Delta p/p_c' - p_0'$ AND $\Delta p/s_u$

Structure	Settlement at 20 years (mm)	$\dfrac{\Delta p}{p_c' - p_0'}$ (%)	$\dfrac{\Delta p}{s_u}$
Skoger Sparebank	40	37	0.6
Scheitliesgate 1	200	58	0.9
Skistadbygget	c. 230	72	1.0
Konnerudgate 12	250	74	1.6
Turnhallen. Light	360	68	1.8
Konnerudgate 16	400	80	1.8
Danvikgate 3	460	80	1.8
Turnhallen. Heavy	600	123	3.5

times greater than that measured on small samples. An hydraulic oedometer, Rowe and Barden (1966), was developed to enable more reliable measurements of c_v to be made. Sample diameters of up to 250 mm with heights of up to 125 mm can be accommodated, with either vertical or horizontal drainage. The loading is applied hydraulically, and porewater pressures, volume changes and axial deformations can be measured.

In situ permeability measurements coupled with laboratory measurements of compressibility (which are not so sensitive to sample size) give values of c_v in fair agreement with field performance and with the results from large samples tested in the hydraulic oedometer at similar stress levels. Reference can be made to the Proceedings of the Conference on In Situ Investigations in Soils and Rocks, London, 1970, and to the Symposium on Field Instrumentation in Geotechnical Engineering, London, 1973.

Much work has been carried out on theories of three dimensional consolidation and reference can be made to Biot (1941), Gibson and Lumb (1953), Gibson (1961), Cryer (1963), Gibson and McNamee (1963), Davis and Poulos (1965), Raymond (1965), Davis and Poulos (1968), Schiffman, Whitman and Jordan (1969), Davis and Poulos (1972) and Lewin (1973).

Recourse to theories of three dimensional consolidation coupled with reliable determinations of c_v, taking due account of orientation and stress level, is necessary if it should be important to obtain reasonably accurate predictions of rates of settlement.

It is beyond the scope of this report to review advances in consolidation theory and technique. Reference can be made to Barden (1969), Poorooshab (1969) and Scott and Ko (1969).

Secondary settlement

Secondary settlement is generally considered to be the settlement which develops following changes in effective stress. After many years of research work into secondary consolidation no reliable method is yet available for calculating the magnitude and rate of such settlement, for which the necessary soil parameters can be obtained fairly simply, and which takes into account the various factors, for example the principal effective stress ratio, the load increment ratio, temperature and time effects which are known to affect significantly secondary settlement. For this reason if estimates of secondary consolidation are required in practice they are generally based one empirical procedures.

It should be pointed out that, contrary to the view which has sometimes been expressed that secondary settlement is often of little practical consequence so far as structures are concerned, several case records are available which show clearly that in certain circumstances a large part of the observed settlement has occurred after full dissipation of excess porewater pressure, e.g. Foss (1969). Satisfactory prediction of secondary settlement is therefore certainly a matter of practical importance.

When considering secondary settlement it should be noted that two different factors may influence the process. The first is reduction in volume at constant effective stress and the second is the vertical strain resulting from lateral movements in the ground beneath the structure. Terzaghi (1948) pointed out that these two factors may be expected to result in completely different types of settlement. The relative importance of these factors will vary from structure to structure, depending on the stress level, type of clay and the geometry of the problem, and for any given structure will vary with the location of any deforming soil element and with time.

No great accuracy can, therefore, be expected from predictions of secondary settlement and caution should be exercised when attempting to extrapolate the results from one particular investigation to another set of conditions.

Investigations into secondary settlement can be considered to fall into three categories:

(1) Laboratory work.
(2) Empirical approaches based on field and laboratory results.
(3) Theoretical analyses based on rheological models.

Laboratory work

Many workers have carried out various types of laboratory test to investigate various aspects of secondary consolidation behaviour, for example Buisman (1936), Gray (1936), Haefeli and Schaad (1948), Leonards and Girault (1961), Lo (1961), Wahls (1962), Leonards and Altschaeffl (1964), Wu, Resendiz

514

and Neukirchner (1966), Bishop and Lovenbury (1969), Berre and Iverson (1972). Important points to emerge are that:

(1) Organic soils show pronounced secondary effects.
(2) Many soils exhibit a linear relationship between settlement and log time for a considerable period, although this relationship cannot hold indefinitely.
(3) Isotropic consolidation results in less secondary effect than consolidation with no lateral yield.
(4) Secondary settlement is more pronounced at stresses under the pre-consolidation pressure, for small load increment ratios, with increasing temperature, with decrease in the length of the drainage path, and for small factors of safety.
(5) Instability may occur, i.e. the rate of settlement may increase temporarily after a long period of time.

Empirical approaches

Generally these are based on the assumption that secondary settlements can be approximated by a straight line on a settlement versus logarithm of time plot, Buisman (1936), Koppejan (1948), Zeevaert (1957, 1958). While there is much evidence to indicate that such a simple extrapolation cannot in general be expected to give reliable predictions in field problems this approach

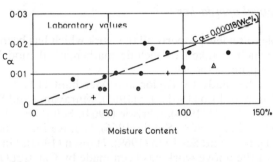

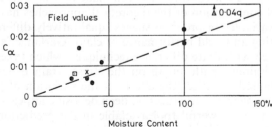

- ● Moran et al (1958)
- ▲ James (1970)
- ▢ Meigh and Corbett (1970)
- + Wilkes (1974)
- × Andersen and Frimann Clausen (1974)

Fig. 7. Values of c_α, the coefficient of secondary consolidation

515

is often adopted in practice. It must be admitted, there is as yet no theoretical solution available which takes into account all the factors which are known to affect secondary compression and in which the relevant soil parameters can be fairly easily obtained.

Use has been made of c_α, the coefficient of secondary consolidation. Values of c_α have been quoted by Moran *et al.* (1958), Wahls (1962), Henkel (1965), James (1970), Meigh and Corbett (1970), and are collected together in *Fig. 7*.

Bjerrum (1967) presented his model to represent 'instant' and 'delayed' compression on a plot of void ratio versus the logarithm of effective vertical pressure, *Fig. 5*, and Foss (1969) applied this approach to predicting the secondary settlements of four structures in Drammen. It must be noted, however, that the predictions were based on the assumptions that

(1) the relative positions of the lines representing field void ratios at differing loading times from 24 hours to 3000 years are available;
(2) if $\Delta p < p'_c - p'_o$, no primary consolidation settlement will occur, and
(3) the clay under a structure will follow these relationships.

While Foss found good agreement between calculated and observed rates of secondary settlement further comprehensive laboratory data for other clays, similar to those which are available for Drammen, and backed up by settlement records of structures on these clays are an urgent requirement.

Theoretical approaches

It is not possible to review properly all the work which has been carried out in this direction which is taken to include mathematical models, rheological models and rate process theory.

Reference can be made to Taylor and Merchant (1940), Taylor (1942), Tan (1957), Gibson and Lo (1961), Lo (1961), Murayama and Shibata (1961), Leonards and Altschaeffl (1964), Christie (1964), Schiffman, Ladd and Chen (1964), Barden (1965 and 1968), Suklje (1966), Wu, Resindiz and Neukirchner (1966), Brinch Hansen and Ses Inan (1969), Hansen (1969), Garlanger (1972).

A summary of the above work has been made by Clayton (1973). Possibly the most useful contribution of the theoretical approaches has been to study the effects of variables in order to suggest qualitatively differences between laboratory testing and field performance of clays. Garlanger's mathematical model, combined with finite difference techniques, which takes into account the effects of drainage path length on the stress–strain behaviour of clays and can be adapted to reflect the effects of the variation of pressure increment that occurs beneath structures on deep deposits, appears at the present time to be the most powerful tool available in the prediction of the time–settlement behaviour of loaded soil in the field.

Long term settlement records

It should be pointed out that although secondary settlements are of significance for many structures there are also a number of cases where settlement

516

records show small or negligible secondary effects. Three illustrative examples are the apartment building At Økernbråten, Oslo, Simons (1963), the trial embankment at Avonmouth, Murray (1971), and the multi-storey buildings in Glasgow, Somerville and Shelton (1972). Typical settlement log–time curves are shown in *Fig. 8*.

At Økernbråten, a nine-storey block of flats founded on strip foundations transmitting 226 kN/m² to the underlying firm to stiff, becoming soft with depth, clays at least 20 m thick, experienced a maximum settlement of 18 mm

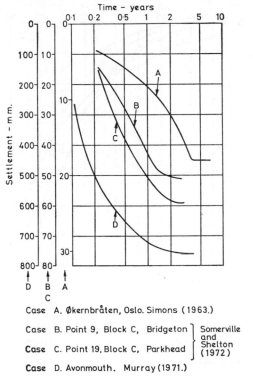

Case A. Økernbråten, Oslo. Simons (1963.)

Case B. Point 9, Block C, Bridgeton ⎤ Somerville
and
Case C. Point 19, Block C, Parkhead ⎦ Shelton (1972)

Case D. Avonmouth. Murray (1971.)

Fig. 8. Settlement log–time curves indicating terminating settlement

and the settlements stopped completely about $3\frac{1}{2}$ years after the end of construction.

At Avonmouth, the square embankment with a maximum height of 9.2 m, a width at the base of 67 m and side slopes of 1 in 2, constructed of pulverised fuel ash of average bulk unit weight 13.4 kN/m³, on highly compressible alluvial soils, about 13 m in thickness, with undrained shear strengths ranging generally from about 40 kN/m² to about 90 kN/m², experienced a maximum settlement of about 780 mm some 1400 days after the start of construction of the embankment, 90% of primary consolidation occurring 300 days after the start of construction. It can be seen that the rate of movement is now quite small and appears to be reducing with time.

At Glasgow, six blocks of fifteen-storey flats were constructed in the Parkhead and Bridgeton districts, which are underlain by deep alluvial deposits of firm laminated clays and silts with deeper deposits of sands, gravels and glacial drifts, followed by productive coal measures. Raft foundations were adopted with net bearing pressures of about 53.5 kN/m². The time settlement curves show very slow rates of settlement only 1 to 2 years after the end of construction, with maximum settlements ranging from 30 to 60 mm.

Clearly there is a need to be able to distinguish between cases where settlement continues over many years, as illustrated in *Fig. 9* for Norwegian and Swedish clays, and cases where the rate of settlement virtually stops a few years after the end of construction, as shown in *Fig. 8*.

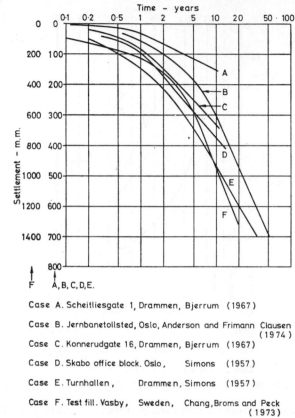

Case A. Scheitliesgate 1, Drammen, Bjerrum (1967)

Case B. Jernbanetollsted, Oslo, Anderson and Frimann Clausen (1974)

Case C. Konnerudgate 16, Drammen, Bjerrum (1967)

Case D. Skabo office block. Oslo, Simons (1957)

Case E. Turnhallen, Drammen, Simons (1957)

Case F. Test fill. Vasby, Sweden, Chang, Broms and Peck (1973)

Fig. 9. Settlement log–time curves indicating non-terminating settlement

The complexity of the problem is shown in *Fig. 10*, which gives the settlement log–time plots for the three well known Chicago structures, the Masonic Temple, the Monadnock Block, and the Auditorium Tower, Skempton, Peck and MacDonald (1955). The behaviour of the Masonic Temple is quite different from that of the other two, more than 90% of the total final settlement having developed after 5 years. The corresponding figures for the Monadnock Block and the Auditorium Tower are 47% and 62%. Furthermore, after 10 years the settlement of the Masonic Temple was virtually complete but for the other two structures settlements were taking place after 30 years.

It may well be that investigations into the geochemistry of the soils, or work on micropaleontology, or electron microscope analysis will be necessary, as well as the determination of engineering properties, in order to be able to classify clays with regard to their long term settlement behaviour.

Settlement observations taken on structures and embankments should be continued over a sufficient period of time for the long term trends to be indicated. To halt observations at a time when excess porewater pressures have just dissipated, for example, may well result in an incomplete picture emerging of the true settlement time performance.

A final point to be noted is that structures subjected to large variations in live load, for example, silos, storage tanks and high structures under wind

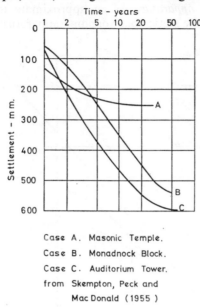

Case A. Masonic Temple.
Case B. Monadnock Block.
Case C. Auditorium Tower.
from Skempton, Peck and
Mac Donald (1955)

Fig. 10. Settlement log–time curves for three structures in Chicago

action, may be expected to experience appreciably larger secondary settlements than would be the case for non-varying loading conditions, Bjerrum (1964 and 1966).

Other methods of predicting deformations

Some other approaches which have been suggested for predicting the deformations of structures are outlined below. As they have not often been applied in practice, their place in settlement analyses could form a topic for discussion.

Centrifugal models. Recent developments in testing centrifugal models indicate that this method of predicting ground displacements may well be a practical tool in the fairly near future. The results obtained by Wroth and Simpson (1973) in predicting the deformations of the trial embankment at King's Lynn, Wilkes (1974), are particularly encouraging. Reference can be made to Endicott (1970), Rowe (1972), Bassett (1973), Bolton *et al.* (1973), Mikase and Takada (1973), Polshin *et al.* (1973).

Janbu's deformation modulus. Settlement calculations are usually based on the parameters C_c or m_v. Janbu (1963 and 1969) proposed basing computations on the tangent modulus, being equal to $d\sigma/d\varepsilon$, as a suitable measure of the compressibility of soils, ranging from sound rock to plastic clays. The tangent modulus depends on stress conditions and stress history and can be considered to represent a resistance against deformation. As the calculation procedure based on the tangent modulus concept is very simple and virgin loading, unloading and reloading conditions can be taken into account, experiences of the use of this approach in practice would be most welcome.

Dutch cone sounding apparatus. As an approximate method to estimate settlement the use of the static cone sounding apparatus has been proposed,

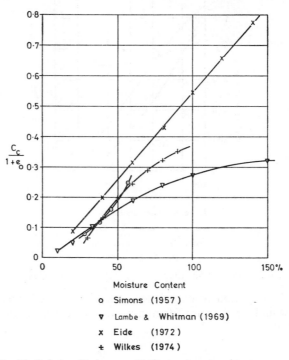

Moisture Content

o Simons   (1957)

▽ Lambe & Whitman (1969)

x Eide   (1972)

± Wilkes   (1974)

Fig. 11. Relationship between $C_c/(1 + e_o)$ and moisture content

Gielly, Lareal and Sanglerat (1970) and Sanglerat (1972), based on correlations between compressibility and the cone resistance. No accurate settlement forecast can be expected, for reasons which have been discussed in this report, but it would seem that the cone resistance can give a useful first approximation of compressibility, Meigh and Corbett (1970).

Correlation of compression index with moisture content. Another first approximation to compressibility for normally consolidated clays is based on correlations between $C_c/(1 + o_o)$ and the moisture content. Various such correlations have been plotted in *Fig. 11*—a fairly narrow band results only for moisture contents less than about 70%.

520

Review of papers presented for Session 2

Andersen and Clausen, Paper II/I, is particularly interesting, not merely because of the length of time (50 years) over which the settlements have been observed. The difficulties which arise when attempting to make a settlement calculation for a structure have been clearly presented, for example the assumptions which have to be made concerning the presence of the short driven displacement piles, the estimate of the net loading actually applied and the allowance to be made for the presence of the surrounding fill, mainly placed before the building was constructed, but partly at the time of construction. It is noteworthy that, although a very high standard of sampling and laboratory technique was achieved, no preconsolidation effect due to delayed compression could be detected. The observed settlement at the present time, which includes undrained deformation, total primary consolidation settlement (since piezometers show no excess porewater pressures), together with some secondary settlement, is considerably smaller (approximately 60%) than the calculated primary consolidation settlement alone, computed without taking into consideration any preconsolidation effect. If calculations are made for the undrained and for the secondary settlements, then the discrepancy between calculated and observed settlement becomes even greater. This case record is somewhat unusual in that the settlements are large and the agreement between calculated and observed settlement is comparatively poor.

Cowley, Haggar and Larnach, Paper II/2, compare calculated with observed settlement for three large cold stores in Grimsby, founded on what are essentially raft foundations on soft alluvial deposits up to about 15 m in thickness. It is interesting to note that appreciably different settlements were observed for geometrically similar observation points, indicating the non-homogeneous nature of the subsoil. A similar observation was recorded by Lambe (1973a). Because of these variations, and also because of the changes in applied loading which have been experienced, it is not possible to draw firm conclusions concerning a correlation between calculated and observed settlements, particularly when the calculations had to be based on samples obtained from standard U-4 cores, which in such soils must result in considerable disturbance. Even so, the predictions were of sufficient accuracy to enable structural design concept decisions to be made and, in spite of the considerable differential settlements which were measured, no distress has been suffered by the structures.

D'Elia and Grisolia, Paper II/3, present a well-documented case record of settlement observations for a rigid foundation of diameter 28 m on a very soft silty clay. The gross loading pressure was 62 kN/m^2 and the net pressure 10 kN/m^2. It was concluded that the *in situ* undrained deformation modulus which could be inferred from the observations was approximately 450 times the undrained shear strength. The authors claim good agreement between the computed consolidation settlements and the settlement observations, on the grounds that at a time three years after the end of construction the calculated settlement, based on Biot's three dimensional theory using the laboratory measured c_v, was 25 mm compared with the observed settlement

521

of 28 mm, the total calculated average consolidation settlement being quoted as 250 mm. A different interpretation could be put forward, namely that the calculated final settlement of 250 mm is an overestimate of the final consolidation settlement which will actually develop, and that the field rate of consolidation is proceeding faster than that indicated by calculations based on laboratory oedometer tests. *Figure 5* in Paper II/3 shows that measured consolidation settlement falls on a straight line when plotted against the square root of time. Even assuming this relationship holds in the future, a further 230 years will be required before the observed settlement equals the calculated settlement. Put another way, three years after the end of construction the observed consolidation settlement is only 10% of the final calculated consolidation settlement. The Reporter suggests that it is more likely that the calculations, based on a normally consolidated clay, overestimate the actual settlement and this view is supported by the small net increase, 10 kN/m², which the foundation applies to the ground. The existence of only a small amount of preconsolidation of the soil would, of course, affect the settlements appreciably.

Jarrett, Stark and Green, Paper II/4, present settlement records taken on 15 structures on the Grangemouth area where the soil conditions consist mainly of soft clayey silts to a depth of about 18 m, underlain by glacial till above rockhead. These records show that a threshold value of the net increase in foundation stress of about 15 kN/m² is indicated, below which only minor settlements, less than 20 mm, occur. Laboratory test results only on U–4 samples were available to the authors at the time of writing the Paper and considerable sample disturbance no doubt occurred. The authors are conducting research orientated tests on block samples and it will be of great interest to compare the threshold value of 15 kN/m² indicated by the field studies with the difference between the preconsolidation pressure and the effective overburden pressure. In this connection, it should be noted that the case records for Konnerudgate 16 and Skoger Sparebank, Bjerrum (1967), show that it is not the net stress increase alone that influences the magnitude of the settlements but the stress increase in relation to $p_c' - p_o'$, which can of course vary within a clay deposit. A regional field study of this type is of the utmost value since it enables correlations to be made with predictions based on laboratory testing.

Liu, Aldrich and Dugan, Paper II/5, draw attention to the practical difficulties which are often, or perhaps generally, encountered when attempting to obtain continuous settlement records on a number of observation points. It is not possible to draw firm conclusions concerning the ratio of computed to observed settlement, partly because the settlement points were installed some $1\frac{1}{2}$ years to 2 years after the start of construction, and partly because only $1\frac{1}{2}$ years of observations are available after the end of construction. During this time, however, very little settlement has been measured. It seems likely that the field consolidation settlement will be less than the computed consolidation settlement. It is interesting to observe that the upper 15 m of the Boston Blue Clay is overconsolidated with values of the overconsolidation ratio greater than $2\frac{1}{2}$, while at lower depths only a very slight degree of overconsolidation is apparent.

Liu and Dugan, Paper II/6, present settlement observations for short term water load tests on three steel storage tanks founded on pads of compacted granular fill bearing directly on soft gray silty clays up to 18 m thick. Fairly good agreement between calculated and observed undrained settlements was obtained and a ratio of undrained deformation modulus to undrained shear strength of about 400 is indicated.

Nordin and Svensson, Paper II/7, present the results of regional studies for two sites, Lidköping (Margretelund) and Vänersborg (Lilla Torpa) in Western Sweden, where settlement observations have been taken on a number of buildings founded on rafts on lightly overconsolidated clays. There are several points of particular interest:

(1) The settlements were measured to 0.01 mm using a dial gauge connected to the structure measuring on to a steel rod driven to rock.
(2) Great care was taken to determine preconsolidation pressures, both from the results of oedometer tests and also using the expression

$$p_c = \frac{s_u}{0.45 \, w_L}$$

At Lidköping, the re-applied foundation loadings were kept to a value 20 kN/m² less than the preconsolidation pressures while at Vänersborg the total loading approached the preconsolidation pressure.
(3) The observed settlements (taken from an early time during construction) were small, not more than 9 mm for Lidköping and not more than 15 mm for Vänersborg, and were virtually complete a few months after the end of construction, at which time no excess porewater pressures were observed. It should be noted that at Lidköping the clay contained very thin seams of silt while at Vänersborg root fibres perforated the clay from top to bottom: it is therefore not surprising that the rates of settlement were so rapid.
(4) The calculated settlements are very much greater than those observed. Typical calculated total settlements are 73 mm at Lidköping, compared with the maximum observed settlement of 9 mm, while the corresponding figures for Vänersborg are 51 mm and 15 mm.

Sanglerat, Girousse and Gielly, Paper II/8, give the settlement observations for a long low residential building founded on a cellular raft consisting of two slabs 0.3 and 0.2 m in thickness separated by a 0.9 m deep gap, sitting on a lean concrete slab 0.5 m thick. The soil conditions vary considerably across the site and soft organic clays of considerable and varying thickness are present. The observed settlements range from 200 mm to 1160 mm; those calculated on the basis of oedometer tests from 70 mm to 1520 mm and those on the basis of Dutch cone soundings, from 40 mm to 660 mm. It is not possible to draw firm conclusions concerning the ratio of calculated to observed settlements because of the heterogeneity of the subsoil and the rather limited amount of geotechnical laboratory data which is available. The fill which was placed at the time of construction has considerably influenced the settlements and it is possible that the fill present before construction started (1.1 m thick at boring S 3) has also had an effect.

523

Vefling, Paper II/9, discusses settlement observations taken on three sugar silos in Italy, two located in Russi and one at Molinella. The structures are supported on driven precast R.C. piles, about 33–36 m long at Russi and some 47 m long at Molinella. At Russi, the subsoil consists of interbedded layers of clays, silts and sands, proved to a depth of 47 m, with somewhat similar conditions at Molinella, although more organic material was found; the borings penetrated to 71 m. The subsoils are overconsolidated to an unknown degree. Of particular interest is the observation that the settlement is increasing linearly with time (one structure at Russi has been observed for more than 10 years) and the author attributes this to fluctuating nature of the live load. Insufficient geotechnical data are available to enable a meaningful comparison to be made between calculated and observed settlements.

Table 5. TWENTY-YEAR SETTLEMENTS GREATER THAN 200 mm

Structure	δ_{20} (mm)	Reference
Skistadbygget, Drammen	c. 230	Engesgaar (1972, 1973)
Konnerudgate 12, Drammen	250	Bjerrum (1967)
Masonic Temple, Chicago	250	Skempton, Peck and MacDonald (1955)
Cold Stores, Grimsby	c. 300	Cowley, Haggar and Larnach (1974)
Silo, Russi	c. 300	Vefling (1974)
Konnerudgate 16, Drammen	400	Bjerrum (1967)
Monadnock Block, Chicago	450	Skempton, Peck and MacDonald (1955)
Danvikgate 3, Drammen	460	Bjerrum (1967)
Skabo Office Block, Oslo	460	Simons (1957)
Jernbanetollsted, Drammen	500	Andersen and Frimann Clausen (1974)
Auditorium Tower, Chicago	540	Skempton, Peck and MacDonald (1955)
Turnhallen, Drammen	600	Simons (1957)
Tower City Hall, Drammen	670	Bjerrum (1967)
Apartment Building, Oslo	c. 700	Hutchinson (1963)
Residential Building, Nantua	c. 1200	Sanglerat, Girousse and Gielly (1974)

Note: δ_{20} = observed or extrapolated maximum settlement at 20 years

An interesting point to emerge is that many structures have been subjected to settlements of considerable magnitude, apparently without experiencing sufficient damage to render them unserviceable. *Table 5* shows observed or extrapolated maximum settlements greater than 200 mm, at an arbitrary time of 20 years, for a number of structures. Oil tanks have not been included in the Table, although many have experienced maximum settlements greater than 200 mm without damage.

Conclusions

(1) Although great advances have been made in various aspects concerning settlement prediction there are still areas where major uncertainties exist and much further work is necessary before reliable, Type A Lambe (1973*a*), settlement predictions can be made.

(2) There is a great need for further regional studies to be carried out, of the kind described by Vargas (1955), Bjerrum (1967), Jarrett, Stark and Green (1974), and Nordin and Svensson (1974). Settlement observa-

tions should be taken over a period long enough to enable long term behaviour to be clarified.

(3) One most significant factor to emerge is the importance of determining accurately the preconsolidation pressure. Since in laboratory tests the test procedure can have a marked effect on the values of p_c' obtained it is important to confirm these values by regional studies.

(4) If it is necessary to obtain reasonable estimates of the rate of settlement then it is vital that the value of c_v used in the computations reflects the permeability of the clay *in situ*. Laboratory oedometer tests on small size samples generally give misleading results. Much better agreement is obtained from large diameter samples or from permeability tests carried out *in situ*. The use of three dimensional consolidation theories may also be necessary.

(5) Sample disturbance, taken to include disturbance when obtaining the sample, in transit to the laboratory, and during handling in the laboratory, is of considerable importance. Disturbance can be expected to result in undrained deformation modulus values being too low, to render less reliable determinations of the preconsolidation pressure and to obscure small strain behaviour.

(6) Secondary settlement is of considerable significance in many cases and no rational method is available for such predictions. Clearly further work is required in this area.

(7) Particularly because of the powerful method of analysis which is now available through the use of finite elements, reliable data concerning stress–strain relationships for soils are vitally necessary. Such data should be obtained not only in the laboratory using the highest possible standards of sampling and testing, but also from the back analysis of field measurements.

Points for discussion

Particular points on which discussion would be most welcome are listed below.

(1) Critical shear stress and threshold value
(2) Sample disturbance
(3) Secondary settlement
(4) Settlement observations
(5) *In situ* testing
(6) Allowable settlements, including differential settlement
(7) Centrifugal models

Acknowledgements

The writer is indebted to Mr. M. A. Huxley of the University of Surrey, and to Soil Mechanics Ltd. for valuable assistance with the preparation of this report.

Aldrich, H. P. (1952). 'Importance of Net Load to the Settlement of Buildings in Boston', Contribution to Soil Mechanics, 1941–1953, *Boston Soc. Civ. Eng.*

Barden, L. (1965). 'Consolidation of Clay with Non-linear Viscosity', *Géotechnique* Vol. 15 345–362

Barden, L. (1969). 'Recent Developments in Consolidation Theory and Technique', *Civ. Eng. and Pub. Works Rev.*, 52–54

Barden, L. (1968). 'Primary and Secondary Consolidation of Clay and Peat, *Géotechnique*, Vol. 18, 1–24

Bassett, R. H. (1973). 'Centrifugal Model Tests of Embankments on Soft Alluvial Foundations', *Proc. 8th Int. Conf. Soil Mech. & Found. Engng.* 2.2, 23–30

Bauer, C. E. A., Scott, J. D. and Shields, D. H. (1973). 'The Deformation Properties of a Clay Crust', *Proc. 8th Int. Conf. Soil Mech. & Found. Engng.* 1.1, 31–38

Berre, T., Schjetne, K. and Sollie, S. (1969). 'Sampling Disturbance of Soft Marine Clays, *Proc.7th Int. Conf. Soil Mech. & Found. Engng.*, Spec. Session 1, 21–24

Berre, T. and Iverson, K. (1972). 'Oedometer Tests with Different Specimen Heights on a Clay Exhibiting Large Secondary Compression', *Géotechnique*, Vol. 23, 1–18

Berre, T. (1973). 'Sammenheng mellom tid, deformasjoner og spenninger for normal-konsoliderte marine leirer', *Norwegian Geotechnical Institute*, Publication No. 97, 1–14

Berre, T. and Bjerrum, L. (1973). 'Shear Strength of Normally Consolidated Clays', *Proc. 8th Int. Conf. Soil Mech. & Found. Engng.*, 1.1, 39–49

Biot, M. A. (1941). 'General Theory of Three Dimensional Consolidation', *J. Appl. Phys.*, Vol. 12, No. 15, 5–164

Bishop, A. W. and Henkel, D. J. (1953). 'Pore Pressure Changes During Shear in Two Undisturbed Clays', *Proc. 3rd Int. Conf. Soil Mech. & Found. Engng.*, 1, 94–99

Bishop, A. W. and Lovenbury, H. T. (1969). 'Creep Characteristics of Two Undisturbed Clays, *Proc. 7th Int. Conf. Soil Mech. & Found. Engng.*, 1, 29–37

Bjerrum, L., Simons, N. E. and Torblaa, I. (1958). 'The Effect of Time on the Shear Strength of a Soft Marine Clay', *Proc. Brussels Conf. 58 on Earth Pressure Problems*, 1, 148–158

Bjerrum, L. and Wu, T. H. (1960). 'Fundamental Shear Strength Properties of the Lilla Edet Clay', *Géotechnique*, Vol. 10, 101–109

Bjerrum, L. (1964). 'Relasjon mellom målte og beregnede setninger av byggverk pa leire og sand', NGF-foredraget 1964, Oslo, *Norwegian Geotechnical Institute*, 92

Bjerrum, L. (1966). 'Secondary Settlements of Structures Subjected to Large Variations in Live Load,' International Union of Theoretical and Applied Mechanics, Rheology and Soil Mechanics Symposium, Grenoble, 460–471

Bjerrum, L. and Eide O. (1966). 'Anvendelse av kompensert Fundamentering i Norge', *Vag och vattenbyggaren*, 4, 170–172

Bjerrum, L. (1967). 'Engineering Geology of Norwegian Normally-consolidated Marine Clays as Related to Settlements of Buildings', 7th Rankine Lecture, *Géotechnique*, Vol. 17, 81–118

Bjerrum, L. (1972a). 'Embankments on Soft Ground', *Proc. A.S.C.E. Specialty Conf. on Performance of Earth and Earth-Supported Structures*, Purdue University, Vol. 2, 1–54, Lafayette, Indiana

Bjerrum, L. (1972b). 'The Effect of Rate of Loading on the p_c' Value Observed in Consolidation Tests on Soft Clay', *Norwegian Geotechnical Institute Publ.*, No. 95

Bjerrum, L. (1973). 'Problems of Soil Mechanics and Construction on Soft Clays and Structurally Unstable Soils (Collapsible, Expansive and Others)', *Proc. 8th Int. Conf. Soil Mech. & Found. Engng.*, 3, 111–159

Bolton, M. D., English, R., Hird, C. C. and Schofield, A. N. (1973). 'Ground Displacements in Centrifugal Models', *Proc. 8th Int. Conf. Soil Mech. & Found Engng.*, 1.1, 65–70

Brinch Hansen, J. and Ses Inan (1969). 'Tests and Formulas Concerning Secondary Consolidation', *Proc. 7th Int. Conf. Soil Mech. & Found Engng.*, 1, 45–53

Brown, J. D. (1969). General Report, *Proc. Bolkesjø Sym. on Shear Strength and Consolidation of Normally Consolidated Clays*, 2–8

Buisman, A. S. K. (1936). 'Results of Long Duration Settlement Tests', *Proc. 1st Int. Conf. Soil Mech. & Found. Engng.*, 1, 100–106

Burland, J. B., Sills, G. C. and Gibson, R. E. (1973). 'A Field and Theoretical Study of the Influence of Non-homogeneity on Settlement', *Proc. 8th Int. Conf. Soil Mech. & Found. Engng.*, 1.3, 39–46

Calhoon, M. L. (1972). Disc. on D'Appolonia, Poulos and Ladd (1971) *J. Soil Mech. & Found. Div. A.S.C.E.*, Vol. 98, SM3, 306–308

Carrier, W. D. and Christian, J. T. (1973). 'Rigid Circular Plate Resting on a Non-homogeneous Half-space', *Géotechnique* Vol. 23, 67–84

Casagrande, A. (1936). 'The Determination of the Pre-consolidation Load and its Practical Significance', *Proc. 1st Int. Conf. Soil Mech. & Found. Engng.*, 3, 60–64

Casagrande, A. and Fadum, R. E. (1942). 'Application of Soil Mechanics in Designing Building Foundations, *A.S.C.E., Proc.*, 68, 1487–1520

Casagrande, L. (1964). 'Effect of Preconsolidation on Settlements', *Int. Conf. Soil Mech. & Found. Engng., A.S.C.E.*, 90, SM5, 349–362

Chang, Y. C. E., Broms, B. and Peck, R. B. (1973). 'Relationship Between the Settlement of Soft Clays and Excess Pore Pressure Due to Imposed Loads', *Proc. 8th Int. Conf. Soil Mech. & Found. Engng.*, 1.1, 93–96

Christie, I. F. (1964). 'A Re-appraisal of Merchant's Contribution to the Theory of Consolidation', *Géotechnique*, Vol. 14, 309–320

Clayton, C. R. I. (1973). 'The Secondary Compression of Clays', *M.Sc. Thesis*, Imperial College

Crawford, C. B. (1964). 'Interpretation of the Consolidation Test', *J. Soil Mech. & Found. Div., A.S.C.E.*, 90, SM5, 87–102

Crawford, C. B. (1965). 'Resistance of Soil Structure to Consolidation', *Canadian Geotechnical J.*, 2, 90–115

Cryer, C. W. (1963). 'A Comparison of the Three Dimensional Theories of Biot and Terzaghi', *J. Mech. Appl. Maths.*, Vol. 16, 401–412

D'Appolonia, D. J., and Lambe, T. W. (1970). 'Method for Predicting Initial Settlement', *J. Soil Mech. & Found Div., A.S.C.E.*, Vol. 96, SM2, 523–544

D'Appolonia, D. J., Poulos, H. G. and Ladd, C. C. (1971). 'Initial Settlement of Structures on Clay', *J. Soil Mech. & Found. Div., A.S.C.E.*, Vol. 97, SM10, 1359–1377

D'Appolonia, D. J. and Lambe, T. W. (1971). 'Floating Foundations for Control of Settlement', *J. Soil Mech. & Found. Div., A.S.C.E.*, Vol. 97, SM6, 899–915

Davis, E. H. and Poulos, H. G. (1965). 'The Analysis of Settlement Under Three-dimensional Conditions', *Sym. Soft Ground Engng.*, Brisbane

Davis, E. H. and Poulos, H. G. (1968). 'The Use of Elastic Theory for Settlement Prediction Under Three-dimensional Conditions', *Géotechnique*, Vol. 18, 67–91

Davis, E. H. and Poulos, H. G. (1972), 'Rate of Settlement Under Two- and Three-dimensional Conditions', *Géotechnique*, Vol. 22, 95–114

Eide, O. (1972). Personal communication

Endicott, L. J. (1970). 'Centrifugal Testing of Soil Models', *Ph.D. Thesis*, Cambridge University

Engesgaar, H. (1970). 'Resultater av to belastningsforsøk på Sundland i Drammen', *Norwegian Geotechnical Institute*, Publ. No. 84, 41–47

Engesgaar, H. (1972). 'Skistadbygget—direkte fundamentering av et høybygg i Drammen sentrum', *Norwegian Geotechnical Institute*, Publ. No. 92, 19–24

Engesgaar, H. (1973). '15-Storey Building on Plastic Clay in Drammen, Norway', *Proc. 8th Int. Conf. Soil Mech. & Found. Engng.*, 1.3, 75–80

Foss, I. (1969). 'Secondary Settlements of Buildings in Drammen, Norway', *Proc. 7th Int. Conf. Soil Mech. & Found Engng.*, 2, 99–106

Frimann Clausen, C. J. (1969). 'Loading Test Mastemyr', *Proc. Bolkesjø Sym. on Shear Strength and Consolidation of Normally Consolidated Clays*, 42–44

Frimann Clausen, C. J. (1970). 'Resultater av et belastningsforsøk på Mastemyr i Oslo', *Norwegian Geotechnical Institute*, Publ. No. 84, 29–40

Garlanger, J. E. (1972). 'The Consolidation of Soils Exhibiting Creep Under Effective Stress', *Géotechnique*, Vol. 22, 71–78

Gibson, R. E. and Lumb, P. (1953). 'Numerical Solutions of Some Problems in the Consolidation of Clay', *J. Instn. Civ. Engrs.*, Pt. 1, 182

Gibson, R. E. and Lo, K. Y. (1961). 'A Theory of Consolidation for Soil Exhibited Secondary Compression', *Acta Polytechnica Scandinavica*, 296

Gibson, R. E. and McNamee, J. (1963). 'A Three-dimensional Problem of the Consolidation of a Semi-infinite Clay Stratum', *Q. J. Mech. Appl. Maths.*, Vol. 16, No. 1, 115–127

Gibson, R. E. (1967). 'Some Results Concerning Displacements and Stresses in a Non-homogeneous Elastic Half-space', *Géotechnique*, Vol. 18, 56–67

Gibson, R. E. and Sills, G. C. (1971). 'Some Results Concerning the Plane Deformation

527

of a Non-homogeneous Elastic Half-space', *Proc. Roscoe Memorial Conf. on Stress Strain Behaviour of Soils*, 564–572

Gielly, J., Lareal, P. and Sanglerat, G. (1970). 'Correlations Between In Situ Penetrometer Tests and the Compressibility Characteristics of Soils', *Proc. Conf. on In Situ Invest. in Soils and Rocks*, 167–172

Golder, H. Q. (1965). State-of-the-Art of Floating Foundations', *J. Soil Mech. & Found. Div., A.S.C.E.*, Vol. 91, SM2, Proc. Paper 4278, 81–88

Gray, H. (1936). 'Progress Report on Research on the Consolidation of Fine Grained Soils, *Proc. 1st Int. Conf. Soil Mech. & Found Engng.*, 2, 138–141

Habibagahi, K. (1973). 'Temperature Effect on Consolidation Behaviour of Overconsolidated Soils', *Proc. 8th Int. Conf. Soil Mech. & Found. Engng.*, 1.1, 159–162

Haefeli, R. and Schaad, W. (1948). 'Time Effects in Connection with Consolidation Tests', *Proc. 2nd Int. Conf. Soil Mech. & Found. Engng.*, 3, 23–29

Hamilton, J. J. and Crawford, C. B. (1960). 'Improved Determination of Preconsolidation Pressure of a Sensitive Clay', *A.S.T.M. Special Tech. Publ.* 254, 254–271

Hansen, B. (1969). 'A Mathematical Model for Creep Phenomena In Clay', Speciality Session No. 12, *7th Int. Conf. Soil Mech. & Found. Engng.*, *Advances in Consolidation Theories for Clays*

Hartlen, J. and Pusch, R. (1973). 'Interpretation of Creep Measurements on Stiff Clay', *Proc. 8th Int. Conf. Soil Mech. & Found. Engng.*, 1.1, 177–180

Henkel, D. J. (1965). 'Problems Associated with the Construction of the Ebute Metta Causeway over Soft Clays in Lagos, Nigeria', *Proc. 6th Int. Conf. Soil Mech. & Found Engng.*, 2, 74–78

Hobbs, N. B. (1971). 'The Menard Pressuremeter—In Situ Loading Tests in Soils and Rocks', *Summer School in Foundation Engineering*, University of Surrey, Notes

Hobbs, N. B. (1974). General Report, Session 4, *Proc. Conf. on Settlement of Structures*, Cambridge

Høeg, K., Andersland, O. B. and Rolfsen, E. N. (1969). 'Undrained Behaviour of Quick Clay under Load Tests at Asrum', *Géotechnique*, Vol. 19, 101–115

Hutchinson, J. N. (1963). 'Settlements in Soft Clay Around a Pumped Excavation in Oslo' *Proc. Euro. Conf. Soil Mech. & Found. Engng.*, 1, 119–126

James, P. M. (1970). 'The Behaviour of a Soft Recent Sediment Under Embankment Loadings', *Quart. J. Eng. Geol.*, Vol. 3, 41–53

Janbu, N., Bjerrum, L. and Kjaernsli, B. (1956). 'Veiledning ved løsning av fundamenteringsoppgaver'. *Norwegian Geotechnical Institute*, Pub. No. 16, 93

Janbu, N. (1963). 'Soil Compressibility as Determined by Oedometer and Triaxial Tests', *Proc. Euro. Conf. Soil Mech. & Found. Engng.*, 1, 19–25

Janbu, N. (1969). 'The Resistance Concept Applied to Deformations of Soils', *Proc. 7th Int. Conf. Soil Mech. & Found. Engng.*, 1, 191–196

Janbu, N. and Senneset, K. (1973). 'Field Compressometer—Principles and Applications', *Proc. 8th Int. Conf. Soil Mech. & Found. Engng.*, 1.1, 191–198

Koppejan, A. W. (1948). 'A Formula Combining the Terzaghi Load-Compression Relationship and the Buisman Secular Time Effect', *Proc. 2nd Int. Conf. Soil Mech. & Found. Engng.*, 3, 32–37

Ladd, C. C. (1964). 'Stress-Strain Modulus of Clay in Undrained Shear', *A.S.C.E. Proc.*, Vol. 90, SM5, 103–132

Ladd, C. C. (1969). 'The Prediction of In Situ Stress–Strain Behaviour of Soft Saturated Clays During Undrained Shear', *Bolkesjø Symp. on Shear Strength and Consolidation of Normally Consolidated Clays*, 14–19, Norwegian Geotechnical Institute, Oslo

Ladd, C. C. (1971). *Settlement Analysis for Cohesive Soils*, M.I.T. Soils Publication, 272

Lambe, T. W. (1964). 'Methods of Estimating Settlement', *J. Soil Mech. & Found Div., A.S.C.E.*, Vol. 90, No. SM5, 43–67

Lambe, T. W. (1967). 'Stress Path Method', *J. Soil Mech. & Found Div., A.S.C.E.*, Vol. 93, SM6, 309–331

Lambe T. W. and Whitman R. V. (1969). *Soil Mechanics*, Wiley

Lambe, T. W. (1973a). 'Predictions in Soil Engineering', *Géotechnique*, Vol. 23, 151–202

Lambe, T. W. (1973b). 'Up-to-date Methods of Investigating the Strength and Deformability of Soils (Laboratory and Field Testing of Soils for their Strength, Deformative and Rheological Properties'). *Proc. 8th Int. Conf, Soil Mech. & Found. Engng.*, 3, 3–43

Leonards, G. A. and Ramiah, B. K. (1960). 'Time Effects in the Consolidation of Clays', *A.S.T.M. Sp. Tech. Publ.*, 254, 116–130

528

Leonards, G. A. and Girault, P. (1961). 'A Study of the One-dimensional Consolidation Test', *Proc. 5th Int. Conf. Soil Mech. & Found. Engng.*, 1, 213–218

Leonards, G. A. and Altschaeffl, A. G. (1964). 'Compressibility of Clay', *J. Soil Mech. & Found. Div.*, *A.S.C.E.*, Vol. 90, SM5, 133–155

Lewin, P. I. (1973). 'Three-Dimensional Anisotropic Consolidation of Clay', *R.I.L.E.M. Sym. Def. and Fail. of Solids*, Cannes

Lo, K. Y. (1961). 'Secondary Compression of Clays', *J. Soil. Mech. & Found. Div.*, *A.S.C.E.*, Vol. 87, SM4, 61–82

Madhloom, A. (1973). 'The Undrained Shear Strength of a Soft Silty Clay from King's Lynn, Norfolk,' *M.Phil. Thesis*, University of Surrey

Marsland, A. (1971). 'Laboratory and In Situ Measurements of Deformation Moduli of London Clay,' *Proc. Sym. on the Interaction of Structure and Foundation*, Birmingham, 7–17

Meigh, A. C. and Corbett, B.O. (1970). 'A Comparison of In Situ Measurements in a Soft Clay with Laboratory Tests and the Settlement of Oil Tanks', *Proc. Conf. In Situ Invest. in Soils and Rocks*, 173–179

Mikase, M. and Takada, N. (1973). 'Significance of Centrifugal Model Tests in Soil Mechanics', *Proc. 8th Int. Conf. Soil Mech. & Found Engng.*, 1.2, 273–278

Moran, Proctor, Mueser and Rutledge (1958). 'Study of Deep Soil Stabilisation by Vertical Sand Drains, ' *Bureau of Yards and Docks, Department of the Navy*, Washington D.C.

Moum, J. and Rosenqvist, I. Th. (1957). 'On the Weathering of Young Marine Clay', *Proc. 4th Int. Conf. Soil Mech. & Found Engng.*, 1, 77–79

Moum, J., Løken, T. and Torrance, J. K. (1971). 'A Geochemical Investigation of the Sensitivity of a Normally Consolidated Clay from Drammen, Norway, *Géotechnique*, Vol. 21, 329–340

Murayama, S. and Shibata, T. (1961). 'Rheological Properties of Clays', *Proc. 5th Int. Conf. Soil Mech. & Found. Engng.*, 1, 269–273

Murray, R. T. (1971). 'Embankments Constructed on Soft Foundations; Settlement Study at Avonmouth', *Crowthorne, Road Research Laboratory Report*, 419

Polshin, D. E., Rudritski, N. Y., Chizhikov, P. G. and Yakovleva, T. G. (1973). 'Centrifugal Model Testing of Foundation Soils of Building Structures', *Proc. 8th Int. Conf. Soil Mech. & Found. Engng.*, 1.3, 203–208

Poorooshab, H. P. (1969). 'Advances in Consolidation Theories for Clays', *Proc. 7th Int. Conf. Soil Mech. & Found Engng.*, 3, 491–497

Raymond, G. P. (1965). 'Rate of Settlement and Dissipation of Pore Water Pressure During Consolidation of Clays', *Ph.D. thesis*, University of London

Raymond, G. P., Townsend, D. L. and Lojkasek, M. J. (1971). 'The Effect of Sampling on the Undrained Soil Properties of a Leda Soil', *Canadian Geotechnical J.*, Vol. 8, 546–557

Rowe, P. W. and Barden, L. (1966). 'A New Consolidation Cell', *Géotechnique*, Vol. 16, 162–170

Rowe, P. W. (1968). 'The Influence of Geological Features of Clay Deposits on the Design and Performance of Sand Drains', *I.C.E. Proc. Sup. Paper* 70585, London

Rowe, P. W. (1972). 'The Relevance of Soil Fabric to Site Investigation Practice', The 12th Rankine Lecture, *Géotechnique*, Vol. 22, 195–300

Rutledge, P. C. (1944). 'Relation of Undisturbed Sampling to Laboratory Testing', *Trans. A.S.C.E.*, Vol. 109

Sanglerat, G. (1972). *The Penetrometer and Soil Exploration*, Elsevier

Schiffman, R. L., Ladd, C. C. and Chen, A. T. F. (1964). 'The Secondary Consolidation of Clay', *Proc. I.U.T. A.M. Symp. on Rheology and Soil Mechanics*, Grenoble, 273–298 (Springer, 1966)

Schiffman, R. L., Whitman, R. V. and Jordan, J. C. (1969). 'Settlement Problem Oriented Computer Language', *J. Soil Mech. & Found Div.*, *A.S.C.E.*, Vol. 96, SM2, 649–669

Schmertmann, J. H. (1953). 'Estimating the True Consolidation Behaviour of Clay from Laboratory Test Results', *Proc. A.S.C.E.*, Vol. 79, Separate No. 311

Schmertmann, J. H. (1955). 'The Undisturbed Consolidation Behaviour of Clay', *A.S.C.E. Trans.*, Vol. 120, Paper 2775, 1201–1233

Scott, R. F. and Ko, H. Y. (1969). 'Stress-deformation and Strength Characteristics', *Proc. 7th Int. Conf. Soil Mech. & Found. Engng.*, State of the Art Volume, 1–47

Simons, N. E. (1957). 'Settlement Studies on Two Structures in Norway', *Proc. 4th Int. Conf. Soil Mech. & Found. Engng.*, 1, 431–436

529

Simons, N. E. (1963). 'Settlement Studies on a Nine-Storey Apartment Building at Økern-bräten, Oslo, Proc. European Conference, Wiesbaden 1, 179–191

Simons, N. E. and Som, N. N. (1970). 'Settlement of Structures on Clay, with Particular Emphasis on London Clays', *C.I.R.I.A. Report* 22, 51

Skempton, A. W., Peck, R. B. and MacDonald, D. H. (1955). 'Settlement Analyses of Six Structures in Chicago and London', *Proc. Inst. Civ. Engrs.* Pt. 1, Vol. 4, No. 4

Skempton, A. W. and Bjerrum, L. (1957). 'A Contribution to the Settlement Analysis of Foundations on Clay', *Géotechnique*, Vol. 7, 168–178

Somerville, S. H. and Shelton, J. C. (1972). 'Observed Settlement of Multi-storey Buildings on Laminated Clays and Silts in Glasgow', *Géotechnique*, Vol. 22, 513–520

Suklje, L. (1969). *Rheological Aspects of Soil Mechanics*, p. 571. London, Wiley

Tan, T. K. (1957). 'Secondary Time Effect and Consolidation of Clays', *Academica Sinica Publ.*, Harbin

Taylor, D. W. and Merchant, W. (1940). 'A Theory of Clay Consolidation Accounting for Secondary Compression', *J. Math. Phys.*, Vol. 19, 167–185

Taylor, D. W. (1942). 'Research on Consolidation of Clays', *Dept. of Civ. and San. Eng.*, Serial 82, MIT

Terzaghi, K. (1936). 'Settlement of Structures', *Proc. 1st Int. Conf. Soil Mech. & Found. Engng.*, 3, 79–87

Terzaghi, K. (1941). 'Undisturbed Clay Samples and Undisturbed Clays', *Boston Society of Civil Engineers J.*, Vol. 28, No. 3, 211–231

Terzaghi, K. (1948). 'Closing Discussion on Foundation Pressure and Settlements of Buildings on Footings and Rafts', *Proc. 2nd Int. Conf. Soil Mech. & Found Engng.*, 6, 118

Vargas, M. (1955). 'Foundation of Structures on Over-consolidated Clay Layers in Saõ Paulo', *Géotechnique*, Vol. 5, 253–266

Wahls, H. E. (1962). 'Analysis of Primary and Secondary Consolidation, *J. Soil Mech. & Found. Div.*, *A.S.C.E.*, Vol. 88, No. SM6, 207–231

Ward, W. H., Burland, J. B. and Gallois, R. W. (1968). 'Geotechnical Assessment of a Site at Mundford, Norfolk, for a large Proton Accelerator', *Géotechnique*, Vol. 18, 399–431

Ward, W. H. (1971). 'Some Field Techniques for Improving Site Investigations and Engineering Design', *Proc. Roscoe Mem. Sym. on Stress Strain Behaviour of Soils*, 676–682

Wilkes, P. F. (1974). 'A Geotechnical Study of a Trial Embankment at King's Lynn', *Ph.D. Thesis*, University of Surrey

Wroth, C. P. and Simpson, B. (1972). 'An Induced Failure of a Trial Embankment', Pt. 2, *Finite Element Computations, Proc. Purdue Conf. on Performance of Earth and Earth Supported Structures*, 1.1., 65–79

Wu, T. H., Resendiz, D. and Neukirchner, R. (1966). 'Analysis of Consolidation by Rate Process Theory', *J. Soil Mech. & Found. Divn.*, *A.S.C.E.*, Vol. 92, SM6, 229–248

Zeevaert, L. (1957). 'Foundation Design and Behaviour of Tower Latino Americana in Mexico City,' *Géotechnique*, Vol. 7, 115–133

Zeevaert, L. (1958). 'Consolidation of Mexico City Volcanic Clay', *Proc. Conf. on Soils for Eng. Purposes*, *A.S.T.M.*, S.T.P. No. 232

Zeevaert, L. (1960). 'Compensated Foundations', *Panamerican Conference on Soil Mechanics and Foundation Engineering*, 1, Mexico 1959, Proc. Vol. 3, 1109–1126

Heavily over-consolidated clays

General Reporter, F. G. Butler,
Geotechnics Division, Ove Arup & Partners

Introduction

Many major towns and cities are underlain by substantial depths of heavily overconsolidated clay. The estimation of settlements of structures on heavily overconsolidated clays has therefore been a subject of particular concern to engineers who contemplate differential settlement rather than bearing capacity as the limiting factor in determining what they can achieve. Thus it might seem reasonable to anticipate that case histories would be plentiful and the response to the Committee's call for such information would be good. This expectation has, however, been only partially realised.

Geological history

As with other soils and rocks, the behaviour of heavily overconsolidated clays is dictated by geological and geomorphological history. The terminology adopted by Morgenstern (1967) is apt in differentiating between the geological history of heavily overconsolidated clays, which fall broadly into two groups.

Firstly, there are marine or lacustrine deposits overconsolidated by the pressure of later sediments which have subsequently been eroded, which Morgenstern termed *simple overconsolidation*, i.e. subjected to a single loading and unloading cycle. These clays are usually relatively uniform in texture and grading and are amenable to some degree to sampling, laboratory or *in situ* testing, and to settlement analysis with a measure of confidence. The London Clay, the Gault Clay, and it seems, the Frankfurt Clay settlement records which are contained in papers to this Conference, are representative of simple overconsolidation.

The second group, which have been subjected to what Morgenstern termed *complex overconsolidation*, are typified by moraine and boulder clays. Their stress history contains major reloading and unloading cycles due to advancing and retreating ice sheets which may be interspersed with periglacial deposits. Because of their heterogeneous nature and the depositional and erosional loading and unloading they have undergone, it is unlikely that empiricism based on local practice will be fully replaced by more definitive methods of settlement analysis.

Methods of settlement analysis

Analytical problems in soil mechanics fall essentially into two groups: those of stability and those of settlement. Terzaghi's published work on consolidation in the period from 1921 to 1925 summarised by Skempton (Terzaghi, 1960) is the starting point of modern soil mechanics.

Terzaghi developed a theory of one-dimensional consolidation with which, using results obtained in the oedometer test, an estimate of consolidation settlement could be made. In cases where the compressible layer is thin relative to the extent of the loaded area or where the compressible layer was contained by layers of sand or rock above and beneath—conditions where lateral strain is negligible—this approach has been shown to be fairly reliable.

The method assumes $\Delta u = \Delta\sigma_v$ and that the relationship between axial compressibility m_v, and vertical effective stress given by the oedometer is applicable. The settlement is given by:

$$\delta_c = \int_o^z m_v . \Delta\sigma_z . \mathrm{d}z$$

However, the more general case, where lateral strain can occur is clearly outside this treatment. In the case of overconsolidated clays the method was suspected of providing a substantial overestimate of settlement.

Skempton and Bjerrum (1957) identified the need for a more generalised approach and proposed a semi-empirical method in which recognition of the nature of the clay and its thickness relative to the size and shape of the loaded area could be allowed for and of which the Terzaghi theory remained a valid particular case. This they achieved by subdivision of the overall settlement into two components, viz. 'immediate' settlement, which was considered to occur in saturated clays due to deformation under no-drainage conditions; 'consolidation' settlement which was controlled by dissipation of pore pressure derived from changes in σ_1, σ_2 and σ_3 rather than by change in vertical stress. This marked an important improvement in the method of estimating settlements; it contained a strong element of logic yet the laboratory testing requirements remained undemanding and the analysis, in computational terms, elementary. It has thus become established as a standard office procedure for the estimation of settlements in overconsolidated clays. However, a structure cannot be built instantaneously, the imposition of the building load can take a significant span even on a primary consolidation time-scale. During construction, immediate primary consolidation and secondary consolidation are occurring simultaneously. Nevertheless, it is convenient to persevere with the notion of instantaneous construction as a means of quantifying the effects of various stages in the construction process.

Immediate settlement

For the purposes of estimating immediate settlement, it is generally assumed that we may resort to elastic theory; to the classical expression for the settlement at the corner of a flexible loaded rectangular area on the surface of an elastic half-space:

$$\rho = \frac{q.B.(1 - v^2).I_\rho}{E} \tag{1}$$

where
q = surface stress
B = breadth of loaded area
v = Poissons Ratio
E = bulk modulus
I_ρ = influence factor

The assumptions implied by this relationship are homogeneity (the soil properties are constant from point to point), linear elasticity (that strain is proportional to stress) and isotropy (its properties are identical in all directions through any point). The half-space is also assumed infinite in extent. Values of the influence factor $I\rho$ are dependent upon the shape of the loaded area and, in the Steinbrenner adaptation (1934), also the thickness of the elastic layer. For saturated clays no volume change occurs so long as there is no dissipation of pore pressure. Therefore in the calculation of immediate settlements $v = 0.5$ is assumed. The search for appropriate values of E_u (the undrained Young's Modulus) has concentrated work upon a variety of laboratory and, latterly, field testing techniques.

Consolidation settlement

In order to compute consolidation settlement, Skempton and Bjerrum proposed that the induced pore pressure due to loading the foundation should be estimated through the use of the pore pressure coefficients A and B where

$$\Delta u = B.[\Delta\sigma_3 + A.(\Delta\sigma_1 - \Delta\sigma_3)]$$

$B = 1$ for saturated clays. The changes in stress $\Delta\sigma_1$, and $\Delta\sigma_3$ at points below the centre of the foundation are derived by elastic theory and the parameter A from observed laboratory tests. In order to retain the practicability of the oedometer test, Skempton and Bjerrum related one-dimensional settlement (ρ_{oed}) to consolidation settlement (ρ_c) by means of the device μ where

$$\rho_c = \mu.\rho_{oed}$$

i.e.
$$\rho_c = \mu \int_o^z m_v.\Delta u.dz$$

The constant μ is a function of a shape of load distribution (breadth, b), depth of compressible layer (z) and pore pressure coefficient A.

The variation of μ with pore pressure coefficient A for various values of z/b and for circular and strip loading were produced in graphical form. Scott (1963) has recalculated and corrected errors relating to strip loading in the original text and the subsequent correction.

Despite the widespread acceptance of the Skempton and Bjerrum method in European foundation practice and the confirmation of its reliability at least in fairly familiar circumstances, the sensitivity of μ to the value of A and the philosophical inconsistency in combining one component based upon elastic theory with another based upon a semi-empirical application of laboratory data prompts the need for a more rational approach.

533

That proposed by Lambe (1964) transports the settlement analysis entirely to the laboratory. He has used the stress-path device to illustrate diagrammatically the shortcomings of the earlier methods of settlement analysis. He demonstrated, by careful triaxial testing, the influence on the behaviour of the soil of the stress history that the soil element has undergone in reaching its final state. The attempted simulation on laboratory specimens of the stress changes on soil elements in the field is, at least in principle, a step in the right direction. A similar line of approach has been advanced independently by Davis and Poulos (1963). However, as a practical means of estimating settlement, the endeavour to find a more rational method merely substitutes the difficulties of identifying 'typical' soil elements and a complex laboratory test programme and analysis for those of estimating E_u and A.

The pursuit of a stress-path technique for estimating settlements particularly in relation to London Clay has been given much impetus by the work of Som (1968), Simons and Som (1969 and 1970). In a very detailed study of the elastic parameters of London Clay, they have examined the extent to which the changes in effective stress in the ground during consolidation depart from the process of one-dimensional consolidation. They propose a correction factor

$$\lambda = \varepsilon_1 / \varepsilon_v$$

where ε_1 = vertical strain and ε_v = volumetric strain, to allow for the effect of lateral strains, their revised equation being

$$\rho_c = \int_o^z \lambda . m_v . \Delta u . \mathrm{d}z$$

The use of this equation still assumes that $m_v \Delta u$ represents the volumetric strain even though it is applied to strain increments which are not one-dimensional. The results of consolidation tests (Som (1968)) with differing values of $\Delta \sigma_3' / \Delta \sigma_1'$ broadly confirm this assumption. Thus m_v is given the wider definition of $\varepsilon_1 / \Delta \sigma_1'$ independent of $\Delta \sigma_3' / \Delta \sigma_1'$, and only a function of stress level. The value of λ has been found experimentally for a variety of stress increment ratios $\Delta \sigma_3' / \Delta \sigma_1'$.

The methods so far considered are illustrated in stress-path terms as in *Fig. 1*. For simplicity let us consider an element of soil on the vertical axis through the centre of a circular loaded area. The *in situ* pressure on the element before applying the surface load is in the vertical direction p_0. If the *in situ* pore pressure is u_0 then the effective stress in the vertical direction is p_0' $(= p_0 - u_0)$ and in the horizontal direction $k_0 p_0'$. In total stress terms, the state of stress of the element is thus represented by the point A, and in effective stress terms by the point A'.

A loading, of intensity q, is then applied, without drainage, causing an increase in vertical stress of $\Delta \sigma_v$ and in horizontal stress of $\Delta \sigma_h$ on the element. From Skempton (1954), assuming a fully saturated clay, the change in pore pressure at the element is given by

$$\Delta u = \Delta \sigma_h + A \left(\Delta \sigma_v - \Delta \sigma_h \right)$$

Immediately after application of the load the effective stresses are therefore

$$(\sigma_v')_{ab} = p_o' + \Delta \sigma_v - \Delta u$$
$$(\sigma_h')_{ab} = k_o p_o' + \Delta \sigma_h - \Delta u$$

and the state of stress in effective stress terms is represented by the point B'. Note that since, for most clays, the value of A is positive and less than 1.0 within the range of stresses encountered in foundation problems the change in pore pressure Δu is greater than the change in horizontal stress. Thus, during the undrained loading the element experiences an increase in vertical effective stress and a decrease in horizontal effective stress. The vertical strain undergone by the element in undrained loading is therefore that associated with the stress path A'–B'.

The element is now allowed to consolidate: in the Skempton and Bjerrum case we assume one-dimensional consolidation. Thus the dissipation of the excess pore pressure Δu results in the stress path B'C', the final state after primary consolidation, in effective stress terms being represented by the point C', and in total stress terms by the point C.

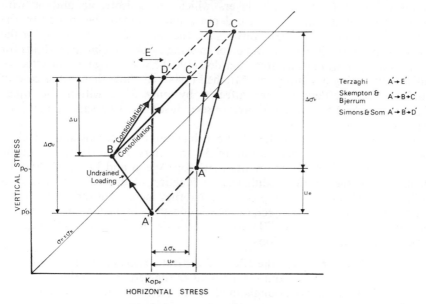

Fig. 1. Stress path settlement analysis

Simons and Som (1970) have taken account of the lateral strain and of the resulting change in stress distribution during consolidation. Thus the true final state of stress of the element is represented by the point D' and the stress path C'–D' is the correct path for the consolidation process.

Referring to *Fig. 1*, since Terzaghi assumes $\Delta u = \Delta\sigma_v$, the stress path followed is A'–E' where E' is not located in terms of horizontal stress. Skempton and Bjerrum ignore lateral strain in the consolidation phase, and follow the stress path A'B'C'. Simons and Som identify the stress path A'B'D'.

The practical application of the Simons and Som method relies upon a simplification of the consolidation properties so that simple oedometer results can be used. This simplification which has been justified by experimental evidence is, strictly in terms of elastic theory, inconsistent with the proposed relationship between λ and $\Delta\sigma_3'/\Delta\sigma_1'$.

535

Settlement prediction by elastic methods

The use of elastic theory to estimate immediate settlements has already been describe as part of the Skempton and Bjerrum method of analysis. A modulus E_u, and influence coefficient $I\rho$, a function of Poissons ratio v, have to be selected for any given loading condition. In estimation of immediate settlements $v = 0.5$ is used for saturated clays. To estimate total settlement it is necessary to use a modulus E' appropriate to drained loading and the v of the mineral skeleton.

The use of Equation 1 implies a homogeneous elastic and isotropic material, which soil is not. In the case of overconsolidated clays, however, total settlements are generally relatively small and there is usually a substantial factor of safety in terms of bearing capacity so that the stress–strain behaviour is sensibly linear. Compressible layers which are anisotropic and/or with non-uniform elastic modulus with depth cannot correctly be treated by the application of Equation 1. However, since the value of Poissons ratio for the mineral skeleton will be low, the error in using a vertical elastic modulus for vertical loading will not be great. Barden (1963) has investigated the effect of anisotropy, using a solution by Mitchell (1900) to show that in the case of surface loads settlements are affected by only 25% within the range $E_h/E_v = 1.0$ to 2.5. It would therefore seem justifiable to assume Equation 1 valid for the present purposes.

Wroth (1971) discusses the elastic behaviour of overconsolidated clays both in terms of undrained and drained parameters. From the results of the sampling at Ashford Common shaft (Bishop, Webb and Lewin (1965)), he deduced that the elastic moduli were a function of pressure and overconsolidation ratio and, more importantly, that they increase linearly with depth. Using the relationships between the elastic constants in the laboratory and the field deduced by Henkel (1971) and in particular for triaxial loading in undrained and drained conditions, where

v_{hv} = Poissons ratio for the effect of vertical strains on horizontal strains
E_v' = Effective stress Young's modulus in the vertical direction
E_h' = Effective stress Young's modulus in the horizontal direction
$R = E_h'/E_v'$
E_{uI} = Undrained Young's modulus for triaxial loading on vertical
      specimens

$$\frac{\Delta\sigma_v'}{\Delta\sigma_h'} = -2\frac{\left(1 - v_{hv}\frac{1+3R}{2}\right)}{R(1-2v_{hv})} = M_I$$

and

$$E_{u_I} = \frac{E_h'}{R}\left(\frac{M_I-1}{M_I-2v_{hv}}\right) \tag{2}$$

Given a simplified C_u–depth profile for any site and assuming a relationship between C_u and E_{uI} and a value of R, it is therefore possible to establish a corresponding profile of E_v' with depth. Commonly, in overconsolidated clay the C_u–depth profile, and thus the E_v' profile, shows a linear increase with depth.

Stress and deformation in a non-homogeneous elastic half-space has been

536

investigated by Gibson (1967) who published solutions for the particular case where the shear modulus G varied linearly with depth and of which the Winkler model was a special case. Gibson, Brown and Andrews (1971) extended this work to examine the case when the elastic medium is of restricted depth and adhering to a rigid base. Brown and Gibson (1972) have since extended Gibson's original work to the analysis of settlement of circular and strip loading on the surface of a half-space with Young's modulus

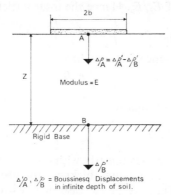

Fig. 2. Steinbrenners adaption of Boussinesq

varying linearly with depth and Poissons ratio of fixed value within the range 0 to 0.5.

Brown and Gibson (1973) have now published the solution to the problem of corner settlement due to a rectangular loading on an isotropic linearly non-homogeneous elastic half-space. However, no definitive analysis has been published which provides influence factors where this compressible layer is of limited depth. An approximate analysis can be undertaken by a simple extrapolation of Steinbrenner (1934). In Steinbrenner the displacement $\Delta\rho_A$ of the point A on the surface of limited depth z is computed from Bousinesq's expressions for the displacement of points A and B ($\Delta\rho_A'$ and $\Delta\rho_B'$) for an infinite depth of soil (*Fig. 2*). Since the actual deflection of point B is zero, $\Delta\rho_A$ is given by

$$\Delta\rho_A = \Delta\rho_A' - \Delta\rho_B'$$

Consider a multi-layered system as in *Fig. 3* and let us calculate the displacements $\Delta\rho_1$, $\Delta\rho_2$, $\Delta\rho_3$, etc., for an infinite depth of soil of modulus E

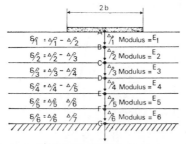

Fig. 3. Adaptation of Steinbrenner to give approximate analysis for variation of E with depth

537

(i.e. as if the whole half-space had the modulus of the upper layer). The settlement in each layer is then given by

$$\delta\rho_1 = \Delta\rho_1 - \Delta\rho_2$$
$$\delta\rho_2 = \Delta\rho_2 - \Delta\rho_3$$
$$\delta\rho_3 = \Delta\rho_3 - \Delta\rho_4 \text{ etc.}$$

But all these settlements are based upon the use of a uniform value of $E = E_I$: to accommodate variations in E with depth we should therefore multiply the nth layer of E_I/E_n. Hence the total settlement $\Delta\rho$ is given by

$$\Delta\rho = \frac{E_1}{E_1}.\delta\rho_1 + \frac{E_1}{E_2}.\delta\rho_2 + \frac{E_1}{E_3}.\delta\rho_3 \dots \text{ etc.} \qquad (3)$$

We actually require influence factors where

$$I_\rho = \frac{\rho/B}{q/E_1}$$

We can therefore rewrite Equation 1

$$I_\rho = \frac{E_1}{E_1}.\delta I_{\rho_1} + \frac{E_1}{E_2}.\delta I_{\rho_2} + \frac{E_1}{E_3}.\delta I_{\rho_3} \dots \text{ etc.} \qquad (4)$$

For the case where E increases linearly with depth the soil can conveniently be divided into finite layers of equal thickness and ascribed an average E for each, E_1, E_2, E_3, etc. The rate of increase of E with depth is expressed in terms of E_0 (at the surface) thus

$$E = E_0 \left(1 + K.\frac{z}{b} \right)$$

and Equation 4 then becomes

$$I_\rho = \frac{E_0}{E_1}.\delta I_{\rho_1} + \frac{E_0}{E_2}.\delta I_{\rho_2} + \frac{E_0}{E_3}.\delta I_{\rho_3} \dots \text{ etc.}$$

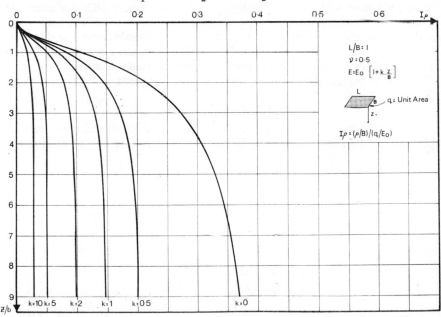

Fig. 4

538

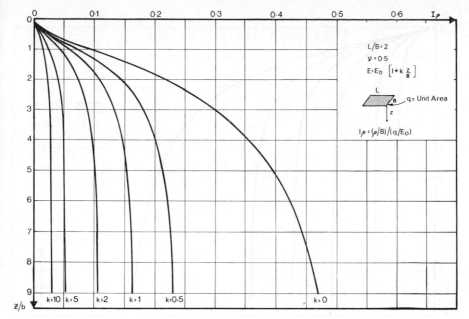

Fig. 5

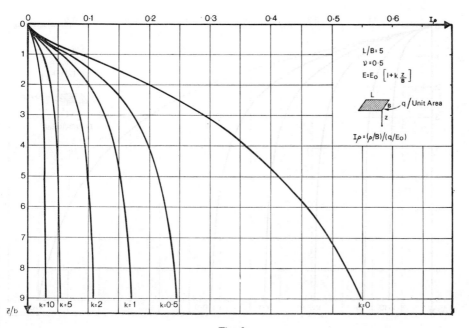

Fig. 6

539

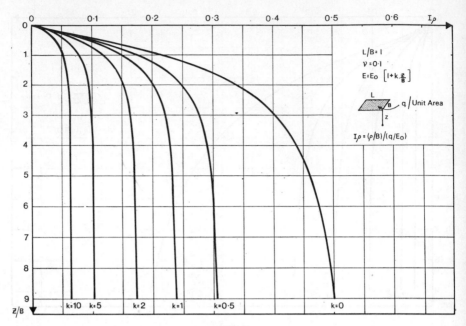

Fig. 7

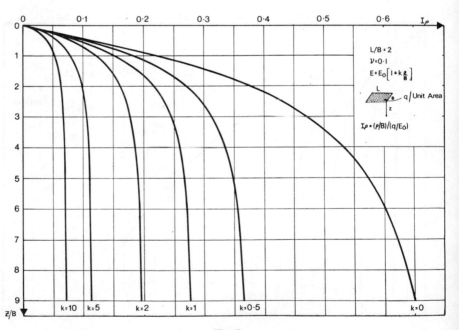

Fig. 8

540

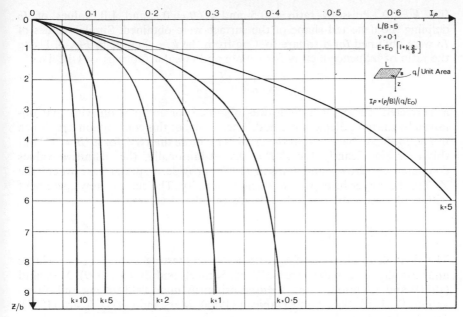

Fig. 9

This expression has been programmed, using a Hewlett Packard computer to provide values of $I\rho$ for various thicknesses of layer (expressed in terms of breadth b) for a range of values of $K [\delta E/\delta(z/b)]$ for rectangular shapes with $L/B = 1.0$, 2.0 and 5.0 and with $v = 0.5$ and 0.1, for the undrained and drained case respectively (*Figs. 4 to 9*).

Since these curves are the result of an approximation it was decided to attempt to assess their accuracy by means of a check against an axi-symmetric case solved by the finite element method. A circular flexible and rough footing radius (B) was modelled, founded on the surface of an elastic half-space of varying depth (Z) with a rigid rough base. With constant E_u

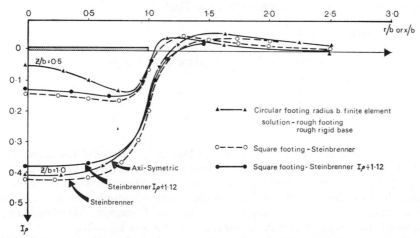

Fig. 10. Comparison between Steinbrenner and axisymmetric finite element analysis, $K = 0$

541

(i.e. $k = 0$), Poissons ratio $= 0.5$ and $Z/B = 0.5$ and 1.0, values of $I\rho$ defining the deflected shape of the surface were obtained. Similar values of $I\rho$ were obtained for a square footing from the $k = 0$ curve in *Fig. 4*. Using the ratio of influence factors for square and circular footings $= 1.12$, i.e.

$$I_{\rho \, \text{circ.}} = \frac{I_{\rho \, \text{square}}}{1.12}$$

a 'corrected' curve for a circle was obtained for comparison with that obtained by the finite element method. The results of this comparison are shown in *Fig. 10*. It can be seen that the error is no more than 8% for $Z/B = 1.0$ but this rises significantly for $Z/B = 0.5$. Additionally, the definitive values computed by Brown and Gibson (1973) provide a special case where the Z/B in the curves in *Figs. 4 to 9* tend to infinity. The curves show agreement within 10% with Brown and Gibson's calculated values.

Estimation of E_u

The direct determination of E_u from laboratory testing is subject to very large and virtually inescapable errors (Ward, Samuels and Butler (1959), Marsland (1971)) and attention is now focused primarily upon field testing and observation (Burland, Butler and Dunican (1966), Marsland (1971), Ward (1971), Marsland (1973)).

Cooling and Skempton (1942) first suggested from laboratory test data and later Skempton and Henkel (1957) substantiated a relationship $E_u = 140 \, C_u$ from a limited number of case histories of settlement in London Clay. More recently Wroth (1971), reconsidering the published data of Bishop, Webb and Lewin (1965) and Webb (1966) derived a relationship $E_u = 150 \, C_u$.

The rapid development of numerical analysis has significantly improved our ability to interpret the results of field observations (Burland (1966), Cole and Burland (1972), Hooper (1973)).

Marsland has reported on the results of an extensive series of tests on the London Clay by a constant-rate-of-penetration testing on large diameter plates at the base of lined and unlined shafts. A summary of some of his results from a site at Hendon is shown in *Table 1*.

Table 1

C_u (kN/m²)	E_u (MN/m²)	E_u/C_u
80	55	690
90	75	830
110	65	600

Earlier plate bearing tests reported by Burland, Butler and Dunican (1966) at Moorfields are shown in *Table 2*. These results indicate far higher ratios of E_u/C_u than have been obtained in the laboratory. However it has been shown that the modulus is very sensitive to the length of time between excavation and testing. For plate loading tests at Ashford Common, Marsland (1973) showed that 10 hr after excavation the modulus had halved.

The impossibility of performing a field test in truly undrained and undisturbed conditions and the difficulties of obtaining representative undisturbed samples for laboratory testing leaves the results of both methods open to question. Certainly from the behaviour of London Clay in the mass a value of E_u/C_u between these two extremes seems more likely.

More recently Hooper (1973) has reported the results of *in situ* measurements of pile load, raft pressure and settlement observations on the Hyde

Table 2

C_u (kN/m²)	E_u (MN/m²)	E_u/C_u
140	100	800
300	200	650

Park Cavalry Barracks. By relating his observations to a finite element analysis of the problem, Hooper established by trial and error a relationship $E_u = 10 + 5.2Z$ where $Z =$ depth below ground surface. Comparing this expression in E_u to the measured values of C_u from the site we obtain the relationship between E_u and C_u tabulated in *Table 3*.

Table 3

C_u (kN/m²)	E_u *(computed)* (MN/m²)	E_u/C_u
170	530	310
256	1050	410
328	1560	480

A back-analysis of the movements recorded in the basement wall of Britannic House (Cole and Burland (1972)) using a finite element mesh and assuming plane-strain conditions showed a value of $E_u/K_{ot} = 295$. Assuming $K_{ot} = 2.0$ and for an average shear strength (U4 samples) of 190 KN/m² gives

$$\frac{E_{uIII}}{C_{u(U4)}} = \frac{2.0 \times 295 \times 1000}{190} = 310$$

Pseudo-plane-strain and triaxial tests on specimens cut from block samples from shafts at Barbican and $U4$ samples from Swiss Cottage (Ove Arup & Partners (1971)) showed the relationship

$$\frac{E_{uIII}}{E_{uI}} = 1.4$$

Hence, from Cole and Burland (1972),

$$\frac{E_{uI}}{C_{uU4}} = \frac{310}{1.4} = 222$$

Some corroboration for this relationship is drawn from the ratio of E_{uI} to undrained shear strength of block samples of 180 obtained in particularly carefully executed sampling and testing at the Building Research Station. In this case, the value of undrained strength of block samples to that of $U4$

543

samples was found to be about 1.3. Thus the ratio of E_{uI} to undrained strength of $U4$ samples would be about $1.3 \times 180 = 234$.

Estimation of E_v'

For the purpose of estimating total settlements we are concerned with the relationship between C_{uU4} and E_v'.
From the foregoing, let us begin by assuming:

$$E_{uI} = 220 \ C_{uU4}$$

Atkinson (1973) from extensive laboratory tests showed $R = 2.0$, $v_{hv} = 0.19$. Hence, from Equation 2,

$$E_{uI} = 1.67 \ E_v'$$

Thus, $$E_v' = \frac{220}{1.67} \ C_{uU4} = 130 \ C_{uU4}$$

This compares with Wroth (1971) who quotes $v_{hv} = 0.12$, from which Henkel (1971) deduces $R = 1.6$ and hence $E_{uI} = 1.58 \ E_v'$ which would give

$$E_v' = \frac{220}{1.58} \ C_{uU4} = 139 \ C_{uU4}$$

Constructional history

In considering case histories of settlement, due account must be taken of the constructional history of the foundation since this may significantly affect its settlement behaviour. For example, let us suppose that a tall building has a two-storey basement and is supported on a piled raft (*Fig. 11*).

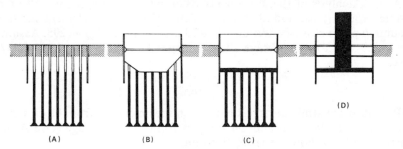

Fig. 11. Construction sequence

Commonly the piles are placed from ground level, with empty bore above raft level as in (a): the excavation process for the piles tends to release lateral pressure which may cause settlement of the site (Burland (1973) and Appendix B) during this stage, and will certainly inhibit heave when the excavation takes place, *Fig. 11* (b). A continuous raft anchored by piles, *Fig. 11* (c), will also inhibit heave effects; the swelling pressures developed under the raft give rise to tension in the piles (Hooper 1973) and the soil is effectively 'pre-stressed' until a net increase in load is applied to the foundations.

Without the piles heave effects would be greater, beginning during stage (b), through stage (c) and into stage (d) until attenuated as a net increase in load is approached. If the basement is much larger in plan than the building above, both initial and long term unloading effects may greatly complicate the settlement behaviour. Thus the response of an element of soil immediately beneath foundation level will be greatly influenced by the sequence of construction which is chosen (*Fig. 12*). The monitored settlement behaviour will

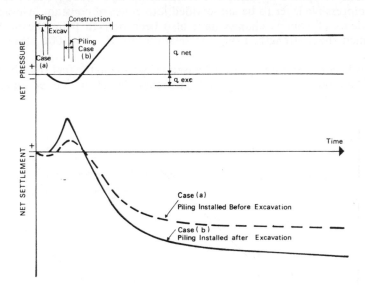

Fig. 12. Foundation settlement: effect of construction sequence

also obviously depend upon the point in time at which the observation points were installed. It follows that it is necessary to examine case histories very closely before applying any form of re-analysis.

The case histories presented to the Conference

The papers which have been submitted in this Session provide us with case histories in a wide variety of overconsolidated clays. The range includes clays with various modes of formation including continuous marine deposition (London Clay, Gault Clay, Woolwich and Reading), glacial action (boulder clay), periglacial deposition (flintz), diluvial clays (Nagoya, Japan).

London, Woolwich and Reading and Gault Clays

The papers by Green and Cocksedge, Hyde and Leach, Morton and Au, and Mould, provide a large and valuable contribution to the available data concerning the settlement effects of London, Woolwich and Reading and Gault Clays. These deposits have been widely investigated and well documented in the literature, and it is therefore appropriate to treat them collectively.

In consideration of the important place which the Skempton and Bjerrum method holds in current practice there is no other logical starting place in considering these case histories. Estimation of the immediate settlements have been computed by using the relationship $E_u = 400C_u$ which requires the compressible layer to be sub-divided into zones of constant 'average' E_u. Coefficients of volume change have also been derived from the profile of undrained shear strength, using the relationship $M_v = 1/100C_u$ (Skempton (1951)).

It has become common practice, based primarily on the fact that the settlement of a rigid circular footing is 20% less than the centre-line settlement of a

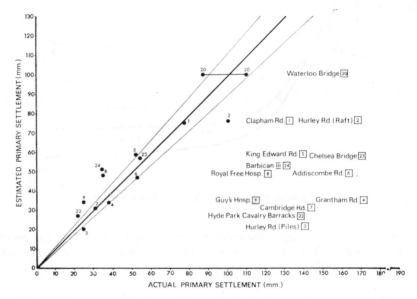

Fig. 13. Comparison between predicted and actual settlement (Skempton and Bjerrum): case histories from London and Gault Clay

circular flexible uniform load, to apply, where appropriate, a factor of 0.8 to the computed centre-line settlement to provide an estimate of the mean settlement of a rigid foundation. Although the justification for this procedure is open to debate, the settlements computed by Skempton and Bjerrum's method have been so factored in an attempt to exemplify current practice. Any more elaborate device to make allowance for structural stiffness would in any event be an intrusion into the subject matter of Session V. A depth correction has been applied (Janbu, Bjerrum and Kjaensli (1956)), wherever appropriate. Thirteen of the case histories presented to the Conference have been analysed and a comparison between predicted and measured settlement is shown in Fig. 13. The range of error obtained is between +60% and −12% which can be regarded as a fairly satisfactory result.

It should be noted however that in conventional application lower values of E_u are often used and no allowance is made for increasing E_u with depth.

In view of the interest in the measurement of elastic properties of over-consolidated clays it is considered justifiable to test the elastic method against the fairly extensive and demonstrably consistent record of case histories which has been accumulated. For this purpose it was assumed that $E_v' = 130C_{uU4}$.

The use of the curves in *Figs. 4 to 9* is illustrated by application to a case history from Paper III/6 by Morton and Au, that of the tower block at Clapham Road, and appears as Appendix A to this Report.

Comparisons have been made not only for the examples quoted in the papers to this Conference but also for a large number of case histories of

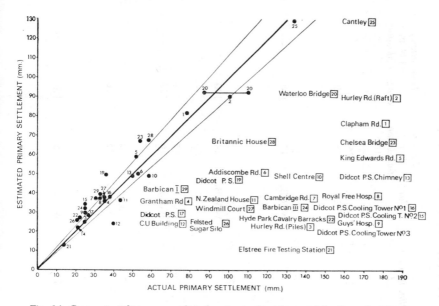

Fig. 14. *Comparison between predicted and actual settlement (elastic analysis): case histories from London and Gault Clay*

structures founded on London, Gault, Oxford, and Woolwich and Reading Clays. A summary of the sources of these case histories and the calculated and observed settlements appears in *Table 4*.

In cases where demolition and excavation has preceded a net increase in load, a similar procedure to that above has been used but the curves for $v = 0.5$ (undrained shear) and the corresponding values of E_u ($= 1.67E_v'$) have been used to estimate the movements. The curves for $v = 0.1$ and corresponding values of E_v' have been used for settlements due to net loading. Where piled foundations have been employed a spread at 1 in 4 down the pile shafts has been assumed, to define a hypothetical raft at pile base level. Settlements have been computed for this hypothetical raft. The result of these comparisons is shown in *Fig. 14*. A good measure of agreement is evident.

Table 4. CASE HISTORIES OF SETTLEMENT—LONDON, WOOLWICH AND READING AND GAULT CLAYS

No.	Site	Location	Type of foundation	Dimensions (m) B	L	D	Clay	Thickness below foundation (m)	Foundation press (kN/m²) Gross	Unloading	Net	Primary Settlement Obs. (mm)	Calc. (mm)	Remarks
1	Clapham Road (Morton & Au)	Lambeth	Surface raft	17	28	1.5	L.C. W.R.B.	25 15	190	—	190	79	82	Raft founded on 7 m gravel $\bar{N} = 46$. Estimated settlement in gravel = 11 mm
2	Hurley Road (Morton & Au)	Lambeth	Surface raft	17	26	1.5	L.C. W.R.B.	24 14	190	—	190	100	90	Raft founded on 5 m gravel $\bar{N} = 80$. Estimated settlement in gravel = 10 mm
3	Hurley Road (Morton & Au)	Lambeth	Large dia. U/R piles	17	26	19	L.C. W.R.B.	7 14	190	—	190	25	25	
4	Grantham Road (Morton & Au)	Lambeth	Large dia. U/R piles	17	26	19	L.C. W.R.B.	15 16	190	—	190	36	35	
5	King Edward's Road (Morton & Au)	Hackney	Surface raft	20	27	1.7	L.C. W.R.B.	4 15	210	—	210	52	59	Raft founded on 3 m gravel $\bar{N} = 90$. Estimated settlement in gravel = 3 mm
6	Addiscombe Road (Morton & Au)	Croydon	Basement raft	14	14	11	W.R.B.	5	446	220	226	53	50	Estimate of settlement 9 mm due to reloading + 41 mm due to net loading

No.	Reference	Location	Foundation type	B	L	D	Founding stratum									Remarks
7	Cambridge Road (Morton & Au)	Waltham Forest	Piled raft	16	31	16	L.C. W.R.B.	10	25	—	194	194	31	37	37	
8	Royal Free Hospital Morton & Au	Hampstead	Raft	see text of reference			L.C. W.R.B.	40	14	250	255	5	34	37		Sloping site. Assumed average cut of 12.5 m for estimate
9	Guy's Hospital (Mould)	Southwark	Piled raft in basement	30	30	24.5	L.C. W.R.B.	5	15	170	430	260	25	32		Settlement points fixed after 20% of loading in place
10	Shell Centre (Green & Cocksedge)	Lambeth	Piled raft in basement	105	55	37	L.C. W.R.B.	13	12	172	294	122	59	49		Approx. 18% of gross loading applied before monitoring began. Estimate consists of 9 mm due to reloading and 40 mm due to net loading
11	N. Zealand House (Green & Cocksedge)	City of Westminster	Piled raft in basement	19	19	28	L.C. W.R.B.	about 60		175	180	5	44	36		
12	C.U. Building (Green & Cocksedge)	City of London	Piled raft in basement	33	42	43	W.R.B. (under piles)	16		280	370	40	24			Piled raft partially loaded before main raft constructed
13	Power station chimney (Hyde & Leach)	Didcot, Oxon	Piled raft	36 dia.		18	Gault	47		—	155	155	50	49		
14	Power station cooling (Hyde & Leach)	Didcot, Oxon	Piled raft	91 dia.		21	Gault	44		—	17.5	17.5	21	22		Piled placed through 1.2 m fill

Table 4—continued

No.	Site	Location	Type of foundation	Dimensions (m) B	L	D	Clay	Thickness below foundation (m)	Foundation press (kN/m²) Gross	Unloading	Net	Primary Settlement Obs. (mm)	Calc. (mm)	Remarks
15	2N cooling tower (Hyde & Leach)	Didcot, Oxon	Piled raft	91 dia.		21	Gault	44	17.5	—	17.5	25	34	Piles placed through 2.1 m fill
16	1N cooling tower (Hyde & Leach)	Didcot, Oxon	Piled raft	91 dia.		21	Gault	44	17.5	—	17.5	38	38	Piles placed through 2.4 m fill
17	Boiler house (Hyde & Leach) B.N.18	Didcot, Oxon.	Footings	9	9	7.5	Gault	57.5			210	28	27	
18	Turbine house (Hyde & Leach) T.N.17	Didcot, Oxon	Footings				Gault	57.5						Settlement incomplete
19	Boiler house (Hyde & Leach) B.N.20	Didcot, Oxon	Footings	10	50	7.5	Gault	57.5			210	35	49	
20	Waterloo Bridge (Skempton, Peck & MacDonald (1955)) (Buckton & Cuerel (1943)) (Cooling (1948))	London	Footings	8.3	36	7	L.C. W.R.B.	} 36	470	182	288	(1) 59 (2) 87 (3) 110 (4) 127	92.5	Pier 4—swelled due to wetting. Estimate consists of 29 mm reloading settlement plus 63.5 mm net settlement

No.	Structure	Location	Foundation											Remarks
21	Fire Testing Station (Skempton, Peck & MacDonald (1955))	Elstree	Footings	1.52	3.04	2.1	L.C.	15.0	86	—	86	14	13	
22	Cavalry Barracks (Hooper 1973)	Hyde Park, London	Piled raft in basement	30	30	34	L.C. W.R.B.	28 20	196			22	27	
23	Chelsea Bridge (Skempton, Peck & MacDonald (1955)) (Buckton & Fereday (1938))	London	Footings	8.5	32	9.5	L.C. W.R.B.	18 15	182			54	67	
24	Tower Block III (Ove Arup & Partners Job No. 1023)	Barbican, City of London	Piled raft	30	30	20	L.C. W.R.B.	3 12	390	—	390	33	32	
25	Sugar silos (Parker & Bayliss (1971))	Cantley Norfolk	Piled raft	19.5	88	11	L.C. W.R.B.	}47.0	290	—	290	134	129	6 loadings assumed equivalent to 2 yr continuous loading ($U = 0.66$)
26	Sugar silos (Parker & Bayliss (1971))	Felstead, Essex	Piled raft	21	21	29	L.C. W.R.B.	}29.0	340	—	340	21	26	3 loadings assumed equivalent 1 yr continuous loading ($U = 0.47$)

Table 4—continued

No.	Site	Location	Type of foundation	Dimensions (m) B	L	L	Clay	Thickness below foundation (m)	Foundation press (kN/m²) Gross	Unloading	Net	Primary Settlement Obs. (mm)	Calc. (mm)	Remarks
27	Windmill Court (Greenfield (1971, 1974))	Kilburn	Bored piles	17	47	18	L.C. W.R.B.	}73.0	160	—	160	38	37	
28	Britannic House (Ove Arup & Partners, Job No. 1405)	City of London	Raft in extensive basement	26.8	64	17.5	L.C. W.R.B.	13.0 13.0	450	350	100	60	68	
29	Tower Block I (Ove Arup & Partners, Job. No. 1023)	Barbican, City of London	Piled raft	20	34	20	L.C. W.R.B.	5.0 13.0	580	—	580	34	38	

Unloading due to excavation

Unloading—short term. As noted above, estimates of rebound effects due to unloading and movement due to reloading have been made by elastic analysis using the curves for $v = 0.5$. Several case histories have been used from the published literature. A summary of the data is given in *Table 5* and

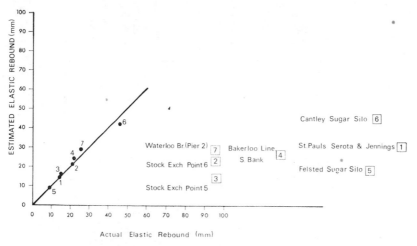

Fig. 15. Comparison between predicted and actual elastic heave (elastic analysis)

the comparison between estimated and observed movement in *Fig. 15*. A good measure of agreement is evident.

Unloading—long term. Occasionally long term or permanent unloading is a feature of foundation problems. Observations of long term heave effects are, however, rare, and for this reason the paper by May is particularly valuable. It is unfortunate that the observations of heave were not begun until the basement was virtually complete since it becomes very debatable where the movement datum should really be. May has plotted the movement against the root of time, and, seemingly by using March 1968 as an origin, extrapolated back to a zero time ordinate representing November 1967, the date of completion of the basement structure. This then takes no account of elastic and long term heave during the preceding 18 months of excavation and construction. May claims that the pattern of heave is generally as would be predicted by one-dimensional consolidation, for which he used $m_v = 0.035$ m^2/MN.

If *Fig. 5* of the paper is replotted to a root time basis but with zero time at June 1966, the commencement of construction, the total heave is about 116 mm at Point 5 (*Fig. 16*). The estimated 'undrained' heave of this excavation, by the elastic method described, is 60 mm at Point 5. Most of this movement would, of course, be 'excavated' during construction, but its effects would be realised on any services or other installations immediately beneath the reduced level of the basement. If it could be assumed that heave takes place at the same rate as settlement it could be argued that some 60% of the

Table 5. CASE HISTORIES OF 'ELASTIC' HEAVE

No.	Site	Location	Source	Dimensions (m)			Clay	Thickness below reduced level (m)	Movement		Remarks
				B	L	D			Obs.	Est.	
1	St. Paul's	City of London	Serota & Jennings	Complex			L.C. W.R.B.	}52	13.0	15.0	Excavation approximated to 64 m square × 5.3 m deep
2	Stock Exchange Point 6	City of London	Ove Arup & Partners	Irregular			L.C. W.R.B.	27.0 15.0	21.0	21.0	Movement registered 3.7 m below reduced level
3	Stock Exchange Point 5	City of London	Ove Arup & Partners	Irregular			L.C. W.R.B.	27.0 15.0	15.0	14.5	Movement registered 12.9 m below reduced level
4	Bakerloo Line	South Bank	W. H. Ward	110	200	12.2	L.C. W.R.B.	}32.0	24.0	27.0	Movement registered 5 m below reduced level
5	Sugar silos	Felstead, Essex	Parker & Bayliss	21	21	29	L.C. W.R.B.	}29.0	9.0	9.5	Elastic movement due to emptying of silo
6	Sugar silos	Cantley, Norfolk	Parker & Bayliss	19.5	88	11	L.C. W.R.B.	}47.0	46	42	Elastic movement due to emptying of silo
7	Waterloo Bridge	Pier 2	Buckton & Cuerel	8.3	36	7	L.C. W.R.B.	}36.0	25.0	29.0	Movement registered 5 m below reduced level

movement after 5 years would have taken place by the end of construction (i.e. when movement observations began; Morton and Au). If this is so then the total movement from June 1966 to November 1972 would be (90 + 60 = 150 mm, i.e. (150 − 116) = 34 mm of 'long term' heave also took place during construction.

A tentative estimate of drained unloading by using $v = 0.1$ and the elastic method used earlier for settlement suggests that a total heave at the Horseferry Road site would be of the order of 180 mm. However, if as seems possible, a major part of this movement took place during construction the authors' conclusion that 20–30 mm of heave remains before stability is reached would appear reasonable.

More observations of heave are required as movements on the scale derived from elastic estimates could present amongst the most difficult design and detailing problems. For example, differential movement between a

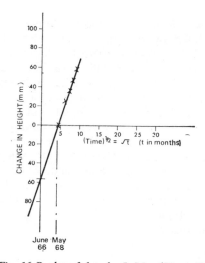

Fig. 16 Replot of data by J. May (Paper III/5)

podium substructure (with possibly a net unloading) and a basement structure supporting a tower above (with a net increase in load) may result in settlement beneath the tower while the basement and podium structure surrounding it rises.

The use of double-skinned basement floors allows the clay beneath freedom to heave and movements of the order of those quoted above have been measured. Where rigid structures on piles have been installed to prevent heave effects a measure of success seems apparent. Elastic heave takes place so rapidly that a containing structure cannot be built quickly enough to make any significant effect. The movements recorded in the Bakerloo tunnels during the excavation of the basement for the Shell Centre (Ward 1961) are typical in this respect, but the long term movement has been very small (Ward 1972) compared with estimates and experience for such an extensive area of unloading. This suggests that the special containing structure built over the tunnels has been at least partially successful in inhibiting long term heave effects.

Boulder clays and glacial tills

Two papers, that by Breth and Chamboisse describing the settlement of a nuclear reactor at Gundremmingen in S. Germany, and that by Vefling describing the settlement of sugar silos at Gørlev, Denmark, are concerned with glacial materials. Because of the difficulties of determining the properties of these deposits in the laboratory, it may be argued that relatively simple robust and universally applicable *in situ* techniques are necessary. The Standard Penetration Test (Stroud, 1974) and the plate bearing test seem to offer the most likely approach.

Paper III/2, Breth and Chamboisse, concerns 'flintz', a tertiary fresh-water deposit, overconsolidated by subsequent deposition. The site is certainly within the periglacial area associated with the Alpine glaciation and during the Pleistocene (Zeuner 1959) and will have been subjected to climatic oscillations, probably causing chemical weathering in temperate times, loess deposition, mechanical weathering and aggradation during glacial phases and possibly even ice during periods of exceptional advance.

The 'flintz' is of low plasticity, seldom containing more than 20% of colloidal material and could perhaps be more accurately described as a heavily overconsolidated sandy silt. The sandy nature suggests that a Poissons ratio of 0.3 for effective stress may be justified. Assuming, from the S.P.T. results $C_u = 7N$, and assuming $v = 0.3$ and $R = 1.6$, $E_{u1} = 1.4 E_v'$.

Applying the same argument as above, by 'elastic' analysis, a mean settlement of 145 mm is obtained, which compares favourably with the mean settlement of 153 mm observed.

Paper III/11, Vefling describes the settlement behaviour of two sugar silos founded on stiff to very hard moraine clay. The clay is heavily overconsolidated and variable in grading including sand and sand lenses. The borings on the site were limited to a depth of 15 m.

The site, about 90 km west of Copenhagen, is on a glacial plain which in the Pleistocene period would have been heavily glaciated during several advances and retreats of ice from the Scandinavian glaciation (Zeuner 1959). Layers of boulder clay and inter-glacial horizons, perhaps less compact ablation till in parts, are to be expected to great depth.

Where the moraine clays are locally more homogeneous, tests have shown (Jacobsen 1970) that the undrained shear strength C_u can be expressed in terms of the void ratio thus:

$$C_u = \exp{(0.77 \times e^{-1 \cdot 2})} \ \text{ton/m}^2$$

However, this expression is rather sensitive to e below about e = 0.35.

An elastic settlement analysis of the settlement of these silos has been attempted with limited success as the results show (*Table 6*).

Vefling's records show that there is a net increase in settlement with each loading/unloading cycle which he suggests may be a secondary consolidation phenomenon similar to the examples cited by Bjerrum (1968).

Table 6

	Estimate	Actual Silo 1	Actual Silo 2
End of construction (assumed undrained)	9.5	5.0	3.0
End of first filling (assume $U = 0.7$ as quoted by Vefling)	31.0	18.0	24.0
Total drained settlement assuming constant fully loaded conditions	40.5	—	—
Actual settlement after 13 fillings		26.0	38.0
Elastic rebound on emptying	18.0	8.0	8.0

Other case histories of settlement of structures on glacial tills have been recorded (De Jong and Harris 1970; Crawford and Burn 1962). These have also been examined by elastic settlement analysis using published data on Canadian glacial till by Klohn (1965), Hanna (1955) and Trow (1955). In several cases these records relate to deposits consisting of more than one definable layer of differing geological age and consistency. Settlements have been computed by elastic means by superposition of E_v'—depth diagrams as shown in *Fig. 17*.

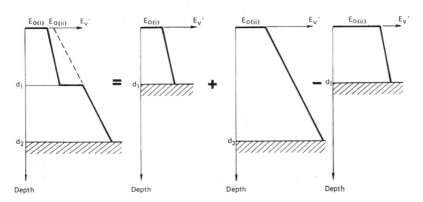

Fig. 17. Superposition of E_v–depth diagrams

The principal data concerning these case histories is summarised in *Table 7* and the comparison between estimates and actual recorded settlements illustrated in *Fig. 18*. These results show an encouraging measure of agreement. It is possible that further correlation of S.P.T. results will provide further confirmation that the elastic method facilitates reliable estimates of settlement even in the case of complex overconsolidation.

557

Table 7. CASE HISTORIES OF SETTLEMENT—GLACIAL DEPOSITS

No.	Site	Location	Type of foundation	Dimensions (m) B	L	D	Clay	Thickness below foundation (m)	Foundation Press. Gross	Unloading	Net	Primary Settlement Obs.	Calc.	Remarks
1	Nuclear reactor (Breth & Chamboisse)	Gundremmingen, Germany	Raft	31.5	31.5	7.5	'flintz'	72.5+	466	166	290	153	145	
2	C.N. Tower (De Jong & Harris)	Edmonton	Footings	40	61	8	Glacial till Saskatchewan sands	23	540	168	372	32	40.5	
3	Avord Arms (De Jong and Harris)	Edmonton	Piles	17.0 Podium 38.0	Overall 58.0 64.0	5.0	Glacial till Saskatche sands	26	343 80		343 80	32	39	
4	Mt. Sinai Hospital (Crawford & Burn)	Toronto	Raft	18.0	70.0	5.0	Sunnybrook till Interglacial till	12.0 30.0+	171	86	85	15.0	17.5	Observations begun after 53 kN/m² applied, i.e. 33 kN/m² reloading and 85 kN/m² net loading observed
5	Silo 1 (Vefleq)	Gøfleq	Raft	22.0	22.0	4.0	Moraine clay		(full) (empty)		250 26	18.0	31.0	
6	Silo 2 (Vefleq)	Gøfleq	Raft	22.0	22.0	4.0	Moraine clay		(full) (empty)		250 26	24.0	31.0	

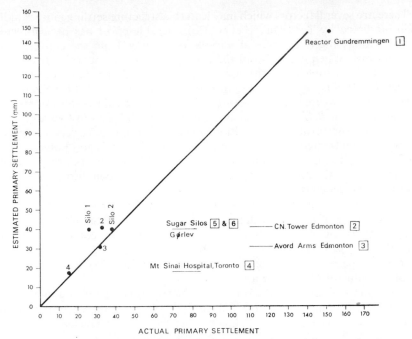

*Fig. 18. Comparison between predicted and actual settlement (elastic analysis)
glacial deposits*

Rate of primary consolidation

The rate at which primary consolidation takes place is dependent upon the rate at which the excess pore water pressure induced by the building load can dissipate, so that the structure is supported entirely by the soil skeleton. The increase in effective stress in the soil corresponds to a decrease in volume which in turn is controlled by the rate at which the pore water can escape from the voids. The rate of settlement of the structure associated with the decrease in volume of the soil is controlled by the finite time taken for the pore water to percolate through the clay mass to the drainage surfaces.

Terzaghi considered the relatively simple case of a uniform load over a very large area so that the conditions of drainage and consolidation were one-dimensional. Under these conditions immediate settlement is negligible because of lateral confinement. In practice a structure takes a relatively lengthy period to construct and the load cannot be considered to be applied in a unique instant. Terzaghi showed that an adequate approximation could be obtained by superimposing the settlements due to a series of discreet loads applied at short intervals. The problem of linear rate of loading has been solved analytically by Schiffman (1958) and the results are applied in graphical form for different rates of loading.

A simplified approach to the analysis of linearly increasing load is to consider the whole load as being applied half way through the period of construction. This leads to a prediction of more rapid pore pressure dissipation than Schiffman's rigorous treatment, but is tolerably accurate for $T_v \leqslant 0.6$.

There are several factors which may lead to a structure settling more rapidly than predicted by Terzaghi's theory. Considerable effort has been directed towards analysing the effects of various departures from the conventional assumptions. When the width of the foundation B is comparable to, or less than, the depth of clay Z the assumption of one-dimensional drainage is unrealistic. If B is much larger than Z the primary settlement time is only a function of Z whereas for a deep clay stratum the rate is almost entirely dependent upon B. The rigorous analysis of consolidation in two and three dimensions was formulated by Biot (1941, 1955) and a variety of special cases has been treated by Gibson and Lumb (1953), Davis and Poulos (1968, 1972) and Gibson, Schiffman and Pu (1970). These results are primarily concerned with the analytical realism of the boundary conditions, but relate only to homogeneous soils.

Most real soil deposits exhibit some change of their properties with depth and in the case of uniformly deposited heavily consolidated clays, the compressibility is found to vary inversely with the average confining pressure (Wroth, 1971). There is generally, although not always, a corresponding decrease in permeability which in the case of London Clay leads to a fairly constant value of C_v with depth.

Schiffman and Gibson (1964) have considered this type of inhomogeneity and have shown that the relative high compressibility of the upper part of the London Clay leads to a much higher rate ($N \times 4$) of consolidation than predicted by Terzaghi's theory for the same value of C_v.

Of the case histories presented to this Conference, two are of particular interest when considered in the light of Schiffman and Gibson's analysis. The blocks at Clapham Road and Hurley Road (Block II) are both founded in the gravel above the London Clay. The 95% consolidation time for both structures is 1.5 yr (after allowing for linear loading during construction) and the structure can be considered to be equivalent to rafts of width 24 m and 20 m respectively at the top of the London Clay. The depths of London Clay (over Woolwich and Reading) are 24 m and 20 m respectively and assuming that the interface with the Woolwich and Reading provides free drainage the drainage path lengths are 12 m and 10 m. If it is further assumed that drainage is one-dimensional and $C_v = 0.3$ (see Morton and Au, Table 2)

Clapham Road

$$t_{95\%} = \frac{12^2 \times 0.9}{0.3} = 430 \text{ yr}$$

Hurley Road

$$t_{95\%} = \frac{10^2 \times 0.9}{0.3} = 300 \text{ yr}$$

There is clearly a large discrepancy between these estimates and the observed time of 1.5 yr. The effect of geometry is not large in these cases as shown by the three-dimensional analysis of Davis and Poulos (1968) whose value of T_v for 95% consolidation differs by only 11% from the one-dimensional value of 0.9. Even allowing for inhomogeneity by the use of Schiffman and Gibson's results, the rate of settlement is still about 60 times larger than expected.

It is interesting to consider the observed rate of settlement in the field as a means of determining the value of C_v appropriate to London Clay in the mass.

560

Rowe (1972) has discussed the difference between laboratory and field C_v in terms of soil fabric. The presence of a permeable fabric consisting of silt and sand inclusions (either on bedding planes or in fissures) is said to provide increased drainage for the individual beds of clay. Such inclusions are often spaced at between 50–100 mm but in order to contribute to the overall drainage under a footing they must form an interconnected network reaching a major permeable stratum. Given these conditions one would expect that the settlement of a footing, which is large in comparison with the spacing of the permeable fabric, will be significantly faster than predicted from laboratory value of C_v.

The effect of footing size is illustrated by considering the rate of settlement of the rafts at Clapham Road and Hurley Road, with that of the foundations at Elstree Fire Testing Station (breadth 1.5 m, drainage path 2.4 m) and Waterloo Bridge (breadth 8.0 m, drainage path 7.5 m) (*Table 8*)

Table 8

	Breadth (m)	Length of drainage path (m)	C_v(field)/C_v(lab.)
Clapham Road	24	12	60
Hurley Road	20	10	60
Waterloo Bridge	8	7.5	10
Elstree	1.5	2.4	2.5

The comparatively small ratio of C_v(field)/C_v(lab.) at Elstree implies that the drainage path of 2.5 m is not much greater than the critical length at which the permeable fabric would cease to influence the rate of consolidation.

The length of this critical drainage path is still some 25–50 times greater than the spacing of permeable fabric in London Clay noted by Rowe (1972) at Ardleigh. This suggests that only the larger features of the permeable fabric affect the overall rate of consolidation.

Secondary consolidation

The study of secondary consolidation of heavily overconsolidated clay has received little attention primarily because of the inherently formidable difficulties which attend such long term testing in the laboratory and the diminution of interest and frequent practical difficulties of access which accompanies long term settlement measurements in the field.

Secondary consolidation is generally relatively small, compared with that of normally-consolidated clays and is seldom a source of problems. Creep of the building structure is able to find accommodation for the small movements and the time scale during which they occur. Exceptions exist however. The results of two long term settlement studies reported by Dawson and Simpson (1948) formed the basis of Bjerrum's (1966) review of secondary settlements. His basic conclusion was that large secondary settlements can be expected in structures with large variations in live load, even in soils where the secondary settlements, in general, are assumed to be small. This point

561

has been referred to by Vefling (Paper III/11) in connection with the settlement behaviour of the silos at Gørlev, although his evidence does not yet seem fully conclusive.

Bishop (1966) initiated a series of creep tests on London Clay and Pisa Clay: the results of these tests were reported by Bishop and Lovenbury (1969) and indicate the limited applicability of simple logarithmic or power laws relating strain and time, the absence of secondary creep and a marked instability of strain rate which appears to be associated with a modification in soil structure. Unless further long term cyclic loading tests are initiated, case histories must be tested against Bjerrum's conclusions. The evidence must also be modelled from the rheological standpoint.

Frankfurt Clay

The settlement behaviour of buildings on Frankfurt Clay has been very comprehensively documented in Paper III/1 by Breth and Amann. Frankfurt Clay is a moderately plastic tertiary clay which has been overconsolidated by about 150 m of overlying sediments, but not by ice. The overlying sediments were subsequently eroded by the Main river during the Quaternary period. The Frankfurt Clay contains a slightly higher colloidal content (50%) than is general in London Clay but in its characteristics is very similar, liquid limit 75%, plastic index 50%, natural moisture content 30–32%. Other references to the engineering properties of Frankfurt Clay are given by Ertel (1967) and Lenssink and Muller-Kirchambauer (1967).

The settlement records of eight buildings on Frankfurt Clay are given in *Table 2* to the Paper. To compare the records they have been normalised, by applying a factor to the actual settlement, to account for plan shape, reducing each to the equivalent settlement of a square foundation using Steinbrenner's curves (assuming uniform homogeneous elastic compressible material). Assuming that the settlement is given by a relationship

$$\rho_{net} = q_{net} \times B \times \left[\frac{(1 - v^2)}{E_v'} \cdot I_\rho \right]$$

and a plot of ρ_{net} against $q_{net} \times B$ is shown in *Fig. 19*. Some scatter is evident but a trend is clear, with

$$\frac{E_v'}{I_\rho(1 - v^2)} \simeq 38\,400 \text{ kN/m}^2.$$

The authors also describe the monitoring of a 70 m × 70 m, 24 m deep excavation in which observation points were installed from −19 m level. During the reduction of the excavation to −21 m level an elastic heave of about 24 mm was recorded. Using the same expression

$$\frac{E_u}{I_\rho(1 - v^2)} = \frac{40 \times 70 \times 1000}{24} = 116\,000 \text{ kN/m}^2$$

therefore we can obtain

$$\frac{E_u}{E_v'} = \frac{116\,000 \times (1 - 0.5^2)}{38\,400\,(1 - 0.1^2)} = 2.3$$

The Frankfurt Clay is calcareous and contains frequent bands of limestone and sand. In the borehole log 15% of the strata in the upper 40 m consists of interbedded limestone and sand layers. It seems highly probable that on this account the Frankfurt Clay would be highly anisotropic. If we assume $\nu_{hv} = 0.12$ and $R = 3.0$, from Henkel (1971), $E_u/E_v' = 2.3$, which is consistent, may be coincidentally, with the result found above.

The observations obtained from the AfE building are of particular interest since the distribution of settlement within the layers beneath have been measured (*Figs. 5 and 6*, Paper III/1). The settlement is seen to be concentrated in the layers immediately beneath the foundation and dies away rapidly with

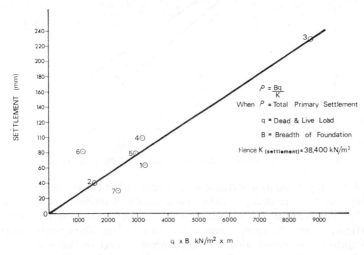

Fig. 19. Settlement on Frankfurt Clay

depth. The settlement at $Z = B/2$ is only slightly greater than 10% of the settlement of the raft itself.

Let us assume that the net settlement of the AfE building is 125 mm and express the settlement at various depths as a percentage of the net settlement of the building and express the depth at which it occurs as a function of the half-breadth, with $\nu = 0.1$ (*Table 9*).

Table 9

Layer	Z/B	Cumulative % of total settlement	$4 \times I\rho$ (for $k = 0$)	$\Delta I\rho$	$\Delta\rho\%$	$E_n/E_1 (= \frac{\Delta\rho_n}{\Delta\rho_1})$
1	0.2	46	0.18	0.18	46	46/66 = 1.0
2	0.4	66	0.36	0.18	20	46/20 = 2.3
3	0.6	77	0.54	0.18	11	46/11 = 4.2
4	0.8	84	0.71	0.17	7	46/7 = 6.5
5	1.0	89	0.86	0.15	5	46/5 = 9.2
6	1.2	91.5			2.5	46/2.5 =

The plot of $Z/B \times \rho/\rho$ total (%) and $Z/B \times E_n/E_1$ are shown in *Fig. 20*

563

A rapidly and almost linearly increasing modulus with depth is indicated. Taken at their face value these results imply a best-fit $K \ (= \delta E / \delta (z/B))$ of 9 to 11.

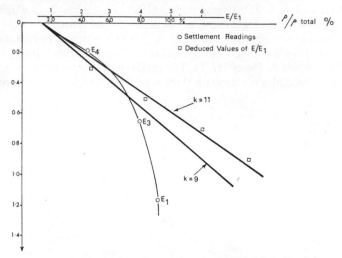

Fig. 20. Re-analysis of Settlement records, AfE Building, Frankfurt

This is not regarded as a realistic assessment of the elastic moduli of Frankfurt Clay since the re-analysis takes no account of anisotropic effects. These would be significant and would have the effect of concentrating both heave and settlement into the layers nearest the foundation. The opening of fissures due to unloading would also be more pronounced in these layers. These results form a valuable contribution to our knowledge of heave and settlement effects under working conditions.

Other case histories

Several interesting papers have been submitted to the Conference which deal with particular problems, geologically or constructionally which need to be reviewed, individually.

Paper III/10, Ueshita, Matsui, Ohoka and Nagase, describes an extension to an existing building in Nagoya, Japan, in which an extensive investigation of the soil properties of diluvial (Pleistocene ?) strata was carried out to establish that the new works need not incorporate deep foundations like those of the original building, yet settlements would be within tolerable limits. This case history is one of few where datum points were installed before construction began so that a continuous record of ground and structure movement is available. A close agreement is shown between estimated and measured rebound recompression and net settlement. However, unlike Breth *et al.*, the authors were unfortunately unable to establish their datum points at different depths in order to provide confirmation of the contributions of the various layers. The method by which the moduli of the layers had been determined is not described in detail but the predictions of movement arising from the non-

564

cohesive layers quoted seem rather high having regard to the density indicated by the S.P.T. results. For example, let us assume that the change in stress during unloading and loading is uniform with depth throughout the strata investigated, then since

$$\rho = f\left(\frac{l}{E}\right)$$

where l = thickness of each layer and E = modulus of each layer, the rebound and recompression in each layer should be a function of l/E. Using the moduli quoted, and assuming three sandy gravel layers of thickness, 6.4, 5.0 and 5.0 m we obtain the data given in *Table 10*.

Table 10

1 Layer	2 Material	3 Thickness (mm) l	4 $\bar{E}$	5 l/E	6 $\dfrac{l/E\,(\%)}{\Sigma l/E}$	7 Estimated rebound (*Ueshita et al.*)	8 (%)
1	Clay	3000	35 800	0.084	39	0.35	16
2	Sand	3000	121 000	0.025	12	0.33	15
3	Clay 2	2600	35 800	0.073	34	0.98	44
4	Gravel 1	6400	350 000	0.018			
5	Gravel 2	5000	610 000	0.008	15	0.55	25
6	Gravel 3	5000	860 000	0.005			
				0.213	100%	2.21	100%

These assumptions would tend to overestimate the contribution of the lower layers. Nevertheless the distribution of heave within the layers as predicted in the Paper shows an even greater proportion at depth (compare columns 6 and 8 of *Table 10*).

An approximate elastic analysis has been attempted, the moduli used being drawn from a tentative correlation with S.P.T. results (Ove Arup and Partners, 1974 and Stroud, 1974). As the graphs of $Ip \times z/B$ are known to be suspect for very small values of z/B, layers 1, 2 and 3 were treated as a single

layer with a modulus given by $E_u = \dfrac{\Sigma l}{\Sigma l/E} = 30\,000$ kN/m².

Layers 4, 5 and 6 were similarly assumed to have an overall modulus of 600 000 kN/m² (taken from Ushita et al. in the absence of any other information). This very approximate analysis suggests that a heave of 17 mm will take place in layers 1, 2 and 3 and 3 mm in layers 4, 5 and 6, a total of 20 mm in all, compared with a recorded heave of 21 mm.

Paper III/8, Samuels and Cheney, provides us with the complete case history of structural damage due to swelling of a clay subsoil resulting from tree removal. This well-known phenomenon is troublesome in many places where the more plastic clays are exposed at the ground surface. The paper contains a useful statement for practising engineers' methods which can be used to estimate the amount of swelling and the pressure likely to be generated beneath foundations if attempts are made to resist it.

Paper III/9, Thomas and Fischer, describing their monitoring of St. Paul's Cathedral attempts to do justice to one of the most remarkably painstaking and exhaustive settlement studies on record, which has now been going on for about 50 years.

The general soil succession which is quoted is a relatively familiar one, but despite the numerous boreholes and pits known to have been made in this area, variations in thickness and level of the brickearth, the sands and gravels, the London Clay and the Woolwich and Reading beds are not recorded. For a building of such size in plan (150 m × 76 m) evidence from surrounding sites indicates that such geological variations exist.

The ground water conditions beneath the cathedral are passingly referred to; in particular the effects of water extraction from the chalk causing under-drainage and hence consolidation of the London Clay are dismissed as being very small. It is somewhat difficult to reconcile the authors' rough calculations of the consolidation of London Clay, during the period 1923–1971, of 2.5 mm with Grace and Wilson's (1942) prediction of about 50 mm. It is noteworthy that Green and Cocksedge in Paper III/3 claim from their observations at the Shell Centre that the rate of regional surface settlement there is of the order of 3 mm/yr, twice that suggested by Wilson and Grace, although a significant part of this may be due to consolidation of recent deposits surcharged by fill.

More information about the position and exact nature of the datum point used in this survey would be welcome. From the details given, the datum appears to terminate about 40 ft into the London Clay. It would therefore be anticipated that it too would be subjected to the effects of consolidation of the London Clay and may not be reliable (*Fig. 20*). Consequently it would seem that the settlements presented in *Fig. 3* are primarily of value as differential settlements. Hopefully a more definitive statement of the soil, and in particular the ground water conditions beneath the cathedral will be provided at this or some future occasion.

Paper III/12, Ward. Finally, we have an interesting and well-documented account of the settlement performance which may result from the intervention of what he euphemistically terms 'typical' workmanship! It is important to emphasise that the settlement characteristics of heavily over-consolidated clays are very much a function of workmanship. As an example compare records of Pier 2 (87 mm) and Pier 4 (127 mm) of Waterloo Bridge resulting from exposure to wetting because the blinding concrete at Pier 4 was too thin. Ward (1967) summarises the problem most graphically when describing a London Clay face in a tunnel. 'One does not have to wait more than a few minutes to realise that the whole face is changing before your eyes.'

It is fitting to leave the last word to Sir Marc Isambard Brunel who noted in his diary 'it is evident that what is wanted is that the ground should be kept pressed'.

Acknowledgements

Previously unpublished observations of heave and settlement are presented by kind permission of Mr. F. G. Greenfield of the Chief Engineer's Department, London Borough of Brent, Dr. W. A. Ward and Dr. J. B. Burland of the Building Research Establishment, the British Petroleum Company, the Barbican Committee of the Corporation of the City of London, the Council of the Stock Exchange and Ove Arup and Partners.

The author would like to acknowledge the enthusiasm, encouragement and assistance of his colleagues of the Geotechnics Division of Ove Arup and Partners of whom Dr. D. L. Borin, Dr. M. A. Stroud, Dr. J. A. Lord and Dr. J. M. Markland merit particular recognition.

APPENDIX A

Example of elastic analysis—Clapham Road (Morton and Au)

The plan shape of the building approximates to a rectangle 17 m × 28 m. The buildings exerts a ground pressure of 190 kN/m² (see Fig. A1). The soil

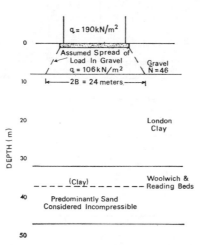

Fig. A1 Example. Clapham Road, Lambeth (from Morton & Au)

succession consists of 7 m of gravel ($N = 46$ blows/ft) overlying 24 m of London Clay and 15 m of Woolwich and Reading beds (10 m considered incompressible).

Settlement in gravel

Assume $\bar{E} = \bar{N} \times 2000 = 46 \times 2000 = 92\,000$ kN/m². Assuming 1 in 2 spread in gravel, average increase in stress in gravel layer = 150 kN/m².

$$\rho_{\text{gravel}} = \frac{150 \times 7 \text{ m} \times 1000}{92\,000} = 11 \text{ mm}$$

Settlement in clay

The building load on the gravel/clay interface will be spread over an area of 24 m × 35 m. Therefore

567

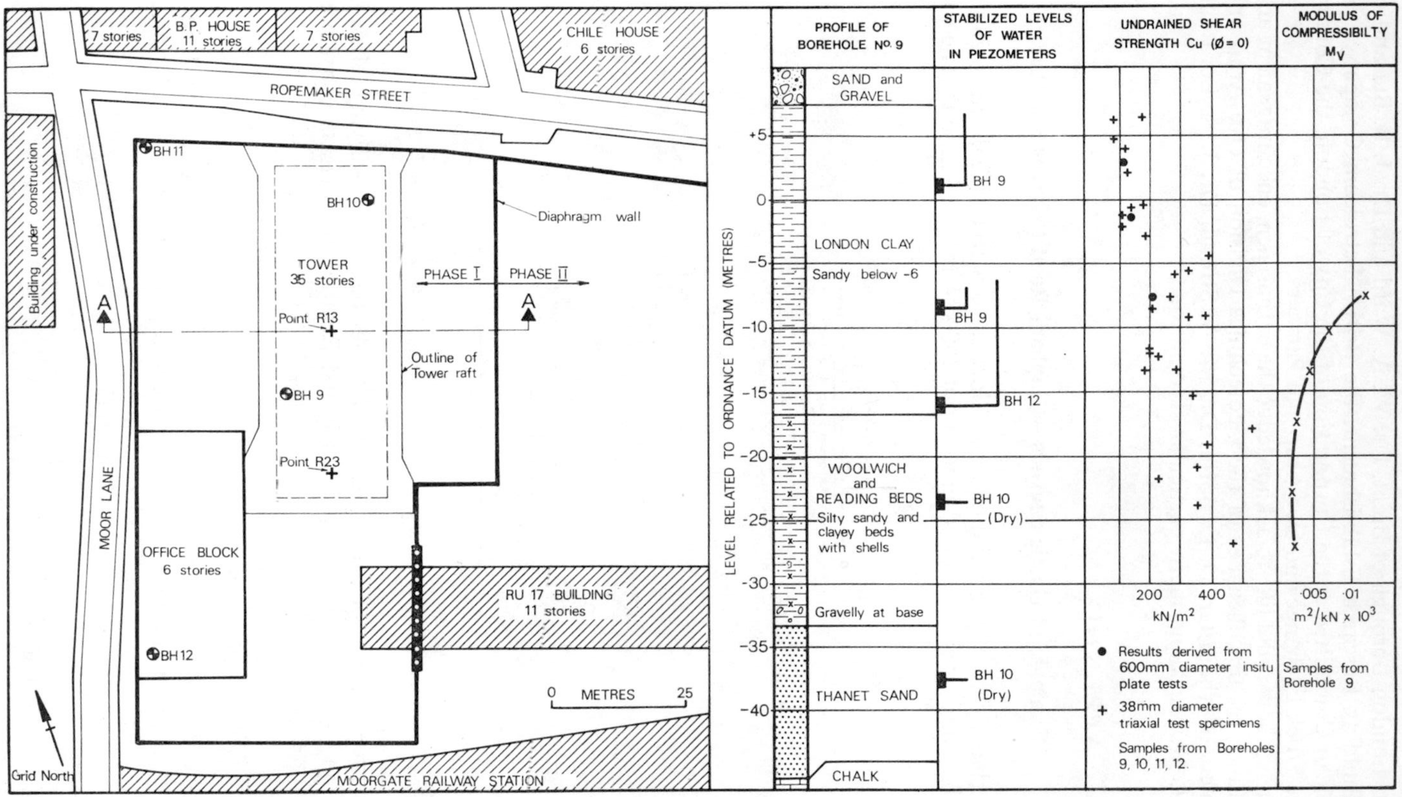

Fig. B1 Britannic House plan and soil profile

$$\text{stress on clay} = 190 \times \frac{17}{24} \times \frac{28}{35} = 106 \text{ kN/m}^2$$

For the centre of the loaded area

$$b = \frac{24}{2} = 12 \text{ m}, \quad l = \frac{35}{2} = 17.5 \text{ m}, \quad \frac{l}{b} = \frac{17.5}{12} = 1.5, \quad \frac{z}{b} = \frac{29}{12} = 2.4$$

$$C_{uo} = 1900 \text{ lb/ft}^2 = 91 \text{ kN/m}^2, \quad E_0 = 11\ 800 \text{ kN/m}^2$$
$$C_{u12} = 4500 \text{ lb/ft}^2 = 217 \text{ kN/m}^2, \quad E_{12} = 27\ 200 \text{ kN/m}^2$$

$$k = \left(\frac{27\ 200}{11\ 800}\right) - 1 = (2.3 - 1) = 1.3$$

$$\therefore I_\rho = 0.207 \text{ (Interpolated from } Figs.\ 7\ and\ 8)$$
$$I_\rho = 0.207 \times 4 \text{ (for } I_\rho \text{ of centre point)}$$
$$= 0.83$$

$$\rho = \frac{106 \times 12 \times 0.83}{11\ 800} = 89 \text{ mm}$$

Depth correction (Fox)

$$\frac{D}{B} = \frac{7}{24} = 0.3, \quad I_D = 0.92$$

$$\therefore \rho = 0.92 \times 89 = 82 \text{ mm}$$

N.B. No factor for the rigidity of the structure is applied in this case. The primary settlement of the structure is actually approximately 90 mm, of which 11 mm (see above) is estimated to be in the gravel. Hence the actual settlement in clay $= 90 - 11 = 79$ mm (compared with the estimate of 82 mm).

APPENDIX B

Additional case histories

References to several additional case histories which have been examined (Ove Arup and Partners, Jobs 1023, 1405, 1642) are included in the text. The relevant details of these are as follows.

Britannic House. The headquarters building of B.P. Trading Ltd. occupies a site between Ropemaker Street and Moorgate Railway Station and is known as Britannic House. The structure consists of a 35-storey tower with an adjoining 2-storey podium and a 6-storey block, surmounting an extensive 18 m deep basement (*Fig. B1*).

The tower is founded on a raft within the basement area. All the remaining structure is founded on large diameter piles which, together with the diaphragm wall of the basement, were constructed before major excavation works began (*Fig. B2*). The sequence of construction has been described elsewhere (Cole and Burland, 1972) and the soil sequence and properties are also shown in *Fig. B1*.

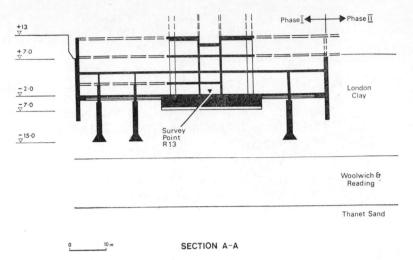

SECTION A-A

Fig. B2 Britannic House

The site had been occupied previously by offices up to six storeys high, and generally with single-storey basements. These buildings had been totally destroyed by bombs during the war and the site reduced to basement level for about 20 years before the new construction began. The tower raft is 70 m long and 27 m wide and imposes an average gross pressure of 450 kN/m². The unloading due to excavation is 350 kN/m² so that the overall net loading under the tower raft is 100 kN/m².

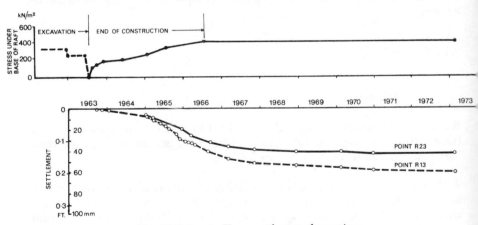

Fig. B3 Britannic House settlement observations

The results of precise levelling on settlement points on the raft and the surmounting structure are shown in *Fig. B3*.

Barbican I and III. Three similar 42-storey residential tower blocks will form part of the redevelopment of the Barbican for the Corporation of the City of London. Two of these have now been completed. The general arrangements of each are shown in *Figs. B4 and B5*. The soil succession under each tower is also shown.

Because of a lesser thickness of London Clay beneath Block III and occasional water bearing sands at the base, the length of the piling was limited. It therefore became necessary to spread the piled foundation over a somewhat larger plan area than in the case of Block I. Neither block has a basement, the average net increase in load at raft level is estimated to be 580 kN/m² in the case of Barbican I and 390 kN/m² in the case of Barbican III.

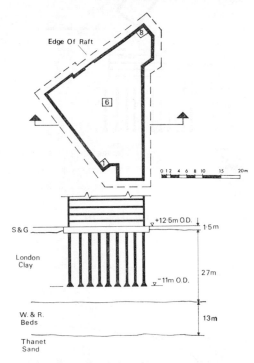

Fig. B4 Barbican I

Load–time settlement curves prepared from precise levelling related to a deep benchmark founded in the Upper Chalk, are shown in *Figs. B6 and B7*.

The Stock Exchange, London. The rebuilding of the Stock Exchange in the City of London has been described by Dunican and Martin (1970). Part of the first phase consists of an office tower of 26 storeys, a plan and general arrangement of the foundations is shown in *Fig. B8*. The tower is founded on a basement raft 3 m thick on 0.6 m diameter piles at 2.6 m centres and effectively 23 m long, but bored from existing basement level, i.e. with 11 m of empty boring.

Twin tube railway lines of the Central Line run near the northern boundary of the basement and the foundation design and sequence of construction was particularly directed towards limiting the movement and deformation of these tunnels. As part of the site control, nine heave points were installed in three groups of three at levels −23 m, −13.7 m and −4.6 m o.d., i.e. 22 m,

571

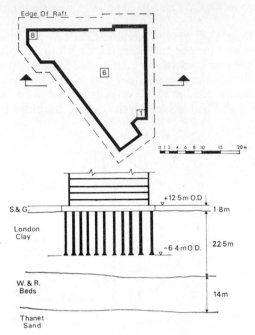

Fig. B5 Barbican III

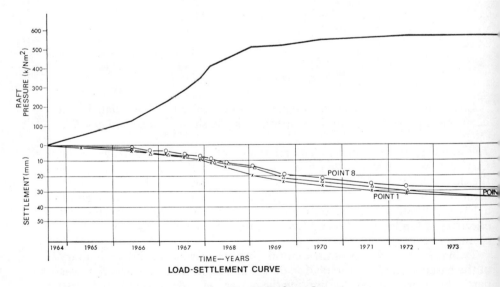

LOAD-SETTLEMENT CURVE

Fig. B6 Barbican I

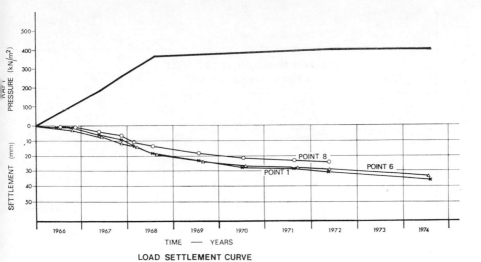

Fig. B7 Barbican III

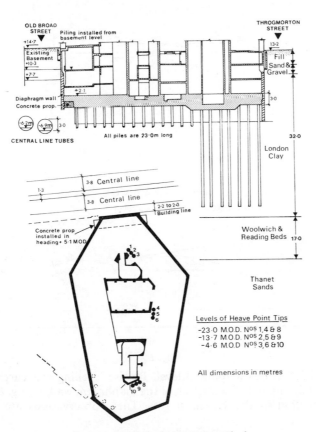

Fig. B8 Stock Exchange Tower Block

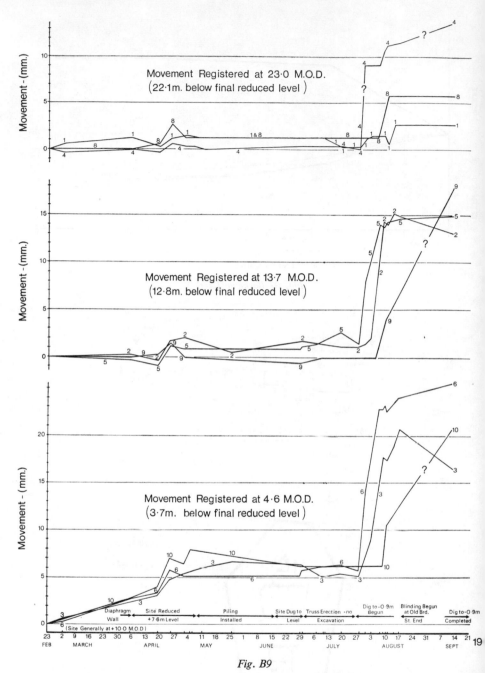

Fig. B9

12.7 m and 3.6 m below reduced level. Precise levelling was carried out on these points during the construction of the diaphragm wall, piling, excavation and construction of the basement raft. These observations are shown in *Fig. B9*.

Referring again to the effects of construction sequence these movements deserve further comment. Prior to the first observations on the heave points in late February 1967, demolition of the existing buildings, which although only five or six storeys high, were of massive construction and therefore relatively heavy, probably equivalent to about 270 kN/m^2. This would have led to undrained and long term heave, most of which was not recorded. However, an upward movement of about 3 mm is recorded on all three heave points between the 23rd February and 6th April 1967, when the diaphragm wall was constructed. It is likely that this is some residual movement due to demolition of the substructure (which in places was up to 17 ft thick) rather than the effect of installing the wall.

The reduction of the site from +10.0 m o.d. to +7.6 m o.d. would, by 'elastic' methods of analysis, produce a heave of 5 mm at a level of −4.6 m o.d. However, only 3 mm was recorded prior to the commencement of the piling, near Point 6.

A cessation of heave and/or a settlement (of only 1 or 2 mm) was evident immediately piling commenced and the excavation of a further 1.8 m of the basement was undertaken without perceptible heave. The reduction of the site from +5.8 m o.d. to −0.8 m would produce, by elastic methods of analysis, an undrained heave of 15 mm at a depth of −4.6 m o.d.; 18 mm was recorded at Point 6.

The effect of installing the piling before excavation and the rapid placing of the raft can be seen to be one of reducing the vertical movements to those of, at most, purely elastic rebound. Movements of the Central Line tunnel were reported by Turner (1970).

REFERENCES

Atkinson, J. H. (1973). 'The Deformation of Undisturbed London Clay', PhD Thesis University of London
Barden, L. (1963). 'Stresses and Displacements in a Cross-anisotropic Soil', *Géotechnique*, Vol. 13, 198–210
Biot, M. A. (1941). 'General Theory of Three-Dimensional Consolidation', *J. Appl. Physics*, Vol. 12, No. 2, 155–164
Biot, M. A. (1955). 'Theory of Elasticity and Consolidation for a Porous Anisotropic Solid', *J. Appl. Physics*, Vol. 26, No. 2, 182–185
Bishop, A. W. (1954). 'The Use of Pore-Pressure Coefficients In Practice', *Géotechnique*, Vol. 4, Pt. 4, 148
Bishop, A. W., Webb, D. C. and Lewin, P. I. (1965). 'Undisturbed Samples of London Clay from the Ashford Common Shaft; Strength–Effective Stress Relationships', *Géotechnique*, Vol. 15, No. 1, 1–31
Bishop, A. W. (1966). 'The Strength of Soils as Engineering Materials', Sixth Rankine Lecture, *Géotechnique*, Vol. 16, No. 2, 91–128
Bishop, A. W. and Lovenbury, H. T. (1969). 'Creep Characteristics of Two Undisturbed Clays', *Proc. 7th Int. Conf. Soil Mech. & Found. Engng.*, Vol. I, 29–37
Bjerrum, L. (1968). 'Secondary Settlements of Structures Subjected to Large Variations in Live Load', *Norwegian Geotechn. Inst.*, No. 730
Breth, H., Amann, P. (1974. 'Time-Settlement and Settlement-Distribution with Depth in Frankfurt Clay', *Proc. Conf. Settlement of Structures*, Cambridge
Breth, H., Chamboisse, G. (1974). 'Settlement Behaviour of a Nuclear Reactor', *Proc. Conf. Settlement of Structures*, Cambridge
Brown, P. T. and Gibson, L. (1972). 'Surface Settlement of a Deep Stratum Whose Modulus Increases Linearly with Depth', *Canadian Geotech. J.*, Vol. 9, No. 467–476
Brown, P. T. and Gibson, R. E. (1973). 'Rectangular Loads on Inhomogeneous Elastic Soil' (Technical Note). *Proc. A.S.C.E.*, SM 10, October, 917–920

Buckton, E. J. and Cuerel, J. (1943). 'The New Waterloo Bridge', *Proc. Instn. Civ. Engrs.*, Vol. 20, 145

Buckton, E. J. and Fereday, H. S. (1938). 'The Reconstruction of Chelsea Bridge', *J. Instn. Civ. Engrs.*, Vol. 7, 383

Burland, J. B., Butler, F. G. and Dunican, P. (1966). 'The Behaviour and Design of Large Diameter Bored Piles in Stiff Clay', *Proc. Symp. on Large Bored Piles, Instn. Civ. Engrs.*

Burland, J. B. (1973) Private communication

Cole, K. W. and Burland, J. B. (1972). 'Observations of Retaining Wall Movements Associated with a Large Excavation', *Proc. 5th European Conf. Soil Mech. & Found. Engng.*, Madrid, Vol. 1, 445–453

Cooling, L. F. and Skempton, A. W. (1942). 'A Laboratory Study of London Clay', *J. Instn. Civ. Engrs.*, Vol. 17, 251

Cooling, L. F. (1948). 'Settlement Analysis of Waterloo Bridge', *Proc. 2nd Conf. Soil Mech. & Found. Engng. (a)*, Vol. 2, 130

Cooling, L. F. (1948). 'Settlement Observations on Four Grain Silos, *Proc. 2nd. Int. Conf. Soil Mech. & Found. Engng. (b)*, Vol. 2, 135

Cooling, L. F. and Gibson, R. E. (1955). 'Settlement Studies on Structures in England' *Conf. Correlation of Stresses, Instn. Civ. Engrs.*, London, 295–317

Crawford, C. B. and Burn, K. N. (1962). 'Settlement Studies on the Mt. Sinai Hospital, Toronto', *Engineering J.*, December, 31–37

Davis, E. H. and Poulos, H. G. (1968). 'Settlement Prediction under Three-Dimensional Conditions', *Géotechnique*, Vol. 18, No. 1, 67–91

Davis, E. H. and Poulos, H. G. (1972). 'Rate of Settlement under Two and Three-Dimensional Conditions', *Géotechnique*, Vol. 22, No. 1, 95–114

Dawson, R. F. (1947). 'Settlement Studies on the San Jacinto Monument', *Proc. 7th Texas Conf. on Soil. Mech. & Found. Engng.*, Austin, Texas

Dawson, R. F. and Simpson, W. E. (1948). 'Settlement Records of Structures in the Texas Gulf Area', *2nd Int. Conf. Soil Mech. & Found. Engng.*, Rotterdam, Vol. 5, 125–129

De Jong, J. and Harris, M. C. (1971). 'Settlement of Two Multistorey Buildings in Edmonton', *Canadian Geotech. J.*, Vol. 8, No. 2, 217–235

Dunican, P. and Martin, J. (1970). 'The Rebuilding of the Stock Exchange', *Structural Engineer*, Vol. 48, No. 4

Ertel, W. (1967). 'Determination of the Initial Strength of Stiff Tertiary Frankfurt Clay', *Proc. European Conf. Soil Mech. & Found. Engng.*, Oslo, Vol. 1, 109–111

Gibson, R. E. and Lumb (1953). 'Numerical Solutions of Some Problems in the Consolidation of Clay', *Proc. Instn. Civ. Engrs.*, Vol. 2, 182–198

Gibson, R. E. (1967). 'Some Results Concerning Displacement and Stresses in a Non-Homogeneous Elastic Half-Space', *Géotechnique*, Vol. 17, 58–67

Gibson, R. E., Schiffman, R. C. and Pu, S. L. (1970). 'Plane Strain and Axially Symmetric Consolidation of a Clay Layer on a Smooth Impervious Base', *Q.J. Mech. appl. Math*, Vol. XXIII, Pt. 4, 505–520

Gibson, R. E., Brown P. T. and Andrews, M. R. F. 'Some Results Concerning Displacements in a Non-Homogeneous Elastic Layer', *J. appl. Math. Physics*, Vol. 22, 855–864

Green, P. A. (1971). 'Some Aspects of the Foundation Design of the Commercial Union Building', *Midland Soil Mech. & Found. Engng. Soc. Proc. Symp. Interaction of Structure & Foundations*, Birmingham, 118–130

Green, P. A., Cocksedge, J. E. (1974). 'The Settlement Behaviour of Three Tall Buildings in London', *Proc. Conf. on Settlement of Structures*, Cambridge

Greenfield, F. G. (1971). 'Early Settlement of Tall Buildings Founded on Piles in London Clay', *Conf. on Behaviour of Piles, Instn. Civ. Engrs.*, London, 71–78

Greenfield, F. G. (1974). Private communication

Hanna, T. (1955). Discussion on Paper by Klohn, E. J., *Canadian Geotech. J.*, Vol. II, No. 2, 129–131

Henkel, D. J. (1971). 'The Relevance of Laboratory Measured Parameters in Field Studies', *Proc. Roscoe Memorial Symp. on Stress–Strain Behaviour of Soils*, Cambridge, 669–675

Hooper, J. A. (1973). 'Observations on the Behaviour of a Piled Raft Foundation on London Clay', *Proc. Instn. Civ. Engrs.*, Vol. 58, Pt. 1

Hyde, R. B., Leach, B. A. (1974). 'Settlement at Didcot Power Station', *Proc. Conf. Settlement of Structures*, Cambridge

576

Jacobsen, M. (1970). 'Strength and Deformation Properties of Preconsolidated Moraine Clay', Bulletin No. 27, *Danish Geot. Inst.*

Klohn, E. J. (1955). 'The Elastic Properties of a Dense Glacial Till Deposit', *Canadian Geotech. J.*, Vol. II, No. 2, 116–128

Lambe, T. W. (1964). 'Methods of Estimating Settlement', *J. of Soil Mech. & Found. Div., A.S.C.E.*, Vol. 90, No. SM5, 43–67

Lenssink, H. and Muller-Kirchambauner, H. (1967). 'Determination of the Shear Strength of Sliding Planes Caused by Geological Features', *Proc. European Conf. Soil Mech. & Found. Engng.*, Oslo, Vol. 1, 131–137

MacDonald, D. H. and Skempton, A. W. (1955). 'A Survey of Comparisons Between Calculated and Observed Settlements of Structures on Clay', *Conf. Correlation of Stresses, Inst. Civ. Engrs.*, London, 318–337

Marsland, A. (1971). 'The Shear Strength of Stiff Fissured Clays', *Proc. Roscoe Memorial Symp. on Stress–Strain Behaviour of Soils*, Cambridge, 59–68

Marsland, A. (1971). 'The Use of *In Situ* Tests in a Study of the Effects of Fissures on the Properties of Stiff Fissured Clays', *Proc. 1st Australian and New Zealand Conf. on Geomechanics*, Melbourne, Vol 1, 180–189

May, J. (1974). 'Heave of a Deep Basement in the London Clay', *Proc. Conf. Settlement of Structures*, Cambridge

Measor, E. O. and Williams, G. M. (1962). 'Features in the Design and Construction of the Shell Centre, London', *Proc. Instn. Civ. Engrs.*, Vol. 21, 475–502

Mitchell, J. H. (1900). 'The Stress Distribution in an Aeolotropic Solid with an Infinite Plan Boundary', *Proc. London Math. Soc.*, Vol. 32, 247–258

Morgenstern, N. (1967). 'Shear Strength of Stiff Clay', *Proc. European Conf. Soil Mech. & Found. Engng.*, Oslo, Vol. II, 59–69

Morton, K., Au, E. (1974). 'Settlement Observations on Eight Structures in London', *Proc. Conf. Settlement af Structures*, Cambridge

Mould, G. (1974). 'Guy's Hospital Tower Block, London—Settlement Records', *Proc. Conf. Settlement of Structures*, Cambridge

Ove Arup and Partners (1971). 'The Design of Excavation and Retaining Works, Barbican Arts Centre'. Internal Report Job 1023 (unpublished)

Ove Arup and Partners (1971). 'The Effects of Construction of the King Reach Development on the Waterloo and City Railway Tunnels'. Internal Report Job 3387 (unpublished)

Ove Arup and Partners (1971). 'Report on the Testing of London Clay in Active and Passive States under Conditions of Axially Symmetric Loading and of Plane Strain' (unpublished)

Ove Arup and Partners (1974). 'Interpretation of the Standard Penetration Test', unpublished

Parker, A. S. and Bayliss, F. V. (1971). 'The Settlement Behaviour of a Group of Large Silos on Piled Foundations', *Conf. on the Behaviour of Piles, Inst. Civ. Engrs.*, London, 59–69

Rowe, P. W. (1972). 'The Relevance of Soil Fabric to Site Investigation Practice', Twelfth Rankine Lecture, *Géotechnique*, Vol. 12, No. 2, 195–300

Samuels, S. G. and Cheney, J. E. (1974). 'Long-Term Heave of a Building on Clay Due to Tree Removal', *Proc. Conf. on Settlement of Structures*, Cambridge

Schiffman, R. L. (1958). 'Consolidation of Soil Under Time-Dependent Loading and Varying Permeability', Annual Meeting, *Highway Research Board, Proc. 37*

Schiffman, R. L. and Gibson, D. E. (1964). 'Consolidation of Non-Homogeneous Clay Layers', *Proc. A.S.C.E.*, Vol. 90, No. SM5, 1–30

Scott, R. F. (1963). *Principles of Soil Mechanics*, Addison-Wesley, Mass., 280–287

Serota, S. and Jennings, R. A. J. (1959). 'The Elastic Heave of the Bottom of Excavations *Géotechnique*, Vol. 9, 62–72

Simons, N. and Som, N. N. (1969). 'The Influence of Lateral Stresses on the Stress Deformation Characteristics of London Clay', *Proc. 7th Int. Congress Soil Mech. & Found. Engng.*, Mexico, Vol. I, 369–377

Simons, N. and Som, N. N. (1970). 'Settlement of Structures on Clay with Particular Emphasis on London Clay', *C.I.R.I.A. Report*, No. 22

Skempton, A. W. (1971). 'The Bearing Capacity of Clays' *Building Research Congress*, Vol. I, 180

Skempton, A. W. 'The Pore-Water Parameters A & B', *Géotechnique*, Vol. 4, No. 3, 143–147

Skempton, A. W. Peck, R. and MacDonald, D. H. (1955). 'Settlement Analyses of Six Structures in Chicago and London'. *Proc. Instn. Civ. Engrs.*, Vol. 53, 525–544

Skempton, A. W., and Bjerrum, L. (1957). 'A Contribution to the Settlement Analysis of Foundations on Clay', *Géotechnique*, Vol. 7, No. 4, 168–178

Skempton, A. W. (1959). Correspondence, *Géotechnique*, Vol. 9, No. 3, 145–146

Skempton, A. W. (1969). 'The Consolidation of Clays by Gravitational Compaction', *Q.J. Geol. Soc.*, London, Vol. 125, 373–411

Skempton, A. W. and Henkel, D. J. (1957). 'Tests on London Clay from Deep Borings at Paddington, Victoria and South Bank', *Proc. 4th Int. Soil Mech. & Found. Engng.*, London, Vol. 1, 100

Som, N. N. (1968). 'The Effect of Stress-Path on the Deformation and Consolidation of London Clay', PhD Thesis, University of London

Steinbrenner, W. (1934). 'Tafeln zur Setzungsberechnung', *Die Strasse*, Vol. 1, 121–124

Stroud, M. A. (1974). 'The Standard Penetration Test in Insensitive Clays and Soft Rocks', *European Symp. on Penetration Testing, Swedish Geotech. Soc.*, Stockholm

Terzaghi, K. (1925). *Erdbanmechanik*, Denticke, Vienna

Terzaghi, K. (Edited by Skempton, A. W.) (1960). *From Theory to Practice in Soil Mechanics*, John Wiley, New York

Thomas, J. B. and Fischer, A. W. (1974). 'St. Paul's Cathedral—Measurement of Settlements and Movements of the Structure', *Proc. Conf. Settlement of Structures*, Cambridge

Trow, W. A. (1955). Discussion on Paper by Klohn, E. J., *Canadian Geotech. J.*, Vol. II, No. 2, 132–139

Turner, F. S. P. (1970). Discussion on Paper by Dunican and Martin (1970)

Ueshita, K., Matsui, K., Ohoka, T. and Nagase, S. (1974). 'Displacement of Over-Consolidated Clay/Sand Soil During Construction of a Building', *Proc. Conf. Settlement of Structures*, Cambridge

Vefling, G. (1974). 'Settlements of Sugar Silos on Moraine Clay in Gørlev, Denmark', *Proc. Conf. Settlement of Structures*, Cambridge

Ward, W. H., Samuels, S. G. and Butler, M. E. (1959). 'Further Studies of the Properties of London Clay', *Géotechnique*, Vol. 9, No. 2, 33–38

Ward, W. H. (1967). Discussion on Shear Strength of Stiff Clay, *Proc. European Conf. Soil Mech. & Found. Engng.*, Oslo, Vol. 2, 140–141

Ward, W. H. (1961). 'Displacements and Strains in Tunnels Beneath a Large Excavation', *Proc. 5th Int. Conf. Soil Mech. & Found Engng.*, Paris, Vol II, 749–753

Ward, W. H. (1971). 'Some Field Techniques for Improving Site Investigation and Engineering Design', *Proc. Roscoe Memorial Symp. on Stress–Strain Behaviour of Soils*, Cambridge, 676–682

Ward, W. H. (1970). Private communication

Ward, W. H. and Burland, J. B. (1973). 'The Use of Ground Strain Measurements in Civil Engineering', *Philosophical Transactions of the Royal Society*

Ward, E. H. (1974). 'Settlements of a Brick Dwelling House on Heavy Clay (1951–73)', *Proc. Conf. Settlement of Structures*, Cambridge

Webb, D. L. (1966). 'The Mechanical Properties of Undisturbed Samples of London Clay and Pierre Shale', PhD Thesis, University of London

Wilson, G. and Grace, H. (1942). 'The Settlement of London Clay Due to Under-Drainage of the London Clay', *J. Instn. Civ. Engrs.*, Vol. 19, 100–127

Wroth, P. (1971). 'Some Aspects of the Elastic Behaviour of Overconsolidated Clay', *Proc. Roscoe Memorial Symp. on Stress–Strain Behaviour of Soils*, Cambridge, 347–354

Zeuner, F. E. (1959). *The Pleistocene Period*, Hutchinson, London

Rocks

Factors affecting the prediction of settlement of structures on rock: With particular reference to the Chalk and Trias

General Reporter: N. B. Hobbs,
Rock Mechanics Limited

Introduction

The estimation of settlement involves two basic questions:
 (1) the magnitude and distribution of the stresses set up in the rock-mass by the structure, and
 (2) the mechanical properties of the rock mass in depth and extent relevant to immediate and long term behaviour.

Other factors such as the effects of excavation, blasting, softening and seepage forces are also important and their proper consideration requires experience and judgment. It is necessary during analysis to strike the correct balance between these three factors bearing in mind the type of structure, its rigidity (which may tend to increase as construction progresses), the construction procedure, and the ground conditions. For example, if the conditions are such that disturbance to the supporting ground cannot be avoided then the importance of the basic questions diminishes. The first question, although great importance is attached to it, generally plays a somewhat subordinate role in as much as any discrepancy arising from one application of the elastic theory rather than another or a simple method in preference to a complex one is far outweighed by our inability to determine the deformability of rock masses with a comparable degree of accuracy, particularly of those rocks which are difficult to recover undamaged from the core barrel. The Chalk and Keuper, the subjects of four papers, particularly in their fractured and weathered states belong in this category.

Lack of space precludes a proper review of even the basic questions. It is proposed therefore to look briefly at the probable errors associated with the use of the elastic theory, the laboratory moduli of the Chalk and Triassic rocks with some remarks on testing, the effects of fracture on the deformability of rock masses with particular reference to the Chalk, concluding with some comments on the two methods of settlement prediction in current use.

Stress distribution and elastic displacements

Errors and accuracy in analysis

The problem of the distribution of stresses beneath loaded areas is a classical and evidently engaging one (some 1000 papers have been written on this and associated subjects since Boussinesq) which has received increasing attention following the advent of the finite element method and the computer. Distributions have long been available for the stresses in homogeneous, linear, elastic, isotropic half-spaces beneath uniformly loaded areas, and recently attention has been turned to the more difficult cases involving non-homogeneity, anisotropy, non-linearity together with rigid loading where frequently only numerical solutions are possible.

It has long been recognised that rock even in an apparently intact state, is not an elastic material. Stress–strain tests on small specimens on many rocks show non-linearity, hysteresis, permanent strain and poor reproducibility, effects which are due to cracks and to a lesser extent pores in the rock fabric. Rock masses contain fractures, fissures, faults, shear zones and bedding planes, features which are generally referred to as discontinuities and which magnify, depending upon their nature and extent, the departures from elastic behaviour observed in the apparently intact rock. In spite of their elastic nature rock masses are nevertheless generally treated as isotropic, elastic continua both for the reduction of test results and for the estimation of settlements, an assumption which conveniently permits the use of the theory of elasticity. The question of the errors likely to be incurred with various assumptions associated with elastic theory and departures from the simple elasticity commonly assumed is briefly discussed below.

The effect of Poissons ratio in a homogeneous material. The vertical strain at any point in a medium having elastic constants E and υ is given by the expression

$$e_z = \frac{1}{E}\left\{\Delta\sigma_z - \upsilon\left(\Delta\sigma_x + \Delta\sigma_y\right)\right\} \tag{1}$$

where the horizontal stress increments $\Delta\sigma_x$ and $\Delta\sigma_y$, clearly reduce the strain due to the vertical stress increment $\Delta\sigma_z$. It is fairly common practice where the modulus varies with depth to summate the vertical strains directly from the classical vertical stress distributions, a procedure which excludes the term $(\Delta\sigma_x + \Delta\sigma_y)$ and results in an over-estimate of settlement. Elastic theory expressions for settlement automatically include the effect of the horizontal stresses, thus the settlement S of a rigid plate diameter B under a stress q is given by

$$S = \frac{\pi}{4}.qB\frac{(1-\upsilon^2)}{E} \tag{2}$$

580

Ignoring the horizontal stresses, that is putting $\upsilon = 0$,

$$S_0 = \frac{\pi}{4} q B \frac{1}{E}$$

The error in assuming $\upsilon = 0$ is thus υ^2, and in rocks is unlikely to exceed 10%.

Anisotropy

The effects of transverse anisotropy on stress distributions and displacements have been studied numerically by Gerrard and Harrison (1970), Milovic (1972) and in closed form by Barden (1963) and Nyak (1973). In bedded rocks modulus ratios (parallel to normal) of up to 8 have been reported. It appears that the use of the modulus normal to the bedding with the assumption of isotropy will lead to over-estimates of settlement with errors approaching 25% for ratios up to 4. The limit may be approximated by assuming that no lateral strain occurs in the rock mass (Westergaard, 1938). This results in the vertical stresses being reduced by up to 35% compared with the isotropic case, the effect decreasing with increasing depth. In some rock masses such as those with up-ended bedding the vertical modulus could be higher than the horizontal modulus, and thus using the vertical modulus would result in an under-estimate of settlement. Where this condition is suspected an upper bound of settlement may be calculated by using $\upsilon = 0$ with the vertical modulus.

Roughness of the foundation base

The well known published solutions generally refer to smooth bases. In the case of a homogeneous half-space the consequence of ignoring the roughness is a small over-estimate of settlement, negligible for values of $\upsilon > 0.3$ but increasing for values of $\upsilon < 0.3$ to a maximum of about 10% at $\upsilon = 0$. Recent work by Popova (1972) suggests that on thin layers the over-estimate of settlement in ignoring roughness is likely to be very large—up to 50% for $z = 0.5R$ and $\upsilon = 0$.

Averaging the settlement of a uniformly loaded flexible area

Exact solutions are not available for the surface settlement of rigid foundations for shapes other than circular or long strip. However, it is known that the average settlement $\bar{S}$ of a uniformily loaded circular area on a homogeneous half-space is not appreciably different from the settlement S^* on a smooth based rigid circular footing. For this case Boussinesq (1885) has shewn that

$$\frac{\bar{S}}{S^*} = \frac{32}{3\pi^2}$$

This suggests that in many cases $\bar{S}$ may be a reasonable approximation of S^* giving rise to errors of 10% or less (Gibson 1974). The effects of flexural rigidity of circular smooth rafts on differential settlements have been discussed by Brown (1969). Rafts on homogeneous rocks cannot generally be regarded as settling uniformly.

Effect of embeddment

Fox's curves for the reduction in settlement are concerned with an ideal case of loading at depth in a homogeneous material having $\upsilon = 0.5$. The maximum reduction is 50%. Burland (1970a) following a comment by N. B. Hobbs has produced curves of reduction factor against depth in an unlined shaft for three values of υ, 0.49, 0.25 and zero, for which the maximum factors are 0.86, 0.8 and 0.75 respectively. Indiscriminate use of Fox's curves can lead to large underestimates of settlement and modulus.

Non-linearity

Rocks containing cracks frequently exhibit non-linear stress-strain characteristics, the Bunter sandstone, closely studied by Morgenstern and Phukan (1966, 1968 and 1969), being a good example. It is customary to use an appropriate secant modulus from either laboratory compression tests or plate tests in the calculation of settlements, but Hobbs (1973) has shown that this can lead to large errors when scaling the foundation settlement directly from

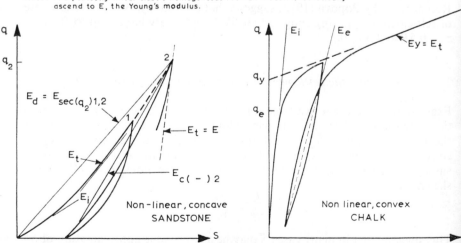

E   The unloading modulus, ideally the Youngs modulus (Walsh and Brace, 1966)
E_i The initial modulus, a tangent
E_e The so-called elastic modulus on reloading, a tangent, secant or chord.
$E_t(\ \ )$ The tangent modulus at any particular stress or range of stress, (   ).
E_d The deformation modulus on first loading, generally a secant,
$E_{sec(\ \ )n}$ The secant modulus to stress (   ) for the nth cycle.
$E_{c(-)n}$ The chord modulus over stress range (-) for the nth cycle

Note : In a perfectly linear elastic homogeneous rock all moduli theoretically ascend to E, the Young's modulus.

Fig. 1. Typical plate loading tests and modulus definitions

plate tests, the extent depending upon the degree of non-linearity. In a particular study of a plate test on the Bunter Sandstone the error in scaling to a 60 ft diameter reactor foundation amounted to 40% compared with 7% using the secant modulus from laboratory tests. The estimates based on secant moduli are generally conservative provided the rock is stiffening under stress, i.e. the stress–strain curve is concave upwards.

A second type of non-linearity is that of the yielding type frequently observed in plate loading tests on fractured and weathered rocks, as reported for example by Lake and Simons (1974) on the Basingstoke Chalk and by Lord and Nash (1974) on the Keuper Marl. *Figure 1* shows typical plate test results on Bunter Sandstone and Chalk and defines the various moduli in current use. It should be noted that non-linear field results do not necessarily imply non-linear characteristics on intact specimens.

Non-homogeneity refers to the steady increase of modulus with depth, that is where

$$E = E_0 + kz,$$

E_0 being the surface modulus and k the rate of increase in E with depth. This characteristic has been reported in the chalk at Mundford by Burland and Lord (1970) and also occurs in the Keuper associated with a gradual decrease in weathering; *Figure 15* is an example of this type of profile obtained with the pressuremeter. The performance of a foundation on such a material will clearly depend on the ratio E_0/k; the limits being infinity, homogeneity and zero, the form of non-homogeneity studied by Gibson and Sills (1971) and by Brown and Gibson (1972). Carrier and Christian (1973) in a numerical study have produced tables and curves which enable the settlement of a rigid circular foundation to be rapidly calculated for any value of E_0 and k. These curves are extremely useful for studies with plates and footings of various sizes, and applied for instance to Lake and Simon's results presented to this conference give values of $E_0 = 97$ MN/m^2 and $k = 19$ MN/m^3 for the grade III Upper Chalk at Basingstoke. Brown and Gibson (1974) have considered the same case for rectangular loaded areas.

The effect of non-homogeneity depends not only on the ratio E_0/k, but also on the diameter of the test plate or footing, and thus serious over-estimates of settlement will be made by using the average value of E determined by Equation 2 from the results of a plate test if the rock mass is strongly non-homogeneous and this factor is not taken into account. It is not possible to indicate ranges of error since the problem can only be studied numerically. The effect however is to cause the contributions to settlement to be concentrated beneath the plate on the Bunter sandstone, as reportded by Moore and Jones (1974) in their paper to this conference. *Fig. 2* shows the influence values for the mean surface settlement S of a uniformly loaded area on a homogeneous semi-infinite medium plotted against depth in terms of the radius R for various values of υ. The heavy line however refers to the non-homogeneous case for $\upsilon = 0.3$ and

$$E = E_0 (1 + z/2R).$$

Comparing the influence value at $z = 6R$ with that for the homogeneous case reveals a discrepancy of over 50%. The curve indicates that about 90% of the settlement of a one metre diameter plate would occur within one diameter of depth, which is the figure reported by Moore and Jones for the Bunter sandstone at Daresbury. In this case the modulus characteristic however is not wholly an independent one, since Moore (1973) has obtained results from the same test which show that as the stress on the plate increases the distribution of vertical strain approaches that for a homogeneous rock; a circumstance which may be due to the effects of bedding immediately beneath the

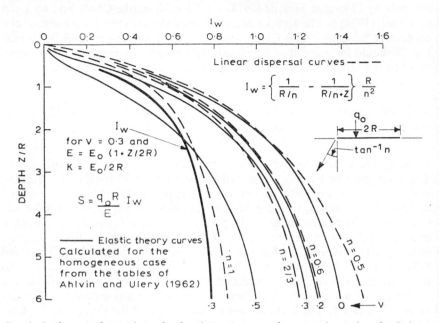

Fig. 2. *Settlement of a rigid circular foundation (mean settlement under uniform load) due to compression within a depth Z in a semi-infinite elastic medium*

plate. Some Bunter sandstones even in the intact state are non-linear, whence variation of modulus with depth can be induced by overburden pressure and by applied pressure (Hobbs, 1973). Thus we may talk of *induced inhomogeneity* in contrast with stress independent non-homogeneity. Problems involving induced inhomogeneity can only be studied numerically.

A rigid base underlying an elastic layer will result in under-estimates of settlement if the effect of the rigid base on the stresses in the layer is not taken into account. This however, it must be admitted, is more easily said than done; because not only is the behaviour sensitive to v, but the influence factors for surface displacement given in the literature show such startling variation that great caution is necessary here. *Figure 3* gives six interpretations for the settlement of a rigid circular area or mean settlement of a uniformly loaded area for $v = 0.5$, and *Fig. 4* some results for lower values of v, appropriate to rock. Burland (1974) has estimated the modulus of the chalk immediately beneath the building at Reading from the settlement observations, considering a rigid

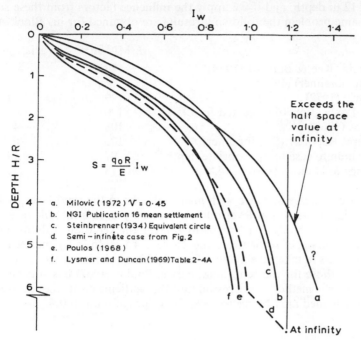

Fig. 3. Settlement of a rigid circular foundation on an elastic layer of thickness H. Poisson's ratio 0.5

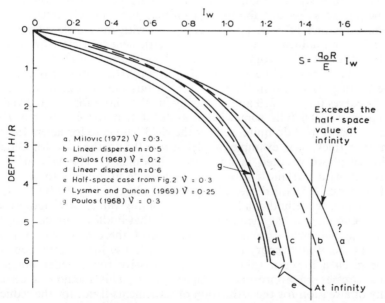

Fig. 4. Settlement of a rigid circular foundation on an elastic layer of thickness H. Poisson's ratio 0.2 to 0.3

585

base at 12 m depth, and if we apply the influence factors from these sources to the same problem the following results are obtained for his Block A.

	E MN/m^2	Variation %
Burland, Kee & Burford (1974) (Steinbrenner) (1934)	113	—
Milovic (1972)	129	+14
NGI Publication 16 (adjusted for υ)	120	+6
Poulos (1968)	108	−4
General Reporter (Steinbrenner)	106	−6
Semi-infinite case to appropriate depth	97	−14
Lysmer & Duncan (1969) 2–4A	78	−31

The discrepancies increase with greater depths to the rigid layer, but after about 3 diameters would level out. An anomaly occurs in Milovic's results which suggest that the settlement in a layer thicker than 2 diameters will exceed the half-space settlement, and Table 2–4A given in Lysmer & Duncan (1969) appears to take no account of the rigid base, the influence factors being lower than those for the semi-infinite case. Poulos (1967) has suggested that Steinbrenner's method underestimates the settlement at the centre of a uniformly loaded area by up to 15% for values of υ up to 0.4.

Simplified methods. The errors briefly considered above all arise within the bounds of the elastic theory, that is to say making one assumption when another would be nearer the mark; whereas rock, as we discover from our tests, is not an elastic material. It is worthwhile therefore seeing how a simplified method of stress distribution, which may more closely reflect the discontinuous nature of rock masses, compares with the elastic distributions. *Figure 2* shows influence factors of mean settlement of a circular area prepared from the elastic theory for various values of υ and also four curves prepared on the basis of a linear dispersal of mean vertical stress from the edge of the footing, as shown in the figure for various values of n. It will be seen that the curves for $n = 2/3$ and 0.6 neatly bracket those for $\upsilon = 0.2$ and 0.3. The curve $n = 0.5$ falls somewhat outside of that for $\upsilon = 0$, but *Fig. 4* indicates that it may do quite well for the rigid base case for layers less than 1 diameter thick. For example, applying $n = 0.5$ to the case for Block A considered above resulted in an average modulus of 110 MN/m^2 which agrees well with Burland's value. The best single fit appears to be $n = 0.6$ which is very close to the dispersal angle of 30° suggested by the Boston Code.

The simplified method can be used with varying values of E, requiring no tables or curves, and has the particular merit that it allows greater scope for engineering judgment through consideration of the clastic nature of rock than does the elastic theory (Trollope, 1968). Loose highly fractured rocks can be given a lower n value, and more massive rocks under horizontal ground stresses an appropriately higher value. This method is certainly worthy of attention for the estimation of settlements if not for the reduction of test results.

Deformability

The deformability of a rock mass depends upon three factors:
  (1) the deformability of the so-called intact rock contained within the discontinuities
  (2) the stiffness, frequency and incidence of the discontinuities
  (3) time

While the deformability of the intact rock and the effect of time can be measured, generally without difficulty, by tests on small specimens, that of a rock mass can only be determined by field tests suitably scaled to the fracture spacing or from observations on structures.

Intact rock

Intact rocks and other materials can be most conveniently compared by plotting compressive strength q_u against modulus E as proposed by Deere and Miller (1966). Lithologically similar materials tend to fall into diagonally distributed groups, in which large variations in strength and modulus are

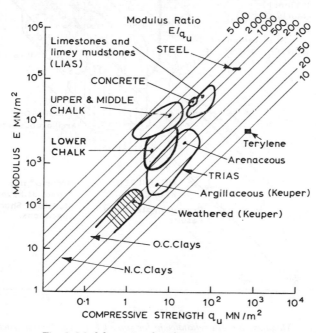

Fig. 5. Modulus ratio of civil engineering materials

contained within relatively narrow ranges of the modulus ratio $M = E/q_u$ Deere, in his more fundamental study used the tangent modulus at $\frac{1}{2}q_u$, but the modulus used here is the secant modulus based on customary bearing pressures rather than on the rock strength.

The rocks and soils fall into well defined groups, *Fig. 5*; the Chalk and the Trias extending over nearly 3 orders of modulus, and the weathered Keuper Marl over an additional order entering the realm of O–C clays which in turn

587

give way to the N–C clays. The calcitic rocks are exceptional in two respects, their extremely high modulus ratio and the variation of this ratio with strength and clay content. The high modulus ratio of the Chalk (Upper and Middle) arises from the nature of its porous structure, its somewhat isolated position in the chart being due as much to its low strength as to its high

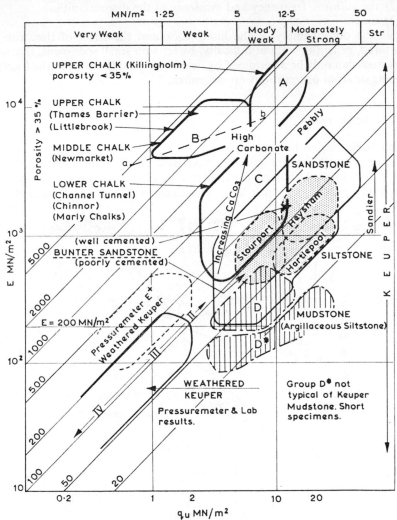

Fig. 6. Modulus strength relationship for some Chalk and Triassic rocks

modulus. Pores, as distinct from cracks, considerably reduce the strength of a material with relatively little effect on deformability. Cracks, on the other hand influence both the strength and modulus, the modulus being particularly affected by low stress levels, Walsh and Brace (1966).

The rocks of these groups based on some 500 tests from selected sites are shown in *Fig. 6** in more detail, the envelopes enclosing the greater majority

* These and other results have been taken from a paper under preparation on foundations on weak rocks.

of the results in each group. Results from other localities may expand the envelope areas, but generally they will tend to fall close to, if not within, the approximate lithological area, provided the rock specimens are tested in their natural condition by the correct procedures. Naturally saturated rocks should not be allowed to dry out as this generally increases the modulus and to a lesser extent the strength, Colback (1965). The results in *Fig. 6* are based on compression tests with strain gauge measurements on rock cores of various diameters ranging from 60 mm (*NX*) to 150 mm (*ZF*), with the exception of one group of tests on Keuper Mudstone marked *D*, and on the highly deformable weathered Keuper where measurement was across the end caps. A considerable amount of the data on the weathered Keuper obtained with the Menard pressuremeter (Hobbs and Dixon, 1970), and the rock strength classification by the Geological Society (1970) has been included in *Fig. 6* for

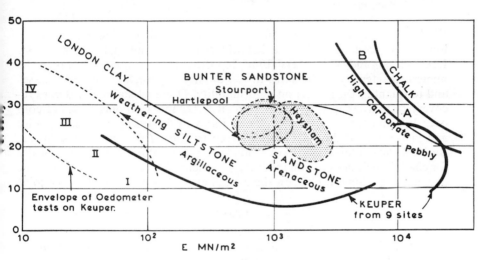

Fig. 7. Modulus porosity relationship for some Chalk and Triassic rocks

convenience. Further light is thrown on the nature and relationship of these rock groups to each other by comparing porosity with modulus as in *Fig. 7* where the London Clay has been included as a reference.

It is not possible to discuss these results in detail, they largely speak for themselves, but the following points should be noted:

(1) the connexion between calcium carbonate and modulus, which is admirably brought out, not only by the isolation of the Upper and Middle Chalk, but also by the results from the Lower Chalk, a high carbonate content corresponding with a high modulus and vice versa.

(2) the reasonably well defined relationship between porosity, strength and modulus in high carbonate Chalk compared with the Triassic sandstones of comparable porosity where no well defined relationship exists.

(3) the central position of the Bunter Sandstone in both figures, and the comparative uniformity of this rock over three widely separated sites.

(4) the extremely wide range in strength and modulus covered by the Keuper, both parameters increasing as the rock becomes more arenaceous.

(5) the low lower bound modulus ratio of the unweathered Keuper com-

pared with the weathered. This is probably due to the effects of stress relief and disturbance on the agglomerations of secondary peds (Chandler, Birch and Davis 1965) which comprise the mudstones and siltstones.

(6) weathering appears to have little effect on the modulus ratio of the Keuper Marl. There is some evidence that the pressuremeter modulus E_p in weathered Keuper Marl roughly corresponds with the modulus from triaxial tests on drill cored specimens after saturation and consolidation under the overburden pressure. This suggests that weathered argillaceous rocks properly belong in the realm of soil rather than rock mechanics.

Rock masses

Rock quality

While the influence of discontinuities on the behaviour of rock masses has long been recognised, vide the ICOLD proceedings and CP No. 4 (1954), Deere (1964) and with others (1966) reported the first attempt to relate the numerical intensity of fractures to the quality of unweathered rock masses and to quantify their effect on deformability. Deere uses a modified core recovery ratio, rock quality designation (R.Q.D.) in which only pieces of core over 100 mm long are counted, as an index of fracturing and rock quality and has proposed the classification given in *Table 1* for hard rocks. Ward, Burland and Gallois (1968) have produced a classification, *Table 2*, specifically for chalk based, not only on the fracture spacing *in situ*, but also on the joint opening and filling. R.Q.D. on the other hand takes no account of the joint

Table 1

Quality classification	R.Q.D. %	Fracture frequency per m	Velocity Index V_F^2/V_L^2	Mass Factor j
Very poor	0–25	15	0–0.2	0.2
Poor	25–50	15–8	0.2–0.4	0.2
Fair	50–75	8–5	0.4–0.6	0.2–0.5
Good	75–90	5–1	0.6–0.8	0.5–0.8
Excellent	90–100	1	0.8–1.0	0.8–1.0

(Deere and others, 1966)   (Coon and Merritt, 1970)

V_F the wave velocity in the field, V_L the velocity in the laboratory

Table 2. CLASSIFICATION OF CHALK (Ward, Burland and Gallois, 1968)

Grade	Brief description
V	Structureless remoulded chalk containing lumps of intact chalk
IV	Rubbly partly weathered chalk with bedding and jointing. Joints 10–60 mm apart, open up to 20 mm and often infilled with soft remoulded chalk and fragments
III	Rubbly to blocky unweathered chalk. Joints 60–200 mm apart, open up to 3 mm and sometimes infilled with fragments
II	Blocky medium hard chalk. Joints more than 200 mm apart and closed
I	As for Grade II but hard and brittle

opening and condition, a further drawback being that with fracture spacings greater than 100 mm the quality is excellent irrespective of the actual spacing. This particular difficulty can be overcome by using the fracture spacing or frequency rather than the R.Q.D. in rock quality studies, a requirement that is becoming common in the U.K. together with the determination of the point load strength of the cores (Broch and Franklin, 1972).

The rock mass factor j

The term rock mass factor is conveniently defined as the ratio of the deformability of the rock mass within any readily identifiable lithological and structural component to that of the deformability of the intact rock comprising the component and is intended to indicate in a simple manner the quality of the rock mass. It reflects the effect of discontinuities on the expected performance of the intact rock. The value of j will depend upon the method of measuring or assessing the deformability of the rock mass, and the value beneath an actual foundation will not necessarily be the same as that determined from even a large-scale field test particularly on the more massive rocks. The assessment of the rock quality in extent and depth requires exploration, testing and judgment and is important since the determination of the modulus profile is the first step in a settlement analysis. It is not proposed to discuss the methods of exploration here, but generally the plate loading test, suitably scaled to the fracture spacing and with the strain measured within the rock mass by means of a deep hole located through the centre of the plate as at Churchill Falls (Benson, Murphy and McCreath, 1970) and as at Daresbury (Moore, 1974), is regarded as the best method of determining the appropriate field modulus, although it is not always either practicable or economically feasible to do so. Deere (1966) presented test results based on large plates from a number of dam sites in which the reduction factor (mass factor j) was plotted against the R.Q.D. The envelopes of Deere's results have been replotted against fracture frequency in *Fig. 8* together with some reworked results, also on large plates, from three dam sites in Tasmania reported by Boughton (1968) in which an attempt was made to express the fracture condition as a joint weathering index, J.W.I.

The results from four Chalk sites are given in *Fig. 9*. All are based on large diameter plate tests with the exception of the Killingholme site where the pressure meter was used. The Littlebrook tests were carried out in an open shaft, which probably accounts for the excessively low results. The fracture frequencies have been related to Burland's Classification for convenience. Two points should be noted in both *Figs. 8* and *9*.

(1) the extremely rapid drop in j-value with the decrease in fracture spacing, followed by a gradual decline as the intensity of fracturing increases
(2) the very wide scatter which is due to the variation in openness, condition and filling (if any) in the fractures; the cleaner and tighter the fractures, the flatter and higher the curves. Boughton's results emphasise this point.

It will be seen from these figures and from the theoretical relationships given below that the greatest difficulties occur with fracture spacings within the

591

range of 10 to 50 cm approximately, in as much as small variations in fracture spacing and condition result in exceptionally large changes in the j-value. The role of water injection tests between packers in drillholes in the assessment of rock quality and deformability will be readily appreciated. High hydraulic conductivity in relation to the fracture frequency means correspondingly low j-values and vice versa. In simple joint systems, it is theoretically possible to relate joint opening to hydraulic conductivity, Louis and Maini (1970), but the application of the possibility awaits further study. Knill (1970) and

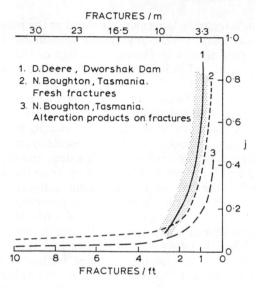

Fig. 8. Variation of mass factor with fracture frequency. Strong rock

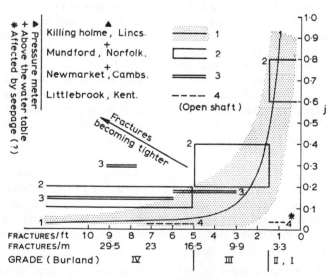

Fig. 9. Variation of mass factor and fracture frequency. Chalk

592

Bellier, Bernede, Bollo (1964) have reported broad correlations between hydraulic conductivity and the fracture index (defined by Knill as the ratio of the field wave velocity to the wave velocity of the intact rock) and seismic velocity respectively, both of which may be regarded as indications of rock mass quality. However, problems of interpretation limit quantification, e.g. the effect of moisture content on laboratory velocity and amplitude difference.

The high deformability of the lower grades of chalk is emphasised by these results, as is the difficulty of its determination. Reducing the modulus of the chalk in *Fig. 6* by 10 to 20 times puts these materials in a category little better than the more deformable Keuper and with a mass modulus ratio typical of the Trias. These chalks when conventionally investigated appear in a considerably worse light, their strength and rigidity being affected to such an extent that, as Burland (1974) has shown, incorrect judgments of their true condition can be easily made. This and other related questions have also been touched upon by Hobbs (1970a) in the discussion on the use of the S.P.T. in the estimation of the modulus of chalk by Wakeling (1970).

Walsh and Brace (1966) in a review on the elasticity of rock have given expressions for modified elastic constants derived from consideration of the effects of pores (more or less spherical cavities which do not close under load) and cracks (sharp narrow cavities which can be closed). Laboratory tests on intact specimens have generally confirmed the theoretical results, prompting the suggestion that joints and fractures may be regarded indicatively as cracks in larger masses of rock. Two limiting cases are considered; initially before loading with the cracks open, and finally with the cracks closed, changing conditions which produce the familiar non-linear stress–strain curve. Their relationships expressed as mass factors for fractures of mean length c in a volume v are given in *Fig. 10** where they have been plotted for both layered and block fractures taking μ as 0.65. The coefficient of internal friction, μ, enters the process since under stress the crack surface slide over each other, a cause of hysteresis on unloading. The block model usually is more typical of rock masses than the layered model.

Waldorf, Veltrop and Curtis (1963) suggested that partial contact between blocks would account for the low field modulus of apparently rigid rocks, since the stresses between the blocks would be transmitted by contact areas varying in size, shape and height. Pursuing this suggestion a very simple ideal model can be made in which uniform contacts are regarded as miniature plate loading tests. The resulting expression for j is easily obtained and is given in *Fig. 11* where, for various values of the diameter b and intensity per unit area n of the contacts, curves of j against the fracture frequency $1/f$ (layered) have been plotted taking v^2 as 0.1. It will be observed that as the total contact area, $\pi b^2 n/4$, per unit area increases the curves become flatter and less acute, a circumstance which is consistent with decreasing permeability. The remarkable similarity between the curves of *Fig. 10 and 11* and the field curves of *Fig. 8 and 9* suggests that even apparently tight fractures are not in perfect contact. Benson (loc. cit.) considers that alteration products occur on apparently tight joints. Both theoretical approaches account for the

---

* A. Malone (1968) has given a theoretical relationship between the velocity index V_F^2/V_L^2 and the fracture intensity which produces a series of curves similar to those in Figs. 8 and 11, thus justifying theoretically the association of the velocity index with the mass factor j. See below and Table 1.

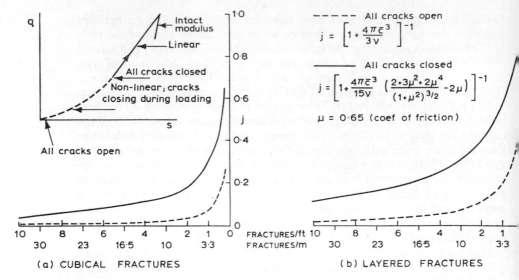

Fig. 10. Variation of mass factor with fracture frequency. (Walsh and Brace 1966)

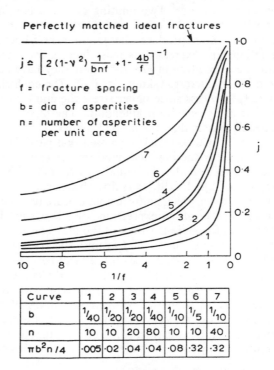

Fig. 11. Variation of mass factor with fracture frequency. Contact model

594

non-linearity of rock masses in that the j-value is clearly stress dependent, thus even apparently uniform rock masses may be expected to be inhomogeneous, that is fractures at depth will be 'stiffer' than similar fractures nearer the surface, a point to be borne in mind when assessing deformability.

Yet a further simple model envisages the fractures having a stiffness K (force per unit volume) attributable to poor contact in fresh rock, to the resistance of cracks to closure, to surface alteration products, infilling or to thin compressible layers where $K = (m_v . t)^{-1}$. If the fracture or layer spacing is f,

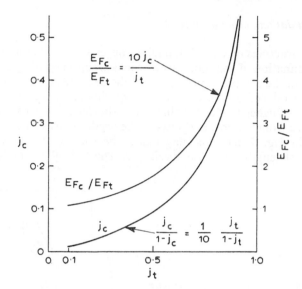

Fig. 12. Variation of j_c (Chalk) and E_{Fc}/E_{Ft} with j_t (Trias) for $E_c = 10\ E_t$

then using a similar approach to the 'contact' model the mass factor is found to be

$$ j = \left(1 + \frac{E}{Kf}\right)^{-1} \tag{3} $$

in rock having an intact modulus E, an expression which will produce similar curves to those given above for different values of K. This equation is interesting because, containing the intact modulus E, it shows that in very rigid rocks (high E) the fractures assume a much more important role in the deformability of the mass than in less rigid rocks. This is demonstrated theoretically in *Fig. 12* where the mass factors j_c and j_t are plotted against each other for two rock masses having comparable fracture spacings and fracture stiffnesses and in which the intact modulus $E_c = 10E_t$, the typical ratio for Chalk to Trias. The mass factor j_c falls from about 0.5 to 0.1 of j_t as the fracture stiffness decreases. In view of its higher E value however (1 order), the mass modulus of the Chalk E_{fc} is never less than that of the corresponding Trias mass modulus E_{ft} as can be seen from *Fig. 12*, and generally it is considerably higher. These observations only apply to the better grades of chalk since the fracture stiffnesses in the poorer chalks are very much lower than those likely to be found in weaker Triassic rocks. Nevertheless it would seem that chalk may deserve a better reputation than hitherto. The papers by Lake and Simons (1974) and Burland (1974) support this view. The greater tendency of calcitic rocks to creep should not be overlooked and thus in the longer term, all other things being equal, chalk masses may tend to settle rather more than other rocks of comparable strength.

Modulus measurement

Uniaxial and oedometer tests

The effect of fractures on overall modulus suggests a possible source of error in the determination of the modulus of rocks when measuring strain across the end caps in the compression test and in the oedometer, since fractures are artificially introduced at the ends of the specimen. Longer test specimens would produce higher modulus values than shorter specimens, and there is some evidence from the group *D* tests in the argillaceous siltstones from one site shown in *Fig. 6* that this may indeed be so. *Fig. 13* shows modulus plotted against length for these specimens. The fall off in modulus in the

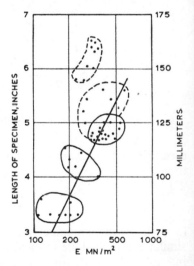

Fig. 13. Modulus against length of Specimen Keuper Mudstone. Tested across end caps

150 mm specimens is not understood, but may be due to the increasing effect of natural weaknesses as the specimens get bigger, rather as in soil testing, or to the greater difficulty in obtaining plane surfaces at the larger diameters.

The oedometer is particularly susceptible to these effects as the 'fracture frequency' is many times higher than in the triaxial test, equivalent to 50 per metre in the standard oedometer. The report to CIRIA on the compressibility of the Keuper Marl by Davis (1970) indicated a very broad relationship between compressibility and water content for the various weathering zones based on an admittedly limited number of oedometer tests. The envelope covering these results and others from the files of Soil Mechanics Limited is plotted in *Fig.* 7 using the reciprocal of m_v as the modulus. Agreement between the envelopes is not good for Zones I & II, *Table 3,* but improves rapidly with increase in weathering. It would seem therefore that testing un-weathered and lightly weathered Keuper Marl in the oedometer leads to excessively high compressibilities, due partly to the proportionally greater disturbance in the preparation of these specimens compared with the very much longer specimens for modulus testing, and partly to the question of specimen thickness. For instance, using the curve for layered fractures in the

Table 3. WEATHERING SCHEME FOR KEUPER MARL (after A. W. Skempton and A. G. Davis) (Chandler, 1969)

Zone		Description	Notes
Fully weathered	IVb	Matrix only	Can be confused with soli-fluction or drift deposits, but contains no pebbles. Plastic, slightly silty clay. May be fissured
Partially weathered	IVa	Matrix with occasional clay-stone pellets, less than ⅛ in. diameter but more usually coarse sand size	Little or no trace of original (zone 1) structure, although clay may be fissured. Lower permeability than underlying layers
	III	Matrix with frequent lithorelicts up to 1 in. As weathering pro-gresses lithorelicts become less angular	Water content of matrix greater than that of lithorelicts
	II	Angular blocks of unweathered marl with virtually no matrix	Spheroidal weathering. Matrix starting to encroach along joints; first indications of chemical weathering
Un-weathered	I	Mudstone (often fissured)	Water content varies due to depositional variations

closed state from *Fig. 10,* a fracture frequency of 50 per metre gives a *j*-value of about 0.1. We note from *Fig. 7* that the oedometer modulus values of the Zone 1 Keuper specimens are about 10% (1 order) of the undrained values obtained by strain gauge measurements on siltstones of the same porosity. While the two tests are not strictly comparable, comparisons of this sort are not altogether inadmissible particularly when the discrepancies are so large. It would seem on both theoretical and practical grounds that testing relatively rigid rocks by making measurements across end caps will lead to large under-estimates of modulus. When the intact material is highly deformable as in highly weathered rocks the end cap factor loses its significance.

597

Rate of testing

In commercial practice standard tests taking 10 to 20 minutes per cycle are regarded as giving the drained modulus on porous coarser grained rocks. The fine grained Triassic siltstones however do not drain rapidly and it is necessary to extend the test in the working stress range over a period of hours to attain an apparently drained condition, that is to say a modulus which is not further reduced by extending the test period. Some 20 tests on Keuper siltstones have indicated apparently drained to undrained modulus ratios varying from 0.65 to 1 with an average of about 0.8. Slow uniaxial tests enable drained modulus and Poisson's ratio values to be determined and thus avoid the problem of a separate determination of the consolidation contribution to the settlement. Creep testing is another matter but is beyond the scope of this report.

Settlement prediction—the extrapolation method

This method is based on loading tests with one or a number of plates at the proposed foundation level and requires a good description of the ground beneath the structure accompanied by either a qualitative or quantitative profile of some relevant mechanical property. This is essential to justify the application of the method to the site. In sands and weathered rocks the S.P.T. is most generally used. Lake and Simons and Lord and Nash have used this procedure on the Chalk and weathered Keuper respectively. In rocks, ideally such information would comprise lithology, weathering, strength, fracture spacing and permeability along the drill hole.

Scaling the results directly from the plate tests to the foundation or obtaining the modulus from the tests and using this in the calculation of settlement amount in fact to the same thing and imply homogeneity and linearity in the rock mass. More generally it is sometimes assumed that

$$\frac{S_f}{S_p} = \left(\frac{B_f}{B_p}\right)^\alpha \tag{4}$$

where α is determined from tests with different size plates. *Figure 14* indicates however that except at values of α approaching 1 for relatively small foundations Equation 4 does not adequately model the condition of steadily increasing modulus with depth, the only condition apart from homogeneity in which this procedure can be realistically applied. The heavy lines in the figure are based on the modulus profile

$$E = E_0 + kz \tag{5}$$

for various ratios of the surface modulus E_0 to the increment term k, ranging from the completely non-homogeneous case, $E_0 = 0$, to the completely homogeneous case, $k = 0$, which corresponds to $\alpha = 1$ in Equation 4. These curves have been prepared from results given by Carrier and Christian (1973) and enable plate tests to be rationally used in extrapolation in those cases where the ground has been shown by some other method to be reasonably uniform in depth, i.e. uniform E or uniform k.

Lake and Simon's results, *Fig. 14*, suggest that the chalk at their Basingstoke site is very nearly homogeneous. Using the large plate as a datum $E_0/k = 5$ m and the modulus of the grade III Chalk works out at

$$E = 97 + 19z \text{ MN/m}^2, \quad \text{say} \quad 100 + 20z \text{ MN/m}^2 \tag{6}$$

compared with an equivalent E_e of 135 MN/m². These results are discussed further below. By contrast the modulus of the grade IV Chalk at Mundford immediately after loading increases much more rapidly according approximately to the relationship

$$E = 360 + 147z \text{ MN/m}^2$$

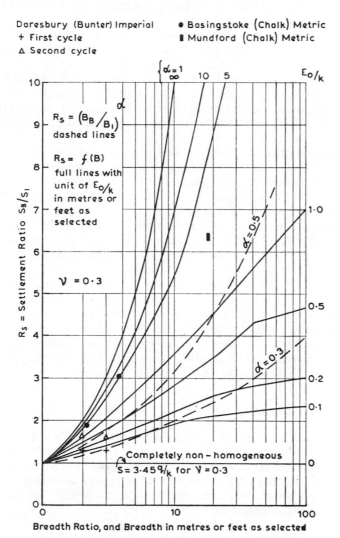

Fig. 14. *Extrapolation of plate bearing tests and foundation performance on a non-homogeneous half-space*

599

Keuper Marl

Lord and Nash (1974) have attempted a similar procedure in the Keuper at Loughborough, but the results given in their *Fig. 6* indicate that extrapolation from plate tests on the weathered Keuper can result in considerable underestimates of settlement. As it is unlikely that the ground becomes generally weaker with depth the reason for the large settlements of the foundations must lie in the effects of excavation beneath the water table, relief, uplift and consolidation. The Keuper frequently contains thin bands of more sandy and thus more permeable material as well as sandstone layers, and water pressure trapped in these layers can result in imperceptible though significant uplift and weakening. The upper sandstone layer at the Loughborough site lies some 10–15 m beneath the site and may not necessarily have been the cause of the unexpectedly high settlement.

Limestone bands in the Lias mudstone can also cause uplift and slight dislocation of the mudstone rock if the water pressure resulting from a deep excavation is not relieved—as has recently occurred beneath a reactor foundation (Haydon 1973). It would seem that the extrapolation approach requires unusually careful exploration, particularly regarding ground water, and that adequate allowance should be made for the effects of construction procedures if settlements are to be assessed with reasonable accuracy, particularly in the Keuper and Lias.

The question of consolidation in weathered Keuper is important, and it is possible that had some allowance been made for consolidation better agreement with measured settlements would have been obtained. Deeper drainage layers which would not affect the plate tests could have a considerable influence on the rate of settlement of the structure. By contrast at three motorway bridges the measured settlements were considerably less than the estimated settlements (CIRIA, 1965), and in a later estimate somewhat lower in the case of one bridge at Loughborough (Davis 1970). It was found here that the estimated consolidation component was twice the immediate component. Consolidation is thus an important factor and should be considered even if the settlement of the plate occurs rapidly. Davis reported that 90% of the settlement occurred within 10 minutes, but it seems that this could have been "delayed elastic" compression rather than consolidation. Certainly Davis found it necessary to include consolidation in the estimate in order to obtain reasonable agreement (12%) with the measured settlement. The drained modulus from the observations on this bridge worked out at about 90 MN/m².

Bunter Sandstone

Moore and Jones (1974) have not reported the first series of plate tests at Daresbury in detail as they considered the tests severely underestimated the modulus of the Bunter sandstone, but the results (Soil Mechanics Ltd 1964), based on a 1 ft plate, have been plotted in *Fig. 14* following the method proposed by Carrier. While there is considerable scatter the modulus expressions for the extremes are

$$E = 55 + 1820z \text{ MN/m}^2 \tag{7}$$

and

$$E = 255 + 840z \text{ MN/m}^2 \tag{8}$$

The estimated mean settlement of a circular area equivalent to the area of the experimental hall from these expressions works out at 0.21 and 0.33 mm respectively, compared with an estimated corner settlement of 5 mm based upon the same plate tests but treating the rock as homogeneous. The measured settlement at the centre of the area is just under 2 mm. There is thus a range of error of -1.8 to $+3$ mm, that is a 900% under estimate to a 150% over estimate, ignoring the differences between mean, centre and corner. With discrepancies of this magnitude insistence on theoretical accuracy in stress distribution is trivial.

The large plate test (Moore, 1974) with vertical strain measurement at depth can be analysed by taking the non-Boussinesq distribution of vertical strain into account. Very approximately the modulus expression for this condition at a stress level of 1500 kN/m⁵ works out at

$$E = 635 + 700z \text{ MN/m}^2 \tag{9}$$

leading to a mean settlement of the equivalent area of 0.3 mm, very close to the values deduced from Equations 7 and 8. Admittedly the conditions at Daresbury are not simple, nevertheless it seems that the non-homogeneous approach grossly under-estimates the settlement as the homogeneous assumption has over-estimated it. Evidently the modulus does not increase as rapidly as the above equations suggest, for Moore has fitted to the observations at Daresbury an E_0 of zero with a k value of only 78 kN/m³ for 13 m, thereafter falling to zero. This is somewhat arbitrary, since E_0 cannot be zero in a cemented rock, as proved by the first series of plate tests where the plates did not all have the same coefficients of subgrade reaction, *Fig. 14*, a theoretical requirement of a purely non-homogeneous medium. Moore accounted for the larger settlement of the experimental hall by making the rock homogeneous below 13 m. No doubt some other combination of E_0 and k would also satisfy the observations. A further complication is that the plate test result refers to a stress level 15 times higher than the prototype loading. Nevertheless the Daresbury results highlight the difficulties in predicting settlements from plate tests on these sandstones even when the load-settlement curves are apparently linear. Large hysteresis occurred on unloading, and perhaps this should serve as a warning of difficulties ahead in these deceptively uniform rocks.

Evidently care is necessary in applying the non-homogeneous approach to stress dependent rocks, i.e. rocks susceptible to 'induced' non-homogeneity as distinct from the 'natural' non-homogeneity considered by Gibson and by Carrier. Carrying out single size plate tests and pressuremeter tests at different depths and also laboratory tests on samples from different depths would readily indicate whether any non-homogeneity is induced or natural. In the Bunter at Heysham k values of 35 to 55 MN/m³ and 20 to 35 MN/m³ were obtained from pressuremeter tests and laboratory tests respectively indicating a modest degree of natural non-homogeneity in this non-linear sandstone. The problems of interpreting plate tests on the Bunter sandstone at Heysham have been discussed by Hobbs (1973).

The modulus profile method

The second and preferable approach requires the assessment, if not the determination, of the variation of the modulus with depth. The settlement can then be rapidly determined by the usual summation procedure, Equation 1, using the appropriate distribution of vertical stress from either the classical results or from the simplified linear dispersal method. It can be fairly assumed that for engineering purposes the vertical stresses will not be significantly affected by the departures found in nature from the classical assumptions of linearity, isotropy and homogeneity (Hobbs, 1973).

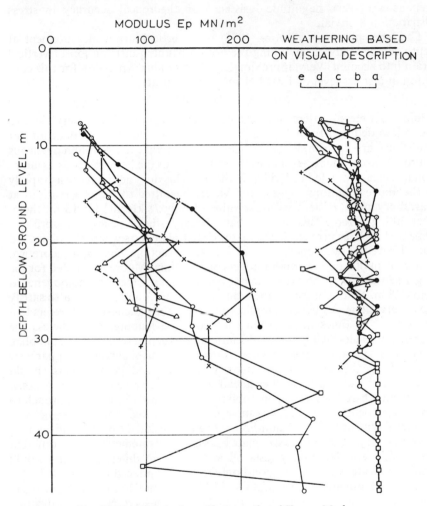

Fig. 15. Modulus–depth profile in weathered Keuper Marl

The profile method does not, like the extrapolation method, require the modulus to follow a uniform trend which can be detected by plates of various diameters. Provided the rock is not excessively fractured and is not significantly affected by the effects of stress relief the intact modulus profile can be determined by testing small specimens in the laboratory. To obtain the field modulus profile it is necessary to reduce the intact modulus by the mass factor j assessed from the condition of the core; but this is more easily said than done since the rock, particularly if laminated, can be fractured by drilling, and in any case the openness of natural fractures cannot be judged in the core box. Packer tests can be of great assistance in providing a qualitative guide to the rock condition in those cases where the rock cannot be physically examined *in situ*. The modulus at various depths can be measured directly by plate tests above the water table and by pressuremeter tests either above or below the water table. Plate tests carried out at depth below the water table without dewatering by deep wells will result in serious underestimates of the modulus (Hobbs, 1970a). The investigation of the Mundford site (Burland and Lord, 1970) is a good example of the modulus profile method applied to chalk. *Figure 15* gives an example of a modulus profile determined by pressuremeter in weathered Keuper Marl.

A method involving acoustic logging of boreholes is receiving increasing attention (Geyer and Myung, 1970). Knill (1970) has proposed that the ratio of the longitudinal wave velocity to the wave velocity of the corresponding core specimen determined in the laboratory may be regarded as an index of fracturing. Coon and Merritt (1970) have gone further than this and propose virtually a one to one correspondence between the velocity ratio and the mass factor *j*, *Table 1*. The difficulty here is that the wave velocity is a function of pressure, degree of saturation in the pore space and the presence or absence of water in the fissures. Nevertheless development of this method should be pursued since it would mitigate the problem of assessing the mass factor of excessively fractured core.

None of the papers in Session 4 has been directly based upon the profile method, although Burland assessed average values for the modulus of the low-grade Chalk supporting a building at Reading from standard penetration tests. The values were confirmed by the observations on the raft foundations. It would be of interest to conclude the discussion on the Chalk by briefly comparing the results from Basingstoke with those from Reading.

Chalk

The S.P.T. values at Reading indicate throughout the 12 m bed a fairly steady improvement from grade VI to grade IV Chalk—clearly a non-homogeneous condition which cannot be quantified from the settlement observations alone. It is reasonable to assume, however, since the stress level is low, that the k value will be lower than that at Basingstoke. According a value of 10 MN/m³ to k, the modulus expression to produce the observed settlement under Block A works out at about

$$E = 65 + 10z \text{ MN/m}^2 \tag{10}$$

having a mean value of 125 MN/m². This expression, although somewhat arbitrary, gives a more realistic picture of the ground than the equivalent homogeneous modulus E_e of 113 MN/m². Had the stress level been higher E_0 would have been smaller, k higher, and E_e smaller. Using this expression modulus values for the grades of Chalk indicated in Burland's *Fig. 2* can be estimated. These indicative results are given in *Table 4* together with values for the Basingstoke Chalk (Lake and Simons) deduced from Equation 6,

603

Table 4. SUMMARY OF RESULTS IN UPPER CHALK

Site	*READING (Block A)* *(High Water Table)*				*BASINGSTOKE* *(Low Water Table)*			
Equivalent modulus (MN/m²)	113 (Authors, 125 (General Report))				100 (Authors), 135 (General Report)			
Applied stress (MN/m²)	0.057				0.365			
Modulus expression (MN/m²)	$E = 65 + 10z$				$E = 100 + 20z$			
Chalk grade	VI	V	V	IV	IV	III	III	II
Depth below foundation (m)	1	3.5	6.5	10	0	2	4	6
Average stress increment, q_a (MN/m²)	0.05	0.037	0.027	0.02	0.365	0.122	0.061	0.0365
Deduced modulus E, (MN/m²)	75	100	130	165	100	140	180	220
'Guessed' intact, Modulus (see *Fig. 6*) (MN/m²)	2000	2500	2500	3000	3500	3500	3500	3500
Mass factor *j*, purely indicative	0.04	0.04	0.05	0.05	0.03	0.04	0.05	0.06
E/q_a	1500	2600	4700	8000	280	1100	3000	6000

While it may appear from these results that the grade III Chalk at Basingstoke is little better than the grade IV to VI Chalk at Reading, this circumstance is in fact entirely due to the much higher foundation pressure used at Basingstoke, 0.365 MN/m² as against 0.057. Even so it must be admitted that the deduced values for the grade III Chalk are low, particularly at depth where the stress increments are small. The modulus of the grade III Chalk at Mundford under comparable stress levels is 10 times higher than at Basingstoke. Lake, unlike Burland, made no allowance for initial bedding down of the foundations; allowance for loosening of the formation would result in higher estimated modulus values at depth.

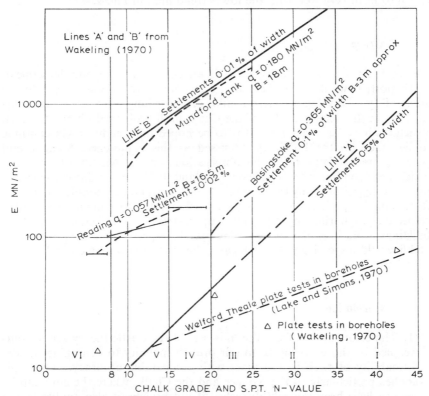

Fig. 16. *Deduced field modulus against Chalk grade and degree of settlement*

These results and those for Mundford are plotted in *Fig. 16* in the form originally given by Wakeling (1970). While the moduli lie considerably above the line 'A' it should be remembered that, being stress-dependent, a particular modulus cannot be generally associated with a particular grade of chalk. It is also quite clear that the modulus profile of chalk cannot be determined by carrying out small-diameter plate tests in boreholes, particularly below the water table.

An important object in plate testing is to establish the initial yield point of the Chalk q_e and the intercept q_y (*Fig. 1*), since it appears that the settlements will be small provided the foundation pressure does not approach too closely to q_y, determined with a large plate. Beyond this stress settlements will tend to increase rapidly, and conversely below this stress level settlements will be smaller. The buildings at Basingstoke and Reading on footings and rafts respectively are in no way regarded as exceptional, although the groundwater conditions were very different. The average modulus for the building on piles at Basingstoke deduced using over simplified assumptions (Kee, Parker and Wehrle, 1974) is very similar to that obtained by Lake and Simons under comparable stresses. Piles may not have been necessary under this 12-storey building.

Values of the mass factor j have been estimated using 'guessed' values of the intact modulus, *Table 4*. Although the values appear to be low they will be found to fit in very well with the lower trend lines in *Fig. 9*.

Creep

Although discussion on this aspect of Chalk behaviour is beyond the scope of this report, it would nevertheless be appropriate to draw attention to the advisability of making a generous allowance for long term settlements particularly in the weaker grades of the Chalk until prolonged observations on structures have been reported. It is to be hoped that readings will continue to be made on the buildings discussed at this conference. Some useful pointers are available from the observations on the Mundford test tank which were continued for one year, (Burland 1973). If the linear trends are projected logarithmically to 10 000 days (say 30 years) the modulus reduction factors in the grade III and IV Chalk work out at very roughly 0.5 and 0.2 respectively. Ignoring any dependence upon stress level the settlements of the buildings at Basingstoke and Reading may be expected to increase two-fold and five-fold respectively. In both cases the long-term settlement will be less than 1 in.

Conclusions

(1) The errors arising from the 'mis-use' of the elastic theory are far outweighed by those arising from our present inability to determine the deformability of rock in the mass with comparable accuracy. While this applies particularly to the Keuper and the Chalk, where the core can be seen to have been severely disturbed if not broken, it also applies to the Bunter and to other rocks where the core may frequently be recovered in solid sticks.

(2) Hitherto, rocks have always been regarded as homogeneous for the purpose of reducing the results of plate loading tests even when the ground is known to be not so. This can lead to significant overestimates of settlement, as can the assumption of linearity in non-linear rocks. Methods are now readily available for the rational analysis of loading tests carried out in series and for the estimation of the settlements in non-homogeneous rock masses.

(3) The Chalk is an exceptional rock in as much as it possesses an extremely high modulus ratio in the pure state. Increasing clay content, as in the Lower Chalk, causes a rapid reduction in modulus with relatively little

change in strength. Generally the Upper and Middle Chalks are one order higher in laboratory modulus than the Triassic rocks although they have comparable strengths.

(4) Chalk masses, like the Keuper and Bunter, are non-homogeneous, the extent showing considerable variation between the few cases so far investigated on the Middle and Upper Chalk. The non-homogeneity is not wholly natural or intrinsic, but in the poorer grades is partly induced by the magnitude and extent of the applied load.

(5) Foundations on Chalk so designed that the applied pressure does not too closely approach the intercept q_y determined from a large plate loading test are unlikely to settle excessively even in the longer term. The poorer grades of Chalk are far less deformable than hitherto suspected.

(6) Surface plate loading tests appear to be a useful method for investigating the deformability of Chalk masses above the water table where the modulus increases moderately with depth. Where the grade improves sharply deep tests may be necessary to assess the modulus profile. Determination of the deformability of chalk below the water table, particularly when flints are present, remains a serious problem.

(7) The assessment of the deformability of the Keuper Marl and Siltstone presents one of the most pressing problems in foundation engineering today. Plate loading tests can be misleading and it may be necessary to allow for consolidation when estimating settlements from these tests. The fresh rock, although recovered in an apparently undisturbed state, may be so disturbed by the effects of stress relief on the ped-structure that the modulus is under-estimated in compression tests.

(8) Generally the modulus (and j-value) of the Bunter Sandstone is sufficiently high to ensure that settlement is unlikely to be a problem except in the case of very special structures where close estimates of settlement are required.

Acknowledgments

Acknowledgment is made to the Central Electricity Generating Board, to the Department of the Environment, to the Tees and Hartlepools Harbour Authority, to Allott and Lomax, Consulting Engineers, Mott Hay and Anderson, Consulting Engineers, to Rendel, Palmer and Tritton, Consulting Engineers and to Sir William Halcrow and Partners, Consulting Engineers, also to George Wimpey Limited, Dredging Investigations Limited and to Soil Mechanics Limited for permission to refer to the results published in this report.

REFERENCES

Ahlvin, R. G. and Ulery, H. H. (1962). 'Tabulated Values for Determining the Complete Pattern of Stresses, Strains and Deflections Beneath a Uniform Circular Load on a Homogeneous Half-space', *H.R.B. Bull.*, 342

Barden, L. (1963). 'Stresses and Displacements in a Cross-anisotropic Soil', *Geotechnique*, London, Vol. 13, 198–210

Bellier, Bernede, J. and Bollo (1964). 'La Deformabilite des Massifs Rocheux et Comparison des Resultats', *Proc. 8th I.C.O.L.D., Q28*, Edinburgh

Benson, R. P. Murphy, D. K. and McCreath, D. R. (1970). 'Modulus Testing of Rock at the Churchill Falls Underground Powerhouse, Labrador', *A.S.T.M., STP* 477: 89–116

Boughton, N. O. (1968). 'Correlation of Measured Foundation Modulus with In situ Rock Properties', *Proc. Int. Symp. Rock Mech.*, Madrid; Blume

Boussinesq, J. (1885). *Applications des Potentials*, p. 329, Paris

Broche, E. and Franklin, J. A. (1972). 'The Point-load Strength Test', *Int. J. Rock Mech. Min. Sci.*, Vol. 9, 669–697

Brown, P. T. (1969). 'Uniformly Loaded Rafts on Elastic Foundations', *Geotechnique*, London, Vol. 19, 3: 399–404

Brown, P. T. and Gibson, R. E. (1972). 'Surface Settlement of a Deep Elastic Stratum Whose Modulus Increases Linearly with Depth', *Geotech. J.*, Vol. 9, 467–476

Brown, P. T. and Gibson, R. E. (1974). 'Rectangular Loads on Inhomogeneous Elastic Soil'. To be published by *A.S.C.E.*

Burland, J. B. and Lord, J. A. (1970). 'The Load Deformation Behaviour of Middle Chalk at Mundford, Norfolk: A Comparison Between Full-scale Performance and In Situ and Laboratory Measurements'. *Proc. Conf. In Situ Investigations in Soils and Rocks*, London, British Geotechnical Society

Burland, J. B. and Hobbs, N. B. (1970a). Discussion on Papers in Session A, *Proc. Conf. In Situ Investigations in Soils and Rocks*, London, p. 62, British Geotechnical Society

Burland, J. B. (1973). Private Communication

Burland, J. B., Sills, G. C. and Gibson, R. E. (1973). 'A Field and Theoretical Study of the Influence of Non-homogeneity on Settlement', *Proc. 8th Conf. Int. Soc. Soil Mech. Found. Engng*, Moscow

Burland, J. B., Kee, R. and Burford, D. (1974). 'Short-term Settlement of a Five-storey Building on Soft Chalk. *Proc. Conf. Settlement of Structures*, Cambridge

Carrier, W. D. and Christian, J. T. (1973). 'Rigid Circular Plate Resting on a Non-homogeneous Elastic Half Space', *Geotechnique*, London, Vol. 23, 67–84

CHANDLER, R. J. (1969). 'The Effect of Weathering on the Shear Strength Properties of Keuper Marl', *Geotechnique*, London, Vol. 21, No. 3, 321–334

Chandler, R. J. (1967). 'Shear Strength Properties of Keuper Marl', *Ph.D. Thesis*, University of Birmingham.

Chandler, R. J., Birch N. and Davis, A. G. (1965). 'Engineering Properties of Keuper Marl', *Res. Rep. 13, C.I.R.I.A.*, London

C.I.R.I.A. (1965). 'Engineering Properties of Keuper Marl', *Research Report 13*, London

Colback, P. S. B., (1965). 'Influence of Moisture Content on the Compressive Strength of Rock', *Proc. 3rd Canadian Symp. Rock Mech.*

Coon, R. F. and Merritt, A. H. (1970). 'Predicting In Situ Modulus of Deformation Using Rock Quality Indexes', *A.S.T.M.*, STP 477, 154–173

Davis, A. G. (1970a). 'The Compressibility of Keuper Marl', *Research Report*, London

Davis, A. G. (1970). 'The analysis of the observed settlement of a multi-storey block of flats at Highgate, Birmingham', *Research Report, C.I.R.I.A.*, London

Deere, D. U. and Miller, R. P. (1966). Classification and Index Properties for Intact Rock,' *Technical Report* AFWL-TR-65-116, U.S.A.

Deere, D. U., Hendron, A. J., Patton, F. D. and Cording, E. J. (1966). 'Design of Surface and Near Surface Construction in Rock', *Proc. 8th Symp. Rock Mech.*, Minnesota (A.I.M.E., 1967)

Deere, D. U. (1964). 'Technical Description of Rock Cores for Engineering Purposes', *Rock Mech. and Engng Geol.* I, 1, 17–22

Geological Society (1970). 'Logging of Rock Cores for Engineering Purposes', *Q. J. Engng Geol.*, London, Vol. 3, No. 1

Gerrard, C. M. and Harrison, W. J. (1970). 'The Effect of Inclined Planar Fabric Features on the Behaviour of a Loaded Rock Mass', *Proc. 2nd Conf. Int. Soc. Rock Mech.*, Beograd, Vol. 2, 20

Geyer, R. L. and Myung, J. I. (1970). The 3-D Velocity Log for the *In Situ* Determination of the Elastic Moduli of Rocks', *12th Symp. Rock Mech.*, Univ. of Missouri, Rolla

Gibson, R. E. and Sills, G. C. (1971). 'Some Results Concerning the Plane Deformation of a Non-Homogeneous Elastic Half-space', *Proc. Roscoe Memorial Symp. on Stress-Strain Behaviour of Soils*, Cambridge

Gibson, R. E. (1974). Private communication

Haydon, R. E. V. (1973). Private communication

Hobbs, N. B. and Dixon, J. C. (1970). 'In Situ Testing for Bridge Foundations in Devonian

608

Marl', *Proc. Conf. In Situ Investigations in Soils and Rocks*, London, British Geotechnical Society

Hobbs, N. B. (1970a). Discussion on Papers in Session A. *Proc. Conf. In Situ Investigations in Soils and Rocks*, London, British Geotechnical Society

Hobbs, N. B. (1973). 'Effects of Non-linearity on the Prediction of Settlements of Foundations on Rock', *Q. J. Engng Geol.*, London, Vol 6, No 2, 153–168

Kee, R. Parker, A. J. and Wehrle, J. E. C. (1974). 'Settlement of a 12-Storey Building on Piled Foundations in Chalk at Basingstoke', *Proc. Conf. Settlement of Structures*, Cambridge, British Geotechnical Society

Knill, J. L. (1970). 'The Application of the Seismic Method to the Prediction of the Grout Take in Rock', *Proc. Conf. In Situ Investigations in Soils and Rocks*, London, British Geotechnical Society

Lake, L. M. and Simons, N. E. (1974). 'Some Observation on the Settlement of a Four-storey Building Founded in Chalk at Basingstoke, Hampshire', *Proc. Conf. Settlement of Structures*, Cambridge, British Geotechnical Society

Lake, L. M. and Simons, N. E. (1970). 'Investigations into the Engineering Properties of Chalk at Welford Theale, Berkshire', *Proc. Conf. In Situ Investigation in Soils and Rocks*, London, British Geotechnical Society

Lord, J. A. and Nash, D. F. T. (1974). 'Settlement Studies of Two Structures of Keuper Marl', *Proc. Conf. Settlement of Structures*, Cambridge, British Geotechnical Society

Louis, C. and Maini, Y. N. (1970). 'Determination of *In Situ* Hydraulic Parameters in Jointed Rock', *Proc. 2nd Int. Conf. Rock Mech*, Beograd

Lysmer, J. and Duncan, J. M. (1969). 'Stresses and Deflections in Foundations and Pavements', *Dept. of Civ. Engng. Univ. of California*, Berkeley

Malone, A. W. (1968). 'Elastic Wave Measurement in Rock Engineering', *Ph. D. Thesis*, Imperial College, London

Meigh, A. C., Skipp, B. O. and Hobbs, N. B. (1973). 'Field and Laboratory Creep Tests on Weak Rocks', *Proc. 8th Int. Conf. Soil Mech. Found. Engng*, Moscow

Milovic, D. M. (1972). Stresses and Displacements in an Anistropic Layer due to a Rigid Circular Foundation, *Geotechnique*, London, Vol. 22, No. 1, 169–173.

Moore, J. F. A. and Jones, C. W. (1974). '*In Situ* Deformation of Bunter Sandstone', *Proc. Conf. Settlement of Structures*, Cambridge, British Geotechnical Society

Moore, J. F. A. (1973). Private communication

Morgenstern, N. R. and Phukan, A. L. T. (1966). Non-linear Deformation of a Sandstone' *Proc. 1st Cong. Int. Rock Mech.*, Lisbon, Vol. 1, 543–548

Morgenstern, N. R. and Phukan, A. L. T., (1968). 'Stresses and Displacements in a Homogeneous Non-linear Foundation', *Proc. Int. Symp. Rock Mechanics*, Madrid, Blume, 313–320

Morgenstern, N. R. and Phukan, A. L. T., (1969). 'Non-linear Stress-strain relations for a Homogeneous Sandstone', *Int. J. Rock Mech. & Min. Sci.*, Vol. 6, 127–142.

Popova, O. V. (1972). 'Stress and displacement distributions in a homogeneous half-space below a circular foundation'. *Soil Mech. & Found. Engng*, Vol. 9, No. 2, 86–89, Consultants Bureau, New York

Poulos, H. G. (1968). 'The Behaviour of a Rigid Circular Plate Resting on a Finite Elastic Layer', *Trans. Inst. Engrs Australia*, 213–410

Poulos, H. G. (1967). 'Stresses and Displacements in an Elastic Layer Underlain by a Rough Rigid Base', *Geotechnique, London*, Vol. 17, No. 4, 378–410

Nyak, M. (1973). 'Elastic Settlement of a Cross-anisotropic Medium Under Axi-Symmetric Loading', *Soils and Foundations*, Tokyo, Vol. 13, No. 2, 83–90

Rocha, M., Borges, J. F. and Serafim, L. (1955). General Report, *Symposium on the Observation of Structures*, R.I.L.E.M., Lisbon

Steinbrenner, W. (1934). See Terzaghi, K. *Theoretical Soil Mechanics* (1947), 423–427, Wiley, New York

Trollope, D. H. (1968). 'The Mechanics of Discontinua', Contrib. *Rock Mechanics in Engineering Practice*, Stagg and Zienkiewiez, Wiley, New York

Wakeling, T. R. M. (1970). 'A Comparison of the Results of Standard Site Investigation in Middle Chalk at Mundford', *Proc. Conf. In Situ Investigations in Soils and Rocks*, London, British Geotechnical Society

Waldorf, W. A., Veltrop, J. A. and Curtis, J. J. (1963). 'Foundation Modulus Tests for Karadj Arch Dam', *A.S.C.E.*, SM 4, Vol. 89, Pt. 1, 91–126

609

Walsh, J. B. and Brace, W. F., (1966). 'Elasticity of Rock: A Review of Some Recent Theoretical Studies', *Rock Mech. and Engng Geol.*, Vol. 4, No. 4

Ward, W. H., Burland, J. B. and Gallois, R. W. (1968). 'Geotechnical Assessment of a Site at Mundford, Norfolk, for a Large Proton Accelerator', *Geotechnique*, London, Vol. 18, 399–431

Westergaard, H. M. (1938). 'A Problem of Elasticity Suggested by a Problem in Soil Mechanics', Contrib. *Mechanics of Solids, Timoshenko, 60th Anniversary*, New York, Macmillan

610

Settlement of Buildings and Associated Damage

General Reporters, J. B. Burland, *Building Research Station, Watford*
C. P. Wroth, *University Engineering Department, Cambridge*

Introduction

Session V of this Conference is concerned with allowable settlements of structures including damage and soil/structure interaction. Bjerrum (1963) pointed out that the engineer has basically two problems in considering design in relation to settlement. In the first place he has to evaluate the allowable differential settlements that he believes the building can withstand and secondly, he has to predict what total and differential settlements can be expected. This Paper is concerned with both these questions.

Compared with the literature on the prediction of settlement the question of allowable settlements and the influence of settlement on the performance and serviceability of structures has received little attention. This is remarkable when it is considered that large sums of money are spent on soils investigations aimed at assessing probable settlement, and that the foundations of many large structures are designed specifically to limit total and differential settlement.

The situation seems to be the same in the design of the structure itself. The approach in structural design, encouraged by the new Unified Code, is to design the various structural members so as to satisfy certain, often arbitrary, load factors. Very little attention is paid to the all-important question of serviceability apart from applying rather crude rules of thumb which are not usually related to the actual fabric of the building. It is seldom that the foundation engineer or the structural engineer takes an overall view of the performance of a building including its foundations, its structural members and its fabric and yet it is the total structure that the client must pay for and use. An outstanding example of co-operation between foundation engineer, structural engineer and architect is the CLASP system (Corsortium of Local Authorities Special Programme) which was evolved to cope with mining subsidence.

The problem of allowable settlements and soil/structure interaction forms a part of the much wider problem of serviceability and structural interaction. There are many obvious reasons why so little progress has been made on this global problem. Some of them are:

(1) Serviceability is very subjective and depends both on the function of the building and the reaction of the users.

(2) Buildings vary so much one from another, both in broad concept and in detail, that it is very difficult to lay down general guidelines as to allowable movements.

(3) Buildings, including foundations, seldom perform as designed because construction materials display different properties from those assumed in design. Moreover, a 'total' analysis including the ground and the cladding would be impossibly complex and would still contain a number of questionable assumptions.

(4) As well as depending on loading and settlement, movements in buildings can be attributed to a number of factors such as creep, shrinkage and temperature. There is as yet little quantitative understanding of these factors and there is a lack of careful measurements of the performance of actual buildings.

There is a tendency amongst foundation engineers to believe that the movements of the foundations are the major cause of distress in buildings and that by controlling these the satisfactory performance of the building is guaranteed. A recent conference on Movement in Buildings* clearly demonstrates that this is very far from being true. Even a casual glance through the proceedings of this conference suggests that engineers are in no better position to calculate relative movements of structural members under working conditions than they are for calculating settlements. Many cases are quoted of damage to finishes which result from movements of the structural members rather than the foundations. Moreover, the problem of movement in buildings is becoming more important because of the modern trend towards longer spans, higher permissible stresses, greater brittleness of walls and facing materials and larger non-structural units.

Another aspect of the problem which engineers may overlook is that a certain amount of cracking is unavoidable if the building is to be economic (Peck *et al.*, 1956). It is said that it is impossible to build a structure that does not crack due to shrinkage, creep, etc. Little (1969) has estimated that in the case of one particular type of building the cost of preventing any cracking could exceed 10% of the total building cost.

It is interesting that in the above conference numerous examples are quoted of simple design and construction expedients which permit the accommodation of movement without damage and it appears that the majority of these are relatively inexpensive. It could be argued that effort should be placed in developing and applying better design and construction details rather than in attempting to control serviceability using traditional methods. It may well be the case, but if building practice is to be improved it is necessary to develop a clearer understanding of the relationship between movement and damage in various types of structure and of methods of estimating such movements.

The first part of the paper is concerned with allowable distortions and a new approach is suggested based on the idea that the initiation of visible cracks is associated with a critical tensile strain. The second part of the Paper deals with the problem of settlements and the various factors that influence them such as soil properties and soil/structure interaction.

* Design for Movement in Buildings—Concrete Society, 1969.

PART 1—DAMAGE DUE TO FOUNDATION MOVEMENT

Previous work on allowable settlements of buildings

Perhaps the two best known studies leading to recommendations on allowable settlements of structures are those of Skempton and MacDonald (1956) and Polshin and Tokar (1957).

Skempton and MacDonald summarised settlement and damage observations on 98 buildings, 40 of which showed signs of damage. The study was limited to traditional steel and reinforced concrete frame buildings and to structures with load-bearing walls. Skempton and MacDonald used as their criterion for damage the ratio of the differential settlement δ and the distance l between two points after eliminating the influence of tilt of the building. The ratio δ/l was defined as the 'angular distortion'. It was concluded that the limiting value of δ/l to cause cracking in walls and partitions is 1/300 and that values of δ/l greater than 1/150 would cause structural damage. Skempton and MacDonald recommended that 'angular distortions' in excess of 1/500 should be avoided and if it was particularly desired to avoid any settlement damage this figure might well be increased to 1/1000. Subsequently Bjerrum (1963) supplemented these recommendations by relating the magnitude of δ/l to various types of damage.

Skempton and MacDonald also attempted to establish correlations between the greatest 'angular distortion' and maximum settlement and maximum differential settlement. This aspect of the work attracted some criticism (Terzaghi, 1956) on the grounds that the relationship between the above quantities is largely dependent upon the uniformity of the ground and the distribution of loads over the area occupied by the structure. Skempton and MacDonald themselves emphasised the need for judgement and experience in particular cases.

Polshin and Tokar (1957) discussed the question of allowable deformations and settlements and defined three criteria:

(1) 'slope', measured as the difference of settlement of two adjacent supports relative to the distance between them
(2) 'relative deflection', comprising the ratio of deflection to the length of the deflected part
(3) average settlement under the building

The limiting values of the above three quantities adopted by the 1955 Building Code of the USSR were then listed. Of particular interest is the fact that frame structures were treated separately from continuous load-bearing structures. Recommended maximum slopes vary from 1/500 for steel and concrete frame infilled structures to 1/200 where there is no infill or no danger of damage to the cladding. These values are clearly in line with Skempton and MacDonald's recommendations. However, much stricter criteria were laid down for load-bearing brick buildings. For ratios of length (between panels) to height less than three the maximum relative deflections are 0.3×10^{-3} and 0.4×10^{-3} (1/3300 and 1/2500) for such buildings on sand and soft clay respectively. For L/H ratios greater than 5 the corresponding relative deflections are 0.5×10^{-3} and 0.7×10^{-3} respectively.

It is interesting that Meyerhof (1956) also treated framed panels and load-bearing brick walls separately. He recommended limiting angular distortions of 1/250 for open frames, 1/500 for infilled frames and 1/1000 for load-bearing walls or continuous brick claddings. It should be noted that a maximum angular distortion of 1/1000 corresponds to a 'relative deflection' of about 1/2500. Therefore Meyerhof's recommendations are closely in line with those of Polshin and Tokar.

In their paper Polshin and Tokar made use of two concepts that are of fundamental importance. The first is the length to height ratio of a building and the second is that of a limiting tensile strain before cracking. For brick-work the limiting value is assumed to be 0.05%. A theoretical limiting relationship between relative deflection and L/H was presented which described the limit of cracking of brick walls and which was shown to be in good agreement with a number of cracked and uncracked buildings.

2. Description of ground and foundation movement

Terzaghi (1935) pointed out that the complete description of the settlement of a structure requires a large number of observation points so that detailed contours and profiles of foundation movement can be plotted. Detailed graphical presentation of observations becomes cumbersome when correlating a number of studies and it is necessary to categorise the various types of movement that can occur.

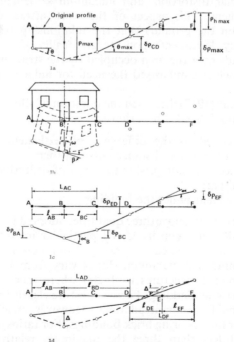

Fig. 1. (a) Definitions of settlement ρ, relative settlement δρ and rotation θ. (b)Definitions of tilt ω and relative rotation (angular distortion) β. (c) Definition of angular strain α. (d) Definition of relative deflection (sag or hog) Δ and deflection ratio (sagging or hogging) Δ/L

614

A study of the literature on allowable settlements has revealed a wide variety of symbols and terminology describing foundation movements—much of it very confusing. For example, the term 'angular distortion' has been used to describe at least four different modes of deformation.

In order to tackle the problem of allowable settlements and criteria of damage successfully it is necessary to have a clear and consistent set of definitions describing the types of movements and deformations experienced by foundations. It is very important that the terms should in no way prejudice concepts about the behaviour of the associated superstructure since this will depend on a large number of other factors such as size, details of construction, materials, time, etc. The following list of definitions and symbols is offered as a basis for discussion. In presenting these it is assumed that the settlement of a number of discrete points is known (see *Fig. 1(a)*). However, the details of the foundation and structure are (deliberately) not specified and the precise deformed shape between the observation points is not necessarily known.

Suggested definitions and symbols for the deformation of foundations

(1) A change of length equal to δL over a length L gives rise to a strain $\varepsilon = \delta L / L$. A *shortening* of $-\delta L$ over a length L gives rise to a *compressive* strain $\varepsilon = -\delta L / L$.

(2) *Settlement* (see *Fig. 1(a)*) is denoted by the symbol ρ and implies that the displacement is downwards. If the displacement is upwards it is termed *heave* and denoted by ρ_h.

(3) *Differential* or *relative settlement* (or heave) is denoted by $\delta\rho$ (or $\delta\rho_h$). In *Fig. 1(a)* the settlement of C relative to D is denoted $\delta\rho_{CD}$ and is taken as positive. (The settlement of D relative to C is denoted by $\delta\rho_{DC}$ which equals $-\delta\rho_{CD}$.) Maximum differential settlement is denoted by $\delta\rho_{max}$.

(4) *Rotation* is denoted by θ (see *Fig. 1(a)*) and is used to describe the change in gradient of the straight line joining two reference points embedded in the foundation or ground.

(5) *Tilt* is denoted by ω and normally describes the rigid body rotation of the whole superstructure or of a well-defined part of it. Normally it is not possible to ascertain the tilt unless details of the superstructure and its behaviour are known. Even then it can be difficult when the structure itself flexes. *Figure 1(b)* shows diagrammatically the tilt ω of a building overlying points ABC. Skempton and MacDonald (1956–Appendix I) give examples of the calculation of tilt.

(6) *Relative rotation* is denoted by β and describes the rotation of the straight line joining two reference points relative to the tilt (see *Fig. 1(b)*). Note that the 'angular distortion' defined by Skempton is identical to the relative rotation β.

(7) *Angular strain* is denoted by α. From *Fig. 1(c)* it can be seen that the angular strain at B is given by:

$$\alpha_B = \frac{\delta\rho_{BA}}{l_{AB}} + \frac{\delta\rho_{BC}}{l_{BC}}$$

Angular strain is positive if it produces sag or upward concavity, as at B in *Fig. 1(c)*, and negative if it produces hog or downward concavity as at E.

Angular strain is useful for predicting crack widths in buildings in which movement occurs at existing cracks or lines of weakness. Note that if the deformed profile between the three reference points ABC is smooth the average curvature is given by $2\alpha_B/L_{AC}$.

(8) *Relative deflection* (relative sag or relative hog) is denoted by Δ (see *Fig. 1(d)*) and is the maximum displacement relative to the straight line connecting two reference points a distance L apart. Relative sag produces upward concavity (as at B) for which Δ is positive. Relative hog produces downward concavity (as at E) for which Δ is negative.

(9) *Deflection ratio* (sagging ratio or hogging ratio) is denoted by Δ/L (see *Fig. 1(d)*). The sign convention is the same as in (8). Note that when $l_{AB} = l_{BD}$ or the deformed profile is approximately circular $\alpha = 4\Delta/L_{AD}$.

No doubt a number of additions could be made to the above list. For example, no attempt has been made to define three-dimensional behaviour such as warping. However, it is thought that the definitions should be adequate to describe most types of 'in plane' deformation.

The above list of symbols and definitions relates to foundation and ground movements. Description of the behaviour of the superstructure has not been attempted since standard terminology and sign conventions in structural engineering are widely used and understood.

3. The concept of critical tensile strain

Past work has aimed at establishing criteria of serviceability by relating observed deformation to damage. For example, the work of Skempton and MacDonald, of Polshin and Tokar and of Bjerrum all result in limiting criteria for allowable foundation deformations. The same approach is adopted in structural engineering. Mayer (1966) has carried out an extensive survey of partition failures and related these to deflection and deflection/span ratio. Beeby and Miles (1969) have made proposals for the control of deflections of structural members.

Deflection criteria based on empirical observations, though of undoubted value to the engineer engaged in day-to-day design, do not throw much light on the fundamental causes of damage. Moreover, they may have the effect of stultifying progress in the development of positive methods of design and construction aimed at minimising the damage caused by movement.

There is a need for a fundamental study of the problem of damage due to movement. In this section a particularly simple approach is outlined in the hope that it will stimulate thought and further work. We take as our starting point the fact that damage due to movement is almost always confined to the cladding and finishes rather than the structural members. Apart from a few notable exceptions buildings will usually become unserviceable before there is a danger of structural collapse.

Most damage manifests itself as cracking which results from tensile strain. Following the work of Polshin and Tokar (1957) it is assumed that the *onset* of visible cracking in a given material may well be associated with a limiting or 'critical' tensile strain ε_{crit}. It must be emphasised that we are not concerned here with collapse but with visible damage. This critical tensile strain is in no way related to that strain at which loss of tensile strength occurs. It is

616

interesting to note that a number of British Standards set limits on permissible shrinkage and expansion of bricks and concrete products. Materials which fall outside these limits often give rise to visible cracking.

For a number of years the Building Research Station has been studying the stiffness and strength of brickwork and infilled frames. The most recent work on the racking of infilled frames has been published by Mainstone (1972) and Mainstone and Weeks (1972) who describe some full-scale tests on frames with brick infills. It was observed that visible cracking first occurred at average diagonal tensile strains ranging between 0.081% and 0.137% with a mean value of 0.115%. These results are in good agreement with earlier work in which it was shown that the shear distortions at the onset of cracking for hollow-tile block and clinker block infills, as well as brickwork, varied from 0.22 to 0.33% which correspond to average principal tensile strains of approximately 0.11 to 0.16% respectively (assuming no volume change).

Fig. 2. Cracking of a brick wall due to bending strains and diagonal strains

The above figures are based on the measurement of boundary displacements. The detailed behaviour of the frame and infill is complex and does not conform to simple uniform shear. Detailed observations on one of the frames indicate that the local critical tensile strains are somewhat lower than the average strains and are probably in the range of 0.05% to 0.10%.

In addition to the work on racking of infilled frames the behaviour of brick walls has been under study at the Building Research Station for many years. Burhouse (1969) has published the results of some of this work. The purpose of the work was to study the composite action between brick walls and supporting beams. Several walls of varying span and height were tested and in the majority of cases the brickwork cracked at the centre of the span prior to collapse. *Fig. 2* is a photograph of one of the walls after failure and vertical cracking at the centre is clearly visible. The span to depth ratio of the walls varied from 1.2 to 3.0 and in all tests a building paper joint was provided between the brickwork and the supporting beam. Tensile strain at the onset of visible cracking varied from 0.038% to about 0.06%. These results agree with the suggestion of Polshin and Tokar (1957) that visible cracking of brick walls occurs at a tensile strain of 0.05%. Although the stress conditions at the lower edge of a brick wall in bending must differ significantly

617

from those set up in a panel subjected to racking it is evident that the critical tensile strains at the onset of visible cracking do not differ appreciably for the two modes of deformation.

It is of interest to note that the onset of cracking of the reinforced concrete support beams for the above tests took place at a value of tensile strain of about 0.035%. Base *et al.* (1966) have studied the cracking of reinforced concrete beams in great detail using a wide variety of concrete strengths and types of reinforcement. A close study of their results shows that for the complete range of tests there is a well-defined relationship between mean crack width w and average tensile strain ε over a 1.1 m length. The ratio w/ε lies between 65 mm and 115 mm with a mean of approximately 90 mm. It was also shown that the maximum crack width is approximately equal to $2w$. No information is given about the strain at onset of cracking but the precise determination is, in any case, rather subjective. If we take a figure of $2w = 0.1$ mm as being the width of crack likely to be first discerned by the casual observer we deduce that the average tensile strain at the onset of visible cracking is 0.05%. Though approximate this figure is in good agreement with the value of 0.035% deduced from Burhouse's results.

At this stage it should again be emphasised that we are concerned only with strains likely to cause visible cracking in walls and panels whose stability does not depend on the tensile strength of the material composing them. There is a great deal of evidence to suggest that tensile 'failure' (i.e. loss of tensile strength) occurs at much smaller values of tensile strain than those causing visible cracking.

It is interesting to note that the critical tensile strains associated with visible cracking of walls and panels under load are of the same order as the permissible values of shrinkage strain laid down in the various British Standards (0.03 to 0.09%).

Finally, it should be emphasised that the onset of visible cracks does not necessarily represent a limit of serviceability. Provided the cracking is controlled, as in a reinforced concrete beam, it may be quite acceptable to allow deformation to continue to well beyond the initiation of cracking. Nevertheless, a criterion based on the initiation of visible cracking may serve as a useful reference point.

4. Application of critical tensile strain to the initial cracking of simple beams

The consequences of adopting a fundamental approach to cracking can best be illustrated by applying the concept of critical tensile strain to a simple structure such as a uniform, weightless, elastic beam of length L, height H and unit thickness. The beam may be thought of as representing a building—see *Fig. 3(a)*. It is recognised that real structures are normally very much more complex but the study of a simple beam does help to illustrate a number of important features.

We will assume that the deflected shape of the soffit of the beam is known. The problem is to calculate the tensile strains in the beam and hence define the criteria for initial cracking. It is clear that little can be said about the performance of the beam unless we know its mode of deformation. Two possible extreme modes of deformation are bending only (see *Fig. 3(b)*) or shearing only (see *Fig. 3(c)*). It is clear that in the case of bending, initial cracking will occur as a result of direct tensile strain at the extreme fibre whereas for shear, cracking will result from diagonal tensile strain. In fact, both modes of deformation are likely to occur simultaneously but initially we will consider the two separately.

Bending only. If the deflected shape of the soffit is circular then the radius of curvature $R = L^2/8\Delta$. Now if pure bending is occurring then the bending

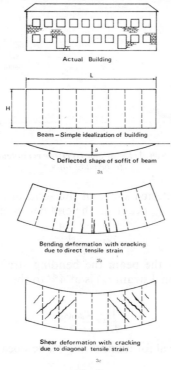

Actual Building

Beam – Simple idealization of building

Deflected shape of soffit of beam

3a

Bending deformation with cracking due to direct tensile strain

3b

Shear deformation with cracking due to diagonal tensile strain

3c

Fig. 3

strain $\varepsilon_b = y/R$ where y is the distance from the neutral axis. When the neutral axis is in the middle of the beam the maximum bending strain $\varepsilon_{b(\text{max})}$ occurs at $y = H/2$. Hence

$$\frac{\Delta}{L} = \frac{L}{4H} \cdot \varepsilon_{b(\text{max})} \tag{1}$$

The equivalent expression for a uniformly loaded beam is

$$\frac{\Delta}{L} = \frac{L}{4.8H} \cdot \varepsilon_{b(\text{max})} \tag{2}$$

and for a central point load

$$\frac{\Delta}{L} = \frac{L}{6H} \cdot \varepsilon_{b(\text{max})} \tag{3}$$

619

For any given critical value of tensile strain $\varepsilon_{\text{crit}}$ Equations 1 to 3 define limiting relationships between Δ/L and L/H for the three types of loading considered. These relationships are shown in *Fig. 4* in terms of $\Delta/L.\varepsilon_{\text{crit}}$ against L/H. Two important conclusions can be drawn from these results:

(1) The limiting relationship between Δ/L and L/H is not very sensitive to the form of load distribution.
(2) For deformation involving bending only the limiting value of Δ/L is directly proportional to L/H.

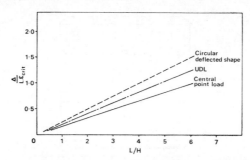

Fig. 4. Relationships between $\Delta/L\varepsilon_{\text{crit}}$ and L/H for rectangular beams deflecting due to bending only

Shearing only. If the deflected shape of the soffit of the beam is circular the maximum shear strain γ_{max} occurs at the end and is given by

$$\gamma_{\text{max}} = 4\frac{\Delta}{L}$$

Since at any section of the beam the bending strain is zero the principal strain (termed the diagonal strain ε_d) is at 45° to the neutral axis and is equal to $\gamma/2$. Hence

$$\frac{\Delta}{L} = \frac{\varepsilon_{d(\text{max})}}{2} \tag{4}$$

The expression is identical for a uniformly loaded beam, but for a beam with a central point load

$$\frac{\Delta}{L} = \varepsilon_{d(\text{max})} \tag{5}$$

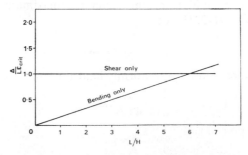

Fig. 5. Comparison between criterion of cracking for rectangular beams deflecting due to bending only and due to shear only

620

In contrast to the case of pure bending the limiting value of Δ/L is independent of L/H and somewhat more dependent on the load distribution. In practice, however, diagonal cracking occurs some distance from the ends of beams and if the diagonal strain at the quarter span points is considered, the limiting relationship between Δ/L and $\varepsilon_{d(max)}$ is given by Equation 5 for all three cases.

The case just discussed is that of a very simple structure. It is clear, however, that even for a simple beam the criterion for initial cracking in terms of Δ/L can vary between wide limits depending upon the mode of deformation. *Figure 5* shows the criteria for shear only and bending only for a beam with a central point load. It can be seen that for $L/H < 6$ bending is much more critical than shear.

Combined bending and shear. For illustrative purposes we will concern ourselves only with the case of a beam with a central point load. Other loading cases give similar results. Timoshenko (1955) gives the expression for the

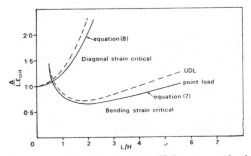

Fig. 6. *Relationship between $\Delta/L\varepsilon_{crit}$ and L/H for rectangular beams deflecting due to combined bending and shear–neutral axis in the middle*

total central deflection of a centrally loaded beam of unit thickness flexing in both shear and bending as

$$\Delta = \frac{PL^3}{48EI}\left\{1 + \frac{18I}{L^2H}\frac{E}{G}\right\} \tag{6}$$

For an isotropic elastic material $E/G = 2\,(1 + v)$ where v is Poisson's ratio; for $v = 0.3$ (say), $E/G = 2.6$. When the neutral axis is in the middle Equation (6) may be written in terms of the maximum extreme fibre strain $\varepsilon_{b(max)}$ as follows:

$$\frac{\Delta}{L} = \left\{0.167\frac{L}{H} + 0.65\frac{H}{L}\right\}\varepsilon_{b(max)} \tag{7}$$

Similarly for maximum diagonal strain $\varepsilon_{d(max)}$ Equation 6 becomes

$$\frac{\Delta}{L} = \left\{0.25\frac{L^2}{H^2} + 1\right\}\varepsilon_{d(max)} \tag{8}$$

Equations 7 and 8 have been plotted in *Fig. 6* as full lines. The broken lines are the equivalent expressions for a uniformly loaded beam (considering the diagonal strain at the quarter span points). As before it is evident that, for a given value of critical tensile strain ε_{crit}, the criteria for initial cracking are not sensitive to the type of loading.

A very important conclusion emerges from *Fig. 6*. It is evident that the minimum limiting value of $\Delta/L\varepsilon_{\text{crit}}$ for diagonal strain is 1.0 and this increases sharply as L/H increases. However, the limiting value of $\Delta/L\varepsilon_{\text{crit}}$ for direct strain in bending decreases as L/H increases from zero reaching a minimum of 0.66 at $L/H = 2.0$ and increasing gradually thereafter. *Fig. 6* demonstrates that, for the case of a simple uniform isotropic beam flexing in combined shear and bending, the direct strain in bending is much more critical than the diagonal strain for values of $L/H > 0.6$. For example, at a value of $L/H = 2.0$ the limiting value of $\Delta/L\varepsilon_{\text{crit}}$ for direct bending strain is one-third that for diagonal strain.

In practice the foundations of many buildings offer a significant restraint to their deformations, as would be the case with a reinforced concrete raft. In these circumstances it may be more realistic to take the neutral axis at the

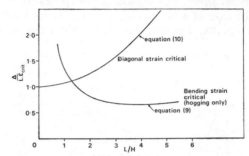

Fig. 7. Relationship between $\Delta/L\varepsilon_{\text{crit}}$ and L/H for rectangular beams deflecting due to combined bending and shear–neutral axis at one edge

lower extreme fibre of the simple beam. The equations for limiting Δ/L due to bending strain and diagonal strain become respectively

$$\frac{\Delta}{L} = \left\{0.083\frac{L}{H} + 1.3\frac{H}{L}\right\}\varepsilon_{b(\text{max})} \tag{9}$$

and

$$\frac{\Delta}{L} = \left\{0.064\frac{L^2}{H^2} + 1\right\}\varepsilon_{d(\text{max})} \tag{10}$$

Equation 9 only applies for hogging since direct tensile strain would be zero for sagging. The above equations are plotted in *Fig. 7* and it can be seen that the only essential difference between the curves in *Figs. 6 and 7* is that for any given value of $\Delta/L\varepsilon_{\text{crit}}$ the value of L/H in *Fig. 7* is twice that in *Fig. 6*.

Influence of the ratio E/G. Up till now it has been assumed that the ratio E/G is given by the simple isotropic theory of elasticity ($= 2(1 + \nu)$). The quantity E may be thought of as a measure of the longitudinal stiffness of the beam and G the stiffness of the beam in shear. In practice the 'structure' may be built in such a way that the equivalent ratio E/G in the model beam does not conform to the simple isotropic elastic relationship. For example, it may be designed in such a way that it has relatively little shear stiffness due, perhaps, to a number of openings in it. Alternatively, the structure may be very stiff in shear with little longitudinal stiffness as might be the case for a wall made of precast concrete units held together with dowels.

In *Fig. 8* are plotted the limiting relationships between $\Delta/L\varepsilon_{\mathrm{crit}}$ and L/H for values of E/G equal to 0.5, 2.6 (isotropic) and 12.5 with the neutral axis at $H/2$. It can be seen that the most desirable structure from the point of view of minimising tensile strain for a given value of Δ/L is one that is relatively flexible in shear, i.e. a large value of E/G. When the structure is stiff in shear relative to its horizontal stiffness the direct strains due to bending dominate and the limiting relationship approaches that given by Equation 3. The curve labelled (4) in *Fig. 8* corresponds to $E/G = 0.5$ for a beam with neutral axis at the lower edge. As for Equation 10 curve (4) is only valid for hogging.

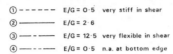

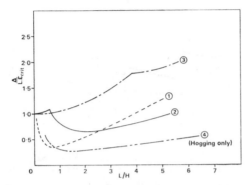

Fig. 8. *The influence of E/G on the relationship between $\Delta/L\varepsilon_{\mathrm{crit}}$ and L/H for rectangular beams*

Criteria for the cracking of simple beams. This section has been concerned with the initial cracking of simple beams. It is evident that even for such a simple type of structure composed of a material with a known value of critical tensile strain the value of Δ/L causing visible cracking can vary through wide limits depending chiefly upon the mode of deformation, the relative shear to tensile stiffness and the geometry. The behaviour of real structures will clearly be much more complicated. Nevertheless, the results of the analysis may be of some value in assessing the performance of real buildings and can be summarised as follows:

(1) Before even an approximate estimate of the limiting relative deflection can be made it is necessary to know the following: the critical tensile strains of the building materials and finishes, the length to height ratio of the structure (L/H), the approximate ratio between the equivalent longitudinal stiffness and the shear stiffness (E/G) and the degree of tensile restraint offered by other parts of the structure and/or the foundations.

(2) When the structure has relatively low stiffness in shear or a significant degree of tensile restraint (e.g. frame buildings, reinforced load-bearing walls) cracking due to diagonal tensile strain will be the

623

limiting factor. For a simple beam Equation 10 provides a conservative limiting relationship for these conditions and is plotted in *Fig. 9(a)*, curve (1).

(3) When the structure has little or no tensile resistance (e.g. traditional brick or masonry buildings) it is probable that cracking due to tensile bending strains will be the limiting factor, particularly when L/H is greater than 2. The curve labelled (2) in *Fig. 9(a)* represents an approximate lower limit for a simple beam with neutral axis at the centre.

(4) When the structure is of traditional brick or masonry without tensile restraint and it is subjected to hogging, cracking due to direct strain

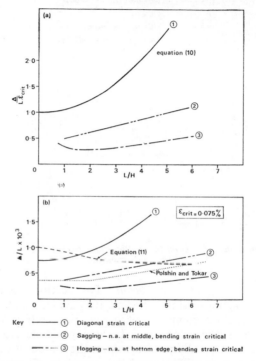

Fig. 9. (a) *Criteria for the onset of visible cracking of rectangular beams.*
(b) *Comparisons between the criteria of Figure 9(a), with* $\varepsilon_{crit} = 0.075\%$, *and other damage criteria*

in bending is likely to occur at very low values of Δ/L. This is illustrated for the case of a simple beam by curve (3) in *Fig. 9(a)* which was derived for a low value of E/G (low tensile stiffness) and the neutral axis at the lower edge.

It must be emphasised that the case of a simple uniform beam has been chosen so as to illustrate a possible method of assessing when damage might occur. No doubt the analysis would be more realistic if plate theory were used to calculate the maximum tensile strains. More complex structures involving openings and changes of section could be studied using numerical methods such as the finite element method. Without attempting such sophisticated calculations the validity of the basic approach may be assessed by comparing the criteria derived from the simple beam analysis with existing criteria of damage.

Comparison with other criteria of damage. In Section 3 it was concluded that the critical tensile strain for brickwork and other materials appears to lie in the range 0.05 to 0.10%. A value of 0.075% will be adopted for present purposes and curves 1, 2 and 3 in *Fig. 9* have been re-plotted in *Fig. 9(b)* on axes of Δ/L against L/H with $\varepsilon_{crit} = 0.075\%$.

Polshin and Tokar (1957) have suggested a limiting relationship between Δ/L and L/H for plane brick walls which they state is based on a limiting tensile strain of 0.05%. The relationship is shown in *Fig. 9(b)* as a dotted line and can be seen to lie a little below line (2) as would be expected. Presumably Polshin and Tokar based their relationship on a similar analysis to the one outlined in this section but they have not extended it to deal with diagonal strain or hogging.

Skempton and MacDonald (1956) suggested that the angular distortion β should be used as the criterion for damage and recommended $\beta = 0.002$ as a suitable limiting value. Bjerrum (1963) subsequently endorsed this value as being a safe limit for buildings where cracking is not permissible. This criterion appears to have been evolved primarily for infilled framed structures where shearing will predominate but it has been widely applied to other types of structure.

In applying the criterion to the present case of a simple beam the maximum rotation is at the supports. For the case of an elastic beam deforming in shear as well as in bending and the neutral axis in the middle it can be shown that the maximum angular distortion is related to Δ/L by the expression

$$\frac{\Delta}{L} = \frac{\beta}{3} \times \left[\frac{1 + 3.9 \, (H/L)^2}{1 + 2.6 \, (H/L)^2} \right] \tag{11}$$

The equivalent expression for the neutral axis at the lower face does not differ significantly.

Equation 11 is plotted as the dashed line in *Fig. 9(b)* with $\beta = 0.002$ which corresponds to the limiting value of $\delta/l = 1/500$ recommended by Skempton and MacDonald (1956). It can be seen that for values of L/H less than 2.5, Equation 10 with $\varepsilon_{crit} = 0.075\%$ (line 1) is in reasonable agreement with Equation 11 which is based on Skempton and MacDonald's criterion. However, for values of L/H greater than 2.5, Equation 11 gives significantly lower limiting values of Δ/L than Equation 10. It should be noted that for L/H less than 4, line (2), which is based on limiting bending strains, gives much lower limiting values of Δ/L than Skempton and MacDonald's criterion.

It is worth noting that Meyerhof (1956), while accepting the value of $\beta = 0.002$ for framed buildings, recommends a limiting value of Δ/L of approximately 0.4×10^{-3} for continuous brick cladding and load-bearing walls. This value is in reasonable agreement with the simple sagging beam theory (line (2)) and Polshin and Tokar's relationship for $L/H < 3.5$. A notable feature in *Fig. 9(b)* is that line (3), which applies to hogging of load-bearing walls, predicts that cracking occurs at significantly smaller values of Δ/L than predicted by either Equation 11 or Polshin and Tokar's relationship.

625

In summary it appears that the simple criteria for initial cracking of beams as summarised in *Fig. 9(a)* are in good agreement with various existing criteria for brickwork; and further that they give a more comprehensive coverage of materials, method of construction and mode of deformation than any other single criterion. The next step is to compare the various criteria with some observed case records but first the behaviour subsequent to the onset of cracking will be discussed briefly.

5. Behaviour subsequent to initial cracking

So far we have considered only deformations that give rise to initial cracking. In this section we will consider briefly the behaviour subsequent to the onset of cracking. Provided the cracking is contained by some form of tensile restraint it is possible that significantly larger deformations can be tolerated than those giving rise to initial cracking. Reinforced concrete beams and slabs are good examples where this type of behaviour occurs and guidelines are laid down for the maximum acceptable width of crack for various applications.

Similar ideas can probably be developed for the cracking of partitions, cladding, etc. It will be necessary to develop criteria for serviceability based on such factors as maximum crack width, frequency of cracks, surface texture and rate of deformation.

The behaviour of even simple structures subsequent to initial cracking is very complex and it will not be easy to lay down general guidelines. A basic question seems to be: Can a crack propagate unrestrained as deformation continues and will movement concentrate across a crack once it has formed?

Cases where the propagation of initial cracks may be fairly well controlled are framed structures with panel walls, and reinforced load-bearing structures. Continuous, unreinforced load-bearing structures founded on reinforced strip footings or rafts may also fall into this category. However, care must be taken about possible planes of weakness. Ward (1956) drew attention to a classic case of a brick building founded on continuous strip footings which had undergone settlement giving rise to sagging. Slip had occurred along the bitumen damp course resulting in extensive cracking due to bending in the overlying brickwork. No amount of tensile reinforcement in the footing would have prevented or controlled such cracking. It is interesting to note that no cracking occurred in the underlying brickwork which had a much larger L/H value.

An important mode of deformation which does not appear to have received the attention it deserves is that of hogging of traditional brick or masonry structures. Once a crack forms at the top of the building there is nothing to prevent it from propagating downwards to the foundations. Thereafter the structure behaves as two or more separate units and very wide cracks can develop. The essential difference between hogging and sagging can be illustrated quite simply by the following example.

Figures 10(a) and (10c) show the damage to model brick walls which have been subjected to angular strain giving rise to sagging. The foundations offer considerable restraint and the deformation occurs as fairly uniform shear distortion without obvious continuous cracks. If, however, the walls are subjected to the same amount of angular strain but giving rise to hogging (see

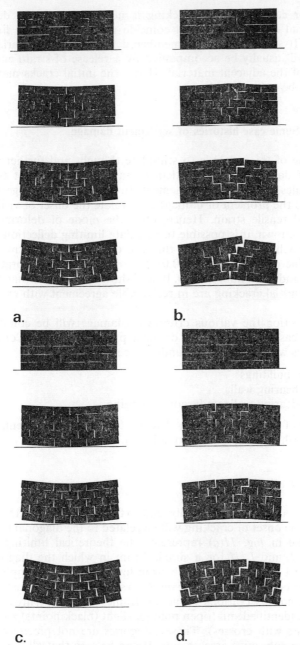

a. b.

c. d.

Fig. 10. The cracking of model brick walls during sagging and hogging

Figs. 10(*b*) and 10(*d*)) it is evident that the damage is much more severe. Continuous cracks run through the full depth of the wall causing severe disruption and the two ends of the wall tend to behave as independent units. Pryke (1974) also gives a striking illustration of the propagation of cracks through a building due to loss of support from the ground on one side.

Hence for cases in which cracking is not restrained, the deformations causing initial cracking may well coincide with the desirable limits of serviceability. It should also be remembered that when a material cracks in tension it will usually be accompanied by a release of strain energy due to unloading of the adjacent material. Hence the initial cracks may form very rapidly and be quite sizeable.

6. Some case histories of settlement damage

The purpose of this paper is to stimulate thought and further work on the problem of damage due to settlement rather than attempt to formulate empirical rules for allowable settlement. In Section 3 it is suggested that one approach to the problem of damage is to relate the onset of visible cracking to a critical tensile strain. Hence, once the mode of deformation of the structure is known, it is possible to calculate limiting deflections. In Section 4 the approach is illustrated for the case of a simple uniform elastic beam using a value of critical strain (0.075%) obtained from laboratory tests on brick walls and brick infill frames. It was shown (*Fig. 9(b)*) that the resulting criteria for initial cracking are in reasonable agreement with existing criteria of damage.

In this section the various criteria of damage will be compared with a number of case records—many of them presented to this Conference. The observations will be considered under three headings:

(1) frame buildings
(2) load-bearing walls
(3) load-bearing walls undergoing hogging.

The values of relative deflection Δ relate to total displacements rather than those occurring subsequent to completion.

Frame buildings

This type of building will generally tend to flex in shear rather than in bending and hence the onset of cracking will be related to the diagonal tensile strains. The full line in *Fig. 11(a)* represents the theoretical limiting relationship between Δ/L and L/H for a simple beam in which the diagonal strain is critical (line 1 in *Fig. 9(b)*). The broken line represents Skempton and MacDonald's criterion for a simple beam. Also shown in *Fig. 11(a)* are some results obtained from observations on actual frame buildings. Three categories of damage are identified: nil (open points), slight (black points) and substantial (black points with crosses). These categories are not precisely defined but are based on subjective assessments. It can be seen that all the observations relating to substantial damage give values of Δ/L lying well above the two limiting relationships whereas the four observations for slight damage lie very close to them.

The results for $L/H < 1$ are of particular interest since they all relate to high-rise buildings constructed from cast *in situ* reinforced concrete and having a variety of types of cladding. Vargas and Silva (1973) have analysed the results of settlement observations on 26 tall buildings, some extensively damaged, and concluded that damage is generally in accordance with Skempton and MacDonald's proposals. Breth and Amann (1974) have listed the observed values of relative deflection Δ for eight tall buildings on Frankfurt Clay—all of them undamaged. The maximum value of Δ/L is 0.565×10^{-3}.

The results in *Fig. 11(a)* are for $L/H \leqslant 3$ for which the two criteria are, for practical purposes, identical. At larger values of L/H the criterion based on critical tensile diagonal strain gives much higher permissible values of Δ/L than Skempton and MacDonald's criterion. Observations on frame buildings

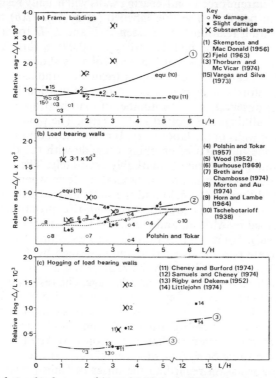

Fig. 11. Relationship between Δ/L and L/H for buildings showing various degrees of damage. The numbered curves refer to the theoretical criteria shown in Figure 9(b) — $\varepsilon_{crit} = 0.075\%$

with $L/H > 3$ are required in order to determine which of the two criteria is the better.

It was concluded by Skempton and MacDonald (1956) that the foundations of open frame structures can undergo angular distortion of as much as 1/150 without damage. In a paper to the Conference Webb describes the performance of a two-storey reinforced concrete open frame building which was specially designed to accept large differential settlements. The relative settlement between two of the columns gave rise to a relative rotation (or angular distortion) of 1/177 causing slight cracking in a beam connecting the two columns. At another point in the building a relative rotation of 1/172 occurred without damage. These results are in good agreement with Skempton and MacDonald's conclusions. The span to depth ratio of the cracked

629

beam was 24.4 giving a theoretical maximum tensile strain of about 0.035% which is in excellent agreement with the values of ε_{crit} (0.03 to 0.05%) for concrete suggested in Section 3.

Load-bearing walls

In Section 4 it was concluded, in agreement with Polshin and Tokar (1957), that cracking of unreinforced load-bearing walls will usually result from direct tensile strains in bending. The line labelled (2) in *Fig. 11(b)* represents the appropriate limiting relationship between Δ/L and L/H for a simple beam with the neutral axis in the middle and $\varepsilon_{crit} = 0.075\%$ (line 2 in *Fig. 9(b)*). The relationship proposed by Polshin and Tokar is shown as a dotted line. Skempton and MacDonald's criterion is frequently applied to load-bearing walls and is therefore also shown in *Fig. 11(b)* as Equation 11.

The results of observations on a number of load-bearing walls and precast concrete panel buildings are plotted in *Fig. 11(b)*. The points marked 'L' are for tests on brick walls in the laboratory. It can be seen that the two relationships which define the onset of cracking coincide with the division between damaged and undamaged walls remarkably accurately. The two points lying below the theoretical criteria are for laboratory tests where loading is generally more rapid than in the field and the walls may well be more brittle.

The points numbered (8) are for two high-rise precast concrete panel buildings described by Morton and Au at this Conference (Cambridge Road and Hurley Road). Point number (3) relates to a 4-storey block of flats also of precast concrete panel construction and described by Thorburn and Mc-Vicar at this Conference. Although the relative sag is quite large for two of these buildings there is no record of any damage. The results suggest that precast concrete panel buildings may be no more sensitive to differential settlement than traditional load-bearing wall structures although many more observations are required before firm conclusions can be drawn.

An important observation is that Skempton and MacDonald's criterion overestimates the value of Δ/L at which cracking first occurs. There is not a great deal of information regarding the onset of 'substantial' damage. The point numbered (9) is for a building on the M.I.T. campus which was reported by Horn and Lambe (1964) to have a substantial amount of cracking both inside and outside the building. They also note that the relative sag of 1/1600 is considerably less than that normally given as the maximum permissible.

Buildings undergoing hogging

As mentioned in Section 5 hogging is an important mode of behaviour which appears to have received relatively little attention in the past. However, four papers have been presented to this Conference which deal with hogging and provide a large amount of extremely valuable information. The papers are by Cheney and Burford, Samuels and Cheney, Littlejohn, and Breth and Chambosse.

In Section 4 it was shown that when a structure of traditional unreinforced brick or masonry is subjected to hogging, cracking due to direct tensile strain in bending is likely to occur at very low values of Δ/L. Moreover, it was demonstrated in Section 5 that cracks are likely to propagate rapidly during hogging resulting in substantial damage at values of Δ/L which give rise to little or no damage during sagging. The chain dotted line labelled (3) in *Fig. 11(c)* is the theoretical limiting relationship for the initial cracking of a simple beam.

Also shown in *Fig. 11(c)* are results from the observations presented in the four papers referred to previously together with some others. The points numbered (11) refer to the 3-storey brick building described by Cheney and Burford and have been obtained from their *Fig. 1*. At a value of $\Delta/L = 0.154 \times 10^{-3}$ hair cracks appeared and at $\Delta/L = 0.21 \times 10^{-3}$ the cracks were as wide as 2.5 mm. When Δ/L had reached a value of 0.58×10^{-3} complaints from the occupants were such that the damage must be classed as 'substantial'.

Littlejohn describes some important experiments on the performance of brick walls subject to mining subsidence. The walls had exceptionally high L/H ratios of between 12.5 and 17. The observations showed that the brick walls underwent significant hogging which was maintained after the passage of the subsidence wave. The foundations were of unreinforced concrete which exhibited hairline cracks at very small tensile strains (0.01%) and may well have been initiated by shrinkage. As the hogging in wall number (1) developed during the passage of the subsidence wave, the cracks extended rapidly through the brickwork due to the increase in curvature and at a hogging ratio of $1/1390 \ (= 0.72 \times 10^{-3})$ the full depth of the wall was cracked. At a hogging ratio of $1/920 \ (= 1.08 \times 10^{-3})$ the damage was classed as 'severe'. The behaviour reported by Littlejohn is complicated by the presence of direct strains in the ground as well as the differential settlements. Nevertheless, the observations, together with the others plotted in *Fig. 11(c)*, appear to be broadly in agreement with the predictions of the cracking of simple beams undergoing hogging.

In Section 5 it was demonstrated by means of a model brick wall undergoing hogging that once cracking has occurred there is a tendency for the structure to split into separate units with the movement being concentrated across the cracks. Littlejohn has observed this behaviour in his experiments, and Breth gives some excellent case records of this type of behaviour resulting from surface subsidence due to tunnelling. In these circumstances a knowledge of the angular strain experienced by the foundations permits a reasonable assessment of the likely width of cracks. For example, building III described by Breth and Chambosse underwent a tilt of between 1/210 and 1/130 relative to an adjoining building giving an angular strain α of between -0.48 and -0.77%. The total height of the building H from founding level to roof top was 13 m. Assuming movement occurs only at cracks the total width of the cracks at the top of the building is calculated to be 60–100 mm ($H \times \alpha$) which compares well with the total measured width of the cracks of 60–80 mm.

7. Discussion and conclusions to Part I

The safe limit for angular distortion (relative rotation—see Section 2) of 1/500 proposed by Skempton and MacDonald (1956) has been shown to be satisfactory for framed buildings of both traditional and modern construction. However, it appears that this criterion is not satisfactory for buildings with load-bearing walls and damage has been reported for very much smaller angular distortions. For this type of building, including those constructed from precast reinforced concrete panels, the criterion proposed by Polshin and Tokar (1957) which limits the value of Δ/L in relation to L/H appears to be much more satisfactory. However, even this criterion is not adequate for load-bearing-wall buildings undergoing hogging. Observations show that in such cases cracking occurs at values of Δ/L which are half those given by Polshin and Tokar.

In Section 3 it was suggested that a fundamental understanding of the causes of damage is required and that one approach is to relate the onset of cracking to a critical tensile strain. The application of this approach has been illustrated for the very simple case of a uniform rectangular beam. It may be argued that this case is too simple to be representative of real buildings. However, its use has been both instructive and surprisingly successful in predicting the cracking of various types of building. Not only does the approach account for the difference in allowable movements for framed and load-bearing-wall buildings but also for the greater susceptibility to damage of buildings subjected to hogging.

The great advantage of the concept of critical tensile strain is that it can be applied to complex structures employing well established analytical techniques with the emphasis being placed on deformation and strain rather than on stress. One can envisage that for important or unusual buildings, in addition to analysing the strength and stability of the structure a deformation and strain analysis will be carried out using a simple idealisation of the whole building, taking account of foundations, partitions and finishes. This approach will be discussed in greater detail in Part II of the paper. Such an approach makes explicit the fact that damage can be controlled not only by limiting settlements or stiffening the structure but by thoughtful design of cladding and partition fixings.

Various papers to this Conference have referred to buildings that have undergone hogging. This mode of deformation can result from a number of causes such as:

(1) swelling or shrinkage of the subsoil
(2) mining subsidence
(3) subsidence due to tunnelling
(4) loss of support during underpinning
(5) drag down by adjacent buildings
(6) ground water lowering
(7) movements around excavations

Clearly the performance of buildings undergoing hogging is of importance. The fact that both theoretical considerations and observations on actual buildings indicate that the magnitude of the movements required to cause damage is very small, particularly in the absence of any tensile restraint, makes the problem even more important. Whereas it is possible to design buildings to accommodate settlements due to their own weight hogging often occurs in existing buildings as a result of nearby construction. The results presented in this paper suggest that deformation giving rise to hogging of load-bearing-wall structures must be very carefully controlled if damage is to be avoided. The effects of excavations are particularly significant since these not only give rise to hogging of the surrounding ground but direct tensile strain as well.

PART 2—INTERACTION BETWEEN STRUCTURE AND UNDERLYING GROUND

8. Main factors affecting interaction

This section of the Report is concerned with the problem of interaction between a structure and the underlying ground on which it is founded, both during construction and subsequently during service. Interaction is a very complex problem because it is the combination of several different effects (some of which are time-dependent). None of these effects is truly linear, so that it is not possible to consider any one separately from the others, and then to superpose them without introducing errors and approximations. These factors include:

(1) the immediate settlements caused by each increment of load as the structure grows
(2) the long-term consolidation settlements (both primary and secondary) which overlap with the immediate settlements, and of which a major proportion may occur during construction
(3) the changing stiffness of the structure as building progresses
(4) the redistribution of loads and stresses within the structure due to differential settlement

But in order to gain some understanding of these problems and to identify the salient features it is necessary to isolate the above factors and to consider them separately in as simple a manner as possible, while remaining aware of the approximations that have been introduced.

The question of the proportion of immediate to long term settlement is considered first, and at some length, because of its important influence on the amount of differential settlement that occurs. The problems of estimating settlement and differential settlement are illustrated in the following section by the example of a stiff circular raft on stiff clay; the details of the raft and ground have been chosen specifically to be as similar as possible to one of the case histories presented to this Conference by Morton and Au (1974). The final section contains a brief discussion of the questions of the changing stiffness of a building during construction, and the redistribution of loads within it due to differential settlement.

9. Proportion of immediate to long term settlement

In the paper by Skempton and MacDonald (1956) an important distinction was made between major deformation and damage to the main structural frame of the building, and the deformation and damage to the finishes, i.e. cladding, in-fill panels, plasterwork. It was shown that major structural damage was rare, and that most damage could be classified as architectural, which may or may not impair the serviceability of the building. Structural damage is due to the deformations resulting from *both* the immediate and long term settlement of the ground, whereas damage to the sensitive finishes is predominantly caused by the long term settlement occurring *after* completion. (There may be a small amount of immediate settlement if the finishes are fixed before all the dead load is on the foundation.) It is therefore important to establish, if possible, the proportion of total settlement that will occur before the finishes are applied to the building: and if the design will allow it, to arrange for this proportion to be as large as possible.

This proportion depends on the properties of the ground and will only be of concern when the soil is fine-grained with a low coefficient of permeability. This means that the problem of differential settlement and hence of architectural damage is unlikely to occur in buildings on sand unless there are special circumstances such as collapse of the sand due to wetting or vibration. A clear distinction must be made between:

(1) overconsolidated clays which for reasonably small stress changes behave essentially as an elastic material (even though the elastic properties may be non-homogeneous and anisotropic)
(2) normally consolidated clays which do not.

A simple approach to assessing the proportion of settlements on these two types of clay is outlined in the following sections.

Uniform circular load on overconsolidated clay

We start by considering the simplest possible case of a uniform circular load (of radius a and intensity q per unit area) applied vertically to the horizontal ground surface as shown in *Fig. 12*. The ground is assumed to be of infinite

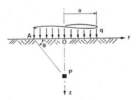

Fig. 12. Uniform circular load on elastic half-space

extent and to be overconsolidated clay which behaves as a perfectly elastic material.

We need to distinguish between the responses of the clay in two extreme conditions:

634

(1) the immediate undrained case related to changes of total stresses and associated with appropriate values of Young's modulus E_u and Poisson's ratio v_u $(= \frac{1}{2})$

(2) the final, fully drained case related to changes of effective stresses with relevant elastic constants E' and v'.

The more fundamental elastic constant, the shear modulus G is defined in terms of stress differences and must therefore be independent of pore pressure; for an elastic material it must be the same for undrained and fully drained cases, so that

$$\frac{E_u}{1 + v_u} = 2G_u = 2G' = \frac{E'}{1 + v'} \tag{12}$$

The settlement ρ of the central point O on the ground surface is given by

$$\rho = \frac{2(1 - v^2)qa}{E} = \frac{(1 - v)qa}{G'} \tag{13}$$

see, for example, Timoshenko and Goodier (1951). It should be noted that the settlement for a fixed value of pressure q is directly proportional to the size of the loaded area; this will be referred to at greater length later.

Directly from Equation 13 we have the immediate settlement

$$\rho_u = \frac{2(1 - v_u^2)qa}{E_u} = \frac{(1 - v_u)qa}{G'} \tag{14}$$

and the total (long term) settlement

$$\rho_t = \frac{2(1 - v'^2)qa}{E'} = \frac{(1 - v')qa}{G'} \tag{15}$$

Hence the proportion of immediate to total settlement is

$$\frac{\rho_u}{\rho_t} = \frac{1 - v_u}{1 - v'} = \frac{1}{2(1 - v')} \tag{16}$$

since $v_u = \frac{1}{2}$. It is significant that this ratio is dependent only on the value of v'; and not very sensitive to the actual value of v' within the restricted range 0.10 to 0.33 relevant to overconsolidated clay, as indicated in *Fig. 13*.

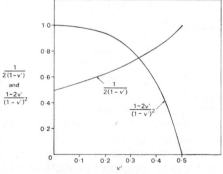

Fig. 13. *Variation of the factors $1/2(1 - v')$ and $(1 - 2v')/(1 - v')^2$ with v'*

The limited experimental evidence to date suggests that v' is a constant for a given clay independent of the overconsolidation ratio, Wroth (1972a); and in addition Calladine (1973) has suggested a possible relationship between v' and Plasticity Index, based on work by Wroth (1972b).

It is believed that the likely range of values for v' is 0.1–0.33 so that Equation 16 suggests that the ratio ρ_u/ρ_t lies in the range 0.55 to 0.75. Moreover, the values of this ratio are unlikely to be very different for other and more realistic loading conditions, and so this simple analysis suggests (assuming no consolidation occurs during the construction period) that the immediate settlement of a foundation on stiff clay will be of the order of 55 to 75% of the total settlement. In Section 10, a similar calculation is carried out on the basis that stiff clay is a non-homogeneous elastic material with moduli that increase linearly with depth. The results indicate that the likely values of ρ_u/ρ_t are reduced, and in particular for rafts on London Clay the suggested range is from 0.35 to 0.55. An assessment of the influence of the stiffness of a raft is also made in Section 10.

Simons and Som (1970) analysed 12 case records of settlement of major structures on overconsolidated clay and quoted a range of values for ρ_u/ρ_t from 0.315 to 0.735 with an average of 0.575. Morton and Au (1974) have studied 8 case records of buildings on London Clay (5 rafts, 3 piled foundations) and quote a range of 0.4 to 0.82 with an average of 0.63. The settlements of these recent buildings are not yet complete, so that the values quoted for ρ_u/ρ_t are slight overestimates. Both sets of records support the findings of the simple elastic analysis.

An alternative approach to the estimate of the total settlement is to integrate the compression for all elements down the centre-line beneath point O in *Fig. 12* assuming that they undergo no lateral strain and consolidate one-dimensionally. This is the settlement obtained from the classical interpretation of the oedometer test, and is denoted by ρ_{oed}.

For the condition of one-dimensional consolidation of a perfectly elastic material the coefficient of volume compressibility

$$m_{\text{vc}} = \frac{(1 - 2v')(1 + v')}{E'(1 - v')} = \frac{(1 - 2v')}{2G'(1 - v')} \tag{17}$$

Using this result it can be shown that

$$\rho_{\text{oed}} = \frac{(1 - 2v')}{G'(1 - v')}qa \tag{18}$$

and further, from Equation 15

$$\frac{\rho_{\text{oed}}}{\rho_t} = \frac{(1 - 2v')}{(1 - v')^2} \tag{19}$$

The latter ratio is plotted in *Fig. 13* and for $v' < 0.24$ it is greater than 0.9, i.e. the settlement ρ_{oed} calculated from the simple one-dimensional approach is within 10% of the correct value ρ_t (assuming, of course, that the appropriate value of m_{vc} has been adopted). The accurate measurement of m_{vc} is very important, and this is commented on later. It is likely that for other more realistic boundary conditions, such as a clay stratum of finite depth, the simple approach would lead to acceptable predictions and could be corrected by using Equation 19 to give a better indication of ρ_t.

The surprising accuracy of the 'one-dimensional' calculation can be attributed to the fact that the major part of the settlement is contributed by the heavily stressed elements just beneath the loaded area. The cumulative compression from O to P is plotted as a proportion of the total settlement ρ_{oed}, against depth as curve (a) in *Fig. 14*. We see that 60% of the settlement is due to the elements above a depth $z = 1.5a$, and 80% due to elements above a depth $z = 4a$. In a real situation where m_{vc} is not constant but decreases with depth, this effect will be even more marked. Also included in *Fig. 14* is the example discussed later of a circular raft on London Clay which is assumed to have a linear increase of Young's modulus with depth. In this case, 80% of the settlement is due to the soil above a depth $z = 1.25a$.

The simple estimates of settlement discussed above and based on perfect elasticity, clearly must be applied with caution, because certain heavily stressed elements close to the applied load may have yielded and undergone

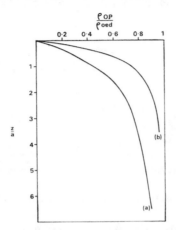

Fig. 14. Cumulative compression down centre line as proportion of total settlement for (a) uniform circular load on homogeneous half space (b) a circular raft on non-homogeneous half space

plastic deformations, and hence have invalidated the assumption of an elastic stress distribution. It is essential therefore to consider the actual stress paths that apply to such elements. This is done in the following section.

Uniform circular load on normally consolidated clay

The case of a uniform circular load acting on the surface of a normally or lightly overconsolidated clay is a more complicated problem to tackle analytically. Much of the clay beneath the load will have yielded, deformed plastically and undergone further virgin consolidation.

For ease of argument let us consider an element on the centre-line which has experienced an immediate increment of *effective* stress T'U' in *Fig. 15* where the point U' is on the current yield envelope. At this instant the element is experiencing a substantial excess pore pressure and, during the subsequent consolidation, the effective stress path will be of the form of U'V'. (The final equilibrium point V' will depend on the final distribution of the total stress in the ground.)

The behaviour of this element is compared with that of a similar element in an overconsolidated clay which is assumed to undergo identical increments of

637

effective stress $P'Q'$, $Q'R'$. If the two elements have the same elastic properties then the immediate responses of both elements will be identical with the same immediate settlements. However, the volume changes, and consequently the settlements during consolidation of the two elements, will be very different. The volume change of the yielding element as it follows the path $U'V'$ is controlled by the compression index C_c associated with normal consolidation. In contrast, the non-yielding element experiences a much smaller volume change associated with $Q'R'$ and controlled by the swelling index C_s.

For most clays the ratio C_c/C_s lies in the range 2 to 5. Adopting, say, a value of 4 for a typical clay then the consolidation settlement for the yielding

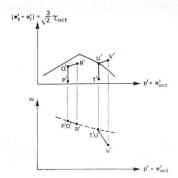

Fig. 15. Stress paths for elements under centre of uniform load

element will be 4 times that of the elastic element. Therefore for the yielding element

$$\frac{\rho_u}{\rho_t} = \frac{\rho_u}{\rho_u + \rho_{U'V'}} = \frac{\rho_u}{\rho_u + 4\rho_{Q'R'}}$$

But from Equation 16

$$\frac{\rho_u}{\rho_{Q'R'}} = 0.65 \text{ say}$$

then

$$\frac{\rho_u}{\rho_t} = \frac{0.65}{0.65 + 4} = 0.14 \tag{20}$$

This analysis shows that there is a major reduction in the *ratio* of immediate to total settlement for an element that yields. For any particular footing only a localised region of soil will yield so that the ratio for the *footing* will not fall as low as 0.14, but it will be the integrated result of those elements which yield (and make the major contribution to settlement) and those which do not. Yield may also occur during the undrained response of the soil, and since this possibility complicates the issue further it is not worth attempting to be more precise.

However, the review of nine case histories of buildings on normally consolidated clays by Simons and Som (1970) give values for ρ_u/ρ_t ranging from 0.077 to 0.212 with an average of 0.156. These values are close to the simple prediction above, and are in marked contrast to those relevant to overconsolidated clays quoted in the previous section. They clearly show that the problems of differential settlement will be much more severe for buildings founded on normally consolidated clays. As is well recognised by designers, the problem is doubly worse—greater total settlement and greater proportion of settlement occurring after the end of construction.

We can conclude from this study that

(1) there is a much smaller proportion of immediate to total settlement for structures founded on normally or lightly overconsolidated clay than those on overconsolidated clay, and that the problems of differential settlement are likely to be much more severe
(2) the proportion of immediate to total settlement for structures founded on overconsolidated clay can be estimated reasonably accurately from simple elastic theory given the relevant values of the elastic parameters.

10. Effect of soil properties on settlement predictions

In the last section the importance and likely magnitude of the proportion of settlement that will occur after the end of construction was discussed. An example was taken of the simplest possible form of the problem—a uniform circular loading on an elastic half-space—but the general findings should be equally valid for complicated structures with various combinations of foundations.

In this section the question of soil properties themselves will be considered in relation to the prediction of settlement of a given foundation. Until the advent of computers, nearly all settlement calculations were based on assuming (1) a Boussinesq stress distribution and (2) that the distribution of vertical total stress did not change during consolidation.

Finite element computations allow more realistic models of soil behaviour and more realistic boundary conditions to be adopted for any particular problem. The main advances have been in the representation of overconsolidated clays or dense sands as elastic materials with non-homogeneity or anisotropy or both. The form of non-homogeneity has generally consisted of a linear increase of elastic moduli with depth, and the type of anisotropy has been that of symmetry about the vertical axis, i.e. cross or orthotropic anisotropy.

Influence of non-homogeneity of elastic moduli

Recent work both in the field and in the laboratory has confirmed (1) the dependence of elastic moduli on the ambient level of effective stress and (2) that the values of moduli measured however carefully in the laboratory are substantially smaller than the corresponding values measured directly or indirectly *in situ* (see for example Marsland (1972), Atkinson (1974)).

Some typical variations of values of undrained Young's modulus with depth for a variety of clays are given in *Fig. 16*. It should be emphasised that all have been taken either directly from *in situ* tests or from back-analysis of field measurements; none has come from laboratory tests. The results for

London Clay are from plate tests at Chelsea by Marsland (1972) and a line selected by Hooper (1973) from a survey of various results; those for Boston Blue Clay and for four separate profiles at Canvey Island are from special pressuremeter tests by Hughes (1974). (For the two sets of results on soft clays, which are not elastic in their behaviour, tangent moduli have been taken at the start of the stress–strain curves.)

Values of moduli for soft rocks are quoted by Hobbs (1974) in his General Report to Session 4 of this Conference. He points out the difficulties of

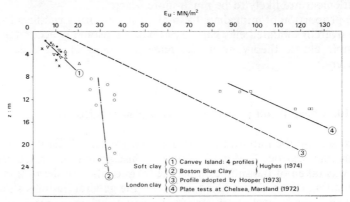

Fig. 16. Variation with depth of undrained moduli for clays from in situ tests

measuring the moduli accurately. Results of pressuremeter tests on sand are reported by Winter (1974).

Important papers have been published recently on the effects of this form of non-homogeneity on stress distributions and settlements. Theoretical contributions include Gibson (1967), Gibson, Brown and Andrews (1971), Gibson and Sills (1972), Brown and Gibson (1972), Awojobi and Gibson

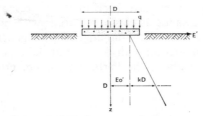

Fig. 17. Problem of smooth rigid plate on non-homogeneous half-space considered by Carrier and Christian

(1973) and Gibson (1974), and finite element computations have been carried out by Smith (1970) and Carrier and Christian (1973).

Particularly valuable is the last-named paper. In this a parametric study has been reported of the variation of soil properties on the settlement and stress distribution caused by a smooth rigid circular plate. The problem is indicated in *Fig. 17*, and for ease of reference the symbols chosen by Carrier and Christian have been adopted here.* The plate is of diameter D, subject

---

*With the minor exception that primes have been added to the symbol E to emphasise that it must be the relevant Young's modulus in terms of *effective* stresses.

640

to a vertical load with average contact stress q; the soil is considered to be a non-homogeneous elastic half-space with

$$E' = E_0' + kz \quad \text{and} \quad \nu = \text{constant} \tag{21}$$

Although this particular set of results will provide no information about *differential* settlement, it gives a clear indication of the effect on total settlement of the non-homogeneity of the ground.

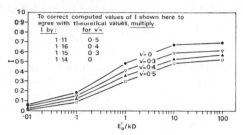

Fig. 18. *Influence factors for the settlement of a smooth rigid circular plate on a non-homogeneous elastic half-space (after Carrier and Christian)*

The main results of the finite element computations are presented in *Fig. 18* in dimensionless form. The influence factor I defined by

$$\frac{\rho}{D} = I\frac{q}{E_0'} \tag{22}$$

is plotted against E_0'/kD. It should be noted

(1) that the latter parameter is a measure of how rapidly E' increases with depth
(2) that for a given soil the parameter will change with the size of the plate
(3) that the parameter is plotted to a logarithmic scale
(4) that the homogeneous case is given by $E_0'/kD \rightarrow \infty$.

The main implications of these results can best be appreciated by considering a specific case of a stiff circular raft founded directly on London Clay, assumed to have the properties in *Fig. 16*. Adopting the values used by Hooper (1973) (see *Fig. 16*) in his successful finite element computations for matching the measured performance of a piled-raft foundation, we have

$$\nu' = 0.1, \ \text{E}' = 0.75E_u = 0.75 \ (10 + 5.2z) \ \text{MN/m}^2$$

where z is measured in metres. Choosing an imaginary raft of 21 m diameter with $q = 210 \ \text{kN/m}^2$ gives the same area and loading* as that of the raft supporting the flats, Hurley Road Block II, reported by Morton and Au (1974). The settlement performances of these two rafts will be compared in some detail. We have

$$\frac{E_0'}{kD} = \frac{7.5}{0.75 \times 5.2 \times 21} = 0.09$$

so that from *Fig. 18*, $I \simeq 0.15$. But this value has to be corrected by a factor of 1.14 to allow for the effect of a finite depth of soil assumed in the finite

* The difference in behaviour between an essentially square raft and a circular one of equal area is considered to be negligible; the loading is the gross pressure applied to the raft, as any heave due to excavation has not been included in the settlement observations

element computations. Therefore the computed total settlement for the imaginary raft is

$$\rho_t = 1.14\frac{IqD}{E_0'} = \frac{1.14 \times 0.15 \times 210 \times 21}{7.5 \times 10^3}\text{m} = 100 \text{ mm} \qquad (23)$$

which compares very favourably with the observed total settlement of the real raft in *Fig. 5* of Morton and Au (1974). However, if the clay were assumed to be homogeneous the total settlement would be predicted by the formula

$$\rho_t = \frac{q\pi D(1 - v'^2)}{4E'} \qquad (24)$$

The selection of a suitable average value of Young's modulus is required; if we consider the classical case of one-dimensional consolidation, the plot of *Fig. 14* shows that half of the central settlement would be due to the soil above a depth $z = 1.2a$, and the other half to the soil below that depth. We might proceed therefore by selecting

$$E' = 7.5 + 3.9 \times (1.2 \times 10.5) = 55.5 \text{ MN/m}^2 \qquad (25)$$

Substituting into Equation 24 gives $\rho_t = 60$ mm which is an underestimate by a factor of 2. This suggests that a prediction of settlement based on a homogeneous model of the soil could be misleading, depending on the choice of the equivalent value of Young's modulus for use in Equation 24.

For the non-homogeneous case if the diameter of the raft is increased say by a factor of 2 the parameter E_0'/kD is halved but the influence factor is reduced by only about one-sixth. The predicted settlement will be relatively insensitive to I, but will be almost directly proportional to the selected diameter for a given intensity of loading.

The proportion of immediate to total settlement can also be deduced from Carrier and Christian's tabulated results. Since the drained and undrained moduli are in constant proportion regardless of depth, the parameter E_0/kD is the same for both cases (for a given value of D), but will lead to different influence factors I' and I_u according to $v = v'$ and $v = \frac{1}{2}$ respectively. Hence

$$\frac{\rho_u}{\rho_t} = \frac{I_u DqE_0'}{I' DqE_{u_0}} = \frac{1 + v'}{1 + v_u} \cdot \frac{I_u}{I'} = \frac{1 + v'}{1.5} \frac{I_u}{I'} \qquad (26)$$

For a given value of E_0'/kD the corresponding influence factors have been taken from Carrier and Christian's table and substituted in Equation 26 to give the curves plotted in *Fig. 19*. The corresponding curve for a uniform circular load on a homogeneous half-space taken from *Fig. 13* is included and as might be expected is indistinguishable from the case of a rigid plate on a homogeneous half-space given by $E_0'/kD = \infty$.

However, there is a significant reduction in the ratio of immediate to total settlement as the value of E_0'/kD decreases. For example, for the imaginary raft of 21 m diameter on London Clay $E_0'/kD = 0.09$ and $v' = 0.1$ so that $\rho_u/\rho_t = 0.34$. Had the clay been considered to be homogeneous the ratio would have been 0.55. The value quoted by Morton and Au for the real raft is 0.63. However, the figure is not directly comparable as the top 5 m of ground is gravel and this will not have contributed to the long term settlement; the observed ratio would have been smaller had the real raft been founded directly on the London Clay. In addition some of the consolidation settlement will have occurred during the time of construction so that the observed ρ_u is larger than the comparable value deduced from elastic theory.

Carrier and Christian carried out similar calculations to those discussed here for a rough rigid plate as opposed to a smooth one, and have concluded for all practical purposes that the roughness of a plate or raft has a negligible effect and can be discounted.

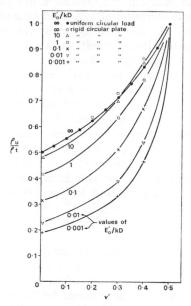

Fig. 19. Ratio of immediate to total settlements of a smooth rigid circular plate on an elastic half-space for different degrees of non-homogeneity

Influence of anisotropy

The influence of transverse anisotropy on the stress distributions and displacements within an elastic half-space caused by some surface loading, appears not to have been studied very extensively. King and Chandrasekaran (1974) have studied the case of a two-bay continuous frame supported by a single raft founded on London Clay which was represented as a non-homogeneous elastic material. By means of finite element computations they have considered the influence of anisotropy, and they have shown for that particular case that the anisotropy reduces the computed settlements of the raft to a minor extent. Closed form solutions for various loadings on anisotropic half-spaces have been obtained by Gerrard and Wardle (1973a, 1973b).

The limited evidence suggests that the only significant influence of transverse anisotropy is on the distribution of horizontal stresses. This is not likely to be important in predictions of settlement caused by vertical loading; but it will be important in the design of retaining walls and for excavations where horizontal stresses and displacements are the major concern.

This section has so far concentrated exclusively on assessing the effect of the properties of overconsolidated clays on predictions of settlement. As was discussed in the last section, and is evident from the General Report by Simons (1974), the estimate of settlements on normally consolidated soils is a much more complex problem. An indication has already been given of the consequences of some elements of soil yielding and undergoing virgin consolidation. However, most so-called 'normally consolidated' clays are in fact lightly overconsolidated, Parry (1968), so that for relatively small increments of load, only a very small region, if any, of the soil yields and hence the primary settlement may be adequately predicted on the basis of an elastic computation. The secondary consolidation may present a problem.

For heavy loads on soft clays, such as those resulting from motorway embankments or large storage tanks for bulk liquids, an elastic analysis will be inappropriate. Several attempts have been made to obtain complete solutions to the load-settlement response of structures with single foundations on normally consolidated clay, by means of finite element computations using complicated elastic-plastic models of soil behaviour. Circular loads have been studied by Hoëg, Andersland and Rolfsen (1969), and strip loads by Hoëg, Christian and Whitman (1968), by D'Appolonia, Lambe and Poulos (1971), by Naylor and Zienkiewicz (1972), by Burland (1972) and by Wroth and Simpson (1972).

Some encouragement can be gained from the fact that the classical method of calculation of settlement based on the assumption of one-dimensional consolidation with the *appropriate* value of m_v or C_c, may be surprisingly accurate. Close agreement has been obtained by Burland (1969) with observations on a model strip footing, and by Naylor and Zienkiewicz (1972) with finite element computations using a critical state model based on the same simple soil parameters.

11. Estimate of differential settlement

There appears to be very little in the literature that deals specifically with differential settlement, that is the relative *displacements* of a single foundation. Account has been taken of the relative stiffness of a raft compared to that of the underlying soil, but the published work concentrates on the *stress distribution* of the contact pressure on the raft. This information is, of course, important from the point of view of designing the raft itself, but gives no direct indication or insight into the relative displacements that it will suffer. Consequently it is not usually possible to estimate readily the value of Δ/L and check the likelihood of damage by using the criteria of *Fig. 11*.

Implicit in such publications is the assumption that there is no lateral variation in the properties of the underlying soil strata, so that no tilt of the structure is expected. This means that in essence no distinction is made between differential settlement $\delta\rho$, and the relative deflection Δ of a structure (see the definitions in Section 2). But it is the latter that is liable to cause damage to the structure.

This assumption is also adopted in the following remarks, and it is important that for any specific case its validity should be questioned.

Brown (1969a and b) has published solutions for a uniformly loaded circular raft resting on a homogeneous elastic layer of different depths. We can use his charts to estimate the relative deflection for our imaginary raft if we are prepared to accept that London Clay acts as a homogeneous elastic material.

We first have to assess the relative stiffness of the raft, defined by Brown as

$$K = \frac{E_r(1 - v_f{}^2)}{E_f} \left(\frac{t}{a}\right)^3 \tag{27}$$

where t is the thickness of the raft, a its radius and the suffices r and f refer to raft and foundation respectively. Taking $t = 0.91$ m, $a = 10.5$ m as for the Hurley Road flats and $E_r = 25 \times 10^3$ MN/m² for the concrete the properties of the raft are well defined. However, the choice of appropriate parameters for the soil is more difficult. We have already seen that selection of a single average value of Young's modulus for the clay, Equation 25, is open to question.

With hindsight we can adopt values of E_u and E' that give the observed maximum settlements of the real raft, say 68 mm and 120 mm, i.e. $E_u \simeq 55$ MN/m² and $E' \simeq 28.4$ MN/m². Substituting these in Equation 27 with the respective values of Poisson's ratio of 0.5 and 0.1 leads to $K_u = 0.222$ and $K' = 0.566$. Reading off the chart in Brown (1969a) gives the maximum relative deflections of $\Delta_u = 15$ mm and $\Delta_t = 23$ mm, compared with the observed values of 6.3 and 10 mm. The proportions of relative deflection to maximum settlement are

$$\frac{\Delta_u}{\rho_u} = \frac{15}{68} = 0.22 \quad \text{and} \quad \frac{\Delta_t}{\rho_t} = \frac{23}{120} = 0.19$$

These computations give some idea of likely values, but apply only to a homogeneous soil. To date, there do not seem to be any reported computations of differential settlements (or relative deflections) of rafts of varying stiffnesses on non-homogeneous soils. There is clearly a need for a parametric survey along the lines conducted by Carrier and Christian, but for various values of raft stiffness.

In order to assess the influence of non-homogeneity on differential settlements, finite element computations have been carried out at the Building

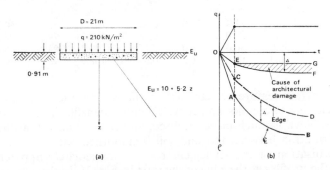

Fig. 20. Details of raft chosen for computations

Research Station for the imaginary raft depicted in *Fig. 20*. Three cases have been considered:

(1) a raft of zero stiffness
(2) a raft of the same stiffness as that of the Hurley Road flats
(3) a raft 5 times as stiff as case (ii)

This last case is included to assess the effect of an increase of raft stiffness due to its interaction with the superstructure. The contribution to extra stiffness from the superstructure will not be proportional to $(t'/a)^3$ where t' is the 'equivalent' depth of raft and superstructure, and it is impossible to assess its magnitude. The equivalent shear stiffness, as discussed in Section 4, will be more relevant because of the small value of L/H. The factor of 5 selected above is therefore arbitrary.

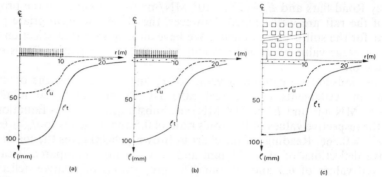

Fig. 21. *Surface settlement profiles for three different stiffnesses of circular raft on London Clay*

The computed profiles of the immediate and total settlements the ground surface are plotted in *Fig. 21*, and the key values tabulated below.

Table 1: COMPUTED AND OBSERVED SETTLEMENTS OF RAFT ON LONDON CLAY

	Immediate				*Long-term*			
	C.L. ρ (mm)	Edge ρ (mm)	Δ_u (mm)	Δ_u/ρ_u	C.L. ρ (mm)	Edge ρ (mm)	Δ_t (mm)	Δ_t/ρ_t
Case (1)	44.2	21.2	23.0	0.52	103.9	53.4	50.5	0.49
Case (2)	43.4	27.0	16.4	0.38	102.0	79.8	22.2	0.22
Case (3)	36.0	30.0	6.0	0.17	90.7	84.4	6.4	0.07
Observed	68.3	62.0	6.3	0.09	108.9	98.9	10.0	0.09

A very important feature brought out by the computations is that the non-homogeneity has two major and complementary benefits. The first is that the 'settlement bowl' is much more localised, and the second is that the contact stress distribution between raft and soil is smoothed out.

The localisation of the settlement bowl is immediately apparent in the 'kinks' in the profiles at the edge of the rafts in *Fig. 21*. In the extreme case of

646

a soil with zero modulus at the surface (i.e. $E_0' = 0$) Gibson (1967) has shown that the immediate surface settlement outside the loaded area is zero everywhere. An alternative way of considering this is to integrate the compression down the centre line. This has been done for the long term settlement of case (2) and plotted as curve (b) in *Fig. 14*, where it can be seen that the settlement is contributed by the layers of soil very close to the surface—much more so than for the uniform circular load on a homogeneous soil represented by curve (a). A major consequence of this finding is that accurate measurements of the soil properties, and particularly of m_{vc}, must be concentrated near the surface of the ground.

The reduction in the size of the zone of influence of a single footing due to non-homogeneity of the soil has the important consequence that it will markedly reduce the effect that one footing has on the behaviour of a neighbouring footing or nearby building. However, *immediately* adjacent to a footing the profile of surface settlement shows a greater curvature than for a non-homogeneous soil; and this effect increases with raft stiffness. In a qualitative way, the stiffer the raft and the more non-homogeneous the soil, the closer the deformations of the soil become to a simple punch failure. The likelihood of causing damage to existing structures is increased very close to the raft, but is substantially decreased with distance.

In this section we have concentrated on the topic of differential settlement of single foundations in stiff clays. In the last section it was shown that it is difficult to solve completely the boundary value problem of a footing on normally consolidated clay, and compute all the boundary stresses and displacements. This difficulty will apply equally to the prediction of differential settlements. It is an area of foundation engineering that needs urgent attention—both theoretically and experimentally from high quality case records.

We can conclude for overconsolidated clays that the influence of non-homogeneity of the soil properties (especially Young's modulus) is important, since ρ, ρ_u/ρ_t, Δ/L and the profile of surface settlement are directly affected. We now have the numerical techniques with which to handle the non-homogeneity. However, although a start has been made on the compilation of charts to aid the designer, much more work along these lines could usefully be done.

There is some evidence to suggest that the ratio of relative deflection to maximum settlement does not alter significantly during consolidation, as pointed out by Morton and Au. Although this finding would not help during the design stage, it could prove useful as a check on the likely future behaviour of a major structure at the end of construction, at a time when remedial measures might still be taken.

12. Buildings with complex foundations

So far we have been concerned only with the behaviour of a structure founded on a single raft foundation. Even for this rather restricted range of cases, it has been shown that the prediction of differential settlement, both in the short and long term, is very difficult.

In reality, a structure will not behave monolithically as a single beam (as was assumed in the first part of this report), but even if a satisfactory model

647

of its behaviour can be found, this model will have to take account of the changing stiffness of the structure, as building proceeds and while the immediate settlement is occurring. The effect of changing stiffness during construction has been considered by De Jong (1971)* who studied in detail the settlements of four multi-storey structures in Edmonton, Alberta. The need for including this effect in any proper solution of soil-structure interaction was emphasised, but this does require a large and involved finite element computation.

In many cases the foundation will not consist of a single raft, but may be formed by piles, piled-raft or a series of discrete foundations which interact with a continuous superstructure. Such cases can now be handled on a rational basis by suitable finite element computations, like those reported to this Conference by Naylor and Hooper (1974).

The case of a structure built on separate footings raises extra complications. If the ground can reasonably be assumed to have uniform properties laterally then the characteristic behaviour of each separate footing, when considered in isolation, should be the same. Assuming that this behaviour can be treated as quasi-elastic then the settlement ρ of any footing will be directly proportional both to the intensity of loading q and to some representative dimension such as the diameter d (cf. Equations 18, 22 and 24), i.e.

$$\rho = k_1 q d \qquad (28)$$

where k_1 is some constant (for all the footings at one particular site). For each footing the value of q will be proportional to P/d^2 where P is the applied load, so that we can write

$$\rho = \frac{k_2 P}{d} \qquad (29)$$

where k_2 will be constant, if the footings are all the same shape.

Equation 29 relates to individual isolated footings and does not take account of interaction between adjacent footings. Interaction can be significant when the ground is homogeneous or when there is a compressible stratum at some depth beneath the founding level. However, when the stiffness of the ground increases with depth, the surface settlement profile is concentrated around each footing thereby considerably reducing interaction (see Section 11).

In order not to induce undesirable locked-in movements and stresses in the continuous superstructure, it would seem advisable to design each footing to settle by the same amount. It follows from Equation 29 that the size of the footing should be proportional to the vertical *load* it is designed to carry, and *not* the average applied stress. This recommendation has been made by Levy and Morton (1974). It has the consequence that in order to achieve equal settlements, internal foundations carrying greater loads, and hence of greater size and lower pressures, might themselves have to be more conservatively designed than the external foundations.

Finite element computations for a framed structure built on separate foundations have been published by Larnach and Wood (1972), and Larnach (1972) has also pointed out that different *rates* of consolidation of the ground, under separate footings, even with the same compressibility will lead to permanent differential settlement and hence to locked-in movement and stresses in the superstructure.

* Also published by De Jong and Harris (1971).

It is our belief, however, that such complex computations will seldom be warranted because of the major uncertainties in

(1) the equivalent stiffness of the superstructure
(2) the varying stiffness during construction (the pattern of which may radically change during the vagaries of the contractor's actual programme of work)
(3) the properties of the ground.

It would seem better practice in most cases to adopt a simple design philosophy, and then to monitor carefully the actual behaviour of the structure during the time that settlements occur, having made provision for remedial measures should these prove necessary.

13. Summary and concluding remarks

Damage due to settlement of buildings is usually confined to the cladding and finishes. It is therefore the relative deflections and rotations that occur subsequent to the application of the finishes which are important (see *Fig. 20(b)*). As a first approximation the settlement that can give rise to damage is equal to $\rho_t - \rho_u$ and the ratio of immediate to total settlement (ρ_u/ρ_t) is therefore an important quantity. For heavily overconsolidated clays ρ_u/ρ_t is usually greater than 0.4 and observed values average about 0.6. On the other hand, for normally and lightly overconsolidated clays ρ_u/ρ_t will generally be less than 0.2. The low values of ρ_u/ρ_t coupled with larger total settlements makes the problems of design for normally consolidated clays much more severe than for overconsolidated clays.

Theoretical studies suggest that the classical method of estimating total settlement based on the assumption of one-dimensional compression is surprisingly accurate both for normally consolidated and heavily overconsolidated clays. Accurate estimates of settlement are much more dependent, therefore, on correct measurement of simple parameters such as m_v and C_c than on sophisticated calculations using complex stress–strain laws.

In the past much effort has been devoted to calculating stress-distributions on rafts and in structures due to interaction with the soil. Such studies may be important to the design of the structural members but are not very useful for assessing architectural damage (Skempton and MacDonald 1956; Fjeld 1963). Much more emphasis needs to be placed on estimating relative displacements with the stiffness of the building being accounted for in simple global terms. In Section 12 the approach is illustrated using a simple example of a circular raft (see *Fig. 21*). Naylor and Hooper (1974) have used a similar approach in their paper for predicting the behaviour of a piled raft.

Non-homogeneity in the form of increasing stiffness with depth is important since ρ, ρ_u/ρ_t, Δ/L and the profile of surface settlement are significantly affected. There is a need for the compilation of charts which take account of non-homogeneity of the ground and building stiffness to aid the designer in estimating the above quantities.

A new approach to the assessment of likely settlement damage has been outlined which is based on the assumption that the onset of visible cracking is associated with a critical tensile strain ε_{crit}. For brickwork and blockwork, set in cement mortar, average values of ε_{crit} lie in the range of 0.05 to 0.1% and for reinforced concrete ε_{crit} appears to lie between 0.03 and 0.05%.

The concept of critical tensile strain can be used to estimate the relative deflections at which a given structure will show visible cracking. In Section 4 the application of this concept has been illustrated for the simple case of a uniform rectangular beam. This study indicates that, for a given value of ε_{crit}, the magnitude of $\varDelta/L$ is significantly dependent on

(1) the length to height ratio
(2) the relative stiffness in shear and in bending
(3) the degree of tensile restraint built into the structure and its foundations
(4) the mode of deformation.

The results of a large number of case records, many of them reported to this Conference, have been presented in *Fig. 11*. The results show that the allowable limit of angular distortion of 1/500 proposed by Skempton and MacDonald (1956) is satisfactory for frame buildings of both traditional and modern construction, but may be conservative for values of $L/H > 3$. However, the above criterion is unsafe for load-bearing walls with $L/H < 4$ and the criterion proposed by Polshin and Tokar (1957), which limits the value of $\varDelta/L$ in relation to L/H, appears to be more appropriate. Even this criterion is not adequate for load-bearing walls undergoing hogging since cracking has been observed to occur at values of $\varDelta/L$ which are half those recommended by Polshin and Tokar. It is of interest to note that the criteria for the cracking of simple beams based on a value of $\varepsilon_{crit} = 0.075\%$ are in good agreement with the observations plotted in *Fig. 11*.

The observations on the cracking of load-bearing walls undergoing hogging are of particular importance since the movements are so small. Such structures, which include most old buildings, are particularly vulnerable to damage from nearby construction which gives rise to hogging of the surrounding ground surface, such as excavations, tunnels or heavy buildings.

In the Introduction it was stressed that damage due to settlement was only one aspect of the much broader problem of serviceability. The problem of coping with differential settlement, as with creep, shrinkage and structural deflections, may be solved best by designing the building to accommodate movements rather than resist them. This can be achieved only if the problem is recognised at an early stage in the design and if there is cooperation between foundation engineer, structural engineer and architect. The papers to this Conference contain a number of interesting examples of buildings designed to accommodate movement.

Cowley *et al.* (1974) describe cold stores designed to accommodate large differential settlements. The success of the scheme can be judged by the fact that angular distortions as large as 1/125 were experienced without damage. Webb (1974) describes the performance of a two-storey reinforced concrete structure designed to accommodate large settlements. The structure was pre-loaded by storing building materials in it prior to fixing the cladding and finishes. In spite of large initial differential settlements subsequent settlement has been uniform. A feature of particular interest is that two columns which had settled excessively were raised periodically by jacking, after cutting with a thermic lance. Provision for jacking may well be economic in certain cases.

The above two cases are examples of very flexible buildings. In contrast, Sanglerat *et al.* (1974) give details of a building which underwent total settlements of between 200 mm and 1800 mm without damage. This was achieved by the use of a stiff cellular raft and by a number of separation joints, one of which opened up by about 100 mm. Thorburn and McVicar (1974) used joints in a long low-rise building in which considerable stiffness was achieved by using precast concrete panels.

The theme of this Report has been to show that the prediction of differential settlements and of the susceptibility of buildings to damage is a very complex problem. In general terms we consider that at this stage in the development of foundation engineering, design philosophy should be concentrated on a simple approach, that this approach should be based on assessing the global stiffness of a structure and then estimating the relative deflections that will result from its interaction with the ground. The likely relative deflections can then be compared with the criteria of *Fig. 11* and if large enough to suggest that damage will occur, either the design of the structure or of the foundations can be modified, or provision can be made for remedial measures to be taken if and when they are required.

It is apparent that there is an urgent need for case histories which include detailed observations of the performance of the structure and its finishes as well as the foundation movements. Regional studies of the type conducted by Morton and Au (1974), Thorburn and McVicar (1974) and Breth and Amman (1974) are of outstanding importance in the design of future buildings in a given locality or on a given stratum.

Acknowledgements

The authors have had helpful discussions with many of their colleagues and are particularly indebted to Mr. G. Mitchell, Mr. G. Weeks and Mr. M. A. Pyle, all from B.R.S., and Dr. G. S. Littlejohn and Dr. I. A. MacLeod. We wish to thank Mr. K. Tarr of B.R.S. who carried out the finite element computations described in Section 11. The Paper is published by permission of the Director of the Building Research Establishment.

REFERENCES

Atkinson, J. H. (1973). Contribution to Discussion, Session IV, *Proc. Symp. on Field Instrumentation*, Butterworths, London

Awojobi, A. O. and Gibson, R. E. (1973). 'Plane Strain and Axially Symmetric Problems of a Linearly Non-homogeneous Elastic Half-space', *Q.J. Mech. & Appl. Maths*, Vol. 26, 285–302

Base, G. D., Read, J. B., Beeby, A. W. and Taylor, H. P. J. (1966). 'An Investigation of the Crack Control Characteristics of Various Types of Bar in Reinforced Concrete Beams', *Cement and Concrete Association, Research Report* 18, Pts. 1 and 2

Beeby, A. W. and Miles, J. R. (1969). 'Proposals for the Control of Deflection in the New Unified Code', *Concrete*, Vol. 3, No. 3, 101–110

Bjerrum, L. (1963). Discussion, *Proc. European Conf. Soil Mech. & Found. Eng.*, Wiesbaden, Vol. III, 135

Breth, H. and Amann, P. (1974). 'Time-Settlement and Settlement-Distribution with Depth in Frankfurt Clay,' *Proc. Conf. Settlement of Structures*, Cambridge, 141–154

Breth, H. and Chambosse, G. (1974). 'Settlement Behaviour of Buildings Above Subway Tunnels in Frankfurt Clay', *Proc. Conf. Settlement of Structures*, Cambridge, 329–336

Brown, P. T. (1969a). 'Numerical Analyses of Uniformly Loaded Circular Rafts on Elastic Layers of Finite Depth,' *Géotechnique*, Vol. 19, 301–306

Brown, P. T. (1969b). 'Numerica l Analyses, of Uniformly Loaded Circular Rafts on Deep Elastic Foundations, *Géotechnique*, Vol. 19, 399–404

Brown, P. T. and Gibson, R. E. (1972). 'Surface Settlement of a Deep Elastic Stratum whose Modulus Increases Linearly with Depth,' *Canadian Geotech. J.*, Vol. 9, 467–476

Burhouse, P. (1969). 'Composite Action Between Brick Panel Walls and their Supporting Beams,' *Proc. Instn. Civ. Engrs.*, Vol. 43, 175–194

Burland, J. B. (1971). 'A Method of Estimating the Pore Pressures and Displacements Beneath Embankments on Soft Natural Clay Deposits', *Proc. Roscoe Memorial Symp.*, Foulis, 505–536

Burland, J. B. (1969). Discussion to Session 2, *Proc. 7th Int. Conf. Soil Mech. & Found. Engng.*, Vol. 3, 248–250

Calladine, C. R. (1973). 'Overconsolidated Clay: a Microstructural View', *Proc. Symp. Plasticity in Soil Mechanics*, Cambridge, 144–158

Carrier, W. D. III and Christian, J. T. (1973). 'Rigid Circular Plate Resting on a Non-homogeneous Elastic Half-space', *Géotechnique*, Vol. 23, 67–84

Cowley, B. E., Haggar, E. G. and Larnach, W. J. (1974). 'A Comparison Between the Observed and Estimated Settlements of Three Large Cold Stores in Grimsby', *Proc. Conf. Settlement of Structures*, Cambridge, 79–90

Cheney, J. E. and Burford, D. (1974). 'Damaging Uplift to a Three-storey Office Block Constructed on a Clay Soil Following the Removal of Trees', *Proc. Conf. Settlement of Structures*, Cambridge, 337–343

D'Appolonia, D. J., Lambe, T. W. and Poulos, H. G. (1971). 'Evaluation of Pore-Pressures Beneath an Embankment', *J. Soil Mech. & Found Div.*, *A.S.C.E.*, Vol. 97, 881–897

DeJong, J. (1971). 'Foundation Displacements of Multi-storey Structures', PhD Thesis, Univ. of Alberta.

DeJong, J. and Harris, M. C. (1971). 'Settlements of Two Multi-storey Buildings in Edmonton, *Canadian Geotech. J.*, Vol. 8, 217–235

Fjeld, S. (1963). 'Settlement Damage to a Concrete-framed Structure', *Proc. European Conf. Soil Mech. & Found Engng.*, Wiesbaden, Vol. 1, 391

Gerrard, C. M. and Wardle, L. J. (1973a). 'Solutions for Point Loads and Generalized Circular Loads Applied to a Cross-anisotropic Half-space', *Div. Appl. Geomechanics*, Technical Paper 13, *C.S.I.R.O.*, Australia

Gerrard, C. M. and Wardle, L. J. (1973b). 'Solutions for Line Loads and Generalized Strip Loads Applied to an Orthorhombic Half-space', *Div. Appl. Geomechanics*, Technical Paper 14, *C.S.I.R.O.*, Australia

Gibson, R. E. (1974). 'The Analytical Method in Soil Mechanics', 14th Rankine Lecture, *Géotechnique*, Vol. 20, No. 2

Gibson, R. E. (1967). 'Some Results Concerning Displacements and Stresses in a Non-homogeneous Elastic Half-space', *Géotechnique*, Vol. 17, 58–67

Gibson, R. E., Brown, P. T. and Andrews, K.R.F. (1971). 'Some Results Concerning Displacements in a Non-homogeneous Elastic Layer', *Z.A.M.P.*, Vol. 22, 855–864

Gibson, R. E. and Sills, G. C. (1971). 'Some Results Concerning the Plane Deformation of a Non-homogeneous Elastic Half-space', *Proc. Roscoe Memorial Symp.*, Foulis, 564–572

Hobbs, N. B. (1974): 'The Prediction of Settlement of Structures on Rock', *Proc. Conf. on Settlement of Structures*, Cambridge

Hoëg, K., Andersland, O. B. and Rolfsen, E. N. (1969). 'Undrained Behaviour of Quick Clay Under Load Tests at Åsrum', *Géotechnique*, Vol. 19, 101–115

Hoëg, K., Christian, J. T. and Whitman, R. T. (1968). 'Settlement of Strip Load on Elastic-plastic Soil,' *J. Soil Mech. & Found. Div.*, *A.S.C.E.*, Vol. 94, 431–445

Hooper, J. A. (1973). 'Observations on the Behaviour of a Piled-raft Foundation on London Clay', *Proc. Inst. Civ. Engrs.*, Vol. 55, 855–877

Horn, H. M., and Lambe, T. W. (1964). 'Settlement of Buildings on the M.I.T. Campus', *Proc. A.S.C.E.*, Vol. 90, SM5, 181–195

Hughes, J. M. O. (1974). 'Result of Tests in Soft Clays with the Camkometer', Private communication

King, G. J. W. and Chandrasekaran, V. S. (1974). 'An Assessment of the Effects of Interaction between a Structure and its Foundation', *Proc. Conf. Settlement of Structures*, Cambridge 368–383

Larnach, W. J. (1972). Contribution to discussion, Session 2. *Proc. Symp. Interaction of Structure & Foundation*, Birmingham, 169–173

Larnach, W. J. and Wood, L. A. (1972). 'The Effect of Soil-structure Interaction on Settlements', *Int. Symp. Computer-aided Structural Design*, Univ. of Warwick

Levy, J. F. and Morton, K. (1974). 'Loading Tests and Settlement Observations on Granular Soils', *Proc. Conf. Settlement of Structures*, Cambridge, 43–52

Little, M. E. R. (1969). Discussion, Session 6, *Proc. Symp. Design for Movement in Buildings*, The Concrete Society, London

Littlejohn, G. S. (1974). 'Observations of Brick Walls Subjected to Mining Subsidence', *Proc. Conf. Settlement of Structures*, Cambridge, 384–393

Mainstone, R. J. (1971). 'On the Stiffness and Strengths of Infilled Frames,' *Proc. Instn. Civ. Engrs.*, Suppl. (IV), Paper 7360S, 57–90

Mainstone, R. J. and Weeks, G. A. (1970). 'The Influence of a Bounding Frame on the Racking Stiffnesses and Strengths of Brick Walls', *S.I.B.M.A.C. Proc.*, 165–171

Marsland, A. (1972). 'Laboratory and in situ Measurements of the Deformation Moduli of London Clay', *Proc. Symp. Interaction of Structure and Foundation*, Birmingham, 7–17

Mayer, H. (1966). 'Bauschäden als Folge der Durchbiegung von Stahlbeton-Bauteilen, Munich, Materialprüfungsamt für das Bauwesen der Technischen Hochschule, München, Report No. 68, 45

Meyerhof, G. G. (1956). Discussion on Paper by Skempton, A. W. and MacDonald, D. H., 'The Allowable Settlements of Buildings,' *Proc. Instn. Civ. Engrs.*, Pt. II, Vol 5, 774

Morton, K. and Au, E. (1974). 'Settlement Observations on Eight Structures in London', *Proc. Conf. Settlement of Structures*, Cambridge, 183–203

Naylor, D. J. and Hooper, J. A. (1974). 'An Effective Stress Finite Element Analysis to Predict the Short and Long Term Behaviour of a Piled-raft Foundation on London Clay', *Proc. Conf. Settlement of Structures*, Cambridge, 394–402

Naylor, D. J. and Zienkiewicz, O. C. (1972). 'Settlement Analysis of a Strip Footing Using a Critical State Model in Conjunction with Finite Elements', *Proc. Symp. Interaction of Structure and Foundation*, Birmingham, 93–107

Parry, R. H. G. (1968). 'Field and Laboratory Behaviour of a Lightly Overconsolidated Clay', *Géotechnique*, Vol. 18, 151–171

Peck, R. B., Deere, D. U. and Capacete, J. L. (1956). Discussion on Paper by Skempton, A. W. and MacDonald, D. H., 'The Allowable Settlements of Buildings', *Proc. Instn. Civ. Engrs.*, Pt. II, Vol. 5, 778

Polshin, D. E. and Tokar, R. A. (1957). 'Maximum Allowable Non-uniform Settlement of Structures', *4th Int. Conf. Soil Mech. & Found. Engng.*, Vol. 1, 402

Pryke, J. F. S. (1974). 'Differential Foundation Movement of Domestic Buildings in South East England—Distribution, Investigation, Causes and Remedies,' *Proc. Conf. Settlement of Structures*, Cambridge, 403–419

Rigby, C. A. and Dekema, (1952). 'Crack-resistant Housing', *Publ. Works of S.A.*, Vol. II, No. 95

Samuels, S. G. and Cheney, J. E. (1974). 'Long-term Heave of a Building on Clay Due to Tree Removal', *Proc. Conf. Settlement of Structures*, Cambridge, 212–220

Sanglerat, G., Girousse, L. and Gielly, J. (1974). 'Unusual Settlements of a Building at Nantua (France),' *Proc. Conf. Settlement of Structures*, Cambridge, 123–131

Simons, N. E. (1974). 'The Settlement of Structures on Normally Consolidated and Lightly Overconsolidated Cohesive Materisls', *Proc. Conf. Settlement of Structures*, Cambridge

Simons, N. E. and Som, N. N. (1970). 'Settlement of Structures on Clay, with Particular Emphasis on London Clay', *C.I.R.I.A.* Report 22

Skempton, A. W. and MacDonald, D. H. (1956). 'Allowable Settlement of Buildings', *Proc. Instn. Civ. Engrs.*, Pt. III, Vol. 5, 727–768

Smith, I. M. (1970). 'A Finite Element Approach to Elastic Soil-structure Interaction', *Canadian Geotech J.*, 7, 95–105

Terzaghi, K. (1935). 'The Actual Factor of Safety in Foundations', *Struct. Eng.*, Vol. 13, 126

Thorburn, S. and McVicar, R. S. L. (1974). 'The Performance of Buildings Founded on River Alluvium', *Proc. Conf. Settlement of Structures*, Cambridge, 425–442

Timoshenko, S. (1957). *Strength of Materials*, Pt. I, van Nostrand, London

Timoshenko, S. and Goodier, J. N. (1951). *Theory of Elasticity*, 2nd edn., McGraw Hill

653

Tschebotarioff, G. P. (1938). 'Settlement Studies of Structures in Egypt', *Proc. A.S.C.E.*, Vol. 64

Vargas, M. and Silva, F. P. (1973). 'Settlement of Tall Buildings in São Paulo and Santos', *Proc. Pan Am. Regional Conf. on Tall Buildings*, Porto Alegre

Ward, W. H. (1956). Discussion on Paper by Skempton, A. W. and MacDonald, D. H., 'The Allowable Settlements of Buildings', *Proc. Instn. Civ. Engrs.*, Pt. III, Vol. 5, 782

Webb, D. L. (1974). 'Observed Settlement and Cracking of a Reinforced Concrete Structure Founded on Clay', *Proc. Conf. Settlement of Structures*, Cambridge, 443–450

Winter, E. (1974). 'Calculated and Measured Settlements of a Mat Foundation in Arlington, Virginia, USA', *Proc. Conf. Settlement of Structures*, Cambridge, 451–459

Wood, R. H. (1952). 'Studies in Composite Construction', Pt. I, *Nat. Building Studies*, Res. Paper No. 13, H.M.S.O.

Wroth, C. P. (1971). 'Some Aspects of The Elastic Behaviour of Overconsolidated Clay', *Proc. Roscoe Memorial Symp.*, Foulis, 347–361

Wroth, C. P. (1972). 'General Theories of Earth Pressure and Deformations', General Report Session 1, *Proc. 5th European Conf. Soil Mech. & Found Engng.*, Vol. 2, 33–52

Wroth, C. P. and Simpson, B. (1972). 'An Induced Failure at a Trial Embankment: Pt. II—Finite Element Computations', *Proc. A.S.C.E. Spec. Conf. Performance of Earth and Earth-Supported Structures*, Vol. 1, Purdue University, 65–79

SESSION I
Granular Materials

Session Chairman: Mr. A. C. Meigh
General Reporter: Professor H. B. Sutherland

DISCUSSION   Page

Fawley tanks

G. E. BRATCHELL, Nachshen Crofts and Leggatt. Professor Sutherland asks how we arrived at the predicted settlement of the fill as 50 mm. In practice, we had decided that we wished to limit the settlement within the compacted fill, to 2 in maximum under an applied load of 2 ton/ft^2. This was to be the criterion for the compaction contract.

The problem was, how much compaction should be carried out in order to keep within this limit? We did not want to be faced with the situation where, having installed the tanks on the new foundations and filled them with water, the settlement and differential settlements turned out to be excessive. It was thus necessary to predict what the centres of compaction treatment should be, and also to monitor progress during compaction.

Initially, we attempted to verify a graph of S.P.T.s versus depth of penetration which was produced by the contractor, the Cementation Company. For this purpose we made references to Terzhagi, and Thorburn, and Gibbs and Holtz, and as you can imagine we got a variety of answers, with a range of about $\pm 50\%$.

The next stage was to carry out vibroflotation tests on the actual foundation at various centres. These were in triangular grids of sides of 7 ft, 8 ft, 8 ft 6 in and 9 ft. All the tests gave proportionately good S.P.T. results so we decided to proceed on 8 ft 6 in centres using a derived graph of S.P.T. versus depth for control purposes.

As soon as enough compacted area was available, a test mound of some 8000 ton was built which was sufficient to create about 1.5 ton/ft^2 pressure through the depth of the fill over a small area under the centre of the mound. The settlements were monitored by steel joists placed in the fill, access being obtained under the test mound through large concrete drainpipes.

The results confirmed that the settlement of the full tank area was likely to be satisfactory, and the work on compaction continued on 8 ft 6 in centres. It is worth listing the reasons for the compaction treatment.

(1) To keep gross vertical settlement to 2 in maximum within the fill.

(2) To reduce differential settlement to 1 in in 30 ft.

(3) To improve the general stability of the fill.

(4) To provide a transition of loading through the compacted fill at the tank periphery, in order to reduce the edge loading on the Barton Clay beneath.

(5) To detect any small patches of dredgings left in, and to remove them.

(6) To detect any clay material in the fill.

(7) To detect any large soft patches which would have to be removed by subsequent excavation.

We have of course full detailed records of the 8000-ton test loading and both of the subsequent hydrotests if anybody would like the details.

Last week we took levels on the perimeters of both tanks to establish how much settlement has taken place in the last 4 years. Both tanks have under-gone a further consistent settlement of about $3\frac{1}{2}$ in at the periphery; no doubt, all due to the Barton Clay beneath.

D. A. GREENWOOD, Cementation Ltd. Usually settlements predicted from *in situ* tests in sandy soils are overestimated—mainly for the following reasons.

Penetration test resistance is due primarily to bearing failure below the penetrometer and is not directly related to settlements at sub-failure stresses comparable with those under a stable footing. Since most natural sands are slightly over-consolidated empirical correlations do not generally relate to truly normally consolidated soils nor to very heavily over-consolidated ones (but are more nearly appropriate to the former condition). This leads to over-estimates of settlement where soil has been pre-loaded.

The modulus E depends on ambient effective stress to which applied stress is a major contributor as pore pressure dissipates rapidly in permeable soils. For confined compression average ambient stress below footings may in-crease by as much as 2 to 4 times respectively for unit loads of 25 to 50 T/m². It follows that E from penetration tests can be at best only appropriate to the unloaded soil.

To avoid these problems a case can be made for screw-plate tests taken throughout the depth of soil to be stressed by a proposed foundation to de-termine E for anticipated values of applied stresses and in appropriate strata for *direct* correlation with the full scale footing. There is much less justification for relating E from screw-plates with N or C_{kd} and then back to estimates of settlement for full sized footings. Such correlations are anyway rather poor.

However, although small screw plates may be geometrically similar to the full scale footing confinement of soils below is not. Thus settlement of small plates is also greater at comparable stresses than that of wide footings on a thin sandy layer in which high ambient stresses and horizontal shear res-traints increase confinement and hence E. This effect becomes more marked as applied stress increases. Hence the effective E is generally under-estimated from all types of *in situ* tests. With this background the gross under-estimates of settlement detailed in Paper I/1, page 3, require explanation.

At Fawley the tank diameters were about eight times the depth of granular material on which the tank rested as a flexible surface load with a sharp discontinuity and local peak stresses of about 200 kN/m² at the edges. For totally confined elastic conditions the average ratio of settlement arising in the fill beneath the centre and perimeter is about 1.44. This may be regarded as an upper limit since confinement at the centre is greater than at the edges where local plastic yielding of the fill surface would reduce the ratio. The Authors suggest that deformation of the Barton Clay allowed increased horizontal strains (and hence centre settlements) in the fill but these deforma-tions are reflected in measured 'dishing' of the tank base from which the average horizontal strain cannot exceed 0.003%: this seems insignificant, whereas any lateral yielding of the alluvium surrounding the fill would tend to increase relative perimeter settlement.

The 'deduced' centre settlement of 60 mm which is fundamental to the quoted ratios of calculated to observed settlements therefore appears ex-aggerated with respect to the measured perimeter settlement of 32 mm.

Furthermore, the calculated settlements may be underestimated because (1) vibroflotation, unlike vibrating rollers, induces an accelerated normal consolidation in permeable soils (vide paragraph 2 above) and (2) the tendency to segregation of wide ranging particle sizes as fill was dumped into 10 m water resulted in large scatter of S.P.T. results with significant numbers of high blow counts as the penetrometers struck coarse particles at depth (*Fig. 1*). It is unlikely that vibroflotation used at the relatively wide probe spacing of 2.6 m would achieve 100% relative density in this essentially sandy

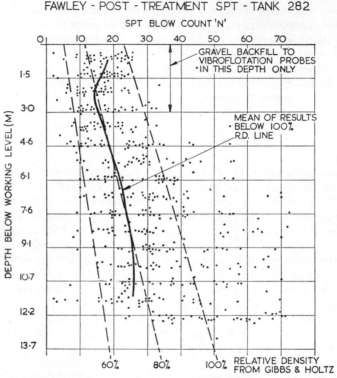

Fig. 1. *Comparison of calculated settlements for Fawley Oil Tanks for different assumed ground conditions*

fill. The average value of S.P.T. (*Fig. 3*, Paper I/1) used to estimate settlement may therefore be fractionally high: this is a problem of interpretation inherent in use of 50 mm diameter penetrometers in soils with gravel content in a compact sandy matrix which controls the soil properties.

I agree therefore with the Author's comment that this was not an ideal case for a comparative study of predictive methods, but it is valuable in highlighting limitations of penetrometers for settlement prediction.

N. E. SIMONS, J. RODRIGUES and P. A. HORNSBY, University of Surrey.   In Paper I/1 page 3, the authors compared the settlements observed in the sandy gravel fill some 10 m thick with those predicted on the basis of the Standard Penetration Test and the Static Cone Test, following various procedures proposed by a number of writers.

It was found that the basic Terzaghi and Peck procedure under-estimated the observed settlement. This was unusual as the Terzaghi and Peck method is bound to lead, in general, to conservative estimates of settlements. The authors suggested that the presence of the underlying Barton Clay and the surrounding alluvium may have allowed greater horizontal strain (and hence vertical settlements) to have taken place in the granular fill than otherwise would have been experienced if the soil conditions had consisted of sandy gravel everywhere. To investigate this possible explanation a finite element analysis was carried out to compare predicted settlements for the actual soil conditions with those calculated assuming a gravel foundation only. The assumptions made in the analysis are given in *Fig. 2*, and the ratio of the computed settlement, in the 10 m of gravel fill, appear in the final column.

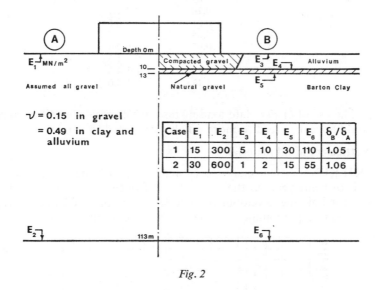

Fig. 2

It can be seen that the presence of the alluvium and the Barton Clay only increases the settlement by 5% or 6%, depending on the soil parameters assumed, and it can be concluded therefore that the fact that the observed settlements are very much larger than those predicted by the standard Terzaghi and Peck procedure, cannot be explained by the influence of the alluvium and the Barton Clay.

A further interesting observation in the Paper was that markedly different computed settlements result from the different methods of settlement prediction which were adopted.

The writers have studied five other case records of structures founded on granular soils, where settlement observations and the results of S.P.T. and Static Cone Tests are available, and have computed the settlements according to the eight methods discussed in the Paper. The results of the computations for the five structures and for the Fawley Tanks are summarised in *Table 1*. It can be seen that while Alpan's method based on the Standard Penetration Test and Schmertmann's method using the Static Cone give the best agreement with observed settlements on average, the ranges of calculated to observed settlements are very wide indeed and hence in any one case, the calculated settlement may be quite different from that actually experienced.

This is perhaps not surprising as neither the S.P.T. nor the Dutch Cone Test measures soil compressibility directly and, in particular, the test results will not indicate whether a granular deposit is normally or over consolidated, and this is one important factor which will influence the settlement which will

Table 1. COMPARISON OF CALCULATED AND OBSERVED SETTLEMENTS FOR SIX STRUCTURES FOUNDED ON GRANULAR SOILS

Method	$\delta_{calc.}/\delta_{obs.}$ average	$\delta_{calc.}/\delta_{obs.}$ range
1	1.89	0.5–3.2
2	0.70	0.2–1.1
3	0.31	0.1–0.6
4	0.63	0.3–1.4
5	0.95	0.1–2.4
6	0.72	0.1–1.3
7	3.22	1.0–4.8
8	1.48	0.2–4.0

develop. The use of *in situ* plate loading tests which measure soil compressibility would appear to lead to more reliable predicted settlements.

Settlement of fill materials

W. M. KILKENNY, University of Newcastle upon Tyne. We have been interested for some years in the restoration of opencast coal sites and their suitability for building development. The strata consists of Upper Carboniferous Measures, predominantly shales and mudstones. Kilkenny (1968)

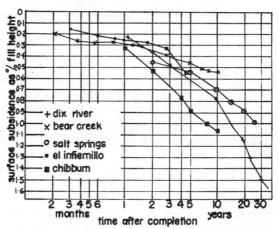

Fig. 3. Settlement–time behaviour of rockfill

reviewed the existing experience on settlement of such fills and the structural precautions which had been adapted for building development. At that time there had been no instrumentation of fills to study settlement and restoration of groundwater levels in a systematic fashion. However, it was possible to show from periodic level surveys undertaken by the Opencast Executive at

Chibburn in Northumberland that the settlement/time behaviour was similar in form to that of other recorded rockfills (*Fig. 3*). Without further experiment it was not possible to come to any conclusion on the moisture movement in these fills. Since that time a series of one-dimensional consolidation tests have been carried out in a Rowe's oedometer on 254 mm diameter specimens

Table 2.

Upper Carboniferous Shale and Mudstone		Initial density (Mg/m³)	Applied vertical load (kN/m²)	α $\delta\varepsilon = (\log t_2 - \log t_1)$
Uniformity coefficient	Moisture condition			
10	Dry	1.80	750	0.12
10	Dry	1.80	428	0.08
10	Saturated, surface dry	1.80	750	0.22
10	Dry	1.60	750	0.14
1.5	Dry	1.55	750	0.16
10	Saturated, surface dry then inundated	1.80	750	0.25
10	Saturated, surface dry then inundated	1.80	214	0.15
In situ observations at Chibburn, Northumberland				0.74

75 mm thick. The specimens consisted of the 38 mm down fraction of shales and mudstones from Widdrington, Northumberland. *Table 2* shows the range of α values recorded for different grading, densities and moisture conditions

These values are rather different to those observed at Chibburn. However, further tests on 250 mm thick specimens confirmed that sample thickness was

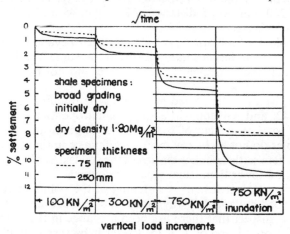

Fig. 4. Effect of sample thickness

critical (*Fig. 4*) as had been proposed in a theoretical approach by Marsal (1965). The high value of particle breakage under loading and inundation characteristic of these deposits was evident. I wish to emphasise the susceptibility of Coal Measure shale and mudstone fills to settlement on saturation either by restoration of groundwater levels or from external sources such as flooding or leakage from services. In this situation light structures such as two-storey residential development, road pavements and pipelines applying only nominal loads to the fill may suffer considerable settlement. Total settlements of 450 mm with differential settlements almost as great have been recorded in the field. While the Laboratory tests are useful in explaining the mechanism of compression they are probably of little value in prediction. Research should now be directed to field experimentation and the investigation of ground treatment methods and structural solutions.

Kilkenny, W. M. (1968). 'Settlement of Restored Opencast Coal Sites and their Suitability for Building Development', Bulletin No 38, Dept. of Civil Engineering, University of Newcastle upon Tyne

Marsal, R. J. (1965). 'Stochastic Processes in the Grain Skeleton of Soils', *Proc. 6th Int. Conf. Soil. Mech. Found. Engng.* Vol. 1, 303

J. E. GUEST, Foundex Ltd.   With reference to Paper I/7 by Penman and Godwin (p. 53) I would like to give some details of the settlement on an opencast coal site which it can be assumed was not loose dumped but received some compaction. The case history is not yet complete and because the contractural and legal problems have not yet been resolved I can give you only the bare bones of the case.

It concerns a 56-acre site which was mined by opencast methods in 1959. A site investigation was carried out in 1970 and the site pronounced fit for a light development. The depth of the fill being 9 m.

A housing estate of 530 dwellings, bungalows and two-storey houses in 85 blocks was planned. An accurate plan of the workings was available and trenches were dug to identify the hidden sides of the excavations. The blocks were sited to avoid sitting on the edge of or across steep slopes. Work has been in progress since 1972. During last year, settlements of up to 200 mm (2% of the depth of the fill) have taken place locally. That is, a differential settlement of 200 mm over a distance of less than 45 m. To date four small areas have been affected.

So far one block of 8 houses has been damaged beyond repair. We know that similar settlement has occurred in a drain run some 2 m deep away from dwellings so that the settlement is deep seated and not confined to loaded areas.

The curves given for compacted material in the Building Research Station Digest No. 142, and Kilkenny in a study of an opencast coal site in 1968 show 75% of total settlement of 1–3% of the fill height to be complete in 5 years. Boreholes do not show any marked difference in the strata except that in one bore, where settlement has occurred, the soil is described as fragmented clay/shale with a lower than usual moisture content.

If the cause of the settlement is softening of the points of contact in the shale fill described in Paper I/7 why should it take place after 14 years. There is no reason to suspect that there has been a change in the water table.

This is a site on a slope of 1 in 15. Could it be that rainfall has run off rather than sinking in, and the digging of drain trenches across the slope has let water through to dry fragmented shale areas which were not fully compacted so inducing severe settlement? Is this an isolated case for opencast coal sites? Should all such sites be investigated in minute detail and by what method can we identify the settlement prone areas and at what cost? Or must a percentage of damaged structures be accepted as inevitable when developing opencast coal sites?

Plate bearing and other in situ tests

L. M. LAKE Mott, Hay & Anderson.   I wish to draw attention to two matters. The first concerns the Standard Penetration Test and the second some aspects of the nature of granular soils, pertinent to their behaviour.

It has been shown that there are many factors which influence the result of a dynamic standard penetration test. One very important and complex factor is the precise technique used to form and advance the borehole, especially the relative size of the lining tubes and 'shell', which is not yet specified in the Code of Practice. However, of particular relevance to the symposium in which test results and observational data from various points of the world are being correlated, is the effect of borehole diameter. In the Americas and also some European countries, it is common to carry out S.P.T.'s in nominal 75 mm diameter boreholes whilst U.K. practice typically involves boreholes of 125 to 250 mm diameter. Indeed, S.P.T.'s have been used in pile boreholes of much greater diameter, as a means of verifying the quality of the ground before concreting. The writer has found reductions in S.P.T. values of the order of 25% to 50% in moving from 125 mm to 200 mm diameter shell and auger borings. A pro rata adjustment to the initial length of test that is normally neglected, i.e. the first 150 mm, which is roughly equivalent to two borehole diameters in the original Raymond test, may provide a suitable basis for correcting the results to a common base. Meanwhile, those publishing data should ensure all the relevant boring data is recorded whilst those referring to the papers should be mindful of the significance of these effects.

So often in the U.K. the granular material which confronts the engineer comprises very strong siliceous material; many have experience of no other. However, this is not a prerequisite of granular soil, for example, calcareous and shell sands are widespread around the world. The latter may often be very weakly cemented, at particle contacts, frequently in an irregular manner. Further, the climatic conditions under which they most commonly occur is also conducive to the deposition of relatively soluble cementing agents, e.g. gypsum. Removal of the cement by leaching may then result in the formation of a metastable soil. The response of granular soils composed of weak particles to static or dynamic stress, whether due to excavation processes, to foundation loading or *in situ* testing, may be quite different from that of strong, inert siliceous granular soil. Caution is therefore advised, in extrapolating the data from one granular soil type to another of different composition.

J. F. LEVY, Greater London Council.   The question may have arisen regarding our use (Paper I/6, p. 43,) of 1 ft square plates on sites where wide footings or even a raft were to be provided as a foundation. The explanation lies in the fact that most of the tests were undertaken in order to settle differences over the proposed design bearing pressures. Since these disputes often arose after work had begun on site, time was an important factor and this led to the use of small plates which could be loaded with a relatively small quantity of easily moved kentledge. The validity of the results in relation to the actual foundation was, of course, another matter.

Where more time was available, larger plates or a series of plates were used. Attempts to obtain reliable settlement readings on the structures themselves were often frustrated by forgetfulness, lack of access or lack of continuity. Improved arrangements are likely to result in worthwhile settlement readings being obtained from Site M and subsequent sites.

Professor Sutherland has pointed out that on the two sites where different size plates were used, the Terzaghi and Peck relationship underestimated the actual settlements of the larger plates when predicted from the measured settlement of the smallest plate. Since this also supports d'Appolonia's findings, it may be that plate tests are of doubtful value unless carried out on plates at least 1 m square so as to overcome the scale effect.

Although, as Professor Sutherland has indicated, *Fig. 6* of the Paper cannot be used for the prediction of the actual settlement of a particular building, it was an attempt to draw, from available empirical data, an envelope for the probable range of settlement to which spread footings are likely to be subjected. It was felt that this could be of use in attempts to limit differential settlements in those cases where a more detailed analysis of estimated settlement is not, or cannot be, made. However, caution must be exercised in using these values.

K. MORTON, Ove Arup & Partners.   It can be seen from our Paper (I/6, p. 43) that at Stratford Bus Station construction had just been completed in March 1973. We were able to take a further set of readings recently and the additional information is given in *Table 3*.

Table 3. SETTLEMENT OBSERVATIONS, STRATFORD BUS STATION (further set of readings to add to *Table 2* on page 49)

13.3.74	5.9	7.4	9.9	8.7	10.0	11.9	12.9	12.9	—	15.5*	occupied

*Excavations for subway adjacent to footing.

I should like to reply to Professor Sutherland's comment in his State of the Art that our *Fig. 6* cannot be used in the prediction of settlement of a particular foundation. This was never the intention and the figure merely gives a range of values based on the results of plate tests taken mainly in the London Terrace Gravels. However, it is possible to use the limits shown to assess the maximum differential settlements of structures founded in those conditions and to design for the predicted distortions.

In order to investigate the validity of this approach, I have made a similar diagram (*Fig. 5*) based on the results of settlements of structures reported to this Session and from other sources (see *Table 4*). *Fig. 5* shows that there is very little detail available for pad foundations but suggests that the limits of settlement/foundation width are well within the scatter obtained from plate bearing test results.

There is a significant number of case histories for raft foundations and these suggest that

(1) the ratio of settlement/foundation width is significantly smaller for raft foundations

(2) the range of values of settlement/foundation width for a particular structure is relatively small.

J. A. LORD, Ove Arup & Partners.   I should like to sound a note of warning with regard to placing too much reliance on high S.P.T. blow counts in sand

Table 4. RELATIONSHIP BETWEEN APPLIED PRESSURE AND SETTLEMENT/FOUNDATION WIDTH FOR STRUCTURES ON GRANULAR SOILS

Ref. no.	Pressure (kN/m²)	N value	Footing width B (m)	Range of settlement (ς mm)	ς/B (%)	Reference
1	500	40	4	10–15	0.25–0.38	Birmingham Arts & Commerce Building
2a	300	45	2	6–10	0.30–0.50	Stratford Bus Station
2b	300	45	3	10–13	0.33–0.43	Morton and Levy, I/6, 1974
3	188	25	79	32–63	0.04–0.08	Fawley Tanks Bratchell. Leggatt and Simons, I/1, 1974
4	425	50	60	37–46	0.06–0.08	Reactor on Rhine Breth and Chambosse, I/2, 1974
5	289	50	55	23–62	0.04–0.11	Dungeness 'B' Dunn, I/3, 1974
6	161	20	20	8–14	0.04–0.07	Queens University. Stuart and Graham, I/8, 1974
7a	316	30	3	3.0–15.0*	0.10–0.50	D'Appolonia, D'Appolonia
	316	30	4	11.5–16.5*	0.29–0.41	and Brisette, *Proc. Am.*
	316	30	7	13.0–19.0*	0.19–0.28	*Soc. civ. Engrs.* Vol. 94, May 1968
8	132	14	12	13–23	0.11–0.19	Tanks, Kansas City
9	169	20	37	0–30	0–0.08	Tanks, Iowa Baker, *Am. Soc. civ. Engrs.* Proc., Vol. 91, 1965
10	193	10	13	17–25	0.13–0.19	Buildings, Lagos. Grimes and Cantlay, *Struct. Engineer*, Feb. 1965
11	240	15	22	19–25	0.08–0.11	Thyssen Bldg, Dusseldorf
12	210	15	16	16–20	0.10–0.12	Ministry Bldg, Dusseldorf
13	160	10	20.5	6.5–9.5	0.31–0.46	Chimney, Cologne
14	270	25	14.5	14–22	0.09–0.15	Chimney, Duisberg
15	240	40	33	40–48	0.12–0.15	Reactor, Julich
16	150	40	15	13–19	0.08–0.12	Building, Aachen. Schultze, E. *Euro. Conf. on Soil Mech. Founding Engrg.*, Vol. 1, 1963
17a	220	50	20	16–22	0.08–0.11	Block 1, Poplar
	220	50	20	29–32	0.14–0.16	Block 3, Poplar. Morton and Levy, I/6, 1974

* Estimated from 'average' results quoted in paper.

and gravel strata as being indicative of high allowable bearing pressures over the *whole* site. In other words, to restate the obvious, local and sometimes significant variations in soil conditions exist.

An example of this occurred at a brewery in East London which Arup Associates are rebuilding. As part of the scheme the shell of the old Brewers House was to be retained. Redesign of the interior imposed additional foundation loads which would have overstressed the firm brown sandy clay on which the original structure was founded. Consequently, it was necessary to transfer these loads on to the gravel by underpinning.

A plan of the area (*Fig. 6*) shows the locations of the two nearest boreholes,

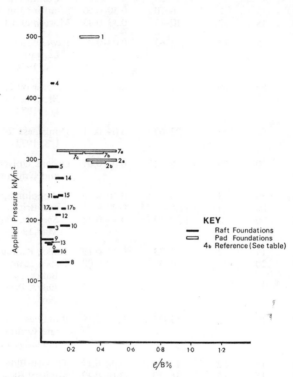

Fig. 5. Relationship between applied pressure and settlement/width of foundation for structures on granular soils

which because of existing structures and other restraints at the time of the site investigation could not be placed closer to the Brewers House. The strata encountered in the top 6 m of each borehole are shown in *Fig. 7*. In both a dense sandy gravel was encountered with cone penetration test blow-counts of between 44 and 75 blows/ft (average 60). The ground water table encountered in the sandy gravel was at a level of +7.7 m A.O.D.

Although calculations based on the procedure recommended by Terzaghi and Peck (1967) indicated that the sandy gravel could sustain an allowable bearing pressure of over 400 kN/m², confirmation of this by plate bearing tests was required. Loading tests on the sandy gravel were carried out in two pits (located as shown in *Fig. 6*) using a plate 300 mm square and jacking

666

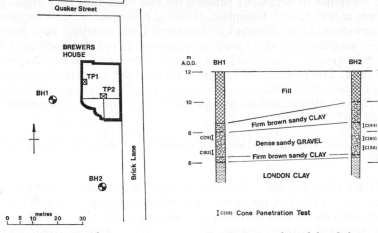

Fig. 6. Site plan   Fig. 7. Section through boreholes

against the underside of the old foundations. The load settlement curves for
the two tests are shown in *Fig. 8*. The plate in the test in pit 1 settled only
about 7 mm under an applied stress of 1000 kN/m², as might be anticipated
from the penetration test results, whereas in pit 2 the settlements became
excessive at a stress of 550 kN/m². (The results of these two tests agree closely
with the upper and lower bounds shown by *Fig. 6* of Paper I/6, page 52.)

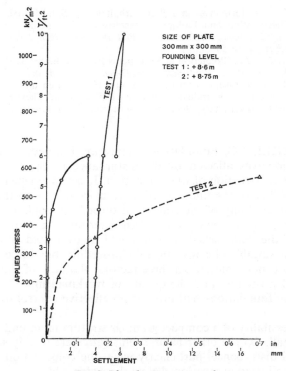

Fig. 8. *Plate bearing test results*

The difference in behaviour between the two plate tests only about 10 m apart was at first difficult to explain. However subsequent excavation beneath the plates showed that whereas the plate in pit 1 was underlain by at least 1 m of dense sand and gravel, in pit 2 the plate was founded on only 0.25 m of sandy gravel which in turn overlay at least a further 0.65 m of loose to medium dense sand. The presence of the loose sandy layer, completely undetected during the site investigation, was confirmed along the whole eastern perimeter of the Brewers House, and in consequence the allowable bearing pressure was reduced to 275 kN/m².

Terzaghi, K. and Peck, R. B. (1967). *Soil Mechanics in Engineering Practice*, 2nd Edn, Wiley.

E. SCHULTZE, Aachen.   The evaluation of 48 settlement observations of foundations on sand by means of a multiple correlation for which standard penetration tests or static sounding tests were available, showed a dependence of the measured ultimate settlement on the contact pressure, the number of blows of the S.P.T. or the point pressure, the foundation depth and the width of the foundation for a correlation coefficient $r = 0.79$ (Schultze and Sherif, 1973). Investigations were further extended to 25 constructions on silt (Sherif, 1973). The ratio ρ_{calc}/ρ_{meas} was found to lie always between the values 0.6 and 1.4.

The applicability of the load plate test for determining the compressibility of sand was also investigated in detail. (Schormann, 1973).

Schormann (1973). 'Verformungs- und Bruchverhalten von Sand in axialsymmetrischen Versuchen' (Deformation and Failure Characteristics of Sand in Axially Symmetric Tests); with English foreword and summary, *Bull. Inst. Soil Mech. Found. Engng*, Techn. University Aachen, VGB 59
Schultze and Sherif (1973). 'Prediction of Settlements from Evaluated Settlement Observations for Sand. *Proc. 8th Int. Conf. Soil Mech. Found. Engng*, Moskwa, Vol. 1.3, 225–230
Sherif (1973). 'Setzungsmessungen an Industrie- und Hochbauten und Ihre Auswertung' (Settlement Observations on Industrial and Tall Buildings and Their Evaluation), with English foreword and summary, *Bull. Inst. Soil Mech., Found. Engng*, Techn. University Aachen, VGB 57

T. R. M. WAKELING, Foundation Engineering Limited.   The settlement of a foundation was affected by the disturbance to the bearing stratum caused by the construction of the foundation. It became dependent on both the compressibility of the undisturbed stratum, and also on the extent and modified compressibility of the zone of disturbed material. The term 'compressibility' refers to the parameter which determines settlement. The compressibility of the undisturbed stratum could be estimated with reasonable accuracy by a suitable site investigation procedure but the constructional disturbance depended mostly on three factors; the type of foundation, the constructional procedure and the quality of workmanship. Quite often the designer of the foundations had little or no effective control over these last two points.

The compressibility of a compact granular stratum in an undisturbed condition was small. On the other hand, such strata were readily susceptible to constructional disturbance, particularly when the construction of the foundations involved deep excavations below the water table.

Disturbance caused a reduction in density and a considerable increase in compressibility. Thus, the settlement caused by constructional disturbance could be an appreciable part of the actual settlement of foundations on compact granular strata. Since constructional disturbance tended to vary at random from site to site, it became impossible to predict settlements with any useful accuracy unless the effect of disturbance could be taken into account. Case records often gave little or no information concerning the constructional procedure or quality of workmanship and it therefore became difficult to compare the observed settlements with those predicted only from the compressibility of the undisturbed strata.

A similar effect occurred with plate bearing tests where disturbance caused by the excavation, and inaccurate bedding of the plate, could lead to observed settlements which were not completely representative of the compressibility of the undisturbed stratum. The test results shown in *Fig. 3* of the Review Paper showed an increased scatter as the strata became more dense, say for S.P.T. values in excess of 20. Possibly much of this scatter reflected the increasing effect of disturbance and imperfections in the testing technique.

Building Codes throughout the world give presumed bearing values for foundations on various types of ground. For compact granular strata these values often appeared to be conservative when compared with allowable bearing pressures determined from conventional settlement analyses. In addition, for granular strata, most Codes reduced the presumed bearing values by 50% where the ground water table was at or above the base of the foundation and allowed no increase for deep foundations. The validity of these two criteria could be criticised on theoretical grounds and a number of proposed modifications have been discussed by the General Reporter. On the other hand, with deep foundations below ground water, an inappropriate construction procedure and poor workmanship could cause severe disturbance. The presumed bearing values quoted in Codes were derived from experience extending back over a considerable number of years and the apparently low values probably represented an allowance for possible constructional disturbance. In projects where any disturbance can effectively be controlled, it might well be feasible to exceed the normally adopted presumed values.

In conclusion, the observed settlements for foundations on compact granular soils have generally been small. However, since a considerable proportion of this settlement might be due to constructional disturbance, any correlation between observed and predicted settlements was likely to show a considerable variation unless allowance could be made for this factor.

G. A. LEONARDS, Purdue University.   I would like to point out that the scaling factor for different footing sizes is strongly influenced by preloading. The residual lateral stress left after the preload has been removed produces a higher K_0, which value decreases very rapidly with depth as the overburden pressure increases. The effect of the higher K_0 on the load-settlement relationship for a very small plate is therefore very pronounced, whereas for a large footing it may be quite small. This causes the ratio of the settlement of the larger footing to that of, say, a 1 ft × 1 ft plate to increase with (a) the amount of preloading, (b) the width of the footing, and (c) the magnitude of the contact pressure.

P. W. ROWE, Manchester University.   The importance of stress history of a site appears to have been overlooked. Let us form two identical beds of loose sand and then increase the density of each by the same amount using vibration for Bed A and pressure for Bed B. These processes form different particle contact geometries. Bed A is normally consolidated and Bed B is overconsolidated but since both have the same relative density, both would exhibit approximately the same failure strength, cone resistance and S.P.T. value. Deformation to failure replaces the previous local particle contact geometry by one which is a function of the relative density and the test method. However, when subjected to a plate or foundation loading at a safe bearing pressure remote from failure Bed A would follow a virgin stress-strain path whereas Bed B would follow a much steeper reloading path. Thus for the same relative density, cone resistance or S.P.T. value, the settlement at safe bearing stress of heavily overconsolidated sands is very much less than for normally consolidated sand because the particle contact geometry or fabric is different. Therefore *in situ* plate loading tests coupled with a knowledge of the strata at depth and the geological history of the site are essential prerequisites for settlement prediction.

G. LOWTHER, Foundation Engineering Limited.   I am disappointed that insufficient emphasis has been placed by the General Reporter and previous speakers on the variability of standard penetration test procedures. One consulting engineer friend of mine told me of a site investigation he had had carried out abroad where apparently the S.P.T. hammer did not weigh 140 lb and the free fall had not been 30 in. The standard penetration test is a standard that is not a standard unless carried out by a trip monkey. For this reason we in Foundation Engineering Limited always use trip monkeys for all our standard penetration tests, both at home and abroad.

E. MARANHA das NEVES, Laboratório Nacional de Engenharia Civil, Lisboa.   Crestuma dam will be constructed on the river Douro within a few kilometers of its mouth. The project of this hydroelectric scheme includes piers which will be founded at a depth of 17.5 m in the alluvial soil and will

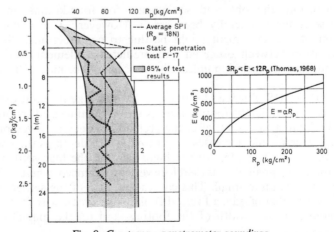

Fig. 9. Crestuma—penetrometer soundings

support the gates and the slabs. It was therefore important to know the deformability of the alluvial material below the foundation level. It essentially consists of medium and coarse sands with gravel.

A large number of static and dynamic penetration tests were carried out. In *Fig. 9* Curves 1 and 2 are the boundaries of the zone corresponding to 85% of the tests results. As can be seen, for depths concerned by the foundation, $R_p \simeq 100$ kgf/cm^2 which, following a classic interpretation, gives to that soil

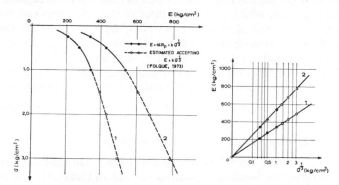

Fig. 10. Crestuma—penetrometer soundings

a value of E ranging from 150 to 300 kgf/cm^2. This value is practically constant with depth. Nevertheless, although R_p may be constant, we cannot accept E to be so, that is, to behave like a quantity independent of the state of stress. Folque (1973), by adopting Thomas's correlation (1968) represented in *Fig. 9*, and by assuming E to evolve in proportion to the cube root of the mean stress, has proposed a method that makes it possible to determine compressibility of sands below the critical depth.

On basis of function $E = \alpha R_p$, *Fig. 10* presents the Curves 1 and 2 of the deformability versus overburden pressure down to the critical depth. In the

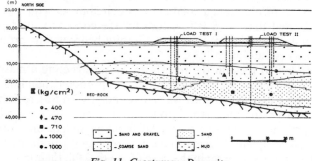

Fig. 11. Crestuma—Dam site

same figure we can see that by plotting these curves under the form $E = f(\sigma^{1/3})$ other curves passing through the origin will be obtained.

In such a way we shall be able to extrapolate for stresses higher than those corresponding to the critical depth, and to obtain the corresponding deformation moduli.

In the present case and for the depths considered, values of E ranging from

500 to 800 kgf/cm² were obtained, which are fairly above those given by the classic methods.

In order to have another approach to study deformabilities of the soil, large load tests (30 × 30) and (45 × 30) m² were carried out. In those tests measurements of the settlement of formations below the foundation level were made. The measuring devices consisted of upright steel rods, connected to plates installed within boreholes at the required depth. Before removing the tube from the hole, the rods were enveloped in plastic sleeves capable of high vertical deformation, which prevented later contact between soil and rods.

Settlements measured after loading (= 0.35 kgf/cm²) quickly stabilized. *Figure 11* shows a cross-section of the northern zone of the alluvial valley together with the load test site and location of measuring devices. As can be inferred from the corresponding order of magnitude, results obtained point to deformabilities that confirm the values derived through the proposed method.

Folque, J. (1973). 'Compressibilidade de Areias Determinada por Ensaios de Penetraçaõ', *Geotecnia*, Vol. 6, 19–27

Test embankment and large raft foundations

J. C. GRAY, W. & C. French (Construction) Ltd.   On the Island of Mahe, the main island of the Seychelles group in the Indian Ocean, there is a narrow coastal plain formed from a mixture of coral debris blown inshore and quartz sands washed down from the granite massive. The granite hills rise steeply from the shore line to a height of 3000 ft in the centre of the island. The coastal deposits are very loose and with the exception of some localized cementing can be considered as normally consolidated especially on their seaward edge. There is very little flat land on the island and consequently an area of the foreshore between the limits of high and low tide, adjacent to an existing causeway, was investigated for reclaimation and subsequent development of a three-storey office and shop complex.

The foreshore was found to consist essentially of loose coral debris to 48 ft with a band of hard coral between 22.5 and 27 ft which probably varied in thickness and continuity. Ten feet of sandy silt was proved below the coral deposits. No test results were available for the site but subsequent work in the area has indicated that the S.P.T. values for the coral and silt deposits vary between 4 and 20 averaging about 10. Granite bedrock was probably 65 ft below the original surface.

It was decided to reclaim the foreshore by constructing a masonary wall round the perimeter, a rockfill raft under the proposed building and earth filling elsewhere. The rockfill raft was surrounded with fine rock material to act as a filter layer and an earthfill preload was placed on it to induce settlement. The preload was about twice the building weight and amounted to 7700 ton when complete. It was placed in two phases as indicated in *Fig. 12*. Rod and plate gauges were set in the base of the rockfill, their movement being observed with normal levelling techniques. The relative movement is indicated in the lower portion of *Fig. 12*.

The time settlement record is recorded in *Fig. 13* under the loading diagram. This clearly indicates that the majority of the movement took place within a few days of placing the load and that gauges 2 and 4 moved twice the amount of 10. It was subsequently established that some filling had been placed in the area around gauges 10 and 6 prior to the placement of those gauges.

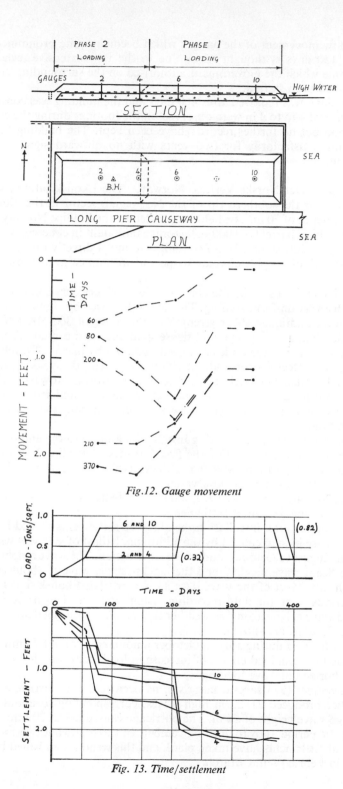

Fig.12. Gauge movement

Fig. 13. Time/settlement

673

The slow movement of the gauges which becomes quite pronounced after the first 150 days is thought to be due to the very extensive reclaimation programme which the Government authorised on the surrounding area after this test had been started.

The preload was removed after 400 days and the building was constructed, movements of $\frac{1}{2}$ in to $\frac{3}{4}$ in were observed on the footings during the construction period but no further records have been kept. The building had been performing satisfactorily for two years with no outward signs of further movement.

IVAR FOSS, Det Norske Veritas, Norway. The Ekofisk oil and gas field is located near the southern tip of the Norwegian sector in the North Sea, at a water depth of 70 m. The field is operated by the Phillips Norway Group.

A one million barrel (160 000 m³) storage tank built in concrete was placed at Ekofisk on 30th June 1973. The tank is founded directly on the sea floor on a base slab of roughly circular shape and with an equivalent diameter of 93 m.

The soil conditions at the site consist of 26 m of very dense, fine to medium sand with a very uniform grading. The sand is followed by about 50 m of hard clay with an undrained shear strength of the order of 300–800 kN/m². At greater depth alternating layers of dense sand and hard clay are found.

The settlements of the tank have been observed by means of levelling to a pile-supported steel platform about 200 m away. Strictly speaking these are differential settlements with respect to this platform. However, the steel platform was installed about one year before the tank, and settlement of the steel platform is not expected to be significant during the period under consideration.

The observed settlements during installation and up to the end of January 1974 are shown in *Fig. 14*. For the first 500 MN of loading bedding errors confused the observations and the settlements for this load have been estimated from the subsequent observations.

During 24 h, 4th to 5th July 1973, 1000 MN of ballast water was pumped into the tank, causing 52 mm of settlements.

For the period after installation the observed settlements should be related to the increase in load caused by the additional ballast of 400 MN which was placed in August to October, and to the wave action. The wave heights shown are 'significant wave heights' and the highest single wave is about twice the height shown. Most of the wave data are recorded at Ekofisk, but for a few short periods when the data recording system was out of operation visually observed wave heights from the weather ship *Famita*, positioned about 100 km North of Ekofisk are shown.

The settlements during July to October amounted to about 75 mm and are caused by the combined effects of increase in load, consolidation of the clay and the horizontal wave loading.

In November 1973 several large storms occurred. The hurricane on 19th November produced 21–22 m high waves, corresponding to about 90% of the design wave height of 24 m. The hurricane caused about 50 mm of settlement in the period 16th to 20th November. In the following two months no additional settlements have taken place, and this trend is confirmed by observations in February and March.

The settlements during the hurricane most likely have taken place in the sand layer. As a consequence of the large horizontal wave force, the dense sand has consolidated under the combined action of vertical and shear stresses. If it is assumed that most of the settlement took place during the peak storm period which had a duration of about 12 h, considerable pore pressures must have developed under the tank. Assuming radial drainage and a permeability coefficient of about 2×10^{-6} m/s as measured in laboratory tests, excess pore pressures of 150–200 kN/m² are calculated for the central part of the tank.

The results of a preliminary settlement analysis, based on the observations,

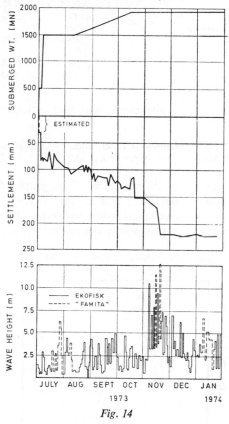

Fig. 14

are given in *Table 5*. For comparison, two different assumptions regarding soil conditions have been made. In the first instance, the foundation soil is considered as a homogeneous, elastic half space. In the second instance, all settlements are assumed to take place in the sand. The deformation modulus is given as a secant modulus (Janbu, 1963), defined as follows:

$$M = mp_a\left(\frac{p'}{p_a}\right)^{1-a}$$

where m = modulus number, p_a = reference stress = 100 kN/m², p' = effective stress, and a = stress exponent = 0.5 for sand. The correct interpretation probably is intermediate between these assumptions.

675

The deformation parameters found in this back calculation appear reasonable. The observed settlements to date also compare favourably with advance predictions made by Norwegian Geotechnical Institute, amounting to about 200 mm. In conclusion, the performance of the tank during the first

Table 5

Period		4–5 *July* 1973	30 *June* 1973 to 1 *Feb.* 1974
Net load	(kN/m²)	135	260
Obs. settlements	(mm)	52	230
Soil modulus (from obs. setlm.)			
1) Hom. elastic soil, E	(MN/m²)	150	65
2) All settlement in sand:			
Average modulus	(MN/m²)	65	32
Modulus number	(m)	400	200

winter of service has been satisfactory, and the settlements close to those anticipated.

Janbu, N. (1963). 'Soil Compressibility as Determined by Oedometer and Triaxial Tests, *Proc. Eur. Conf. Soil Mech. Found. Engng*, Wiesbaden, Vol. 1, 19–25

C. S. DUNN, University of Birmingham. The graph of load–settlement for the centre of the raft shown in *Fig. 6* of my Paper clearly shows two phases of settlement. Up to the preconsolidation pressure, the compressibility is small and above this, the compressibility is significantly greater. The geology

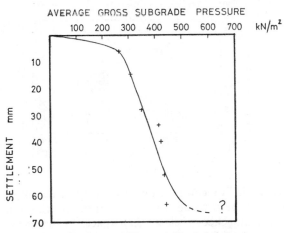

Fig. 15. Settlement of Reactor No. 1, 'A' Station

of Dungeness promontory suggests that the beach deposits must be normally consolidated and the settlement behaviour corroborates this. Settlement records for the reactor rafts of the first station are few and of doubtful accuracy but a load–settlement graph for reactor No. 1 of this station shown in *Fig. 15* also indicates two-phase settlement. In carrying out an evaluation of sand raft interaction it is suggested that the soil model should be assumed to possess two-phase elasticity in which an initially high value of E be used for average subgrade pressures up to preconsolidation pressure and a lower value of E be used for pressures above this.

The restoration of the ground water table had little effect on settlement presumably because of the counteracting effects of uplift of the foundation concurrent with decrease in effective stress in the ground.

There was the advantage of a 'full scale load test' in predicting settlement but rather than assume that the modulus of subgrade reaction K for the second station would be the same as that of the first it might have been more prudent to calculate K for the second station by multiplying the K for the first station by the ratio

$$\frac{\text{Average slope of pressure/settlement for Station B}}{\text{Average slope of pressure/settlement for Station A}}$$

taken from *Fig. 5*. This would have made allowance for the fact that the sand under the second station was more compressible than that under the first.

June 1966 was taken as the zero time for plotting settlements because this

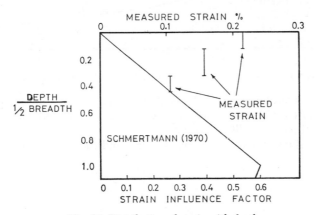

Fig. 16. *Distribution of strain with depth*

was the start of the loading phase and because of doubts about the interpretation of measurements of plate movement during excavation. Plate 5 in BH 2 rose 30.2 mm during excavation indicating a substantial heave compared with a settlement of 22.4 mm on reloading to overburden pressure. However, plates above this rose by smaller amounts suggesting that the superficial layers of sand actually compressed. A much more likely explanation is that the top plates settled within the loose backfill in the boreholes. There was considerable vibration due to vibroflotation on an adjacent area during this period.

Prediction of settlement using Schmertmann's method, accepting the idealised I_z distribution and calculating E at various depths, by multiplying cone resistance by two, gave an answer of 18 mm for an applied net pressure of 83 kN/m². This compares with a prediction of 58 mm by DeBeer and Martens and a measured settlement at BH.2 of 34 mm.

Schmertmann's method seriously underestimates the vertical strain distribution immediately under the raft as shown in *Fig. 16*. This is due to (1) the low Poisson's ratio of about 0.2 for sand instead of the higher values considered by Schmertmann and (2) the important redistribution of strain which occurs when there is a 'rigid' layer at a relatively shallow depth.

H. BRETH and G. CHAMBOSSE, Technische Hochschule Darmstadt. In our Paper (p. 10) observations were reported of the settlement of a nuclear reactor (called A) at Biblis in Germany built on sand and gravel. Since then

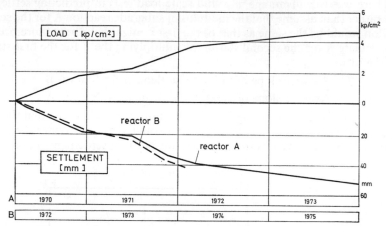

Fig. 17. Average soil pressure and settlement of point $X/R = 0.845$

the constructions of a second indentical reactor (B) of the same size and weight has been started, 100 m from the first. More details and new results in addition to those in the paper will be presented.

Figure 17 shows the average soil pressure and the settlement of reactor A

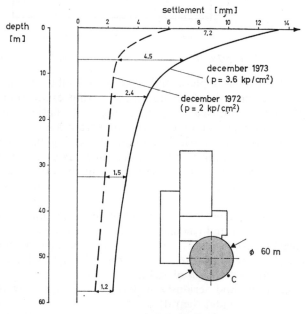

Fig. 18. Distribution of settlements underneath Reactor B, Biblis

678

and B at a 'representative point'. At this point the settlement depression of a flexible foundation is equal to the settlement depression of a rigid foundation.

Figure 18 shows the distribution of settlements with depth underneath reactor B of Point C situated 1.8 m outside the foundation raft. The settlement of Point C was measured at a depth of 7, 15, 33 and 57 m. The increase of settlements between December 1972 and December 1973 shows that the settlements mainly occur in the soil above a depth of 30 m. The lower settlements are due to a big silt layer below 60 m.

The compression of the sand layers between the particular measuring points is shown in *Fig. 19*. The compression of one year is plotted as a result of an increasing stress of 1.6 kp/cm². At a depth of 30 m the compression is only one tenth of the compression in the foundation area.

Figure 20 shows the measured settlement of reactor A and B. The calculated

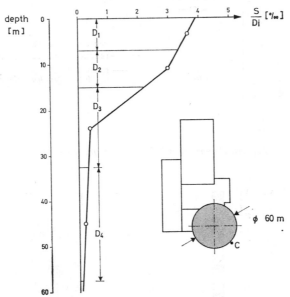

Fig. 19. Compression of subsoil (%). Reactor B, Biblis, 1972–73 $\Delta p = 1.6$ kp/cm²

settlement of reactor A is also plotted. The settlements were calculated using the theory of Boussinesq for a rigid foundation. Young's modulus for dense sand was estimated to be 2500 kp/cm² and to increase with depth to 5000 kp/cm². For the silt layer underneath 60 m a constant Young's modulus of 1000 kp/cm² was estimated.

The settlement distribution with depth of measured and calculated settlements is shown in *Fig. 21*. The two curves for the calculated settlements differ in the Young's modulus only for the silt layer. 80% of the settlements measured were due to the soil down to a depth of 60 m.

In the Institute of Soil Mechanics and Foundation Engineering at the Technische Hochschule Darmstadt, triaxial tests on dense sand at low stress levels have been carried out. At present the geological stress history of the subsoil in Biblis is simulated to investigate the strain conditions and the stress ratio which might have led to the prestress in the sand before construction of

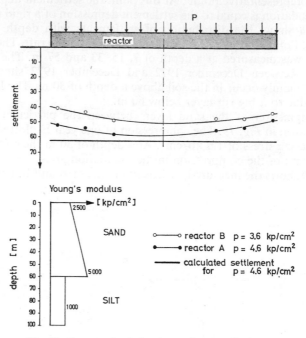

Fig. 20. Compression index for settlement calculation

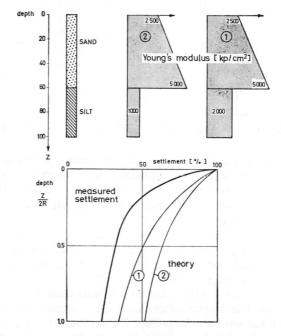

Fig. 21. Comparisons between measured and calculated settlements—Reactor Biblis

the reactors. We hope to find out how the stress strain behaviour under anisotropic loading of soil differs from the stress strain behaviour of a pre-stressed soil. With the results of these tests the settlement of the reactor will be calculated using the finite element method, to get a better understanding of the settlement behaviour.

D. H. CORNFORTH.   Five of the eight papers presented in this session of the Conference describe settlements measured on large raft foundations. In the paper by Dunn (p. 14), the author used the data available from the pre-viously observed settlements of reactor base A to provide the information for predicting the settlements below the new reactor base B. Bratchell, Leggatt and Simons (p. 3) compared their observed settlements with various design methods originally developed for isolated spread footings, and found that the correlations were poor. However, raft foundations are usually wide relative to the depth of compressible soil and approximate to the condition for one-

Table 6

Site & location of settlement	Period estimated relative (dry) density	Calculated settlement (mm)	Observed settlement (mm)	Ratio of calculated: Observed
Dungeness, Paper I/3 (BH2)	Week 51 45/75% —403	41	57	0.72
Fawley, Paper I/1 (centre of tank)	Final 50/80%	20	60 (deduced)	0.33
Biblis, Paper I/2 (average values)	To Feb. 60/110% 1973	46 (R.D.D. = 85%)	42	1.09

dimensional consolidation. An alternative method is, therefore, to treat the problem as a conventional settlement computation. The main problem is that it is difficult to obtain reliable measurements of sand compressibility. The oedometer test is unsuitable because, in addition to the difficulty of measuring relative density, the compression of the apparatus itself is of the same order as the compressibility of sand.

I have recently analysed the K_3 consolidation data from my research work on Brasted Sand. The test specimens were 100 mm diameter, 200 mm high and these dimensions allow the relative density to be reliably measured. Consistent data was obtained and a set of curves representing relative dry densities of 10, 30, 50, and 70% have been drawn. Further details of this work are contained in a short paper which has been submitted for publication.

Brasted Sand is a uniform fine to medium river sand which is inorganic and thought to be composed of quartz. Since the compressibility properties of different sand deposits are probably not too variable—this is an implicit assumption in the existing design charts—I have used my data to calculate the settlements beneath the raft foundations at Dungeness, Fawley and Biblis (Table 6). In the first two cases, the S.P.T. data given in the papers has been plotted on the S.P.T. versus depth relationship suggested by Peck and Bazaraa (1969) to estimate the relative densities in the foundation. At Biblis, the cone penetration resistances are available and the relationship $q_c = 4N$ has been used (see comments by the General Reporter) to convert the data to blow counts and hence relative density.

Bearing in mind the uncertainties about relative density, the correlations between observed and calculated settlements are reasonably good at both Dungeness and Biblis, especially since the observed settlement at Dungeness appears to include some contribution from the underlying Hastings Beds (see *Fig. 8*, page 20).

The correlation for Fawley is poor but there are several possible reasons to account for larger vertical settlements than predicted, such as lateral yield towards the alluvium, imperfect contact between the tank base and foundation, over-estimate of the relative density due to the gravel stones increasing the S.P.T. blow count, etc.

SESSION II
Normally Consolidated and Lightly Over-consolidated Cohesive Materials

Session Chairman: Professor R. E. Gibson
General Reporter: Dr. N. E. Simons

DISCUSSION

Embankments and centrifuge tests

R. H. BASSETT, King's College, London. At the BGS Conference on Field Instrumentation in London last June I presented the results of a centrifugal model test of a road embankment section constructed on soft clay (Bassett 1973b). The model was arranged to represent the 8 m high approach embankment to the River Ouse crossing on the A64 at York and *Figs. 1 and 2* reproduce some of the displacement data predicted by the model. The

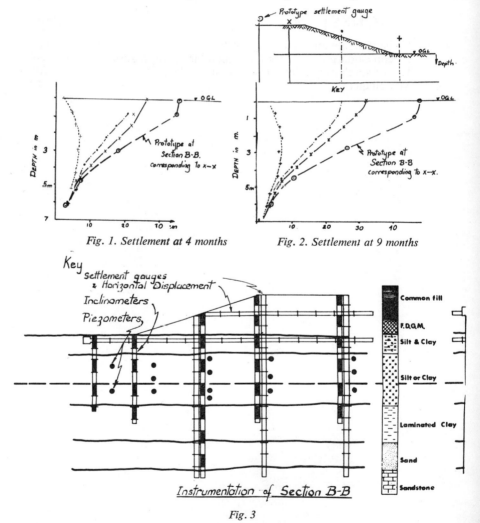

Fig. 1. Settlement *at 4 months*   Fig. 2. Settlement *at 9 months*

Fig. 3

prototype was instrumented as shown in *Fig. 3* and some prototype data can now be presented.

Construction commenced on May 1, 1973 and was carried out in two distinct phases:

(1) 5 weeks to raise the embankment to $4\frac{1}{2}$ m high. This was followed by a 14 week pause.

(2) The embankment was completed to 8 m high in a further 8 weeks.

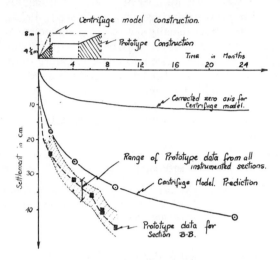

Fig. 4 (a). Settlement–time relationship

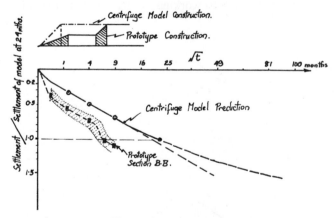

Fig. 4 (b). Settlement–time relationship

This construction sequence differed considerably from the model test which assumed rapid construction to complete height in 4 weeks.

However, comparison at the current time, viz. nine months after commencement, has proved interesting. *Figures 4(a) and (b)* are plots of the settlement of the original ground surface under the full height section (the line marked X in *Fig. 2*) against t and √t. Superimposed on *Fig. 4* are the prototype results for the vertical displacement of the central settlement gauge in the main

685

instrumented section (BB) and limit reading taken from a number of similar locations. The prototype data shows a markedly larger settlement than predicted by the centrifugal model. However, examination of *Fig. 4(b)* suggests that the discrepancy occurs due to an instantaneous compression during the loading processes. The subsequent consolidation settlement being accurately reproduced by the centrifugal model (this is deduced from the fact that the displacement $\sqrt{t}$ curves are parallel).

The author has carried out a number of other centrifugal tests which have been compared with prototype construction (Bassett, 1973a, Bassett and Beasley, 1971) and these have usually shown close agreement with prototype performance. The discrepancy in these York models requires careful consideration as it suggests an instantaneous compressive response in the lightly over-consolidated soils which may have been destroyed by the sampling or handling techniques used in the model making.

The prototype will be continuously monitored and with my colleagues from the North Eastern Road Construction Unit I hope in the near future to present this data in full.

Bassett R. H. and Beasley H. W. (1971). 'Centrifugal Model Tests of the Flood Protection Embankments for the Centinel Mist Project', Report to USAF and Department of the Environment

Bassett, R. H. (1973a). 'Centrifugal Model Tests of Embankments on Soft Alluvial Foundations', *Proc. 8th Int. Conf. Soil Mech. Found. Engng, Moscow*, Vol. 45, 23–30

Bassett, R. H. (1973b). *Field Instrumentation in Geotechnical Engineering*, 687–694, Butterworths

L. J. ENDICOTT, G. Maunsell & Partners.   This contribution is concerned with the use of centrifugal models to predict the settlement of structures founded on cohesive materials. Whilst working as a Research Student at Cambridge

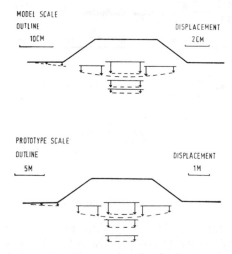

Fig. 5. Predicted and observed settlements of trial embankment, King's Lynn at 265 days

University I was able to use the centrifuge facility, developed by Avgherinos (1969) to test small models, about 250 mm high, of real structures about 10 m high. The results I wish to discuss are of a model test of the Trial Embankment at King's Lynn. The prototype structure was a well instrumented test embankment about 4 m high, founded on about 6 m of compressible, very soft, estuarine deposits of clay, peat and peaty clay. The trial embankment has been described by Wilkes (1970) who made available some of the results of measurements taken on site.

The model was built as a plane section from large bulk samples obtained from the site. Photographs taken of the model during the test were used to record the displacements within the plane section. The test was conducted at 56 gravities. A linear scale of 56 was adopted and a time scale of 56^2 was assumed. A comparison between vertical settlement after 265 days as predicted from the model and as measured on site at some of the settlement points is shown in *Fig. 5*. The plots of settlement against time for four levels on a vertical line below the embankment are shown in *Fig. 6*. It can be seen from

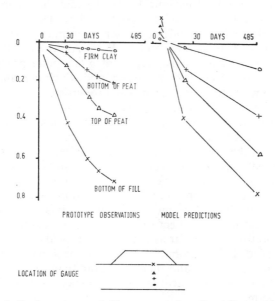

Fig. 6. Settlements recorded by gauge and predicted from model test

the two figures that there is good overall agreement between the predictions from the model test and the observations on site. A more detailed report of the model test with a critical discussion on the results obtained is given by the writer, Endicott (1970).

Avgherinos, P. J. (1969). 'Centrifugal Testing of Models Made of Soil', Ph.D. Thesis, University of Cambridge
Endicott, L. J. (1970). 'Centrifugal Testing of Soil Models', Ph.D. Thesis, University of Cambridge
Wilkes, P. F. (1970). 'The Installation of Piezometers in Small Diameter Boreholes', *Geotechnique* Vol. 20, No. 3, 330–333.

J. M. McKENNA. My discussion is concerned with the relationship between primary and secondary settlement and the use of centrifugal models to predict field rates of settlement.

The conventional method of calculating the amount and rate of settlement of buildings (or embankments) on soft ground is to consider that settlement is made up of three separate components, immediate, primary and secondary, the last only starting when primary settlement has ended after all the excess pore water pressures have dissipated. This approach is permissible when the drainage paths in the clay are short and the time taken for the excess pore water pressures to dissipate is fairly quick, say one year. However, when the time for pore pressure dissipation is longer than this, there is now sufficient laboratory and field evidence to show that secondary settlement also occurs in the primary phase while the pore water pressures are dissipating, and therefore the conventional approach may be inaccurate. Although this is allowed for in some analytical studies (Christie, 1964), the rheological models employed lack sufficient flexibility to embrace a range of clay types.

In considering this problem, the laboratory work of Berre and Iversen (1972) at the Norwegian Geotechnical Institute and the theoretical analysis of these results by Garlanger (1972) are outstanding. The method of analysis involves a numerical solution, and a computer programme is therefore essential. In association with R. E. Gibson and K. R. F. Andrews of King's College, London, I hope shortly to publish details of a computer programme which involves an extension of Garlanger's theory and which can analyse a multilayer soil profile with stress dependent soil properties.

I was interested in R. H. Bassett's comparison of the actual rate of settlement of the York Trial Embankment with the prediction based on one centrifugal model test. This showed that the rate of settlement in the field was much faster than that in the laboratory. I am not surprised by this result, for it is my experience that alluvial soil profiles are never sufficiently homogeneous, even after dividing them into three or four layers, for one sample from each layer to be representative of the actual field conditions. As with all laboratory tests on relatively small samples, more than one result is necessary to establish average values, and in this respect the centrifugal model test is no different from other laboratory tests as far as predicting the field coefficient of consolidation is concerned. This is probably the most difficult parameter to measure accurately in the laboratory, and therefore more than one centrifugal model test is essential if a reasonably accurate prediction of the rate of settlement in the field is to be made.

Berre, T. and Iversen, K. (1972). 'Oedometer Tests with Different Specimen Heights on a Clay Exhibiting Large Secondary Compression, *Geotechnique* Vol. 22, 53–70

Christie, I. F. (1964). 'A Re-appraisal of Merchant's Contribution to the Theory of Consolidation,' *Geotechnique* Vol. 14, 309–320

Garlanger, J. E. (1972). 'The Consolidation of Soils Exhibiting Creep Under Constant Effective Stress,' *Geotechnique*, Vol. 22, 71–78

R. H. BASSETT, King's College, London.   In reply to Endicott's contribution on centrifugal test data, I would completely endorse his observation that remarkably good predictions are made. Both he (Endicott, 1972) and I (Bassett, 1973 and Beasley, 1974) have succeeded in obtaining excellent predictions of prototype behaviour. I feel the York data I presented earlier in this session is important as I did not intend to be critical of the method but rather to show that there can still be inexplicable results. To do this gives the engineering fraternity a more balanced picture of centrifugal testing, not just the outstanding successes.

The York data will be presented in full at a later date. The pattern of combined consolidation and shear which is being shown in the displacements occurring under the slopes of the embankment is very close to the model.

Consideration of the problem of combined consolidation and shear leads me to endorse Wroth's contribution concerning the explanation behind the preconsolidation characteristic being considered by some as a critical normal stress and by others as a critical shear stress. I believe it is important for both students and experienced engineers to stop thinking of shear deformations and consolidation as separable independent processes. This is particularly important in the case of normally or lightly overconsolidated clays. Shear and consolidation are part of the complex interaction behaviour pattern of soils for which the critical state concept (Roscoe, Schofield and Wroth, 1958; and

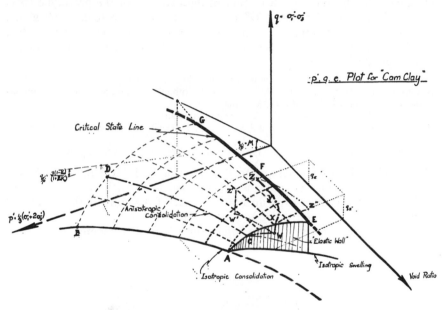

Fig. 7. $e' - p' - q'$ envelope for Cam Clay

Schofield and Wroth, 1968) provides a good representation and I make no apology for reproducing the $e:p'$ diagram for a 'Cam Clay' in Fig. 7. In this figure, p' is a normal stress parameter $\sigma'_1 + \sigma'_2 + \sigma'_3/3$, which for radial symmetry $= 1/3\,(\sigma'_1 + 2\sigma'_3)$, q is a shear stress parameter (which for radial symmetry) $= \sigma'_1 - \sigma'_3$, and e is void ratio. The isotropic consolidation line AB (for $\sigma'_1 = \sigma'_2 = \sigma'_3$) represents the only situation in which shear stresses are zero. The one-dimensional K_0 consolidation line $C–D$, which represents the standard oedometer and the normal field situation, is a consolidation relationship between void ratio and normal stress (p') which is constrained to lie in the plane defined by the shear stress (q)/normal stress (p) relationship

$$q/p' = \frac{3(1 - K_0)}{(1 + 2K_0)}.$$

Even the ultimate yield condition or critical state line (EFG) is a consolidation characteristic at the limiting stress ratio $q/p' = M$.

The response path $WXYZ$ is typical of the 'state path' (or void ratio/shear stress/normal stress relationship) of a soil element in the foundations below the sloping section of an embankment and is a complex inter-relationship of consolidation and shear deformation. Only with caution should it be considered the same as the independent processes:

(1) Consolidation W to W' followed by no volume shear $W'-Z'-Z$ or

(2) No volume change shear WXZ'' followed by consolidation $Z''-Z'$ (this has apparently exceeded the critical failure conditions).

The complex real inter-relationship is very difficult to analyse numerically. The centrifugal model test technique is, in contrast, ideal as it should provide an almost identical analogue of the prototype without requiring the 'state path' to be predetermined. If correctly made the centrifuge model will automatically produce the displacement data to $1/G$ of the prototype (G is the gravitational factor) in $1/G^2$ of the prototype time (e.g. at 60 g, i.e. $G = 60$. 24 hours running in a centrifuge represents 10 years in a prototype).

It is in this interesting inter-relationship field that the writer is pursuing the centrifuge method.

Endicott, L. J. (1972). Cooling prize paper, Cambridge University

Bassett, R. H. (1973). 'Centrifugal Model Tests of Embankments on Soft Alluvial Foundations', *Proc. 8th Conf. S.M.F.E., Moscow*

Beasley, D. H. (1974). Ph.D thesis, Cambridge University

Roscoe, K. H., Schofield, A. N. and Wroth, C. P. (1958). 'On the Yielding of Soils'. *Geotechnique*, Vol. 8, 22–53

Schofield, A. N. and Wroth, C. P. (1968). *Critical State Soil Mechanics*, McGraw-Hill

M. JAMIOLKOWSKI, Associate Professor of Soil Mechanics, Technical University of Turin. I would like to be associated with G. Barla and L. Garassino in describing experimental evidence of the contribution of lateral confined plastic flow and creep to settlement.

We would like to refer to the behaviour of an instrumented 3.5 m high

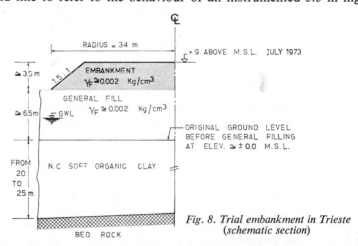

Fig. 8. Trial embankment in Trieste (schematic section)

circular trial embankment built in Trieste over a deposit of a very soft NC organic clay of high plasticity. The geometry of the problem and solid profile may be deduced from *Fig. 8*. In *Fig. 9* the location and type of the adopted field instrumentation are shown.

The construction history of the site was as follows: (1) early 1969 a general

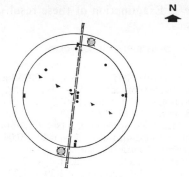

■ SETTLEMENT PLATES
● PNEUMATIC PIEZOMETERS
▲ BRS SETTLEMENT MERCURY GAUGES
⊕ VERTICAL INCLINOMETER TUBES
═══ HORIZONTAL INCLINOMETER TUBE

Fig. 9. Trial embankment in Trieste—scheme of foundation instrumentation

fill was dumped over the whole area in very uncontrolled geotechnical conditions: (2) July to November 1972 extensive *in situ* and laboratory soil investigations were carried out: (3) March 1973 small diameter (= 10 cm) displacement sand drains (like 'Sandwicks', see Dustidar *et al.* (1969)) were installed in equilateral triangular array with a 2 m spacing: (4) July 1973 a trial embankment was placed on the sand drains area.

The soil investigation revealed the conditions summarized in *Fig. 10(a)*,

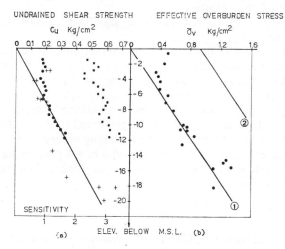

● FIELD VANE TESTS
+ LABORATORY TESTS
● SENSITIVITY AS OBTAINED FROM THE FIELD VANE TESTS

① EFFECTIVE OVERBURDEN STRESS PRIOR TO GENERAL FILLING
② EFFECTIVE OVERBURDEN STRESS SUPPOSING FULL CONSOLIDATION OF THE SOFT CLAY UNDER THE WEIGHT OF GENERAL FILL
● MAXIMUM CONSOLIDATION PRESSURE FROM THE OEDOMETER TESTS

Fig. 10. (a) Undrained shear strength; (b) stress history

691

Fig. 10(b) and *Table 1*. Examination of these results leads to the following statements:

Table 1. GEOTECHNICAL PROPERTIES OF THE SOFT CLAY

Bulk density 0.00065 $< \gamma < 0.00075$ kg/cm^3
Natural water content $45 < W_n < 55\%$
Liquid limit $70 < LL < 80\%$
Plastic index $45 < PI < 70\%$
Undrained shear strength (C_u) (see *Fig. 10(a)*)
Virgin compression ratio $CR = 0.20 \pm 0.04$
Coefficient of consolidation:
   in vertical direction $c_v = 2.11 \pm 1.27 \times 10^{-4}$ cm^2/s
   in horizontal direction $c_h = 3.30 \pm 1.23 \times 10^{-4}$ cm^2/s
Coefficient of earth pressure at rest $K_o = 0.5 \div 0.56$

Undrained deformation modulus, for $\dfrac{\sigma_1 - \sigma_3}{2C_u} = 0.5*$

$200 . c_u < E_{u5o} < 300 \ C_u$

---

* Stress–strain curves from CK_oU triaxial compression tests and plane strain active compression tests are of hyperbolic shape.

(1) When the trial embankment was placed the soft clay stratum was practically completely unconsolidated under the pressure imposed by the general fill.

(2) The undrained shear strength was dependent only upon the effective consolidation stress due to the self weight of the soft clay.

(3) There was no appreciable anisotropy within the soft clay with respect to the coefficient of consolidation.

(4) The soft clay layer does not exhibit any significant aging effect, therefore it can be classified as a young clay.

The observed vertical settlement of the contact surface between fill and embankment, and the horizontal displacement of the soft clay on the periphery of the embankment, both measured by means of Digitilt servo-accelerometer inclinometers, are shown in *Fig. 11* and *Fig. 12*.

A preliminary examination of the above mentioned measurements (the complete results of this case record will be presented to the 1st North Baltic Conference scheduled for 1975) allows the following observations to be made:

(1) Lateral displacements of the soft clay which took place in the undrained conditions correspond to the so called 'immediate settlement' larger than those predicted on the basis of

$$\frac{E_u}{c_u} = f \left(\frac{\sigma_1 - \sigma_3}{2c_u} \right)$$

relationship obtained from CK_oU triaxial compression tests even introducing the correction for confined plastic flow as recommended by d'Appolonia *et al.* (1971). In the tests the specimens were reconsolidated under conditions corresponding to the effective geostatic pressure existing prior to the placement of the trial embankment.

(2) During consolidation the lateral displacement is still growing indicating that an appreciable amount of creep is taking place parallel with the dissipation of the pore pressure. The vertical settlement due to this creep could not be taken into account with existing methods for settlement calculations.

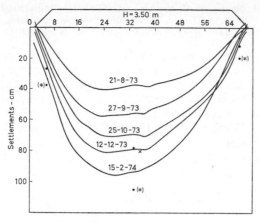

(*) Values as measured on April 4. 1974
    inclinometer results not yet elaborated
— Horizontal inclinometer torpedo
× B.R.S. Mercury settlement gauges
• Settlement plate

Fig. 11. Vertical settlement of trial embankment measured at top of the general filling

(3) The observed rate of pore pressure dissipation was even smaller than predicted on the basis of consolidation coefficients determined from laboratory tests on relatively undisturbed specimens of small dimensions using conventional 1-D consolidation theory. This fact suggests that when the safety factor against shear failure is low, (in this case $F_s = 1.1$ to 1.3 depending

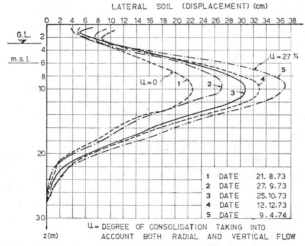

1	DATE	21. 8.73
2	DATE	27. 9.73
3	DATE	25.10.73
4	DATE	12.12.73
5	DATE	9. 4.74

U = DEGREE OF CONSOLIDATION TAKING INTO
ACCOUNT BOTH RADIAL AND VERTICAL FLOW

Fig. 12. Lateral displacement as measured in the northern inclinometer tube

upon assumption made about fill shear strength), the creep phenomena may influence the pore pressure dissipation (Zaretzky, 1967) or/and generate some kind of 'strain induced pore pressure'.

Dastidar, A. G., Gupta, S. and Gaaosh, T. K. (1969). 'Application of Sand Wick in Housing Project', *Proc. VIIth Int. Conf. Soil Mech. Found. Engng*, Mexico City

693

D'Appolonia, D. I., Poulos, H. G. and Ladd, C. C. (1971). 'Initial Settlement of Structure on Clay', *J. Soil Mech. and Found. Engng*, Div., *Am. Soc. civ. Engs*, Vol. 97, SM-10
Zaretzky, J. K. (1967). *The Theory of Soil Consolidation*, Nauka-Moscow (in Russian)

S. HANSBO, Chalmers Institute of Technology, Göteborg. In addition to consolidation, viscous elasto-plastic flow also contributes to settlements. We made a plate loading test on a very soft plastic clay near Stockholm and measured the horizontal movement of the clay. From this we concluded that half the settlement was due to flow of the clay in a horizontal direction and half due to consolidation. This half agreed with consolidation theory.

This has also been confirmed in larger field structures, e.g. under a road embankment where—after a loading time of 2000 days—about 35% of the total settlement was due to lateral flow and 65% due to consolidation.

E. SCHULTZE, Aachen. The statements by Hansbo have been confirmed by our tests with strip footings of 0.2 m width on silt having a water content between 21 and 23 % (Altes, 1974). After the excavation strong lateral deformations showed beneath the footings up to depths of about $z/b = 1.0$ to 1.5. Therefore, the pre-conditions of the usual settlement calculations no longer existed. For the rest, similar conditions were found for sand (Schormann, 1973).

Altes (1974). 'Formanderungen und Bruch von Schluff unter schmalen Streifenfundamenten' (Deformations and failure of silt below narrow strip footings), Dr. Thesis, Techn. University, Aachen
Schormann (1973). 'Verformungs- und Bruchverhalten von Sand in axialsymmetrischen Versuchen' (Deformation and failure characteristics of sand in axially symmetric tests). With English foreword and summary, *Bull. Inst. Soil Mech. Found. Engng*, Techn. University, Aachen, VGB 59

Preconsolidation pressure and secondary compression

C. R. I. CLAYTON, Ground Engineering Limited. Dr. Simons has asked me to say a few words on secondary compression. I have certain misgivings about this, since I believe it may be necessary to give up various long-held ideas about the consolidation process, and in particular the application of the principle of effective stress in consolidation problems in order to understand the problem. Since Terzaghi published his theory in 1923 there appears to have been a basic misunderstanding among some engineers of the compression problem.

Terzaghi developed his consolidation theory while attempting to find a quick way of determining the permeability of clays. He was, therefore, primarily interested in the passage of water through the voids in the tested material, rather than in the reaction of the soil structure to changes in loading. Implicit in his equations was the idea that the soil condition would remain constant under constant effective stress, or that the deformation of the soil skeleton without the retarding effect of pore water would always occur at a very much faster rate than the dissipation of pore pressures.

That his theory has proved such a useful tool over the past fifty years bears witness to the frequency with which soils occur to which it is approximately applicable. However, there have been frequent reports of the inability of the Terzaghi theory to imitate the time–settlement behaviour of foundations. Generally, where settlement has continued after the supposed dissipation of pore pressures this has been labelled 'secondary compression', the implication being that secondary consolidation is separate from primary. In fact it is now well appreciated that the compression process in soils is a finely balanced reaction between the time-deformation behaviour of the soil, and the time-dependence of the dissipation of pore water pressure.

While for many clays a large proportion of compression takes place during the dissipation of pore pressures, this is not true for soils where the grain size (and hence permeability) is not small enough to cause measurable excess pore water pressures and retard skeleton compression. Bearing this in mind, it is hardly surprising that large secondary settlements have been reported for a large number of soils. These have included quick clays, organic clays, peats, normally consolidated clays, silts, and overconsolidated clays.

Furthermore, even relatively normal soils can show large secondary settlements, when deposited in thin layers between free draining soils, as have been reported at Mexico City. The dissipation of pore pressures in this case occurs rapidly, and the time dependancy of settlement is largely determined by the soil structure.

It therefore appears that the principle of effective stress is largely a 'red-herring' in consolidation. This is simply illustrated by considering two laboratory consolidation tests. Firstly if we consider a load increment in the oedometer test, we assume that the instant after load application there is a uniform pore pressure throughout the clay. However, near the free draining boundaries drainage occurs virtually instantly. Consolidation theory assumes that the skeleton can compress instantly, but since this is not the case, we can no longer mimic the time compression curve even in an oedometer. Secondly, consider a triaxial consolidation test. Here we can measure pore pressures as well as volume change. If a consolidation stage is started, and pore pressures are dissipating, when drainage is then prevented we would expect to have constant conditions. However, if all soils creep under constant effective stress we can expect to obtain an increase in pore pressure with time and hence a decrease in volume or a decrease in effective stress.

If the time dependency of the soil skeleton is an important part of the compression process we obviously require information about it, as much in the short term as in the long term. However a brief review of the tools available to the engineer for design purposes shows that the most useful method of prediction available, developed by Garlanger, is based on the model of skeleton compression proposed by Buisman at the 1st International Conference on Soil Mechanics and Foundation Engineering in 1936. This says that compression is linear when plotted against the logarithm of time. This more or less sums up the value of much of the research into secondary compression in the intervening period. Since 1936 a large number of papers have been published, as categorized by Dr. Simons. Virtually all these papers have been concerned with long-term effects only, and until recently laboratory equipment of sufficient precision and durability has not been available.

Much effort has been made to model the results of these tests, using either mathematical or rheological models, or rate-process theory, but these models do not seem to provide reasonable fits when compared with other sets of data.

In conclusion I would like to say that it seems to me that skeleton compression is important both during, as well as after, dissipation of pore pressures. This being so we require to investigate more fully the time dependency of skeleton compression in the short term. The results of such investigation may well lead to a better assessment of the sorts of soils that may show secondary effects.

G. SÄLLFORS, Chalmers Institute of Technology, Gothenberg.   From the General Report we can see that p_c' is an extremely important parameter. However, Dr. Simons does not mention any of the new oedometer routines, such as CRS (constant rate of strain test) where the sample is continuously deformed at a constant rate of strain. This is a most powerful method and it has many advantages. It can be readily automated and the computer takes care of all calculations and plotting of curves.

Figure 13 shows the stress-strain relationship for a soft clay from the western part of Sweden. Important is the distinct and well defined part of the curve around the p_c'-value. In *Fig. 13* there is also a plot of pore-pressure

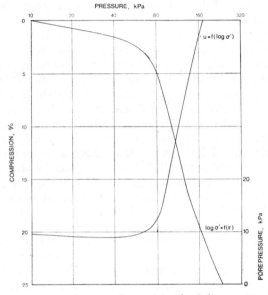

Fig. 13. Pressure–compression curve for specimen loaded at constant rate of strain

(measured at the undrained bottom of the sample) versus log σ'. The pore pressure is fairly constant during the early stage of the test. However, as the preconsolidation pressure is exceeded the pore pressure increases rather rapidly, indicating a breakdown in the structure. If, during a test like this, the horizontal pressure is measured, a stress-path can be plotted, see *Fig. 14.*

The test results plotted by the computer are indicated by crosses. The curve consists in principle of two straight lines. Line 1 represents a stage where the shear stress is built up rather rapidly. The breaking point between the two lines 1–2 represents a critical shear stress and a further increase in shear-stress above this value is followed by large deformations, line 2.

The vertical pressure at the breaking point corresponds very well with the preconsolidation pressure. This indicates that the critical shear stress is a *fundamental* parameter, more so than the vertical pressure p_c'.

It is well known that the p_c'-value exhibits a time dependency. If CRS-tests are run with different rates of strain on similar samples the result can be as shown in *Fig. 15*. The faster the test is run the higher the p_c'-value, which is consistent with earlier investigations. Bjerrum explained these phenomena using the concept of delayed secondary compressions. I would like to make a slightly different approach.

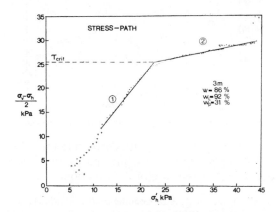

Fig. 14. Stress-path for specimen loaded at constant rate of strain

We know from earlier investigations at Chalmers University of Technology (Torstensson, 1973) that the undrained shear strength, determined by vane tests, is time dependent; the longer the time to failure, the lower the shear strength. If τ_{crit} is evaluated from CRS-tests, performed at different rates of strain similar results will be obtained, e.g. such as shown in *Fig. 16*.

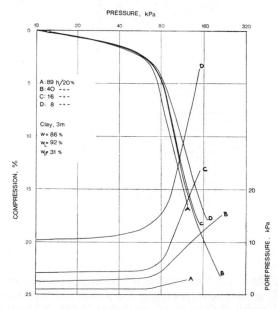

Fig. 15. Pressure–compression curve for specimens loaded at different rates of strain

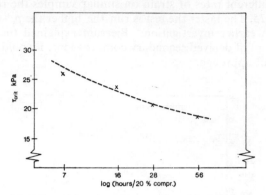

Fig. 16. Effect of duration of test on the critical shear-strength

It is quite clear that τ_{crit} decreases with decreasing rate of strain. Thus, the time dependency of the p_c' could be explained by the fact that the critical shear-stress is dependent of the rate of deformation.

Bjerrum, L. (1972). 'The Effect of Rate of Loading on the p_c-value Observed in Consolidation Tests of Soft Clays,' contribution to discussion, Session III (*Am. Soc. civ. Engs. Specialty Conf.*, Purdue)
Torstensson, B. A. (1973). 'Friction Piles Driven in Soft Clay. A field study.' Thesis, Chalmers University of Technology, Göteborg

H. SHIELDS, University of Ottawa. A new town is proposed for an area south-east of Ottawa, Canada in a region of soft, lightly overconsolidated,

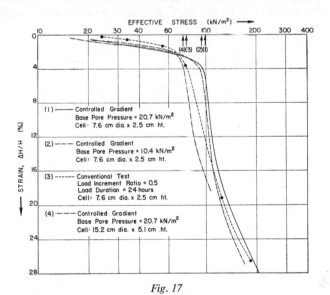

Fig. 17

sensitive clay. The correct prediction of the preconsolidation pressure, p_c, is of considerable economic importance. Consolidation tests have been carried out at the University of Ottawa on specimens cut from block samples. Standard Rowe cells were used and the tests were run as Controlled Gradient Consolidation (CGC) tests without back pressure.

Well-defined strain–log pressure curves result from these tests since results are obtained continuously. Curves 1 and 2 in *Fig. 17* illustrate typical CGC test results. Conventional, incremental consolidation tests on the other hand are poorly defined even with a reasonable load increment ratio–see Curve 3. Comparing Curves 1, 2 and 3, one immediately notices the much

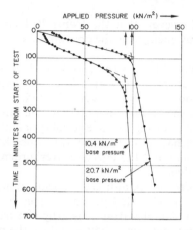

Fig. 18. *Pressure–time curves from controlled gradient consolidation tests*

steeper virgin compression curves from the CGC tests. The p_c value from the CGC tests is considerably higher (using the Casagrande construction in all cases). The small difference between Curves 1 and 2 is due to a difference in pore pressure gradient and, therefore, the duration of the test.

An advantage of the Controlled Gradient test is that an independent measure of p_c can be had from pressure versus elapsed time plots (*Fig. 18*). A

Table 2. COMPARISON OF RESULTS

Type of test	Base pore-pressure (kN/m²)	Diameter (cm)	Preconsolidation pressure Casagrande (kN/m²)	Rate change (kN/m²)
Controlled gradient	10.4	7.6	96	93
Controlled gradient	20.7	7.6	99	98
Controlled gradient	20.7	15.2	74	75
Conventional		7.6	76	

comparison between p_c Casagrande and p_c rate change for typical tests is given in *Table 2*. The results are identical for practical purposes. However, all is not as rosy as it seems. Curves 1, 2 and 3 of *Fig. 17* are from 7.6 cm diameter cells. Curve 4 is a typical result from a 15.2 cm cell. It is noted that the larger diameter cells give results which are considerably to the left of the smaller diameter results. Curve 4 is equally as well defined as Curves 1 and 2, of course, and shows the near vertical virgin line. However, the p_c from Curve 4 now happens to be similar to the p_c from the conventional test, Curve 3.

Curves 1 and 4 should be identical. This means there must be a fault in the Rowe cells—most probably in the comparative stiffness in the diaphragms between the 7.6 cm and 15.2 cm diameter cells. Tests are planned on a 25.4 cm diameter Rowe cell to help resolve this matter of size effect. In the meantime, it is hoped that anyone having a similar experience with the Rowe cells and soft clay will contact the discusser—particularly if they have discovered the fault.

The fact remains, however, that the Controlled Gradient Consolidation test is a marked improvement over the conventional test.

P. M. JARRETT, Royal Military College of Canada.   The General Reporter noted that in our paper (page 99) a threshold value for the net increase in foundation stress of about 15 kN/m² was indicated, below which only minor settlement occurred in the 15 buildings studied. He raised the question as to whether any more recent testing of better quality samples than the originally available U-4 samples had indicated any difference between the measured preconsolidation pressure and the effective overburden pressure, in support of such a threshold value.

The answer at this time is no, although investigations are still proceeding. Consolidation tests on thin walled piston samples of soil from the site in question and on block samples from a nearby site have indicated preconsolidation pressures less than, or at best equal to, the present effective overburden pressures. Furthermore, testing to investigate the effects of rate of loading have shown it to be negligible. There is nothing therefore in the laboratory testing programme to suggest that the soil is exhibiting a delayed compression type of behaviour as proposed by Bjerrum (1967). If reliance is placed on the field behaviour then the probability exists that although good quality samples have now been used, good is still not good enough for this soil.

The General Reporter presents in *Fig. 6* a graph of p_c'/p_0' against plasticity index, after Bjerrum (1972). If this chart is applied to the presented Grangemouth results for Layer 2, the major seat of the settlements, then one obtains for the 35% plasticity index of Layer 2 a p_c'/p_0' value of 1.6. For a value of p_0' at the centre of the layer of 33 kN/m² this gives a p_c' of 53 kN/m² and a $(p_c' - p_0')$ of 20 kN/m². This predicted difference from Bjerrum's plasticity index chart gives a good measure of agreement with the field threshold value of 15 kN/m² especially when it is noted that any useful threshold value from the Bjerrum approach is only a fraction of the difference $(p_c' - p_0')$.

A deliberate effort was made in Paper II/4 to present data in a simple form without the burden of too many assumptions. It was mentioned in general discussion that it would be preferable to present data in a non-dimensional form. This approach whilst having quite obvious advantages does lead to the necessity for making what are at this stage arbitrary assumptions as to the load distribution within layers, the amount of settlement in the individual layers and the overall compressible thickness. Sufficient information is contained within the paper to allow readers to develop their own assumptions. As an example of where such assumptions lead, for Layer 2, if a Boussinesq distribution of stress increase is used for the centre of the layer and it is assumed almost arbitrarily that 60% of the overall settlement occurs in that layer then from the amended version of *Fig. 4* in the Paper a value of C_c of about 0.8 is obtained which is considerably in excess of that obtained in laboratory testing. The continuing investigation will show how close such assumptions are to the truth.

Bjerrum, L. (1967). 'Engineering Geology of Norwegian Normally Consolidated Marine Clays as Related to Settlements of Buildings', *Geotechnique*, Vol. 17, 81–118
Bjerrum, L. (1972). 'Embankments on Soft Ground', *Proc. Am. Soc. civ. Engs Speciality Conf. Performance of Earth and Earth-Supported Structures*, Purdue University, Vol. 2, 1–54

E. SCHULTZE, Aachen. In contrast to the normally consolidated *clays* dealt with in the reports, we mainly have to do in the Rheineland with normally consolidated *silts*. Here the precompression found in the oedometer

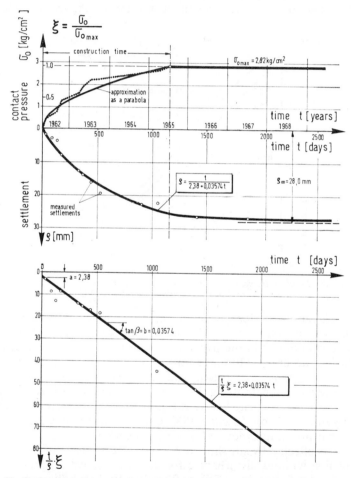

Fig. 19. *Evaluation of time–settlement–load curves by a modified hyperbolic function*

test usually deviates more or less from the overburden pressure (Schultze, 1970, Kahl, 1972). The same follows from an evaluation of settlement observations according to the following method.

Christow (1968) has shown that the time–settlement curve at a constant loading can be represented by a hyperbola, and by transforming its coordinate system, a straight line the slope of which allows the prediction of the final settlement with a high reliability. Sherif (1973) showed that for an irregular increase of load within the construction period, a straight time–settlement curve can be achieved in the hyperbolic coordinate system by using a correction factor (*Fig. 19*). The difficulty in such an evaluation, however, is that in most cases it is not possible to obtain for the observed time-settlement-curves the appertaining time-load curve during the construction period. Generally structural engineers are reluctant to make the necessary intermediate calculations of the loading of the foundations. It has therefore been investigated, whether one can obtain with sufficient accuracy a straight time–settlement curve in the hyperbolic scale by substituting the time-course within the construction period by a parabola without knowing the time-course of loading.

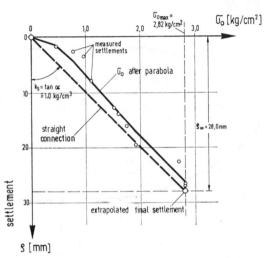

Fig. 20. *Load–settlement curve for calculation of the modulus of subgrade reaction and of the modulus of compressibility*

The investigations performed by means of examples for which the exact course of the time-load curve was known, show that a straight time-settlement curve is obtained in the hyperbolic coordinate system by this method with sufficient accuracy. Moreover on the basis of such representation, the valuable relationship between load and settlement during the construction period can be derived permitting the modulus of the subgrade reaction of the construction to be calculated (*Fig. 20*) and from this the modulus of compressibility of the subsoil (Schultze 1963 and 1964). Here, however, the break indicates that precompression exists in rare cases only.

Creep tests effected by silt showed in the case of *D*-tests in the log *t*-scale a straight line without disturbance; for cup-tests, however, after a certain time it showed a flutter, i.e. a failure (Schultze 1971, Schultze and Schmidt-Schleicher, 1973).

Christow (1968). 'Beitrag zur praktischen Setzungsberechnung und Auswertung von Zeitsetzungslinien' (Contribution for Practical Settlement Calculation and Evaluation of Time–Settlement Curves), *Proc. Donau Eur. Conf.*, Wien, 6

Kahl (1972). 'Geologie und Bodenmechanik des Rheinischen Schluffs' (Geology and Soil Mechanics of the Rheine-silt), with English foreword and summary, *Bull. Inst. Soil Mech. Found. Engng*, Techn. University, Aachen, VGB 54, 155

Schultze (1963). 'Probleme bei der Auswertung von Setzungsmessungen' (Problems in the Evaluation of Settlement Observations), *Bull. Inst. Soil Mech. Found. Engng*, Techn. University, Aachen, VGB 27

Schultze (1964). 'Beispiele für Setzungsbeobachtungen in bindigen und nichtbindigen Böden,' *Bull. Inst. Soil Mech. Found. Engng*, Techn. University, Aachen, VGB 29

Schultze (1970). 'Vorbeanspruchung auf Ton' (Initial Loading of Clay), Discussion, *Proc. Conf. Soil Mech. Found. Engng*, Dusseldorf, 537

Schultz (1971). 'Essais de fluage sur des sols normalement compactés' (Creep Tests in Normally Consolidated Soils), *J. Nat. Comité Français de Mécanique des Sols*, Paris, Comportement des sols avant la rupture, 20

Schultze and Schmidt-Schleicher (1973). 'Die Kriecheigenschaften von Schluff' (The Creep Properties of Silt), *Scient. Bull. Nordrhein-Westfalen*, Nr. 2268, West Germany Publishing House, Opladen

Sherif (1973). 'Setzungsmessungen an Industrie- und Hochbauten und ihre Auswertung' (Settlement Observations of Industrial and Tall buildings and their Evaluation). With English foreword and summary, *Bull. Inst. Soil Mech. Found. Engng*, Techn. University, Aachen, VGB 57

C-J. F. CLAUSEN, Norwegian Geotechnical Institute. In his General Report, Dr. Simons had a direct question concerning our paper, relating to *Fig. 5* on page 76. He pointed out that his test result could have been expected from Bjerrum's work to show a value of about 1.4 for the ratio p_c/p_0. Instead the result has shown a value less than 1.0. There are probably two reasons for this:

(1) This sample was taken from a depth of 20 m, and we all know that taking good samples in soft material from depths greater than, say 10 to 15 m is very difficult. We probably cannot obtain good samples below these depths.

(2) It is important to know the geological history of the site. The building was constructed in 1920, but 25 to 30 years earlier a fill 2.5 m thick was placed over the site. During construction an additional 0.5 m of fill was placed. This fill thickness of 3.0 m corresponds to a load of 6 tonne/m^2 which has destroyed the p_c value that had developed in the clay during a few thousand years while it enjoyed constant effective stresses. At the end of construction this clay was a truly normally consolidated young clay and the gain in strength resulting from secondary consolidation has only gone on from 1920 until today. This second explanation is probably the more important cause of the observed low value.

G. A. LEONARDS, Purdue University. The preconsolidation pressure appropriately determined in the oedometer test, where the time required to dissipate excess pore pressures may be measured in minutes, or at most a few hours, will agree tolerably well with the preconsolidation pressure at the end of pore pressure dissipation in the field.

I realise that this is a very contentious statement, particularly in view of the fact that sizeable settlements have been measured when values of $\Delta p/(p_c - p_0)$ were much less than one. I submit that there may be other reasons why this occurs:

(1) Wroth has drawn attention to the importance of the stress path: I would add the importance of the strain path in influencing the way a lightly overconsolidated clay will respond when stressed. There need be no change in the macroscopic shear stress in order to observe a preconsolidation pressure

in the stress change—volume change phenomenon. One may, of course, apply a hydrostatic state of stress and still observe a stress level at which there is a sharp break in the apparent compressibility. There is, however, on either a stress path or a strain path, a point at which the structure of the clay will break down and when this happens you will observe a preconsolidation pressure effect. It may not be generally recognised that the pressure at which this occurs, particularly with a sensitive clay, can be influenced by the size of the increment of load applied.

(2) One may also cause a critical shear strain by making an excavation in weak soil of sufficient depth to cause, because the factor of safety will be low, sufficient remoulding leading to relatively large subsequent settlements.

(3) If a sample of apparently uniform clay is sliced into thin layers, say, 1 to 2 mm thick, and the water content of each layer is measured, the scatter and range of water content values will be very large. If the preconsolidation pressure could be obtained for each of these thin layers, a similar large range of values could be expected. Thus, any given pressure may exceed the preconsolidation values for some number of thin layers and can lead to unexpected settlements although the *apparent* overall value of $\Delta p/(p_c - p_0)$ is small.

There are, therefore, a variety of reasons why one gets different apparent relationships between the settlement and p_c, p_0 and Δp, which some may choose to interpret as a time rate effect on the value of preconsolidation pressure.

P. SMART, Glasgow University. Some care should be taken in using Bjerrum's relationship between p_c'/p_0' and PI (Simons, General Report, *Fig. 6*), because both p_c' and p_0' depend on geological history. This is evident from Bjerrum's plot of e and log p (*ibid.*, *Fig. 5*), which shows clearly that p_c'/p_0' depends on the period of secondary consolidation, at least for the materials to which the data relate.

Care is also required in using the relationship $\Delta_p/p_c' - p_0'$, because, for a normally consolidated clay, p_c' and p_0' are, by definition, equal.

P. M. JARRETT, Royal Military College of Canada. Professor Leonards questioned the time dependency of the preconsolidation pressure of sensitive clays. He suggested that the shock loading of differently sized loading increments could produce similar effects to time dependent loading and might indeed have been a factor in tests purporting to prove the time dependency. Whilst shock loading can be a factor in the behaviour of such clays, the time dependency of the value of preconsolidation pressure has been shown by Crawford (1965) using constant rate of strain consolidation testing and Jarrett (1967) using constant rate of loading testing. In both these examples completely uniform and continuous build up of strain or load respectively was applied with no shock factor involved.

Crawford, C. B., (1965). 'The Resistance of Soil Structure to Consolidation' *Can. Geotech. J.* Vol. 2, May

Jarrett, P. M. (1967). 'Time-dependent Consolidation of a Sensitive Clay,' *Materials Research and Standards* (ASTM) Vol. 7, No. 7

G. A. LEONARDS, Purdue University. In reply to Jarrett's remarks about the time-dependency of the preconsolidation pressure, I must ask (1) what constitutes sufficient evidence and (2) what is the definition of preconsolidation pressure. In my comments on time effects on the preconsolidation pressure I stated that the consolidation strains at the end of pore pressure dissipation in the laboratory sample, which may occur in minutes, would be comparable with the consolidation strains in the field at the end of the consolidation process, if there was one-dimensional compression with no lateral strains.

It is true that if a consolidation test is made by whatever method—constant rate of strain, incremental loading, constant gradient—deviations can be obtained as a result of the rate of loading. These deviations, however, are due to strains occurring at essentially zero excess of pore pressure, so that the critical strain causing a breakdown in soil structure can occur at a smaller stress level.

Use of surcharge, piles, etc., to limit settlement

Z. HROCH, Soil Mechanics Division, Stavební geologie, Prague.   I wish to describe work we have done to reduce settlements when we had to move a Gothic Church.

The town of Most in North Bohemia is situated above very rich coal deposits. It was decided that the coal seam in this territory should be exploited by opencast mining and the town of Most should be demolished with the exception of the Late Gothic Church of the Virgin Mary. After some consideration a decision was made to move the church along a curved track to the vicinity of another Early Gothic church and a medieval hospital (*Fig. 21*). The width of the church is 30 m and the length is 60 m. The building,

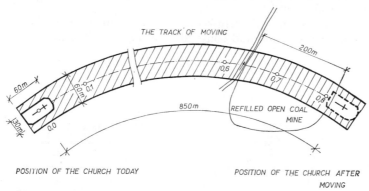

Fig. 21. Scheme for moving the Church

including the stiffening steel structure will be supported on 53 transfer cars mounted on four tracks. Each of the hydraulic cylinders mounted on the transfer cars will have the load bearing capacity of 500 Mp and a piston stroke of 270 mm.

The transfer will follow a circular track of length 850 m. The scheduled period of transfer is the summer 1975. On the basis of field investigations it was recognized that an old opencast mine is situated in the line of transfer (*Fig. 22*).

The bottom of the old opencast mine is about 33 m below ground level. The length of the part of the track crossing the opencast mine is 200 m, i.e. almost one fourth of the total distance. The remaining space after mining was refilled by clay overburden, coal dust, brick rubble and ashes. The age of this loose and heterogeneous filling is about 30 years.

It is possible to conceive that the settlement caused by the excess load from the moving church may be higher than the piston stroke of the hydraulic cylinders. Therefore it was decided before transfering the church to compress the filling by a pressure equal to the excess load from the moving church

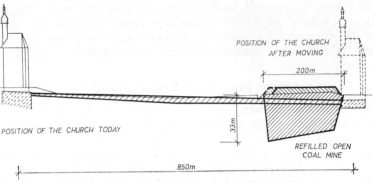

POSITION OF THE CHURCH AFTER MOVING

200m

POSITION OF THE CHURCH TODAY

33m

REFILLED OPEN COAL MINE

850m

Fig. 22. Section through track

(*Fig. 23*). A consolidation embankment of 5 m height was constructed and three pair of measuring points were installed. In every pair there was one measuring point situated on the natural ground and one point at a depth of 6 m below the natural ground level.

The measurement of settlement lasted 17 months and covered the period of embankment construction and consolidation (*Fig. 24*). The measurement of

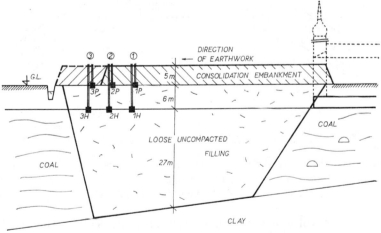

DIRECTION OF EARTHWORK

G.L.

CONSOLIDATION EMBANKMENT

5m

3P 2P 1P

6m

3H 2H 1H

COAL

LOOSE UNCOMPACTED

FILLING

COAL

27m

CLAY

Fig. 23. Section through refilled coalmine

elastic heave lasted 6 months and covered the period of removing the embankment and the resulting heave.

From measurement results we see that the total settlement at ground level reached (Bench mark No. 1 P) 98 cm and at a depth of 6 m below ground level

706

(Bench mark No. 1 H) 76 cm. The elastic heave after removing the consolidation embankment was 12% of the total settlement.

We anticipate that by these precautions the settlement of track subgrade caused by the moving church will be minimized to the elastic part of the settlement caused by the consolidation embankment and will remain within the piston stroke range of the hydraulic cylinders.

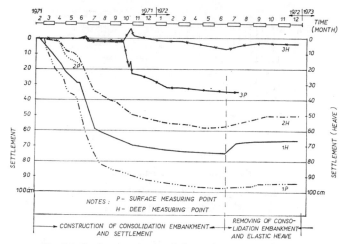

Fig. 24. Settlement–time and elastic heave–time diagram

S. HANSBO, Chalmers Institute of Technology, Gothenberg.   I would like to present to you three case records, all from Gothenburg, on settlements of buildings founded on friction piles—or, with Swedish terminology, 'cohesion piles'–in extraordinarily deep, normally consolidated high plastic clay.

The first case, Quay shed No. 107 in the free harbour, was built during 1944–1946. The basement has an area of 3560 m². The building is founded on a mat, 1.2 m in thickness, resting on 1412 spliced timber piles 26 m in length (pile spacing 1.6 m except for the uttermost pile rows where the pile spacing is 1.3 m). The net pressure increase on the subsoil at the foundation level is 26 kN/m². The depth of the clay layer is about 90 m. The geotechnical data, and the observed settlements, are shown in *Fig. 25*.

The second case, Quay shed No. 183 in Lindholmshamnen, was built during 1946–1949. The basement has an area of 3050 m². The building is founded on a mat, 1.25 m in thickness, resting on 1336 piles 30 m in length (the same pile spacing as in Quay shed No. 107). The net pressure increase on the subsoil at the foundation level is 30 kN/m². The depth of the clay layer is 95 m. The geotechnical data of the subsoil and the observed settlements are given in *Fig. 26*.

The agreement between observed and calculated settlements in these two cases is found to be good if the theoretical calculation of the settlements is based on the assumption (1) that the building load is acting at a fictitious foundation level located one third of the pile length above the pile tip, (2) that the buildings behave like rigid bodies, (3) that the fictitious foundation level is fully drained and (4) that pore water flow takes place only in the vertical direction (one-dimensional consolidation).

707

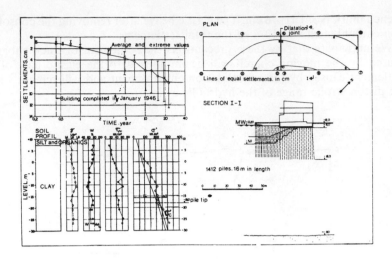

γ   density

w,w$_L$ water content, liquid limit

τ_{fu} undrained shear strength

σ'   effective vertical pressure

σ_c' acc. to oedometer tests

Fig. 25

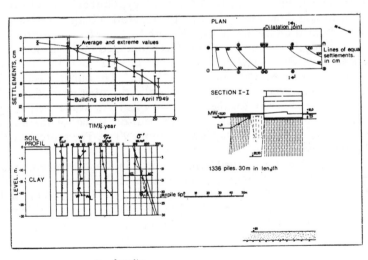

γ   density

w,w$_L$ water content, liquid limit

τ_{fu} undrained shear strength

σ'   effective vertical pressure

σ_c' acc. to oedometer tests

Fig. 26

The third and perhaps most interesting case, Östra Nordstaden, with a total, common basement area of about 53 000 m², is an example of compensated foundation.

The depth of the clay layer varies in this case between 49 m in the lower left corner, *Fig. 27*, to 92 m in the three other corners. The clay has a water content of about 60–80% and a plasticity index of about 65–70%. The shearing strength varies from about 20 kN/m² at the clay surface to about 70 kN/m² at 35 m depth. The fully compensated buildings in this area are founded on a common mat on timber piles, 18–20 m in length, except for the building in Zone 4 where spliced timber piles 26 m in length were used. The latter building is placed on piles concentrated below the columns in a conventional way. The other buildings, however, are resting on more or less evenly distributed piles. The number and distribution of piles were determined in order that the building load be carried partly by the piles (pile load equal to half failure

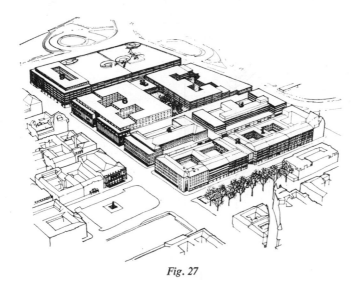

Fig. 27

load) and partly by contact pressure between mat and subsoil (contact pressure maximum 50 kN/m²). In some cases, the pore water pressure under the mat is regulated in order to obtain full compensation (where the buildings are too heavy) or to avoid uplift (where the buildings are too light). The differential settlements obtained are less than 1 : 1000, *Fig. 28*. The settlements at different depths are given in *Fig. 29*. As can be seen, the additional settlements at measuring points 5 and 7 by the load of the building are caused mainly by relative compression of the clay below half the depth of the piles. Pile spacing in this case is 2.4 m. At measuring point 6 the additional settlement due to the load of the building is caused mainly by relative compression of the clay above half the depth of the piles. The pile spacing in this case is 4.8 m.

The conclusion of these case records is that a mat foundation on evenly distributed cohesion piles can be a sound and safe solution of foundation problems in the case of extraordinarily deep, soft, normally consolidated clay layers.

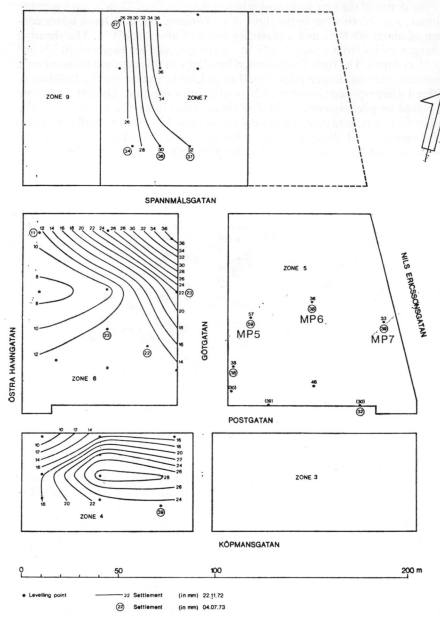

ZONE 9

ZONE 7

SPANNMÅLSGATAN

ÖSTRA HAMNGATAN

ZONE 6

GÖTGATAN

ZONE 5

MP6

MP5

MP7

NILS ERICSSONSGATAN

POSTGATAN

ZONE 4

ZONE 3

KÖPMANSGATAN

0   50   100   200 m

• Levelling point ——————22 Settlement (in mm) 22.11.72
㉒ Settlement (in mm) 04.07.73

Fig. 28

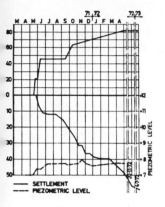
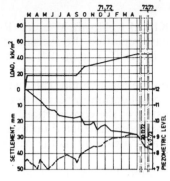
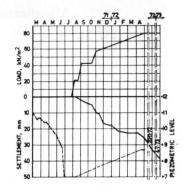

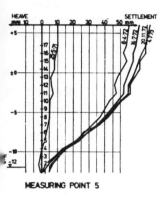
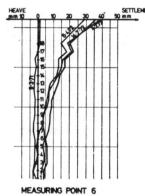
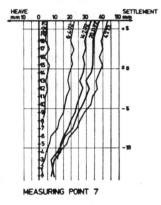

— SETTLEMENT
- - - PIEZOMETRIC LEVEL

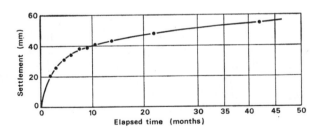

MEASURING POINT 5

MEASURING POINT 6

MEASURING POINT 7

Fig. 29

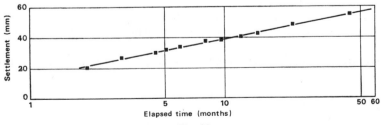

Fig. 30

Miscellaneous

M. M. H. AL-SHAIKH-ALI, North Western Road Construction Unit, Cheshire sub-unit. I have replotted the average of settlement points which appeared in *Table 1* of Paper II/3 (page 94) on a natural and logarithmic scale as shown in *Fig. 30*. It appears that the settlement points in this case fall on a straight line on the logarithmic scale, which suggests that creep is predominent and that consolidation appeared to cease at the end of the 12-month period after the start of the construction.

The total settlement has been calculated up to 833 years and found to be 113 mm and not 250 mm as suggested by the Authors.

G. LOWTHER, Foundation Engineering Limited. The measurement of pore pressures will be effected by the chemical composition of the fluid in the pore pressure measuring system. If this is not identical to that in the pores of the soil there will be osmotic pressures which will prevent accurate measurement of pore pressures. The diffusion process should not be considered to be isolated from the measuring system. This would apply particularly when the pore fluid was saline or low effective pressures were being measured, in the field and in the laboratory.

G. SANGLERAT, Lyon. All engineers who accept the validity of calculated settlements in sand with S.P.T. values must accept the validity of the method based on the static cone penetrometer. For clay it is more difficult. We wish to clarify our aim so as to leave no ambiguity regarding the purpose of our research on the compressibility of clays as estimated by the cone resistance of the static penetrometer.

It is not the intention to present a precise method of calculating total differential settlements using the static cone resistance R_p, but only a practical means to quickly determine in everyday routine work the range of settlements so as to give a rigid appraisal of the difficulties which could be expected from settlements.

The rule of thumb, which I proposed several years ago, has been proven many times, which is when the R_p of the Dutch Cone moving at 2 cm/s is more than 12 bars, and if footings are properly designed, no harmful settlements need be anticipated. But, if R_p is less than 12 bars, it is absolutely necessary to recover a sample to determine the water content. If the water content is low, say less than 30%, problems will be minor.

If the water content is high, the total of differential settlements will be high. They would require additional investigation by means of laboratory testing.

Of course, the penetrometer diagrams give immediately the thickness of compressible layers and their variations, so we strongly believe that in practice the penetrometer helps in putting settlement problems in the right perspective.

We are still waiting for someone to submit results which would prove us wrong.

Heavily Over-Consolidated Cohesive Materials

Session Chairman: Professor A. W. Bishop
General Reporter: Mr. F .G. Butler

DISCUSSION

Regional settlement

P. A. GREEN, Scot Wilson Kirkpatrick and Partners. We all talk rather loosely about differential settlement and total settlement, but what do we mean by total settlement. It is the value that we all try to calculate and strictly it is the relative shortening of the distance from the point being considered to the centre of mass of the earth. However, in practice what is normally measured is the settlement in relation to an arbitrary point in the earth's crust. In looking at case histories, including settlement surveys which I have done myself, I find that in general the settlement of a building is measured with respect to one of three types of datum:

(1) A deep datum founded in what is considered to be uncompressible strata. (In London this is usually the Chalk at depth well below the more compressible clays. If some near-surface regional settlements are taking place, these will also be measured with the building settlements. From the limited evidence available, it looks as though the regional settlements in London due to consolidation of the upper strata may be significant when a comparison is being made between predicted and measured settlements.)

(2) Surface monuments to measure settlements of the general surface of the ground adjacent to the building.

(3) Bench marks, either Ordnance Survey bench marks or pins, established on buildings with shallow foundations or on garden walls. (In this case settlement of the bench mark will include regional settlement and the settlement of the structure on which it is founded.)

The behaviour of these types of datum in the London area can be illustrated by two case histories.

(1) On the south bank of the Thames in the area of the Shell Centre, there is a building with a 2-storey basement which is not founded on piles. Part of the building has 9 storeys and part 12 storeys above ground level. Special datums were installed, one founded in the Chalk and an adjacent shallow one. These were sited 3 m from the building. I will refer to points 1 and 6 and 3 on the building and a point X on the building adjacent to the special datums (*Table 1*). Nearby there were two Ordnance Survey bench marks, one on an old railway viaduct and one on the Thames embankment wall. Measured settlements of the shallow datum, points on the building and the Ordnance Survey bench marks are given in Table 1.

If we make the basic assumption that there is regional settlement taking place, it appears to be at a rate of about 2.7 mm per year, and can be used to correct the annual settlement figures for the building. With this correction, the shallow datum is settling at a rate of 1.3 mm per year compared with a rate of 2.2 mm per year for point X on the building only 3 m away. This suggests that the shallow datum is within the 'cone' of settlement caused by the weight of the building.

It is of interest to note that the assessed rate of regional settlement of 2.7 mm per year is about twice the rate predicted by Wilson and Grace (1942).

(2) New Zealand House was built at the bottom of Haymarket adjacent to Pall Mall in London and settlements were measured in relation to five Ordnance Survey bench marks along Pall Mall:—

(1) on a building later demolished
(2) on the wall of the Reform Club
(3) on the United Services Club
(4) on a bank, and
(5) on a boundary wall outside the National Gallery.

Bench mark (4) was used as the main datum for measuring the settlements of the building, but it was continually checked against the movements of the other bench marks. During the period 1960 to 1973 bench marks (2), (3) and (5) moved up and down relative to (4) by as much as 20 mm up and 10 mm down. In addition, it seems that these three bench marks are gradually coming up with respect to (4) at an average rate of about 1 mm per year. This may explain why New Zealand House also now appears to be coming up: the bank is clearly settling by a small amount.

Table 1

Position	Period	Settlement (mm)	Settlement per year (mm)	Settlement/yr excluding regional[4] settlement (mm)
Shallow datum	1963–73	40[1]	4.0	1.3
BM on railway arch	1954–70	40[2]	2.6	−0.1
BM on embankment	1954–70	43[2]	2.7	0
Points 1 and 6	1961–71	37[1]	3.7	1.0
Point 3	1961–71	85[1]	8.5	5.8
Point X	1961–71	49[3]	4.9	2.2

(1) Relative to the special deep datum.
(2) Information from Ordnance Survey.
(3) Calculated assuming a linear settlement between points 1 and 3 which are on either side of point X.
(4) Assuming regional settlement of 2.7 mm/yr assessed from the behaviour of the Ordnance Survey bench marks.

These two cases illustrate the difficulty of defining exactly the settlement of a building on a heavily overconsolidated clay in which the 'total' settlement is relatively small when compared with other effects that may be present.

K. MORTON, Ove Arup & Partners.   I should like to make a few comments on Mr. Green's observations on the effects of regional settlements on bench-marks.

In our Paper, Mr. Au and I have presented eight case histories. For these we adopted benchmarks of the type (3) discussed by Green, that is bench-marks on existing structures or roads adjacent to the building under observa-tion. If one does this, two principles should be used:

(1) The benchmarks should be sufficiently far away to be outside the zone of influence of the settlement of the structure;

(2) Three or more benchmarks should be used for each to provide a check against differential regional settlement and possible loss of individual points.

Having made these provisions, I would suggest that this type of benchmark is the most appropriate to measure the settlement of a structure relative to the

715

adjacent ground, as this is what counts when considering the capability of the structure to resist settlement. Deep benchmarks are of assistance primarily if one is interested in assessing regional settlements of a particular stratum.

Heave

W. H. WARD, Building Research Station. Heave measured at the Shell Centre on the South Bank of the Thames described by Ward and Burland (1973), may be compared with that reported by May (p. 177) for a site on the opposite side of the river. At Shell, heave has continued since 1959 and by 1970 had reached 28 mm and is still going on. Values given by May for heave between 1968 and 1972 show 52 mm at a comparable position in the basement. What are the reasons for this difference? Incidentally you will notice a kink in the middle of the heave curve plotted from the figures given by May in *Table 1*, on p. 181.

At Shell, the excavation area we are considering is about 100 m × 300 m causing a reduction of vertical loading of about 150 kN/m². The last 2 m of excavation in London Clay was removed and replaced by concrete in one day. The excavation procedure was described by Ward (1961). The long-term uplift is being measured from the Bakerloo line which passes about 1.4 m below the underside of the raft of the basement. The May basement is about 70 m × 70 m. On both sites there is filling of early Victorian age, about 3 m thick. On the Shell site, the previous structures were two-storey houses, while on the May site, there used to be a fairly substantial building. On the Shell site there is 11 m of excavation so that the raft is resting just below the surface of the London Clay. There is also 11 m of excavation on the May site, and half the site is on a wedge of gravel while the rest is on London Clay. There is about 30 m of London Clay under both sites. There is a drainage system under the Shell basement to prevent water pressure form developing, but on the May basement some water pressures have been measured. I remember the May area during the 1928 floods and know that it is low-lying and subject to flooding from the river causing a rise of water pressures in the gravel. Superficially it looks as though the change in effective stress caused by the two excavations is similar, but I wonder if there is a difference in the stress history or the lithology of the London Clay which can explain the different amounts and rates of heave at the two sites.

Ward, W. H. (1961). 'Displacements and Strains in Tunnels Beneath a Large Excavation,' *Proc. 5th Int. Conf. Soil Mech.*, Vol. 2, Dunod, Paris, 749–753

Ward, W. H. and Burland, J. B. (1973). 'The Use of Ground Strain Measurements in Civil Engineering,' *Phil. Trans. R. Soc.*, Lond., A274, 421–428; and Building Research Station Current Papers, CP 13/73

J. MAY, Department of the Environment, Property Services Agency. Four further sets of readings of the survey marks have been taken between the time of preparation of my Paper (p. 177) and the date of the Conference. These readings are shown in *Table 2*. If these readings are plotted on the graph of *Fig. 5* of the paper the curves now show, if viewed optimistically, a tendency to follow the predicted heave pattern showing that the maximum heave is being approached.

Measurements have been taken of the ground water pressures beneath the slab at the two points P1 and P2 of *Fig. 2* of the Paper. The Bourdon gauges were abandoned in favour of ordinary stand pipes and the readings obtained are shown in *Table 3* in this discussion. There is a difference of approximately 1 m head of water between the pressure at P1 and P2, the lowest pressure being where the London Clay is immediately beneath the slab. In no case has the pressure recorded been greater than the level of 3.4 m above the top of the slab derived from the original site investigation and there is no evidence of the water pressure ever being higher than the recorded values shown at least over the period the stand pipes have been in place.

The benchmark from which all readings are taken is on a large government office block some 100 m from the site. This is checked periodically from

Table 2

Date of reading	Survey marks										
	1	2	3	4	5	6	7	8	9	10	11
	(accumulative change in height in millimetres—upward changes positive)										
Apr. 73	46.6	50.9	57.4	59.8	58.9	49.4	33.4	52.0	40.8	41.5	37.5
July. 73	47.4	52.2	58.7	61.0	60.3	49.9	33.3	53.8	42.3	41.9	37.6
Oct. 73	53.5	55.4	62.4	65.1	64.0	53.0	36.2	57.7	45.6	44.8	40.9
Feb. 74	53.7	56.1	63.2	66.0	65.0	53.9	36.7	58.7	46.6	45.5	41.8
	measurements accurate to 0.1 mm										

two other ordnance survey benchmarks on other buildings. The original benchmark was damaged and had to be replaced in October 1970 by the present datum. This change coincides with the apparent drop in level of all the survey marks of approximately 2 mm at this time and remarked upon by Ward. Since the error in some earlier readings prior to October 1970 could not be checked these results have been shown exactly as recorded.

Level readings were taken using first a Watts Autoset and then over the last 2 years a Wild N3. The levels are only relative to the surrounding area and

Table 3. WATER LEVELS AT POSITIONS P1 AND P2 (see *Fig. 2* of paper)

Levels recorded in metres above bottom slab level		
Date	P1	P2
19 Nov. 1973	2.04	—
10 Dec. 1973	*	3.23
18 Dec. 1973	*	3.06
7 Jan. 1974	*	3.11
17 Jan. 1974	2.04	3.00
31 Jan. 1974	2.22	3.08
20 Feb. 1974	2.04	†
11 Mar. 1974	2.14	3.04

* Leak in standpipe joint.
† Water level below sight glass (i.e. less than 2.95 m above bottom slab level).

are not absolute taking no account of any overall settlements in the London Clay.

The construction programme of the basement was as follows: (1) access to site June 1966, (2) diaphragm wall commenced July 1966, (3) Diaphragm wall completed October 1966, (4) Excavation commenced in September 1966 and continued in stages until November 1967.

P. R. VAUGHAN, Imperial College. I wish to speak about pore pressure changes during the settlement of structures, which is a subject which receives surprisingly little attention. Pore pressure measurements are mentioned in only three of the papers to this Symposium. The coefficient of consolidation can range, according to the *in situ* structure of the clay, from values close to those measured in laboratory tests on small samples to values several orders of magnitude larger. Thus, for a typical building on clay, the time for consolidation may vary from several years to several hundred years. Recent observations of the rate of swelling of the blue London Clay (Vaughan and Walbancke, 1973) indicate that the field value of c_v may be no more than four times the laboratory value for small samples. The shape of a typical time-settlement curve is ill-conditioned for the estimation of the degree of primary settlement which has occurred. It seems to be impossible to analyse and use field settlement observations for foreward prediction unless the pore pressure is measured and the amount of primary consolidation monitored directly.

Settlement observations of structures are often made difficult by instability of the bench mark to which they are referred. It seems more logical to measure the compression of the clay beneath the structure, for instance in the manner suggested by Burland, Moore and Smith (1972). It would be quite feasible to combine an instrument of this type with a multi point piezometer, installed in a grout which acts as both seal and piezometer filter (Vaughan, 1969) Such an approach would allow both vertical compressions and pore pressure changes to be monitored at different levels in the foundation, at the cost of only one borehole and instrument head. In this way the progress of primary consolidation could be monitored in such a way that accurate foreward prediction could be made.

Burland, J. B., Moore, J. F. A. and Smith, P. D. K. (1972). 'A Simple and Precise Bore-hole Extensometer,' *Geotechnique* Vol. 22, No. 1, 174–177

Vaughan, P. R. (1969). 'A Note on Sealing Piezometers in Boreholes,' *Geotechnique*, Vol. 19, No. 3, 405–413

Vaughan, P. R. and Walbancke, H. J. (1973). 'Pore Pressure Changes and the Delayed Failure of Cutting Slopes in Overconsolidated Clay,' *Geotechnique*, Vol. 22, No. 4, 531–539

Settlements associated with piles

J. F. A. MOORE, Building Research Station. I would like to give an example of settlement measured by the Building Research Station adjacent to bored piles and referred to by the General Reporter. Two magnet extensometers (Burland, Moore, and Smith, 1972) were installed 42 m deep just into the Woolwich and Reading beds, each at the centre of a square of side 8 m. Augered piles 2.1 m diameter were constructed, at each corner of these squares, 30 m deep through 10 m of sand and gravel overlying the London Clay. A liner through the gravel was sealed into the London Clay to exclude water. A loose liner extending down to the top of the 4.5 m diameter bell remained in the shaft which was concreted up to 16 m below ground level.

At the stage when 4 piles of one square and one of the second square had been completed settlements at the top of the London Clay of 24 mm and 6 mm respectively were recorded. On completion of the remaining 3 piles of the second square the associated settlement had increased to 20 mm.

These movements were clearly related to the piling operation, the resulting reduction in lateral ground stresses and the associated expansion of the clay adjacent to the pile shafts. While this explanation appears reasonable the quantitative prediction of movements of such a magnitude has not yet been achieved. Clearly such movements could create problems for structures close to substantially unsupported large diameter pile shafts.

Burland, J. B., Moore, J. F. A. and Smith, P. D. K. (1972). 'A Simple and Precise Borehole Extensometer, *Geotechnique*, Vol. 22, No. 1, 174–177

M. J. LLOYD AND M. M. H. AL-SHAIKH-ALI, North Western Road Construction Unit, Cheshire Sub-unit. The 150 m long viaduct constitutes part of the M56 motorway which was opened to traffic in February 1971. The sub-structures consist of two abutments and six piers for each carriageway. Each structure is founded on a group of piles 61 cm diameter and 20 m long. The piles were cast *in situ*. The strata on which this viaduct was founded is as shown in *Fig. 1* which is a typical borehole record of the area.

The viaduct was designed on groups of piles sitting on the sand layer. The

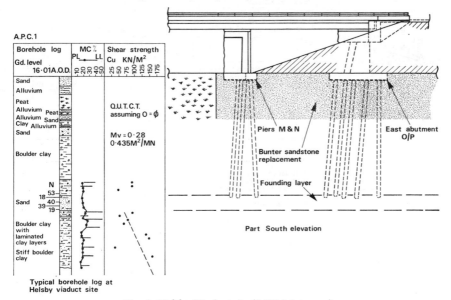

Fig. 1. Helsby Viaduct site (M56 Motorway)

average load on each pile for piers fell between the values 41–65 ton, and that for the abutment 63–69 ton. The settlement calculations took into consideration the pressure of the overburden, and the cohesion of the clay underlying the peat which was estimated to be between 38–48 KN/m².

The pressure distribution was carried out by the Boussinesq method having regard to the distribution of the pressure at the toe of the piles at the effective area at the base of the end bearing piles (Tomlinson). The formula used for the settlement calculation was

$$S_t = 0.5M_v\sigma_z HK_D$$

When depth factor K_d was in this case 0.74, σ_z at the base of the piles was, in most cases, calculated to be 55 KN/m².

Due to the granular nature of the boulder clay and the uncertainty of the drainage paths, it was assumed that settlement would take place soon after construction. The total settlement predicted at the design stage is plotted in *Fig. 2*.

The measurement of the elastic settlement was not fully monitored during construction and therefore the settlements recorded in *Fig 3* were mainly consolidation and creep.

The prediction of the settlement based on the laboratory test results, together with the recorded settlement is plotted to natural and logarithmic scale (see *Figs. 3 and 4*). The latter shows most points fall on a straight line

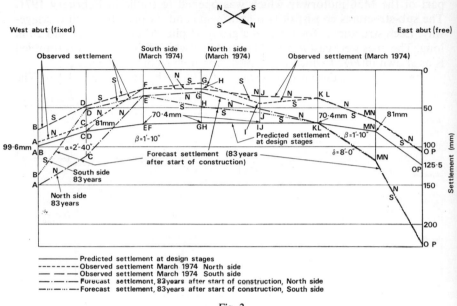

Predicted settlement at design stages
Observed settlement March 1974 North side
Observed settlement March 1974 South side
Forecast settlement, 83 years after start of construction, North side
Forecast settlement, 83 years after start of construction, South side

Fig. 2

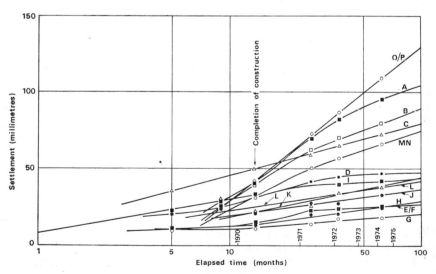

Fig. 3

for the two years from the opening of the Motorway. This phenomenon suggests that creep is predominant over consolidation.

The prediction of the total settlement, including creep, was calculated from the cycle of log time to a period of 83 years, the values of which have been superimposed on *Fig. 2*. It can be seen that the settlement for each substructure under this viaduct (predicted, observed, and postulated from creep graphs) have one thing in common—they generally follow the shape of the graph but not necessarily the value, and that the ratio of the elastic to total settlement p_u/p_t vary between 0.25–0.45. The observed settlement indicated that there is substantial differential settlement within the two halves of each abutment and pier, particularly at the West side.

More importantly, some piers, such as *KL* and *EF*, (see *Fig. 4*) showed a

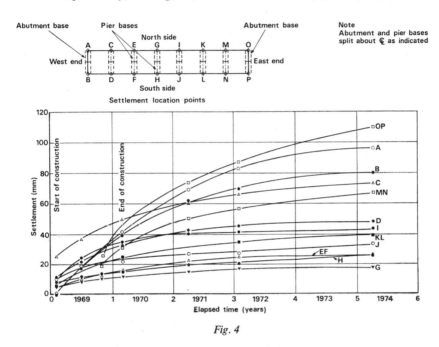

Fig. 4

forward displacement (tilt) relative to each other, some of which measured about 15 mm and 39 mm at the top of the joint; others exhibited small horizontal displacement along the line of the pier. The excessive and the differential settlement, particularly at the west side is thought to be partly due to negative skin friction of the buntersandstone which might have been underestimated. Furthermore, as the carriageway is at present settling at a rate faster than the viaduct, the alluvial clay layer might have contributed additional negative skin friction. One of the reasons for the tilt may be due to side effect of the bunter sandstone in the sand replacement zone.

It is the opinion of the writers that field experimentation should be used to forecast the behaviour of such complex strata. The variability of the boulder clay in Cheshire merits further research work. Creep in boulder clay should always be allowed for in the analysis of settlement. The views expressed are those of the writers and do not necessarily conform to those of their employer.

H. BRETH, Technische Hochschule, Darmstadt. The main point of the General Reports seems to be the evaluation of the settlement data presented. Therefore I want to deal with the interpretation of settlement readings beneath the AfE-Building (given in our Paper, p. 145) in consideration of the questions raised in the discussion.

Using a non-linear stress–strain relationship the finite-element method has been applied to investigate the difference between the observed settlement distribution and prediction by elastic theory. The quadratic foundation has been

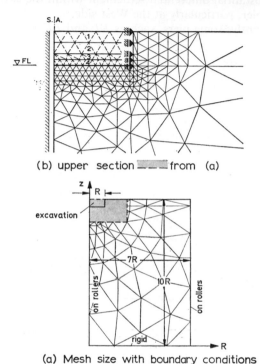

(b) upper section ▬▬ from (a)

(a) Mesh size with boundary conditions

Fig. 5. Finite element representation of the subsoil, AfE building

converted into a circle of equal area and an axisymmetric mesh has been employed as shown in *Fig. 5*. Using the incremental method, the excavation process was first simulated in four steps. The initial state of stress has been computed before as

$$\sigma_z = \gamma . z \text{ and } \sigma_r = \sigma\theta = K_0 . \sigma_z.$$

In the second part of the computation, steps of uniform settlement were brought up to the foundation level due to the observed rigid behaviour of the foundation. Different Young's moduli were used for primary loading and unloading and reloading, in this case with respect to the magnitude of the stress deviation and the related stress history of each element (see *Fig. 6*). The Young's modulus for primary loading has been determined from the slope of normal triaxial tests following a hyperbolic stress–strain relation suggested by Duncan and Chang. In our case specimens of $1\frac{1}{2}$ in diameter

722

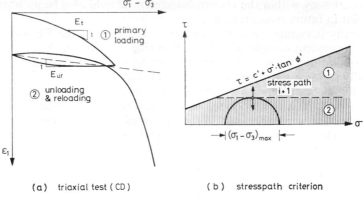

(a) triaxial test (CD)   (b) stresspath criterion

Fig. 6. Stress–strain relation from triaxial tests of Frankfurt Clay

were isotropically reconsolidated at different cell pressures up to 5 kp/cm². They were sheared slowly with a constant rate of strain allowing full drainage. For unloading and reloading a constant value of the Young's modulus was used with respect to the slope of the hysteresis in corresponding triaxial tests.

Figure 7 shows the computed settlement distribution beneath the foundation, using this assumption, in comparison with the measured data. From the nonlinear computation we get a settlement-distribution, which show

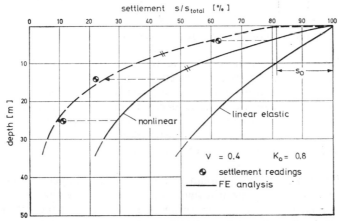

Fig. 7. Analysis of measured settlement distribution by non-linear finite element computation

more settlement occurring immediately below the foundation than does the linear elastic case. Although the settlement readings are not matched, the tendency and the slope of the curvature are equal. The observed settlement distribution would agree with the computed values if we assume an additional portion of the settlement in the uppermost part of about 20% of the total settlement. This phenomenon might be understood as a 'surface effect' not to be explained by the tested stress–strain behaviour. Unfortunately no measurements had been made in this zone. So we propose that the compression in this part—say within the uppermost meter—should also be an important viewpoint in future measurements.

Naturally it cannot be our intention to recommend the FE-method for normal settlement computation in engineering practice. With the investigations pointed out we want to improve our methods for settlement analysis especially for a better design of foundation slabs.

Mr. Butler has given a valuable suggestion for settlement analysis in his report using a Young's modulus increasing with depth, which is similar to our mode of proceeding related to the measured settlement distribution. For Frankfurt clay we usually compute the final settlement by considering the

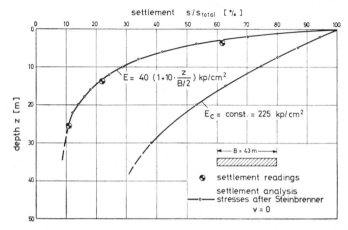

Fig. 8. Settlement distribution for Young's modulus constant increasing with depth

foundation pressure as the dead load reduced by the excavation pressure. *Figure 8* shows the comparison between the measured data and the computed settlement distribution using this assumption. Knowing the total settlement the time settlement behaviour can be derived from *Fig. 1* of our Paper (p. 145).

As mentioned in our Paper, Amann has worked upon the evaluation of these measurements. For further information see his thesis which will be published in the current reports of our Institute.

G. MOULD, R. Travers Morgan & Partners.   My Paper (p. 204) was, of course, prepared many months ago, and I wish to take this opportunity to present the latest level readings, bringing *Fig. 6* in the paper up to date to March 1974. There has been, in fact, no significant change over the past year. The latest check (29 March 1974) on the level of the benchmark on the nearby building in relation to the deep datum also shows no change which is beyond the limits of human error in reading. If anything, a rise in level of 0.9 mm may be indicated.

Mr. Butler has asked for additional information from contributors to his case histories, in particular in relation to actual construction procedures of foundation work. In the case of Guy's Tower, the procedure was first to excavate approximately 3 m from the surface in open cut, the main purpose being to remove basements and foundations of nineteenth century warehouses which had occupied the site. The perimeter diaphragm wall, and the bored piling, were both installed from this reduced level, the piling having some 4.5 m of empty bore. This case is therefore more or less in accordance with *Fig. 11* in Butler's General Report.

T. OHOKA, Nikken Sekkei, Ltd., Japan.   Further details relating to our paper (p. 234) have been given in a paper by the authors (1973).

The authors tried to estimate the rebound and the recompression of over-consolidated clay layers by use of the consolidation test, but the estimated values were higher than the measured values. Although the reason has not been made clear, the difficulty of undisturbed sampling of heavily consolidated clay may play an important part.

In the authors' case, the estimated value by the theory of elasticity was used to give the recompression value of underlying layers, because the pressure in question was about one third of the maximum previous consolidation pressure, and the estimated value by the theory of elasticity agreed with the measured value for the rebound. In this computation, the authors used Steinbrenner's method (1934) and Holl's relation (1940).

Assuming that Poisson's ratio of the gravelly sand and underlying layers was 0.2, the modulus of elasticity E was estimated to be proportional with the third power of depth z according to Holl's relation $E = \alpha z^3$.

In general, the relation $E = \alpha z^n$ where $n = 0.5$–1.0 is often seen and used. On the other hand, $E = \alpha z^3$ is seldom seen. But, in this case, the computed value based on $E = \alpha z^3$ was comparable with the measured value. The authors are still trying to find the true relation between E and z in such a case using the lateral loading test in boreholes.

It is now one and a half years after completion of this building and a 3 mm settlement has been observed during this period. This settlement may be caused mainly by the increased live load and is not serious.

The relative deflection of this raft foundation is shown by the three symbols in the recompression process in *Fig. 7* (p. 240).

Steinbrenner, W. (1934). 'Tafeln zur Setzungsberechnung,' *Die Strasse*, Vol. 1, 121–124.
Ueshita, K., Matsui, K., Ohoka, T. and Nagase, S. (1973). 'An Example of Rationalization of Building Foundation by Measuring of Ground Behaviour', *J. Jap. Soc. Soil Mech. Found. Engng.*, Vol. 13, No. 3, 87–95 (in Japanese)
Holl, D. L. (1940). 'Stress Transmission in Earth', *Proc. Highw. Res. Bd*, Vol. 20

A. W. FISHER, Freeman Fox & Partners.   In relation to our paper on St. Paul's Cathedral (p. 221), Mr. Butler in his report regretted that so many questions had been left unanswered, but he has, to some extent, answered these himself by stating that we are concerned only with differential settlements. With regard to the movement of the principal datum we agree that it may be moving but the effect of this will only be to retard or accelerate the apparent rate of settlement but will not affect the differential settlement values. Consideration had been given to the installation of a deep level datum but the accuracy of this is less than the 0.0001 ft, achieved in levelling.

It would be presumptuous of me to comment on the soil conditions since this is Professor Skempton's province, but I can assure you that we have a plethora of information from the many boreholes which have been sunk over the years. We are, in fact, in the process of re-plotting the levels of the clay and water surfaces and the depths of water, the last two having been affected

725

by the construction of many new buildings around the Cathedral and also the urban development of Hampstead to the north.

We are indeed fortunate that successive Deans and Chapters have had the foresight to continue the measurements and also that the vast majority of reference points are intact and accessible. It is difficult to persuade clients that the costs involved are worthwhile, but I would echo the plea of many speakers for the early installation and retention of such reference points; so many of the authors refer to the fact that measurements ceased because these had become damaged or lost. I do not suggest that the sophisticated equipment we have at St. Paul's be provided, but at least we can get away from the aptly named glass tell-tale type of installation which only tells us if a crack has moved and not by how much or in which direction. The careful pre-planning of the design and installation of reference points will prove amply rewarding.

I would also urge the need for continuous or at least regular observations—so often the measurement lapses for one reason or another during the very period that something happens. In addition errors due to instrumentation, human frailty, seasonal variations, etc., are minimised. The information so obtained will be inestimable value for future investigation and research.

J. A. LORD Ove Arup & Partners.　The paper (p. 221) by Thomas and Fisher on St. Paul's Cathedral is of great interest; in particular the precision of their measurement survey over such long duration is most commendable. However, I am a little concerned as to the reliability of the benchmark to which they

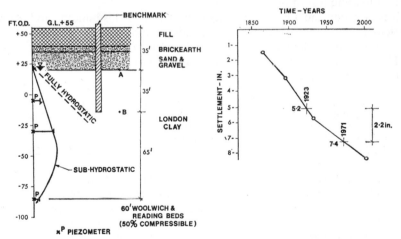

Fig. 9. Geological and hydrostatic profile in the vicinity of St. Paul's

Fig. 10. Settlement of surface of London Clay in the vicinity of St. Paul's due to underdrainage (Wilson and Grace, 1942)

have referred all their settlement measurements. Unfortunately the information provided by the authors as to the soil succession beneath the Cathedral and the hydrostatic conditions existing in the London Clay are very scant. The soil conditions and ground levels at a site investigated by us immediately to the north of St. Paul's Cathedral appear to be similar and are shown in Fig. 9. The Woolwich and Reading Beds underlying the London Clay are approximately 60 ft thick (Imperial units have been used in this contribution so as to conform with the Paper by Thomas & Fisher) in the area and piezometers installed in the clay for a period of about 4 years exhibit the piezometric heads shown. These confirm that water pressures in the clay are substantially sub-hydrostatic, as might be expected as a result of the underdrainage of London over the past 100 years due to pumping from the chalk.

Wilson and Grace (1942) have considered the settlement of the surface of London Clay as a result of consolidation due to underdrainage. Their findings for the clay surface in the vicinity of St. Paul's are shown in *Fig. 10*. If correct during the period 1923–71 (the period of Thomas and Fisher settlement observations), the surface of the London Clay, represented by point A in *Fig. 9*, will have settled about 2.2 in. The benchmark used by Thomas and Fisher is 70 ft deep and so extends about 35 ft into the London Clay (see *Fig. 9*). If it is assumed that 50% of the Woolwich and Reading Beds are compressible (i.e. 30 ft), as a first approximation the settlement of point B, the bottom of the benchmark, will be proportional to the amount of compressible material beneath it as compared with point A.

$$\rho_B = \frac{\text{Thickness of compressible material beneath B}}{\text{Thickness of compressible material beneath A}} \times \rho_A$$

$$= \frac{(65 + 30)}{(100 + 30)} \times 2.2 = 1.61 \text{ in.}$$

(This assumes that the material is of the same compressibility—in practice the compressibility of the material below B will be less than that above, so that ρ_B will be an overestimate.)

Therefore during the period 1923–71, the surface of the London Clay will have settled at least 0.59 in ($= \rho_A - \rho_B$) relative to the bottom of the benchmark. This relative movement will have induced some negative skin friction on the benchmark so causing an indeterminate amount of dragdown.

Settlements of St. Paul's presented by Thomas and Fisher in their *Fig. 3* for the period 1923–71 vary between 0.3 in (choir) and 0.4 in (dome piers). These values are of the same order of magnitude as the movement of the benchmark relative to the surface of the clay as presented above. The settlements presented by Thomas and Fisher may therefore be valueless as *total* settlements as they neither represent movements relative to the surface of the clay nor to a fixed point below the consolidating strata. However the measured settlements are of value as *differential* settlements, for example, by considering settlement of the dome piers, transepts and choir relative to the nave. Similarly as movement of the bench mark would be time dependent, the concept of settlement versus time is only of value for differential settlements, such as presented in the writer's *Fig. 11*. (The settlements of the nave after 1938 as plotted in Thomas and Fisher's *Fig. 3* are in error, being 0.10 in greater than those actually observed (Fisher (1974)). Plotted in this way it can be seen that relative settlements have been very small over the period of observation. Three points should be noted: (1) the relative settlement of nave and choir is negligible, only 0.01 in. up to 1971, as might be expected of parts of the structure of similar size and loading, (2) the settlement of the transepts relative to the nave is 0.06 in. and (3) the settlement of the dome piers relative to the nave is 0.13 in.

In addition to the precise settlement measurements discussed above, Thomas and Fisher quote (p. 231) a differential settlement from when construction began to 1900 of 2 to 3 in between the east, north and west dome piers and the walls of the nave and choir. Such observations combined with latterday measurements provide an excellent example of a long term settlement survey.

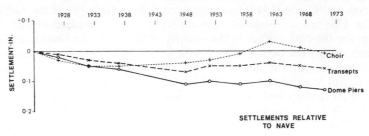

Fig. 11. Average settlements relative to the nave

An attempt has been made to predict such settlements using of necessity simple soil parameters in the absence of more detailed information.

The Cathedral is founded on the Brickearth at bearing pressures of 5 ton/ft² (dome piers) and $2\frac{1}{2}$ ton/ft² (nave and choir walls (Fisher (1974)). In the gravel, a 2:1 distribution of stress has been assumed, so that the area of the top of the London Clay stressed by the weight of the dome (supported on the piers and bastions) may be approximated to a circular annulus of 200 ft external diameter and 60 ft internal diameter, as shown in *Fig. 12*. The be-

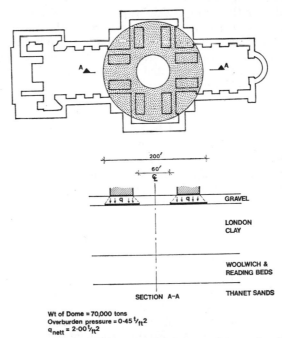

Fig. 12. St. Paul's Cathedral showing approximation of Dome foundations to a circular annulus

haviour of the inner perimeter of the annulus is considered as being analogous to that of the dome piers and the outer perimeter to the transept. Consolidation data for the Brickearth is very scant—assumed values, based on results from King's College Hospital site and confirmed by Fisher (1974) are presented in *Table 3*. The properties of the London Clay from a site immediately

Table 3. SOIL PARAMETERS

	Depth (ft)		c_v (ft^2/year)
Brickearth	5	$m_v = 0.01$ ft^2/ton	40*
Gravel	15–20	$E = 1000$ T/ft^2	
London Clay	100	$m_v = 0.003$ ft^2/ton	5
Woolwich & Reading Beds	60†	$m_v = 0.0015$ ft^2/ton	7

* From Kings College Hospital site
† 30 assumed compressible.

to the north of St. Paul's, are shown in *Figs. 13(a) and (b)*. The parameters assumed are listed in *Table 3*.

The relative settlements in 1900 quoted by Thomas and Fisher would almost certainly be based on the differential movements measured on readily identifiable features in the Cathedral such as string courses in the stonework. Such string courses would be above nave floor level—therefore settlements due to loads applied by the foundations and the crypt piers would have taken place in the Brickearth before the string course was laid. This initial loading

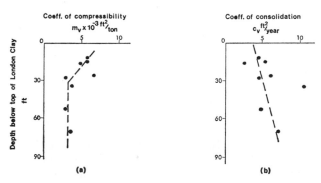

Fig. 13. Variation of consolidation parameters with depth for London Clay

is estimated to be 1 ton/ft^2. The contribution of the various strata towards estimated settlements for the period 1680 (the assumed completion date of the foundations and crypt) to 1900 is tabulated in *Table 4* based on the soil parameters listed in *Table 3*. The settlements of the dome piers and transept relative to the choir are 3.0 and 2.0 in respectively, which compare very favourably with Thomas and Fisher's quoted 2–3 in. It should be noted that contrary to Thomas and Fisher's assertion, the London Clay has contributed more to settlement than the Brickearth.

By 1923 consolidation of the Brickearth would have been completed—so

Table 4. ESTIMATED SETTLEMENTS (c. 1680*–1900)

Stress increment	Dome piers $1T/ft^2$ to $5T/ft^2$	Transept	Choir $1T/ft^2$ to $2\frac{1}{2}T/ft^2$
Brickearth	2.4 in	2.4 in	0.9 in
Gravel	0.4 in	0.4 in	0.4 in
London Clay	3.2 in	2.1 in	1.7 in
Woolwich & Reading Beds	0.4 in	0.4 in	0.4 in
Total	6.4 in	5.4 in	3.4 in
Settlement relative to Choir	3.0 in	2.0 in	

* Completion of foundations and crypt.

that over the period 1923–71 settlement would be due to continued consolidation of the London Clay due to the applied building loads and the effects of underdrainage (the latter is assumed to be uniform beneath the Cathedral). The estimated settlements over this period are presented in *Table 5*. The settlement of the dome piers and transepts relative to the choir are calculated to be 0.11 and 0.03 in respectively, which again compare favourably with the

Table 5. ESTIMATED SETTLEMENTS (1923–71)

Settlement due to	Dome piers (in)	Transept (in)	Choir (in)
Consolidation of L.C.	0.23	0.15	0.12
Under-Drainage	2.2	2.2	2.2
Settlement relative to Choir	0.11	0.03	—

For $cv = 5$ ft²/year   1923 U = 75.5%
                          1971 U = 80.5%

respective measured values of 0.13 and 0.06 in. Thus despite simple and often crude approximations, estimates can be made of the behaviour of the Cathedral which agree reasonably with that observed.

Fisher, A. W. (1974). Personal Communication.
Wilson, G. and Grace, H. (1942). 'The Settlement of London Due to Underdrainage of the London Clay', *J. Inst. civ. Engs*, Vol. 19, 100–127

A. P. K. TATE, Howard Humphreys & Sons, formally E. W. H. Gifford & Partners.   Ratios of 60% and 25% for end of construction (50-year life) and differential maximum settlements respectively were given in the paper (p. 183) by Morton and Au. Data for such ratios were obviously very valuable at both design stage and for check monitoring during construction with the possibility of subsequently modifying structural junctions, should settlement behaviour during construction not accord with design assumptions. However, as shown in the preceding more detailed discussion (p. 201) variations to such ratios can be considerable and must be related to the specific conditions appertaining. Further site observations, particularly in different regions, with possible variations in characteristics of the overconsolidated clays were essential.

An example for possible addition to Table IV of the General Report was a 20-storey tower block, on the outskirts of Southampton, with an outcrop of London Clay on the site overlaid by about 4 m of river gravel and made ground. On the basis of an initial single bore hole 18 m deep (proving the London Clay to such depth) a cullular raft of 4.5 m overall depth (forming a basement area for plant) was selected to form a very rigid foundation.

A subsequent, more detailed site investigation was undertaken once the location of the tower had been fixed. Bore holes 7.5 m deep at each of the four corners were used to obtain samples for conventional oedometer testing at depths of 4.5 m (just below the proposed foundation level) and 7.5 m. Duplicate testing of all samples for settlement characteristics was undertaken in an attempt to separate testing variation from potential differential effects.

In terms of the ratio (difference from mean)/(mean value), scatter for m_v was of the order 5–10% for duplicate samples and 15% between different corner

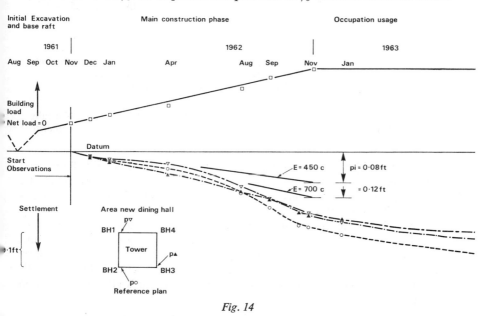

Fig. 14

locations. Values of m_v measured from samples from the lower depth were 65% of that found from samples at foundation level.

Scatter of values for the time coefficient (C_v) was more marked, particularly between locations (3 out of 4 within 20% variation and 7 out of the 8 duplicate sets had variations under 10%). Perusal of such test results did not however suggest that any valid estimation of differential settlement could be made.

The overall observed settlements (see *Fig. 14*) from time of replacement of excavation by approximate equivalent basement load gave differential— maximum settlement ratios of 13% and 12.5% at end of construction and one year later.

Comparative values for the consolidation component (i.e. total less immediate) settlement to relate to test values are dependent on interpolation of observations to derive the immediate settlement. Assuming this occurred when construction was complete (Nov. 1962) the maximum value from the shape of the observed settlements would appear to be 36.5 mm corresponding to $E = 700c$. If this were uniform then differential/maximum settlement ratios of 40% and 25% would apply to the consolidation component. An interpolation of immediate settlement based on $E = 450c$ gave more compatible comparative values of such ratios, namely 20% and 18%.

The example is summarised in *Table 6* in relation to Table IV of the General Report.

Table 6

Site	Location	Type	Dimensions (m)			Clay
			B	L	D	L.C.
South Stoneham	Southampton, Hants	Rigid basement raft	15	16.2	4.5	

Thickness below foundation (m)	Gross (kN/m²)	Foundation Press Unload (kN/m²)	Net (kN/m²)	Primary settlement Obs. (mm)	Calc. (mm)	Remark
20+ (estimate 50)	210	75	135	49*	55	Not fully complete

J. H. ATKINSON, Cambridge University. In their paper King and Chandrasekaran (p. 368) have adopted values for the anisotropic elastic parameters for London Clay of $v_1' = 0.156$, $v_2' = 0.12$ and $n' = 1.6$ obtained by Wroth (1971) and Henkel (1971). In obtaining a value for v_1' Henkel (1971) made the assumption that v_1' could be equated to the mean of v_2' and v_3' where $v_3' = n'v_2'$ is the strain in the axis of symmetry due to unit strain in the plane of symmetry. The calculations made by Wroth (1971), by King and Chandrasekaran and also by Butler in his General Report to this Session have accepted this assumption for v_1'. Henkel's (1971) assumption leads to

$$v_1' = \tfrac{1}{2}v_2'(1 + n')$$

for which $v_1' > v_2'$ for $n' > 1.0$. On the other hand from an investigation of the results of drained cylindrical compression tests on samples of undisturbed London clay from the Barbican Arts Centre (Atkinson, 1973), from Ashford Common (Webb, 1966) and from High Ongar (Som, 1968) an empirical relationship

$$v_1' = v_2'(2 - n')$$

was obtained by Atkinson (1973) for which $v_1' < v_2'$ for $n' > 1.0$. These two relationships between the elastic parameters, one assumed and the other empirical, agree at $n' = 1.0$ when $v_1' = v_2'$ but otherwise they depart significantly.

It is not immediately obvious whether, for undisturbed London Clay, v_1' will be greater than, less than, or indeed different from v_2' for $n' > 1.0$. The relationship between the undrained parameters is however quite specific and may be obtained from King and Chandrasekaran's Equation 9.

$$v_1 = 1 - \tfrac{1}{2}n, \; v_2 = \tfrac{1}{2}$$

732

for which $v_1 < v_2$ for $n > 1.0$. Clearly $n' > 1.0$ for $n > 1.0$ but this does not necessarily mean that v_1' will be less than v_2' for $n' > 1.0$; nevertheless the evidence of laboratory tests on undisturbed London Clay does not seem to support Henkel's (1971) assumption for calculating v_1' from v_2' and n'.

Furthermore, although King and Chandrasekaran allowed non-homogeneity in the soil moduli for undisturbed London clay they have assumed homogeneity in the values of v_1', v_2' and n'. Wroth (1971) investigating Webb's (1966) data found a unique value for v_2' for samples from Ashford Common but Atkinson (1973) using data from Ashford Common (Webb 1966) and High Ongar (Som 1968) found evidence to suggest a dependence of v_2' on mean normal stress level and a possible dependence of n' on overconsolidation ratio.

The kind of studies undertaken by King and Chandrasekaran and also by Gibson (1974) and others to investigate the effects of non-homogeneity and anisotropy on the behaviour of structures are to be welcomed but the writer would question whether 'the properties of clay beds at Ashford Common with regard to inhomogeneity and transverse isotropy have been established—' (p. 379–380). In particular the assumption made by Henkel (1971) should not be allowed to assume the mantle of an established material property for London Clay.

Atkinson, J. H. (1973). 'The Deformation of Undisturbed London Clay', Ph.D. Thesis, London University
Gibson, R. E. (1974). 'The 14th Rankine Lecture: The Analytical Method in Soil Mechanics', *Geotechnique*, Vol. 24, No. 2, 115–140
Henkel, D. J. (1971). 'The Relevance of Laboratory Measured Parameters in Field Studies', *Proc. Roscoe Mem. Symp.*, 669–675
Som, N. N. (1968). 'The Effect of Stress Path on the Deformation and Consolidation of London Clay', Ph.D. Thesis, London University
Webb, D. L. (1966). 'The Mechanical Properties of Undisturbed Samples of London Clay and Pierre Shale', Ph.D. Thesis, London University
Wroth, C. P. (1971). 'Some Aspects of the Elastic Behaviour of Overconsolidated Clay', *Proc. Roscoe Mem. Symp.*, 347–361

A. SULLIVAN, McClelland Engineers, Ltd.   Papers to this session report settlement data on some 12 tall buildings resting on London or Woolwich and Reading Clays, and four case histories are given in Appendix B of the General Report. Only 6 of the buildings are supported on raft foundations and the remaining 10 buildings are carried on drilled shafts or drilled-and-under-reamed caissons. Breth and Amman (p. 141) report settlements of eight multi-storey buildings on rafts underlain by Tertiary Frankfurt Clay.

Observed settlements of the raft supported buildings are plotted against net bearing pressure in Fig. 15. The upper two points on the London curve are from similar structures of 13 m in width that are underlain by 25 m of London clay and 10 m of Woolwich and Reading clay overlying Thanet sand. The lower point is from the 27-m-wide Brittanic House resting on 16 m of clay. The Woolwich and Reading curve is indicative of movements of 20 m and 27 m rafts resting on 15 m and 5 m, respectively, of the gravelly basal portion of the Woolwich and Reading beds. The Frankfurt curve is developed from observations on rafts ranging from 19 to 32 m in width that are generally underlain by more than 100 m of Tertiary clay.

Included in *Fig. 15* is a curve for the Beaumont clay underlying Houston, Texas, that was developed some years ago from settlement observations of rafts in the order of 40 m in width. The Beaumont clay formation consists of poorly-bedded plastic clay interbedded with silt and sand lentils and some more-or-less continuous sand layers. The clays were overconsolidated by desiccation when the sea level was more than 120 m below its present level during the late Wisconsin glacial stage. It is standard practice in the Houston area to support tall buildings on raft foundations resting directly on the Beaumont clay. The tallest building is the 52-storey One Shell Plaza that is

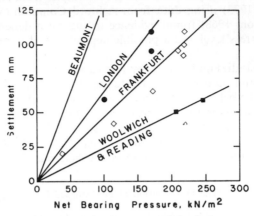

Fig. 15. Rafts on over consolidated clay

supported on a 53 m by 70 m mat about 18 m below street level producing a net bearing pressure of 33 kN/m².

Curves of the type shown in *Fig. 15* are extremely useful in preliminary project planning and in preparing comparative cost estimates. Foundation design of tall buildings underlain by strong overconsolidated clays is generally governed by settlement requirements rather than bearing capacity. For satis-factory building performance, differential settlement are usually limited to 25% of total settlement. To meet this settlement criteria, net bearing pressure on the Beaumont clay is usually restricted to 60 kN/m² for raft foundations. By applying the same differential settlement criteria to the London curve, a net bearing pressure of 130 kN/m² could tentatively be used for preliminary comparative design studies of tall buildings on London clay.

B. A. LEACH, Allott & Lomax.   I would like to present some further infor-mation regarding the settlement of the main buildings at Didcot Power Station. As we stated in our Paper (p. 169), the problems associated with monitoring the movements of such a complex structure have been found to be immense and of 57 measuring points either installed or planned to be in-stalled, we have continuous readings on only a handful. Readings on other points have given a nearly complete record but with discontinuities in the results that have made plotting the deflected shape of the building difficult. However, we have surveyed the steelwork at the firing floor level, some 12m above the ground floor, which has yielded a good indication of the deflected shape of the building as it stands at the present time. Using this information we have been able to go back through our results and verify to a large degree the validity of many of the readings and to sort out anomalies within them.

As a result of this we have traced as best we can the build-up of deformation of the structure.

Figure 16 shows contours of equal settlement as recorded in May 1968. At this stage foundation work was complete to ground level over the total extent of the structure, steelwork was complete to Line 27, superstructure to Line 3 (wall cladding to Line 7) and plant erection was in progress in the areas of Units 1 and 2. In the boilerhouse plant erection was 70% complete in Unit 1 and 30% complete in Unit 2. In the turbine house plant erection was only 40% complete in Unit 1 and 15% complete in Unit 2. It will be seen that

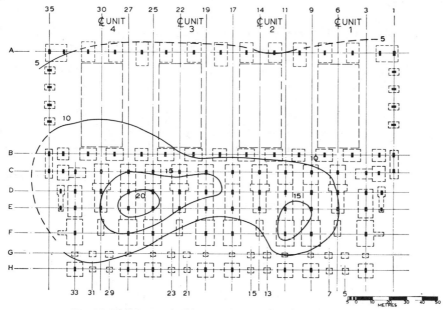

Fig. 16. Didcot main building—contours of settlement, May 1968

the settlement pattern is emerging as predicted and that a maximum of just over 20 mm settlement had been observed.

Moving on to March 1970 (*Fig. 17*) the developing pattern is seen. Although settlements are only some 7 mm greater than at two years previously, a much more developed bowl of settlement occurs over the heavily loaded boiler house area. At this stage steelwork and superstructure were complete and plant erection was in progress in all four units. In the boiler house Units 1 and 2 were complete, No. 3 70% complete and No. 4 50% complete. In the turbine house Unit 1 was virtually complete, Unit 4 25% complete and the other units at intermediate stages of erection.

Figure 18 shows the situation in September 1973 and is based on the latest set of readings which we have processed. The structure is now fully loaded and settlements of up to 37 mm have been observed.

The settlements observed thus far are in fact only some 50% of those predicted for 30 years at the design stage and some 66% of the maximum which can be predicted from extrapolating the curves of settlement presented as *Fig. 6* of our Paper.

735

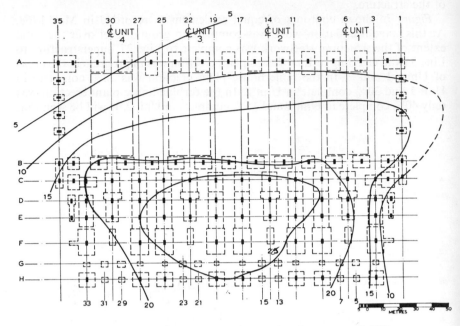

Fig. 17. Didcot main building—contours of settlement, March 1970

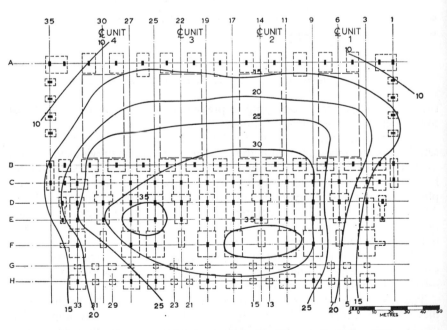

Fig. 18. Didcot main building—contours of settlement, September 1973

736

The maximum slopes between column bases in the building at present are about 1/1300. If the readings are projected forward 30 years we can predict maximum slopes in the building of about 1:850 which compares with the 1:830 predicted at the design stage.

The building at the present time shows no distress due to differential movement and we are confident that it will be able satisfactorily to absorb the greater differentials to be expected in the future.

General problems

P. R. RAUKILOR, Wanstock Engineering Co. Ltd.   I am going to describe the use of an instrument which may find application to some of the interesting settlement studies upon which many people here are engaged. The instrument is a tilt meter designed with industrial field application in mind. The idea started when, researching into mining subsidence at Salford University, we needed an instrument to measure subsidence tilting on buildings. The electronics were developed by Dr. Jones in the Department of Electrical Engineering at Manchester University.

The instrument comprises an electronic box and numerous virtually disposable tilt-sensing heads. The basic working concept is that one electronic box is used to monitor many remote sensing heads which, because of their

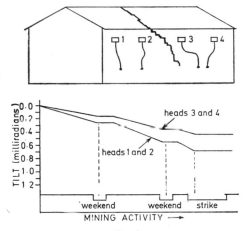

Fig. 1

low price, can be permanently built into structures and effectively 'written-off'. Such places as on high beams, underwater, beneath poured concrete, come to mind as extreme cases of inaccessible tilt-reading locations. In general, however, the heads have been used in positions where, after several months, they have been recovered for further use elsewhere.

The instrument has been found to have good long-term readout stability, and requires no current supply for the sensing head on site. It is accurate to 1/100 m rad (0.0006°) or expressed as gradient, to a tilt of 1 mm in 100 m. It has a digital readout and is specially designed to be used by a non-technical operator.

With a conventional electronic level, one cannot obtain perfect repeatability of setting; readings can be scattered, and one has to wait for the sensing head to reach ambient temperature. With fixed heads, they can be outside or inside a building and still be read immediately upon arrival at a site, even during the winter. This makes reading quick and inexpensive.

K. Wardell and Partners sponsored the development of this instrument at Manchester University, and have themselves used it on sites for purposes such as checking that shallow mine grouting did not disturb adjacent buildings.

Figure 1 shows diagramatically a building cracking as a result of mining subsidence. It might be argued that in the case of a new building damage was due to inadequate foundation design resulting in differential settlement. Frequent reading of tilt heads could produce a graph as shown in *Fig. 1*, which clearly ties in the damage with periods of mine working.

The frequent acquisition of tilt data to supplement periodic levelling and to define in detail periods of quiescence and activity, could lead to a more detailed knowledge of the processes of settlement. The installation of numerous permanent sensing heads in major structures might prove to be an inexpensive method of monitoring their settlement activity over a considerable period of time and by virtue of their sensitivity and frequency of reading could give warning of incipient failure of major structural units.

S. HANSBO, Chalmers Institute of Technology and Consulting Civil Engineer, A. B. Jacobson and Widmark. According to the Swedish regulations the permissible average ground pressure with reference to footings and end-bearing bored piles on rock is restricted to a maximum value of 8 MN/m².

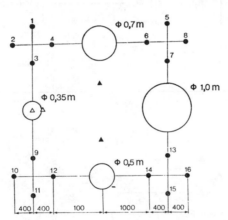

●	TIE BACK. LENGTH OF TIE IN mm, 15500 ODD TIE NUMBER
	11500 EVEN TIE NUMBER
▲	CORE DRILLING BEFORE LOADING TEST
△	CORE DRILLING AFTER LOADING TEST
■	ROCK SOUNDING

Fig. 2

As a matter of fact only 2 MN/m² is allowed on rock in connection with bridge structures.

In order to investigate the deformation characteristics of rock, and its bearing capacity, plate loading tests were carried out in the field on severely cracked, slaty gneiss of the worst kind that could be found in the Gothenburg region. The plates, or rather the footings, which were made of concrete, cast in steel tubes, with diameters 0.35, 0.50, 0.75 and 1.0 m were placed directly on the rock surface which had a dip of about 20°, *Fig. 2.* The load was applied by means of a heavy steel beam which was anchored in the bedrock, *Fig. 3.* The loading capacity in this case was 10 MN which gives an average ground pressure of 103 MN/m² for the smallest footing and 13 MN/m² for the biggest footing.

The results of the loading tests are given in *Fig. 4.* It is obvious that the deformations of gneiss—even if severely cracked—in this case are of elastic nature. We also find that the E-modulus increases with increasing pressure.

ELEVATION

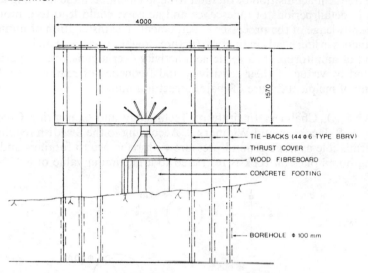

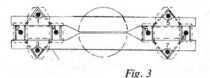

Fig. 3

This increase is obviously due to the cracks which close with increasing load. The hysteresis effect is especially pronounced for the smallest footing, probably due to small, local fractures by which the various rock elements are brought together. The load/settlement curves for the three smallest footings are fairly similar and almost linear, even at the highest pressure, *Fig. 5.*

It is obvious that the loading capacity was not enough to produce failure—not even of this severely cracked rock—although the ground pressure for the smallest footing is 13 times greater than the maximum permissible pressure on good solid gneiss or granite rock. In no case did the settlements exceeded 8 mm. The E-modulus of the rock is about 1000 to 35000 MN/m².

The time dependent settlements were also studied at certain loads, *Fig. 6.* The creep deformation—as expected—is directly proportional to the logarithm

742

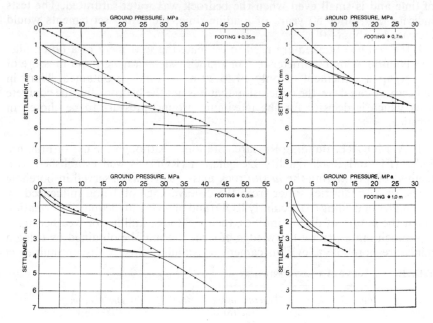

Fig. 4

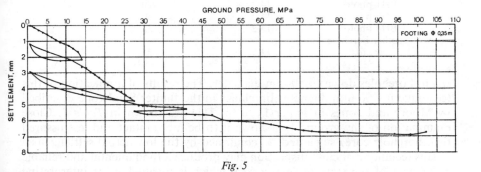

Fig. 5

Fig. 6

of time and is small even when the bedrock was water-saturated. The tests indicate that, after 50 years of loading time, the initial settlements would increase by about 30%.

If the results obtained are used to estimate the settlement of a bored pile, or a column, with a diameter of 5 m and with an average ground pressure of 5 MN/m^2 a settlement of only 10 to 20 mm would be obtained. Thus, in our case we can draw the conclusion that it is not the bedrock but the concrete itself which is decisive for the maximum ground pressure to be chosen in practice.

J. F. A. MOORE, Building Research Station. Although the use of Fracture Index (FI) and Rock Quality Designation (RQD) is well established as a method of quantifying the *in situ* rock mass quality as observed in borehole cores, these quantities suffer from the disadvantage of being influenced by the quality of drilling (particularly RQD) and of giving inadequate information about the nature of discontinuities, their tightness, infilling or weathering.

A much more precise knowledge of structure may be obtained by using a miniature borehole television camera to examine directly discontinuities *in*

Table 1. ASSESSMENT OF DISCONTINUITIES

Record	Fracture index (joint/m)	Rock Quality Designation (% length $\geqslant$ 100 mm)
CCTV1	7.3	81
CCTV2	4.2	77
BH1 photo	14.2	20
BH2 photo	14.2	30
BH4 photo	13.2	33
other borehole records	14.5	35

situ in boreholes down to 50 mm diameter. The results of application of this technique in 100 mm diameter boreholes in the Bunter Sandstone described by Moore and Jones in Paper IV/6 (p. 311) are shown in *Table 1*. Close examination of the ground has shown it to be more intact and less fissured than borehole cores suggested, as confirmed by the low actual settlements.

This technique of close inspection may produce a fundamental and reliable assessment of the state of *in situ* rock which is invaluable for interpreting test results and in predicting settlement behaviour.

J. L. JUSTO, University of Seville. I want to supplement the results presented in our Paper (p. 266) because the observations have been continued after submission of the paper.

Up to now settlement measurements have been made on 141 piers, and some of them have been continued for more than two years. The maximum settlement is still 5 mm. The most important field results for gypsum rock are summarized in *Table 2*. For very stiff clay ($c_u = kN/m^2$), the corresponding field results are shown in *Table 3*.

For clays the terminology employed by Butler in his General Report (Session 3) has been used.

It has been confirmed that settlements in gypsum rock are time-dependent.

Table 2

Foundation type	Immediate settlement	Final settlement (mm)	Average E (MN/m^2)
Footings	negligible	1	500
Point bearing piles	negligible	2	800

For piles, a further reason may be added to the list indicated in the paper: relaxation of skin friction and load transfer to the point. Total stabilization of settlements took place after about one year.

It is interesting to notice that the ratio of either the undrained or the drained field modulus to the undrained strength for Spanish clays is much higher than the values indicated by Butler in the General Report for Session 3.

Table 3

Foundation type	E_u/c_u	E_v'/c_u	Average E_v' (MN/m^2)
Floating piles	1000	700	100

T. K. CHAPLIN, University of Birmingham. First, may I express my thanks to Mr. Hobbs for such a thorough review of settlement predictions for structures on rock.

It is desirable to distinguish between settlements due to shear strains of a cemented rock, e.g. a sandstone, and those due to closure of fissures and microfissures. Though Terzaghi's law of effective stress is familiar to all, it is only a special case of superposition and one can equally well *superpose shear stresses* carried by different systems, so far as we know.

The cementitious material in a cemented granular material creeps steadily at constant stress, and tends to transfer shear stress to the granular structure. In *Fig. 7* of my paper to the Montreal Conference (Chaplin, 1965) I showed the possible influence of testing rate on the stress-strain curve of a cemented granular material. A typical granular structure at or near constant volume seems to have a nearly parabolic stress–strain curve. At extremely slow rates of shear strain one would expect a typical cemetitious filling to exert relatively little shearing stress; with a fairly high mean stress (to avoid bond failure) it might very well be able to keep the granular structure at nearly constant volume and thereby make it carry most of the shear stress.

Carey (1953) emphasised the vastly different time scales for different minerals, which he expressed as 'rheidity': one would like to have much more information about time effects from real rock cores. My interest in the creep of cementing minerals came from seeing the noticeably concave-upwards stress-strain curves obtained some forty years ago at Leeds University by Wood and Evans on sandstone and concrete, incrementally loaded and unloaded over several weeks.

A significant example of fissures apparently closing and opening reversibly was given by Terry and Morgans (1958), who loaded a cube of Barnsley Hards

745

coal in compression. Below a critical load the stress–strain curve was closely parabolic and at higher loads was linear; on unloading there was negligible hysteresis.

My final point concerns settlement analysis, and applies equally to rocks and soils. Where the strata under one footing are different from those under some other footing, the conventional settlement analysis would give a ground flexibility matrix which is *not* symmetrical about the leading diagonal. This error is easily avoidable, and the following improvement is particularly advised where settlements outside a loaded area are calculated, as for soil-structure interaction.

Where the calculated settlement Δ_{AB} of footing A due to a unit load at B differs from Δ_{BA}, the settlement at B due to the *same* load at A, the average $\frac{1}{2}(\Delta_{AB} + \Delta_{BA})$ should be used for both. Then the ground flexibility matrix becomes symmetrical before it is inverted to give the ground stiffness matrix for addition to the structural stiffness matrix before settlements are calculated. Where Δ_{AB} and Δ_{BA} are grossly different, the geometric or harmonic mean might well be even better.

Carey, S. W. (1953). 'The Rheid Concept in Geotectonics', *J. Geol. Soc. Aust.* Vol. 1, 67–117

Chaplin, T. K. (1965). 'A Fundamental Stress–Strain Pattern in Granular Materials Sheared with Small or no Volume Change', *Proc. 6th Int. Conf. Soil Mech. Found. Engng.* (Montreal) Vol. I, 193–197.

Terry, N. B. and Morgans, W. T. A. (1958). 'Studies of the Rheological Behaviour of Coal', *Mechanical Properties of Non-Metallic Brittle Materials* (Edited by W. H. Walton), 239–256 Butterworths, London

Problems associated with Chalk

J. B. BURLAND, Building Research Station. In this contribution I will present the up-to-date observations on the settlement of the building on Chalk

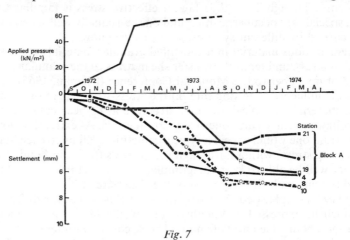

Fig. 7

which is described in our Paper IV/1 (p. 259) and then make some remarks on the creep of Chalk.

The observations presented in the paper relate to the short term settlements. *Figure 7* shows that additional settlement has occurred since completion of the building, however it would be misleading to attribute the major part of

746

this to creep. The marked increase in settlement of a few of the points (8, 10 and 19) is probably due to extensive groundworks that took place around the building subsequent to completion. Over the last 6 months of observation the time-dependent settlement has been generally less than 1 mm. The values of equivalent elastic modulus of the Chalk nine months after completion are 78 MN/m² and 87 MN/m² for blocks A and B respectively. In view of the low rate of settlement it seems unlikely that the long-term moduli will lie much below these values.

The General Reporter has referred to the work carried out by BRS on the Chalk at Mundford. Most of the published data from the tank loading test relate to the short-term loading. However, a long-term test was carried out in which the creep of the Chalk underlying the tank was observed over a period of a year. *Figure 8* shows the relationship between time and vertical strain for three levels beneath the tank. The Grade II Chalk between levels 3 and 4 showed no creep and the Grade III Chalk between levels 2 and 3 showed a small amount of creep which terminated within 120 days. The Grade IV Chalk

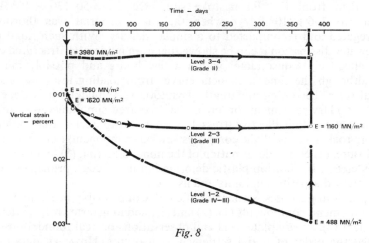

Fig. 8

between levels 1 and 2 showed significant creep which was by no means complete when the tank was finally unloaded. This result seems to be the only well documented case of creep on Chalk and it is unfortunate that the test had to be terminated. It should be noted that loading of the tank took place in only two days. Most buildings take upwards of six months to complete and much of the creep observed at Mundford would probably have been 'built out'.

On the basis of the Mundford results it would seem prudent to assume for the present that the ratios between the long-term and short-term moduli for Grades V, IV and III Chalk are 0.2, 0.3 and 0.5 respectively. However, it should be emphasised that these figures only apply when the foundation pressures are less than the yield pressure q_y (see *Fig. 1* of the General Report). The results of plate tests at Mundford suggest that at pressures greater than q_y the creep may be expressed as a proportion of the immediate settlement by means of a parameter

$$R = \frac{\text{settlement 1 log cycle of time}}{\text{immediate settlement}}$$

which was found to lie between 10 and 15% for a wide range of pressures. It is evident that more settlement observations are required before any definitive conclusions can be reached about the creep of Chalk.

L. M. LAKE, Mott, Hay and Anderson. Both Burland and Hobbs referred to the yield stress level (q_y) indicated by plate bearing tests; the available data shows q_y to vary between about 350 and 400 kN/m^2 for chalk. In this connection it will be noted that the 4-storey buildings described in our Paper IV/4 (p. 283) was supported on spread foundations designed to a pressure of 430 kN/m^2 whilst the 20-storey block was carried on a raft designed to 650 kN/m^2. The latter figure accords with the Code of Practice CP 2004 for shallow foundations in sound chalk. It is clearly not the bearing capacity of virgin chalk which is in question but the settlement which may be induced and some recent work suggests that it is long-term movement that may be important.

Laboratory triaxial dissipation tests carried out on small intact specimens of chalk taken from the Basingstoke site gave $C_v = 50–130 \times 10^3$ ft^2/yr reducing to $2–6 \times 10^3$ ft^2/yr when the same material was thoroughly disaggregated and recompacted to a similar density. With such rapid rates of porewater dissipation it can be shown that even moderately fractured chalk would effectively 'consolidate' within a few hours and possibly minutes. Thus, although the time/settlement curves from loading tests may have a classical form, becoming apparently asymtotic very quickly, it is clear that some 'secondary settlement' or 'creep' has occurred simultaneously; that is, settlement at constant effective stress. It is helpful to appreciate that settlement of a soft rock such as chalk comprises several components: (1) fissure and joint closure, (2) elastic deformation of the intact material, (3) 'consolidation' and (4) 'creep'. In addition plastic deformation may occur, principally at the highly stressed asperities in joints and fissures.

For practical purposes, settlement of chalk can probably be simplified into initial settlement, components (1), (2) and (3), and long term creep settlement, component (4). *In situ* plate tests and observations on real foundations have indicated the order of initial settlement that occurs. However, data on the creep behaviour of real foundations has yet to be obtained and laboratory tests provide the only guidance. Uniaxial and oedometer creep tests on small specimens indicate continued creep movement which is linear on a log–time scale; the longest known tests have continued for several months. Whether such tests can be realistically extrapolated to real situations is doubtful but is clearly in need of further research.

T. R. M. WAKELING, Foundation Engineering Limited. Specimens were obtained from rotary core boreholes in the Upper Chalk at Woolwich, London, and three intact specimens were tested separately in unconfined compression. The three specimens were then stacked to form a composite jointed specimen and tested in unconfined compression between steel end caps. Allowing for the compressibility of the end caps, it was assumed that the composite specimen simulated the effect of three joints. The results are given in *Fig. 9*. The stress strain curve for the composite specimen shows an initial bedding down, followed by a steeper part that is repeated on cyclical loading. As a first approximation it might be assumed that the initial compressibility simulates the effect of loose joints and the steeper cyclical part, the compressibility of tight joints.

Using the moduli of elasticity (*E*) measured on the composite specimen, a tentative indication of the effect of joint spacing can be obtained. *Figure 10*

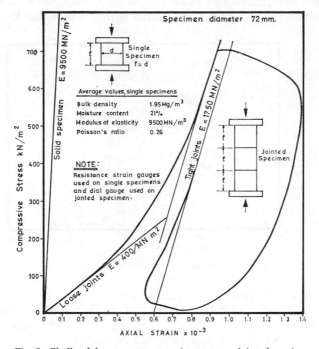

Fig. 9. Chalk—laboratory compression tests on jointed specimen

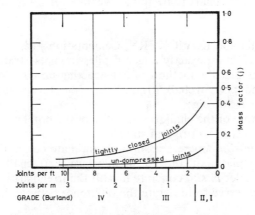

Fig. 10. Chalk—variation of mass factor and joint frequency from laboratory tests

shows the derived curves for mass factor versus joint frequency. It will be noted that these curves show comparable trends to the field results (*Fig. 9* of Review Paper), although there is no clear indication of the significance of the laboratory 'loose' and 'tight' joint condition. The laboratory results can also be extended to give curves for $\bar{E}$ versus standard penetration test value using an empirical classification of chalk quality, see *Fig. 11*, (Wakeling 1970, Hobbs 1970a). The derived curves are roughly comparable with field results (*Fig. 16* of Review Paper).

The papers and discussions presented to the Conference have suggested that settlements in chalk can be predicted with a reasonable level of con-

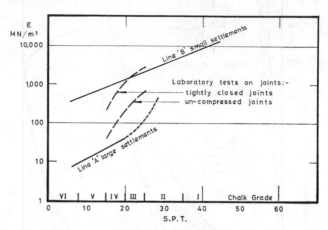

Fig. 11. Chalk—elastic modulus E

fidence. However, the predicted settlements are based on the mass properties of chalk and are only applicable when the foundation is large in comparison with the joint spacing; otherwise the properties of the joints themselves become influential. As a rough guide it could be considered that the minimum foundation dimension should be at least four times the joint spacing for the mass properties of the chalk to be applicable. (See references at end of General Report, p. 607.)

Z. J. SLIWINSKI and K. VICKERY, Cementation Ltd.   Our contribution refers to bored or more exactly augered piles in chalk strata. From our experience it seems that the method of constructing bored piles is particularly suitable for foundations in chalk strata. The reasons for this may be tentatively attributed to the following:

(1) The structure of the chalk on the sides of the borehole is not destroyed and remains almost in its original state.

(2) The cast *in situ* concrete assures an intimate connection between the borehole and the pile shaft. Indeed the interface shaft/chalk follows the surface of the borehole: *in situ* concrete fills all the irregularities.

Both these points serve to illustrate the difference between bored and driven piles.

Cementation Ltd., has successfully constructed over 1000 piles in chalk (600–1200 mm diameter) for foundation of multi-storey buildings, retaining walls, bridges, etc. Unfortunately we do not have records of settlement of structures founded on these piles but, in genral, the following characteristics have been observed: (a) that complete stabilisation takes place rapidly and is completed during construction of the superstructure, at least within the limits of dead load, and (b) that settlements are very moderate and easily tolerated by the superstructure.

The assessment of bearing capacity may be stated as follows: (a) visual classification of samples (if available), (b) regional characteristics, and (c) empirical relationship between S.P.T. and bearing capacity which can be assessed approximately by previous experience.

The method as will be appreciated requires both experience and critical engineering judgement.

In order to check the assessment it is often considered advisable to carry

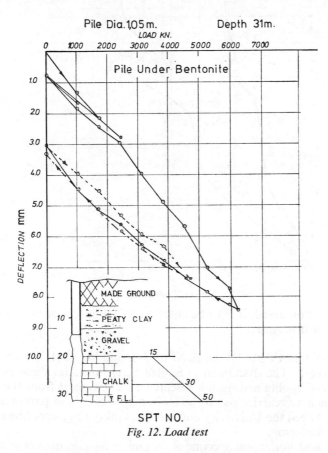

Fig. 12. Load test

out a full scale load test to 1.5 or 2 times working load. From a summary of such tests that we have carried out it may be observed: (a) that the settlements are quite small—a few millimetres at working load, and (b) that the applied working stress is consecutively increased which means that higher loads are acceptable.

Simple test loads provide useful information for a given site—but in the case of superimposed layers above chalk, the load on the chalk itself cannot be easily deduced. More sophisticated tests with piles equipped with strain gauges and extensometers are much more informative. An example of such a test is shown on *Fig. 12* and *Fig. 13*. High frictional resistance mobilised in

751

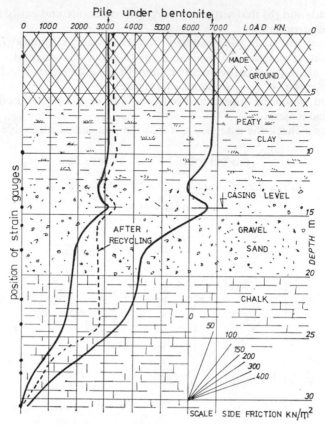

Fig. 13. Load profiles

chalk may be observed. It is to be noted that the pile was excavated and concreted under bentonite.

N. M. AL-MOHAMMADI, N. E. SIMONS, and M. A. HUXLEY, University of Surrey. The discussion presented herein includes some of the preliminary results obtained during research on chalk marl from the Channel Tunnel which was carried out at the University of Surrey as part of a research contract between the University and the Consulting Engineers Messrs. Mott, Hay, and Anderson.

Vertical and horizontal specimens 76 mm × 38 mm diameter were consolidated isotropically in the triaxial cell to the estimated *in situ* vertical effective stress using a back pressure of 758 kN/m^2. The specimens were sheared by increasing the vertical stress.

The average values of 6 horizontal specimens and 3 vertical specimens taken from the same depth show that the average undrained shear strength for vertical and horizontal specimens is approximately the same, while the tangent modulus in the horizontal direction is 1.4 times the tangent modulus in the vertical direction. Therefore assuming isotropy in such material will lead to errors in horizontal and vertical stress distributions and thus an over estimate of settlement.

150 mm × 76 mm diameter specimens from land borings were tested in uniaxial compression in the triaxial cell. Lateral deformations were measured with a special stainless steel spring fitted with 4 ERS gauges and attached to the middle of the sample. Typical stress-strain curves with cycles of loading and unloading are shown in *Fig. 14*. A clear characteristic of all the stress-strain curves is that upon start of loading all curves show a low value of the initial modulus (E_i). At higher stress levels the curves tend to be steeper and more linear, but without approaching truly linear elastic behaviour. It seemed possible that the early curved part of the stress–strain curve was due

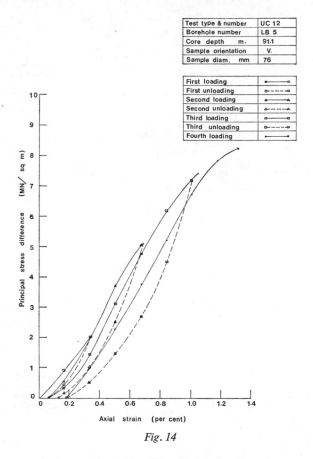

Test type & number	UC 12
Borehole number	LB 5
Core depth m.	91.1
Sample orientation	V.
Sample diam. mm	76

First loading	•———o
First unloading	o-----o
Second loading	▲———▲
Second unloading	▲-- --▲
Third loading	□———□
Third unloading	□-----□
Fourth loading	•———•

Fig. 14

to bedding, but during cycles of loading and unloading the same behaviour occurred which indicates that this is mainly due to closure of fissures.

It appears that there is no constant value for the moduli or for Poisson's ratio, being stress level dependent as can be seen from the plots of secant modulus (E_s) and Poisson's ratio vs stress levels, i.e.

$$\frac{(\sigma_1 - \sigma_3)}{(\sigma_1 - \sigma_3)_f} \times 100$$

in *Figs. 15 and 16*. From these plots it is clear that the value of the secant modulus is low at low stress levels and increases until it reaches a high value

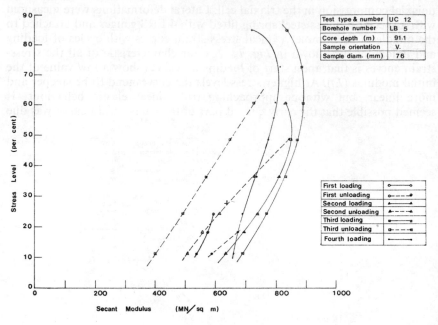

Fig. 15

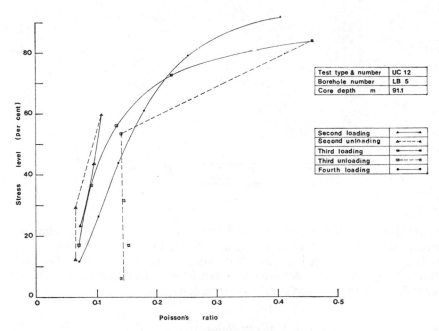

Fig. 16. Radial distribution of settlement, pressure and pile load
(alternative treatment of uplift forces)

at a stress level of about 65–70%, beyond which there is no or very little increase in its value (see *Fig. 15*). A typical value of the secant modulus (E_s) taken at stress level of 50% is 680 to 880 MN/m² while the tangent modulus (E_t) for the approximately straight part of the stress-strain curve during loading cycles is 980 MN/m². Poisson's ratio is taken as the slope of the lateral strain vs axial strain curve, has low values (0.1 or less) at low stress levels and increases steadily with increasing stress level until it reaches a value near 0.5 at failure (see *Fig. 16*). Typical values for the Poisson's ratio at stress level 50% are 0.1–0.3.

76 mm diameter × 150 mm specimens were consolidated under an effective stress of 345 kN/m² (50 psi) using a back pressure of 758 kN/m² (110 psi). Stainless steel inserts 6.4 mm diameter × 9.5 mm were fixed into the specimen

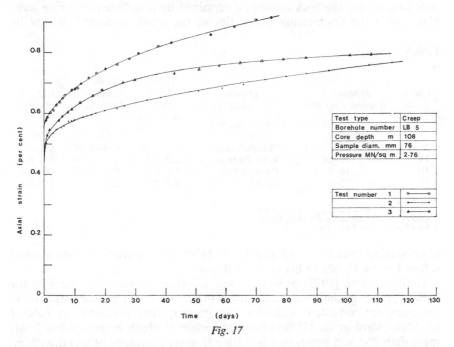

Test type	Creep
Borehole number	LB 5
Core depth m	108
Sample diam. mm	76
Pressure MN/sq m	2·76

Test number	1	○——○
	2	•———
	3	▲——▲

Fig. 17

by 'plastic padding'; these inserts were fixed at 120° at a height of 25.4 mm from both ends of a specimen. After consolidation in the triaxial cell, samples were removed quickly, enclosed in a copper sheath and sealed completely to prevent any loss of moisture, leaving the stainless steel inserts projecting outside the copper sheath for axial strain measurement with demec gauges.

The specimens were mounted on a special loading frame and subjected to a constant compressive stress of 2758 kN/m² (400 psi) applied in one increment. This stress represents a stress level of 80–85% of the maximum compressive strength of the control test. The plots of axial strain vs time are presented in *Fig. 17*. 0.5–0.6% axial strain occurred almost instantaneously upon application of the stress. Then the specimens continued to show a considerable amount of axial compression.

Although it is too early to come to a definite conclusion about the numerical value of creep from this limited number of tests, it can be said that it has a considerable effect on settlement. According to the results of these tests, creep may account for as much as 40% of the total settlement at high stress levels.

J. A. LORD, Ove Arup and Partners. I should like to draw the General Reporter's attention to the paper by Grainger, McCann and Gallois (1973). In this the authors have applied seismic refraction techniques to study the fracturing of the Middle Chalk at Mundford, Norfolk. They found that the seismic velocity increases with depth in clearly defined steps, and that these steps broadly correspond with some of the grades of chalk in the engineering classification adopted by the geotechnical survey (Ward, Burland and Gallois, 1968). To the writer this seems a unique instance in which it has proved possible to compare the rock quality, determined by a surface-measuring technique, with the engineering properties of the chalk measured *in situ* by

Table 5

Chalk grade	Fracture frequency per m	Quality class	Wave* velocity in field (km/s)	Velocity index $(V_F{}^2/V_L{}^2)$	E/E_{lab}† $(=j)$
V	—	(Structureless)	0.65–0.75	0.06–0.08	0.1
IV	100–16	Very Poor	1.0–1.2	0.14–0.20	0.1–0.2
III	16–5	Poor to Fair	1.6–1.8	0.35–0.45	0.2–0.4
II	5–1	Good	2.2–2.3	0.6–0.8	0.6–0.8

Computed Velocity in Laboratory V_L = 2.7 km/s
* After Grainger, McCann and Gallois (1973)
† After Burland and Lord (1969)

plate-bearing tests (Burland and Lord, 1969)—the modulus profile method referred to by Hobbs in his General Report.

Grainger *et al.* (1973) present seismic velocities measured in the field for various grades of chalk—these are listed in *Table 5*. It can be shown that these velocities are consistent with the fracture frequency presented in *Table 1* (p. 590). Ward *et al.* (1968) classified Grade II chalk as possessing joints more than 200 mm apart, that is, with a fracture frequency of less than 5/m. Assuming a velocity index $(V_F{}^2/V_L{}^2)$ of 0.7 corresponding to this quality of chalk, the velocity in the laboratory V_L is calculated to be 2.7 km/s. Applying this value to Grades III to V chalk the velocity indices are as shown in *Table 5* (in which the joint spacing and rock quality are also tabulated) and plotted graphically in *Fig. 22*. The shape of the curve in *Fig. 18* is similar to that of *Fig. 9* of the General Report. If the rock mass factor is defined as $j = E/E_{\text{lab}}$ (where E is Young's modulus for the rock mass and E_{lab} relates to the intact material as measured in the laboratory) then estimates of j for the various grades of chalk can be made from *Table 5* of Burland and Lord (1969). These values have been included in *Table 5* of this discussion. Thus for chalk at one particular site, there appears to be a direct correlation between the rock mass factor j and the velocity index $V_F{}^2/V_L{}^2$. However, Grainger *et al.* (1973) stress that correlation of seismic velocities with fractured rock grades at one site can only be extrapolated to another site if the intact material is identical

756

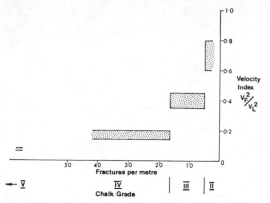

Fig. 18. Variation of velocity index with fracture frequency in middle chalk

at both sites. They illustrate this at the Mundford site by comparing seismic results of lithologically different chalks—in the lower Upper Chalk and upper Middle Chalk (classified as Grade II and I in engineering terms) the presence of a large number of flints, often technically shattered, appeared to reduce the seismic velocities to those characteristic of Grade III chalk elsewhere in the Middle Chalk.

Burland, J. B. and Lord, J. A. (1969). 'The Load Deformation Behaviour of Middle Chalk at Mundford, Norfolk', *Proc. Conf. In situ Investigations in Soils and Rocks*, British Geotechnical Society, London

Grainger, P., McCann, D. M. and Gallois, R. W. (1973). 'The Application of the Seismic Refraction Technique to the Study of the Fracturing of the Middle Chalk at Mundford, Norfolk', *Geotechnique*, Vol. 23, No. 2, 219–232

Ward, W. H., Burland, J. B. and Gallois, R. W. (1968). 'Geotechnical Assessment of a Site at Mundford, Norfolk, for a Large Proton Accelerator', *Geotechnique*, Vol. 18, 399–431

Problems associated with Marl

A. G. DAVIS, University of Birmingham. The authors of Paper IV/5 (p. 292) are to be congratulated for their valuable contribution to the analysis of the settlement of structures on Keuper Marl. Of particular interest are the deductions of apparent modulus E for relatively unweathered marls from pile tests at Nottingham.

The writer recently participated in pile testing for the Clarence Road Bridge over the River Taff in Cardiff. Two bored cast *in situ* test piles, each of 760 mm shaft diameter, were founded through river gravel into Zone II Keuper Marl. Test Pile No. 1 (TP 1) was constructed normally, while Test pile No. 2 (TP 2) was formed with a 300 mm void under its toe. Both piles were double sleeved down to the top of the marl to prevent skin friction due to the river gravel. In both cases, the piles penetrated 4 m into the marl. A full description of the site and construction process is given in the New Civil Engineer of 21 February, 1974.

TP1 was loaded to 550 tonf, then completely unloaded and reloaded to 550 tonf. TP 2 was loaded to 270 tonf, at which value it proved impossible to keep a constant load on the jack. This was taken as the maximum value of skin friction mobilised by the test pile in the marl.

Values of pile load at equal settlements for the two test piles are given in

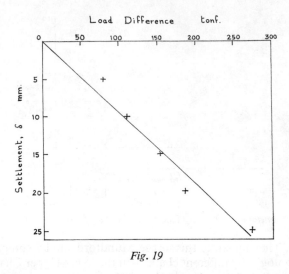

Fig. 19

Table 5, together with the pile load differences. The latter are inferred to be that proportion of the total pile load on TP 1 which is carried by base resistance alone. *Figure 19* shows the test pile load difference for equal settlement plotted against settlement. The relationship is reasonably linear for a settlement range, S between 5 and 25 mm, and corresponds to a load/settlement ratio of 10.5 tonf/mm.

Taking the expression for apparent modulus given in *Table 3*, p. 304, the above load/settlement ratio is equivalent to an apparent modulus of between

Table 5

Settlement (mm)	Load on TP1 (tonf)	Load on TP2 (tonf)	Load difference TP1 − TP2 (tonf)
5	200	120	80
10	280	170	110
15	360	207	153
20	415	230	185
25	520	247	273

1.26×10^5 and 1.38×10^5 kN/m² for values of μ between 0.3 and 0. These agree very satisfactorily with the results presented for the Nottingham site.

The ratio $\dfrac{E}{q_u}$ is approximately 100 for these pile tests, which falls in the Zone II bracket in *Fig. 6* in the General Report to this session. However, it should be emphasised that it is possible to have consolidation settlement at high stress levels under pile toes in the marl. This was noticed at the Clarence Road Bridge, where two load increments were left on TP 1 for over 12 h, and typical time-settlement curves were recorded with values for 90% consolidation of approximately 200 min. This corresponds to values for the coefficient of consolidation, c_v of approximately 7×10^{-3} m²/min and a modulus of volume change, $m_v = 8.1 \times 10^{-3}$ m^2/MN for a stress increment at the toe of 3000 kN/m^2. This is equivalent to an apparent modulus at the toe $E = 1.24 \times 10^5$ kN/m^2. The increase in toe stress was calculated by subtracting the increased shaft load from the total load increment of 40 tonf. The former was estimated by measuring the increased load on TP 2 over the equivalent settlement range of 6.2 to 8.2 mm. This gave an increased shaft load of 26.7 tonf, or an increased toe load for TP 1 of 13.3 tonf. Given the relatively rapid dissipation of pore pressure, it would appear desirable to maintain each load increment for pile tests in Zone I and Zone II Keuper Marl for a minimum of 4 h at stress levels greater than 1000 kN/m^2 at the pile toe. An estimate of likely skin friction would have to be made to calculate the actual toe stress level. In this way, a lower bound, fully drained apparent modulus could be derived from pile tests.

Settlement measurements are being taken on the Clarence Road Bridge, which is under construction at the present moment.

D. H. CORNFORTH. Present site investigation practice for Keuper Marl can be briefly summarized as follows: The highly weathered marl usually encountered at the surface is bored and sampled by shell and auger methods to a depth of about 10 to 20 m, below which the marl is hard enough to be cored by rotary methods. This means that the settlements of shallow foundations are computed from laboratory data on soils recovered by the 100 mm diameter (U4) sampler tube. The sampler is usually driven into the stiff, fissured, and brittle marl by about 100 to 150 blows of a jarring link. The end area ratio of the cutting shoe is 25–27% and considerable sampling disturbance is evident from the results obtained recently at a bridge site near Warwick. At this site, laboratory data from 100 mm diameter (U4) tube samples in the upper 5 m of the Grade III/IV marl (Chandler Classification, see p. 597 of the General Report) have been compared with block samples taken from test pits at depths of around 1.3 m.

The stress-strain curves of undrained triaxial tests show that the U4 samples failed at axial strains of 6 to 10% whereas the block samples failed at axial strains of only 2 to 4%. Since the average strength of the block samples was about twice that of the U4 samples, the effect of sampling disturbance on the undrained modulus E_u was even more pronounced, amounting to a factor of 4 on average. However, there are other implications to these results, which have been plotted against water content on a semi-logarithmic plot, *Fig. 20*. This plot shows that the U4 samples showed no discernible trend towards increases in modulus with decreasing water content, as might be expected. It suggests that sampling disturbance may increase in severity as the soil structure becomes stiffer so that the modulus, at a particular sampling depth, may reach a limiting value. The two sets of results diverge below a water content of about 34%, which is the mean plastic limit of the marl. Another implication of the data is that it could be inaccurate to try to 'correct' the results obtained from U4 samples by using a fixed ratio for sample disturbance effects. At increasing depth below ground, the ratio is likely to increase.

For consolidation settlement studies, the results of 6 oedometer tests on block samples compared with 7 tests on U4 samples showed that, on average, the compressibility of the former was only about one-half of the latter. A more surprising set of results was obtained from triaxial dissipation tests which showed that specimens cut from block samples had 100 to 1000 times faster rates of primary consolidation than the specimens cut from U4 samples. This suggests that the fundamental structure of the marl, known to consist of agglomerations of clay particles, has been altered by the sampling method.

The above results, together with the previous studies of sampling disturbance on London Clay by the Building Research Station clearly demonstrate that the intact properties of stiff clays and soft rocks are very severely affected by our present sampling methods. The Pitcher Sampler is probably the best sampler available for such soils at the present time. In view of the increasing sophistication of laboratory and analytical procedures, it is to be hoped that increasing attention will be paid to improving field sampling techniques in the immediate future.

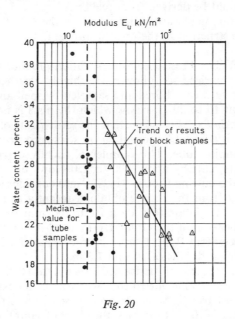

Fig. 20

M. M. H. AL-SHAIKH-ALI, North Western Road Construction Unit, Cheshire Sub-unit.   In Paper IV/5 (p. 292) the Authors stated that creep in Keuper Marl is very slight. With pressures as low as 300 kN/m² one would not expect the creep to be very pronounced, and using the Authors own phraseology, 'this pressure has been masked by the incidence of the negative skin friction on the shaft of the pile'. Furthermore the Authors did not appear to have carried out prolonged pile tests to measure the creep directly associated with Keuper Marl.

Two of the test piles which were executed on the M56 Motorway were kept under constant load (1800–200 kN) for over 165 h. (See 'Procceedings of Symposium on the interaction of Structure and Foundation', Birmingham University 1971.) The material immediately underneath the pile was Zone II (as in Cheshire, Zone I can only be found at a very deep level). From the test pile the settlement points appear to be on a straight line (*Fig. 21*) when plotted on the logarithm of time. This suggests that creep is imminent but very minimal under this pressure.

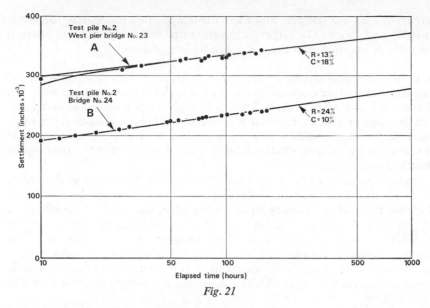

Fig. 21

The creep in Keuper Marl was fully investigated and it confirms the view expressed by others that when the stresses are excessive, say over 60% of the ultimate, the creep does not necessarily have a linear relationship with log of time. *Fig. 22* showed that when the stresses are as high as 28 kg/cm² the creep was proportional to the logarithm of time under constant load and temperature, 'the stress in this case was over 50% of the ultimate'.

When the material is mechanically or thermally activated and, provided that the activation energy is large enough to surmount the energy barriers, the material would creep. In *Fig. 21* (of the above mentioned Symposium) the

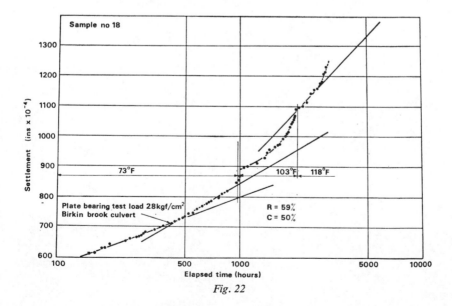

Fig. 22

761

activation used was thermal, and when the temperature was raised from 70°F to 103°F and then to 118°F there was substantial increase in the rate of creep per cycle of log time, and the creep was no longer linear.

A. C. MEIGH, Soil Mechanics Ltd.   I wish to emphasise the need for careful identification and description of the materials that we are dealing with. We mustn't talk about The Keuper Marl as though it was a unique material and had the same properties wherever it is found. Within the Keuper series there are to be found materials with a range of compressive strength or modulus of several orders of magnitude, and they can occur on one site, within quite short distances of each other.

Similar remarks may be made about London Clay. While tens of thousands of bored piles have been put into The London Clay in the London area, there are parts of the country where the London Clay will hardly allow the construction of a bored pile, because the hole will collapse, and where it is impossible to underream because of the intensity and configuration of the fissures, sometimes complicated by the presence of clay stones and with difficult ground water conditions. To avoid such difficulties we must carefully describe the material we are dealing with, and not simply describe it as The London Clay.

SESSION V

Allowable and Differential Settlements of Structures, Including Damage and Soil-Structure Interaction

Session Chairman: Professor A. W. Skempton
General Reporters: Dr. J. B. Burland and
Dr. C. P. Wroth

DISCUSSION   Page

Examples of damage

G. S. LITTLEJOHN, University of Aberdeen. I wish to illustrate some simple observations of the cracking of brickwork with particular reference to formation, propagation and influence on subsequent structural behaviour. My Paper V/8 (p. 384) on wall behaviour describes some detailed monitoring of movements at the foundation/soil interface and these observations have proved useful when studying damage.

The walls, situated at stations H, I and J on the dynamic subsidence line (*Fig. 2*, p. 386,) were damaged initially when subjected to convex ground curvature, with its accompanying tensile ground strain. Since the latter is usually the main cause of damage it is important to understand how tensile ground strains create forces on a foundation, because it is these forces which create structural strains and possibly damage. *Figure 1* shows a simple footing subjected to uniform tensile ground strain. Up to the point of cracking the foundation acts as a rigid unit in the horizontal direction so that the slip of the ground past the footing at a point, distance X from the neutral axis is simply equal to $\varepsilon_g X$ and the maximum slip is $\varepsilon_g L/2$ (*Fig. 1(c)*). Making some simplifying assumptions that pushing half the footing over the ground gives the required relationship between shear resistance and slip at the interface (*Fig. 2(a)*) then the subgrade forces acting on the footing (*Fig. 2(d)*) may be estimated enabling simple calculation of the structural forces (*Fig. 2(e)*). It should be noted that the maximum tensile force in the footing occurs at the position of zero slip, and field observations (*Fig. 8*, p. 390) on the 15.8 m long $\times$ 1.2 m high brick wall show that the first crack formed at 9 m whilst at a later stage, a hairline crack occurred at 5 m. Both cracks formed at positions of zero slip and these results therefore confirm that the main cause of damage was tensile ground strain. It is interesting to note that the first crack occurred at a maximum differential settlement of 8.6 mm over a length of 15.8 m, and a hogging ratio of 1/6000, i.e. the same order of movement that might be expected in ground adjacent to large open excavations or tunnels (see Paper V/2, p. 329). However whilst measurement of ground strain is common in mining areas, this parameter is seldom measured by civil engineers and bearing in mind that a tensile ground strain of only 0.02% was sufficient in this case to create aesthetic damage, it is recommended that more attention should be paid to monitoring movements at the foundation/soil interface with particular reference to ground strains.

With regard to the propagation of cracks, initially Crack 1 extended only 0.66 m up the wall. The widths monitored as the mine face advanced are shown in *Figs. 2(a) and (b)* and it is noteworthy that even the trained eye could easily miss crack widths of less than 0.05 mm in brickwork; the width when a crack first becomes discernable clearly depends on surface roughness as well as the skill of the observer. *Figure 2(b)* shows that with increasing ground hogging curvature the crack extended upwards to the top of the wall and subsequent widening was due to the combined effects of curvature and tensile strain, curvature gradually becoming more dominant as the movements progressed. It is considered that the reduction in width of the crack (*Fig. 2(b)*) was influenced by the interaction of the bricks in the damage zone, high contact stresses being mobilised as the curvature increased thereby giving rise to significant horizontal shear resistance. As a result this crack acted as a 'stiff

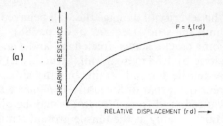

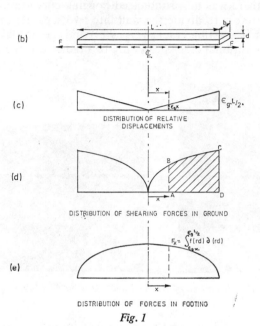

$$F_x = \int_{\epsilon_g x}^{\epsilon_g \frac{L}{2}} f(rd) \, \partial \, (rd)$$

Fig. 1

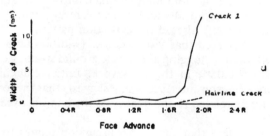

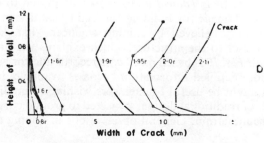

Fig. 2. Crack movements—Structure 1

hinge', and in spite of the severity of damage the wall behaved as a composite structural unit throughout the study (see *Fig. 5*, on p. 388).

On the other hand some cracks act as 'free edges' and *Fig. 3* shows cracks 1 and 1(a) on a 31.2 m long by 1.83 m high wall (Structure 2) where again the damage was caused by tensile ground strain, but the subsequent mode of deformation was different from that of Structure 1. *Figure 4* shows the crack widths as the mine advanced, and from *Fig. 4(b)* it may be observed that the main crack 1 widened fairly evenly as the tensile ground strain increased but on this occasion there was no pronounced hogging curvature. The effect of this crack was therefore to divide the wall into two separate units. The subsequent behaviour of existing cracks (*Fig. 4(c)*) and the probability, severity and influence of additional cracks can only be predicted if the existence and

Fig. 3. Pattern of cracks 1 and 1 (a), Structure 2

time of origin of the dual behaviour above is taken into account in the analysis. Subsequent cracking occurred at more or less central positions on the two structural units and severely altered the deflection pattern.

In general it was observed that visible cracks (width = 0.05–0.10 mm) in the brickwork, subjected to hogging and tensile ground strain, were monitored at strains of 0.02 to 0.05% in the brickwork. Initial cracking invariably occurred at the brick/mortar interface which suggests that the tensile strength of brickwork is closely related to the quality of the bond between the mortar and the bricks.

Finally bearing in mind that the above discussion relates to simple walls and that the real problem concerns buildings, it is noteworthy that King and Orchard in (1959) (see their *Table 1* and *Fig. 10*) attempted to relate ground strain to severity of damage for buildings of traditional design (N.B. at this time it was commonly believed by mining engineers that ground strains were transferred direct to the structure, which explains the first column of their *Table 1*). Their *Fig. 10* is based on 17 case records and for guide purposes only and until more detailed information becomes available, it is suggested that these curves might be useful to engineers wishing to predict damage to existing structures of traditional design, subjected to ground curvature with its accompanying ground strain. General observations in mining areas indicate

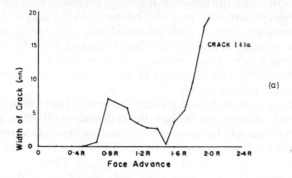

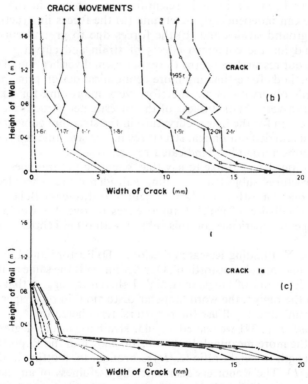

Fig. 4. Crack movements, Structure 2

that aesthetic damage, i.e. cracking of internal plaster and sticking of doors and windows, tends to occur when tensile and compressive ground strains exceed 0.04% and 0.08%, respectively.

King, H. J. and Orchard, R. J. (1959). 'Ground Movement in the Exploitation of Coal Seams', *Colliery Guardian*, Vol. 198, 471–477

C. M. KAWULOK, The Institute for Building Technics, Gliwice, Poland. I would like to present some remarks with reference to the note by Littlejohn describing observations of brick walls subjected to mining subsidence.

First of all the problem of building in mining areas of Poland concerns the most industrialized region of our country, of small area (about 3000 km^2) but very densely covered with buildings of various structural types, and inhabited by about 2.5 million people.

Current coal extraction in Poland amounts to nearly 150 million ton a year of which about 50% is extracted beneath those densely built areas. However, this percentage will steadily increase. This extraction involves various forms and sizes of ground deformations. Horizontal strains of the order of 0.6%, curvatures of 0.16% and slopes of 1.0% are typical. There often occurs a necessity to erect new buildings in places that have bigger ground deformations reaching 1.5 to 2 times the above values. There are also cases of structures built in the area of discontinuous ground deformations due to old workings and faults.

There have been considerable scientific achievements in this subject in Poland. We can independently determine (a) the forces in structures due to horizontal ground strains and (b) the forces due to ground curvature. We are now studying the combined effects of strain and curvature. The main direction of our research concerns the experimental verification of known engineering methods for estimating more economical design principles.

Littlejohn's observations are therefore very interesting for us. One may suppose that a very small stiffness of the investigated structure could be an important reason for the bigger curvature in the structure compared with the ground. Tests carried out on real structures in our scientific centres as well as in other centres abroad have indicated reverse results.

In conclusion, I would like to say that scientific programmes referring to building structures subjected to mining subsidence are scheduled and performed in Poland mostly at the Department of Institute for Building Technics in the city of Gliwice, of which I am a representative. We would appreciate any exchange of experiences on this subject with other centres abroad.

J. E. CHENEY, Building Research Station. D. Burford and I, in our Paper on p. 337 define 'angular distortion'. Our definition is the same as the General Reporters' definition of 'angular strain'. I therefore suggest that to be consistent with the report, the word 'angular distortion' in our paper be read as 'angular strain', and I will use the reporters' term hereafter.

In our *Table 1*, p. 341 we related angular strain to various degrees of cracking. I note the more meaningful relationship shown in the Reporters' *Fig. 11* where the length to height ratio (L/H) is brought in, plotting it against relative hog (Δ/L). The Reporters also note the usefulness of angular strain in predicting crack widths where movement occurs at existing cracks or at weakness lines. I appreciate the possibly unjustified over-simplification of trying to associate angular strain assumptions derived from levelling with diffused cracking of a building. However we have measured the cracks on a building and attempted to relate them to angular strain. The investigation was restricted to the external brickwork as decorative repairs to internal wall finishes over the years prevented useful information being obtained.

In order to measure the cracks we hired a mobile hoist. The device came with a driver who could place us with speed and skill before the brickwork at any point we desired (the hire charges of about £40 a day are a fraction of the cost of scaffolding and the hoist is quicker and safer than ladder work). We measured the cracks using a pocket magnifier manufactured by Granticules Limited. It is equipped with a glass graticule having etched scales graduated to 0.1 mm. The device is used with the glass scale in contact with the brickwork.

It was decided to record the horizontal element of the cracks at four horizontal sections of the building. These sections were at the brick course above DPC and at a chosen course at each of the levels: first floor, second floor and below the eaves. *Figure 5* shows the west elevation of the building with the movements in level plotted below. It clearly shows the crack pattern radiating

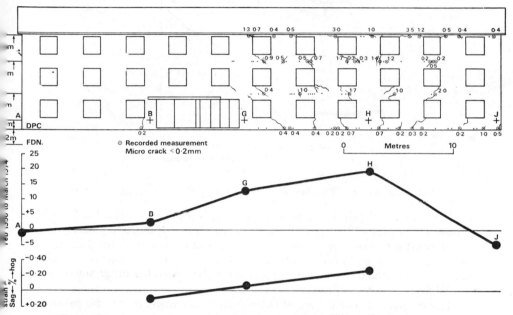

Fig. 5. *West elevation—horizontal component of crack widths* (mm) *at March 1974*

from the hogged part of the foundations. The numbers on the elevation show the horizontal width of those cracks which measured at least 0.2 mm. We also noted the presence of cracks below this size. They are shown as dots on the figure. Although it is the horizontal component of the cracks that are recorded, it is reasonable to compare their dimensions with crack visibility. This is because the cracks usually resolve into zig-zags in the brickwork, there being near vertical cracks in the bricks themselves or between them and the vertical mortar joints. Thus the maximum width of the cracks is of the same order as the measured horizontal component.

The actual crack tracks shown on *Fig. 5* are not obtained by observation from the hoist, but are visible by reasonable deliberate inspection from the ground, or in the case of those above first floor by 8 × 30 binoculars. It will be seen that generally cracks below 0.2 mm in brickwork are not easily visible. I suggest that cracks below 0.5 mm in brickwork are not really noticed by the user and it is only when cracks reach 1.0 mm that comments are made. J. F. S. Pryke (Paper V/10 p. 406) intimates that owners do not actually complain before 3.0 mm is reached. I think they had noticed the 1.0 mm cracks and had watched them grow for some time before committing themselves to a course of action likely to prove relatively expensive.

For the hog length *G–H–J Fig. 6* shows the angular strain at *H* drawn from foundation level, and the integrated horizontal width of cracks of 0.2 mm and greater, over the length *G–H–J* at the four observed heights. On this basis about 100% of angular strain appears as observed cracks at dpc reducing to about 35% at eaves. Possibly this could be explained by a greater distribution as micro-strain in the brickwork at higher levels, but I am more inclined to

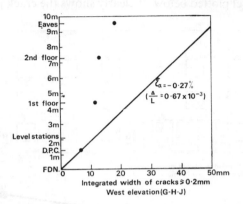

Fig. 6. Integrated crack widths at various heights—related to angular strain at H

believe that the depth of the horizontal strain is some way below the foundations. If the plot of the aggregated crack width is extrapolated linearly in a downward direction, zero horizontal strain is indicated at about 3 m below foundation level and on this basis 25% of the angular strain shows as visible cracking. A similar picture emerges from dealing with the other side of the buildings.

In conclusion I would suggest there is scope for many more case histories to be studied. The difficulty is to find buildings for early instrumentation which are expected to crack. Offers of such buildings will be welcomed by the authors! It would also be useful to measure horizontal strains in the foundations and in the ground below on suitable sites.

C. J. F. JONES, West Yorkshire Metropolitan County Council. It appears that there are two main causes of the settlement of structures; (1) natural settlement due to foundaton conditions, and (2) settlements which are man made. As far as it can be seen, settlement which results in structural damage is usually caused by the latter. The ground strains, tilts and settlements resulting from the mining of minerals being the prime cause.

There is no such thing as allowable settlement. The nature of the structure involved determines its behaviour. If the settlement is sufficient to cause damage or loss of serviceability then that structure is the wrong one for the site. It may of course be that settlement could not have been anticipated at the time of construction, or that the costs of taking the necessary precautions are excessive. However, this is not always the case and the success of the CLASP system of building medium sized structures illustrates what can be accomplished. It is possible to argue that CLASP structures are special structures, however they are better described as representing a satisfactory solution to a structural problem.

A variety of structural forms have been used in building the M1 motorway. What is not immediately obvious is that many of the structures in Derbyshire and Yorkshire are capable of accommodating considerable differential settlements. These structures did not cost more than those at the southern end of the motorway and their appearance generally meets with approval. Indeed the low torsion decks adopted in the old West Riding of Yorkshire, initially to cater for mining subsidence, are now used generally, whether there is subsidence or not, as they have proved easier to construct and less costly.

In the Author's opinion, the success of the designs for the bridges in areas of mining subsidence in Yorkshire are due to the fact that the design of the structure as well as the design of the foundations and the interpretation of the soils information were all the responsibility of individual design teams. As the General Reporters point out this was the general philosophy behind the development of CLASP.

R. J. MAINSTONE, Building Research Station. The identification and use of a critical tensile strain ε_{crit} for predicting the distortions at which visible cracking of the superstructure might occur is an attractive suggestion of the General Reporters. Clearly it holds out the promise of a powerful tool for the determination of allowable relative settlements. Having unwittingly

Table 1. VALUES OF ε_{crit} FROM BRS RACKING TESTS OF BRICK WALLS AND FRAME-INFILLS

Test arrangement and principal-stress pattern in infill or wall	Measured change in diagonal length at first diagonal crack	Approximate diagonal strain pattern in infill or wall	Approximate value of ε_{crit} at centre of infill or wall
A	−0.04% (stiff frame)		between +0.01% and +0.02%
B	−0.125% (normal frame)		between +0.02% and +0.03%
C	+0.004% (no frame)		between +0.01% and +0.03%

furnished some of the basic data through tests carried out for quite a different purpose, I am very happy to accept Dr. Burland's invitation to comment.

I have only one major reservation about the present development of the idea in Sections 4 and 5 of the General Report. I feel that, in by-passing all the analytical awkwardnesses and complexities of the true strain distributions in brittle walls and infills both prior to cracking and during the early stages of crack development, the Reporters have inevitably sacrificed some of the potential generality that is the chief merit of the basic concept. The approximations made are of kinds that cannot be equally valid for all practical situations.

Table 1 illustrates, for instance, typical internal strain patterns up to initial cracking for walls and infills undergoing predominantly shearing distortion. (*A* simulates a normally framed infill surrounded by other infills; *B* and *C* exhibit similar strain patterns but the changes in diagonal length were measured, for convenience, on different diagonals.) The physically significant values of ε_{crit} are all substantially less than the averaged values quoted in the Report. Likewise, a re-examination of the tests reported by Burhouse will show values within the range 0.01% to 0.03% for the strain at which the first crack formed, the strains were thereby re-distributed, and the stiffness markedly fell. The larger values quoted in the Report are here averaged over an indeterminate length at a slightly later stage of crack development.

In principle, there is no reason why more precise analyses should not be substituted for those in Section 5 for the determination of $\Delta/L\varepsilon_{crit}$ and coupled with revised values of ε_{crit} relevant to initial cracking. But it is more difficult to see how the values of Δ/L so derived should be increased to make them more appropriate to unsightly visible cracking. Certainly a greater increase

Table 2

	Speedwell Court North London	Abbotsford Farm Derbyshire	H. M. Prison Stafford
Foundation details	Concrete piers and simply supported beams	Unreinforced concrete strip cast in trench 0.75 m deep × 0.4 m wide	1 m wide brick strip
Type and cause of ground movement	Swelling of London Clay following tree removal prior to construction	Settlement of poorly compacted backfill to opencast coal pit	Settlement due to extraction of salt by brine pumping
Rate of ground movement (m/year)	0.02	0.05–0.10 (with seasonal variations)	0.10
Mode of deformation	Hogging	Hogging at edge of settlement bowl Sagging across settlement bowl	
Cause of cracking	Direct tensile strain	Direct tensile strain at edge of bowl Diagonal tensile strain across bowl	

would be admissible in a case (such as *A* or, to a lesser extent, *B* in *Table 1*) where continued distortion would tend to initiate new cracks rather than widen the first one, than in a case (such as *C*) where it would all be accommodated by widening this first crack.

The fairly innocuous character of some kinds of cracking as compared with others should also clearly be taken into account in arriving at a final criterion. The Reporters have, indeed, rightly pointed out that even large distortions can be accommodated without unsightly cracking by appropriate detailing. What they have failed to emphasise is that such detailing will also lead to a more flexible superstructure—to one that will distort more if set on a given foundation. Perhaps this is so obvious that it hardly needs stating, but it does have an important converse: a superstructure not so detailed will usually be considerably stiffer and will thus distort less and, to that extent, be less at risk than might otherwise be expected. The task of the designer is to strike the right balance in particular circumstances. The Reporters have taken a notable step forward in helping him to do so.

772

D. W. HIGHT, Scott Wilson Kirkpatrick and Partners.   Details are tabulated (see *Table 2*) of three examples of damage to traditional brick structures. The types of ground movement and patterns of cracking observed are illustrated in *Figs. 7, 8 and 9*; these can be seen to be consistent and the crack patterns more or less confirm those predicted in the General Report for both hogging and sagging modes of deformation.

A direct comparison between the case records is justified since, although the cause of ground movement differs, the rate of movement, the construction details and the restraints provided by the foundations are similar. In *Fig. 7* the bending deformation has produced no relative vertical movement at the cracks but greater relative horizontal movement at the top of the building than at the bottom. The hogging ratio at which the cracks became visible

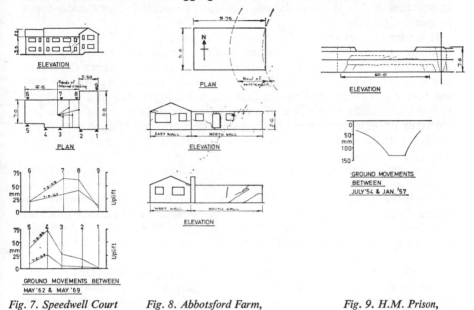

Fig. 7. Speedwell Court   Fig. 8. Abbotsford Farm,   Fig. 9. H.M. Prison,
North London   Derbyshire   Stafford

has been estimated as 0.15×10^{-3}. In *Figs. 8 and 9* the structures have been affected by bowls of settlement and the resulting crack patterns are similar, that observed in the prison wall being an unfolded version of that observed in the bungalow at Abbotsford Farm. In both, the lower part of the wall has fallen away from the upper part which is behaving as an arch; on each side of the sagging section vertical cracks wider at the top than at the bottom have formed.

At Abbotsford Farm the sagging ratio at which the diagonal cracks became visible has been estimated as 0.7×10^{-3}; these cracks developed after the vertical cracks but before the horizontal cracks.

Analytical methods

W. J. LARNACH, University of Bristol.   In his general report Mr. Butler reminded us that we calculate settlements, by whatever method, for a purpose, and not as an end in itself. In my view estimates are used for two main purposes. First, as a design decision aid. Here both total and differential settlements need to be evaluated in order to permit rational decisions to be taken about the type of foundation to be adopted, and there are significant economic overtones. It may well be that conventional non-interactive methods cause the wrong or doubtful decisions to be taken, because the estimated differential settlements between, say, adjacent footings in a proposed pad foundation solution will almost certainly be considerably overestimated, giving rise to angular distortions approaching or exceeding accepted damage limits. Hence decisions to adopt other solutions, e.g. rafts or piles may be made, leading to needless expense.

The second use is that of a structural analysis and detailing aid, and this has been almost entirely ignored in this conference. By this I mean the estimation of stress resultants which occur within the structure due to differential movements within the foundation, and this is essential information for a proper design.

It is very evident that settlement estimates for these specific purposes can be made only by using an interactive analysis. In our paper (p. 460) Wood and myself give one facet of an interactive analysis which has been developed to a state in which many of the effects mentioned in discussion can be accounted for: e.g. elastic solutions in layered, unhomogeneous and non-linear systems, and using either drained or undrained parameters; time dependent solutions using pore-pressure dissipation; the modelling of change in frame stiffness with construction time; or, more conventionally, the use of the simple Terzaghi soil model.

Such analyses provide a useful adjunct in design, and I thought that the General Reporters adopted a rather dismissive attitude to these attempts. Of course there are deficiencies in these methods and in the input information in regard to both soils and structures, but my view is that bounded solutions (i.e. those involving likely ranges of important parameters) are easily possible, and can give guidelines within which engineering judgements can be more effectively exercised.

I think I agree with Breth that complete finite element solutions may not be suitable for all cases, and in our work there is the possibility, already referred to, of using simple soil models based on compressibilities. An interactive analysis using even this unsophisticated approach must produce more realistic estimates of differential settlements than current practice. However there is a great need for fully documented case records (including the measurement of strains within the structure) so that theoretical treatments, of whatever complexity, can be tested against experience.

G. J. W. KING, Liverpool University. As the General Reporters have pointed out most research workers who have considered structure/soil interaction have concentrated on its effects on contact pressure distributions under rafts and the forces in structures and not published values of differential settlement. Indeed, this is true of the paper by Chandrasekaran and myself (p. 368). It is therefore of interest to consider the influence of structure/raft interaction on the immediate settlement of the two bay continuous frame which we studied (*Fig. 4*, p. 374).

For a value of $q = 1000$ lbf/ft^2 the values of settlement at the centre and edge of the raft obtained from both independent and interactive analyses, for a range of raft thickness, are shown in *Table 3*.

For the given loading the raft thickness would usually be of the order of 9 to 12 in and for this thickness it is clearly essential to consider the stiffness of the superstructure when predicting differential settlements. It is also obvious that differential settlements are not related to overall settlements since only the former are significantly dependent on the stiffnesses of superstructure and raft.

The variations of differential settlement with raft thickness for independent and interactive analyses can be used to consider the General Reporter's suggestion that the stiffness of the superstructure might be accounted for empirically by adjusting the stiffness of the raft. For corresponding differential settlements a 9 in thick raft would have to be considered as almost 16 in thick in an independent analysis. Even if there was some empirical way of guessing such a value for this or any other structure, and I don't believe there

Table 3

Raft thickness	Settlements from interactive analysis (in)			Settlements from independent analysis (in)		
(in)	Centre	Edge	Difference	Centre	Edge	Difference
6	4.1501	4.0163	0.1338	4.4493	3.7980	0.6513
9	4.1073	4.0250	0.0823	4.2325	3.9292	0.3033
12	4.0870	4.0314	0.0556	4.1398	3.9907	0.1491
24	4.0605	4.0446	0.0159	4.0621	4.0426	0.0195
36	4.0544	4.0487	0.0057	4.0544	4.0486	0.0058

is, it would not be of any use to the design engineer. Reference to *Figs. 6 and 7* on pp. 375 and 376 shows that the bending moments in the structure and the raft would be quite different. Further, for a multi-span frame only the differential settlement for one of the spans could be modelled and the overall displacement patterns of the rafts would be quite different. It really doesn't make sense to exclude the structure from any analysis in which it is intended to calculate the deflections in that structure.

The structure considered in the paper is admittedly an idealised plane structure which serves simply to illustrate the importance of structure/soil interaction. Our more recent work has been concerned with the analysis of rafted space frames on overconsolidated inhomogeneous clays (elastic) and we have evolved techniques so that this can be done economically with respect to computer storage and time. The modern engineer is trained to make use of the computer and the preparation of data for a well-documented programme is becoming almost as easy as using design charts and in relation to the problem of differential settlement is far more reliable.

There are admittedly problems yet to be solved in relation to the stiffness of structural claddings, the variable stiffness and partial settlement during construction and deformation moduli for soils, but we must not bury our heads in the soil and ignore the very important influence of structure/soil interaction.

M. J. KEEDWELL, Lanchester Polytechnic. With reference to analytical methods of assessing settlement the following rheological equations may be of interest. These equations were originally proposed by the writer in 1971 and relate to soils tested in the triaxial cell with the cell pressure held constant. They show the effects of stress level, time, and stress history on the behaviour of isotropic cohesive soils.

Effect of stress level. If time effects are ignored the equations may be written in the form:

$$\left.\begin{aligned} E_q &= \frac{1}{K}\, e^{-\alpha q} \\[2mm] \varepsilon &= \frac{K}{\alpha}\, e^{\alpha q} + C \\[2mm] q &= \frac{1}{\alpha}\log_e\left[\frac{\alpha}{K}(\varepsilon - C)\right] \end{aligned}\right\} \text{provided } 0.9q_F > q > 0.3q_F$$

where E_q is the elastic modulus, ε = axial strain, q = deviator stress level, q_F = failure deviator stress, K, α = soil constants at given void ratio, and C = constant of integration.

As an example of their application, the equations might be used, respectively, in the 'variable stiffness', 'initial strain', and 'initial stress' methods of

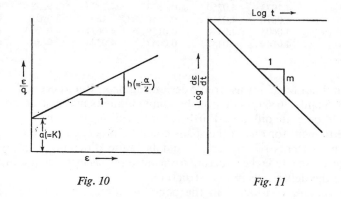

Fig. 10   *Fig. 11*

non-linear stress analysis using finite elements. The values of K and α may be obtained from a plot of the logarithm of E_q versus q (see *Fig. 14*) or, alternatively, since

$$\alpha \simeq 2b \qquad K \simeq a \qquad C \simeq \frac{9}{2b}$$

where a and b are constants in the equation

$$\frac{\varepsilon}{q} = a + b\varepsilon$$

(Kondner, 1963) the α and K values could also be evaluated from the transformed stress-strain curve shown in *Fig. 10*.

Effect of time. The more general form of the stiffness equation is

$$E_q^{-1} = G.e^{\alpha p}\dot{q}^{m-1}$$

776

where $\dot{q} = \dfrac{\mathrm{d}q}{\mathrm{d}t}$ (and $G\dot{q}^{m-1} = K$, when $m = 1$). The same theory also yields the creep equation

$$\frac{\mathrm{d}\varepsilon}{\mathrm{d}t} = H.\, \mathrm{e}^{\alpha q}\, t^{-m}, \qquad \left[\dot{q} = 0,\ H = G\left(\frac{\alpha}{1-m}\right)^{-m}\right]$$

The parameters α and m are common to both equations. Usually m is close to unity so that when $m \neq 1$, E_q is slightly sensitive to the rate of stress change $\dot{q}$. The creep equation predicts a linear relationship between (1) the logarithm of strain rate and the logarithm of time (*Fig. 11*), and (2) the logarithm of strain rate at a given time in any test and the stress level q (*Fig. 12*).

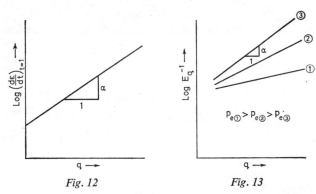

Fig. 12   Fig. 13

Effect of stress-history. Only the effect of the consolidation pressure at which the samples were prepared has so far been considered. The effect of consolidation pressure p_e on the values of the constants in the equation

$$E_q^{-1} = K\, \mathrm{e}^{\alpha q}$$

is shown diagrammatically in *Fig. 13*. It has been found that the constants in the dimensionless equation

$$E'q^{-1} = K'\, \mathrm{e}^{\alpha_c q'}, \qquad \left(q' = q/p_e, \qquad Eq' = \frac{\mathrm{d}q'}{\mathrm{d}\varepsilon}\right)$$

are not dependent on the consolidation pressure (see *Fig. 14*).

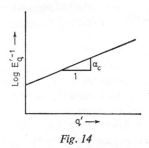

Fig. 14

Keedwell, M. J. (1971). 'The Rheology of Clays', *Proc. Symp. Interaction of Structure and Foundation*, The Midland Society of Soil Mechanics and Foundation Engineering, University of Birmingham

Kondner, R. L. (1963). 'Hyperbolic Stress–Strain Response: Cohesive Soils,' *Proc. Am. Soc. civ. Engrs*, Vol. 89, SMI, 115–143

V. S. CHANDRASEKARAN, The University of Liverpool. The General Reporters have compared settlements obtained using oedometer results with those using elastic analysis. This is based on the assumption that the settlements by the former approach are calculated according to the equation

$$\rho_{oed} = \int \sigma_z m_v \, dz. \tag{1}$$

They have shown that these values are less than those obtained from elastic analysis and for values of v' less than 0.24 to be within 10%. They therefore suggest that the settlements could be corrected according to Equation 19 of the General Report.

Another procedure using oedometer results is based on the equation

$$\rho_t' = \rho_i + \rho_c \tag{2}$$

in which

$$\rho_c = \int u_i m_v \, dz.$$

It is interesting to compare the values obtained from this equation with those

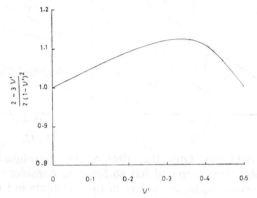

Fig. 15. Variation of factor $\dfrac{(2 - 3v')}{2(1 - v')}$ with v'

from the elastic analysis. If ρ_i and u_i are calculated assuming the medium to be a homogeneous elastic half-space, Equation 2 yields

$$\rho_t' = \frac{2 - 3v'}{2(1 - v')^2} \, \rho_t \tag{3}$$

The factor $(2 - 3v')/2(1 - v')^2$ is plotted against v' in *Fig. 15*. It is seen that ρ_t' is greater than ρ_t and the maximum difference is less than 12·5% for the entire range of v'.

For nonhomogeneous media the General Reporter of the session for heavily overconsolidated clays has outlined a procedure for calculating immediate settlements. This procedure assumes a stress distribution corresponding to Boussinesq's equations, but uses variable modulus with depth for calculating deformations in the soil medium.

A similar procedure for calculating consolidation settlements would be to use initial pore pressure values corresponding to a homogeneous elastic half-space and using variable coefficients m_v for calculating deformations in the soil medium. Then the consolidation settlement would be

$$\rho_c = \sum u_i m_v \, \Delta H \tag{4}$$

This procedure will be restricted to overconsolidated clays.

778

Some useful expressions for calculating initial pore pressures in a homogeneous elastic half-space due to surface loads over flexible areas are given in Equations 5, 6 and 7. At any point due to concentrated load Q

$$u_i = \frac{Qz}{2\pi} \frac{1}{(r^2 + z^2)^{3/2}}$$ (5)

in which r = radial distance of the point from the load and z = depth of the point below the surface.

Below the centre of a circular load of intensity of q

$$u_i = q\left(1 - \frac{z}{\sqrt{(a^2 + z^2)}}\right)$$ (6)

in which a = radius of the circular load and z = depth of the point below the surface.

Below the corner of a rectangular load of intensity q

$$u_i = \frac{q}{2\pi} \tan^{-1}\left(\frac{m^2 n^2}{m^2 + n^2 + 1}\right)$$ (7)

in which $m = B/Z$, $n = L/Z$, B = breadth of the loaded area, L = length of the loaded area and z = depth of the point below the surface.

It is not suggested that the computed overall settlements should be reduced according to Equation 3 since the overestimate would compensate to some extent for errors introduced by estimating initial stresses and pore pressure from the homogeneous half-space model.

D. J. NAYLOR, University College of Swansea. I wish to make five general points concerning the role of the finite element method in soil structure interaction. I have in mind applications where the finite element representation of the soil dominates, the structure being represented in a simplified form.

Direct or indirect application. The f.e.m. is probably most frequently applied directly to particular problems which can be represented in 2 dimensions, either by plane strain or axi-symmetric meshes. I think there is a growing case for indirect applications where the f.e.m. is used to produce design charts. The General Reporters have drawn attention to the work by Carrier and Christian (1973) in this connection.

Another example is found in the paper by Lefebre, Duncan and Wilson (1973) who carried out both 2- and 3-dimensional analyses of a layered fill dam in a V-shaped valley and present results making it possible for the engineer to assess the error inherent in a 2-dimensional finite element approximation.

There is scope for work on these lines in the context of foundations. For example, it could be valuable in assessing the error inherent in the assumption of axi-symmetry used by Breth in his Paper, p. 141, and in the Paper by Hooper and myself p. 394.

Sensitivity studies. The f.e.m. is well suited to estimating the output error (displacements, stresses) associated with input error. Professor Gibson in his recent Rankine lecture (Gibson, 1974) gives an extreme example of sensitivity. The settlement at the surface of a material with E proportional to depth and subject to distributed loading changes from finite to infinite if Poisson's ratio is reduced marginally below 0.5.

Another example concerns the distribution of the applied load. A proposed design for a pile supported bridge pier consisted of a centre row of vertical piles and outer rows of raking piles. The centre lines of all three rows intersected above the pier base. The pile batter was chosen so that the intersection point was at the same level as the resultant horizontal design load. The design was very rigid under these conditions. However, a small change in the assumed position of the resultant leads to very large displacements and rotations of the pier. This design was not adopted.

Lastly, the sensitivity to changes in geometry can readily be assessed by the f.e.m.

Sophisticated models. Sufficient is now known about soil behaviour to justify in some situations a more elaborate model than linear elastic isotropic. The following types can currently be used with the f.e.m.: (1) transversely isotropic linear elastic (5 constants), (2) isotropic variable elastic (4 or 5 constants), (3) elastic-plastic critical state (2 or more constants), and (4) no tension (2 constants).

Use of these models becomes justifiable when significant yielding occurs in part of the soil mass (e.g. at the edge of a rigid footing).

Sequential analyses. This technique provides a means of modelling the construction sequence. It has particular potential where major geometric and/or stiffness changes occur during construction. An example is given in the Paper by Hooper and myself (V/9, p. 394).

Back analyses. In my opinion soil tests are unsatisfactory as a means of obtaining the constants defining the soil stiffness. It is much better to find them by back analysis of an instrumented prototype, either one similar to the structure under consideration or the structure itself in its early construction stages. In the latter case provision for later design change must be made.

Carrier, W. D. III, and Christian, J. T. (1973). 'Rigid Circular Plate Resting on a Non-homogeneous Elastic Half-space', *Geotechnique*, Vol. 23, 67–84

Gibson, R. E. (1974). 'The 14th Rankine Lecture—the Analytical Method in Soil Mechanics', *Geotechnique*, Vol. 24, No. 2, 115–140

Lefebre, G., Duncan, J. M. and Wilson, E. L. (1973). 'Three-dimensional Finite Element Analyses of Dams', *Am. Soc. civ. Engrs*, Vol. 99, SM7, 495–507

J. A. HOOPER, Ove Arup and Partners.   In the effective stress analysis of a piled-raft foundation described in Paper V/9 (p. 394), it was assumed that the entire uplift force generated by the removal of soil was mobilized before the raft was cast. This led to some rather unexpected results; particularly those concerning the predicted increases in pore pressure near the base of the raft during consolidation.

An alternative approach, which represents a second bound to the problem, is to assume that the full uplift force is mobilized after the casting of the raft, and the results of such an analysis are given in *Fig. 16*. The general effect is that pile loads and total raft settlements are decreased and that raft contact pressures are increased, relative to the corresponding computed values shown in *Fig. 2*, p. 399. Furthermore, the effects of soil consolidation on the load

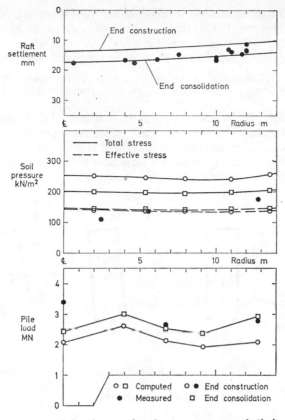

*Fig. 16. Radial distribution of settlement, pressure and pile load
(alternative treatment of uplift forces)*

sharing at the base of the raft are quite different in the two analyses. It is evident from *Fig. 3*, for example, that during consolidation the computed pore pressures decrease in the vicinity of the raft, which in turn gives rise to decreases in total pressure on the raft base and increases in pile loads. The

Table 4. PERCENTAGE DISTRIBUTION OF SUPPORT AT BASE OF RAFT (alternative treatment of uplift forces)

	End of construction	End of consolidation
Pore water	23	12
Soil skeleton	30	30
Piles	47	58

resulting distribution of computed vertical support at the base of the raft is given in *Table 4*.

Comparison of measured and computed values of load and raft settlement (*Figs. 2 and 3*) indicates that, during the construction of the foundation, considerable uplift forces were mobilized after the raft was cast. In this

781

connection, it may be recalled from an earlier total stress analysis of the same foundation (Hooper, 1973a) that a mobilized uplift force of approximately 70% of the net weight of material removed on completion of the raft was found to give the closest agreement between measured and computed results.

Hooper, J. A. (1973a). 'Observations on the Behaviour of a Piled-Raft Foundation on London Clay', *Proc. Inst. civ. Engrs*, Vol. 55, Pt. 2, December, 855–877

E. SCHULTZE, Technische Hochschule, Aachen.   In the case of tall and slender structures such as smoke-stacks and staircase towers the danger exists that due to a casual inclination an additional moment is produced enlarging the inclination causing an additional increment of the moment (*Fig. 17*). For this a safety factor against instability exists (Puwein, 1955; Schultze and

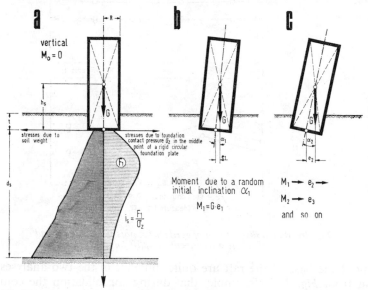

Fig. 17. *Effect of increasing inclination due to displacement of centre of gravity*

Muhs, 1967; Habib and Puyo, 1970; Schultze, 1972). The safety factor against instability is F_s, Equation 1 of *Fig. 18*.

In the case of $F_s > 1$, the inclination approaches a final value, possibly very slowly, otherwise there is a collapse. In this case the behaviour of the construction has a similarity with creep. The nature of the time-course of such an inclination has therefore been investigated in the two cases: creep and no creep. For this equations have been derived (Equation 2 of *Fig. 18*).

Settlement functions for primary settlements with an ultimate value, are hyperbolic, according to Christow (1968), i.e.

$$f(t) = \frac{bt}{a + bt} \quad \text{(no creep)}$$

or a more exact function according to the laws of clay mechanics in the form of a series (Terzaghi and Fröhlich, 1936), *Fig. 19*, which in the simplest case can be terminated after the first term

$$f(t) = 1 - e^{-\beta_1 t} \quad \text{(no creep)}$$

For a creep function without an ultimate value of settlement one can use (*Fig. 20*)

$$f(t) = \ln t$$

By means of the following equations (Schultze, 1972) the time-settlement curve for an increase of settlement with time can be obtained for an arbitrary

$$F_s = \frac{R^3 \cdot E}{G \cdot h_s \cdot i} \qquad \text{(1)}$$

i = inclination coefficient due to Fischer (1965, p. 39), depening on ν, d_s/R

$$s = \frac{i_s \cdot \bar{\sigma}_0}{E} \cdot f(t) = K_1 \cdot \bar{\sigma}_0 \cdot f(t) \qquad \text{(2)}$$

i_s = settlement coefficient of a circular plate depening on d_s/R, ν, observed point and the contact pressure distribution, practically equal to the stress area under the observed point due to a unit loading.

$\bar{\sigma}_0$ = uniformly distributed contact pressure, constant with time

E = elastic modulus, modulus of compressibility E_s

$f(t)$ = function for the dependence of the settlement on time

$$\bar{\sigma}_0(\vartheta,\varrho) = \frac{M}{R^3 \cdot \pi} \cdot m \quad , \quad \varrho = r/R \qquad \text{(3)}$$

m = coefficient for the contact pressure distribution

$M = \dfrac{R^3 \cdot \pi}{m_r} \cdot \bar{\sigma}_{or} = K_2 \cdot \bar{\sigma}_{or}$

m_r = m at the edge by $\varrho = 1$

$\tan \alpha = \dfrac{M \cdot i}{R^3 \cdot E} = K_3 \cdot M$

$\sin \alpha = \Delta s / R$

$\sin \alpha = \dfrac{i_s \cdot m_r}{R^4 \cdot E} \cdot M \cdot f(t) = K_4 \cdot M \cdot f(t)$

Increase of the settlement due to an increase of the contact pressure in the time $\Delta t = t_2 - t_1$

$ds_{21} = K_5 \cdot d\bar{\sigma}_{01} \cdot f(t_2 - t_1)$

$d \sin \alpha_{21} = K_4 \cdot d M_1 \cdot f(t_2 - t_1)$

$\sin \alpha_2 = K_4 \cdot f(t_2) \cdot M_0 + \displaystyle\int_{t_1=0}^{t_1=t_2} d \sin \alpha_{21}$

$\qquad = K_4 \left[f(t_2) M_0 + \displaystyle\int_{t_1=0}^{t_1=t_2} d M_1 f(t_2 - t_1) \right]$

$\dfrac{M_2 - M_0}{\sin \alpha_2} = \dfrac{M_1 - M_0}{\sin \alpha_1} = G \cdot h_s$

$$M_2 - M_0 = \frac{G \cdot h_s \cdot i_s \cdot m_r}{R^4 \cdot E \cdot \pi} \left[f(t_2) M_0 + \int_{t_1=0}^{t_1=t_2} d M_1 \cdot f(t_2 - t_1) \right]$$

Fig. 18. Equations for time–settlement moment curves

function even in the case of a self-produced inclination. For a circular area Equation 3 of *Fig. 18*.

By means of this equation the increase of the moments and thus of the inclinations can be calculated step by step beginning with $M_1 = M_0$ at time $t = 0$. M_1 always results as M_2 from the preceding step.

According to the function used one obtains time-inclination curves with (Equation 3, *Fig. 19*) or without (Equation 4, *Fig. 20*) ultimate value.

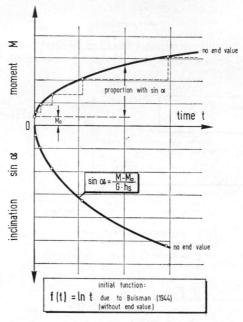

Fig. 19. *Stability of exclusively primary settlements (consolidation) and n ⩾ 1*

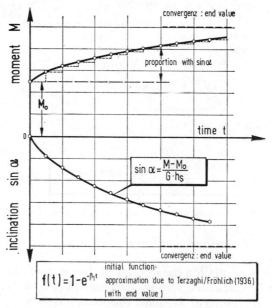

Fig. 20. *Instability due to secondary settlements (creep)*

Christow (1968). 'Beitrag zur praktischen Setzungsberechnung und Auswertung von Zeitsetzungslinien' (Contribution for Practical Settlement Calculation and Evaluation of Time-Settlement Curves), *Proc. Donau Eur. Conf.* Wien, 6

Fischer (1965). *Beispiele zur Bodenmechanik. Ansätze mit Formeln, Tafeln und Schaubildern* (Examples in Soil Mechanics. Statements with formulas, tables and figures). Berlin/München, Wilh. Ernst & Son

Habib and Puyo (1970). 'Stabilité des fondations des constructions de grande hauteur', *Ann. Inst. Techn. Batiment et Trav. Publs*, Paris 275, Serie: *Sols et Fondations* 77, 118

Puwein (1955). 'Baugrundknickung' (Collapse of foundation soil), *Öst. Bauz.* Vol. 10, 116

Schultze (1972). 'Das Problem des Turmes von Pisa' (The problem of the tower of Pisa), *Mededelingen van de Koninklijke Academie voor Wetenschappen, Letteren en Schone Kunsten van Belgie, Jaargang* XXXIV, No. 7, Brüssel, p. 12

Schultze and Muhs (1967). *Bodenuntersuchungen für Ingenieurbauten* (Soil Investigations for Engineering Structures), Springer, 677

Terzaghi and Fröhlich (1936). *Theorie der Setzung von Tonschichten* (Theory of the Settlement of Clay Layers), Leipzig/Wien, Deuticke

M. B. JAMIOLKOWSKI, Technica University of Turin. Some participants have posed questions on the choice and use of a simple linear elastic model in soil interaction studies. Up to recent years problems of this type have been generally solved by assuming that the deformability of the soil in contact with the foundation could be approximated by means of the well known Boussinesq or Winkler models. Unfortunately those two approaches, which cannot be considered an equivalent or alternative, give the designer quite different results with particular reference to the following aspects of the problem:

(1) The surface deflection outside the loaded area is overestimated or neglected according to whether Boussinesq or Winkler theory is used; it follows that, when investigating the reciprocal influence between adjacent foundations, both models cannot be used.

(2) The observed shape of the surface deflection beneath the foundation is always intermediate between those predicted by the Winkler and Boussinesq models.

(3) The observed variation of the centre settlement with increasing dimensions of the loaded area gives patterns which are intermediate between those predicted using Winkler and Boussinesq models: see *Fig. 21* and Basaara (1967), Bjerrum and Eggestad (1963) and Terzaghi and Peck (1967).

(4) The use of the Boussinesq model in soil-foundation interaction studies generally overestimates the positive bending moments and underestimates the negative bending moments; this is particularly true in the case of rigid foundations. The opposite statement applies when the Winkler model is used.

In the writer's opinion the best approach to soil-foundation interaction problems, when one remains within the limits of linear elastic behaviour, is represented by the assumption that the soil behaves as an isotropic heterogeneous elastic half-space whose modulus of elasticity increases with depth

$$E(z) = E_0 + E_n \cdot z^m$$

where $E(z)$ = modulus of elasticity at the depth z below the upper limit of the halfspace, E_0 = modulus of elasticity at $z = 0$, E_n = coefficient which gives the variation of the elasticity modulus with depth, and m = modulus exponent.

The recent works by Gibson (1967), Gibson and Sills (1971), Awajobi and Gibson (1973), Brown and Gibson (1972), Burland *et al.* (1973), Carrier and Christian (1973), Klein (1971), Durajev and Klein (1971). Belloni and

Jamiolkowski (1973) on the stresses and deformations in the isotropic-heterogeneous elastic half-space have shown very clearly that both Boussinesq and Winkler models represent two extreme limits of the possible behaviour of deep deposits of relatively uniform soils.

Also the use of the heterogeneous half-space model in soil-foundation, interaction problems confirmed that (1) stresses and deflections of foundation members, when calculated by using heterogeneous halfspace model are closer to the observed ones, and (2) the results obtained by using the above mentioned model are always intermediate between those given by the Winkler and Boussinesq theories.

A good approximation of the surface deflection of the heterogeneous elastic

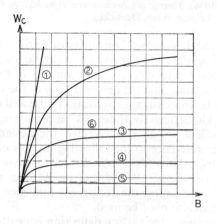

① BOUSSINESQ HOMOGENEOUS ISOTROPIC ELASTIC
   HALF SPACE B → ∞

②÷⑤ HETEROGENEOUS ISOTROPIC ELASTIC HALF SPACE
   WITH DECREASING VALUES OF B [$B_5 < B_2$], OR
   REPNIKOV MODEL WITH DIFFERENT E_s/k^* RATIOS

⑥ WINKLER SPRING MODEL

(*) E_s = MODULUS OF ELASTICITY ⎫ PARAMETERS OF
    k = WINKLER SPRING CONSTANT ⎰ THE REPNIKOV
                                     MODEL

Fig. 21. Influence of foundation width B on settlement W_c at centre of loaded area.

half-space can be obtained by using the Repnikov model in which the Boussinesq elastic isotropic and homogeneous half-space is reinforced internally with the Winkler springs. The application of this model in solving soil-structure interaction studies, made by Repnikov (1969), Repnikov and Rachimov (1971), Schultze (1970), and Barla *et al.* (1974), proved that the combination in parallel of Winkler and Boussinesq models allows one to reproduce more realistically the true soil surface deflection obtaining results which are in agreement with those predicted on the basis of the heterogeneous half-space theory.

In view of the above comments the following criteria can be followed in the choice of a linear-elastic model for a simplified calculation of the differential settlements and when solving problems in the field of soil-structure interaction:

786

(1) For a degree of heterogeneity $\beta \geq 10$ the Boussinesq elastic isotropic homogeneous half-space is to be adopted.

(2) When $\beta \leq 0.01$ the use of the Winkler spring model can represent a good approximation of surface deflection under the foundations.

(3) In all other cases, when $0.01 < \beta < 10$, the heterogeneous isotropic elastic half-space or the equivalent Repnikov model (only with respect to the surface deflection) have to be used.

β = coefficient of the degree of the heterogeneity defined for a half-space whose modulus increases linearly with depth according to the following formula and B = width of the rectangular area or diameter of the circular loaded area.

$$\beta = \frac{E_0}{E_n B}$$

Bazaara, A. R. S. (1967). 'Use of the S.P.T. test for Estimating Settlements of Shallow Foundations on Sand', Ph.D. Thesis, University of Illinois

Bjerrum, L. and Eggestad, A. (1963). 'Interpretation of Loading Tests on Sand', *Proc. Eur. Conf. Soil Mech. and Found. Engng*, Wiesbaden, Vol. 1, 199–203

Terzaghi, K. and Peck, R. (1967). *Soil Mechanics in Engineering Practice*, 2nd Edn, Wiley

Gibson, R. E. (1967). 'Some Results Concerning Displacements and Stresses in a Non-homogeneous Elastic Half-space', *Geotechnique*, Vol. XVII, No. 4, 58–67

Gibson, R. E. and Sills, G. C. (1971). 'Some Results Concerning the Plane Deformation of a Non-homogeneous Elastic Half-space', *Proc. Roscoe Memorial Symp. Stress–Strain behaviour of Soils*, G. T. Foulis & Co. Ltd, 564–572

Awajobi, A. O. and Gibson, R. E. (1973). 'Plane Strain and Axially Symmetric Problems of a Linearly Non-homogeneous Elastic Half-space', *Q. Jl. Mech. Appl. Math.*

Brown, P. T. and Gibson, R. E. (1972). 'Surface Settlement of a Deep Elastic Stratum Whose Modulus Increases Linearly With Depth', *Can. Geotech. J.*, Vol. 9, No. 4, 467–476

Burland, J. B., Gibson, R. E. and Sills, G. C. (1973). 'A Field and Theoretical Study of the Influence of Non-homogeneity on settlements', *Proc. VIII Int. Symp. Soil Mech. Found. Engng*, Moscow. Vol. 1.3 39–46

Carrier, D. W. and Christian, J. T. (1973). 'Rigid Circular Plate Resting on Non-homogeneous Elastic half-space', *Geotechnique*, Vol. XXIII, No. 1, 67–84

Klein, G. K. (1956). 'Study of the Influence of Heterogeneity, Discontinuity of Deformations, and Soil Properties on Soil-Structure Interaction' (in Russian), *Sbornik Trudov Moskovskij Iniz. Strait. Instit.*, No. 14

Klein, G. K. and Durajev, A. E. (1971). 'Influence of the Soil Modulus Increase With Depth on the Behaviour of Beams Resting on Soil' (in Russian), *Hydraulic Constructions*, No. 7

Belloni, L. and Jamiolkowski, M. (1973). 'The Heterogeneous Elastic Half-space as a Model for the Foundation Soils' (in Italian), *Rivista Italiana di Geotecnica*, No. 4

Schultze, E. (1970). 'Die Kombination von Bettungszahlung-Steifenzahl-fahren', *Mitt. V.G.B. Tech. Hochschule Aachen*, Heft 48

Repnikov, L. N. (1969). 'Calculations of Beams Resting on a Soil with Deformation Properties Intermediate Between Those Predicted on the Basis of the Winkler and the Boussinesq Models (in Russian), *Osnov. Fund. i Mekanika Gruntov*, No. 6

Repnikov, L. N. and Rachimov, S. (1971). 'Method for the Calculation of Deformations of a Medium Having Characteristics Intermediate Between Those Predicted on the Basis of the Winkler and the Boussinesq Models' (in Russian), *Osnov. Fund. i Mekanika Gruntov*, No. 4

Barla, G., Belloni, L., Jamiolkowski, M. and Pasqualini, E. (1974). 'On the Behaviour of Circular Plates Resting on Repnikov Foundation' (in Italian), *Rivista Italiana di Geotecnica*, No. 1.

Storage Tanks

L. S. WILLS, British Petroleum Co. Ltd.   A number of papers have been presented concerning differential settlement of oil storage tanks and the effect this has on the tank structure. With the advent of large diameter tanks and higher steel stresses it is of the utmost importance to decide whether previous settlement criteria adopted for the smaller diameter tanks can be scaled up. It has rightly been pointed out in a number of papers that differential settlement relative to the tilt plane, that is distortional settlement, causes horizontal distortion of the shell of floating roof tanks. In the case of the smaller diameter tanks when distortion reaches a limit where the floating roof malfunctions it is relatively simple to packup the tank shell. This takes on much greater significance with large diameter tanks and the practice of repacking around the shell and say 2 to 3 m under the base plates must be considered very carefully otherwise increased stresses can be induced in the area of the annular plates and the bottom to shell plate weld where stresses are high under operating conditions.

In view of these worries, my Company commissioned Imperial College to undertake a series of tests on a model tank which was subjected to varying patterns of differential settlement. This was a scaled down model of a floating roof tank built at Rotterdam, which had been fitted previously with strain gauges and where stresses under varying loading conditions were known. The scale factors of the model were chosen to reproduce the strains in the full sized tank. The tests on the model tank have only been completed recently and the results are being analysed and compared with settlement of various operating tanks. To date the results are encouraging.

In the model various prescribed displacements were applied to the foundation under the base to simulate distortional settlements from the tilt plane likely to be experienced in practice. These settlements were in the form of a sine curve of wave length $2l$ which were denoted by parameters n and Δs where n equals circumference/$2l$ and Δs equals twice the amplitude of the sine curve ($\Delta s =$ crest to trough height).

The tests proved that, at the points of maximum distortional settlement of the base from the tilt plane, the deflection at the top of the shell was directly proportional to the settlement up to different limits for different values of n. For n equals 2, 3 and 4 these limits were not reached in the tests but for n equals 6, 8 and 12 they were reached at progressively lower values of Δs. Non linearity indicated bridging of the tank shell and was associated with increases in strains in the tank shell which were also measured.

From the analysis of the results I consider that criteria can be developed for tolerable vertical distortion of the shell from the tilt plane. These criteria will take account of ovality and possibly the order of stresses. Our analysis is showing that these relationships depend on n, height of tank, diameter/height ratio and construction tolerances of which most of the papers presented do not take fully into account. Our work indicates that no simple solution involving only Δs or relating Δs only to arc length is adequate.

Our work is continuing and I am hopeful that this will lead to a more accurate method of determining permissible distortional settlements. If this is possible it will be a step in the right direction but still only half the story. The prediction of possible settlements due to the subsoils is all important to the designer when deciding on the maximum tank sizes and any necessary ground treatment. De Beer (1969) has pointed out the importance of the interaction between structure and soil and between soil variations under the area of the tank. Our analysis of the structure indicates that it is over short

788

arc lengths that soil/structure interaction is important. Soils at depth will effect long arc lengths and therefore variations are of lesser importance. It follows that as well as studying settlements to the full depth of loading it is important to assess the proportion attributable to the shallow soils and to determine from this whether it is advisable to carry out soil improvement in this area.

Although it will still be difficult to design economically in marginal cases we feel we can now go a long way in defining permissible distortional settlements from the tilt plane, and predict when remedial action will be necessary to avoid malfunction of the roof.

De Beer, E. (1969). 'Foundation Problems of Petroleum Tanks', *Annales de l'Institut Belge du Petrole*, Vol. 6, 25–40

H. A. CRAIG, British Petroleum Co. Ltd.   Greenwood and Belloni *et al.* have referred to a paper by de Beer (1969). Greenwood includes in his table of records (p. 366) the settlements for de Beer's Tank B. *Figure 22(b)* is the

(a)

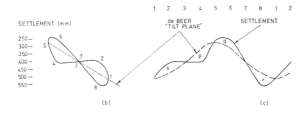

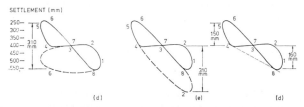

Fig. 22. de Beer (1969), Tank B settlement records, D = 76.2 cm, H = 16.46 cm: Settlements at water height 13.4 m

total settlement, as described by de Beer, measured at the points around the shell shown in *Fig. 22(a)*. *Figure 22(c)* is the development of the settlement (full line) and of the 'tilt plane' assumed by de Beer (dashed line). De Beer considered the possible effects on the structure of the displacements from the 'tilt plane' (vertical distortion) between points 3, 4 and 5, denoted by area (*p*) in *Fig. 22(c)*.

From *Fig. 22(b)* it is evident that this tank has folded on either side of a diameter through points 4 and 8. It is also clear that tilt planes other than de Beer's can be assumed, for example as indicated in *Figs. 22(d), (e) and (f)*. In *Fig. 22(f)* maximum vertical distortion of the base is noted to be at two points diametrically opposed, each equal to 150 mm and 160 mm and the distortion at each taking place over $\frac{1}{2}$ the perimeter. The effects of these two displacements on the structure should be added to give the total effect which is equal to the effect of a single displacement over $\frac{1}{2}$ the perimeter of the pattern and magnitude shown in *Fig. 22(d) and (e)*.

The 'tilt plane' is thus only an arbitrary plane which may be chosen for convenience to allow definition of the vertical distortion. Unless particular planes, such as in *Fig. 22(d) or (e)*, are used, the effects of vertical distortions at more than one point on the shell must be added together. Criteria of the type proposed by Greenwood, by Sullivan and Nowicki, and by Belloni *et al.* will thus only be valid in particular circumstances, since they appear to be based on consideration of only one section of the shell perimeter. Other factors to be considered have been listed by Wills in the discussion to this session. *Table 5* shows a comparison of the true distortion of de Beer's 'Tank B' with some of the criteria proposed.

Table 5

Source:	Suggested maximum criterion		'Tank B'
Sullivan and Nowicki	Distortion	= 45 mm	310 mm
Greenwood	*t*	= 60 mm	310 mm
Belloni *et al.*	Δs/Arc length	= 0.22%	0.52%

'Tank B' tolerated large vertical distortion because of the nature of the settlement pattern ($n = 2$). For other patterns, e.g. $n = 4$, the critical value would be less.

D. A. GREENWOOD, Cementation Ltd. All the papers in this Session concerning tanks correlate observed settlements with damage to the tank, lack of damage, or action taken to prevent potential damage triggered by some pre-determined concept of limits of tolerable differential settlement. There is a remarkable similarity in their conclusions (derived from large numbers of tank sites) about the magnitude of permissible deviation from the mean tilted plane at the base of the tank shell: this despite necessarily crude assessments. It is equally clear that action to forestall potential damage has sometimes been taken unnecessarily early. Specifications of tolerable differentials vary widely. Perhaps naturally enough those from tank manufacturers tend to be tightest but the major oil companies have long known that much wider tolerance is possible. The confused situation reflects the difficulties of stress evaluation for tanks.

The contribution of the two previous speakers is therefore extremely valuable because, in contrast to the papers submitted for this session they

790

attempt to define settlement limits from consideration of strains developed in the tank shell, and furthermore define how they may vary for different modes of deformation. Until their study has been fully developed it is only possible to make simple observations on its potential influence.

As it appears that shell deformation with many nodes is more severe on the tank than with few nodes the role of the granular foundation pad is important. It is implicit that design details should allow local plastic deformation of the pad surface beneath the sole plate of the shell. If many nodes tend to develop the related short arcs of the sole plate will then be stiff enough to cause local bearing failure in the sand, allowing the tank to relax to a condition of minimum stress, i.e. fewer nodes. Appropriate details would include a relatively narrow sole plate and a topping of sand below the asphalted surface of the pad as in Sullivan and Nowicki's *Fig. 3* on p. 422.

It follows that any correction by jacking numerous points of the shell is a critical operation. Also concrete or other semi-rigid ring beams are perhaps best avoided unless they are on a rigid foundation since when they crack they tend to produce sharp angular distortions of the shell over the cracks.

With correct pad detail the main concern is with deformations of the underlying soil. As many tank farms are built on recent filling this is often the most difficult problem, especially where fill covers former water courses cut through stiff crusts to a lower soft alluvium or buries disused embankments. For typical tank sizes both conditions occur at a scale sufficient to cause severe two node deformation—sagging or hogging. It is perhaps comforting to know from Mr. Craig's discussion that greater tolerance may well be associated with two node deformation.

Finally, a comment on bottom plate distortion. When tank pads rest on soft soils the radially outward plastic movement of the soil from below the tank perimeter always results in maximum settlements occurring other than at the centre. Measurement simply at the centre point can therefore serve only as an index giving a useful practical guide for comparing similar situations. However, data presented to this conference confirm that where tanks rest on sandy soils of thickness about $\frac{1}{10}$th tank diameter or more the bottom 'dishes' to give maximum settlement at the centre, even where soft soils exist below.

A. LEGGATT, Nachshen, Crofts and Leggatt. Settlement is a difficult enough problem for people such as ourselves here at this conference, but to many outsiders it is akin to witchcraft or an Act of God—depending on which side of the insurance policy you are. From my own personal experience I can quote two extremes of the spectrum of appreciation of settlement.

One was an architect with whom I had a somewhat short professional relationship, who when I pointed out that some settlement might occur in the building which we were designing, said 'I have never had settlement in any of my buildings so far and I don't intend to start now'. At the other extreme was a factory owner near Stoke-on-Trent in a mining subsidence area who said almost blithely that he would not worry about broken windows or tiles falling off—'you get to live with that sort of thing round here—just make sure that the building does not collapse'. Regrettably the average lies much nearer the case of the architect.

Often the conception of the design ignores settlement, the engineer finds later that the soil is not exactly solid rock and is given the challenge to design a foundation which will cause little or no settlement. It is at this point that the engineer has a special responsibility. It is sometimes thought that the difficult challenge is to solve a stringent technical requirement. It is not! The real challenge is to help ones client or employer towards the correct total solution. Design concepts often get solidified far too early and it is difficult then for the engineer to re-open them, especially in the esoteric field of foundation design. It is much easier to let things be and press on with the solution of a difficult technical problem which in itself can often be a lot of fun and usually remunerative. High costs of such foundations can often be sanctioned easily by the formula 'abnormal soil conditions'. This is engineering for engineering's sake. Technology will get—is getting—a bad name in this respect.

Engineers have a duty to look beyond the solution of the immediate problem presented to them and Sir John Baker's opening remarks about plastic structural design putting settlement engineers out of business, although intended light-heartedly, deserves to be given some deep contemplation. To design a foundation to minimise settlement may be clever, but it could be a dereliction of the engineers true function.

Harking back to my Paper (p. 3) right at the very beginning of the conference, it could be said that I have not been practising what I am now preaching. The foundations which we provided for the Oil Tanks at Fawley Refinery cost at least as much as the tanks themselves and if one adds in the former and the less successful attempts by others, together with the consequent costs, the final price paid for the foundations to these unfortunate tanks amounts to between 5 and 10 times their actual cost. Maybe we should try to design tanks which will take differential settlement in their stride.

Design and construction

I. A. MACLEOD, Paisley College of Technology and G. S. LITTLEJOHN, University of Aberdeen.   We would first of all like to talk about the activities of a special study group of the Institution of Structural Engineers on Structure/Soil interaction. This study group was set up some three years ago by Mr. Sam Thorburn of Thorburn and Partners, Glasgow, and has received encouraging support from people all over Britain. The result of the work of this group is a State of the Art report on Structure/Soil Interaction which is now available from the Institution of Structural Engineers.

The reaction of structural engineers to the question of structure/soil interaction has been interesting, since some engineers do not feel that structure/soil interaction is a problem, whilst others consider that it is most important. Of the order of 60 people were involved in writing the report, all of whom felt that structure/soil interaction is a problem and that something ought to be done about it. The question is—what is to be done? Before trying to briefly answer this question, we would like to discuss what constitutes a structure/soil interaction problem. A structure/soil interaction problem is one where it is best not to consider the behaviour of the structure, the foundation and the soil independently, but where to achieve a realistic design, the interaction between them should be considered. A very good example of this is given by Crouser, Schuster and Sack (p. 344) where the stiffness of silos is shown to be important in assessing the distribution of load below the foundation. Similar examples occur in shear wall buildings supported on rafts. These are two examples of problems where the degree of interaction between the structure and the foundation is high. There is of course a wide spectrum of interaction in structural problems, but it is most important that the critical problems be identified. To achieve this we offer two suggestions.

Firstly, structural engineers, foundation engineers and soils engineers should get together more to discuss their mutual problems. We feel that Thorburn's study group has been a step in this direction, and although this conference has been biased towards the geotechnical side, many of the papers do deal with interaction problems. We would like to draw your attention to a conference on the Performance of Building Structures, sponsored by the University of Glasgow, Paisley College of Technology, and the Institution of Structural Engineers to be held in Glasgow in April 1976. It is intended that this conference will be complementary to the present one in that more emphasis will be given to the structural side than to the geotechnical side.

Table 6(a). SUPERSTRUCTURE

Type of structure	Material	Cladding*	Geometry
Frame	Steel ⎰braced ⎱unbraced Concrete Timber	Type 1 Type 2 Type 3	varies
Wall	Brickwork Blockwork Precast Concrete *In-situ* Concrete		varies

*CLADDING TYPES
Type 1—minimal structural action
Type 2—cladding would not normally be considered in strength calculations but could have significant stiffening effect against settlement.
Type 3—would normally be considered in strength calculations.

Further information can be obtained from Dr. D. Green, University of Glasgow.

Secondly, although a lot of the information presented at this conference forms very valuable case records, much more monitoring work is required on the real behaviour of building and soils. We feel that a significant improvement in design could result from a concentrated programme of monitoring, and in this connection we support the recommendations by the General Reporters concerning types of measurements, with particular regard to the concept of critical tensile strain, the use of relative deflections and the important distinction which should be made between sag and hog when assessing deformation thresholds in relation to degree of damage. These suggestions, when adopted, will greatly facilitate comparisons of case records and in time lead to a better understanding of the subjects. However there is also a need to classify building types and damage.

The special study group has some tentative suggestions for simple categories of superstructure and substructure, indicated by the *Tables 6(a)*, (*b*) *and* (*c*).

With regard to a better classification of damage *Table 6(d)* is proposed for consideration. It is noteworthy that crack widths (*w*) less than 0.1 mm are

Table 6(b). SUPERSTRUCTURE

Type of foundation	Rigidity
strip/pad thin raft (thickness < 450 mm)	flexible
deep beam/grillage/box deep raft (thickness > 600 mm) piled foundation† a) short piles (D/B < 0.5) b) long piles (D/B > 1.0)	stiff

† Depth of 'reinforced' block of soil (D) should be related to dimensions of the building, e.g. width (B)

considered negligible which is in good agreement with the General Reporters who judge that 0.1 mm is the width of crack likely to be first discerned by a casual observer. Pryke (p. 406) states that cracks seem to worry people when $w = 3$ mm, which again fits in quite well and would be classified as 'appreciable'.

In general it is recommended that simple settlement points referred to a stable datum should be installed in all buildings as a matter of construction procedure. The cost of this is small, and if trouble arises with the structure the settlement can be measured and a valuable case history is thus created.

Table 6 (c). SUPERSTRUCTURE INCLUDING ASPECT RATIO

Main type:		Frame						Wall		
Material		Steel			Concrete		Timber	Brick	Concrete	
Cladding		1	2	3	1	2	3		Block	In-situ	Precast
Aspect Ratio (height/width)	> 10										
	10–5										
	5–2										
	2–1										
	1–0.5										
	0.5–0.2										
	0.2–0.1										
	< 0.1										

Table 6(d). CLASSIFICATION OF STRUCTURAL DAMAGE

Classification of damage	Description of typical damage
Negligible	Hair line cracks in plaster, crack width $\not> 0.1$ mm
Very slight	Hair line cracks in plaster, perhaps isolated fracture in building not visible outside, crack width $\not> 0.5$ mm
Slight	Several slight fractures, width $\not> 2$ mm, showing inside building, doors and windows may stick slightly, repairs to decoration probably necessary.
Appreciable	Slight fractures showing on outside of building (or one main fracture, width $\not> 6$ mm), doors and windows sticking
Severe	Open fractures, width $\not> 15$ mm, requiring rebonding and allowing weather into structure, window and door frames distorted, walls leaning or bulging noticeably, some loss of bearing in beams
Very severe	Open fractures, width > 15 mm, and repairs involving partial or complete rebuilding, roof and floor beams lose bearing, walls lean badly and require to be shored up

W. H. WARD, Building Research Station. Leggatt had said earlier that building design conception never involved the ground. Traditionally this occurred with buildings in mining subsidence areas, and often led to the use of extremely expensive foundations—sometimes heavier than the building and its contents—to stiffen and strengthen an unsuitable structure.

On one rare and exciting occasion in the mid 1950s when three architects, Donald Gibson, Dan Lacey and Henry Swain came to my office, I had the opportunity to introduce the basic and important foundation requirements into a building design from its conception. These architects had recently joined the Nottinghamshire County Council to provide new schools in a region with a large incidence of mining subsidence. My proposals were novel and radical, and I was fortunate in finding a group of sympathetic and enterprising architects who were prepared to put so much effort into the development of what is now called the CLASP system.

In a particular mining area the transient strains and distortions occurring at the ground surface can be defined fairly closely, as a result of many measurements by mining engineers of cause and effect. The building problem is therefore one of accepting these basic data and designing a structure to cope with them. The design principles we followed have been given by Lacey and Swain (1957), but as they are not widely known to foundation engineers it is worth repeating them in outline. The structure resists only the horizontal ground strains, particularly the tensile ones which are mainly responsible for damage to the traditional brick and masonry buildings (An Interdepartmental Committee, 1951), but is allowed to follow the ground distortions in the vertical planes. The first requirement is met by making the floors horizontal diaphragms, which maintain their dimensions and shape in plan, but have frequent cross joints that allow them to bend in vertical planes. The principal diaphragm is the ground floor slab which forms a surface foundation. It consists of 130 mm of concrete, reinforced at its mid-plane to take tensions arising from the ground tearing apart beneath it in any direction. The tensile force in the slab is kept small by using a light building, while the horizontal friction coefficient between the slab and the ground is kept small and reproducible by using a continuous interface of plastic sheeting overlying a rolled layer of sand or fine granular filling. In practice light precast concrete foundation pads are placed first and the whole superstructure is erected on them. These are later integrated to form the floor.

The second requirement is met basically by making the superstructure a pin-jointed 3 m bay framework—a mechanism—with a diagonal brace in two bays orthogonal to each other to stop the wind blowing it over. In practice, for large structures telescopic diagonal braces are used in several bays; these are pre-loaded with springs so that they extend or compress under the weight of the building as the ground subsides, but are not deflected by the wind. The various types of cladding are effectively pin-jointed to the framework. Overlapping rather than butt joints are a feature of many details. Staircases, doors and windows can all distort within specific limits.

Since 1956 the CLASP building system has been widely used for schools and many other buildings in both mining and non-mining areas. The total cost of CLASP construction to date is about £230 million in the U.K. (2085 buildings) and about £92 million abroad. The total current expenditure is about £50 million per annum. Swain (1974) has recently made a survey of the successful performance of the system in Nottinghamshire schools. Between 1957 and 1971, 269 schools were built; 70 of these have been subjected to the influence of mining subsidence. Among these 70 schools there were 128 separate exposures to mining subsidence, some of the schools being undermined four times. The total repair costs attributed to mining were only £3300, which is less than one tenth the annual expenditure due to wilful and accidental damage. The greatest saving however has been in the initial cost of the structure itself. When CLASP was first developed it was usual for the Ministry of Education to permit some $7\frac{1}{2}\%$ extra cost on new schools in mining areas. This has never been claimed for CLASP buildings.

An Interdepartmental Committee (1951). 'Mining Subsidence Effects on Small Houses' *National Building Studies*, Special Report No. 12, HMSO
Lacey, W. D. and Swain, H. T. (1957). 'Design for Mining Subsidence', *Architects' J.*, Vol. 126, 557–570
Swain, H. T. (1974). *Architects' J.* (in press)

M. JAMIOLKOWSKI, Technical University of Turin.
In association with my co-authors I would like to give the results of the controlled water tests which are now completed on tanks Nos. 1 and 4. Details of the tanks are given in our Paper, p. 323.

The Ravenna tank has undergone a very large total settlement as shown by *Fig. 23*. During a year of service the following values have been reached:

Settlement in the center; $s_c = 142$ cm
Settlement along the perimeter; $s_{p\ max} = 78.5$ cm, $s_{p\ min} = 62.5$ cm
Maximum differential settlement along the perimeter defined as:

$$\Delta = \frac{s_{pi} - s_{pk}}{l_{ik}} = 0.00215$$

Notwithstanding this unusually high settlement the water test has been completely successful from both an economical and technical point of view.

The apposite conclusion can be reached from experience gained during the water test on the Ancona tank (*Figs. 24, 25 and 26*) where the total settlements were relatively small ($s_c \simeq 30$ cm, $19 < s_p < 25$ cm) and the differential settlement along the perimeter was of this some order of magnitude ($\Delta \simeq 0.0028$) as for the Ravenna tank. It was necessary to jack up a small part of the perimeter of the tank between points 11 and 13 where the shell bridged over the soil for a distance of 2 to 3 m in order to reduce the ovality of the upper edge of the shell (see *Fig. 26*) and assure the safety of the floating roof.

These two examples and other author's experience permit the following

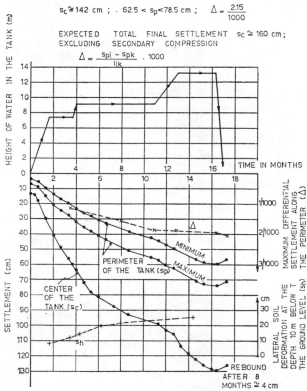

Fig. 23. *Porto Corsini (Ravenna)—50 000* m³ *floating roof tank one year after end of water test*

observations concerning the state of the art in the field of large petroleum tank foundations to be made:

(1) Foundation instrumentation allows very careful control of the tank behaviour during the water tests.

(2) With this provision the danger of general and/or local shear failure can always be avoided.

(3) The problem of allowable differential settlements along the tank perimeter still remains the main one. There is a need to define in a rational way the distortional settlements caused by the effect of differential settlement on the tank structure.

797

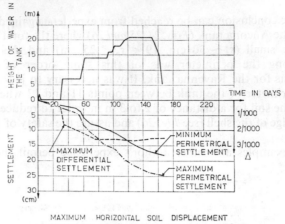

MAXIMUM HORIZONTAL SOIL DISPLACEMENT
MEASURED 10 m BELOW G.L.: s_{hmax} = 5.8 cm

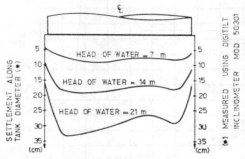

Fig. 24. Falconara (Ancona)—160 000 m³ floating-roof tank

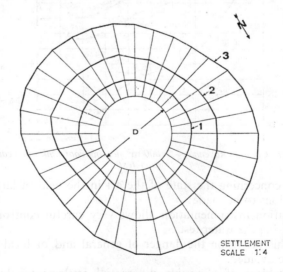

NUMBER	HEAD OF WATER	SETTLEMENT BELOW TANK CENTER
1	7.00 m	8.0 cm
2	14.00 m	17.5 cm
3	21.00 m	29.0 cm

*Fig. 25. Folconara (Ancona)—tank settlement measured with 32 datum points on
the tank perimeter—tank diameter D = 96 m*

(4) Settlements due to the confined plastic flow and/or creep may be partially governed by measuring vertical and lateral soil displacements and adjusting the loaing programme accordingly.

(5) Further development of the controlled water test technique requires a rational definition of the allowable settlement limits in relation to the additional stresses induced into the shell, particularly by distortional settlements along the tank perimeter.

(6) Existing limits of Δ ($0.0035 < \Delta_{max\ all} < 0.0050$), generally mentioned

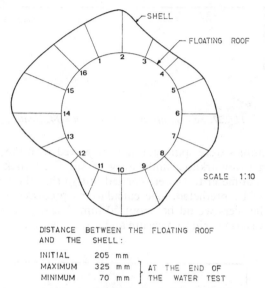

DISTANCE BETWEEN THE FLOATING ROOF
AND THE SHELL:

INITIAL   205 mm
MAXIMUM   325 mm ⎤ AT THE END OF
MINIMUM   70 mm ⎦ THE WATER TEST

Fig. 26. Falconara (Ancona)—160 000 m³. Ovalisation of the top of the shell, July 10, 1973. Head of water 21.0 m

as allowable values, were developed on an empirical basis and are not always on the safe side.

(7) The rational approach to the problem of the allowable differential settlements of a large steel tank will require the preparation of a computer programme to evaluate the additional stresses in the tank shell induced by the differential settlements measured during the water tests. In such a programme the large strains and post-buckling behaviour of the shell should be taken into account.

L. J. ENDICOTT, G. Maunsell and Partners.   This contribution is concerned with the allowable settlements of a steel box girder viaduct. The structure comprises 17 continuous spans totalling more than 500 m in length. The structural elements are four parallel box girders, 1.2 m square working compositely with a 250 mm reinforced concrete deck. The ground conditions are as follows. The site is covered with recent alluvial deposits of the River Neath Saltings in Glamorgan. There is a layer of dense gravel about 8 m below ground level overlying preglacial silt of variable thickness. At the west end of the structure the dense gravel is about 5 m thick and the silt is more than 40 m thick. Towards the east end the dense gravel was found to be in excess

799

of 30 m thick. It was calculated that the dense gravel was capable of providing an adequate bearing for driven piles but the extra load would lead to settlement of the underlying silt. The questions were how much settlement would result, how quickly would it be substantially complete and could the structure accommodate it.

Material from the boreholes was examined carefully in the laboratory. Undisturbed samples were taken from the soft material below the gravel

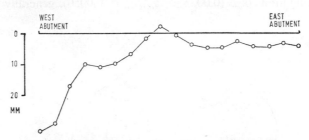

Fig. 27. Settlements at piers. Structure 14, Neath

layer. Many samples were split open and inspected with the aid of a microscope revealing a remarkable uniformity of silt sized rock flour material. Conventional oedometer tests were carried out so that the magnitude of the settlement could be predicted. The calculations predicted that the greatest settlement of the piers would be about 50 mm. During the tests the rate of consolidation was too rapid to measure and it was believed that the settlement

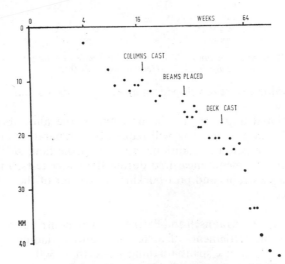

Fig. 28. Settlement–time plot. Pier No. 1, Structure 14, Neath

on site would take place fairly rapidly. It was decided that such settlements would be acceptable if certain features were incorporated in the design of the structure.

Calculations were performed to determine the maximum differential settlements of the piers that could be accommodated by the structure without

800

exceeding the requirements of the Merrison Rules. Typical allowable settlements are 28 mm at an abutment, 25 mm at any one pier or 24 mm at alternate piers. The ratios of settlement to span being 1:1110, 1:1245, 1:1295. It was anticipated that these allowable settlements might be exceeded and a scheme was prepared for jacking the box girders over the piers. The bearing diaphragms to the box girders were checked for jacking forces. In order to accommodate the settlements at the anchorage abutment, a hinged link device was incorporated which would withstand the longitudinal anchorage forces during the jacking of the box girders.

In order to check on the possible need for jacking, the settlements of the piers and abutments were recorded from the completion of the box girders on the adjoining spans. The differential settlements were then calculated for each pier. Results are shown in *Fig. 27*. It can be seen that, at the time of measurement, the allowable settlements have not been exceeded and there is, as yet, no need to implement the jacking. The plot of settlement against time for one of the piers is shown in *Fig. 28*, showing that the total settlement is approaching 50 mm and the rate of settlement is rapidly diminishing. The structure is now substantially complete but not yet open to traffic.

R. W. COOKE, Building Research Station.   In their General Report for Session V, Burland and Wroth suggest that to equalise as far as possible the settlements of the individual footings supporting a structure the dimensions of each footing should be proportioned to the vertical load it is to carry and not related to an allowable average pressure. It seems relevant to consider whether similar reasoning should be applied to the design of footings on single piles or very small pile clusters.

The suggestion made by Burland and Wroth can be applied directly to bearing piles and to the loads reaching the base of piles in deep cohesive soils. However, in these soils, unless the base is enlarged, about 90% of the load applied to a pile is transferred to the soil from the pile shaft. Thus, in the case of straight shafted piles in clay it is the shaft:soil characteristic which determines the settlement.

To minimise differential settlements between pile caps one might expect to be able to reduce the settlement of some of the piles by increasing their diameter, by lengthening them, or by increasing the number of piles in the cluster. Increasing the number of piles brings us into group action problems which I do not want to discuss here. However, it may be useful to consider the effects of varying the diameter D and the length L of a friction pile. Poulos and Davis (1968) have shown theoretically that in a homogeneous elastic soil of modulus E the settlement ρ of a pile carrying a load P is given by

$$\rho = \frac{P}{LE} I$$

where I_p is an influence factor depending on the pile dimensions and Poisson's ratio for the soil. The term D does not appear directly but for Poisson's ratio $0.5 I_p$ varies only from 1.3 to 1.8 as L/D varies over the common range from 10 to 25, with a mean value of 1.55. The relationship between settlement and load is not, therefore, very sensitive to D.

Recent work at BRS on the stress transfer from loaded friction piles in London Clay has tended to confirm the validity of the Poulos and Davis equation. Thus, if it is desired to reduce the settlement of some piles in order to minimise differential movements this can be done by increasing pile lengths but probably not by increasing the diameter.

To check whether this conclusion is valid results of loading tests from piles of various diameters on the same site are necessary. Such data is rare but the work done by Whitaker and myself in London Clay appeared relevant. Piles of approximately 0.6, 0.75 and 0.9 m diameter had been tested but, unfortunately for the present discussion, the lengths also had been varied. For this reason, in order to provide additional comparable data, the load settlement curves for piles less than 15 m long in that investigation have been adjusted. The loads have been scaled up in proportion to both shaft length and

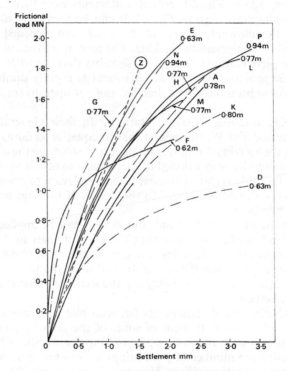

Fig. 29. Shaft load–mean settlement curves for piles having a range of
diameters and length 15 m

the mean shear strength of the clay to correspond to piles unchanged in diameter but 15 m in length. Only frictional loads and mean shaft settlements have been examined and these are plotted for eleven piles in *Fig. 29*. While there is considerable scatter, there is no evidence that the load: settlement curves are related to pile diameter for frictional loads less than about 1 MN.

The mean straight line *OZ* through the lower linear part of these load settlement curves has a slope of 1.18 MN/mm. Using the expression given by Poulos and Davis this corresponds to a mean E value of 120 MN/m² or E/C rather less than 900. These experimental results do therefore support the conclusion that, while increasing the length of a pile designed to carry a given load will decrease the settlement, increasing the diameter is unlikely to. This may be a useful extension for friction piles to the suggestion made by Burland and Wroth for footings near the soil surface.

Poulos, H. G. and Davis, E. H. (1968). 'The Settlement Behaviour of Single Axially Loaded Incompressible Piles and Piers,' *Geotechnique*, Vol. XVIII, No. 3, 351–371
Whitaker, T. and Cooke, R. W. (1966). 'An Investigation of the Shaft and Base Resistances of Large Bored Piles in London Clay', *Symp. Large Bored Piles*, 7–49, *Inst. civ. Engrs.*, London

R. H. BASSETT, King's College London.   In his impressive introduction to this conference Professor Sir John Baker confessed that he had hoped when developing his Plastic Theory to produce a technique which would 'eliminate the problem of differential settlement and put foundation engineers out of business'. I would like to put forward some ideas that could in some respects

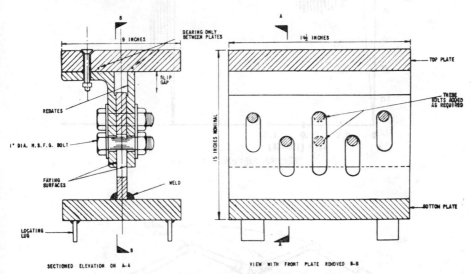

Fig. 30. Typical bolted slip unit

achieve, if not such a drastic aim, at least a method of minimising the uncertainties and complexities which differential settlements cause in rigid structures.

In 1969 the writer faced this type of problem with the very stiff frame structure of a power station for Alcan Limited (Bassett, 1971 and Clark, Bassett and Bradshaw 1973). Experimentation with steel joints (*Fig. 30*) built with elongated bolt holes and friction grip bolts showed that units with repeatable slip characteristics (*Fig. 31*) could be built. The solution to the Alcan project incorporated these units in the foundation piles above the rigid or non-settling areas and this appears to have minimised overstressing in the frame. In retrospect the author feels that considerably more advantage could be taken of the slipping characteristics of the joints by incorporating them directly into the structural frame itself. Utilising the concepts of Professor Baker's plastic theory suitable locations and yielding moments could be chosen for special joints. An example of such an application is shown in *Fig. 32*. The joint units would be set to distort at fixed 'load factors' above working load

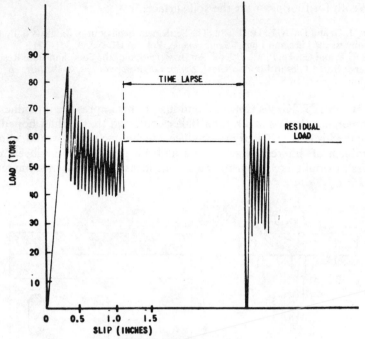

Fig. 31. Slip characteristics

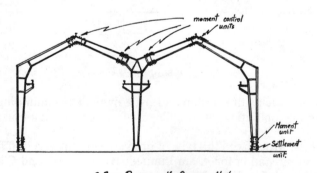

2 Bay Portal with Control Units

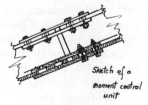

Sketch of a
moment control
unit

Fig. 32

804

but still below the plastic yield of the steel sections employed. The advantages of this approach would be: (1) that the points of displacement and moment compensation would be easily observed, yet (2) only one joint would yield at any instant and the structure would remain a structure and not become a mechanism, but immediately slip was complete the joint would again be a rigid joint and no secondary instabilities would have occurred (e.g. necking, or buckling), and (3) the structure at normal working loads is effectively still stiff or continuous but has the flexibility of a pinned structure to any unpredicted displacements which cause excessive overstressing.

Ward gave graphic examples of flexible or pin jointed structures successfully resisting mining subsidence under schools. Even more important he showed that the difficulties of flexible cladding has been overcome. The author feels that deliberately incorporated slip joints would be no more complex to fabricate or erect than conventionally bolted frames but would endow the stiffer and more economic continuous structure with controlled flexibility. In particular the authors (Clark, Bassett and Bradshaw, 1973) envisaged their use in supporting structures which are very sensitive to small differential displacements, box girders and high space frames in particular.

Bassett, R. B. (1971). Contribution *Proc. Con. Behaviour of Piles, Inst. civ. Engs*, 45–47, 89–90

Clark, P. J., Bassett, R. H. and Bradshaw, J. B. (1973). 'Plate Friction Load Control Devices—Their Application and Potential', *Proc. Inst. civ. Engrs*, December, 947–951

J. ŠKOPEK, Charles University, Prague.   I should like to report on foundation movements due to the groundwater level fluctuation observed during the installation of a turbo-generator in a Shanghai Power Plant (China). A new turbogenerator was installed on an old concrete base close to the River Whampoo, where the difference between the maximum and minimum water level due to the tidal activity was 2 to 3 m.

The concrete foundation block of the turbogenerator rested on driven piles 40 m long. Each pile consisted of three parts jointed with annular rings (*Fig. 33*) The foundation soil consists of a thick layer of normally consolidated soft sandy clay with layers of fine sand with shells. Water content is higher or close to *LL*. During the installation of the generator it was observed that the foundation moved in correspondence with the water-level fluctuations of the river. During high tide, the foundation slowly heaved, and during low tide the foundation sank. The movements of the foundation were measured with very accurate water-level gauges. At the same time, the water-level in the river and that in the standpipes were recorded. A typical result of these measurements is shown in *Fig. 34*. The difference between the water-levels in the river during the period of observation ranged from 2 m to 3 m. In the standpipe S_1 the difference was about 0.5 m while in the standpipe S_2 practically no change was registered. The mean elevation change of point A on the foundation was 1 mm. Point A was nearest to the river. Point B, on the foundation, which was more distant from the river, showed practically no change in elevation. The measurement was carried out for three months during which there was practically no change in the resulting pattern.

An attempt has been made to decrease the movement of the foundation by loading it by an additional 62.5 tonne in excess of the weight of the turbogenerator, but the magnitude of the movement did not change. The foundation

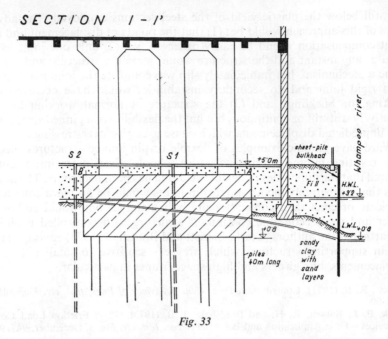

Fig. 33

rests on long piles, but this did not prevent the movement. This is probably due to the fact that the piles act as floating piles built up of freely jointed sections. Nevertheless, it is obvious that the effect of water-level fluctuation reaches to a great depth.

It is well known that lowering the ground-water level increases the effective weight of the soil and surface settlements follow. Conversely, a rising water table brings about a decrease of effective pressure on the underlying layers; due to its elasticity, soil expands and the ground surface heaves up. This rebound has sometimes been observed in connection with the measure-

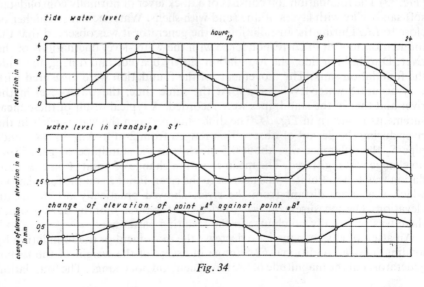

Fig. 34

ment of land subsidence caused by ground-water fluctuation, but only exceptionally in connection with movements of foundations of structures.

D. L. WEBB, D. L. Webb & Associates, Durban, S.A.   At the suggestion of Dr. Burland I would like to describe, briefly, the method employed for the preloading and relevelling of columns as mentioned under 'Performance of the Structure' in my Paper (p. 450).

Preloading comprised bricks, to be used later for the external and internal walls of the building, and sand bags moved by front end loader and generally stacked by hand in such a way as to minimise shear stresses in the beams and slabs. Maximum preload of 37 000 kg was slightly more than the design live load, which included everything except the mass of the reinforced concrete and the foundations themselves, and amounted to about 35% of the total foundation load. The preloads were applied to each column gradually over periods of 3 to 5 days to allow the structure to distort slowly. Another reason for the slow application of preload and, in fact, for the preloading programme as a whole, was to be able to observe the effects of the applied live load and decide whether it would be safe to continue with the structure, or whether it would be necessary to omit the top floor in order to reduce the loads on the foundations.

Preloads were left in position for various periods up to a year, but the additional settlements, as reflected by an acceleration in the time-settlement curves, were virtually complete within about six weeks. The magnitude of the additional settlement was usually 3 to 7 mm in the range 2 to less than 15 mm. The construction programme was planned so that the bricks and sand of the preloading to a particular column were employed for construction in the vicinity of the column. In this way possible adverse effects of cyclic loading were largely obviated.

The thermic lance process proved to be very satisfactory and it was possible to cut and relevel a column 750 × 600 mm in one day. Preformed steel end plates were grouted in position on the cut surfaces of the column, 450 mm above the foundation, and the upper part of the column was then raised by jacking against two steel brackets, one on either side. The brackets were held in position by high tensile steel bolts through the column and short steel supports were provided on either side of each jack to support the column during cutting and in case a jack failed during operation. Spacers, consisting of steel plates between thin lead sheets were inserted between the steel end plates. Subsequent relevelling simply required the insertion of additional spacers. Despite the large rotational strain of the beams and columns, there was no relative displacement of the upper and lower sections of the columns after cutting.

Differential settlements between certain columns before preloading were reaching critical values, but as a result of preloading the columns and cutting and jacking two of them, it was possible to restrict differential settlements to within tolerable limits. Walls, plaster, face brickwork, glass and other finishes have now been in position for 18 months, and although minor regional settlement continues, there has been no distress of either the structure or the finishes.

G. E. BRATCHELL, Nachshen, Crofts & Leggatt.   I would like to draw attention to the fact that settlements in structures cause induced strains, which set up induced stresses. Thus for example, in a reinforced concrete slab lying on the ground, any reinforcement placed in the slab will be strained in accordance with the curvature of that slab. As mentioned by other speakers, the strength of the structure may have some influence on change in the ground deformation pattern, and this will generally be true of large framed and panelled structures, but slabs for example will be relatively weak and will hardly influence the settlement pattern. In the latter cases, it is important to appreciate that adding reinforcement will merely mean that this will be stressed at the same level. Thus the concept of calculating induced bending moments and then providing steel to resist these moments is a commonly found error in design work.

The important point to appreciate is that the stress available in the material to resist the applied loads in the structure, is that remaining after the induced stress has been subtracted from the working stress. In some cases the curvature of the ground from the settlement can be so great that the steel is stressed to its working stress and in such cases adding steel has little or no effect and the design must be reconsidered.

The second point is that differential settlement is a particularly difficult problem when dealing with heavy braced structures. Several cases have been noted by us where the stresses in the bracing due to settlement exceed that due to wind loading. Again, a designer is often faced with the anomaly that the stronger the bracing he uses, the more load it induces under settlement and the design has to be completely reviewed.

The third point concerns rakerpiles. It is recommended that rakerpiles should not be used in foundations where settlement is likely to occur either due to the structure itself or to some associated heavy construction such as local filling. Very large forces indeed can be induced in rakerpiles due to the change of geometry caused by settlement. This point has been the subject of informal discussions of the Piling Group of the Institution of Civil Engineers.

K. STARZEWSKI, Institute of Building Technology, Warsaw. I wish to bring to the attention of the readers a very comprehensive record of settlement studies of 70 structures which was recently published by the Institute of Building Technology (Instytut Techniki Budowlanej) in Warsaw (Bolenski, 1973). The studies were started in 1948 and cover a wide variety of structures (*Table 7*) and soil conditions (*Table 8*).

Another interesting settlement study, worthy of attention, was presented by Laczkowski *et al.* (1973). It involved measurements of settlements (during and after construction) of a grain silo consisting of two large bin houses separated by an elevator tower; settlement readings were also taken during loading and unloading of the silos. The ratios of the calculated (according to PN-59/B-03020) to observed settlements vary between 1.03 and 1.57, depending on the range of loading.

Boleński, M. (1972). 'Osiadanie nowo wznoszonych budowli w zależności od rodzaju podłoża gruntowego' (Settlement of Newly Constructed Structures with Reference to the Ground Conditions), *Inżynieria i Budownictwo*, Vol. 3, 94–102
Wiłum, Z. and Starzewski, K. (1972). *Soil Mechanics in Foundation Engineering*, Intertext Books, London
Laczkowski, A., Młynarek, Z. and Przystański, J. (1973). 'Osiadania silosów zbożowych posadowionych na gruntach piaszczystych' (Settlement of Grain Silos Founded on Sandy Soils), *Inżynieria i Budownictwo*, Vol. 7, 305–7

Table 7

Type of structure	No of structures	Total No	Blocks of flats (multi-storey dwellings) No of storeys						Grain silos	Chimneys	Turbo-generator foundations
			3–4	5	7	9–13	17	20–24			
Brickwork with RC ring beam at every floor level	11	10	9	—	—	1	—	—	—	1	—
Monolithic RC construction	21	10	—	—	1	7	1	1	9	2	—
Monolithic RC block-frame construction	9	—	—	—	—	—	—	—	—	—	9
Monolithic RC frame infilled with brickwork	2	2	1	—	—	1	—	—	—	—	—
Prefabricated RC frame infilled with lightweight concrete	2	2	—	—	—	2	—	—	—	—	—
Prefabricated construction: large block elements	7	7	3	—	—	4	—	—	—	—	—
Prefabricated construction: large slab elements	14	14	—	6	—	8	—	—	—	—	—
Steel frame	4	4	—	—	—	—	—	4	—	—	—
Total	70	49	13	6	1	23	1	5	9	3	9

Table 8

Soil conditions beneath foundations	Number of structures	% of total
Sands	18	28
Glacial (moraine) cohesive soils	23	35
Soils varying considerably with depth and across the site	8	12
Very compressible natural and compacted organic soils underlain by deep deposits of river sands	9	14
Thin deposits of sands underlain by pliocene clays	5	8
Lacustrine sands interbedded with organic soils (peats)	2	3

Implications of the up to date observations are summarized by Bolenski (1972 and 1973): it is suggested that in general the values of stiffness moduli recommended by Polish Standard PN-59/B-03030 (1959) (see also Wiłun and Starzewski, 1972) for evaluation of the calculated allowable total and differential settlements could be increased by factors ranging from 1.3 to 2.0, depending on the type of soil in question—this would narrow the gap between the calculated and observed settlements. It is also suggested that in the classification of the types of soils, silts should be separated from the general groups of cohesive soils, as their compressibility is much smaller than that of the various clays. The observed total and differential settlements suggest that the allowable calculated settlements given in *Table 9* (Wiłun and Starzewski, 1972) can safely be used in determination of the allowable bearing stresses.

Table 9

Class of building and structure	Type of building or structure	Maximum allowable final settlement s_{all} (mm)	Maximum allowable angular distortion calculated for three, collinear, adjoining points or foundations of a structure α_{all}
1	Massive structures of considerable rigidity about horizontal axes founded on rigid mass concrete foundations or cellular or rigid reinforced concrete rafts	150–200 (6 in–8 in)	Maximum differences of settlements at various points of the structure should not cause tilting of the foundation greater than $1/100 \div 1/200$ of the ratio of the smallest dimension of the foundation in plan to the height of the structure
2	Statically determinate structures with actual pin joints (three pinned arches, single-span steel trusses, etc.) and timber structures	100–150 (4 in–6 in)	$\dfrac{1}{100} - \dfrac{1}{200}$
3	Statically indeterminate steel structures and load-bearing brickwork construction with reinforced concrete ring beams at every floor level, with longitudinally reinforced concrete strip foundations and with cross walls of at least 250 mm (9 in) thickness and spaced at not more than 6 m (~20 ft) centres and reinforced concrete frame—structures with columns at less than 6 m centres and founded on strip or raft foundations	80–100 ($3\frac{1}{2}$ in–4 in)	$\dfrac{1}{200} - \dfrac{1}{300}$
4	Structures of class 3 but not satisfying one of the stated conditions and reinforced concrete structures founded on isolated footings	60–80 ($2\frac{1}{2}$ in–$3\frac{1}{2}$ in)	$\dfrac{1}{300} - \dfrac{1}{500}$
5	Prefabricated structures consisting of large slab or block elements	50–60 (2 in–$2\frac{1}{2}$ in)	$\dfrac{1}{500} - \dfrac{1}{700}$

(1) Smaller values quoted relate to public buildings, dwellings, or buildings with structural members or finishes particularly sensitive to differential settlement; larger values relate to taller buildings of considerable rigidity about horizontal axes or to structures which can accept such movements.

(2) In special cases (such as gantry beams, high-pressure boilers, special storage tanks, silos under differential loading, etc.) allowable maximum or differential settlements or both should be taken as specified by service or mechanical engineers or by manufacturers.

Contributors to the Discussion